Advances in SAR: Sensors, Methodologies, and Applications

Advances in SAR: Sensors, Methodologies, and Applications

Special Issue Editors

Timo Balz
Uwe Soergel
Mattia Crespi
Batuhan Osmanoglu

MDPI • Basel • Beijing • Wuhan • Barcelona • Belgrade

MDPI

Special Issue Editors

Timo Balz
Wuhan University
China

Uwe Soergel
University of Stuttgart
Germany

Mattia Crespi
University of Rome "La Sapienza"
Italy

Batuhan Osmanoglu
NASA/GSFC
USA

Editorial Office
MDPI
St. Alban-Anlage 66
Basel, Switzerland

This is a reprint of articles from the Special Issue published online in the open access journal *Remote Sensing* (ISSN 2072-4292) from 2017 to 2018 (available at: http://www.mdpi.com/journal/remotesensing/special_issues/rs_SAR)

For citation purposes, cite each article independently as indicated on the article page online and as indicated below:

LastName, A.A.; LastName, B.B.; LastName, C.C. Article Title. *Journal Name* **Year**, *Article Number*, Page Range.

ISBN 978-3-03897-182-5 (Pbk)
ISBN 978-3-03897-183-2 (PDF)

Contents

About the Special Issue Editors

Timo Balz, was born in Stuttgart, Germany. He received the Diploma degree (Dipl.-Geogr.) in geography and the Doctoral degree (Dr.-Ing.) in aerospace engineering and geodesy from the Universitaet Stuttgart, Stuttgart, in 2001 and 2007, respectively. From fall 2001 to the end of 2007, he was a Research Assistant with the Institute for Photogrammetry, Universitaet Stuttgart. Between 2004 and 2005, he was a Visiting Scholar with Wuhan University, Wuhan, China. From 2008-2010, he has been a Postdoctoral Research Fellow with the State Key Laboratory of Information Engineering in Surveying, Mapping and Remote Sensing (LIESMARS), Wuhan University. From 2010-2015 he was Associate Professor for Radar Remote Sensing with LIESMARS. Since 2015 he is Full Professor with LIESMARS. He serves as Associate Editor for the IEEE Geoscience and Remote Sensing Letters and he is member of the editorial board of Geo-Spatial Information Science. Since 2016 he is Chair of the ISPRS Commission I Working Group on SAR and Microwave Sensing. He has authored and co-authored about 150 scientific articles in journals, books, and conference proceedings. Timo Balz's research interests include surface motion estimation with SAR, data visualization, SAR geodesy, and the use of SAR data to support archaeological prospections.

Uwe Soergel chairs the Institute for Photogrammetry at University of Stuttgart, Germany. He received the Diplomingenieur (M.Sc.) degree in electrical engineering from University of Erlangen-Nuremberg, Germany, in 1997. From fall 1997 to the end of 2005, he was a research associate with the Institute for Optronics and Pattern Recognition (FOM) located in Ettlingen (Germany), which was part of FGAN, a former German research establishment focusing on defense-related studies. At that time, he dealt mainly with pattern recognition of man-made objects from remote sensing imagery, with emphasis on SAR data. In parallel, he earned a PhD in electrical engineering and computer science from the Leibniz Universität Hannover, Germany, in 2003. Prior to his current position starting from 2006 he was first Assistant Professor and later Associate Professor for Radar Remote Sensing and for Radar Remote Sensing and Active Systems, respectively, at Leibniz Universität Hannover. From October 2013 until end of March 2016 he was Full Professor for Remote Sensing and Image Analysis at Technische Universität Darmstadt, Germany.

Mattia Crespi received his degree in Civil Engineering Summa cum Laude in 1987 at Politecnico di Milano (Milan, Italy) and his PhD in Geodesy and Surveying in 1992 at Politecnico di Torino (Turin, Italy) (Supervisor: Em. Prof., DDDr. H.C. Fernando Sansò). He is Professor of Positioning and Geomatics at the University of Rome "La Sapienza" since 2005 (where he was Associate Professor in the same disciplines since 1998), and Senior Fellows of the Sapienza School for Advanced Studies since 2015. He was recipient of the following awards: DLR Special Topic Prize and Audience Award of the European Satellite Navigation Competition in 2010 (VADASE approach); Success Story of the European Satellite Navigation Competition in 2012 (VADASE approach); ESA Certificate for Galileo In-Orbit-Validation in 2014, Accademia Nazionale dei Lincei Prize of the Ministry of Culture for Astronomy, Geodesy and Geophysics in 2018. Mattia Crespi is chair/co-chairman of study groups within the European Association of Remote Sensing Laboratories (EARSeL) and the International Society for Photogrammetry and Remote Sensing (ISPRS), member of EARSeL Bureau as treasurer and vice-president of the Inter-Commission Committee on Theory of the International Association of Geodesy (IAG); he is also the Italian Representative within IAG Council.

Batuhan Osmanoglu, Research Physical Scientist, holds a B.Sc. in telecommunications engineering and a Ph.D. in synthetic aperture radar (SAR) interferometry. He has been a part of Istanbul Technical University, University of Miami, University of Alaska-Fairbanks developing his primary area of expertise in radar remote sensing, and he has worked on applications for observing surface deformation, measuring target velocities, and radar design. Since 2013 he has been working at the NASA Goddard Space Flight Center. He is working on the instrument and algorithm development of P- , L-, X- and Ku-band SAR systems. He is also one of the NASA Disasters Program coordinators at Goddard. He is a member of the IEEE and American Geophysical Union. He chairs the Microwave Remote Sensing workgroup under International Society for Photogrammetry and Remote-Sensing. He takes part in several remote sensing projects, for which he uses various radar imagery collected using domestic and international sensors.

remote sensing

MDPI

Editorial

Editorial for Special Issue "Advances in SAR: Sensors, Methodologies, and Applications"

Timo Balz [1,*], Uwe Sörgel [2], Mattia Crespi [3] and Batuhan Osmanoglu [4]

[1]   State Key Laboratory of Information Engineering in Surveying, Mapping and Remote Sensing, Wuhan University, Wuhan 430079, China
[2]   Institute for Photogrammetry, University of Stuttgart, 70174 Stuttgart, Germany; soergel@ifp.uni-stuttgart.de
[3]   Geodesy and Geomatics Division, University of Roma "La Sapienza", 00184 Rome, Italy; mattia.crespi@uniroma1.it
[4]   NASA/GSFC, Greenbelt, MD 20771, USA; batuhan.osmanoglu@nasa.gov
[*]   Correspondence: balz@whu.edu.cn; Tel.: +86-(0)27-68779960

Received: 31 July 2018; Accepted: 1 August 2018; Published: 6 August 2018

The Special Issue "Advances in SAR: Sensors, Methodologies, and Applications" aims to give an overview of recent advancements in Synthetic Aperture Radar (SAR) remote sensing. SAR remote sensing is a wide field including sensor technologies—the hardware—as well as algorithms and methods—the software. The sensors and algorithms are, however, just a means to an end, with that end being the applications.

In recent years, SAR remote sensing technology has made huge steps forward. With the increase in available sensors and the tremendous growth of available SAR data, access has become easier and SAR has become more relevant. With this increase in data, we have seen recent advancements in SAR, especially regarding sensors and methodologies, but also in terms of newly developed applications.

The articles in this Special Issue focus on these advancements, while also covering different aspects of SAR remote sensing. The Special Issue starts with a paper from Giudici et al. [1] describing the pre-flight experiments during the outdoor performance assessment campaign for the upcoming SAOCOM-1A SAR mission. Corner reflectors are important for various applications in a SAR sensor's lifecycle. Consequently, Garthwaite discusses design considerations for corner reflectors used for deformation monitoring [2]. In an interesting piece on interferences, Monti-Guarnieri et al. [3] present their work on the radio frequency interferences in C-band based on an analysis of the first 8–10 echo measurements per burst to provide a first radio frequency interference map over Europe.

Another important topic is SAR signal processing. Zhang et al. [4] describe a new accelerated back-projection algorithm. A multiple-input, multiple-output (MIMO) video SAR signal processing approach is presented by Kim et al. [5]. Park et al. [6] demonstrate an efficient correction of ground moving targets from SAR SLC images. Bu et al. [7] present a unified algorithm for the calibration of single-pass multi-baseline TomoSAR systems.

SAR interferometry is one of the most important applications in SAR remote sensing. Furthermore, it is the topic of several papers in this Special Issue. Even and Schulz [8] present a detailed review on the deformation analysis with distributed scatterers. Their review offers an excellent starting point to learn more about recent advancements in distributed scatterer (DS) InSAR. Tian et al. [9] present a method for an improved orbital error modeling relevant to various interferometric applications. Besides orbital errors, atmospheric and ionospheric effects are important error sources. To reduce the ionosphere error, Wang et al. [10] present a method based on the Faraday rotation with polarimetric SAR data. Precise DEM generation is one of the main applications of interferometric SAR. Dong et al. [11] demonstrate a multi-baseline InSAR approach using Maximum Likelihood Estimation.

Infrastructure stability surveillance is an important application, especially for high-resolution SAR sensors. Bridges are especially important in this context. The possibility to measure deformation is shown by a case study of the Lupu bridge in Shanghai [12]. Another case study by Neelmeijer et al. [13] shows the deformation around the Toktogul Reservoir in Kyrgyzstan based on ASAR and Sentinel-1 data over two periods from 2004–2009 and 2014–2016.

As an alternative to differential interferometry, surface motion estimation with SAR pixel-tracking can provide precise motion measurements in the range and azimuth directions. Sun et al. [14] demonstrate this by surveying landslides in the Three Gorges Region. Shi et al. [15] take another look at landslides in the Three Gorges Region with Split-Bandwidth Interferometry. On the other hand, Libert et al. [16] use Split-Band Interferometry to assist in phase unwrapping.

Feature detection from SAR images is another major topic in Microwave Remote Sensing. Ghafouri et al. [17] present a method to better estimate IEM (Integral Equation Model) input parameters for multi-frequency SAR data. Di Martino et al. [18] describe the role of resolution for the estimation of fractal dimension maps. Deng et al. [19] give an overview over different methods for the statistical modeling of polarimetric SAR data. Segmenting polarimetric SAR data is important in understanding and classifying SAR data. For high-resolution PolSAR data, Chen et al. [20] demonstrate a multi-feature segmentation technique based on fractal net evolution approach. Tao et al. [21] show a land cover classification method with polarimetric SAR data based on roll-invariant and selected hidden features in the polarimetric rotation domain.

SAR data can be a very good data source for change detection analysis. Braun and Hochschild [22] use this for detecting landscape changes in the African Savannas. Behnamian et al. [23] report the development of a semi-automated surface water detection technique especially suitable for wetlands. Washaya et al. [24] use coherence change detection to identify and monitor damages caused by natural and anthropogenic disasters, including hurricane, forest fire, and earthquake damage detection, in addition to providing an extensive overview of the damages caused by the Syrian War in Aleppo, Raqqa, and Damascus.

Zhai et al. [25] present a multi-layer model for SAR images that is based on multi-scale and multi-feature fusion. Last but not least, Molan et al. [26] introduce a new temporal decorrelation model for L-band data over Alaska, taking the amplitude and snow depth into account.

Funding: This work was supported by the Natural Science Foundation of China under the Grant 61331016.

Conflicts of Interest: The authors declare no conflict of interest.

References

1. Giudici, D.; Monti-Guarnieri, A.; Cuesta Gonzalez, J. Pre-flight SAOCOM-1A SAR performance assessment by outdoor campaign. *Remote Sens.* **2017**, *9*, 729. [CrossRef]
2. Garthwaite, M. On the design of radar corner reflectors for deformation monitoring in multi-frequency InSAR. *Remote Sens.* **2017**, *9*, 648. [CrossRef]
3. Monti-Guarnieri, A.; Giudici, D.; Recchia, A. Identification of C-Band radio frequency interferences from sentinel-1 data. *Remote Sens.* **2017**, *9*, 1183. [CrossRef]
4. Zhang, H.; Tang, J.; Wang, R.; Deng, Y.; Wang, W.; Li, N. An accelerated backprojection algorithm for monostatic and bistatic SAR processing. *Remote Sens.* **2018**, *10*, 140. [CrossRef]
5. Kim, S.; Yu, J.; Jeon, S.-Y.; Dewantari, A.; Ka, M.-H. Signal processing for a multiple-input, multiple-output (MIMO) video synthetic aperture radar (SAR) with beat frequency division frequency-modulated continuous wave (FMCW). *Remote Sens.* **2017**, *9*, 491. [CrossRef]
6. Park, J.-W.; Kim, J.; Won, J.-S. Fast and efficient correction of ground moving targets in a synthetic aperture radar, single-look complex image. *Remote Sens.* **2017**, *9*, 926. [CrossRef]
7. Bu, Y.; Liang, X.; Wang, Y.; Zhang, F.; Li, Y. A unified algorithm for channel imbalance and antenna phase center position calibration of a single-pass multi-baseline TomoSAR system. *Remote Sens.* **2018**, *10*, 456. [CrossRef]
8. Even, M.; Schulz, K. InSAR deformation analysis with distributed scatterers: A review complemented by new advances. *Remote Sens.* **2018**, *10*, 744. [CrossRef]

9. Tian, X.; Malhotra, R.; Xu, B.; Qi, H.; Ma, Y. Modeling orbital error in InSAR interferogram using frequency and spatial domain based methods. *Remote Sens.* **2018**, *10*, 508. [CrossRef]
10. Wang, C.; Chen, L.; Zhao, H.; Lu, Z.; Bian, M.; Zhang, R.; Feng, J. Ionospheric reconstructions using faraday rotation in spaceborne polarimetric SAR data. *Remote Sens.* **2017**, *9*, 1169. [CrossRef]
11. Dong, Y.; Jiang, H.; Zhang, L.; Liao, M. An efficient maximum likelihood estimation approach of multi-baseline SAR interferometry for refined topographic mapping in mountainous areas. *Remote Sens.* **2018**, *10*, 454. [CrossRef]
12. Zhao, J.; Wu, J.; Ding, X.; Wang, M. Elevation extraction and deformation monitoring by multitemporal InSAR of lupu bridg in Shanghai. *Remote Sens.* **2017**, *9*, 897. [CrossRef]
13. Neelmeijer, J.; Schöne, T.; Dill, R.; Klemann, V.; Motagh, M. Ground deformations around the toktogul reservoir, kyrgyzstan, from envisat ASAR and sentinel-1 data-A case study about the impact of atmospheric corrections on InSAR time series. *Remote Sens.* **2018**, *10*, 462. [CrossRef]
14. Sun, L.; Muller, J.-P.; Chen, J. Time series analysis of very slow landslides in the three gorges region through small baseline SAR offset tracking. *Remote Sens.* **2017**, *9*, 1314. [CrossRef]
15. Shi, X.; Jiang, H.; Zhang, L.; Liao, M. Landslide displacement monitoring with split-bandwidth interferometry: A case study of the shuping landslide in the three gorges area. *Remote Sens.* **2017**, *9*, 937. [CrossRef]
16. Libert, L.; Derauw, D.; d'Oreye, N.; Barbier, C.; Orban, A. Split-band interferometry-assisted phase unwrapping for the phase ambiguities correction. *Remote Sens.* **2017**, *9*, 879. [CrossRef]
17. Ghafouri, A.; Amini, J.; Dehmollaian, M.; Kavoosi, M. Better estimated IEM input parameters using random fractal geometry applied on multi-frequency SAR data. *Remote Sens.* **2017**, *9*, 445. [CrossRef]
18. Di Martino, G.; Iodice, A.; Riccio, D.; Ruello, G.; Zinno, I. The role of resolution in the estimation of fractal dimension maps from SAR data. *Remote Sens.* **2018**, *10*, 9. [CrossRef]
19. Deng, X.; López-Martínez, C.; Chen, J.; Han, P. Statistical modeling of polarimetric SAR data: A survey and challenges. *Remote Sens.* **2017**, *9*, 348. [CrossRef]
20. Chen, Q.; Li, L.; Xu, Q.; Yang, S.; Shi, X.; Liu, X. Multi-feature segmentation for high-resolution polarimetric SAR data based on fractal net evolution approach. *Remote Sens.* **2017**, *9*, 570. [CrossRef]
21. Tao, C.; Chen, S.W.; Li, Y.Z.; Xiao, S. PolSAR land cover classification based on roll-invariant and selected hidden polarimetric features in the rotation domain. *Remote Sens.* **2017**, *9*, 660. [CrossRef]
22. Braun, A.; Hochschild, V. A SAR-based index for landscape changes in african savannas. *Remote Sens.* **2017**, *9*, 359. [CrossRef]
23. Behnamian, A.; Banks, S.; White, L.; Brisco, B.; Millard, K.; Pasher, J.; Chen, Z.; Duffe, J.; Bourgeau-Chavez, L.; Battaglia, M. Semi-automated surface water detection with synthetic aperture radar data: A. wetland case study. *Remote Sens.* **2017**, *9*, 1209. [CrossRef]
24. Washaya, P.; Balz, T.; Mohamadi, B. Coherence change-detection with sentinel-1 for natural and anthropogenic disaster monitoring in urban areas. *Remote Sens.* **2018**, *10*, 1026. [CrossRef]
25. Zhai, A.; Wen, X.; Xu, H.; Yuan, L.; Meng, Q. Multi-layer model based on multi-scale and multi-feature fusion for SAR images. *Remote Sens.* **2017**, *9*, 1085. [CrossRef]
26. Eshqi Molan, Y.; Kim, J.-W.; Lu, Z.; Agram, P. L-band temporal coherence assessment and modeling using amplitude and snow depth over interior alaska. *Remote Sens.* **2018**, *10*, 150. [CrossRef]

remote sensing

MDPI

Technical Note

Pre-Flight SAOCOM-1A SAR Performance Assessment by Outdoor Campaign

Davide Giudici [1], Andrea Monti Guarnieri [2] and Juan Pablo Cuesta Gonzalez [3,*

[1]   ARESYS srl, Via Flumendosa 16, 20132 Milan, Italy; davide.giudici@aresys.it
[2]   Politecnico di Milano, Piazza Leonardo da Vinci 32, 20133 Milan, Italy; andrea.montiguarnieri@polimi.it
[3]   CONAE, Avda. Paseo Colon 751, 1063 Buenos Aires, Argentina
*   Correspondence: davide.giudici@aresys.it; Tel.: +39-02-8724-4809

Academic Editors: Timo Balz, Uwe Soergel, Mattia Crespi, Batuhan Osmanoglu, Daniele Riccio
and Prasad S. Thenkabail
Received: 23 May 2017; Accepted: 12 July 2017; Published: 14 July 2017

Abstract: In the present paper, we describe the design, execution, and the results of an outdoor experimental campaign involving the Engineering Model of the first of the two Argentinean L-band Synthetic Aperture Radars (SARs) of the Satélite Argentino de Observación con Microondas (SAOCOM) mission, SAOCOM-1A. The experiment's main objectives were to test the end-to-end SAR operation and to assess the instrument amplitude and phase stability as well as the far-field antenna pattern, through the illumination of a moving target placed several kilometers away from the SAR. The campaign was carried out in Bariloche, Argentina, during June 2016. The experiment was successful, demonstrating an end-to-end readiness of the SAOCOM-SAR functionality in realistic conditions. The results showed an excellent SAR signal quality in terms of amplitude and phase stability.

Keywords: SAR; pre-flight testing; SAR performance

1. Introduction

The forthcoming Argentinean mission Satélite Argentino de Observación con Microondas (SAOCOM) [1] is constituted by two identical Low-Earth-Orbit (LEO) satellites, SAOCOM-1A and -1B, carrying as the main payload an L-band Synthetic Aperture Radar (SAR). The SAOCOM-SAR has an active array antenna with elevation and azimuth steering capability for frequent-revisit, wide-swath, and medium resolution Earth observation.

The mission's target applications [2] require a tight overall amplitude and phase stability performance, which is imposed to a large extent by the instrument critical elements, such as the chirp generator, the distribution network, the Transmit-Receive Modules (TRMs), the clock, and the antenna. The pre-launch assessment and verification of the phase and amplitude performance is paramount in order to limit the calibration and verification efforts to be done in-flight (e.g., during the commissioning phase).

Typically, the instrument elements can be tested on a singular basis, and the antenna far-field radiation pattern can be reconstructed from planar-near-field-scanner (PNFS) measurements of the near field radiation patterns of the elements of the antenna by means of an accurate antenna model [3]. In the latter case, extreme care is needed in the measurement setup to avoid biases in the measurements. The far-field pattern is not available on-ground and the antenna model can be verified only in-flight, exploiting e.g., homogeneous targets for the elevation case and recording transponders for the azimuth. A typical example is the acquisition over the Amazonian rainforest. However, the in-flight data intensity profile depends on a list of other factors (e.g., pointing, time-variant instrument gains, processor gains), also to be calibrated or verified, so that it might be difficult to isolate the antenna contribution from the others, as reported for the case of Sentinel-1 in References [4,5]. Furthermore, the in-flight verification allows a limited verification of the elevation antenna pattern, within the angular range used by the imaging swath.

One way to check the mid- and long-term stability of the system, say from seconds to minutes, is an outdoor experiment where SAR acquisitions are carried out with the full antenna (or a part of it, e.g., one tile) pointed to targets of opportunity or man-made calibrators in the far-field. In addition, the collected power measures for different antenna pointing can be correlated to the theoretical antenna model to obtain a first far-field verification, complementary to what could be obtained with PNFS measurements.

2. Outdoor Experiment Concept, Design, and Simulations

The well-known far field condition on the distance R for an antenna with size L operating at central wavelength λ, is:

$$\frac{\lambda}{L} R \gg L \tag{1}$$

In the case of the SAOCOM-1A SAR instrument, we refer to the SAR antenna, a planar array with the total size along azimuth L equal to 9.94 m. The antenna is made by seven tiles with the size of 1.42 m $\times$ 3.48 m (azimuth $\times$ elevation), and the center frequency is 1.275 GHz ($\lambda = \frac{c}{f}$ is equal to 23.53 cm). The far field distance results are in the order of 40 m. Moreover, the SAR instrument cannot receive echoes returning to the antenna while the chirp transmission is still active. Considering a minimum chirp length of 10 μs, and another 10 μs of guard time between the end of transmission and the start of echo reception, the minimum distance of a "visible" target is 3.3 km.

Compared to the SAR acquisition in space, here the antenna is not moving along an orbit, so there is no synthetic aperture. The area illuminated by the antenna is limited by the range resolution ρ_g and by the real antenna aperture in azimuth $\frac{\lambda}{L_a} R$. Considering a homogeneous reflective scenario (e.g., bare soil or short vegetation) in front of the radar, with reflectivity equal to σ_T^0, the radar cross-section (RCS) of each resolution cell is computed as:

$$\sigma_c = \frac{\rho_g \lambda R}{L_a} \sigma_T^0 \tag{2}$$

Considering the bandwidth range equal to 50 MHz and one tile of the antenna with a length equal to 1.42 m, the resolution cell at the SAOCOM central frequency is as large as 1600 m². Assuming a reflectivity of -15 dB, the obtained RCS is 17 dBsm. (We indicate with dBsm the ratio of the area with respect to 1 m², expressed in logarithmic scale).

This large backscattered power can be considered either as a signal—if stable—or as a noise—if unstable. Water and leaves are highly unstable; however, their contribution is effectively reduced if averaged for seconds. Slower-moving targets could provide long-term noise contributions that affect the measure. We then decided that the observed scenario might not be stable enough for an accurate measurement, so we inserted a trihedral corner reflector in the scene.

Figure 1 shows the schematic representation of the overall concept of the SAOCOM outdoor test (ODT): the SAOCOM SAR antenna points towards a known point target (trihedral corner reflector), moving along the line of sight, with a linear motion, thanks to a linear actuator. The acquired data are downloaded and processed to extract signal characteristics and to assess the phase and gain stability of the instrument.

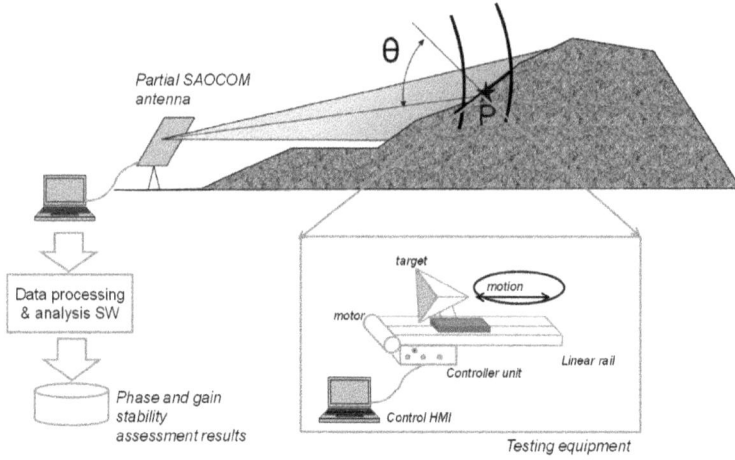

Figure 1. SAOCOM outdoor test (ODT) concept overall view. In the figure, P represents the moving target to be acquired and ϑ the inclination of the Radar Line of Sight with respect to the local normal to the terrain.

The reason for target motion is to effectively limit the size of the corner reflector to reasonable values. The target, moving with controlled and known motion, can be separated from the background and other moving targets provided that the motion is significant with respect to half the wavelength.

The RCS of the trihedral corner reflector with length d can be modeled as:

$$\sigma_x^2 = \frac{4}{3\pi} \frac{d^4}{\lambda^2} \tag{3}$$

In order to distinguish it from the background and to have sufficient accuracy in the phase and amplitude measurement, we ask for a signal to clutter ratio $\frac{\sigma_x^2}{\sigma_c^2}$ of at least 30 dB. With the computed RCS of the clutter as above, this would lead to a target with a size greater than 8 m.

Let us then assume that the corner reflector is mounted on a rail, and moves with time, say of an extent much less than a range resolution cell. The return from successive echoes measured at the target range can be modeled as a mono-dimensional time-variant signal, given by the superposition of the distributed target (or clutter), the signal from the trihedral target, and the noise w, due to thermal and quantization:

$$y(\tau) = \int_{\rho_a} a_c e^{j\phi_{APS}(\xi,\tau)} d\xi + a_s e^{j\phi_{APS}(\xi_0,\tau) + \phi_r(\tau)} + w(\tau) \tag{4}$$

where the first integral is extended to the unfocused azimuth aperture, the resolution ρ_a. In the equation above, τ represents the time and ξ the spatial coordinate. The phase noise term ϕ_{APS} accounts for the propagation within the troposphere and it is discussed later on in the text. The two terms a_c and a_s are the returns, respectively, from the distributed target and the trihedral. The second term is the signal related to the target. The phase term ϕ_r accounts for the motion of the target and represents an equivalent of the "phase history" of the target, to be compensated for ("focused").

In the case of linear motion along the line of sight with velocity v, the phase history corresponds to a linear phase ramp:

$$\phi_r(\tau) = -\frac{4\pi}{\lambda} R(\tau) = -\frac{4\pi}{\lambda} (R_0 + v \cdot \tau) \tag{5}$$

And hence the observed data sequence at the target range R_0 is a sampled sinusoid with frequency:

$$f = -\frac{2}{\lambda}v \tag{6}$$

The term f is the Doppler Frequency created by the target motion. Unlike a spaceborne SAR, in which every target in the scene has its own Doppler, here all the fixed targets appear at Doppler zero except for the one of interest.

Assuming a very good knowledge of the phase ϕ_r, the "focused" target response is retrieved by compensation of the phase on the signal $y(\tau)$, performing what in SAR focusing is called "backprojection":

$$y_F = \frac{1}{N}\sum_i y(\tau_i) \cdot e^{-j\phi_r} \tag{7}$$

In the following, we study the content of $y(\tau)$, making different hypotheses on the disturbing phase noise ϕ_{APS} and on the clutter statistics.

The impact of propagation (the phase noise term ϕ_{APS}), accounts for the additional delay introduced by the non-free space propagation within the troposphere. If the monochromatic approximation holds (a small bandwidth compared to the carrier frequency), the phase term ϕ_{APS} is linearly related to the delay, which in turn depends on the integral of the refraction index along the path.

$$\phi_{APS} = \frac{4\pi}{\lambda}d_{ATM} = \frac{4\pi}{\lambda}\int_{Lp} N(r(\vec{l}))dl \tag{8}$$

where N is the space- and time-variant refractivity index of the atmosphere, a function of the space r and the time of acquisition. The statistics of the atmosphere have been widely studied in the past [6]. Limiting the temporal scope to one day (far longer than one SAR acquisition), we can well consider the Kolmogorov turbulence statistical model of the delay, represented by the power law:

$$E[d(t+\Delta t) - d(t)] = \left(\frac{\Delta t}{\overline{\tau}}\right)^\alpha ad \tag{9}$$

where $\overline{\tau}$ is a temporal constant depending on the local atmospheric conditions (typical values are in the order of hours). The exponent α is also defined as slope, as the function is a line in a log-log space, assuming values experimentally determined to be close to $2/3$. The applicability of the model was verified with experimental ground-based radar measurements in Reference [6]. Numerical simulations were carried out to investigate the impact on the obtained signal-to-clutter ratio due to the atmospheric delay spatial and temporal variation.

We now analyze the clutter contribution (the first term in Equation (4)), assuming sufficient stability of the atmosphere in the time interval of one acquisition. This assumption was verified by analyzing the Doppler spectrum of the acquired clutter echoes (see Section 4). The result of the focusing operation in (7) is a stochastic process with variance depending on the clutter spatial and temporal covariance matrix. This will depend on the scenario (e.g., rocks or vegetation) and on its stability (e.g., due to wind).

To show the principle of the outdoor test, we analytically derive the case of clutter perfectly correlated in time (frozen scene). In this case, the clutter a_c is a constant along time, so its contribution to the focused signal y_F is given by:

$$\sigma_{c,F}^2 = \sigma_c^2 \text{sinc}^2\left(\frac{2}{\lambda}L\right) \tag{10}$$

where L is the total length of the target motion. The relation above shows that the contribution of the clutter to the focused moving target is a cardinal sinc, which depends only on the total motion extent *L*. With a motion extent multiple of $\frac{\lambda}{2}$, the clutter power is (ideally) completely cancelled, as the total signal-to-noise ratio is only thermal-noise limited. In this ideal case, the velocity of the target is completely irrelevant (provided that the radar Pulse repetition Frequency (PRF) is sufficient to sample the sinusoid at $f = \frac{2}{\lambda}v$, which, for typical PRFs in the order of KHz, is practically always true).

3. SAOCOM Outdoor Experiment Setup

The SAOCOM-1A outdoor test (ODT) took place in the Investigación Aplicada (INVAP) facility close to the Bariloche Airport, Patagonia, Argentina. The facility consisted of a radome hosting one tile of the SAOCOM antenna and the SAOCOM electronics, connected to a control and data-processing room.

The moving target was hosted in a radome placed at a distance of 3755 m in the plane in front of the facility.

Figure 2 shows the setup on Google Earth (a) and an aerial view of the area (b). The picture was taken standing on a small hill of approximately 50 m altitude above the large plane. In the foreground, there is the area hosting the shelter with the guard and the power generator, connected to the small radome hosting the moving target. The fixed target was placed on the right, approximately 20 m away from the radome. Far beyond, the facility hosting the SAOCOM with the large radome can be seen.

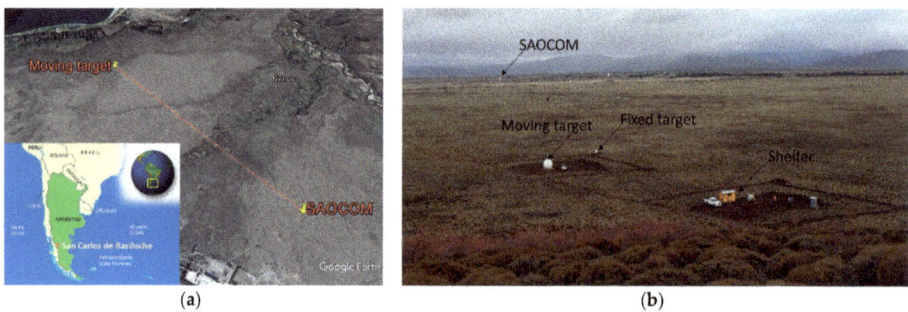

(a)   (b)

Figure 2. SAOCOM-1A outdoor test setup. (a) View on map; (b) aerial photo.

Three trihedral corner reflectors were available during the ODT campaign:

- 75 cm
- 100 cm (visible in Figure 3b)
- 150 cm

The first two corner reflectors could be mounted on the moving actuator, while the large one was placed to be kept immobile at approximately 20 m from the moving target radome.

The Advanced Remote Sensing Systems (ARESYS) corners are trihedral-shaped aluminum reflectors with modular faces that were assembled on-site.

Table 1 below reports the RCS in dB square meters (dBsm) of the corner reflectors at the SAOCOM center wavelength (0.2353 m).

Table 1. Radar Cross Section (RCS) of the corner reflectors.

75 cm Corner RCS	100 cm Corner RCS	150 cm Corner RCS
13.79 dBsm	18.79 dBsm	25.83 dBsm

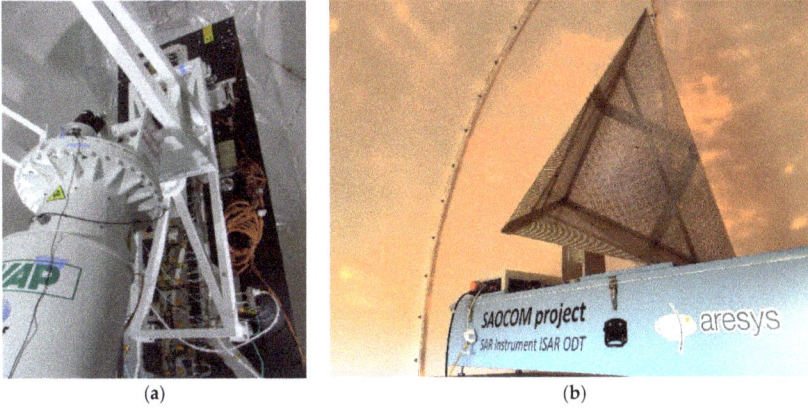

(a) (b)

Figure 3. (a) Radome hosting the SAOCOM antenna tile; (b) Radome hosting the moving target.

The accurate pointing of the corner reflectors mounted on the moving rail was ensured by the alignment of the rail itself with the line of sight, which was carried out by Comisión Nacional de Actividades Espaciales (CONAE) with the use of a differential Global Positioning System (GPS).

The pointing of the large corner reflector was instead performed with the use of a compass and an inclinometer (resulting thus in a coarser pointing). This method was employed because the large corner reflector was thought to have sufficient RCS so as to be left movable around the scene, in order to ease the execution of preliminary visibility tests.

The SAOCOM antenna was one single tile of the total antenna. One tile is composed of 20 elements in the elevation direction and one element in the azimuth direction, with a total size of 3.48 m in elevation and 1.424 m in azimuth. Each TRM was transmitting 50 W with an efficiency of 75%. The excitation coefficients were set according to a Taylor (amplitude only) tapering, in order to shape the side-lobes. The pointing in elevation and azimuth was possible thanks to the mechanical support equipment which could be steered with steps of approximately 1 degree and the knowledge of the pointing was of approximately 0.1 degree. The link budget, accounting for all the gains and losses from the transmitter, through the medium, and to the receiver, and the corresponding signal-to-noise-ratio (SNR) computation, are reported in Table 2 below. The noise figure N_F term is used to define the equivalent noise temperature of the receiver and to then compute the thermal noise power with the classical equation:

$$P_{Noise,Thermal} = K_B T_{eq} B = K_B N_F T_a B \qquad (11)$$

where K_B is the Boltzmann constant, B is the signal bandwidth, and T_a is the ambient temperature (290 K).

The acquisition parameters were set in order to ease the execution of the outdoor test by putting:

- a short chirp to allow close range
- the maximum bandwidth to reduce the resolution
- a reduced sampling window length to avoid unnecessarily large datasets

The main SAOCOM settings are described in Table 3 below.

Table 2. Link budget computation and Signal to Noise Ratio (SNR) computation (ideal propagation media).

Parameter	Value (Linear)	Value (Log-Scale)
Peak power	750.0 W	28.75 dBW
Antenna area transmit (TX)	4.87 m^2	6.88 dBsm
Instrument and antenna TX losses	1.17	0.70 dB
TX path loss (R = 3755 m)	5.64×10^{-9}	−82.48 dB
RCS corner (1-m corner case)	75.66 m^2	18.79 dBsm
Receive (RX) path loss	5.64×10^{-9}	−82.48 dB
Instrument and antenna RX losses	1.17	0.70 dB
Received power	6.4×10^{-12} W	−111.94 dBW
Noise figure	2.14	3.3 dB
Noise power at receiver	4.28×10^{-13} W	−123.69 dBW
SNR raw	14.96	11.75 dB
Number of focused steps	1000	30 dB
SNR focused	14,962	41.75 dB

Table 3. SAOCOM 1A acquisition parameters for outdoor test.

Parameter	Value
Acquisition mode	Stripmap/TOPSAR
Center frequency	1,274,140,000 Hz
Bandwidth	50 MHz
Sampling window start time	23 μs
Chirp duration	11 μs
Sampling window length	20 μs
Acquisition duration	variable 1 min–10 min
Polarization	Quad Pol
Pulse Repetition Frequency (PRF)	4545 Hz
Chirp	Down

In addition to the moving target, a sampling equipment was placed to measure the transmitted chirp and to check the PRF.

4. SAOCOM Outdoor Experiment Results

The raw data, provided by a dedicated implementation of the CONAE User Segment Service (CUSS), called mini-CUSS, were non-Block-Adaptive-Quantizer (BAQ) compressed and modulated (at 30 MHz) data. The processing software had then to perform the following steps:

- digital down conversion
- range compression
- azimuth compression

The digital down conversion step performs the demodulation of the signal to baseband. In the employed software, it also allowed the sub-sampling of the input dataset in order to speed up the processing for a fast analysis of the input data. In fact, the PRF of the SAOCOM is quite high compared to the Doppler content of the illuminated scene. The real input samples are also converted into complex samples during this step.

The range compression performs the matched filter either with an ideal chirp (synthetically generated by the routine itself, or with the chirp replica coming from internal calibration.

It is noted that the chirp replica is close to the ideal chirp, except for an amplitude tapering in the upper part of the chirp (Figure 4a). The effect of the imperfect flatness in frequency is seen in the impulse-response-function (IRF) side-lobes level, which are lower than the ideal level of the perfectly rectangular spectrum of the ideal chirp (Figure 4b). The effect on resolution is of about 4%

resolution loss (see Table 4), which is almost completely recovered when the replica is focused with the ideal chirp.

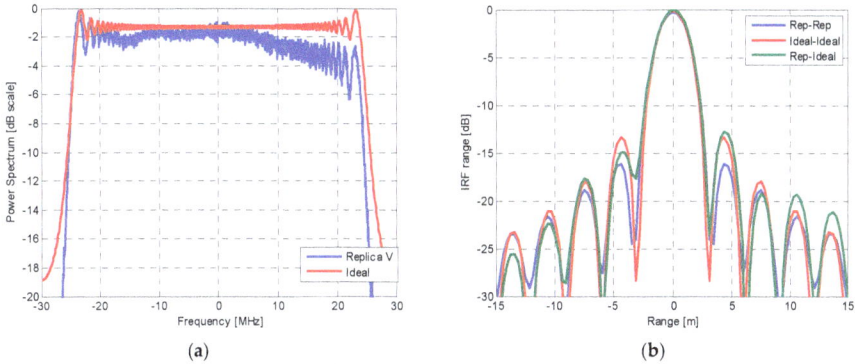

(a)   (b)

Figure 4. (a) Comparison between the ideal chirp and the chirp extracted from internal calibration; (b) Range-compressed Impulse Response Function (IRF) comparison.

Table 4. Range IRF analysis results.

IRF Parameter	Replica-Replica Case	Ideal-Ideal Case	Replica-Ideal Case
Resolution	2.77 m	2.66 m	2.68 m
Peak-to-Side-Lobe ratio	−16.2 dB	−13.3 dB	−14.98 dB

The range power profile of one acquisition is shown in the following figure, for the VV polarization.

The high backscatter from the small hill and three main peaks can be recognized, corresponding to the fixed corner reflector (closest), a small heap of dirt (mid), and the moving target inside the radome (at 3755 m).

The signal to clutter ratio of the latter is approximately 13 dB. Considering the 100-cm corner RCS as in Table 1, we can estimate a clutter RCS in the resolution cell in the order of 5 dBsm. This value is 12 dB lower than the assumption made during the design phase (clutter RCS of 17 dBsm) and reported in Section 2. We can then refine the clutter signal level to −27 dB. The analysis in the Doppler domain (Figure 5b) of the clutter shows a very low impact of the wind, with an estimated stable to time-variant components ratio (also called DC/AC ratio) of 40 dB.

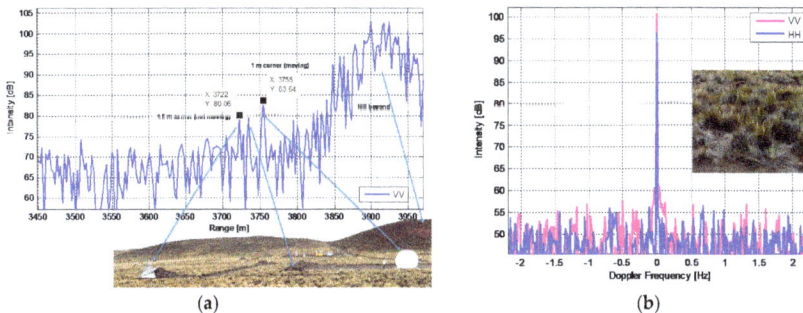

(a)   (b)

Figure 5. (a) Range-compressed data intensity profile and interpretation; (b) Doppler analysis of the clutter.

The effect of the target motion is clearly seen in the range-Doppler map, showing the Doppler spectrum for each range. The two "targets" are the focused responses of the moving target. The bright stripe at the higher range is the backscatter of moving people inside the control shelter.

The azimuth impulse response function of the system can be extracted as a horizontal cut of the range-Doppler map and is shown in Figure 6. The importance of the result below is to have the possibility to analyze in advance the impulse response of the Synthetic Aperture Radar, normally obtained only once the instrument is flying and the data are properly processed on ground.

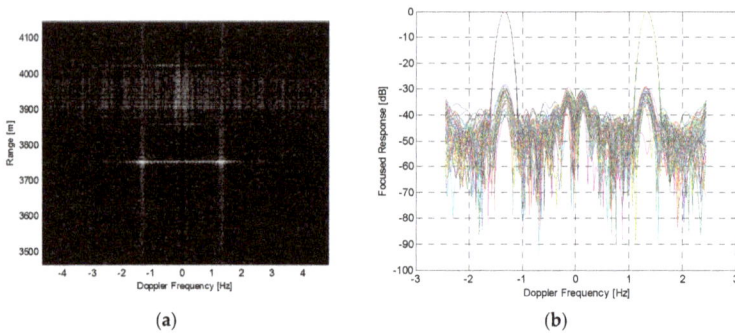

(a)   (b)

Figure 6. (a) Range-Doppler map with the focused moving target clearly visible; (b) Azimuth IRF.

Concentrating now on the moving target range bin, we can assess the end-to-end overall amplitude and phase stability, which will be a combination of the instrument and of the propagation medium stability.

Figure 7 shows, on the left, the reconstructed phase of the signal corresponding to the moving target bin. The expected sawtooth trend can be seen, corresponding to the movement of the corner reflector back and forth during the acquisition. The image in the center shows the corresponding time-variant concentration of the signal energy in the azimuth frequency domain, moving from positive to negative frequencies depending on the direction of motion. The location of the peaks in frequency allows one to estimate with high precision the actual motion velocity and to synthesize an ideal linear motion and the corresponding phase trend. The rightmost plot shows instead the residual phase after linear motion compensation.

By collecting the phase of all the peaks from the acquired 10 min of long data, the amplitude and phase stability results reported in Table 5 were obtained.

Table 5. Amplitude and phase stability results over 10 min.

Parameter	Value (VV)	Value (HH)
Amplitude stability	better than 0.1 dB	better than 0.1 dB
Phase stability	3.9 deg	3.4 deg
Amplitude drift	<0.1 dB/min	<0.1 dB/min
Phase drift	0.19 deg/min	0.35 deg/min

The correctness of the antenna excitation setting and the validation against the theoretical antenna pattern calculation was carried out by repeating the data acquisition with different antenna pointing in elevation thanks to the mechanical steering of the antenna.

Figure 8 below shows an example of the obtained results, where each dot on the plot represents the average power at the moving target range, estimated on one data acquisition, either VV or HH. The dots are superimposed over the theoretical antenna pattern shape, corresponding to the applied tapering on the antenna (Taylor tapering). The good agreement of the measures with the expected pattern can be seen up to the second side-lobe. The agreement is better for positive angles than for negative angles. This can be explained considering the known interaction of the antenna with the ground, introducing a ripple on the whole pattern.

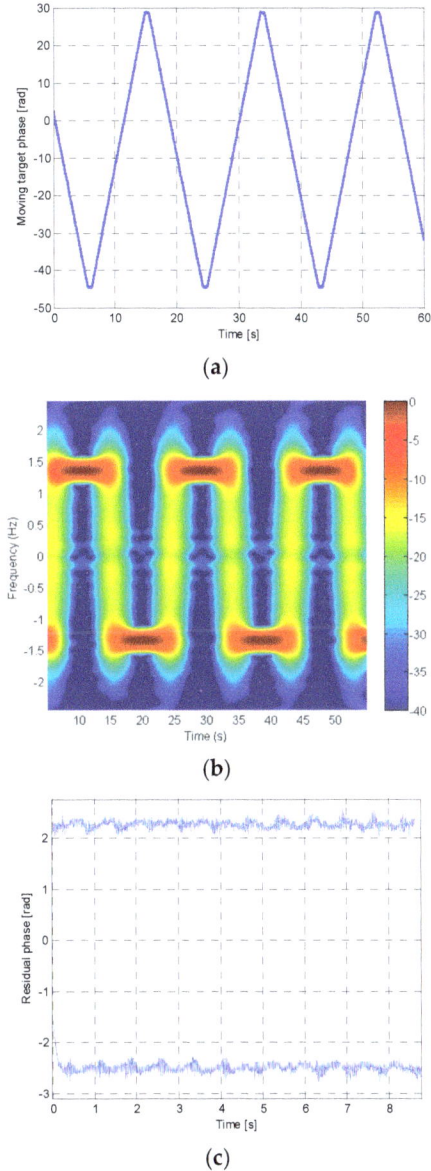

(a)

(b)

(c)

Figure 7. Analysis of the moving target signal. (a) Phase versus time; (b) Doppler frequency versus time (colorscale in dB, normalized to the maximum); (c) Residual phase after linear motion compensation.

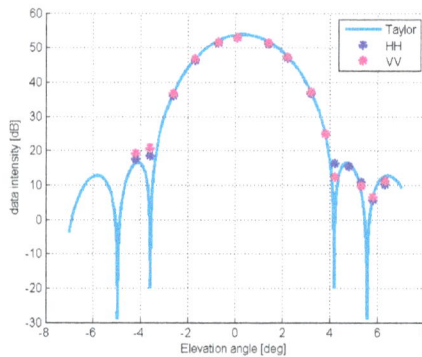

Figure 8. Far-field elevation antenna pattern validation against the theoretical model.

5. Discussion

The execution of the ODT and the processing of the collected data provides an important set of results that give the first significant insights on the overall system performance. First of all, the transmitted signal, after its replica were provided by internal calibration and their similarity to an ideal chirp were checked, showed an almost ideal IRF in range. The compensation of the known phase function, corresponding to the target motion, allowed us to check the phase stability of the system over a long time interval spanning up to 10 min. The analysis over datasets acquired with different azimuth and elevation pointing of the SAOCOM antenna tile allowed the successful validation of the antenna pattern pointing and shape over a large interval including the first and second side-lobes. The main limitations remain the single antenna tile used, thus allowing us to test only the beamforming in the elevation direction, and the target used, which responded only with the co-pol channels. Nevertheless, the collected results allow us to state that the key functionalities of the SAR system are verified, even with the limitations that an on-ground setup unavoidably brings.

6. Conclusions

The SAOCOM-1A outdoor test took place in Bariloche, Patagonia, Argentina, during June 2016.

The main objective of the outdoor test (ODT) was to provide an end-to-end validation of the SAOCOM-SAR functionality, in a realistic condition, where the SAR pulses are radiated by the antenna, reflected by a target, and then received by the antenna and recorded as SAR data. The test setup was made up of two main elements: the SAOCOM SAR engineering model, including one of the seven antenna tiles, and the moving target. It was proven to work as expected throughout the full duration of the tests. Several datasets were successfully acquired through the setup and processed to L0B data.

Overall, the ODT objectives were met and the SAOCOM-1A proved to show an excellent signal quality, both from radiometric and interferometric points of view. The ODT provided the unique occasion to obtain a set of pre-flight far-field measurements. The results can be considered representative for SAOCOM-1A and for SAOCOM-1B as well, so no additional experiments are foreseen. The conclusive and formal verification of the SAOCOM performance is left to the laboratory pre-flight tests and to the commissioning phase.

Acknowledgments: The authors would like to thank CONAE and INVAP for the fruitful cooperation before, during, and after the execution of the outdoor test campaign.

Author Contributions: Andrea Monti Guarnieri originally conceived the experiment. Davide Giudici carried out the feasibility study, coordinated the moving target design and development and processed the acquired SAR data. Juan Pablo Cuesta Gonzalez provided a thorough review of the feasibility study and coordinated the actual setup and execution of the experiment.

Conflicts of Interest: The authors declare no conflict of interest.

References

1. SAOCOM Mission Website. Available online: http://www.conae.gov.ar/index.php/english/satellite-missions/saocom/introduction (accessed on 13 July 2017).
2. SAOCOM Mission Applications. Available online: http://www.conae.gov.ar/index.php/english/satellite-missions/saocom/technical-characteristics (accessed on 13 July 2017).
3. Hees, A.; Koch, P.; Rostan, F.; Huchler, M.; Croci, R.; Østergaard, A. Sentinel-1 Antenna Model Validation: Pattern Prediction vs. PNFS Measurements. In Proceedings of the 2013 IEEE International Symposium on Phased Array Systems and Technology, Waltham, MA, USA, 5–18 October 2013.
4. Miranda, N.; Meadows, P.; Hajduch, G.; Pilgrim, A.; Piantanida, R.; Giudici, D.; Small, D.; Schubert, A.; Husson, R.; Vincent, P.; et al. The Sentinel-1A Instrument and Operational Product Performance Status. In Proceedings of the 2015 IEEE International Geoscience and Remote Sensing Symposium (IGARSS), Milan, Italy, 26–31 July 2015.
5. Schwerdt, M.; Schmidt, K.; Ramon, N.T.; Alfonzo, G.C.; Döring, B.J.; Zink, M.; Prats-Iraola, P. Independent Verification of the Sentinel-1A System Calibration. *IEEE J. Sel. Top. Appl. Earth Obs. Remote Sens.* **2016**, *9*, 994–1007. [CrossRef]
6. Iannini, L.; Guarnieri, A.M. Atmospheric phase screen in ground-based radar: Statistics and compensation. *IEEE Geosci. Remote Sens. Lett.* **2011**, *8*, 537–541. [CrossRef]

remote sensing

MDPI

Article

On the Design of Radar Corner Reflectors for Deformation Monitoring in Multi-Frequency InSAR

Matthew C. Garthwaite

Geodesy and Seismic Monitoring Branch, Geoscience Australia, GPO Box 378, Canberra ACT 2601, Australia;
Matt.Garthwaite@ga.gov.au

Academic Editors: Timo Balz, Uwe Soergel, Mattia Crespi, Batuhan Osmanoglu and Prasad S. Thenkabail
Received: 19 April 2017; Accepted: 21 June 2017; Published: 25 June 2017

Abstract: Trihedral corner reflectors are being increasingly used as point targets in deformation monitoring studies using interferometric synthetic aperture radar (InSAR) techniques. The frequency and size dependence of the corner reflector Radar Cross Section (RCS) means that no single design can perform equally in all the possible imaging modes and radar frequencies available on the currently orbiting Synthetic Aperture Radar (SAR) satellites. Therefore, either a corner reflector design tailored to a specific data type or a compromise design for multiple data types is required. In this paper, I outline the practical and theoretical considerations that need to be made when designing appropriate radar targets, with a focus on supporting multi-frequency SAR data. These considerations are tested by performing field experiments on targets of different size using SAR images from TerraSAR-X, COSMO-SkyMed and RADARSAT-2. Phase noise behaviour in SAR images can be estimated by measuring the Signal-to-Clutter ratio (SCR) in individual SAR images. The measured SCR of a point target is dependent on its RCS performance and the influence of clutter near to the deployed target. The SCR is used as a metric to estimate the expected InSAR displacement error incurred by the design of each target and to validate these observations against theoretical expectations. I find that triangular trihedral corner reflectors as small as 1 m in dimension can achieve a displacement error magnitude of a tenth of a millimetre or less in medium-resolution X-band data. Much larger corner reflectors (2.5 m or greater) are required to achieve the same displacement error magnitude in medium-resolution C-band data. Compromise designs should aim to satisfy the requirements of the lowest SAR frequency to be used, providing that these targets will not saturate the sensor of the highest frequency to be used. Finally, accurate boresight alignment of the corner reflector can be critical to the overall target performance. Alignment accuracies better than $4°$ in azimuth and elevation will incur a minimal impact on the displacement error in X and C-band data.

Keywords: InSAR; persistent scatterers; geodesy; corner reflector; point target; interferometry; Synthetic Aperture Radar; calibration and validation

1. Introduction

The Persistent Scatterer Interferometric Synthetic Aperture Radar (PSInSAR) technique [1–5] has become a popular remote sensing method for monitoring ground or infrastructure displacements induced by wide-ranging natural and anthropogenic phenomena. The technique uses a stack of SAR interferograms and determines the motion history for pixels that are identified to have temporal phase stability (i.e., pixels whose overall response is dominated by a strong back-scatterer). The distribution of these "persistent scatterers" (PS) can be quite dense in urban areas (e.g., several hundred PS/km^2), where there are many man-made angular structures and corners to reflect incident radar energy back to the observing SAR sensor. However, in non-urban areas the distribution of PS may be much sparser, or even non-existent. Consequently, artificial targets are increasingly being deployed in the field to introduce coherent point targets in regions where naturally occurring PS are sparse or non-existent (e.g., [6–14]).

The exact position of naturally occurring PS is generally not known and it is therefore useful to have targets with known position distributed throughout the area of interest that can be used to validate PSInSAR with other geodetic observations. Indeed, there is a growing trend for artificial targets to be considered for permanent deployment in national geodetic reference networks to complement and enable inter-comparison of PSInSAR observations with ground-based geodetic measurements (e.g., from the GNSS, levelling, VLBI, and SLR techniques). Previous validation experiments using artificial targets have found that the accuracy of displacement estimates from PSInSAR analysis is at the millimetre level [15,16]. Recent advances have also seen the development of algorithms for absolute positioning of SAR scatterers using stereo SAR images of targets acquired using multiple imaging geometries [17,18]. Common to all these applications is the need for an artificial target with a geodetically known position that has been designed to have a bright and stable response in SAR imagery.

Despite the increasing and widespread use of artificial targets, the comparison of different target designs and the implications for performance has not been undertaken before in the context of geodetic or geophysical monitoring. The aim of this study is to determine:

- what size of target is suitable for use with the commonly employed SAR frequencies,
- to what extent one size of target can be effectively used across all SAR frequencies, and
- what considerations should be made with respect to manufacturing and long-term or permanent installation of artificial targets.

To answer these questions, I first outline the theoretical considerations around the brightness of targets required to satisfy a certain geodetic tolerance on displacement error. I then discuss aspects of the design of artificial targets and the physical size requirements to meet the required brightness, including some general practical recommendations on target design and installation. Finally, I describe field experiments undertaken to determine the radar response from different target sizes and validate those against the theoretical considerations.

2. Theoretical Considerations

In this section I outline the technical issues that should be considered when designing suitable targets for deformation monitoring with different frequencies and resolutions of SAR data. Firstly, I review the relevant theory around making amplitude and phase measurements from SAR data. I then show how the pixel brightness for a particular SAR frequency and imaging resolution can be derived and used to identify a performance criterion. Following this I discuss how the performance criterion can be distilled into the design of a suitable target.

2.1. Amplitude Measurements

The Radar Cross Section (RCS) of an imaged target is a measure of the size of that target as seen by the imaging radar. The conventional measure of brightness of a distributed target within a SAR image, the backscattering coefficient "Sigma Nought", is equivalent to the RCS (in dBm^2) normalised by the area A of the illuminated resolution cell [19]:

$$\sigma^0 = \frac{<\sigma_n>}{A},$$

(1)

where σ_n is the nth RCS value and angle brackets indicate an ensemble average. The illuminated area projected on the ground is:

$$A = \frac{p_r p_a}{\sin \theta},$$

(2)

where θ is the local incidence angle with respect to the normal to the scattering surface and p_a and p_r are the azimuth and slant range pixel resolutions, respectively. Using these relations, the approximate

RCS of any point target in a SAR image can be estimated. To be of use as a stable phase target, the point target must be visible in the SAR image above the background backscattering level (referred to as the "clutter"). The typically used measure of target visibility in a SAR image is the Signal-to-Clutter Ratio (SCR) [19]:

$$SCR = \frac{\sigma_T}{< \sigma_C >} = \frac{\sigma_T}{< \sigma^0 > A},$$ (3)

where σ_T is the point target RCS and $< \sigma_C >$ is the ensemble average of clutter RCS near to the point target. It is generally understood that to be of use in radiometric calibration, the SCR of an artificial target should be at least 30 dB whilst not being so bright that it saturates the receiving antenna [19,20].

The magnitude of clutter depends on many factors including: terrain type, vegetation density, soil moisture, radar wavelength, incidence angle, polarisation and SAR image resolution. It is therefore important to consider carefully the clutter characteristics near to potential target deployment sites prior to installation. This can be done by a priori analysis of SAR imagery over the area of interest to identify regions with relatively low radar backscatter. Generally, flat cultivated terrain with low vegetation density is ideal. Typical backscatter levels for this type of land cover when considering a range of radar incidence angles is likely to be within the range of −10 dB to −14 dB at C-band [21]. Backscatter levels at C- and X-bands should be broadly similar because the small difference in frequency means that attenuation rates in vegetation will be similar. L-band backscatter measurements of tussocky grassland from an airborne SAR at VV polarisation vary between about −15 dB to −20 dB, with bare soil at the lower end of this range [22].

2.2. Phase Measurements

The complex radar observation at each pixel is the coherent sum of the response from many distributed scatterers located within that pixel. Deformation studies using differential InSAR techniques exploit the phase component of the complex radar signal. The phase for pixels containing distributed scatterers (i.e., those that are un-correlated and where no single scatterer dominates) is unlikely to remain correlated for long periods of time. The PSInSAR technique only exploits those pixels within which there is a dominant scatterer, the so called "PS", exhibiting long-term stable phase characteristics. The phase component from the PS depends on the range (distance) from the target to the SAR sensor whereas the phase due to the other distributed scatterers within the pixel is essentially random (Figure 1).

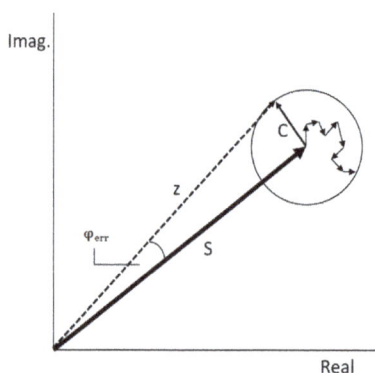

Figure 1. The backscattered signal from a pixel is a complex sum of each scatterer within the pixel, represented here by the vector z. A PS within the pixel is represented by the vector $S = \sigma_T$ and the complex sum of the background clutter by the vector $C = < \sigma_C >$. The angle φ_{err} subtended by z and S is the phase error due to the superposition of the clutter. Redrawn from [23,24].

A simulation of the phase response for a single SAR image pixel comprising many uncorrelated distributed scatterers under the assumption of a complex circular Gaussian statistical model is presented in Figure 2. Under these conditions the amplitude and phase probabilities are approximated by Rayleigh and uniform distributions, respectively [6]. The simulation is repeated twice with the presence of a single PS replacing a distributed scatterer within the pixel. The PS has an amplitude response of about four and eight times the background amplitude in the two further simulations (i.e., an SCR of about 16 and 64, respectively) and a defined phase. What can be seen is that with no PS the phase is uniformly distributed between ±π. The effect of adding the PS is to reduce the level of phase variability in proportion with the amplitude of the PS. As can be seen in Figure 2e, the normal distribution is only a crude approximation to the phase statistics. An important observation is that the phase standard deviation (derived from the best-fitting normal distribution to the phase observations) decreases as the SCR increases (Figure 2f). Below an SCR of about 10 (i.e., ~10 dB), the phase can no longer be adequately approximated by a normal distribution [23]. This is demonstrated in Figure 2f, where below an SCR of 10 the phase standard deviation stagnates, indicating that a uniform distribution of phase values prevails.

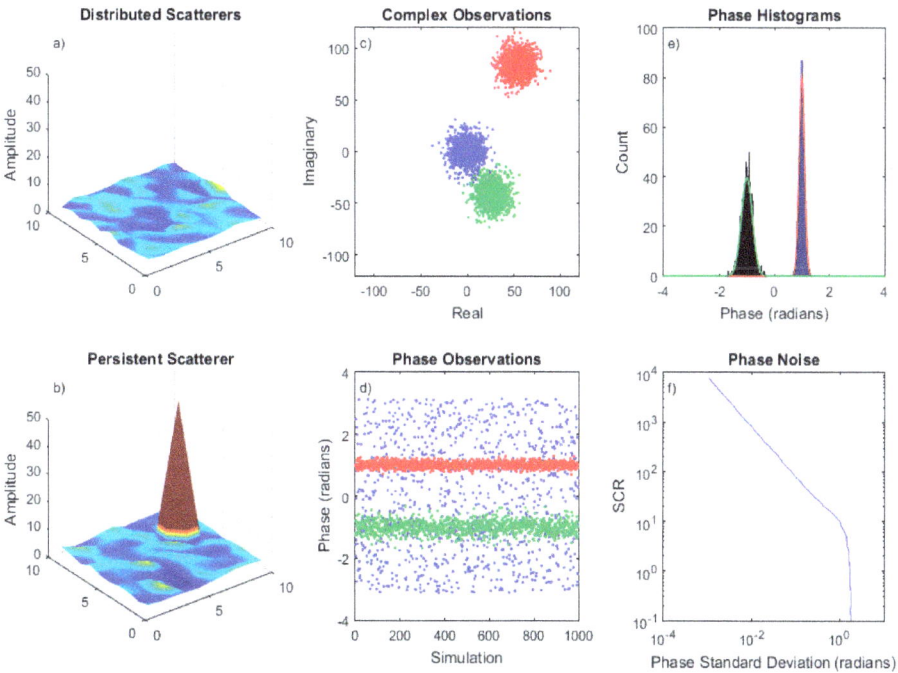

Figure 2. Simulation results of phase response within a pixel. (**a**) Visualisation of a pixel containing 100 distributed scatterers of similar amplitude; (**b**) visualisation of a pixel as before but containing one dominant PS; (**c**) complex observations for 1000 simulations for the situation with no PS (blue), a PS with an SCR of 16 and a phase of −1 radian (green) and a PS with an SCR of 64 and a phase of one radian (red). (**d**) Phase observations for the simulations in (c); (**e**) histograms of the phase for the two PS simulations with the best fitting normal distribution for each; (**f**) the relationship between the SCR and the standard deviation of the phase observations estimated by fitting a normal distribution to the results of 1000 PS simulations.

The effective phase error in radians has been estimated as [7,23,24]:

$$\varphi_{err} = \frac{1}{\sqrt{2 \cdot SCR}}.$$ (4)

Using this simple analytical expression, the estimated phase error can be directly derived from the measured point target SCR in any single SAR image. Furthermore, the SCR and phase error is dependent only on the image resolution and the radar frequency. It is therefore useful as an a priori metric during network design when a large number of SAR images may not yet be available for a particular area of interest and therefore direct interferometric measurements are not feasible. The estimated phase error can be converted to a displacement error in the line of sight (LOS) using the radar wavelength λ:

$$d_{err} = \frac{\varphi_{err} \cdot \lambda}{4\pi}.$$ (5)

The relationship between the SCR and LOS displacement error is plotted in Figure 3 for the radar bands generally used on current SAR missions.

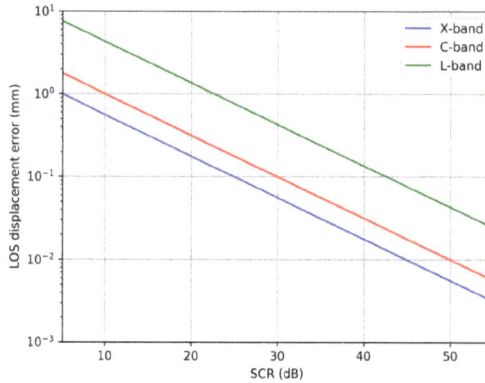

Figure 3. LOS displacement error as a function of SCR for the radar frequencies of interest.

2.3. Target RCS Requirements

By considering the expected background pixel RCS for the SAR imaging mode to be used, an RCS requirement for deployed targets can be derived based on an acceptable level of displacement error (due to random phase error alone). The RCS of a pixel is equal to the product of the illuminated ground range resolution area and the clutter intensity. The required RCS for the target is then found by multiplying the derived pixel RCS with the SCR (in the linear domain; cf. Equation (3)). As an example, I derive approximate values for the RCS of artificial targets that would achieve a nominal LOS displacement error not greater than a tenth of a millimetre in different imaging modes of the currently orbiting SAR sensors (Table 1). A 0.1 mm magnitude of error would require an SCR exceeding 25 dB, 30 dB and 43 dB at X, C, and L-bands, respectively (Figure 3).

For the purpose of classification in this paper, "high-resolution" image modes are those with a ground range resolution area less than 5 m^2, "medium-resolution" are those between 5 m^2 and 100 m^2, and "low-resolution" are those above 100 m^2. Based on the pixel RCS values calculated in Table 1 and the identified SCR requirements, the artificial target must have an RCS up to 31 dBm^2 for medium-resolution X-band SAR imagery, 37 dBm^2 for medium-resolution C-band SAR imagery, 41 dBm^2 for low-resolution C-band SAR imagery (excluding Sentinel-1's Extra Wide Swath mode), and 47 dBm^2 for medium-resolution L-band SAR imagery.

Table 1. Estimates of target RCS and the equivalent triangular trihedral CR required to achieve a LOS displacement error of 0.1 mm for different SAR sensors and imaging modes. Ground range resolution areas are calculated for a 35° local incidence angle using pixel resolution information available from the online sources (retrieved 3 April 2017) listed below the table. Pixel resolutions are for single polarisation SLC products. All calculated values are rounded to one decimal place for clarity.

Band	Sensor	Image Mode	Pixel Resolution (m)			Ground Range Resolution Area (m²)	Clutter (dB)	Pixel RCS (dBm²)	Required SCR (dB)	Required RCS (dBm²)	Equivalent Triangular Trihedral CR Size (m)
			Azimuth	Slant Range	Ground Range						
X (9.65 GHz)	TerraSAR-X [1]	Staring Spotlight	0.24	0.6	1.0	0.3	−10	−16.0	25	9.0	0.2
		High Res Spotlight	1.1	1.2	2.1	2.3	−10	−6.4	25	18.6	0.4
		Stripmap	3.3	1.2	2.1	6.9	−10	−1.6	25	23.4	0.5
		ScanSAR (4 beam)	18.5	1.2	2.1	38.7	−10	5.9	25	30.9	0.7
	COSMO-SkyMed [2]	Spotlight	1.0	1.2	2.0	2.0	−10	−7.0	25	18.0	0.3
		HIMAGE (Stripmap)	3.0	3.5	6.0	18.1	−10	2.6	25	27.6	0.6
		Wid-region (ScanSAR)	16.0	8.1	14.1	225.5	−10	13.5	25	38.5	1.1
C (5.41 GHz)	Sentinel-1 [3]	Stripmap	5.0	5.0	8.7	43.6	−12	4.4	30	34.4	1.2
		Interferometric Wide Swath	20.0	5.0	8.7	174.3	−12	10.4	30	40.4	1.7
		Extra Wide Swath	40.0	20.0	34.9	1394.8	−12	19.4	30	49.4	2.8
	RADARSAT-2 [4]	Spotlight	0.8	1.6	2.8	2.2	−12	−8.5	30	21.5	0.6
		Ultra-Fine	2.8	1.6	2.8	7.8	−12	−3.1	30	26.9	0.8
		Multi-Look Fine	4.6	3.1	5.4	24.9	−12	2.0	30	32.0	1.0
		Fine	7.7	5.2	9.1	69.8	−12	6.4	30	36.4	1.3
		Standard	7.7	9.0	15.7	120.8	−12	8.8	30	38.8	1.5
		Wide	7.7	13.5	23.5	181.2	−12	10.6	30	40.6	1.7
L (1.27 GHz)	ALOS-2 [5]	Spotlight	1.0	3.0	5.2	5.2	−15	−7.8	43	35.2	2.6
		Stripmap Ultra-Fine	3.0	3.0	5.2	15.7	−15	−3.0	43	40.0	3.4
		High-sensitive	4.3	6.0	10.5	45.0	−15	1.5	43	44.5	4.4
		Stripmap Fine	5.3	9.1	15.9	84.1	−15	4.2	43	47.2	5.2
		ScanSAR (28 MHz)	77.7	47.5	82.8	6434.6	−15	23.1	43	66.1	15.3

[1] http://www.intelligence-airbusds.com/files/pmedia/public/r459_9_20140818_tsxx-itd-ma-0009_tsx-productguide_i2.00.pdf; [2] http://www.e-geos.it/products/pdf/csk_product%20handbook.pdf; [3] https://sentinel.esa.int/documents/247904/685163/Sentinel-1_User_Handbook; [4] http://mdacorporation.com/docs/default-source/technical-documents/geospatial-services/52-1238_rs2_product_description.pdf?sfvrsn=10; [5] http://en.alos-pasco.com/alos-2/palsar-2/.

2.4. Target Design

A radar reflector is a passive device that reflects incoming electromagnetic energy directly back to the source of that energy. There are many different types of reflectors, including: spheres, cylinders, dihedrals, trihedrals, flat plates, top hats and bruderhedrals. A trihedral radar reflector (often known as a "corner reflector") facilitates a triple-bounce of the incident radar energy from three mutually orthogonal plates [25]. The RCS pattern of a trihedral has a 3 dB beam-width of approximately 40° [20,26]. This means that a trihedral design is much more forgiving of field alignment errors when compared to other reflector designs. Consequently, trihedral corner reflectors have been used for many years as targets suitable for calibration of SAR images (e.g., [19,27–29]) and they are also the most common target type being deployed for use in PSInSAR analysis of ground deformation.

The shape of the reflecting plates impacts on the RCS magnitude of a trihedral corner reflector. The most commonly used plate shape is the triangle, but square and quarter-circle shaped plates have also been used [10,30]. Of these shapes, the triangle has the lowest RCS for a given size, but has the advantage of being structurally rigid and easy to manufacture. Sarabandi and Tsen-Chieh [31] described 'optimum' corner reflectors with pentagonal-shaped plates, created by trimming the ineffective part of a triangular plate that does not contribute to the RCS pattern. However, this trimming complicates the manufacture process and reduces the overall rigidity of the corner reflector. For these reasons, the focus of this paper is on the triangular trihedral corner reflector, hereafter abbreviated TCR. The theoretical peak RCS value (σ_T) of a TCR (expressed in m^2) is given by [32]:

$$\sigma_T = \frac{4\pi a^4}{3\lambda^2},$$

(6)

where λ is the radar wavelength and a is the length of the non-hypotenuse sides of the right-angled isosceles triangular plate (the inner-leg dimension). Using this relation, the equivalent size of TCR required to meet the RCS requirement of the design tolerance (0.1 mm LOS displacement error) is given in Table 1. Due to the frequency dependence of the RCS response, the design requirement results in non-overlapping size specifications for TCR (Figure 4). At X-band, TCR with an inner-leg dimension of 0.7 m should theoretically be able to achieve the displacement error tolerance for all medium-resolution image modes. At C-band, medium and low-resolution modes require a TCR dimension of between 0.8 m and 1.7 m. At L-band medium-resolution modes require a dimension of 5 m or greater.

It is therefore necessary to make a compromise if more than one SAR frequency is to be exploited using the same target. The compromise should be made at the higher frequency, since it is better for a target to be too bright but still visible rather than too dark and not visible. Due care and consideration should be taken to ensure that the target design will not saturate the signal, particularly at higher frequencies. As an indication of 'safe' target sizes, the German Aerospace Center (DLR) report usage of 3.0 m TCR for calibrating the TerraSAR-X sensor without saturation [28]. Furthermore, DLR have designed a C-band transponder with an RCS of 60.8 dBm^2 for calibration of Sentinel-1 and tested using RADARSAT-2 [33]. This is equivalent to a TCR with an inner-leg dimension of approximately 5.5 m. If a 5.5 m TCR were used at L-band (for example, with ALOS-2 Stripmap Fine data) a 0.1 mm LOS displacement error could be achieved. It is therefore possible to obtain a sub-millimetre LOS displacement error at L-band without saturating the signal at X- or C-band. However, the large size of TCR required to achieve sub-millimetre LOS displacement error at L-band could be impractical for widespread use in geodetic networks.

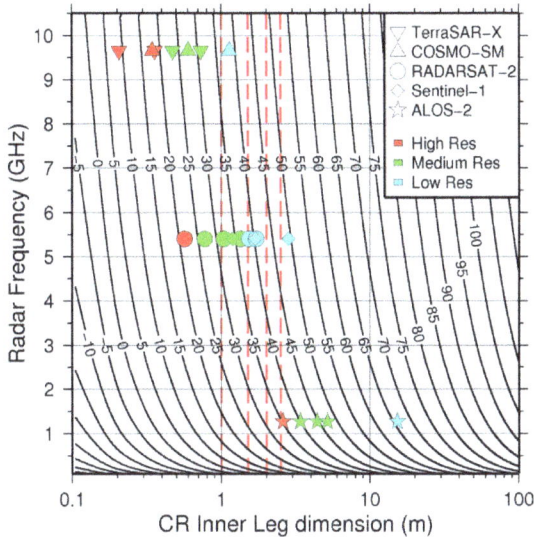

Figure 4. Contour plot of peak theoretical RCS (in units of dBm^2) for TCR at microwave frequencies between L and X-band. Plotted symbols represent the required RCS estimates for the SAR sensors and imaging modes given in Table 1 based on a design tolerance of 0.1 mm LOS displacement error. Red dashed lines indicate the size of TCR manufactured and tested in this study.

3. Manufacturing and Design Considerations

3.1. Losses Due to Manufacturing

There are several factors that can introduce a loss of RCS for a TCR compared to the theoretical values given by Equation (6), including inter-plate orthogonality, plate curvature and surface irregularities. To achieve an RCS accuracy of better than 1 dB with respect to theoretical values, DLR specify the following tolerances on their TCR manufacture process: Inter-plate orthogonality $\leq 0.2°$; Plate curvature ≤ 0.75 mm; Surface irregularities ≤ 0.5 mm [28]. These tolerances apply to X-band, with less stringent tolerances applying to lower frequencies.

The inter-plate orthogonality is the extent to which the three plates form 90° angles at their intersections. Robertson [34] conducted a series of physical experiments to measure the RCS profile of trihedral reflectors when the inter-plate angles are varied from 90°. When only the angle between the two vertical plates is varied, the azimuth profile flattens. Furthermore, the peak RCS is less, with the loss being more severe when the inter-plate angle is less than 90°. Robertson [34] also found that the RCS loss effect of inter-plate angle variation is more severe as the size of corner reflector increases (cf.) [35] and the radar wavelength decreases. Sarabandi and Tsen-Chieh [31] found that for distorted triangular, square and pentagonal plated corner reflectors the relative loss of RCS is between 0.2 dB and 1 dB for an angular deviation of $\pm 1°$ and between 1.3 dB and 2.8 dB for $\pm 2°$. Again, the losses are more severe when the inter-plate angle is acute rather than obtuse. These results indicate the importance of ensuring 90° angular relations between the intersecting plates of a corner reflector during manufacture but also through transportation and installation.

Plate curvature is the deformation of the plate from a perfectly flat plane along its entire length, such as a gradual warp across the plate. In general, RCS loss due to plate curvature is inversely proportional to the radar wavelength and the target size. An RCS loss exceeding 10 dB can result from a 5 mm plate curvature in a 1.0 m trihedral corner reflector at X-band [35].

Plate surface irregularities are the presence of any small-scale feature that causes a deviation from perfect flatness at any given location across the plate. Usage of fasteners such as pop rivets or retaining bolts on any of the plates reflecting surfaces could affect the RCS performance of the target. The RCS loss is proportional to the surface deviation across the plate and inversely proportional to the radar wavelength. A surface feature of only 1 mm deviation could introduce a 1 dB loss at the X-band [35].

3.2. Other Design Features

The material and finish to be used to manufacture the corner reflector plates also needs consideration. Aluminium is commonly used for the construction of plates. Aluminium is generally more costly than steel, but it does not suffer as badly from corrosion and is relatively lightweight. A plain metallic finish should achieve the best radar reflection properties. A thin thermoplastic powder-coat layer may assist in ensuring the longevity of the corner reflector when deployed by resisting oxidation and degradation of the metal but it will introduce RCS losses. Another design feature worth considering is whether to use pre-fabricated mesh sheeting or adding perforations to solid metal sheet. Introducing a large number of holes in the plates has the benefit of allowing quick drainage during heavy rainfall, relieving some of the force applied to the structure by wind, reducing overall weight and promoting self-cleaning of dust and other wind-blown deposits. The addition of holes to the plates will reduce the RCS, with the hole size and spacing both having an impact. To keep RCS losses below 1 dB, the hole diameter must be less than about one-sixth of the radar wavelength (Cheng Anderson, Defence Science and Technology Organisation (DSTO), Pers. Comm. 2012). For utility at X-band, the maximum size of perforation should therefore not exceed 5 mm. Measurements made by DSTO indicate an RCS loss of 0.2 dB and 1.2 dB for mesh samples with 5 mm diameter holes and a ~20% open (non-metallic) area for C and X-band, respectively [36]. Note that using a physical punch to add holes to sheet metal could affect the plate curvature and/or introduce surface irregularities.

Even if mesh or perforations will not be used, it is recommended that several holes are introduced close to the trihedral apex to allow precipitation to drain. Flooding of the corner reflector will introduce an RCS loss at least an order of magnitude greater than that caused by the presence of holes in the plate due to the breakdown of the triple bounce reflection mechanism within the TCR aperture. For example, during the field experiment described in this paper a build-up of dirt blocked the single drainage hole in one TCR prototype, which subsequently caused that TCR to fill with water following a heavy rainfall event. The resulting drop in RCS measured in two X-band COSMO-SkyMed images spanning the rainfall event was 13.2 dBm^2 (±0.69 2-sigma).

3.3. Target Characterisation

As a prelude to the experiments reported here, Geoscience Australia [36] described X and C-band RCS characterisation measurements made at a ground radar reflection range on 12 TCR prototypes of four different designs: 1.0 m solid metal sheet; 1.5 m solid metal sheet, 1.5 m powder-coated solid sheet; 1.5 m with perforated plates (15.7% open area and 5 mm diameter holes). The same twelve TCR prototypes were used in the field experiments reported in this paper, and the different designs correspond to the "type groups" A, B, C and D reported in Table 2. Three of each design were manufactured and characterised to test the consistency of the manufacturing process. The results of these measurements (Figure 5) show that there is good consistency between individual TCR regardless of plate finish, particularly at C-band where 1.5 m TCR have an RCS of 2.0 $\pm$ 0.3 dBm^2 less than theory and 1.0 m TCR are around 1.6 $^{+0.6}/_{-0.3}$ dBm^2 less than theory. At the X-band the RCS of individual TCR is more variable, ranging between 5.0 $^{+1.5}/_{-1.0}$ dBm^2 less than theory for 1.5 m TCR and 3.2 $\pm$ 1.0 dBm^2 less than theory for 1.0 m TCR. Since several distortions of the tested TCR have been documented [36], these results confirm that departures of the TCR from perfect inter-plate orthogonality and plate flatness are less tolerated at shorter radar wavelengths and, in the case of plate curvature, is more severe for smaller targets.

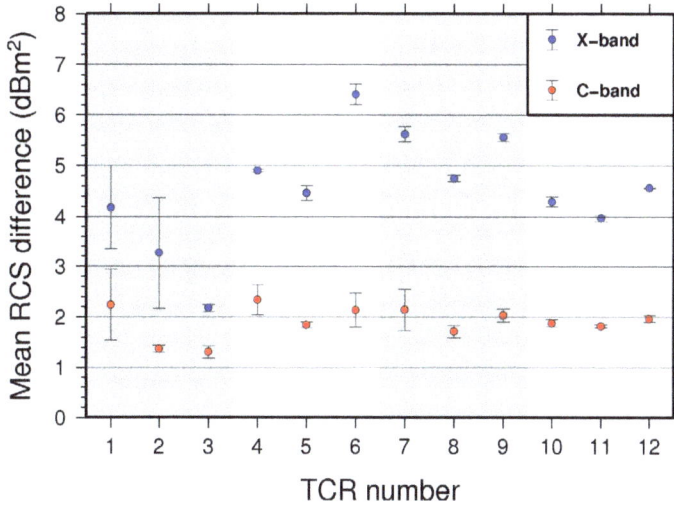

Figure 5. Mean RCS difference (theoretical minus measured) for TCR prototype designs measured at the ground radar reflection range. The mean for each frequency is calculated from four independent measurement combinations (HH/VV polarisation and azimuth/elevation sweep) with error bars indicating the standard 1-sigma error of these measurements. Grey polygons delineate the different type groups of corner reflector design indicated in Table 2. Modified from [36].

Table 2. Details of the 18 TCRs of six type groups deployed at Gunning. Misalignments of the TCR from the optimum boresight orientation for each SAR sensor were only applied for certain image acquisitions (see Table 3) during a secondary experiment described in the text.

TCR Type Group	TCR Size (m)	Plate Finish	Perforations	TCR Number	Misalignment (Degrees)	
					Azimuth	Elevation
A	1.0	Metallic	☒	1	20	0
				2	0	20
				3	0	0
B	1.5	Metallic	☒	4	0	−20
				5	0	0
				6	0	20
C	1.5	Powder-coat	☒	7	0	−10
				8	20	0
				9	20	20
D	1.5	Metallic	☑	10	10	10
				11	0	10
				12	10	0
E	2.0	Metallic	☒	13	0	0
				14	0	20
				15	20	0
F	2.5	Metallic	☒	16	0	0
				17	20	0
				18	0	20

Table 3. SAR acquisitions of the Gunning TCR array.

Acquisition #	Date (UTC)	Time (UTC)	SAR Sensor	TCR Alignment Notes	Stable Clutter Period
1	15 November 2013	19:27:59	TSX	Before TCR deployment	☒
2	7 December 2013	19:27:59	TSX	Average; only 1.0 m and 1.5 m reflectors	☒
3	11 December 2013	7:14:35	CSK-1	Average; only 1.0 m and 1.5 m reflectors	☒
4	14 December 2013	19:18:48	RSAT-2	Average alignment	☑
5	27 December 2013	7:14:31	CSK-1	Average alignment	☑
6	29 December 2013	19:27:58	TSX	Average alignment	☑
7	7 January 2014	19:18:47	RSAT-2	Average alignment	☑
8	9 January 2014	19:27:57	TSX	Average alignment	☑
9	12 January 2014	7:14:23	CSK-1	Average alignment	☑
10	20 January 2014	19:27:58	TSX	TSX	☑
11	28 January 2014	7:14:18	CSK-1	CSK	☑
12	31 January 2014	19:27:57	TSX	RSAT-2	☑
13	31 January 2014	19:18:49	RSAT-2	RSAT-2	☑
15	11 February 2014	19:27:56	TSX	TSX	☑
16	13 February 2014	7:14:12	CSK-1	CSK	☑
17	22 February 2014	19:27:56	TSX	TSX but with misalignment	☒
18	24 February 2014	19:18:44	RSAT-2	RSAT-2 but with misalignment	☒
20	5 March 2014	19:27:57	TSX	RISAT-1 *	☒
21	25 March 2014	7:14:00	CSK-2	CSK but with misalignment	☒
22	10 April 2014	7:13:59	CSK-2	CSK but with misalignment	☒
23	14 April 2014	7:13:57	CSK-4	CSK but with misalignment	☒
24	18 April 2014	7:13:57	CSK-1	CSK but with misalignment	☒

* Several acquisitions were made by the Indian Space Research Organisation for the purpose of calibration and validation of RISAT-1.

4. Field Experiments

In this section I describe field experiments that were conducted using prototype TCR with inner-leg dimensions of 1.0, 1.5, 2.0, and 2.5 m to test whether the theoretically-derived performance (defined by RCS, SCR and LOS displacement error) is achievable in a typical deployment environment in Australia. As indicated in Figure 4, the range of TCR sizes manufactured for this experiment spans the requirements of medium- and low-resolution X- and C-band SAR imaging modes. At the time of this experiment L-band SAR data was not readily available from any sensor (the Japanese ALOS-2 satellite was launched after the conclusion of the field experiment). Correspondingly, L-band is not discussed any further in this paper.

4.1. Test Site

Eighteen TCR prototypes (including the twelve TCR tested at the ground radar reflection range, see Figure 5) were deployed in a temporary array on a grazing property near Gunning, New South Wales, for a period of five months between December 2013 and May 2014 (Figure 6; Table 2). Several factors were considered when choosing installation sites for each TCR, including the flatness of the site and immediate surrounds, the perceived sources of clutter near to the site and the distance from metallic boundary fences. The TCR were also carefully positioned in such a way that the impulse response in the SAR imagery would not overlap with or intersect the response from adjacent TCR sites. The baselines between all TCR deployment sites were 186 m or greater.

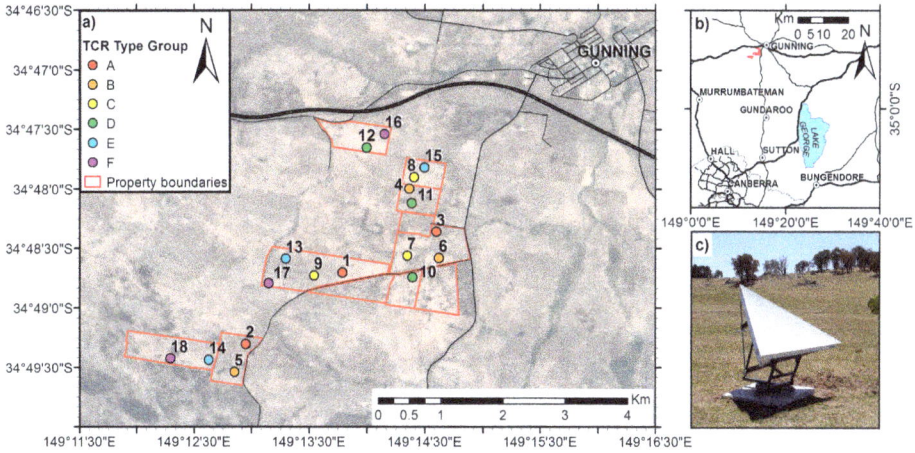

Figure 6. (**a**) Map showing the 18 TCR deployment sites at the Gunning test site coloured by type group (Table 2). The background image is a Landsat-8 RGB-composite optical image acquired on 15 January 2014 with 30 m pixel resolution. Red polygons outline the boundaries of the property available for the experiment. (**b**) Overview map showing the location of the Gunning test site (red polygons) in relation to Canberra, Australia; (**c**) photo of TCR number 7 as deployed at the Gunning test site.

4.2. Target Alignment

The TCR boresight is the vector of the maximum RCS response emanating from the internal intersection of the reflector plates (the apex). From physical optics, the boresight vector for any trihedral target is oriented 45° from the two vertical plates, and elevated $\Psi = tan^{-1}\left(\frac{1}{\sqrt{2}}\right) = 35.26°$ from the baseplate [26]. Misalignment of the TCR boresight with respect to the SAR sensor boresight will incur a loss of RCS. In the field, the TCR must be aligned in azimuth and elevation so that the boresight vector is oriented toward the LOS of the orbiting SAR platform of interest, including consideration of any squint angle of the SAR sensor. The required orientations can be calculated using published orbital information for the SAR platforms of interest, and the deployment position of the target [36]. For a particular orbital pass direction (ascending or descending), the azimuth alignments for the SAR sensors in Table 1 only vary by ~1°, whereas the elevation may vary by much more depending on the incidence angle of the swath and image mode chosen. Typically, the imaging modes have incidence angle ranges of at least 25°. An incidence angle range of 20–45° results in a pixel RCS variation of 3.2 dBm^2 across this range, independent of imaging resolution. Therefore, if the choice of imaging sub-swaths to be used is carefully considered, one target alignment can be used for multiple SAR sensors with only marginal difference in RCS between sensors because of imaging geometry.

Alignments were calculated for the viewing geometry of each SAR sensor, and an average of these three geometries. The boresight of each TCR was re-oriented to these alignments prior to each satellite overpass according to the notes in Table 3.

4.3. SAR Imagery

Twenty-two SAR image acquisitions at X and C-bands were made using the TerraSAR-X (TSX), COSMO-SkyMed (CSK) and RADARSAT-2 (RSAT-2) SAR systems (Table 3). All SAR images used in the analysis are HH polarised and were acquired on descending passes. In total, nine TSX Stripmap mode "SSC" images, nine CSK HIMAGE mode "SCS_B" images and four RSAT-2 Fine mode "SLC" images were acquired for the field experiment (Table 3). The beam modes used (9, 5 and 21, respectively) resulted in average local incidence angles at the TCR deployment sites of 35.22°, 33.31° and 34.90° for

TSX, CSK and RSAT-2, respectively. TSX products were ordered with a −10 dB gain attenuation and RSAT-2 products were ordered with the Calibration-2 look-up table to ensure the dynamic range of the SAR data could accommodate the impulse response of even the largest 2.5 m TCR present in the imaged area. An example of the impulse response from each size of TCR in a TSX image is shown in Figure 7.

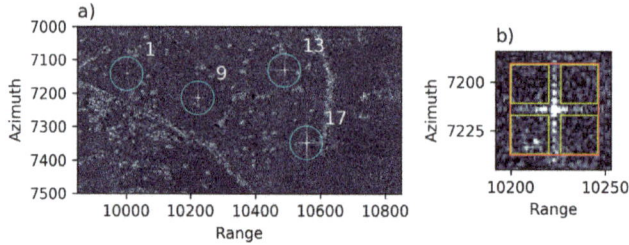

Figure 7. (**a**) Extract of a TSX image acquired on 20140120. The impulse responses for four numbered TCR of different sizes are labelled and highlighted by the cyan circles. (**b**) Zoomed view of the impulse response for TCR 9. The red polygon indicates the *'target window'* and the yellow polygons indicate the four *'clutter regions'* used in the point target analysis of each TCR. The part of the *'target window'* not intersected by the *'clutter regions'* defines a cross region that encompasses the main lobe and side lobe response of the TCR in range and azimuth directions. TSX data is ©DLR.

4.4. Image Processing Methodology

I used the *GAMMA* software [37] to process the received Single Look Complex (SLC) imagery for each SAR sensor before applying the integral method [38] to compute the RCS of each TCR in each image. The integral method is commonly used to determine the absolute calibration factor for SAR imagery by measuring the radar response of targets of known RCS. Since the received SAR imagery is already externally calibrated, the procedure is simply reversed in order to determine the RCS of the TCR. The procedure used is as follows:

1. Read the SLC imagery and convert to Sigma Nought. For TSX and CSK this involves applying the annotated product calibration factor and then scaling the image by $\sin(\theta)$ to get Sigma Nought. For RSAT-2 this involves applying the provided Sigma Nought look-up table.
2. For each SAR sensor, coregister all SLC images to a single master image (chosen to be the earliest acquisition). Verify the co-registration of each image and determine the range (column) and azimuth (row) coordinates of each TCR in the co-registered images.
3. Define a *'target window'* that encompasses the impulse response of the target and four *'clutter regions'* in the quadrants surrounding the side lobe response of the target (Figure 7b).
4. Determine the mean signal clutter in the four *'clutter regions'*. By computing the clutter level as the mean of all pixel values falling within standard-sized windows, a representative view of the actual reflector RCS and SCR is obtained that removes any bias associated with the common practice of manually choosing the location of windows to sample only the lowest clutter in the general surrounds of the target.
5. Calculate the integrated point target energy:

$$E_{CR} = E_n - \left(\frac{N_{CR}}{N_{clt}}\right) * E_{clt},\tag{7}$$

where E_n is the integrated (summed) energy in the *'target window'*, E_{clt} is the total integrated energy in the four *'clutter regions'*, N_{clt} is the number of samples contained within the four *'clutter regions'* and N_{CR} is the number of samples in the *'target window'*.

6. Compute the RCS of the point target by multiplying the integrated point target energy by the area of the ground range resolution cell:

$$\sigma_T = E_{CR} \cdot A. \tag{8}$$

7. Compute the SCR; the ratio between the point target energy corrected for clutter and the average clutter level per pixel:

$$SCR = \frac{E_{CR}}{(E_{clt}/N_{clt})}. \tag{9}$$

8. Compute the phase error (Equation (4)) and convert to LOS displacement error (Equation (5)).

5. Results

In this section I will describe the results of the TCR field experiments in terms of the radar clutter, RCS, SCR and derived LOS displacement errors detected at the deployment sites.

5.1. Clutter Intensity

In general, the average clutter levels at Gunning were between -11 dB and -16 dB for both X- and C-band (Figure 8). The clutter level is roughly the same for both X- and C-bands, consistent with the expectations discussed previously. There is a larger variation in clutter values between TCR sites in the RSAT-2 imagery, which may be because of the coarser pixel resolution resulting in greater speckle variation compared with the higher resolution X-band imagery.

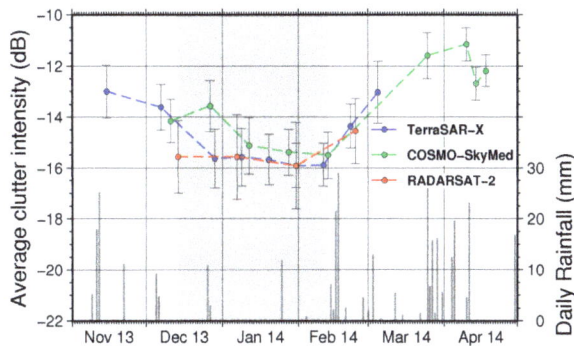

Figure 8. Time series of clutter intensity averaged over the 18 TCR sites in imagery from each SAR sensor. Error bars plot the 2-sigma standard error of the 18 observations for each image. Also plotted in the lower bar chart is the daily rainfall record for Gunning town centre, approximately 3 km away from the TCR array (data obtained from Australian Bureau of Meteorology). The grey polygon indicates a period of relatively stable radar clutter characteristics at Gunning prior to the advent of heavier rainfall events. Image acquisitions falling within this polygon (and indicated in Table 3) are further analysed in terms of RCS, SCR and LOS displacement error in the following sub-sections.

There is a strong correlation between rainfall and trends in clutter intensity for all SAR sensors (Figure 8). Significant rainfall occurred in early November 2013, prior to the installation of TCRs at Gunning. Following this time, a period of mainly dry conditions ensued until February 2014, interspersed by sporadic rainfall events of 1 day duration and around 10 mm or less. During this time, ground conditions at the Gunning test site were observed to become drier, whilst vegetation dried out and the overall volume of biomass reduced. Between 14 to 17 February 2014, ~60 mm of rain fell during a four-day period. Corresponding increases in soil moisture resulted in an increased clutter intensity in imagery from all three SAR sensors. The total increase in clutter following the February rainfall event was about 2–3 dB for TSX with a similar increase inferred for CSK and RSAT-2.

Prior to March 2014, all CSK acquisitions were made using satellite #1 of the constellation. In March and April 2014, four further CSK acquisitions were made using a combination of satellites #1, #2 and #4. Average clutter intensity had a range of ~1.5 dB between these four acquisitions. It is difficult to draw objective conclusions about an inter-constellation comparison from this observation due to the significant amounts of rainfall that fell between adjacent acquisitions that could majorly impact the backscattered signal.

5.2. Radar Cross Section

Figure 9 shows the RCS measured in SAR images for each deployed TCR and a comparison with theoretical RCS values for each size of target. Between mid-December 2013 and mid-February 2014, the background clutter at X and C-bands is consistently low (−16 to −14 dB; Figure 8). Therefore, only images acquired within this time period with relatively stable background clutter were used to calculate a mean RCS (and measurement standard error) for each TCR (see Table 3). The estimated RCS values for many TCR in RSAT-2 images turn out to be greater than theory. This situation is not a possibility, and it highlights that the calibration of RSAT-2, and indeed any SAR system, is performed using different methods and software. It is therefore not likely that equivalency is being compared between the signals from RSAT-2, CSK and TSX. When considering the measured RCS values of these three systems, it appears that CSK is performing the worst because it is the most different to the theoretical RCS values. The relative differences between the three SAR systems are consistent across all TCRs. There is a mean difference of 1.32 dBm^2 (±0.08 2-sigma) between TSX and CSK.

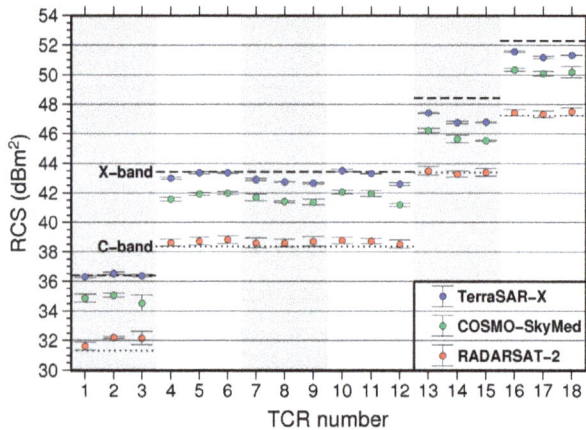

Figure 9. RCS measurements for each TCR at Gunning. RCS estimates are derived by averaging the values measured in 5 TSX images, 4 CSK images and 3 RSAT-2 images occurring during the period of relatively stable clutter (Figure 8). Error bars are the 2-sigma standard errors of these observations. Theoretical RCS values (Equation (6)) for each TCR size are plotted as dashed lines (X-band) and dotted lines (C-band). Background grey polygons are plotted to aid delineation of the different type groups of TCR design described in Table 2.

There are no consistent differences in RCS that can be attributed to the differences in plate finish of the 1.5 m TCRs. Since variations in RCS within each TCR type-group is correlated across different SAR sensors (particularly TSX and CSK) it appears that site-specific conditions are having a larger impact on measured RCS. Variability of RCS within type-groups is again more pronounced at X-band as was the case in the ground radar reflection range measurements (Figure 5).

Generally, the RCS difference between observations and theory reduces in proportion with the TCR size. This is more apparent at X-band; for TSX the difference between smaller and large TCR is on the order of 1.0 to 1.5 dBm^2 and for CSK is on the order of 0.6 to 1.1 dBm^2. The trend is not as obvious in C-band (RSAT-2) RCS estimates. This could be reflective of the fact that departures from inter-plate orthogonality and plate flatness are tolerated less at shorter radar wavelengths. It could also be indicative that for the larger TCR, the *'target window'* used to sample the impulse response is not fully capturing the full extent of side lobes, which are much broader for the 2.0 m and 2.5 m TCR (e.g., Figure 7). Consequently, the RCS measurements for the larger TCR could be adversely biased to be less than the true RCS because of the sampling choice.

5.3. Impact of Misalignment

In a secondary experiment, certain TCR were purposefully misaligned from the optimum calculated boresight for at least one acquisition of each sensor (see Table 2) to measure the RCS loss as a result of these misalignments in consecutive images from each SAR sensor (Figure 10). To ensure consistent inter-constellation signal level, images from the CSK-1 satellite were used. Four TCR, one of each size, were used as 'control' and were not misaligned and so theoretically should exhibit a zero RCS loss. In practice, the difference is not zero due to temporal clutter changes occurring between the two acquisitions. To partially account for this, the standard error of the difference in RCS measurements for these four control TCRs (numbers 3, 5, 13 and 16) were used to derive error bars for the other RCS loss measurements.

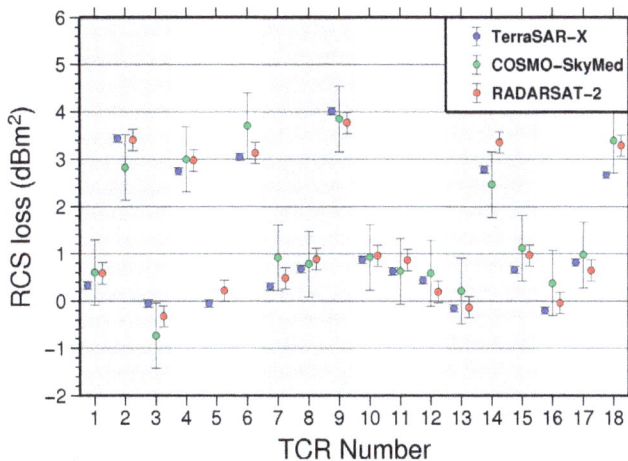

Figure 10. RCS loss due to misalignment of the TCR from the optimum boresight orientation for each SAR sensor. Misalignments for each TCR are given in Table 2. Each RCS loss is calculated by subtracting the image with misalignment from a previous image without misalignment. The RCS of four 'control' TCR (numbers 3, 5, 13 and 16) with no misalignment are used to derive the standard errors for each SAR sensor, which are plotted here as 2-sigma error bars. The CSK measurement for TCR number 5 is discarded from this analysis due to flooding of the TCR as described previously in the text. Background grey polygons are plotted to aid delineation of the different type groups of TCR design described in Table 2.

It is clear from these results that RCS loss is mainly dependent on alignment error rather than TCR size. For instance, all the TCR with an elevation misalignment of 20° (TCR numbers 2, 4, 6, 9, 14 and 18) suffer an RCS loss of 3–4 dBm^2 regardless of size. The level of RCS loss is comparable (within measurement error) in TSX, CSK and RSAT-2 images for all TCR sites, and therefore is not dependent on radar frequency. The other notable observation is that RCS loss is more severe for elevation misalignments than equivalent azimuth misalignments. Furthermore, the greatest RCS loss is seen for TCR 9 which has the worst misalignment (20° in both elevation and azimuth). Figure 11 summarises the impact of these azimuth and elevation misalignments from the optimum boresight direction. In general, the observations at X- and C-band imply that if azimuth and elevation alignment accuracies of 10° are adhered to, the resulting RCS will be within 1 dB of the peak value. Furthermore, a realistic alignment accuracy of a few degrees [36] would result in an RCS loss of less than about 0.2 dB.

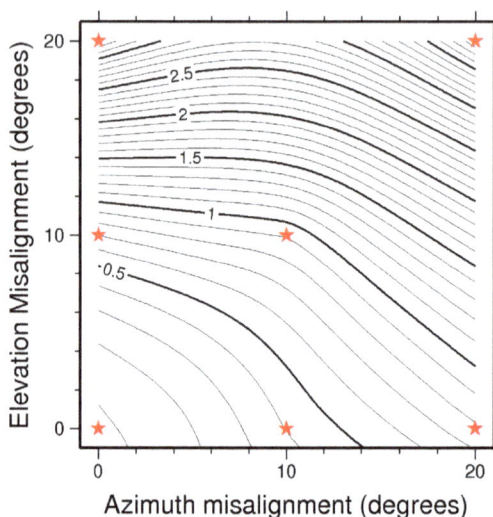

Figure 11. Contour map of RCS loss as a function of azimuth and elevation misalignments. A minimum curvature surface is fitted to the 7 RCS loss measurements from Gunning SAR imagery for the 1.5 m TCR with positive-valued misalignments. The positions in parameter space of the observed data are plotted as red stars. Observed data is taken as the mean of the TSX, CSK and RSAT-2 values. Contour interval is 0.1 dBm^2 with every 0.5 dBm^2 bold and annotated.

5.4. Displacement Error

Figure 12 shows the LOS displacement errors derived directly from measurements of the SCR from SAR imagery at Gunning. As for RCS measurements, only images acquired within the time period with relatively stable background clutter were used to calculate a mean LOS displacement error (and measurement standard error) for each TCR (see Table 3). Generally, we see that the LOS displacement error decreases with TCR size as expected since SCR is proportional to TCR size for a fixed clutter magnitude. The LOS displacement error is also frequency dependent, with C-band having greater displacement errors than X-band for the same TCR. At X-band, all TCR larger than 1.0 m meet the nominal design LOS displacement error criteria of 0.1 mm. The 1.0 m TCRs also meet this criterion in TSX imagery, but not in CSK. In fact, all TCRs have an SCR consistently about 4 dB less in CSK imagery than in TSX imagery. At C-band only the 2.5 m TCRs come close to the design criteria of 0.1 mm, although all TCR larger than 1.0 m have a LOS displacement error less than 0.5 mm. According to the theoretical calculations in Table 1, the 2.5 m TCR should exceed the 0.1 mm LOS displacement error. The fact that it does not highlights that in general the TCR prototypes are not performing as well in real data as in our expectations.

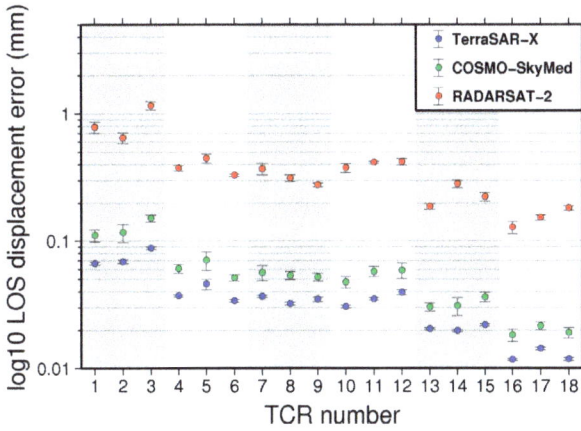

Figure 12. LOS displacement error derived from the SCR measurement for each TCR at Gunning. Each error estimate is derived by averaging the values measured in the five TSX images, four CSK images and three RSAT-2 images highlighted in Figure 8 that correspond to relatively stable clutter characteristics. Error bars are the 2-sigma standard errors of these observations. Background grey polygons are plotted to aid delineation of the different type groups of TCR design described in Table 2.

The magnitude of the 2-sigma standard errors on the LOS displacement error estimates are plotted against TCR size in Figure 13. There is a clear relationship between TCR size and variability of displacement error (and implicitly SCR) with a consistent exponential trend between sensors. This confirms the theoretical phase noise behaviour demonstrated previously in the PS phase simulation (Figure 2) whereby the phase noise becomes less variable as the SCR increases in concert with an increase in target size.

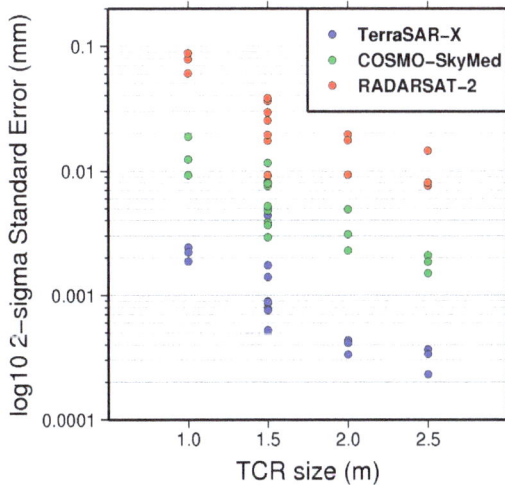

Figure 13. Magnitude of 2-sigma standard errors on the estimated LOS displacement errors (plotted as error bars in Figure 12) grouped according to TCR size.

Figure 14 shows the ratio of the theoretical SCR against observed SCR values derived using Equation (9) for the TCR deployed at Gunning. Theoretical SCR is calculated using the average observed clutter at each TCR in those images highlighted in Figure 8 and Table 3 (corresponding to relatively stable clutter characteristics) and the ground range resolution area derived from the pixel resolution. With an average ratio close to 1, the TCRs are performing close to expectations in TSX images. In CSK this ratio is marginally worse at ~1.2. In RSAT-2 the ratio is much worse at ~2.2. Interestingly the ratio between SCR observations and theory is consistent across all TCRs for TSX and CSK, though in RSAT-2 the ratio is variable, with larger TCR performing better. This indicates that TCR between 1 m and 2.5 m size are adequate for use with TSX Stripmap and CSK HIMAGE imagery if the goal is to achieve a phase noise error equivalent to a tenth of a millimetre displacement in the line of sight. Only the larger TCR are close to achieving that criterion in RSAT-2 Fine imagery.

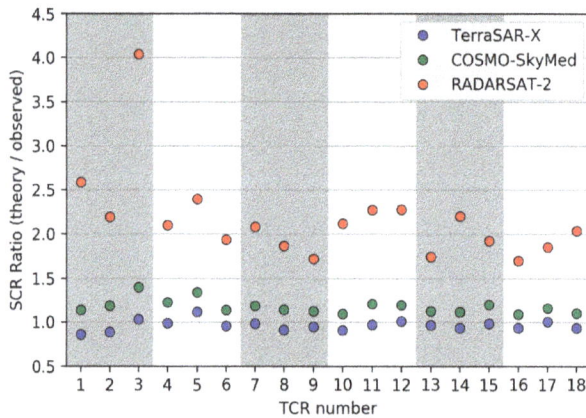

Figure 14. Ratio of theoretical SCR to observed SCR for all TCR deployed at Gunning. Background grey polygons are plotted to aid delineation of the different type groups of TCR design described in Table 2.

6. Discussion

TCR performance is clearly limited by the choice of deployment site. For instance, TCR number 3, a 1.0 m target, was situated in a position relatively close to distributed metallic debris and farm buildings. The impact of this background clutter at the deployment site on SCR performance is self-evident in Figures 12 and 14. Furthermore, the effect of the increased background clutter was compounded in this case by the relatively small size of the target deployed at this site. This highlights the importance of choosing clutter-free sites for deployment of targets to be used as geodetic targets whenever feasibly possible.

The results presented here show that there are differences between the backscattered signal measured by CSK and TSX. The fact that TSX has higher RCS than CSK is consistent with previous work where a mean backscattering difference of 3.15 dB between TSX Stripmap and CSK HIMAGE images was identified [39]. While these inconsistencies may be attributed to the absolute radiometric calibration of the two systems, there are other reasons for at least some component of the observed differences. As reported previously, the average local incidence angle across all the TCRs in all the images is 35.22° in TSX and 33.31° in CSK. This 2.1° difference in incidence angle results in a backscatter difference of 0.2 dB purely due to differences in the viewing geometry, though taking into consideration the difference in spatial resolution of TSX Stripmap and CSK HIMAGE images the deviation could be as much as 1.7 dB. The standard deviation of the local incidence angles infers that the satellite orbital path varies more for CSK (or RSAT-2) in comparison to TSX. The standard deviations across all images used in this experiment are 0.051° for CSK, 0.008° for TSX and 0.044° for RSAT-2.

Another possible explanation for observed discrepancies between SAR sensors is the different processing applied to the raw SAR data during SLC production by the data providers. Interestingly, the extent of side-lobe ringing from TCRs deployed at Gunning is visually more pervasive in CSK images when compared to TSX images. For the purpose of a 'fair test', a fixed *'target window'* size was used to determine the RCS of TCRs across all sensors. As a consequence, it is certainly possible that not all the reflected energy pertaining to the TCR is sampled. This would give rise to a bias in the measured signal, which in the case of CSK data (with the larger side lobes), could manifest as a reduced RCS and SCR measurement.

The ground radar reflection range measurements of twelve of the TCR prototypes deployed at Gunning indicate RCS performance in the -2 to -5 dBm^2 with respect to theoretical values (see Figure 5). It is interesting that the measured performance in the satellite SAR data is better (Figure 9). Several compromises that were made during the ground measurement procedure (documented in [36]) are the likely reason for these discrepancies.

Using the SCR as an a priori proxy for phase error, and therefore LOS displacement error, should be treated with some caution. Ketelaar et al. [23] conducted a validation experiment with five corner reflectors, comparing heights derived from ERS and ENVISAT InSAR analyses with repeated levelling surveys. From this experiment, they found that the SCR-derived phase error is under-estimated by up to four times compared to the levelling measurements. An alternative method for estimating the phase error, commonly used during candidate PS selection in the PSInSAR workflow, is the amplitude dispersion method [1]. Through a simulation exercise, [24] find that the SCR is a more effective estimator of phase error than the amplitude dispersion when the SCR is greater than 9 dB. Below this threshold, both methods are optimistically biased, with the amplitude dispersion being more so. When using the amplitude dispersion method to select candidate PS pixels during PSInSAR analysis, generally at least 20 SAR images are required to achieve an unbiased estimate of the phase error [24]. Therefore, the small number of SAR images acquired with each sensor during the Gunning field experiment precludes conducting a robust interferometric analysis using this data. However, I consider the estimated LOS displacement error as a suitable quantity with which to relatively assess different TCR designs.

Compact, active transponders (as described by [13]) are an alternative type of artificial target to corner reflectors that can be deployed readily in geodetic networks and utilised with C-band sensors such as Sentinel-1 and RSAT-2. The clear advantage of these transponders over passive corner reflectors is that they are consistently compact and therefore less obtrusive in the environment and more easily coupled (due to their small size) to other geodetic monumentation for cross-validation of signals. The RCS of these transponders is restricted to a narrow band of a few hundred MHz since they are designed to target specific SAR sensors. Therefore, the existing transponder designs cannot be used as a single point of reference across multiple frequency bands in the same way as a corner reflector. One other consideration is that transponders are active transmitting devices and therefore require a power supply for long term deployment and an appropriate radio transmission licence to be operated. These may be difficult and expensive to obtain in some jurisdictions.

The Sentinel-1 SAR constellation mission [40] has brought about a new era for SAR remote sensing in which open access to data acquired with blanket global coverage at frequent and regular revisits is becoming a reality. To some extent SAR missions like Sentinel-1 removes the need for radar targets to perform adequately at multiple frequencies, and in this situation it is much easier to design a target towards a specific level of displacement error without the need to make compromises for different SAR frequencies. Generally, this will enable the designed target to be as small as possible. However, in many situations it may still be desirable to use the same set of targets across multiple SAR datasets, for instance to combine high and low-resolution data over a particular area of interest or to validate InSAR measurements derived from data acquired at different frequencies. As shown in this study with respect to TSX and CSK, SAR sensors operating at the same frequency may still perform differently. Using this strategy, Geoscience Australia recently installed a regional geodetic monitoring network in Queensland, Australia that included co-located geodetic survey marks and radar corner reflectors [41]. The triangular trihedral corner reflectors deployed in that network were at least 1.5 m in size in order to be visible across the SAR frequency spectrum (X-, C-, and L-band) and enable the cross-validation of all SAR data being actively acquired over that area of interest.

7. Conclusions

As a result of this experiment, I find that triangular trihedral corner reflectors as small as 1 m inner-leg dimension can achieve a displacement error derived from observations of the signal to clutter ratio below a tenth of a millimetre or less in medium-resolution X-band data (Figure 12). Much larger corner reflectors (2.5 m or greater) are required to achieve the same magnitude of displacement error in medium-resolution C-band data, though displacement errors of less than a millimetre are achievable for corner reflectors of 1.5 m or larger.

I find that the theoretically-derived performance of triangular trihedral corner reflectors between 1 m and 2.5 m inner-leg dimension are broadly achievable in X-band SAR data, though in C-band data all sizes underperform by a factor of at least 2 (Figure 14). Despite the relative high quality of the manufactured corner reflectors used in this study, the expected performance of the targets derived from theoretical considerations, is only fully achieved in TSX data.

Although it is not possible to achieve the same level of displacement error across all SAR frequency bands with a single corner reflector size, it is feasible to use a single design to act as a common reference point across sensors and frequency bands providing that the user recognises that phase noise will increase as the radar frequency decreases. Compromise designs aimed at multiple frequencies should aim to satisfy the requirements of the lowest SAR frequency to be used, providing that these targets will not saturate the sensor of the highest frequency to be used. The choice of target design and size for a particular project will ultimately come down to the choice of an 'acceptable' displacement error. I arbitrarily chose an 'acceptable' displacement error of 0.1 mm herein, as a small fraction of an expected geophysical signal of interest. Relaxing this criterion will enable a smaller target to be used for a given radar frequency.

The most important considerations when manufacturing trihedral corner reflectors are the quality of the materials and the manufacturing processes used. Generally, improved performance can be achieved through better engineering at increased cost. Attention should be made to ensure flatness of the reflector plates and that orthogonality between the three plates is maintained. Corner reflectors must be appropriately designed so that precipitation and debris does not accumulate in the corner reflector aperture. Flooding of the aperture can have a catastrophic impact on the performance of the target.

Deployment sites should be chosen to limit the influence of background clutter wherever feasibly possible. The influence of clutter on the performance is greater as the target gets smaller. Accurate alignment of the target boresight with respect to the SAR sensors of interest should also be carefully considered. Generally, trihedral corner reflectors are lenient to alignment inaccuracies when compared to other target types. A field alignment methodology that makes use of a magnetic sighting compass for azimuth measurement and a digital level for elevation measurement can achieve absolute alignment accuracy of better than 3° [36]. This alignment accuracy should yield an RCS loss of 0.2 dBm^2 or less (Figure 11), which will have a minimal impact on the displacement error at X- and C-bands.

Acknowledgments: This work was undertaken as part of the Australian Geophysical Observing System project, using funds from the AuScope initiative. AuScope Ltd is funded under the National Collaborative Research Infrastructure Strategy (NCRIS), an Australian Commonwealth Government Programme. TerraSAR-X data is copyright of DLR and was provided under science project LAN1499. Four COSMO-SkyMed images acquired in March and April 2014 were supplied by e-GEOS for evaluation. The author would like to thank numerous GA colleagues for their contribution to this work and to Thomas Fuhrmann, John Dawson and three anonymous

Remote Sens. **2017**, *9*, 648

reviewers for improving the manuscript. Mark Williams is also acknowledged for his involvement early in the project. This paper is published with the permission of the CEO, Geoscience Australia.

Conflicts of Interest: The author declares no conflict of interest.

References

1. Ferretti, A.; Prati, C.; Rocca, F. Permanent scatterers in SAR interferometry. *IEEE Trans. Geosci. Remote Sens.* **2001**, *39*, 8–20. [CrossRef]
2. Hooper, A.; Zebker, H.; Segall, P.; Kampes, B. A new method for measuring deformation on volcanoes and other natural terrains using InSAR persistent scatterers. *Geophys. Res. Lett.* **2004**, *31*, L23611. [CrossRef]
3. Kampes, B.M. *Radar Interferometry—Persistent Scatterer Technique*; Springer: Dordrecht, The Netherlands, 2006.
4. Hooper, A.; Bekaert, D.; Spaans, K.; Arıkan, M. Recent advances in SAR interferometry time series analysis for measuring crustal deformation. *Tectonophysics* **2012**, *514–517*, 1–13. [CrossRef]
5. Crosetto, M.; Monserrat, O.; Cuevas-González, M.; Devanthéry, N.; Crippa, B. Persistent scatterer interferometry: A review. *ISPRS J. Photogramm. Remote Sens.* **2016**, *115*, 78–89. [CrossRef]
6. Hanssen, R.F. *Radar Interferometry—Data Interpretation and Error Analysis*; Kluwer Academic Publishers: Dordrecht, The Netherlands, 2001.
7. Ketelaar, V.B.H. *Satellite Radar Interferometry—Subsidence Monitoring Techniques*; Springer: Dordrecht, The Netherlands, 2009.
8. Fu, W.; Guo, H.; Tian, Q.; Guo, X. Landslide monitoring by corner reflectors differential interferometry SAR. *Int. J. Remote Sens.* **2010**, *31*, 6387–6400. [CrossRef]
9. Li, C.; Yin, J.; Zhao, J.; Zhang, G.; Shan, X. The selection of artificial corner reflectors based on RCS analysis. *Acta Geophys.* **2012**, *60*, 43–58. [CrossRef]
10. Qin, Y.; Perissin, D.; Lei, L. The design and experiments on corner reflectors for urban ground deformation monitoring in Hong Kong. *Int. J. Antennas Propag.* **2013**, *2013*, 191685. [CrossRef]
11. Strozzi, T.; Teatini, P.; Tosi, L.; Wegmüller, U.; Werner, C. Land subsidence of natural transitional environments by satellite radar interferometry on artificial reflectors. *J. Geophys. Res. Earth Surf.* **2013**, *118*, 1177–1191. [CrossRef]
12. Caro-Cuenca, M.; Dheenathayalan, P.; Van-Rossum, W.; Hoogeboom, P. Deployment and design of bi-directional corner reflectors for optimal ground motion monitoring using InSAR. In Proceedings of the 10th European Conference on Synthetic Aperture Radar, Berlin, Germany, 3–5 June 2014; pp. 1–4.
13. Mahapatra, P.S.; Samiei-Esfahany, S.; van der Marel, H.; Hanssen, R.F. On the use of transponders as coherent radar targets for SAR interferometry. *IEEE Trans. Geosci. Remote Sens.* **2014**, *52*, 1869–1878. [CrossRef]
14. Singleton, A.; Li, Z.; Hoey, T.; Muller, J.P. Evaluating sub-pixel offset techniques as an alternative to D-InSAR for monitoring episodic landslide movements in vegetated terrain. *Remote Sens. Environ.* **2014**, *147*, 133–144. [CrossRef]
15. Ferretti, A.; Savio, G.; Barzaghi, R.; Borghi, A.; Musazzi, S.; Novali, F.; Prati, C.; Rocca, F. Submillimeter accuracy of InSAR time series: Experimental validation. *IEEE Trans. Geosci. Remote Sens.* **2007**, *45*, 1142–1153. [CrossRef]
16. Marinkovic, P.; Ketelaar, G.; van Leijen, F.; Hanssen, R. InSAR quality control: Analysis of five years of corner reflector time series. In Proceedings of the Fringe 2007 Workshop (ESA SP-649), Frascati, Italy, 26–30 November 2007; pp. 26–30.
17. Dheenathayalan, P.; Small, D.; Schubert, A.; Hanssen, R.F. High-precision positioning of radar scatterers. *J. Geod.* **2016**, *90*, 403–422. [CrossRef]
18. Gisinger, C.; Willberg, M.; Balss, U.; Klügel, T.; Mähler, S.; Pail, R.; Eineder, M. Differential geodetic stereo SAR with TerraSAR-X by exploiting small multi-directional radar reflectors. *J. Geod.* **2017**, *91*, 53–67. [CrossRef]
19. Freeman, A. SAR calibration: An overview. *IEEE Trans. Geosci. Remote Sens.* **1992**, *30*, 1107–1121. [CrossRef]
20. Curlander, J.C.; McDonough, R.N. *Synthetic Aperture Radar Systems and Signal Processing*; John Wiley & Sons, Inc.: Hoboken, NJ, USA, 1991.
21. Skolnik, M.I. *Radar Handbook*; McGraw-Hill: New York, NY, USA, 1970.
22. Dong, Y. *L-Band VV Clutter Analysis for Natural Land in Northern Territory*; DSTO-RR-0254; Defence Science and Technology Organisation: Salisbury, Australia, 2003.

23. Ketelaar, V.B.H.; Marinkovic, P.; Hanssen, R.F. Validation of point scatterer phase statistics in multi-pass InSAR. In Proceedings of the 2004 Envisat and ERS Symposium, Salzburg, Austria, 6–10 September 2004; European Space Agency: Salzburg, Austria, 2005.

24. Adam, N.; Kampes, B.; Eineder, M. Development of a scientific permanent scatterer system: Modications for mixed ERS/Envisat time series. In Proceedings of the 2004 Envisat and ERS Symposium, Salzburg, Austria, 6–10 September 2004; European Space Agenc: Salzburg, Austria, 2005.

25. Knott, E.F. *Radar Cross Section Measurements*; SciTech Publishing, Inc.: Raleigh, NC, USA, 2006.

26. Doerry, A.W.; Brock, B.C. *Radar Cross Section of Triangular Trihedral Reflector with Extended Bottom Plate*; SAND2009-2993; Sandia National Laboratories: Albuquerque, NM, USA, 2009.

27. Bird, P.J.; Keyte, G.E.; Kenward, D.R.D. An experiment for the radiometric calibration of the ERS-1 SAR. *Can. J. Remote Sens.* **1993**, *19*, 232–238. [CrossRef]

28. Döring, B.J.; Schwerdt, M.; Bauer, R. TerraSAR-X calibration ground equipment. In Proceedings of the Wave Propagation in Communication, Microwave Systems and Navigation (WFMN07), Chemnitz, Germany, 4–5 July 2007; pp. 86–90.

29. Shimada, M.; Isoguchi, O.; Tadono, T.; Isono, K. PALSAR radiometric and geometric calibration. *IEEE Trans. Geosci. Remote Sens.* **2009**, *47*, 3915–3932. [CrossRef]

30. Zhou, Y.; Li, C.; Ma, L.; Yang, M.Y.; Liu, Q. Improved trihedral corner reflector for high-precision SAR calibration and validation. In Proceedings of the 2014 IEEE Geoscience and Remote Sensing Symposium, Quebec City, QC, Canada, 13–18 July 2014; pp. 454–457.

31. Sarabandi, K.; Tsen-Chieh, C. Optimum corner reflectors for calibration of imaging radars. *IEEE Trans. Antennas Propag.* **1996**, *44*, 1348–1361. [CrossRef]

32. Ruck, G.T. *Radar Cross Section Handbook*; Plenum Publishing Corporation: New York, NY, USA, 1970; Volume 2.

33. Döring, B.; Schmidt, K.; Jirousek, M.; Rudolf, D.; Reimann, J.; Raab, S.; Antony, J.; Schwerdt, M. Hierarchical bayesian data analysis in radiometric SAR system calibration: A case study on transponder calibration with RADARSAT-2 data. *Remote Sens.* **2013**, *5*, 6667–6690. [CrossRef]

34. Robertson, S.D. Targets for microwave radar navigation. *Bell Syst. Tech. J.* **1947**, *26*, 852–869. [CrossRef]

35. Zink, M.; Kietzmann, H. *Next Generation SAR—External Calibration*; German Aerospace Center (DLR): Köln, Germany, 1995; p. 45.

36. Garthwaite, M.C.; Nancarrow, S.; Hislop, A.; Thankappan, M.; Dawson, J.H.; Lawrie, S. *Design of Radar Corner Reflectors for the Australian Geophysical Observing System*; Geoscience Australia: Canberra, Australia, 2015.

37. Wegmüller, U.; Werner, C. Gamma SAR processor and interferometry software. In Proceedings of the 3rd ERS Scientific Symposium, Florence, Italy, 17–20 March 1997; European Space Agency: Florence, Italy, 1997.

38. Gray, A.L.; Vachon, P.W.; Livingstone, C.E.; Lukowski, T.I. Synthetic aperture radar calibration using reference reflectors. *IEEE Trans. Geosci. Remote Sens.* **1990**, *28*, 374–383. [CrossRef]

39. Pettinato, S.; Santi, E.; Paloscia, S.; Pampaloni, P.; Fontanelli, G. The intercomparison of X-band SAR images from COSMO-SkyMed and TerraSAR-X satellites: Case studies. *Remote Sens.* **2013**, *5*, 2928–2942. [CrossRef]

40. Torres, R.; Snoeij, P.; Geudtner, D.; Bibby, D.; Davidson, M.; Attema, E.; Potin, P.; Rommen, B.; Floury, N.; Brown, M.; et al. GMES Sentinel-1 mission. *Remote Sens. Environ.* **2012**, *120*, 9–24. [CrossRef]

41. Garthwaite, M.C.; Hazelwood, M.; Nancarrow, S.; Hislop, A.; Dawson, J.H. A regional geodetic network to monitor ground surface response to resource extraction in the northern Surat Basin, Queensland. *Aust. J. Earth Sci.* **2015**, *62*, 469–477. [CrossRef]

remote sensing

MDPI

Technical Note

Identification of C-Band Radio Frequency Interferences from Sentinel-1 Data

Andrea Monti-Guarnieri [1,*], Davide Giudici [2] and Andrea Recchia [2]

[1] Dipartimento di Elettronica, Informazione e Bioingegneria, Politecnico di Milano, Piazza Leonardo da Vinci 32, 20133 Milan, Italy

[2] ARESYS srl, Via Flumendosa 16, 20132 Milan, Italy; davide.giudici@aresys.it (D.G.); andrea.recchia@aresys.it (A.R.)

* Correspondence: andrea.montiguarnieri@polimi.it; Tel.: +39-02-2399-3446

Academic Editors: Batuhan Osmanoglu, Timo Balz, Zhong Lu and Prasad S. Thenkabail
Received: 16 September 2017; Accepted: 15 November 2017; Published: 17 November 2017

Abstract: We propose the use of Sentinel-1 Synthetic Aperture Radar (SAR) to provide a continuous and global monitoring of Radio Frequency Interferences (RFI) in C-band. We take advantage of the first 8–10 echo measures at the beginning of each burst, a 50–70 MHz wide bandwidth and a ground beam coverage of ~25 km (azimuth) by 70 km (range). Such observations can be repeated with a frequency better than three days, by considering two satellites and both ascending and descending passes. These measures can be used to qualify the same Sentinel-1 (S1) dataset as well as to monitor the availability and the use of radio frequency spectrum for present and future spaceborne imagers and for policy makers. In the paper we investigate the feasibility and the limits of this approach, and we provide a first Radio Frequency Interference (RFI) map with continental coverage over Europe.

Keywords: SAR; Radio Frequency Interferences; Synthetic Aperture Radar; Geosynchronous SAR

1. Introduction

The availability of radio frequency spectrum is maybe the most critical resource for present and future satellite remote sensing, like microwave passive radiometers or active Earth Exploration Satellite Services (scatterometers, altimeters, wind profilers and SAR [1]). Monitoring the actual spectrum use is currently addressed in L-band [2] and constantly updated using SMOS data [3,4]. An L-band RFI map covering the USA has been made with ALOS data [5,6], while a cube-sat mission for monitoring the whole 6–40 GHz spectrum has been proposed by NASA [7]. In C-band, examples of some RFI monitoring were done by ASCAT scatterometers [8], Envisat and RadarSAT in a dedicated workshop cited in [1], where it was pointed out that the increased exploitation of C-band spectrum for Radio LAN (RLAN) is indeed a major issue for present and future SAR missions.

S1 SAR with its two TOPSAR acquisition modes, the Interferometric Wide Swath (IW) and the Extra Wide Swath (EW), is quite valuable for a frequent and ubiquitous monitoring of C-band RFI for several reasons. The constellation of two satellites [9] is continuously acquiring data over landmasses with a revisit of 1–3 days, thanks to the IW mode wide swath of 250 km with a bandwidth that covers 40–60% of the overall C-band spectrum not open to RLAN (5350–5470 MHz) [1]. Moreover, the burst-mode TOPSAR acquisition, shown in Figure 1 ensures the presence of 8–10 received echoes, in both H and V polarizations, that are unaffected by background scattering, as it will be shown in Section 2, and that can be extracted from S1 raw data. In Section 3 we discuss the method to detect RFI from S1 data, while in Section 4, we show examples of results achieved by processing an entire set of ascending and descending products, covering almost the entire Europe.

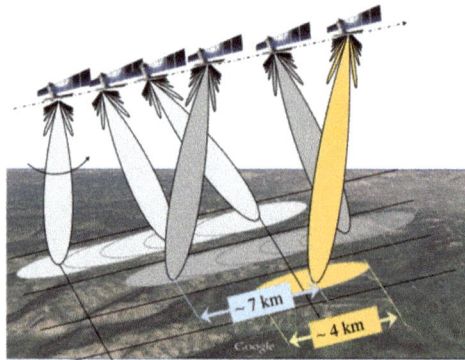

Figure 1. Sketch of TOPSAR acquisition mode, here represented as the three swaths Sentinel-1 IW.

2. Sentinel-1 Acquisition Timeline

We address here Sentinel-1 IW-mode, the default acquisition mode at low and mid latitudes (up to 60°) over still landmasses. At the beginning of each burst, as shown in Figure 1, the antenna beam is pointed backward of −0.6°. During the acquisition it sweeps forward up to +0.6° at the burst end. The system starts acquiring since the transmission of the very first pulse, although no backscatter is foreseen, at least for the two-way sensor-Earth travel-time, which corresponds to 8–10 pulses, that we define "rank" pulses, and are here exploited for RFI monitoring.

The S1 acquisition timeline is made of a fixed pattern of three distinct component. The first part is the preamble, including a number of warmup echoes for instrument stabilization and noise and internal calibration measurements. The central part is the "actual" SAR acquisition and can be several minutes long. For IW mode the adopted timeline is listed in Table 1, the base scheme (duration about 5.4 s) is repeated an integer number of time according to the mission planning. The final part of the data-take is the postamble and includes again internal calibration and noise pulses.

Table 1. IW timeline, cyclically repeated.

Type	Pulse (s)	Rank	PRF (Hz)	Notes
IW1	1409	9	1717.1	
TxCal	12		1717.1	Same pol. of data
TxCal ISO	8		1717.1	H pol
IW2	1548	8	1451.6	
RxCal	20	N/A		No TX
IW3	1410	10	1685.8	
EpdnCal	20	N/A		No TX
IW1	1409	9	1717.1	
TACal	20	N/A		No TX
IW2	1548	8	1451.6	
ApdnCal	20	N/A		No TX
IW3	1410	10	1685.8	
ApdnCal	20	N/A		No TX

The IW data-takes are split into slices of 25 s to ease processing and data dissemination. The slicing procedure is not data driven and the data "cuts" can occur at every time instant within the scheme described in Table 1. At the end of each burst a few calibration pulses are acquired to monitor instrument status during the data-take. Most of the calibration pulses do not foresee any transmission and hence the "rank" echoes (belonging to IW1 and IW3) acquired after such pulses should contain only noise and RFI coming from the Earth, if any. On the other hand, for the second sub-swath, IW2, the case is slightly different, since the calibration pulses, named "TxCal ISO" in Table 1, are effectively transmitted, but only one cycle over two (as shown in the table) and with antenna beam pointing boresight (approximately 30° with respect to Nadir) and no electronical steering. This means that there could be some potential ground backscatter echo in the first rank pulses of IW2, but only every odd cycle.

The quality of the rank echoes is fundamental for the RFI assessment: we expect that S1 is behaving like a receive-only C-band radiometer. Figure 2 shows the rank echoes power gathered over a very long strip of 730 s, covering from North Africa to arctic. The return power from each of the 20,000 pulses is plotted on the right for both V and H polarizations. It is the superposition of a smooth trend, fast ripples, within 1 dB, and sporadic big spikes, due to RFI. The smooth trend is a good representation of thermal noise, and one can appreciate the sudden changes in the land-sea and see-land transitions, due to the different albedo, like in Crete Island, at 7 s, and the Baltic Sea, at 400 s. A zoom of the ripple pattern has been plotted on the right panel in the same figure: one can observe the saw-tooth behavior with a periodicity of two cycles that matches the timeline listed in Table 1. The shape and the periodicity let then think to transients in the receiver gain after calibration. The ±0.5 dB is well acceptable for sensitive RFI monitoring.

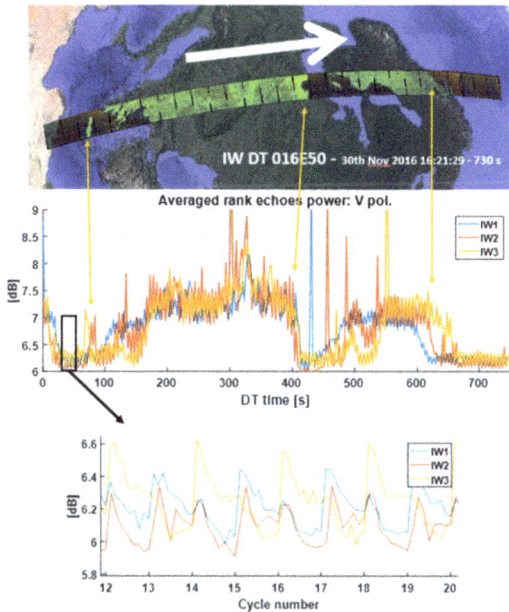

Figure 2. A 730-s long data-take, upper panel, has been used to validate the measure of noise power from rank echoes, mid panel. Notice the changes in the mean power occurring near see/land transitions, as expected from the change in the albedo. A zoom of the power profile, lower panel, shows a saw-tooth behavior with two cycle's periodicity, following the calibration in Table 1.

There is no evidence of a backscatter, not even in the cross-pol echoes in IW2 each two cycles. This result, that comes from the combination of the antenna backward squint (due to TOPSAR), boresight steering, is the key element to enable measures of noise and RFI.

3. RFI Detection and Estimation

The estimation of RFI is carried out according to the method summarized in Figure 3. It comprises a calibration step (on the left in the figure), that runs once in the long term, by exploiting a large amount of data, and an estimation step, on the right, performed on each single data-take.

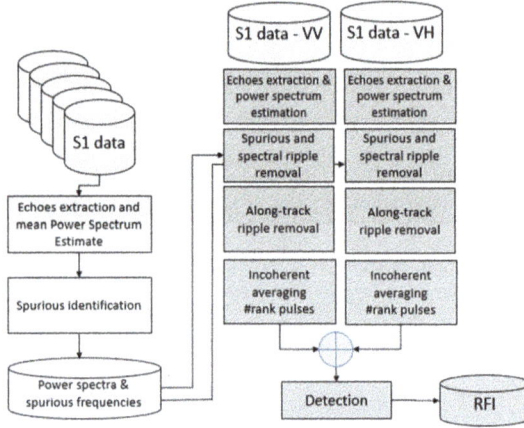

Figure 3. Schematic block diagram for the identification of RFI.

The aim of calibration is the identification of spurious tones and the precise receiver power spectrum profile. In principle, all the combinations of swaths, polarizations, sensors (S1A and S1B) and modes (IW and EW) should be calibrated. In the present study, we focused to IW mode and S1A sensor: the power spectra for the three sub-swaths and the two polarizations are plotted in Figure 4, after averaging over 10,000 bursts in each of the two polarizations, acquired all over the world from November 2016, up to April 2017. S1 demonstrated very good stability in gain, ripples and spurious locations, and this suggest the use of noise data to speed up calibration. A total of 100–200 spectral samples of the whole ~20,000, were found affected by spurious tones, which can be easily detected by a median filter. Once removed, the spectral profiles are estimated and stored.

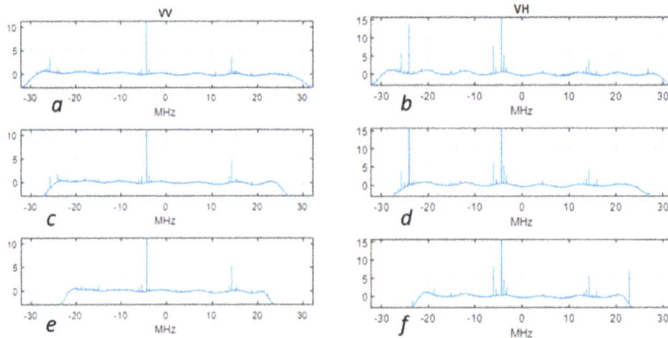

Figure 4. Mean power spectrum density, dB, measured from the "rank" pulses at beginning of each swaths in V (*a,c,e*) and H (*b,d,f*) polarizations, for sub-swaths IW1 (*a,b*), IW2 (*c,d*) and IW3 (*e,f*), averaged on 10,000 bursts. Notice the spurious tones and the ripples.

In the processing step, first the data spectrogram $H(f,t)$ is estimated as the squared amplitude of the Fourier Transform of each echo. The frequencies corresponding to the spurious previously identified are nulled. Ripples are then removed in both frequency and time by modelling the expectation of the spectrogram as:

$$E[H(f)] = H_0(f) \times G(t), \tag{1}$$

where $H_0(f)$ is the mean power spectrum estimated by the calibration step, and $G(t)$ is the along track ripple pattern, in time domain, as the one shown in Figure 2 on the bottom panel. This second term is not stable in time, as it appears in Figure 2 on the middle panel, and it needs to be estimated from the data. The estimate is carried out by first compensating the range spectra, by the inverse profile $1/H_0(f)$, and then computing the spectral median for each time. The median is robust respect to the RFI that can be in the dataset, however, for further improving RFI rejection, a second order polynomial Savitzky-Golay filter [10] is applied to the along-track profile to derive the estimate of $G(t)$.

An example of a data spectrogram prior and after compensating for the frequency and time-domain ripples is shown in Figure 5. The spectrogram has already been averaged in frequency by a multi-look factor of 100, to reduce noise fluctuation, then enhancing the detection of the slightest RFI.

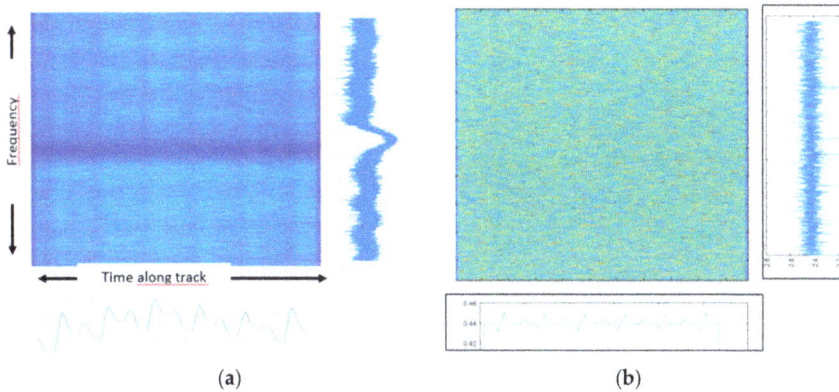

(a)   (b)

Figure 5. Example of a spectrogram of noise data before (**a**) and after (**b**) compensation for the ripples in frequency and time. The residual along-track ripple is less than 0.1 dB.

Detection is performed by a combination of the one tailed Fisher's Z test [6]:

$$Z = \frac{H_f - E[H_f]}{VAR[H_f]}, \tag{2}$$

H_f being the power spectrum, and the use Kullback–Leibler (KL) divergence [11], proposed here:

$$K = \int p(H_f) \log\left(\frac{p(H_f)}{p_{fit}(H_f)}\right) dH_f \tag{3}$$

The first detector, (2), finds those extremal peaks, say isolated RFI with powers much stronger than noise one. The KL based detector, in (3), measures the mean distance between the logarithms of the actual power spectrum density distribution, $p(H_f)$, and the Normal one, p_{fit}. This corresponds to the maximum-likelihood hypothesis testing for non-Normal distributed data [12], and is sensitive to maybe many RFI samples of very low power, that is the case of RLAN [13], distributed RFI, or other factors affecting the raw data quality. An example of identification of low power RFI is shown in Figure 6. The V and H power spectra for the 12 short strips, each formed by the nine rank pulses at the beginning of each burst are shown in a linear scale, normalized with respect to S1 thermal noise floor.

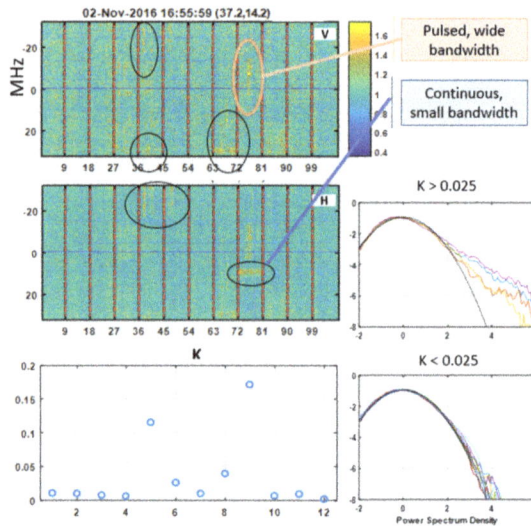

Figure 6. Example of detection of low-power RFI. Top and middle panels on the left: spectrograms for V and H polarizations. Each vertical strip corresponds to the 9 pulses at the beginning of a bursts, all coming from the same sub-swath (IW1). The strongest RFI are encircled. Lower left panel: plot of KL divergence for each burst (*x*-axis define the burst number, and it has been aligned with the spectrograms above). Right panels: histograms, in log scale, and fitting normal PDF, for the bursts with K > 0.025 and K < 0.025.

The multi-looked, whitened spectra are quite homogeneous, and slight deviations may be appreciated, like those encircled. The availability of multiple pulses-per-burst allow to distinguish between pulsed and continuous RFI. As for the detection, the two sets of histograms show that KL divergence was capable to detect those affected by RFI (upper plot) by the others.

4. Sensitivity

The Effective Isotropic Radiated Power (EIRP) of the transmitter causing the RFI contributes to S1 receiver with a power:

$$P_R = \frac{P_{EIRP} \cdot A_S \cdot \eta}{4\pi R^2}, \tag{4}$$

where A_S is the S1 antenna area, η the total losses and R the range. The sensitivity is the RFI power that cause the same contribution as noise at S1 receiver:

$$P_n = K_B \cdot T_S \cdot B_{RFI}, \tag{5}$$

K_B being Boltzmann constant, B_{RFI} the RFI bandwidth, T_S is S1 equivalent noise temperature. We can then define the minimal power P_{EIRPZ} of the RFI that gives the same contribution as thermal noise, by equating (4) and (5):

$$P_{EIRPZ} = \frac{K_B \cdot T_S \cdot B_{RFI} \cdot 4\pi R^2}{A_S \cdot \eta}. \tag{6}$$

We can achieve the same result by assuming that S1 is illuminating a homogenous target with backscatter σ_{NESZ} (where NESZ stands for Noise Equivalent Sigma Zero), by a means of a mean power that is the product between S1 peak power and the duty cycle, and contributing to the RFI bandwidth B_{RFI}:

$$P_{EIRPZ} = P_S \cdot dc \cdot \sigma_{NESZ} \cdot \frac{B_{RFI}}{B_S} \tag{7}$$

The evaluation of the sensitivity from (7) is straightforward, given the S1 peak power, $P_S = 5.2$ kW, the duty cycle, $dc = 9\%$, the bandwidth, $B_S = 50$–70 MHz (depending on the swath, see Figure 4) and the σ_{NESZ} as from [14,15]. If we assume S1 requirement $\sigma_{NESZ} < -22$ dB, and a ratio B_{RFI}/B_S of 100 looks over the 20,000 spectral samples in range, we get from (7) $P_{EIRPZ} = 15$ mW for a single tone RFI. This sensitivity becomes as small as 8 mW if the actual figure for S1 $\sigma_{NESZ} = -25$ dB is assumed [14,15]. In that case, the equivalent noise temperature from (6) and (7):

$$T_s = \frac{P_S \cdot dc \cdot \sigma_{NESZ} \cdot A_S \cdot \eta}{K_B \cdot B_{RF} \cdot 4\pi R^2} \tag{8}$$

results in roughly 800 K, by assuming an equivalent antenna area $A_S = 9.6$ m^2, total losses $\eta = -4.5$ dB and $R = 840$ km [14,15].

The sensitivity computed refers to a single tone RFI, and does not allow to detect such slight RFI in the raw data in time domain. However, for the strongest RFI power exceeds by far that value, say over a factor 20,000/100 = 20, a time domain identification is possible, like the case shown in Figure 7, that refers to the strongest RFI found. The peak power spectrum is 32 dB above the thermal noise level, and the bandwidth is about 5 MHz, that, from (7) evaluated for IW3, gives $P_{EIRP} = 250$ W. Such a high power is still in the dynamic range of S1, from the level of -3 dB to $\sigma_{NESZ}+39$ dB, say with 10 dB margin.

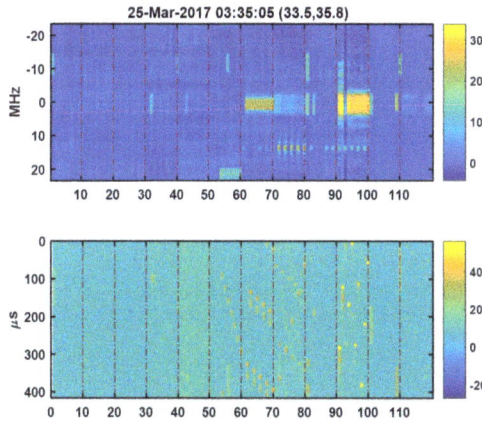

Figure 7. Example of high-power RFI in the spectrogram (**top**); and in the time-along track domain (**bottom**). Color scales are in dB. Horizontal axis is in number of rank pulses that is 10 for IW3.

The method here proposed has been used for estimating RFI all over Europe, in different times—by exploiting ascending and descending passes, and different periods: from November to December 2016 and from February to March 2017. A total of 960 products were considered, spanning 32,000 bursts. In the analysis, we aimed to a very high sensitivity, therefore we discarded the first one-two pulses at the beginning of each burst, since there were still affected by some residual ripples, as shown in Figure 5.

In order to compare the two detectors, we have implemented the Z test in (2) with a very conservative *threshold of 4σ*, which was tuned to the data to exclude false alarms, then we integrated the probability of tails exceeding the threshold:

$$Z = \int_{\mu+4\sigma}^{\infty} p\left(H_f\right) dH_f \tag{9}$$

A comparison between the factor Z and the KL divergence, K, is provided in the bidimensional histogram in Figure 8a, whereas the two values sorted for all the bursts are shown in Figure 8b. In both cases it is possible to notice a two-class behavior, where the thresholds separating the data most likely to be RFI from the good one have been found experimentally from the marginal PDF, plotted on the right, as $\log_{10}(Z_{th}) = -3$ and $\log_{10}(K_{th}) = -1.6$.

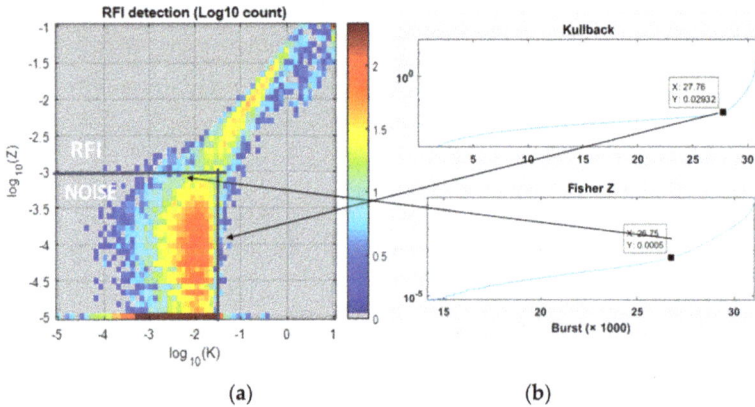

Figure 8. (**a**) Bidimensional histogram, counting the power spectra as function of Kullback-Leibler divergence (K), horizontal, and Fisher figure of merit, (Z), vertical, both in log scale. The two classes, marked as noise and RFI, can be separated better in the 2D than by a single figure of merit, either Z, or K; (**b**) Marginal histograms representing the RFI pdf with respect to Fisher and KL divergence.

In order to appreciate the complementarity between Z and K test, we have plotted a set of histograms exceeding the threshold for K, but not for Z (Figure 9a), and vice versa (Figure 9b). In the total of 32,000 bursts, 3800 were classified RFI according to K, then 12% and 4300 according to Z, 14%, whereas the union of K and Z classified 17% of the bursts.

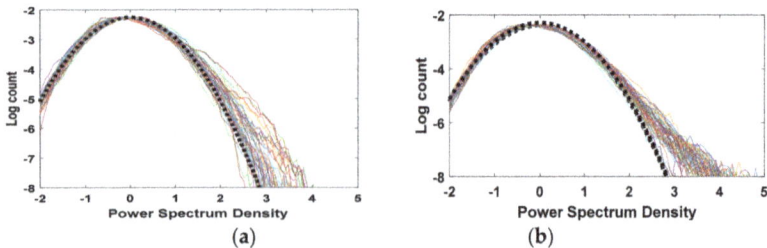

Figure 9. Example of histograms where RFI can be identified only by KL divergence (**a**); and of RFI that can be identified by Fisher Z test (**b**).

This confirm the quality of Z test, however, suggests the use of both (at practically the same cost) if searching for very small RFI, like to RLAN [13].

A map of RFI power has been computed by averaging over the rank echoes and over successive passes, both ascending and descending. The map, shown in Figure 10, reports the highest values measured in the whole spectrum. Detection has been made by assuming a very low threshold in order to visualize the slightest RFI, which would include many false alarms. However, one can appreciate that there is a nearly homogeneous background floor below 0 dBm. The power seems loosely correlated with urban areas, therefore not due to RLAN, whereas there are few occurrences co-located with Radar installations, placed as pinpoints in the map, from those listed in [16].

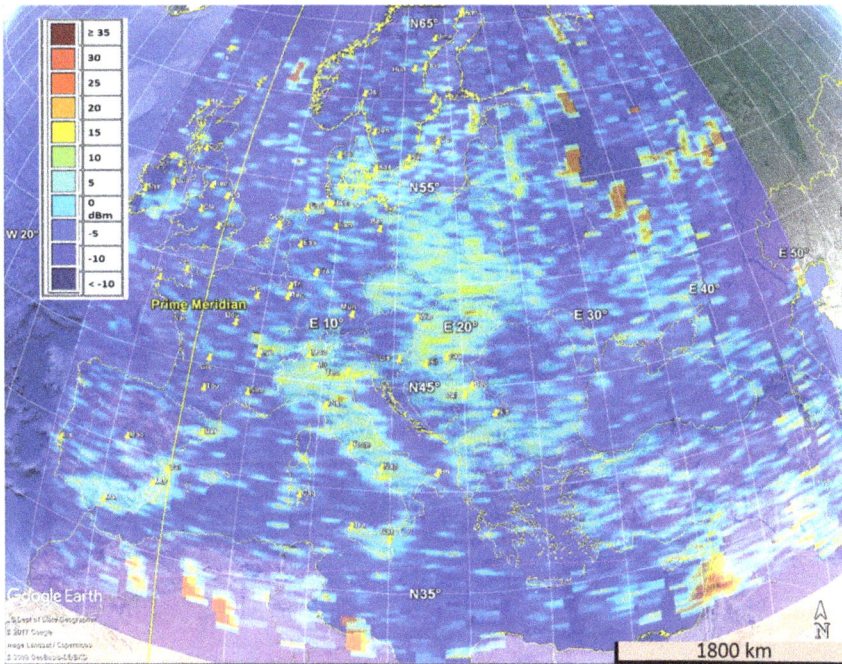

Figure 10. Map of RFI power, averaged in approximately 6 ms. The label superposed marks a part of weather Radar listed in [16].

That power map does not reproduce the spectral occurrence of the RFI. Therefore, we represented in Figure 11, the cumulated histograms of detected RFI power, normalized with respect to S1 noise power, for each frequency bin. The histograms have been normalized so that the figure shows, for each frequency bin, the probability that a certain power level is exceeded. One can notice that, in the majority of the cases, when they are present, the RFI contribute as an additional noise factor less than a figure five, that corresponds to an equivalent temperature of ~4000 K.

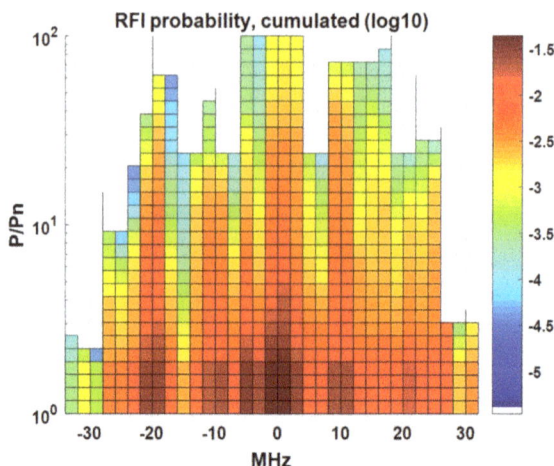

Figure 11. Cumulated histograms of RFI power distributions with frequency, normalized to S1 noise power, estimated by processing data from all-Europe. Vertical scale is probability in $\log_{10}$.

5. Discussion

The main innovation of the proposed method is both in the use of rank-pulses S1 data and in the joint exploitation of two different detectors: Fisher Z test, for the strongest RFI, and KL divergence, sensitive to diffused RFI with low power, like RLAN, that are becoming more and more diffused. The along track squinting of the antenna beam, is a necessary requirement, since it provides rejection to previous transmitted pulses by exploiting the 2D antenna pattern. The method is insofar limited to TOPSAR, and then to S1 (the sole system presently using that mode). Furthermore, the measure is not spanning the whole C-band spectrum, but only up to 40% (in IW1), and it leaves gaps along track, since each swath is sampled for 4 km each 25 km, though near swaths will still be influenced by RFI.

The analysis assumed IW mode data, that exclude RFI detection over polar regions, that are systematically imaged EW mode. Indeed, it could be extended to the EW mode, but the bandwidth would be quite limited, to <20 MHz.

The detection of RFI, carried out with 4σ threshold, would ensure ideally a false alarm rate in the order of $\sim10^{-5}$. However, this applies with respect to thermal noise, whereas the analysis of rank pulses it has shown time-varying behavior due to internal instrument instability. It is expected that detections above threshold are not to be attributed just to RFI, but also to some internal spurious or transients. Nonetheless, the geo-political correlation of the RFI power maps let think that values starting from RFI of 5 dBm are more likely to be attributed to on-ground sources than to the instrument. This seems to be a reliable sensitivity of the system.

6. Conclusions

The paper proposed the use of the first "rank" pulses per burst to derive a very sensitive RFI monitoring from Sentionel-1 SAR. In fact that, thanks to TOPSAR varying squint, those pulses are ideally unaffected by clutter. This has been demonstrated by a careful analysis of S1 acquisition scheme, and checked by processing hundreds of scenes. The achievement of a high sensitivity, in the order of 8 mW of minimal detectable ground power, requires a careful calibration to remove ripples and spurious tones, and frequency domain multi-looking to suppress random fluctuation of noise power.

Detection of RFI has been then approached by exploiting both Fisher Z test, for the strongest RFI, and Kullback-Leibler divergence, sensitive to diffused RFI with low power, like RLAN. The best results have been found by jointly exploiting the two figures of merit.

The power estimation has been applied to a repeat coverage of Europe comprising ascending and descending passes for a total of 32,000 bursts (each of ~20 km along track and 100 km across track), and producing the first map of C-band RFI power. The analysis evidenced many low power sources, spread over the whole bandwidth, and just a few very high power ones, that are active for only part of the time, and on very precise frequency bands.

The force of this approach is the ubiquitous time and space coverage of S1 constellation, which would ensure the capability of producing and updating a world-wide RFI map.

This map could be exploited by policy makers to understand how important is the RFI issues, and to drive the evolution of the RLAN regulation for WIFI.

The production of these map could be done at a negligible computational cost, and in very short time, if rank pulses are made available apart form the whole raw-datasets (that are totally useless for such goal). This upgrade in S1 products is expected for early 2018.

Acknowledgments: The research and the open publication are funded by Italian Space Agency (ASI).

Author Contributions: Andrea Monti-Guarnieri designed and tested the algorithms and wrote the paper. Davide Giudici had the original idea to exploit the rank-burst, revised the paper, and provided information on Sentinel-1 performance. Andrea Recchia wrote the Matlab code to extract rank pulses from Sentinel-1 raw data, the table and schemes of Sentinel-1 calibration, and made the long strip analyses in Figure 2.

Conflicts of Interest: The authors declare no conflict of interest.

References

1. Ulaby, F.; Lang, D.A. Committee on a Survey of the Active Sensing—Uses of the Radio Spectrum Board on Physics and Astronomy. In *A Strategy for Active Remote Sensing Amid Increased Demand for Radio Spectrum*; The National Academy Press: Washington, DC, USA, 2015; ISBN 978-0-309-37305-0.
2. Soldo, Y.; Cabot, F.; Khazaal, A.; Miernecki, M.; Slominska, E.; Fieuzal, R.; Kerr, Y.H. Localization of RFI sources for the SMOS mission: A means for assessing SMOS pointing performances. *IEEE J. Sel. Top. Appl. Earth Obs. Remote Sens.* **2015**, *8*, 617–627. [CrossRef]
3. RFI Monitoring. SMOS Blog at CESBIO. Web Resource. Available online: http://www.cesbio.ups-tlse.fr/SMOS_blog/?page_id=4087 (accessed on 16 September 2017).
4. Soldo, Y.; de Matthaeis, P.; le Vine, D.; Richaume, P. Comparison of L-band RFI from SMOS and Aquarius observations. In Proceedings of the 2nd SMOS Science Conference, ESA-ESAC, Villafranca, Madrid, 25–29 May 2015.
5. Meyer, F.J.; Nicoll, J.; Doulgeris, A.P. Characterization and correction of residual RFI signatures in operationally processed ALOS PALSAR imagery. In Proceedings of the EUSAR 2012: 9th European Conference on Synthetic Aperture Radar, Nuremberg, Germany, 23–26 April 2012; pp. 83–86.
6. Meyer, F.J.; Nicoll, J.B.; Doulgeris, A.P. Correction and Characterization of Radio Frequency Interference Signatures in L-Band Synthetic Aperture Radar Data. *IEEE Trans. Geosci. Remote Sens.* **2013**, *51*, 4961–4972. [CrossRef]
7. Johnson, J.T.; Chen, C.C.; O'Brien, A.; Smith, G.E.; McKelvey, C.; Andrews, M.; Ball, C.; Misra, S.; Brown, S.; Kocz, J.; et al. The CubeSat Radiometer Radio Frequency Interference Technology Validation (CubeRRT) mission. In Proceedings of the 2016 IEEE International Geoscience and Remote Sensing Symposium (IGARSS), Beijing, China, 10–15 July 2016; pp. 299–301.
8. Ticconi, F.; Anderson, C.; Saldana, J.F.; Wilson, J. Analysis of the noise scenario measured by ASCAT. In Proceedings of the 2015 IEEE International Geoscience and Remote Sensing Symposium (IGARSS), Milan, Italy, 26–31 July 2015; pp. 4882–4885.
9. Potin, P.; Rosich, B. Sentinel-1 Mission Status. In Proceedings of the EUSAR 2016: 11th European Conference on Synthetic Aperture Radar, Hamburg, Germany, 5 October 2016; pp. 1–6.
10. Orfanidis, S.J. *Introduction to Signal Processing*; Prentice-Hall: Englewood Cliffs, NJ, USA, 1996.
11. Kullback, S.; Leibler, R.A. On information and sufficiency. *Ann. Math. Stat.* **1951**, *22*, 79–86. [CrossRef]
12. Gray, R.M. *Source Coding Theory*; Kluwer: Boston, MA, USA, 1990.

13. Scott, P.J.; Sydor, J.; Brandão, A.; Yongacoglu, A. Radio Local Area Network (RLAN) and C-Band Weather Radar Interference Studies. In Proceedings of the 32nd AMS Radar Conference on Radar Meteorology, Albuquerque, NM, USA, 24–29 October 2005.

14. European Space Agency (ESA). *Sentinel-1: ESA's Radar Observatory Mission for GMES Operational Services*; ESA SP-1322/1; ESA Communications: Noordwijk, The Netherlands, March 2012.

15. Collecte Localisation Satellites (CLS). *Sentinel-1A and -1B Annual Performance Report 2016*; doc. MPC-0366, DI-MPC-APR, v1.1; European Space Agency (ESA): Paris, France, April 2017.

16. Newsome, D.H. *Weather Radar Networking: COST 73 Project/Final Report*; Springer: Berlin, Germany, October 2013.

remote sensing

MDPI

Article

An Accelerated Backprojection Algorithm for Monostatic and Bistatic SAR Processing

Heng Zhang [1,2], Jiangwen Tang [1,2], Robert Wang [1,*], Yunkai Deng [1], Wei Wang [1] and Ning Li [1]

[1] Department of Space Microwave Remote Sensing Systems, Institute of Electronics, Chinese Academy of Sciences, Beijing 100190, China; caszhmail@163.com (H.Z.); jiangwen@mail.ustc.edu.cn (J.T.); ykdeng@mail.ie.ac.cn (Y.D.); wwang@mail.ie.ac.cn (W.W.); lining_nuaa@163.com (N.L.)

[2] Department of Electronics, Electrical and Communication, University of Chinese Academy of Sciences, Beijing 100049, China

* Correspondence: yuwang@mail.ie.ac.cn; Tel.: +86-10-5888-7166

Received: 27 November 2017; Accepted: 16 January 2018; Published: 18 January 2018

Abstract: The backprojection (BP) algorithm has been applied to every SAR mode due to its great focusing quality and adaptability. However, the BP algorithm suffers from immense computational complexity. To improve the efficiency of the conventional BP algorithm, several fast BP (FBP) algorithms, such as the fast factorization BP (FFBP) and Block_FFBP, have been developed in recent studies. In the derivation of Block_FFBP, range data are divided into blocks, and the upsampling process is performed using an interpolation kernel instead of a fast Fourier transform (FFT), which reduces the processing efficiency. To circumvent these limitations, an accelerated BP algorithm based on Block_FFBP is proposed. In this algorithm, a fixed number of pivots rather than the beam centers is applied to construct the relationship of the propagation time delay between the "new" and "old" subapertures. Partition in the range dimension is avoided, and the range data are processed as a bulk. This accelerated BP algorithm benefits from the integrated range processing scheme and is extended to bistatic SAR processing. In this sense, the proposed algorithm can be referred to simply as MoBulk_FFBP for the monostatic SAR case and BiBulk_FFBP for the bistatic SAR case. Furthermore, for monostatic and azimuth-invariant bistatic SAR cases where the platform runs along a straight trajectory, the slant range mapping can be expressed in a continuous and analytical form. Real data from the spaceborne/stationary bistatic SAR experiment with TerraSAR-X operating in the staring spotlight mode and from the airborne spotlight SAR experiment acquired in 2016 are used to validate the performances of BiBulk_FFBP and MoBulk_FFBP, respectively.

Keywords: accelerated backprojection algorithm; bistatic SAR; monostatic SAR; bulk processing

---

1. Introduction

With ongoing technological progress, synthetic aperture radar (SAR) systems, including multi-mode monostatic SAR and bistatic SAR, are becoming increasingly sophisticated. For monostatic SAR, many effective processing algorithms have been developed [1]. Because of the high efficiency, frequency domain imaging algorithms, such as the Range Doppler [2], chirp scaling [3] and ωK [4] algorithms are widely-applied methods for stripmap SAR data focusing. Their modified versions can focus data from other imaging geometries, such as the SPECAN algorithm presented in [5] for ScanSAR data processing and the extended chirp scaling algorithm proposed in [6] for spotlight SAR. For bistatic SAR, both advantages and disadvantages exist because of the spatial separation between the transmitter and the receiver. On the one hand, this transmitter-receiver separation increases the system design flexibility and makes bistatic SAR a more promising technology for global remote sensing mission applications. On the other hand, it generates increased complexity in the system operation and data processing. Before the general-purpose graphics processing unit (GPGPU) was applied, frequency domain algorithms were developed by researchers for the bistatic SAR data focusing requirement. Due to the diversity of bistatic SAR geometries, the analytical two-dimensional spectrum is difficult to obtain, and the existing frequency domain methods are applicable only to certain environments. Algorithms based on Loffeld's bistatic formula [7], such as the 2D inverse scaled FFT algorithm [8] and the bistatic chirp scaling algorithm [9], can focus the data acquired in azimuth-invariant bistatic geometry where the transmitter and receiver run along different trajectories with identical velocity vectors. The range Doppler algorithm based on the series reversion [10], equivalent velocity approximation and NuSAR[11] can focus data from azimuth-variant bistatic SAR mode where the transmitter and receiver run with different velocity vectors. Nonlinear chirp scaling algorithms [12–14] and the wavenumber-domain algorithm proposed in [15] can focus bistatic SAR data from a hybrid bistatic configuration where the transmitter and receiver are mounted on two very different platforms, such as the spaceborne/stationary bistatic SAR mode.

As a correlation algorithm in the time domain, the backprojection (BP) algorithm can be applied to almost every SAR configuration [16]. However, the immense time cost limits its application. To improve the computational efficiency, two mainstream approaches have been explored. First, parallel computing platforms with incredible computing power, e.g., GPGPUs, have been used to accelerate the progress of the BP algorithm [17]. Second, incremental modifications have been applied to the conventional BP algorithm; the typical products are the fast BP [18] and fast factorized backprojection (FFBP) algorithms [19]. Inspired by the FBP algorithm developed for monostatic SAR data, several bistatic FBP algorithms have been proposed [20–25]. For example, a BiSAR_FBP algorithm was proposed in [23] to focus ultra-wideband-ultra-wide-beam bistatic SAR data. Bistatic FBP algorithms for general bistatic SAR configurations and for one-stationary geometry were proposed in [24,25], respectively.

In [19], the FFBP algorithm was developed based on the theory that the bandwidth is much lower than the sampling rate in an angular coordinate system. However, processing in a polar coordinate system may be cumbersome. To simplify the algorithm process, raw data are partitioned into several blocks in the range direction. This algorithm is hereafter referred to as Block_FFBP. The center of each block is taken as the reference point, and the slant range of other points can be calculated by adding an offset value according to the range of the reference point. Nevertheless, due to the partitioning of range data, the interpolation process of each data block in subaperture summation can only be conducted using an interpolation kernel, which decreases the efficiency. Thus, an accelerated BP algorithm based on uniform rather than partitioned range processing is proposed in this paper. In this algorithm, partitioning of the range data is not needed, and a fixed number of pivots is applied to calculate the slant range of a grid point under both the current aperture and the synthesized one. The range domain data are processed as a bulk, and interpolation can be performed with an FFT, which improves the efficiency of subaperture summation. The derivation of this accelerated BP algorithm in the Cartesian coordinate system is provided for monostatic SAR (MoBulk_FFBP) and bistatic SAR (BiBulk_FFBP). However, although the slant range computation is taken in the Cartesian coordinate system, this approach only simplifies the computation and does not disrupt the sampling theory in the angular frequency domain. Two datasets are used to validate the performance of the proposed algorithms. The first one is from the spaceborne/stationary bistatic SAR system with TerraSAR-X as the transmitter operated in the staring spotlight mode. The second one is from the spotlight experiment with an airborne X-band SAR system operating at a bandwidth of 1200 MHz.

This paper is arranged as follows. Section 2 briefly describes the basic principle of the FBP and describes its disadvantages. In Section 3, derivations of MoBulk_FFBP and BiBulk_FFBP are described in detail. In Section 4, detailed performance analysis of the proposed algorithm, including the error analysis of residual phase, the parallelization consideration, the computational complexity and pivot selection issue, is discussed. In Section 5, monostatic and bistatic images are shown to demonstrate the validity of the algorithm. A comparison of the efficiency between Block_FFBP and the proposed algorithm is also given here. Finally, the conclusion is drawn in Section 6.

2. Description of the Fast BP Algorithm

2.1. Fundamental Concept

In the polar format algorithm [4] where two-dimensional matched filtering is implemented in the phase history domain, the azimuth extension is a reciprocal of the sample spacing in the azimuth frequency domain. Similarly, sample spacing in the azimuth time domain and the azimuth frequency span follow the same principle. According to this time-frequency mapping attribute, the computational complexity can be reduced by data segmentation, which is adopted by the FBP algorithm. If the scene extension is divided into two sub-planes in the azimuth dimension, the corresponding frequency sampling space is consequently extended, which corresponds to azimuth de-sampling. After the aforementioned azimuth partitioning, the new datasets can also be divided until a proper data size is obtained, which is the basic concept of the FFBP algorithm [19]. In general, the computational complexity is $O\left(mN^2 log_m N\right)$ when the factor of the factorization of aperture N is m. After data partitioning, azimuth de-sampling arises, which could cause azimuth spectrum aliasing and ambiguity in the final image. For its sake, subaperture processing is introduced in the FBP algorithm.

2.2. Subaperture Processing

Following the azimuth de-sampling mentioned above, the pulse repetition frequency (PRF) is also decreased. To avoid aliasing, PRF should be larger than the azimuth bandwidth during each stage of factorization. Subaperture summation is performed in this sense, as shown in a schematic diagram in Figure 1. The left image in Figure 1 shows the data synthesis using the full aperture, while the right image illustrates the subaperture processing. Through the summation of two subapertures, two beams pointing in different directions are obtained. Meanwhile, the beamwidth is halved, and the decreased PRF still satisfies the Nyquist sampling constraint. Subaperture processing is used in many FBP algorithms, such as [19,25].

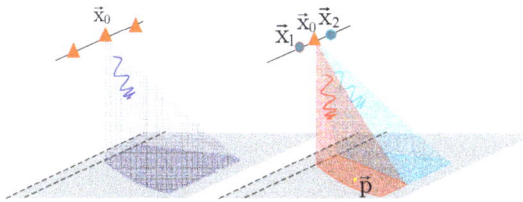

Figure 1. Subaperture summation progress. In the left panel, 'triangle' denotes the location where the platform transmits the signal. In the right panel, $\vec{x_1}$ and $\vec{x_2}$ denote two subapertures and $\vec{x_0}$ denote the new synthesized aperture.

Suppose that the factorization factor is m and that the beam center of a synthesized aperture on an imaging plane is $\vec{p}$. Then, the summation of m subapertures can be written as:

$$s\left(\vec{x}, \vec{p}; \tau\right) = \sum_{i=1}^{m} s\left(\vec{x}_i, \vec{p}; \tau - \Delta\tau_i\right) \exp\left\{-j2\pi f_0 \Delta\tau_i\right\}, \tag{1}$$

where τ is the fast time, $\Delta\tau_i$ is the time difference of signal propagating from subaperture $\vec{x}$ to $\vec{p}$ and from $\vec{x}_i$ to $\vec{p}$,

$$\Delta\tau_i = 2\frac{\left(|\vec{x} - \vec{p}|_2 - |\vec{x}_i - \vec{p}|_2\right)}{c}. \tag{2}$$

where f_0 is the carrier frequency, $\vec{x}_i$ is the location of the original aperture, $\vec{x}$ is the location of the synthesized aperture and $s\left(\vec{x}_i, \vec{p}\right)$ is the echo signal at $\vec{x}_i$.

In this derivation, the summation of subapertures is precise only for the beam center point, for which there is no residual error of slant range. When the scene extends beyond a certain scope, the residual range error will increase, and the summation will deteriorate. To control the maximum error, a partition of the range data according to the multiple beam centers is adopted in Block_FFBP at each stage of subaperture summation. Figure 2 presents an example of the subaperture summation and range data partition in each processing stage. Initially, the number of subapertures is eight. In each stage, the number of subapertures is reduced by a factor of two.

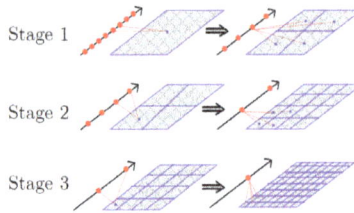

Figure 2. Schematic illustration of the FFBP with eight apertures and three factorization stages. In each processing stage, a common factorization factor of two is used. Narrower "new" beams are formed based on wider "old" beams formed in the previous stage. In the first stage, each of two adjacent subapertures forms a new beam, and 2×2 beams are obtained. In the second stage, repeat the summation operation of Stage 1, and 4×4 beams are obtained.

At each stage of subaperture summation in the conventional FBP algorithm, the steering angle and radius are calculated within the local polar coordinate system where the origin is the new aperture. Then, the coordinate of the target and the slant range to the final aperture are determined using subaperture summation. Meanwhile, in the Block_FFBP, the position of the beam center and the corresponding slant range, as well as the differential range are determined for each range block. The slant range of the other points can then be obtained through linear extrapolation using the range and differential range.

The main limitation of Block_FFBP is that the range data are partitioned according to the beam centers; therefore, in the subaperture summation process, the synthesized data in each block can only be obtained through interpolation. If the range dimension partition could be avoided, range data from the identical beam are maintained as a bulk, and the interpolation can be replaced by an FFT. Furthermore, to ensure the interpolation quality of a margin point for a selected interpolation kernel, a fixed backup allowance of the beam data length should be retained. With an increase in the number of factorization stages, the proportion of the interpolation kernel length throughout the entirety of the beam data increases, which indicates that additional memory space is required.

3. Accelerated BP Algorithm

3.1. Fundamental Concept

In this accelerated BP algorithm, a fixed number of pivots, rather than beam centers whose number is variant in each subaperture factorization and summation stage, are applied. Like the general interpolation process, values at sample points are provided, and the interpolated values at specific query points can be obtained using an interpolation method. Pivots are analogous to the sample points. The correspondence relationship of the propagation time delay from these pivots to current and to synthesized subapertures can be constructed. Afterward, the slant range of the other imaging points can be determined using a conventional interpolation method.

Due to oversampling in the angular coordinate system, the slant range calculations are transferred from the polar coordinate system to the Cartesian coordinate system so that the range calculations are simplified. Therefore, this accelerated BP algorithm can be applied in both bistatic and monostatic SAR configurations.

3.2. Monostatic SAR Case

The left image in Figure 3 shows an example of subaperture summation where the factorization factor is two. The platform runs along the y-axis; the x-axis is the range direction; and the z-axis denotes the height dimension. The new subaperture at $\vec{A}_0 (x', y', z')$ is the summation of subapertures at $\vec{A}_1 (x'_1, y'_1, z'_1)$ and $\vec{A}_2 (x'_2, y'_2, z'_2)$. The number of pivots along the range dimension is n, and a pivot can be denoted as $\vec{p}_i (x_i, y_i, z_i)$. In this accelerated BP algorithm, the number of pivots is fixed during the focusing process, which is different from Block_FFBP, wherein the number of reference points grows with an increase in the number of factorization stages.

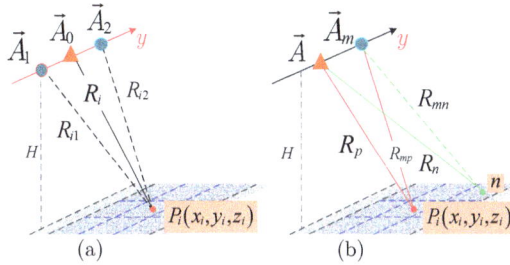

Figure 3. Monostatic SAR case: (**a**) diagram of subaperture summation; (**b**) diagram of error analysis.

In the left panel of Figure 3, the slant ranges R_{i1}, R_{i2} and R_i are the distances between pivot $\vec{p}_i (x_i, y_i, z_i)$ and the apertures $\vec{A}_1 (x'_1, y'_1, z'_1)$, $\vec{A}_2 (x'_2, y'_2, z'_2)$ and $\vec{A}_0 (x', y', z')$, respectively. The corresponding time delay can be expressed as:

$$\tau_{im} = 2\frac{\left|\vec{A}_m - \vec{p}_i\right|_2}{c}, \ (m = 1, \ 2); \ \tau_i = 2\frac{\left|\vec{A}_0 - \vec{p}_i\right|_2}{c} \tag{3}$$

where c is the speed of light. The range-compressed signals belonging to the "old" subapertures $\vec{A}_1 (x'_1, y'_1, z'_1)$ and $\vec{A}_2 (x'_2, y'_2, z'_2)$ are $s_1 (\tau)$ and $s_2 (\tau)$, respectively. Afterward, according to the subaperture summation stage illustrated in Section 2.2, the "new" synthesized subaperture data $s (\tau)$ can be written as:

$$s (\tau_i) = \sum_{m=1}^{2} s_m (\tau_{im}) e^{-j2\pi f_0 (\tau_{im} - \tau_i)}, \ i = 1, \ ..., \ n. \tag{4}$$

To construct the mapping relationship between the echo delay of the "new" and "old" subapertures, an interpolation function $g_m (\tau_i)$ can be used for the delay time pair (τ_i, τ_{im}), which can be expressed as:

$$g_m (\tau_i) = \tau_{im}; \ i = 1, \ 2, \ ..., \ n; \ m = 1, \ 2, \tag{5}$$

where $g_m (\tau_i)$ can be a spline interpolation kernel or a sinc kernel. The subaperture summation can be updated by substituting this interpolation function into Equation (4). The new expression of $s (\tau)$ is

$$s (\tau) = \sum_{m=1}^{2} s (g_m (\tau)) e^{-j2\pi f_0 (g_m (\tau) - \tau)}. \tag{6}$$

Equation (6) is a continuous expression of the subaperture summation process. Hence, partitioning of the range data is favorably avoided, and the complete range data can be processed in bulk. Therefore, the interpolation operation for subaperture data, which is denoted as $s_m(\cdot)$, can be performed with an FFT to improve the overall efficiency [26].

Equation (3) is a general expression for the monostatic two-way propagation delay, and some simplifications can be made to get a more functional version. If the height of the imaging plane is const z_0 and the height of platform is z_h, then $z_i = z = z_0; z'_m = z' = z_h, m = 1, 2$. Equation (3) becomes:

$$\tau_{im} = \frac{2\sqrt{(x'_m - x_i)^2 + (y'_m - y_i)^2 + (z_h - z_0)^2}}{c}, \tag{7a}$$

$$\tau_i = \frac{2\sqrt{(x' - x_i)^2 + (y' - y_i)^2 + (z_h - z_0)^2}}{c}, \tag{7b}$$

In Equation (7b), the range offset $x' - x_i$, between "new" subaperture $\vec{A}_0$ and $\vec{p}_i$, is:

$$\Delta x_i(\tau_i) = -\sqrt{\left(\frac{\tau_i \cdot c}{2}\right)^2 - (y' - y_i)^2 - (z_h - z_0)^2}. \tag{8}$$

Substituting Equation (7b) into Equation (7a), a new expression for the time delay can be obtained,

$$\tau_{im} = \frac{2\sqrt{(x'_m - x' + \Delta x_i(\tau_i))^2 + (y'_m - y_i)^2 + (z_h - z_0)^2}}{c}. \tag{9}$$

Furthermore, when the platform runs along a straight trajectory (i.e., $x'_m = x'$), Equation (9) can be simplified as:

$$\tau_{im} = \sqrt{\tau_i^2 + \frac{4}{c^2}\left[(y'_m - y_i)^2 - (y' - y_i)^2\right]}. \tag{10}$$

Equation (10) is an analytical expression that can determine the relationship between the two propagation delays of the "new" and "old" subapertures. Let $\Delta\epsilon = 4\left[(y'_m - y_i)^2 - (y' - y_i)^2\right]/c^2$; the subaperture summation can be expressed in a continuous form as:

$$s(\tau) = \sum_{m=1}^{2} s\left(\sqrt{\tau^2 + \Delta\epsilon}\right) e^{-j2\pi f_0\left(\sqrt{\tau^2 + \Delta\epsilon} - \tau\right)}. \tag{11}$$

Equation (11) represents a case in which the factorization factor is two, and it can be naturally extended to cases with other factorization factors. It is a more simple and practical subaperture summation approach without pivots that facilitates the computational operation. In Equation (10), the locations of "new" and "old" subapertures are only required to be known, and the pivots are no longer required. In this sense, this functional version of Equation (10) is known as range determination. This makes a further simplification of the process. In practice, with the use of fine motion compensation, an equivalent straight line can be obtained, and MoBulk_FFBP can be applied without pivots.

3.3. Bistatic SAR Case

According to the basic principle of the proposed accelerated BP algorithm, this algorithm can also be applied in the bistatic SAR mode. In this section, the accelerated BP algorithm is provided for two bistatic SAR geometries: the one-stationary bistatic SAR mode and the tandem mode. In the former case (including the spaceborne/stationary and the airborne/stationary cases), only the moving platform contributes to the azimuth modulation, whereas the stationary platform introduces a range offset to the range migration trajectories of targets at the same range [14,27]. Therefore, the subaperture summation can be conducted for the moving platform.

Figure 4 shows the subaperture summation for one-stationary bistatic SAR mode where the factorization factor is set to two. The transmitter runs along the y-axis, and the receiver is fixed. R_{it1}, R_{it2} and R_{it} are the ranges between $\vec{p}_i(x_i, y_i, z_i)$ and the transmitter subapertures $\vec{A}_1(x'_1, y'_1, z'_1)$,

$\vec{A}_2\left(x_2', y_2', z_2'\right)$, $\vec{A}_0\left(x', y', z'\right)$, respectively; while, R_{ir} is the range between the receiver and $\vec{p}_i$. Thus, the echo delay for each subaperture can be written as:

$$\tau_{im} = \frac{R_{itm} + R_{ir}}{c}, \quad (m = 1, 2),$$ (12)

$$\tau_i = \frac{R_{it} + R_{ir}}{c}$$

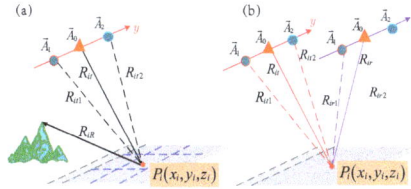

Figure 4. Bistatic SAR cases: (**a**) one-stationary bistatic mode; (**b**) azimuth-invariant bistatic mode.

The data of "new" subaperture $\vec{A}_0$ can also be obtained using Equation (6). For a given range line in the focus plane parallel to the transmitter trajectory, the range to the receiver is different for each grid point. Therefore, it is difficult to obtain an analytical expression for the subaperture summation without deriving a relationship for the location of the pivots.

However, in the azimuth-invariant bistatic SAR mode, this situation may be different [11,28]. The delay times for the "old" and "new" subapertures are:

$$\tau_{im} = \frac{R_{itm} + R_{irm}}{c}, \quad m = 1, 2$$

$$\tau_i = \frac{R_{it} + R_{ir}}{c}$$ (13)

For this mode, let the "new" subapertures of a transmitter and receiver be $\vec{A}_0\left(x_1, y_1, z_1\right)$ and $\vec{A}_0\left(x_2, y_2, z_2\right)$, respectively. Let $R_i = \tau_i \cdot c$,

$$R_{it} = \sqrt{\left(x_1 - x_i\right)^2 + \left(y_1 - y_i\right)^2 + \left(z_1 - z_i\right)^2}$$

$$R_{ir} = \sqrt{\left(x_2 - x_i\right)^2 + \left(y_2 - y_i\right)^2 + \left(z_2 - z_i\right)^2}$$ (14)

For convenience, let $c_1 = \left(y_1 - y_i\right)^2 + \left(z_1 - z_i\right)^2$, $c_2 = \left(y_2 - y_i\right)^2 + \left(z_2 - z_i\right)^2$ and $c_3 = x_1 - x_2$; thus, $x_i\left(\tau_i\right)$ can be solved as

$$x_i\left(\tau_i\right) = \frac{-b + \sqrt{b^2 - a \cdot d}}{2a}$$ (15)

where:

$$a = c_3^2 - \left(c \cdot \tau_i\right)^2,$$

$$b = c_3 \left[c_1 - c_2 + a\right],$$ (16)

$$d = c_1^2 + c_2^2 + a^2 - 2c_1\left(c_2 - a\right) - 2c_2\left[c_3^2 + \left(c \cdot \tau_i\right)^2\right].$$

Therefore, $\tau_{im}\left(\tau_i\right)$ can be obtained by substituting (15) into τ_{im} in Equation (13). Similar to Equation (11), the subaperture summation can also be expressed in a continuous form:

$$s\left(\tau\right) = \sum_{m=1}^{2} s\left(\tau_m\left(\tau\right)\right) e^{-j2\pi f_0\left(\tau_m\left(\tau\right) - \tau\right)}$$ (17)

3.4. Summary of the Algorithm

Figure 5 presents the flowchart of proposed algorithm. The green and cyan blocks represent processing with and without pivots, respectively. Range compression is conducted in the first stage. The number of factorization stages and the factorization factor in each stage are set. For the general case, pivots are required. According to Equation (5), an interpolation function is used to calculate the delay time of the "new" subapertures, which corresponds to the propagation time delay reconstruction in Figure 5. Specifically, when the platform runs along a straight track, pivots are not required, and in the factorization stage, delay time is determined according to Equation (10). After the propagation time delay relationship is established, range data interpolation is performed using an FFT followed by the subaperture summation. When the factorization is done, a conventional BP algorithm is applied to focus the new synthetic data, and a final focused image is obtained.

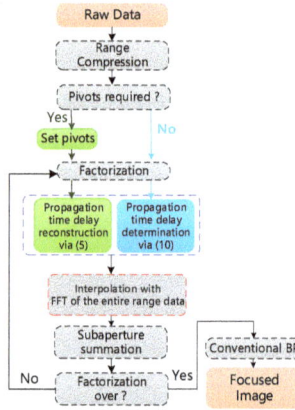

Figure 5. Flowchart of the proposed algorithm. In this flowchart, the MoBulk_FFBP and BiBulk_FFBP are integrated together. Mo, monostatic; Bi, bistatic; FFBP, fast factorization BP.

4. Performance Analysis

4.1. Error Analysis

This section provides the phase error caused by an incorrect slant range when the back-projected data are accumulated during a subaperture summation. Since the proposed algorithm is derived from Block_FFBP and extended to the bistatic SAR case, the error analysis will be conducted through a comparison with Block_FFBP using a numerical method. Moreover, as Block_FFBP was developed for the monostatic SAR case, the error analysis is mainly performed for this case.

In Block_FFBP, the slant range error between the "old" and "new" subapertures causes a phase error, which can affect the focusing. For a certain imaging block, like the right panel in Figure 3, only the delay time of the beam center point $\vec{p}_i$ is correct during the subaperture summation. Assume that $\vec{A}_m$ is the "old" subaperture and $\vec{A}$ is the "new" one. Point $\vec{n}$ is one grid point in the block. The slant ranges from $\vec{p}_i$ to $\vec{A}_m$ and to $\vec{A}$ are R_{mp} and R_p, respectively. Likewise, for point $\vec{n}$, the slant ranges are R_{mn} and R_n, respectively. In Block_FFBP, the slant range of "new" aperture $\vec{A}$ should be mapped to the "old" aperture to perform the subaperture summation. For any point $\vec{n}$, the obtained range is $R'_{mn} = R_n + R_{mp} - R_p$, and the two-way delay time error caused by this operation is:

$$\Delta \tau_{error} = \frac{2}{c} \cdot \left(R_{mn} - R'_{mn} \right) = \frac{2}{c} \cdot \left(R_{mn} - R_n - \left(R_{mp} - R_p \right) \right). \tag{18}$$

Thus, the residual phase error can be written as:

$$\Delta\phi_1 = 2\pi f_0 \cdot \Delta\tau_{error}. \tag{19}$$

In MoBulk_FFBP, the beam center point $\vec{p}$ in the right panel of Figure 3 can also be taken as a pivot. When the subaperture summation is performed using Equation (6), the residual phase error of point $\vec{n}$ can be written as:

$$\Delta\phi_2 = 2\pi f_0 \cdot \left[g_m\left(\frac{2R_n}{c}\right) - \frac{2R_{mn}}{c} \right], \tag{20}$$

where $g_m(\cdot)$ is the interpolation kernel defined by Equation (5). Moreover, when the platform runs along a straight trajectory, pivots are not required, and points that have the same azimuth coordinates as $\vec{p}$ have no phase error. The residual phase error can be expressed as:

$$\Delta\phi_3 = 2\pi f_0 \cdot \left[\sqrt{\left(\frac{2R_n}{c}\right)^2 + \Delta\epsilon} - \frac{2R_{mn}}{c} \right]. \tag{21}$$

Here, a numerical simulation is conducted to intuitively compare these phase errors. An airborne geometry with a straight trajectory, a platform height of 8 km, an off-nadir angle of 40° and a carrier frequency of 9.6 GHz are investigated. The offset range between the "new" and "old" subapertures $|AA_m|$ varies from 1 m–1000 m, and the squint angle varies from 0°–24°. For a certain offset and squint angle, the scene size is set according to the principle that the maximum residual phase error of an imaging point in the scene is $\pi/8$. In each such scene, a 200 × 200 point array is set. Through statistics, the maximum $\Delta\phi_2$ or $\Delta\phi_3$ of these points is obtained. To compare these with $\Delta\phi_1$, the maximum residual phase error is normalized by $\pi/8$. The final result is shown in Figure 6.

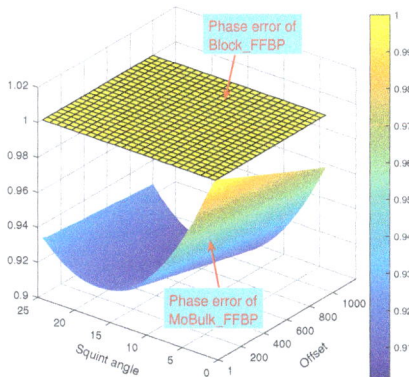

Figure 6. Comparative analysis of the residual phase error in Block_FFBP and MoBulk_FFBP. The top yellow plane indicates the residual phase error of Block_FFBP, which is equal to $\frac{\pi}{8}$ (for a specific offset and squint angel pair, the scene size changes to ensure that the maximum residual phase error is no more than $\frac{\pi}{8}$). The offset axis is the range between the "new" and "old" aperture.

The top yellow plane in Figure 6 is the residual phase error of Block_FFBP, which is normalized by $\pi/8$. A larger offset range between the "new" and "old" subapertures and a smaller residual phase error for the proposed MoBulk_FFBP can be observed. Moreover, under this simulation configuration, the residual phase error decreases with the squint angle, which varies from 0° to 18°. Although the phase error increases from 18°–25° in this experiment, the overall phase error is lower than that of Block_FFBP, which could validate the accuracy of the proposed algorithm.

4.2. Parallelization

As is shown in Figure 5, the factorization is operated sequentially. However, the range data are processed as a bulk set, and the upsampling of this is performed using an FFT in each stage of subaperture summation. In this sense, each stage of the factorization can be executed in parallel. Moreover, after factorization, the conventional BP algorithm can also be executed in parallel, such as with a GPU [29,30] or a multi-thread technical CPU. This will further accelerate the proposed algorithm.

4.3. Computational Complexity

A detailed theoretical derivation of the computational complexity of the proposed algorithm is provided here. Suppose that the aperture length is L and the data size is $N(Az) \times M(Rg)$. "Az" and "Rg" denote the azimuth and range dimension, respectively. In the processing, a factorization of L into K integer factors, corresponding to K processing stages, is established. The reduction in the number of apertures is defined as $l_i, (i = 1, 2, \ldots, K)$, and L can be expressed as:

$$L = l_1 \times l_2 \times l_3 \times \cdots \times l_K. \tag{22}$$

For simplicity, a common factorization throughout all of the stages is used ($l_i = n$ for all i); then, $L = n^K$, and $K = \log_n N$. The number of pivots is set to Q and is fixed at each stage. The aperture length is equal to the azimuth data length such that $L = N$.

In the first stage, the original aperture is split into N/n subapertures, each of which has a length n. To construct the mapping relationship between the echo delays of the "new" and "old" subapertures, an efficient cubic spline interpolation scheme [31] is used. As demonstrated in [31], this spline interpolation is very efficient: $O(Q)$ is used to generate the spline, and $O(\log Q)$ is used to evaluate the spline at a single point, where Q is the number of input data points. Thus, the interpolation computational burden is $\frac{N}{n} \times n \, (Q + M \log Q)$. Next, the interpolation of range data is implemented using upsampling with an FFT, and the overall computational complexity for the FFT and inverse FFT is $\frac{N}{n} \times n \, (M \log M + \alpha M \log \alpha M)$, where α is the upsampling rate. At this stage, the number of beams is n, and the computational burden required to form the beams is:

$$\frac{N}{n} \, (subapertures) \times n \, (subaperture\ points) \times (nM) \, (beams\ samples) \,. \tag{23}$$

Equation (23) can be written as nMN for simplicity. Based on the first factorization stage, the second processing stage forms N/n^2 new beams. Because of the common number of pivots and the unchanged number of range data, the operations for interpolation and upsampling are the same as those in the first stage. The number of operations required to form new beams becomes $\frac{N}{n^2} \times n \times (n^2 M) = nMN$. Therefore, each processing stage has the same number of operations, and the total computational complexity can be written as:

$$[N \, (Q + M \log Q + M \log M + \alpha M \log \alpha M) + nMN] \times \log_n N. \tag{24}$$

4.4. Pivot Selection Issue

During the factorization and subaperture summation process, the accuracy of signal propagation time between the "new" and "old" subapertures could affect the final image quality. In Block_FFBP, the slant range computation error changes across the data block. In the proposed algorithm, the accuracy of slant range can be ensured by the interpolation, which is described in Equation (5). To evaluate the influence of the chosen number of pivots, slant range computation errors between "new" and "old" subapertures are calculated in the condition of different swaths and different number of pivots. Assume that the platform height is 8 km, the look angle is 40° and the subaperture offset is 400 m. The interpolation scheme is a spline function.

In this simulation, the number of pivots changes from 4–64, and pivots are distributed along the range dimension at the same intervals. First, 3000 points are located along the range direction with different range extensions. The average slant computation error is shown in Figure 7a. Second, 3000 points are located along the azimuth dimension with different azimuth extensions. The average slant computation error is shown in Figure 7b.

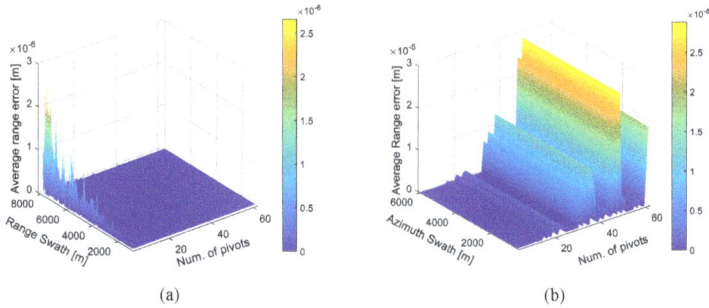

(a)   (b)

Figure 7. Simulation of slant range error caused by different numbers of pivots and (**a**) range swaths and (**b**) azimuth swath.

It can be seen that due to the high accuracy of spline interpolation, the slant range computation is very accurate. The influence of different chosen numbers of pivots is ignorable and varies slightly. In this sense, the choice of pivots is flexible. Please note that in the experiment, the smallest number of pivots is four; this is because the spline interpolation method requires at least four sampling points.

5. Simulation and Real Data Results

In this section, the accuracy and efficiency of the proposed algorithm are validated using point target simulation and real data. The time cost for each stage of the factorization is also taken into consideration. Due to the parallelizability of the processing scheme, a horizontal comparison between different parallel processing strategies is given. For the azimuth-invariant bistatic SAR case, synthesized SAR data are utilized to confirm the performance of the algorithm. Meanwhile, for the spaceborne/stationary bistatic SAR configuration, real data acquired on 31 January, 2015, using TerraSAR-X as an illuminator in the staring spotlight mode, are used for validation.

5.1. Monostatic SAR Case

Airborne monostatic SAR data in spotlight mode with a straight trajectory are investigated here. The simulation parameters are given in Table 1. In this scene, a 5×5 point array is established, and the spacing between each point is 1000 m. The simulated echo signal is focused using the proposed MoBulk_FFBP (both with and without pivots), FFBP and Block_FFBP. All algorithms have four processing stages with a common factorization factor of four for each stage. In MoBulk_FFBP with pivots, the number of pivots is 32, and the interpolation kernel used in this experiment is a spline kernel. In the simulation, the focusing results of the two versions of MoBulk_FFBP are almost identical. The results are shown in Figure 8.

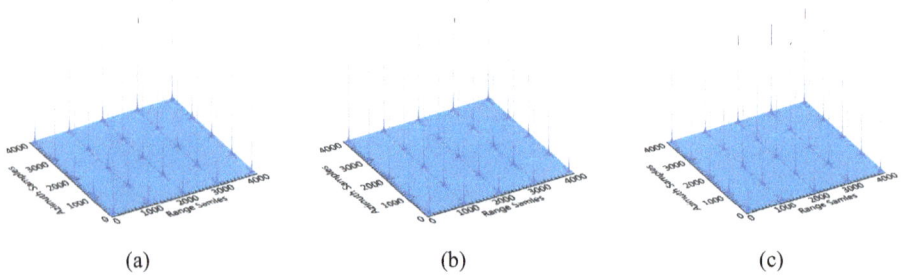

Figure 8. Point array simulation results: (**a**) processed by FFBP; (**b**) processed by Block_FFBP; (**c**) processed by MoBulk_FFBP. A 4 km × 4 km scene containing a 5 × 5 point array is shown.

Table 1. Simulation parameters of the monostatic SAR case. PRF, pulse repetition frequency.

Parameter	Values
Carrier frequency (GHz)	9.6
Bandwidth (MHz)	400
PRF (Hz)	160
Platform height (km)	10
Look angle (°)	4
Velocity (m/s)	120
Azimuth steering angle (°)	±1.62

Figure 8c indicates that the focusing quality of MoBulk_FFBP is adequate. A close look at the upper right target is shown intuitively in Figure 9. By examining the acquired range and azimuth profiles through the peak of the focused target, the impulse-response width (IRW), peak sidelobe ratio (PSLR), integrated sidelobe ratio (ISLR) and signal-to-noise (SNR) are evaluated and summarized in Table 2. According to the system parameters, the theoretical IRWs in the range and azimuth directions are 0.525 and 0.276 m, respectively. The measured range and azimuth IRWs agree well with the theoretical values. The deviations of PSLRs in each profile are within 0.2 dB of the theoretical values of −13.26 dB. Meanwhile, in Figure 9b, the focusing quality in the azimuth direction is decent, while the amplitude of third sidelobe in the range direction indicated by the red arrow is higher than the standard level, which causes the ISLR to deviate from the theoretical value, and the SNR decreases. This ringing effect of pulse response in the range direction is introduced by the data partition. As explained above, the upsampling operation can only be performed through interpolation rather than with an FFT due to the effects of data partitioning. The ringing effect of the interpolation can become increasingly stronger after each stage of the factorization. In the proposed algorithm, range data are processed in bulk rather than in blocks; according to the measured SNRs, it can be seen that the 'focusing depth' of MoBulk_FFBP is higher.

Table 2. Accuracy measurements. IRW, impulse-response width; PSLR, peak sidelobe ratio; ISLR, integrated sidelobe ratio; Az, azimuth; Rg, range.

	IRW (dB)	PSLR (dB)	ISLR (dB)	SNR (dB)
FFBP	(Rg)0.53/(Az)0.28	(Rg)-12.35/(Az)-13.04	(Rg)-9.82/(Az)-9.7	52.61
Block_FFBP	(Rg)0.53/(Az)0.28	(Rg)-12.36/(Az)-13.11	(Rg)-6.9254/(Az)-10.1	50.25
MoBulk_FFBP	(Rg)0.53/(Az)0.28	(Rg)-12.33/(Az)-12.99	(Rg)-10.93/(Az)-9.58	53.39

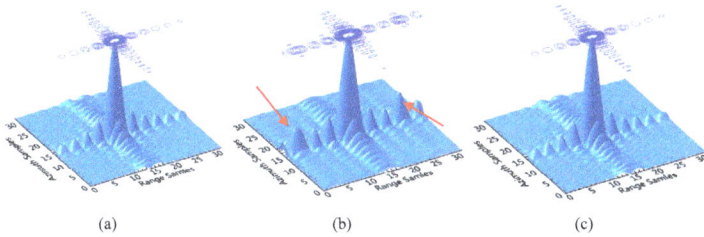

(a) (b) (c)

Figure 9. Magnified image of a point target in the focused results. (**a**) Processing result with FFBP; (**b**) processing result with Block_FFBP. The two red arrows note that the third sidelobe of the point processed using Block_FFBP are incorrect. (**c**) Processing result with MoBulk_FFBP.

To compare the efficiency, the raw data of these points are focused by Block_FFBP and MoBulk_FFBP independently with different factorization stages. Since it has been demonstrated in [19] that Block_FFBP is much more computationally efficient than FFBP, thus the comparison of computation complexity with FFBP is not conducted here. The programs are executed in a single-threaded environment, and the factorization factor is four in each processing stage. The hardware configuration is listed in Table 3. Let K be the number of factorization stages. In Figure 10a, the smaller processing time cost indicates that the processing efficiency of MoBulk_FFBP is generally higher than that of Block_FFBP. This is because the required memory of Block_FFBP at each stage is unstable, which increases the time cost for the factorization and reduces the overall efficiency. The trend shown in Figure 10a indicates that the time cost diminishes as K rises. However, in this experiment, the required processing time increases when K is larger than five. The entire processing scheme incorporates the factorization and BP focusing of the accumulated data. Figure 10b shows the time costs for these two steps for MoBulk_FFBP and Block_FFBP, respectively. First, the time cost for the factorization stage rises as K rises; on the contrary, the time cost decreases for increasing K values for the following residual BP step. Moreover, the time cost for the factorization of MoBulk_FFBP is larger than that of Block_FFBP when K is less than five. This is due to the fact that the interpolation kernel is short and the interpolation computation speed is low for the subaperture summation. When K becomes larger, the time cost for the factorization of Block_FFBP increases dramatically. Then, the time cost for the subsequent BP operation of MoBulk_FFBP is much less than that of Block_FFBP, which benefits from upsampling with an FFT instead of interpolation.

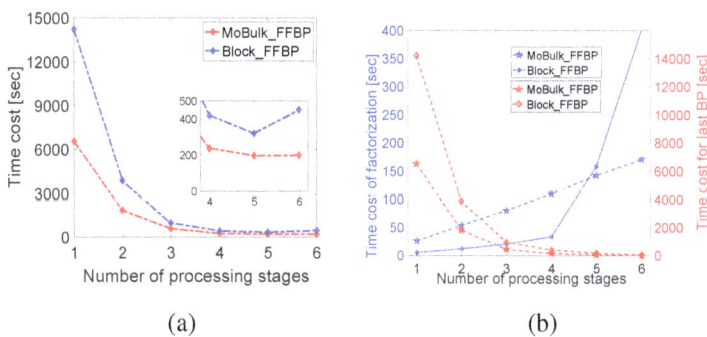

(a) (b)

Figure 10. Comparative analysis of the execution times of Block_FBP and MoBulk_FBP. (**a**) The total processing time when $K = \{1,2,3,4,5,6\}$. (**b**) Time costs of the factorization and residual BP.

Table 3. Hardware configuration for simulation.

Items	Values
CPU	Xeon E5620
Clock speed	2.4 GHz
Memory	192 GB

After the point target simulation, the raw data of synthesized distributed scene containing land and water are generated. Figure 11a–d shows the focusing results, which are processed using BP, MoBulk_FFBP, FFBP and Block_FFBP, respectively. In the red rectangle in each figure, the focusing quality of Block_FFBP is not as good as the others due to the residual phase error induced by the changing range error. The water area in the green rectangle is used for SNR comparison. From Figure 11a–d, the measured SNR is 22.83 dB, 22.78 dB, 22.46 dB and 22.61 dB. although the SNR of each result are similar, the SNR of MoBulk_FFBP is a little better. However, the focusing performance can be validated by the area in the red rectangle.

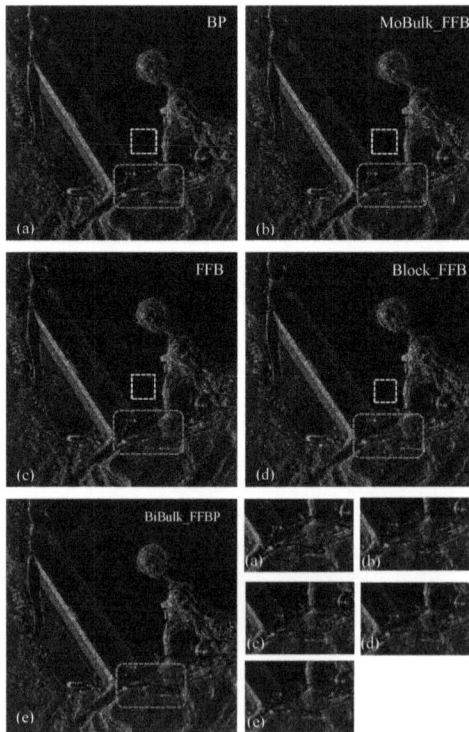

Figure 11. Focusing results of the synthesized distributed scene. Processing result with (**a**) BP, (**b**) MoBulk_FFBP, (**c**) FFBP, (**d**) Block_FFBP and (**e**) BiBulk_FFBP for azimuth-invariant bistatic SAR geometry. The red rectangle is used for focusing performance comparison, and the green rectangle is used for SNR comparison. The five sub-images at bottom-right are the red rectangle areas in (a–e).

Here, the MoBulk_FFBP will be validated using real airborne data, which were acquired in a spotlight SAR experiment taken in Zunhua, China, on June 2016. The radar data were acquired with an airborne wideband SAR operating at a 1200 MHz bandwidth. The carrier frequency was 9.6 GHz, and the PRF was 2000 Hz. The flight altitude was approximately 7500 m, and the track was measured using both a GPS and an INS. The local incidence angle of the scene center is 66°. To apply MoBulk_FFBP, a procedure is designed to focus the data. First, the equivalent velocity is computed using the Doppler centroid frequency, which is calculated using the azimuth cross correlation. Then, motion compensation [32,33] is performed using the platform velocity in the east, north and up directions provided by the INS. After motion compensation, the equivalent velocity is corrected and updated. With the location information provided by the GPS and the obtained equivalent velocity, a straight flight track is fitted. The focusing plane is set equal with the average scene height. Then, the MoBulk_FFBP with or without pivots can be applied to focus the data. The focused monostatic spotlight SAR image is shown in Figure 12. The result indicates that MoBulk_FFBP exhibits a satisfactory processing performance. According to the aforementioned subaperture summation principle, the MoBulk_FFBP was established with an aperture block size of 256, i.e., the 24,576 aperture positions in the echo data were divided into 96 blocks. Four aperture positions are summed in each processing stage until the entire block is processed during the four stages. In this sense, all of the processing stages correspond to an aperture factorization according to 24,576 = $4^4 \times 96$. The subaperture summation is conducted using Equation (8), and the range data upsampling operation is performed in bulk with an FFT.

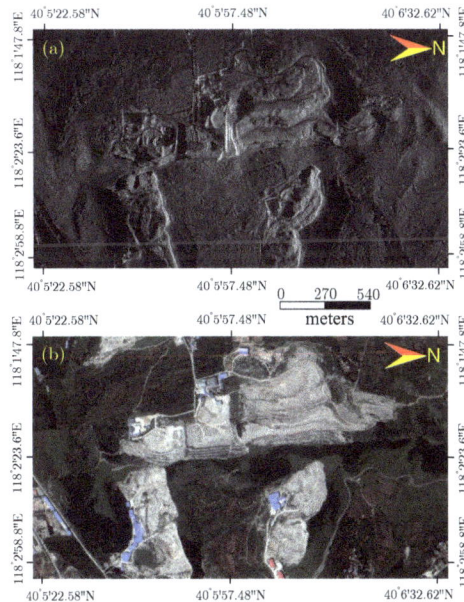

Figure 12. Monostatic spotlight image processed using MoBulk_FBP. (**a**) The monostatic spotlight image; (**b**) the optical image of the imaging area from Google Earth.

5.2. Bistatic SAR Case

To demonstrate the focusing ability of the proposed algorithm for the bistatic SAR case, an azimuth-invariant bistatic spotlight configuration is initially investigated. The transmitter and receiver run along straight trajectories that are separated by 5 m. The system parameters are the same as the point target simulation in the monostatic SAR case. The focused master and slave SAR images are shown in Figure 11b,e, and the result indicates that BiBulk_FFBP can also perform well. Due to the relatively short baseline, the results are very similar, but they demonstrate a slight difference in the marked zone. The factorization factor of BiBulk_FFBP is four, which means that four aperture positions are summed in each processing stage until the entire block is processed, thereby requiring four stages.

Spaceborne/stationary bistatic SAR also use a bistatic SAR configuration that is easily implemented using orbital sensors as coherent transmitters of opportunity with fixed-location receivers. Here, the fast back projection for focusing the one-stationary bistatic SAR data proposed in [25] is used for accuracy comparison. It can handle the synchronization problem and focus the data more efficiently than the BP algorithm, which makes it very appropriate for spaceborne/stationary bistatic SAR processing. A point target simulation based on a spaceborne/stationary bistatic configuration is conducted, and the parameters are given in Table 4.

Table 4. Spaceborne/stationary bistatic SAR simulation parameters.

Parameter	Values
Carrier frequency (GHz)	9.6
Bandwidth (MHz)	150
Sampling rate (MHz)	180
Pulse repetition frequency (Hz)	8000
Synthetic aperture time (s)	1.27
Transmitter center position (km)	$(0, 400, 692.8203)$
Synchronization channel position (m)	$(0, 0, 533)$
Echo channel position (m)	$(0, 0, 533)$
Target for evaluation (m)	$(-320, -9216, 0)$

The results processed by the FBP in [25] and the proposed BiBulk_FFBP are shown in Figure 13. To evaluate the IRW, PSLR and ISLR, the contours of the point at left-up corner are enlarged and shown in Figure 14.

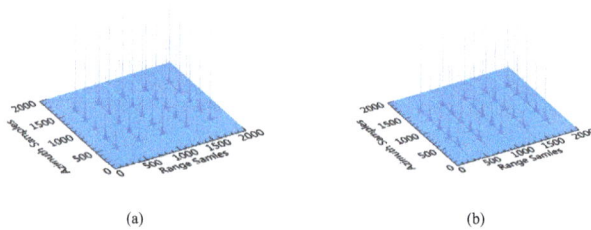

(a)   (b)

Figure 13. Focusing result of point targets. The result is processed with (**a**) BiBulk_FFBP and (**b**) FBP in [25].

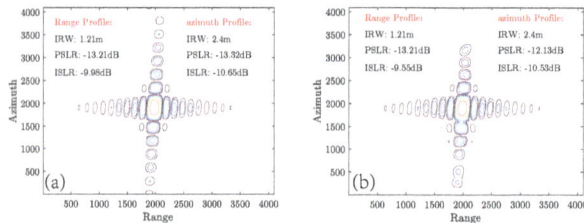

Figure 14. Extended target contours. The result is processed with (**a**) BiBulk_FFBP and (**b**) FBP in [25].

In Figure 14, it can be intuitively found that the focusing quality of BiBulk_FFBP is better that of the FBP in [25]. This is because the series of sub-images is focused at fixed grids, and the nearest interpolation method is used in the sub-images fusion process. Therefore, the imaging quality of the final image is nonuniform. The measured SNR of Figure 13a,b is 58.83 dB and 58.79 dB. From the two-dimensional profiles, it can be seen that two versions of fast BP algorithm have similar performance. However, from the perspective of computational complexity, BiBulk_FFBP is more computational efficient. Equation (24) can be approximated as $nN^2\log_n N$. Comparing with the computational complexity of FBP in [25], $N^{2.5}$, the computational cost of BiBulk_FFBP is lower.

On 31 January 2015, a spaceborne/stationary bistatic SAR (SS-BiSAR) experiment with the transmitter, TerraSAR-X, operating in ST mode was conducted by the Institute of Electronics, Chinese Academy of Sciences (IECAS). More details about the system configuration and the preprocessing of the raw data are provided in [34]. In this experiment, the direct signal was used as the matched filter to perform the range compression. With this method, the time synchronization error, phase synchronization error and tropospheric delay error are eliminated. Ignoring the two-dimensional envelop function, the compressed and synchronized signal in the range frequency domain is:

$$S_{com}\left(\tau, f; \tilde{\mathbf{r}}\right) = e^{-j2\pi f \frac{R_{TR}(\tau; \tilde{\mathbf{r}}) - R_D(\tau)}{c}} \tag{25}$$

where τ is the azimuth time, f is the range frequency, $R_D\left(\tau\right)$ is the direct signal path and $R_{TR}\left(\tau; \tilde{\mathbf{r}}\right)$ denotes the signal propagation range for a target located at $\tilde{\mathbf{r}}$. According to the information given in the TerraSAR-X product file, an XML file, the direct pulse phase history compensation can be performed in the SS-BiSAR coordinate system, after which the range-compressed signal becomes:

$$S_{com}\left(\tau, f; \tilde{\mathbf{r}}\right) = e^{-j2\pi f \frac{R_T(\tau; \tilde{\mathbf{r}}) + R_R(\tilde{\mathbf{r}})}{c}} \tag{26}$$

After performing the direct pulse phase history compensation and an inverse Fourier transformation, the range-compressed signal can be focused using the BiBulk_FFBP. In this experiment, the factorization factor is four, and the four subapertures in each of the four stages are summed into a new one. The focused result is shown in Figure 15. The magnified Areas A and B are also shown in the right panel. Area A is located in the near range of the receiver, from which a high SNR is consequently obtained. Because of the large incidence angle of the receiver, the tree canopies are clearly identified, and the track of an athletic field is easily recognized. Area B has some buildings that were still in construction at the time of data acquisition, and two tower slewing cranes are clearly focused. These details validate the focusing ability of the proposed BiBulk_FFBP.

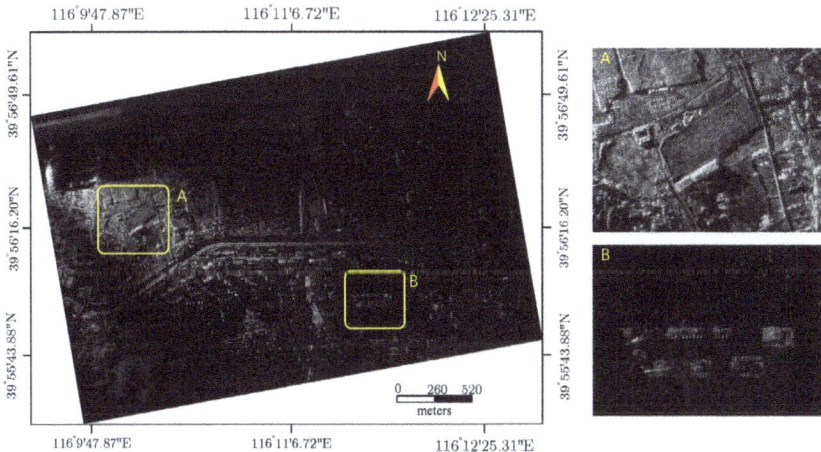

Figure 15. The SS-BiSAR image processed by the proposed algorithm. Area A and B are used for demonstrating the focusing performance. Area A contains an athletic field and some trees. Area B contains some buildings.

As previously discussed, each stage of factorization can be executed in parallel, which further accelerates the processing scheme. To show the acceleration with a multi-thread operation, the factorization and subsequent BP are executed in a single-threaded (ST) and a multi-threaded (MT) environment, respectively. The results are shown in Figure 16. The hardware configuration is shown in Table 3. The computation time of the raw radar data containing 1536 MSampleson an image grid of 144 MPointsis 203.6 min, 306 min and 400.405 min for MT-MT, ST-MT and MT-ST processing pairs, respectively. Therefore, the parallelization of the processing scheme provides a large increase in computational speed compared to the conventional BP algorithm. Moreover, when the program is executed within a GPU platform, the computation speed is even faster.

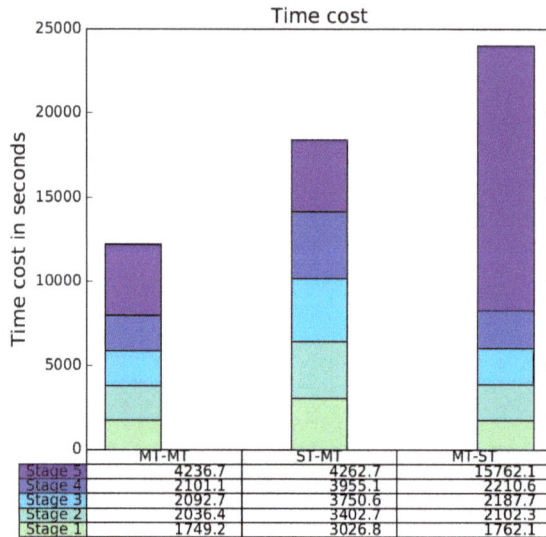

	MT-MT	ST-MT	MT-ST
Stage 5	4236.7	4262.7	15762.1
Stage 4	2101.1	3955.1	2210.6
Stage 3	2092.7	3750.6	2187.7
Stage 2	2036.4	3402.7	2102.3
Stage 1	1749.2	3026.8	1762.1

Figure 16. Time cost of BiBulk_FFBP in different executing environments. Stages 1–4 represent the factorization step, and Stage 5 is the subsequent focusing with BP. (MT: multi-thread; ST: single-thread).

6. Conclusions

In this paper, an accelerated BP algorithm is proposed to focus monostatic and bistatic SAR data. In this algorithm, the range data are processed in bulk rather than through block partitions. This unified range data processing scheme improves the computational efficiency and simplifies the procedure. A fixed number of pivots rather than beam centers is applied to construct the relationship of the propagation time delay between the "new" and "old" subapertures. Moreover, when the trajectory is a straight line, the pivots are not required, and the analytical expression of the subaperture summation can be derived for the monostatic and azimuth-invariant bistatic SAR case. Error analysis shows that the accuracy of the proposed algorithm is verifiable. Since the algorithm is an improved version of Block_FFBP, it satisfies numerical performance standards and retains the processing ability of Block_FFBP. Moreover, it can also focus the bistatic SAR data acquired from the tandem mode and the one-stationary bistatic SAR mode. Simulation data and real radar data validate the performance of the algorithm.

Acknowledgments: This work is funded jointly by the National Key R&D Program of China (2017YFB0502700), the National Ten Thousand Talent Program-Young Top-Notch Talent Program, the National Natural Science Funds for Excellent Young Scholar under Grant 61422113 and National Science Fund under Grant 61701479.

Author Contributions: Heng Zhang designed the algorithm and mainly drafted this manuscript. Jiangwen Tang helped to process the simulated data and assessed the results. Robert Wang and Yunkai Deng directed the research. Robert Wang is the corresponding author and spent much time providing suggestions on the organization of

Remote Sens. **2018**, *10*, 140

the paper. Wei Wang and Ning Li contributed to the revision of this paper and provided insightful comments and suggestions.

Conflicts of Interest: The authors declare no conflict of interest.

References

1. Franceschetti, G.; Lanari, R. *Synthetic Aperture Radar Processing*; CRC Press: Boca Raton, FL, USA, 1999; pp. 32–58.
2. Cumming, I.G.; Bennett, J. Digital processing of SeaSAT SAR data. In Proceedings of the IEEE International Conference on ICASSP '79 Acoustics, Speech, and Signal Processing, Washington, DC, USA, 2–4 April 1979; pp. 710–718.
3. Raney, R.K.; Runge, H.; Bamler, R.; Cumming, I.G.; Wong, F.H. Precision SAR processing using chirp scaling. *IEEE Trans. Geosci. Remote Sens.* **1994**, *32*, 786–799, doi:10.1109/36.298008.
4. Carrara, W.G.; Majewski, R.M.; Goodman, R.S. *Spotlight Synthetic Aperture Radar: Signal Processing Algorithms*; Artech House: Norwood, MA, USA, 1995.
5. Bamler, R.; Eineder, M. ScanSAR processing using standard high precision SAR algorithms. *IEEE Trans. Geosci. Remote Sens.* **1996**, *34*, 212–218, doi:10.1109/36.481905.
6. Mittermayer, J.; Moreira, A. Spotlight SAR processing using the extended chirp scaling algorithm. In Proceedings of the 1997 IEEE International IGARSS'97, Remote Sensing—A Scientific Vision for Sustainable Development, Singapore, 3–8 August 1997; pp. 2021–2023.
7. Loffeld, O.; Nies, H.; Peters, V.; Knedlik, S. Models and useful relations for bistatic SAR processing. *IEEE Trans. Geosci. Remote Sens.* **2004**, *42*, 2031–2038, doi:10.1109/TGRS.2004.835295.
8. Natroshvili, K.; Loffeld, O.; Nies, H.; Ortiz, A.M.; Knedlik, S. Focusing of general bistatic SAR configuration data with 2-D inverse scaled FFT. *IEEE Trans. Geosci. Remote Sens.* **2006**, *44*, 2718–2727, doi:10.1109/TGRS.2006.872725.
9. Wang, R.; Loffeld, O.; Nies, H.; Knedlik, S.; Ender, J.H.G. Chirp-scaling algorithm for bistatic SAR data in the constant-offset configuration. *IEEE Trans. Geosci. Remote Sens.* **2009**, *47*, 952–964, doi:10.1109/TGRS.2008.2006275.
10. Neo, Y.L.; Wong, F.H.; Cumming, I.G. Processing of azimuth-invariant bistatic SAR data using the range Doppler algorithm. *IEEE Trans. Geosci. Remote Sens.* **2008**, *46*, 14–21, doi:10.1109/TGRS.2007.909090.
11. Bamler, R.; Meyer, F.; Liebhart, W. Processing of bistatic SAR data from quasi-stationary configurations. *IEEE Trans. Geosci. Remote Sens.* **2007**, *45*, 3350–3358, doi:10.1109/TGRS.2007.895436.
12. Wong, F.H.; Cumming, I.G.; Neo, Y.L. Focusing bistatic SAR data using the nonlinear chirp scaling algorithm. *IEEE Trans. Geosci. Remote Sens.* **2008**, *46*, 2493–2505, doi:10.1109/TGRS.2008.917599.
13. Qiu, X.; Hu, D.; Ding, C. An improved NLCS algorithm with capability analysis for one-stationary BiSAR. *IEEE Trans. Geosci. Remote Sens.* **2008**, *46*, 3179–3186, doi:10.1109/TGRS.2008.921569.
14. Zeng, T.; Hu, C.; Wu, L.; Liu, L.; Tian, W.; Zhu, M.; Long, T. Extended NLCS algorithm of BiSAR systems with a squinted transmitter and a fixed receiver: Theory and experimental confirmation. *IEEE Trans. Geosci. Remote Sens.* **2013**, *51*, 5019–5030, doi:10.1109/TGRS.2013.2276048.
15. Wang, R.; Loffeld, O.; Nies, H.; Ender, J.H.G. Focusing spaceborne/airborne hybrid bistatic SAR data using wavenumber-domain algorithm. *IEEE Trans. Geosci. Remote Sens.* **2009**, *47*, 2275–2283, doi:10.1109/TGRS.2008.2010852.
16. Tang, J.; Deng, Y.; Wang, R.; Zhao, S.; Li, N.; Wang, W. A weighted backprojection algorithm for azimuth multichannel SAR imaging. *IEEE Geosci. Remote Sens. Lett.* **2016**, *13*, 1265–1269, doi:10.1109/LGRS.2016.2580907.
17. Behner, F.; Reuter, S.; Nies, H.; Loffeld, O. Synchronization and processing in the HITCHHIKER bistatic SAR experiment. *IEEE J. Sel. Top. Appl. Earth Obs. Remote Sens.* **2016**, *9*, 1028–1035, doi:10.1109/JSTARS.2015.2471082.
18. Yegulalp, A.F. Fast backprojection algorithm for synthetic aperture radar. In Proceedings of the The Record of the 1999 IEEE Radar Conference, Waltham, MA, USA, 22 April 1999; pp. 60–65.
19. Ulander, L.M.H.; Hellsten, H.; Stenstrom, G. Synthetic-aperture radar processing using fast factorized back-projection. *IEEE Trans. Geosci. Remote Sens.* **2003**, *39*, 760–776, doi:10.1109/TAES.2003.1238734.
20. Rodriguez-Cassola, M.; Baumgartner, S.V.; Krieger, G.; Moreira, A. Bistatic TerraSAR-X/F-SAR spaceborne/airborne SAR experiment: Description, data processing, and results. *IEEE Trans. Geosci. Remote Sens.* **2010**, *48*, 781–794, doi:10.1109/TGRS.2009.2029984.

21. Duque, S.; Lopez-Dekker, P.; Mallorqui, J.J. Single-pass bistatic SAR interferometry using fixed-receiver configurations: Theory and experimental validation. *IEEE Trans. Geosci. Remote Sens.* **2010**, *48*, 2740–2749, doi:10.1109/TGRS.2010.2041063.

22. Walterscheid, I.; Espeter, T.; Brenner, A.R.; Klare, J.; Ender, J.H.G.; Nies, H.; Wang, R.; Loffeld, O. Bistatic SAR experiments with PAMIR and TerraSAR-x;setup, processing, and image results. *IEEE Trans. Geosci. Remote Sens.* **2010**, *48*, 3268–3279, doi:10.1109/TGRS.2010.2043952.

23. Vu, V.T.; Sjogren, T.K.; Pettersson, M.I. Fast time-domain algorithms for UWB bistatic SAR processing. *IEEE Trans. Aerosp. Electron. Syst.* **2013**, *49*, 1982–1994, doi:10.1109/TAES.2013.6558032.

24. Rodriguez-Cassola, M.; Prats, P.; Krieger, G.; Moreira, A. Efficient time-domain image formation with precise topography accommodation for general bistatic SAR configurations. *IEEE Trans. Aerosp. Electron. Syst.* **2011**, *47*, 2949–2966, doi:10.1109/TAES.2011.6034676.

25. Shao, Y.; Wang, R.; Deng, Y.; Liu, Y.; Chen, R.; Liu, G.; Loffeld, O. Fast backprojection algorithm for bistatic SAR imaging. *IEEE Geosci. Remote Sens. Lett.* **2013**, *10*, 1080–1084, doi:10.1109/LGRS.2012.2230243.

26. Cumming, I.G.; Wong, F.H. *Digital Processing of Synthetic Aperture Radar Data: Algorithms and Implementation*; Artech House: Norwood, MA, USA, 2005; pp. 32–58.

27. Wang, R.; Loffeld, O.; Neo, Y.L.; Nies, H.; Walterscheid, I.; Espeter, T.; Klare, J.; Ender, J.H.G. Focusing bistatic SAR data in airborne/stationary configuration. *IEEE Trans. Geosci. Remote Sens.* **2010**, *48*, 452–465, doi:10.1109/TGRS.2009.2027700.

28. Walterscheid, I.; Ender, J.H.G.; Brenner, A.R.; Loffeld, O. Bistatic SAR processing and experiments. *IEEE Trans. Geosci. Remote Sens.* **2010**, *44*, 2710–2717, doi:10.1109/TGRS.2006.881848.

29. Shi, J.; Long, X.; Zhang, X. Streaming BP for non-linear motion compensation SAR imaging based on GPU. *IEEE J. Sel. Top. Appl. Earth Obs. Remote Sens.* **2013**, *6*, 2035–2050, doi:10.1109/JSTARS.2013.2238891.

30. Di Bisceglie, M.; Di Santo, M.; Galdi, C.; Lanari, R.; Ranaldo, N. Synthetic aperture radar processing with GPGPU. *IEEE Signal. Proc. Mag.* **2010**, *27*, 69–78, doi:10.1109/MSP.2009.935383.

31. Kluge, T. Pricing Swing Options and Other Electricity Derivatives. Ph.D. Dissertation, St Hugh's College, University of Oxford, Oxford, UK, 2006. Available online: http://kluge.in-chemnitz.de/docs/phd/ (accessed on 11 July 2006).

32. Nies, H.; Loffeld, O.; Na, K.; Natroshvili, F. Analysis and focusing of bistatic airborne SAR data. *IEEE Trans. Geosci. Remote Sens.* **2007**, *45*, 3342–3349, doi:10.1109/TGRS.2007.900689.

33. Mao, X.; Zhu, D.; Zhu, Z. Autofocus correction of ape and residual rcm in spotlight SAR polar format imagery. *IEEE Trans. Aerosp. Electron. Syst.* **2013**, *49*, 2693–2706, doi:10.1109/TAES.2013.6621846.

34. Zhang, H.; Deng, Y.; Wang, R.; Li, N.; Zhao, S.; Hong, F.; Wu, L.; Loffeld, O. Spaceborne/stationary bistatic SAR imaging with TerraSAR-X as an illuminator in staring-spotlight modeAnalysis and focusing of bistatic airborne SAR data. *IEEE Trans. Geosci. Remote Sens.* **2007**, *54*, 5203–5214, doi:10.1109/TGRS.2016.2558294.

remote sensing

MDPI

Article

Signal Processing for a Multiple-Input, Multiple-Output (MIMO) Video Synthetic Aperture Radar (SAR) with Beat Frequency Division Frequency-Modulated Continuous Wave (FMCW)

Seok Kim [1,2], Jiwoong Yu [1], Se-Yeon Jeon [1], Aulia Dewantari [1] and Min-Ho Ka [1,*]

[1] School of Integrated Technology, Yonsei University, 21983 Seoul, Korea; skim0511@yonsei.ac.kr (S.K.); jiwoong.yu@yonsei.ac.kr (J.Y.); seyeonjeon@yonsei.ac.kr (S.-Y.J.); dewantariaulia@yonsei.ac.kr (A.D.)
[2] Hanwha Systems, Inc., 17121 Yongin-si, Gyeonggi-do, Korea
* Correspondence: kaminho@yonsei.ac.kr; Tel.: +82-32-749-5840

Academic Editors: Timo Balz, Uwe Soergel, Mattia Crespi, Batuhan Osmanoglu, Zhenhong Li and Prasad S. Thenkabail
Received: 20 April 2017; Accepted: 14 May 2017; Published: 17 May 2017

Abstract: In this paper, we present a novel signal processing method for video synthetic aperture radar (ViSAR) systems, which are suitable for operation in unmanned aerial vehicle (UAV) environments. The technique improves aspects of the system's performance, such as the frame rate and image size of the synthetic aperture radar (SAR) video. The new ViSAR system is based on a frequency-modulated continuous wave (FMCW) SAR structure that is combined with multiple-input multiple-output (MIMO) technology, and multi-channel azimuth processing techniques. FMCW technology is advantageous for use in low cost, small size, and lightweight systems, like small UAVs. MIMO technology is utilized for increasing the equivalent number of receiving channels in the azimuthal direction, and reducing aperture size. This effective increase is achieved using a co-array concept by means of beat frequency division (BFD) FMCW. A multi-channel azimuth processing technique is used for improving the frame rate and image size of SAR video, by suppressing the azimuth ambiguities in the receiving channels. This paper also provides analyses of the frame rate and image size of SAR video of ViSAR systems. The performance of the proposed system is evaluated using an exemplary system. The results of analyses are presented, and their validity is verified using numerical simulations.

Keywords: video synthetic aperture radar; multiple-input multiple-output; multi-channel azimuth processing; frequency-modulated continuous waveform; beat frequency division; polar format algorithm

1. Introduction

Synthetic aperture radar (SAR) technology is an active microwave remote sensing technique, capable of day/night, all-weather operation, used to detect and acquire electromagnetic information about objects without physical contact [1–3]. The enormous potential of SAR—invented by Carl Wiley in 1951—has been evident since the first demonstration of the concept. This potential has facilitated extensive research into SAR technology [4], including avenues for improving its operation. For instance, in the case of conventional single aperture SAR, it is difficult to obtain a high-resolution wide swath image, due to the trade-off between resolution and image size [5]. This problem was solved by adopting multi-channel processing techniques to create a high-resolution wide swath (HRWS) SAR system [6–10]. By adding multiple transmit antennas to the HRWS SAR system using multiple-input multiple-output (MIMO) technology, it is possible to operate a single SAR system in multiple modes, such as HRWS, interferometric SAR (InSAR), and polarimetric SAR (PolSAR). These modes can be operated simultaneously, without performance degradation [11,12]. By applying MIMO and multi-channel azimuth processing techniques to video synthetic aperture radar (ViSAR), improvements to the frame rate and image size of SAR video can be achieved, without compromising other aspects of the system's performance.

ViSAR is a SAR imaging mode which has recently gained increased research interest. In this mode, images can be generated at a much higher rate than in conventional SAR; thus, they can be viewed continuously, just like watching a video [13–15]. Since ViSAR systems can provide high resolution SAR images at a high frame rate regardless of adverse weather conditions, they can be utilized in many day/night, all-weather military and civilian applications, including fire control support [16], as a replacement for electro-optical (EO)/infrared (IR) sensors, which can be only operated in clear weather, land and maritime traffic monitoring, and surveillance [17].

There are two kinds of SAR video synthesis methods associated with ViSAR, full aperture synthesis, which uses a circular SAR mode [13,15], and sub-aperture synthesis, which uses a spotlight SAR mode [14,16]. In the full aperture synthesis mode, an overlapped processing method is typically used between SAR video frame updates to obtain high frame rates, since the synthetic aperture time is very long at relatively low frequencies. Due to its parallel nature and ease of motion compensation, the backprojection algorithm (BPA) is suitable for use with full aperture synthesis. The frame rate in this case can be adjusted easily by changing the overlapping ratio between SAR video frames. Due to the large amount of computation power required, the BPA uses sub-images generated from sub-apertures, which are partial sections of the full aperture. In the sub-aperture synthesis mode, the frame rate can be improved by increasing the transmitted frequency and the radar velocity at a given resolution. When operated as a partial measurement, circular SAR is the same as spotlight SAR. Therefore, on a circular path, SAR video can be generated from the sub-aperture data, using the polar format algorithm (PFA) in spotlight SAR mode. However, as the radar velocity increases, the Doppler bandwidth of the antenna beamwidth increases, thus, azimuth ambiguities occur. These azimuth ambiguities can be suppressed effectively, using multi-channel azimuth processing techniques.

The X band video SAR system (XWEAR) [13], generates SAR video using the BPA on a circular path. The NanoSAR, operating in the X band, and the MiniSAR, operating in the Ku band, can also perform the video SAR function using a circular path. The Defense Advanced Research Projects Agency (DARPA) is currently developing a ViSAR system capable of achieving a high frame rate at an extremely high frequency (233 GHz), to replace EO/IR sensors used for fire control on maneuvering ground targets [15,16]. The Agency for Defense Development (ADD)—a defense research center in the Republic of Korea—is currently in the process of developing a 94-GHz airborne MIMO video SAR system, as a possible concept for surveillance sensors in unmanned aerial vehicles (UAVs).

Frequency-modulated continuous wave (FMCW) technology is suitable for small size, lightweight and low cost systems. Various technologies and algorithms related to FMCW-based SAR systems have been studied in [11,18,19]. Several FMCW SAR systems have historically been used successfully in ice measurement, environment monitoring, and three-dimensional applications [20]. The combination of FMCW technology and ViSAR mode gives birth to a light-weight, cost-effective, high-resolution, active microwave remote sensing instrument, which is suitable for small platforms such as UAVs [20].

In 1994, Paulraj and Kailath patented the use of multiple antennas for both transmission and reception in wireless communications, in order to increase channel capacity [21]. This invention has shown important potential for deployment in various wireless communication applications, as well as for radar sensor technology. The potential applications of this technology have promoted extensive research on the use of multiple antennas. The MIMO radar, the combination of multiple antennas with radar sensors, is currently one of the most interesting topics in radar communities [22]. In this case, the MIMO function is implemented using a coding technique to apply a pseudo-orthogonal waveform. This solution is not suitable for SAR operation, because, unlike the point target case, the received signal consists of distributed targets from large swaths, in the time-frequency domain [23].

Orthogonal frequency division multiplexing (OFDM) chirp [24,25], short-term shift orthogonal (STSO) waveforms [26], and OFDM chirp diverse waveforms [27] have been studied with pulsed MIMO SAR.

MIMO technology can be combined with ViSAR systems using orthogonal waveforms for FMCW radars such as the beat frequency division (BFD) waveform [28], the chirp rate division waveform [28], and the OFDM (or interleaved OFDM, I-OFDM) chirp waveform [29]. This MIMO technology can be used for various objectives, for example, the virtual array can be used to increase the equivalent number of phase centers or receiving channels, for the reduction of the aperture size. When reconfigurable transmitting (Tx) antennas are employed, MIMO technology can also be used for multi-mode operation within a single system [12,25], such as along-track interferometry, ground moving target indication (GMTI), and polarimetric SAR. In this paper, we use MIMO technology to increase the equivalent number of receive channels of a virtual array, and to reduce aperture size, by combining it with a ViSAR system. The stop-and-go approximation used in pulsed SAR is not valid for FMCW SAR. Since an FMCW SAR has a long sweep duration, the motion of the target object within the sweep must be considered [11]. Doppler shift is more sensitive in OFDM chirp waveforms [24,25], than in linear frequency modulation (LFM) [29]. It is therefore difficult to apply Doppler compensation techniques when the sweep duration is long, as in FMCW SAR. This problem can be solved by applying the BFD FMCW, and then compensating the Doppler shift. As such, the proposed MIMO ViSAR system is based on this orthogonal FMCW.

A multi-channel azimuth processing technique is utilized for improving the frame rate and image size of SAR video by suppressing the azimuth ambiguities in the receiving channels. Applying the multi-channel reconstruction algorithm (MCRA) to the ViSAR system allows SAR image formation without Doppler ambiguities, which are due to the simultaneous increase of the Doppler bandwidth and the radar velocity [6–9]. As a result, the frame rate and image size of the SAR video can be increased. There exist several different HRWS algorithms and MCRAs for reconstructing a stationary scene. These include the matrix inversion method [6–9], the orthogonal projection method [9], the maximum signal method [9], the maximum signal-to-ambiguity-plus-noise ratio (SANR) method [9], the minimum mean-square error (MMSE) method [30], and the improved digital beamforming (IDBF) method [10]. In addition, in [17], the GMTI function is implemented by extending the MCRA to the case of imaging moving targets. A method for simultaneously imaging moving and stationary targets is presented in [31]. This paper focuses on the imaging of stationary targets.

When ViSAR is operated on a circular path, the azimuth and the elevation beamwidth are selected to be of similar magnitudes for efficient observation of targets [23]. In this case, the beamwidths can be widened, depending on the requirements of the specific scenario. In the case of a wide beam, it is difficult to apply the PFA, due to the scene size limitation associated with this algorithm. The BPA is advantageous for generating SAR image frames at high frame rates in an overlapping fashion due to its parallel nature. Therefore, the BPA is often applied to ViSAR. However, this algorithm is not suitable for synthesizing SAR images in real time, on small platforms such as UAV, due to its huge computational burden and large memory usage. Conversely, the PFA can be applied when using a narrow beam. In this case, a SAR video can be generated by continuously forming the SAR images from a sub-aperture that is a part of the full circular aperture that has been selected to satisfy the scene size limitation of the PFA.

The PFA, which was developed by Walker [32], is the classical algorithm proposed for raw data processing of spotlight SAR [2]. Although the PFA is significantly faster than the BPA, it approximates a matched filter response, thereby causing image errors in large scenes. Due to these errors, the PFA has a scene size limitation. Since the PFA can be implemented easily for a circular and a linear path, it is possible to operate the SAR system more flexibly. Therefore, video from the proposed MIMO ViSAR system is formed using the PFA.

In this paper, we derive a signal model based on the MIMO radar signal model, and propose a novel signal processing method for ViSAR systems suitable for UAV environment operation, which improves aspects of the system performance, such as frame rate, and image size of the SAR video. The proposed ViSAR system focuses on a sub-aperture synthesis method. The new ViSAR system is based on a FMCW SAR structure that is combined with MIMO and multi-channel azimuth processing techniques, to make use of the characteristics listed above. MIMO technology is utilized for increasing the equivalent number of receiving channels in the azimuthal direction, and reducing the aperture size using a co-array concept with a BFD FMCW. A multi-channel azimuth processing technique is utilized for improving the frame rate and image size of the SAR video, by suppressing the azimuth ambiguities in the receiving channels. This paper also provides analyses of the frame rate and image size of SAR video. The performance of the proposed system is evaluated using an exemplary system comprised of two Tx and two receiving (Rx) channels. The results of analyses are presented, and their validity is verified using numerical simulations.

This paper consists of five sections and is organized as follows. We start with an overview of ViSAR systems and analysis of the frame rate and image size in Section 2. In Section 3, we describe the MIMO signal model and MIMO video SAR processing steps. In Section 4, an exemplary system is designed and the numerical simulation results and performance estimation are given. Finally, conclusions and future opportunities are presented in Section 5.

2. Theory of Video SAR

To gain a better understanding of its operation, the basic concept, the frame rate, and the image size of the ViSAR system are described in detail in this section.

2.1. Overview

ViSAR is a SAR imaging technique in which the radar is operated in spotlight mode on a circular flight path, as shown in Figure 1. Radar data is collected on a region of interest, and images are generated at high frame rates. The ViSAR system can generate electromagnetic SAR images at frame rates similar to conventional video formats, which are typically $2 \sim 5$ Hz or more.

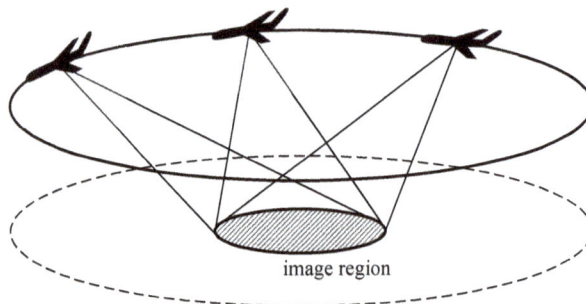

image region

Figure 1. The collection geometry of video SAR on a circular path. SAR: synthetic aperture radar.

Figure 1 shows the ViSAR geometry when it is operated on a circular path. There are two kinds of image formation methods used in ViSAR with circular antenna motion: full synthetic aperture methods [13,15] and sub-aperture methods [14,16]. As the name suggests, in the full synthetic aperture method, SAR images are generated from the full aperture, which corresponds to the complete circular path. In this case, an ultimate resolution of $\frac{\lambda}{4}$ can be obtained. Another advantage of this method is that images are acquired over 360 degrees. If a full aperture image is synthesized every time data is obtained, the frame rate becomes very small. Also, as the BPA is typically used in forming a full aperture image, the computation power required is very large. Therefore, a fast-factorized BPA is used in order to increase the frame rate and reduce the computation load, by factorizing the full synthetic aperture into sub-apertures, and updating the SAR image every time sub-aperture data is obtained. A higher frame rate can be obtained because the SAR image is updated in a sliding window fashion;

the update is performed every time sub-aperture data is modified instead of waiting for a full aperture data update. Conventional ViSAR systems often use this method [13]. Since fast-factorized BPA updates the image using the sliding window method, it has the disadvantage that an afterimage is retained in the SAR image. This is also seen with circular SAR when there is a large amount of overlap between input data. The fast-factorized BPA can achieve a high frame rate, even at relatively low frequencies [15,16].

The sub-aperture method for circular antenna motion is a method for synthesizing SAR images that uses only data obtained from the sub-aperture, which is a fraction of the full aperture. This is the same as operation in spotlight SAR mode, with partial measurement of a circular path. In this case, it is possible to update the SAR image at a high frame rate without an afterimage. However, since the SAR image must be synthesized using only sub-aperture data, the synthetic aperture time must be short to achieve a high frame rate. Therefore, at a given resolution, the transmitted frequency or the speed of the radar should be high. In addition, as only sub-aperture data is used, this technique is less restricted by the flight path than the full aperture method. Therefore, it is suitable for applications such as weapon assignment [15,16]. In this paper, we will focus mainly on the sub-aperture method.

When ViSAR is operated in spotlight SAR mode on a circular path, the direction of the image changes according to the aspect angle of the SAR image. Because of this, stationary targets appear to rotate in the SAR video. In this case, it would be convenient to interpret the SAR images as if they were taken from an optical camera at a fixed point. To do this, a reference angle is defined. SAR images obtained at different aspect angles are subsequently rotated and aligned to this reference. This reference angle is called the cardinal direction up (CDU), as shown in Figure 2. θ_{as} denotes the aspect angle with respect to the cardinal direction.

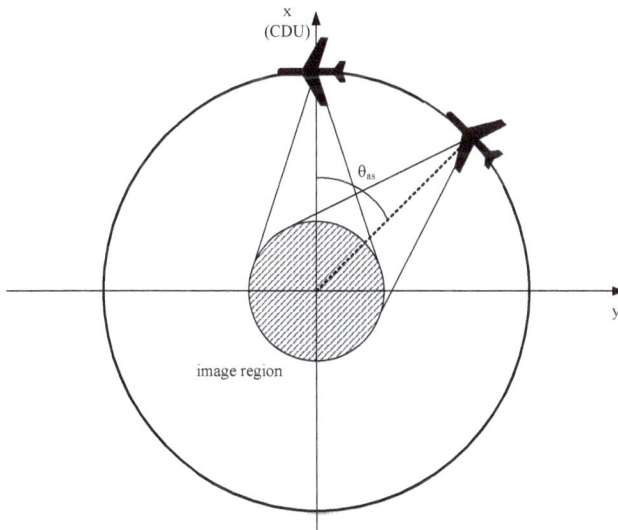

Figure 2. The collection geometry of video SAR on a circular path: top view. CDU: cardinal direction up.

2.2. Frame Rate

The frame rate is one of the most important parameters in the ViSAR system. In ViSAR, the frame rate is proportional to the inverse of the SAR video frame time, which is a fraction of the synthetic aperture time (SAT) and therefore can be expressed as:

$$f_v = \frac{1}{T_f}$$

(1)

where T_f is the SAR video frame time, which is defined as:

$$T_f = F_o T_a \tag{2}$$

where T_a is the SAT and F_o is the overlap ratio between SAR video frames. Note that $F_o = 1$ describes the non-overlap processing condition, which will be applied throughout this paper.

For spotlight SAR mode (or partial measurement of circular SAR) on a circular path, the synthetic aperture time for spotlight SAR can be expressed as [2] Equation (2.21)):

$$T_a = \frac{L}{v} = \frac{\lambda R_a K_a}{v \cdot 2\rho_a \sin(\alpha_{dc})} \tag{3}$$

where L is the synthetic aperture length, λ is the wavelength at the transmitted frequency, R_a is the distance from the antenna phase center (APC) to the scene center, K_a is the beam broadening factor, v is the sensor velocity, ρ_a is the cross-range resolution, and α_{dc} is the antenna cone angle, which is the antenna squint angle and is assumed to be 90 degrees throughout this paper. For the rest of this paper, it is convenient to assume that, in general, the radar is in a broadside condition. This assumption simplifies the subsequent mathematics. Therefore, the frame rate in the spotlight SAR mode can be represented as:

$$f_v = \frac{1}{T_a} = \frac{v \cdot 2\rho_a \sin(\alpha_{dc})}{\lambda R_a K_a} = \frac{v \cdot 2\rho_a \sin(\alpha_{dc})}{c R_a K_a} f_c \tag{4}$$

where f_c is the transmitted frequency, and c is the speed of light. From the above equation, when the resolution and the distance to the observation point are fixed, the radar velocity or transmitted frequency should be increased in order to increase the frame rate of the SAR video. However, the radar velocity is limited by the Doppler bandwidth, which is constrained to avoid azimuth ambiguity.

Figure 3a,b shows an example of the frame rate of the ViSAR system, which is calculated using Equation (4) at $R_a = 1000$ m and $K_a = 1$ with various resolutions and platform velocities. As shown in Figure 3a, the frame rate is 1.003 Hz at 94 GHz in the W band, which is 9.4 times as high as it is in the X band—which includes systems such as Global Hawk or NanoSAR—where the frame rate is 0.107 Hz at 10 GHz, according to Equation (4). Therefore, the frame rate can be increased significantly, by increasing the transmitted frequency. Additionally, a higher radar velocity leads to higher frame rate. For example, two times higher frame rates can be achieved by increasing the platform velocity from 20 m/s to 40 m/s; i.e., frame rates are increased from 1.003 Hz to 2.005 Hz, at 94 GHz. However, in conventional SARs, the radar velocity is limited, to prevent Doppler ambiguities from occurring. At this time, if the MCRA is applied and the number of receiving channels is increased, Doppler ambiguities can be eliminated, even if the Doppler bandwidth increases due to the increase in the radar velocity. Therefore, the radar platform velocity can be increased without increasing Doppler ambiguities, by increasing the number of receiving channels using the MCRA.

In summary, according to Equation (4), the frame rate in the circular path can be increased by increasing the transmitted frequency or the platform velocity while maintaining the azimuth resolution. Pulse repetition frequency (PRF) is proportional to the frame rate, since this is proportional to the platform velocity. As the platform velocity increases, the increased Doppler bandwidth requires an increased PRF. Unfortunately, the increased PRF in FMCW SAR systems causes the transmission time to decrease, which leads to a reduction in detection distance. However, by using a multi-channel reconstruction algorithm, the image rate can be increased without increasing the PRF at a given azimuth resolution, by increasing the platform velocity.

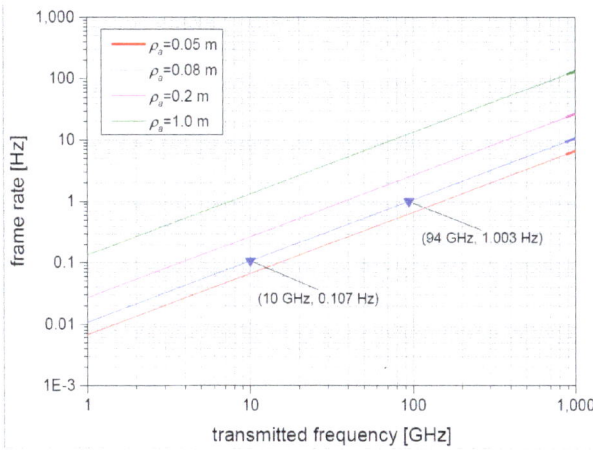

(a)

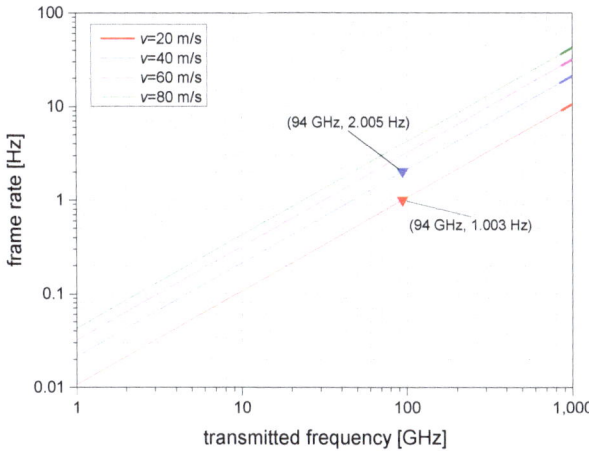

(b)

Figure 3. Frame rate of video SAR on a circular path: (**a**) varying resolution at $v = 20$ m/s; (**b**) varying platform velocity at $\rho_a = 0.08$ m.

By using the spotlight SAR mode for the proposed MIMO ViSAR system, it is possible to generate a SAR video with a high frame rate in the circular path. In the system proposed in this paper, in order to increase the frame rate in the spotlight mode, the transmitted frequency and the radar velocity are increased using the frame rate relation shown in Equation (4). In this case, a high frame rate SAR video can be obtained without using the full aperture synthesis method (BPA), which has very high computational complexity. In particular, the Doppler bandwidth is increased due to the increase in the radar velocity, and the MCRA algorithm is used to suppress Doppler ambiguities.

2.3. Image Size

This section describes the image size of ViSAR. The image size is limited by the Doppler bandwidth of the antenna beamwidth. With spotlight SAR, the image size is determined by the beamwidth. On the other hand, in the case of stripmap SAR, image size is determined by the pulse repetition interval (PRI) in pulsed SAR operation, or by the bandwidth of the dechirp signal in FMCW SAR operation. In the case of a ViSAR mode that is operated using spotlight SAR on a circular path, the received signal is the dechirped azimuth signal. The Doppler bandwidth is limited by scene size or antenna beamwidth as follows ([2], p. 44):

$$B_a \approx \frac{2vW_a \sin \alpha_{dc}}{\lambda R_a}$$

(5)

where $\theta_a \approx \frac{W_a}{R_a}$ approximately corresponds to the antenna beamwidth in azimuth and W_a is the azimuth size of the scene area. The Doppler bandwidth is therefore proportional to both the radar velocity and the image size in the azimuth. According to the Nyquist theorem, the PRF of conventional SAR should be greater than the Doppler bandwidth. Figure 4 shows an example of the relationship between the image size and the Doppler bandwidth described by Equation (5), with $v = 20$ m/s, and $R_a = 1000$ m. Note that, for example, the Doppler frequency of a scatterer located at the radius of 30 m is 376 (=752/2) Hz in this case. In order to maximize the effective image area, the antenna beamwidths in the azimuth and elevation directions should be of the same order [23].

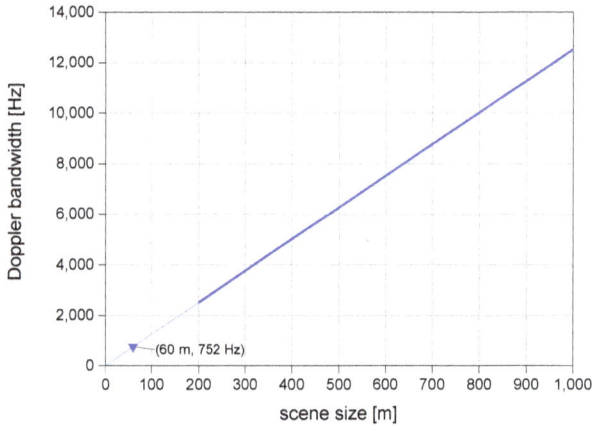

Figure 4. Doppler bandwidth limited by scene size.

When the SAR image is synthesized using the PFA, the image size is limited by wave front curvature, since the algorithm uses the plane-wave approximation. Therefore, these characteristics should be taken into consideration in the design process. The maximum image size can be increased by using wave front curvature compensation methods, as in [33–35]. In PFA, the scene size is limited by range curvature, and residual video phase ([2], Equation (3.130), [36], Equation (B.23)).

(a)   Scene size (diameter) is limited by range curvature as follows:

$$S < 2\rho_a \sqrt{\frac{2R_a}{\lambda}}$$

(6)

(b)   Scene size is limited by residual video phase (RVP) ([36] (Equation (B.26)), when the chirp rate is very high, as follows:

$$S < 2\rho_a \frac{f_c}{\sqrt{\frac{k_r}{\pi}}}$$

(7)

where k_r is the chirp rate in range.

Figure 5 illustrates an example of the scene sizes generated using the PFA, with $R_a = 1000$ m and $f_c = 94$ GHz, constrained by wave front curvature according to Equation (6). Therefore, the antenna beamwidth should be determined prior to operation so that it corresponds to the required image size. The scene size can be converted to beamwidth limit.

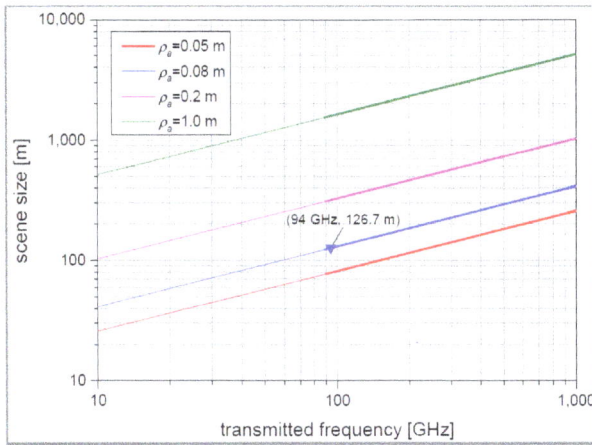

Figure 5. Scene size limitation of polar format algorithm (PFA).

3. Signal Processing

This section describes the signal model and signal processing procedure of the proposed MIMO ViSAR system.

3.1. Geometric Models

The collection geometry of a ViSAR system operating on a circular flight path is shown in Figure 6. The radar moves along a circular flight path at a constant speed, v. R_b is the radius of the flight path, R_a is the slant range between the radar and the scene center, R_z denotes the altitude of the aircraft, θ_a is the azimuth beamwidth of the antenna, θ_g is the grazing angle between the antenna beam axis and the ground plane, and $R_t = S/2$ is the radius of the scene area.

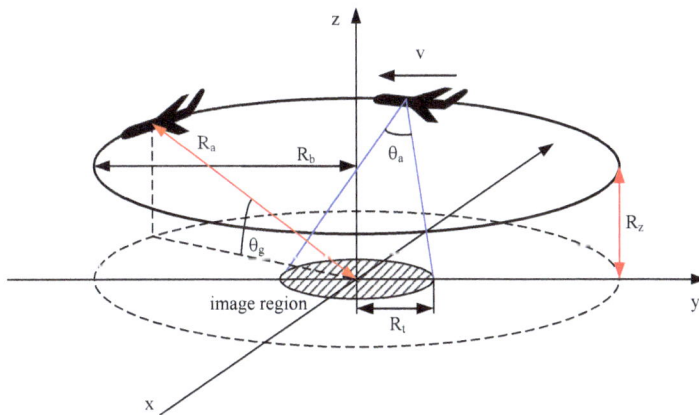

Figure 6. Collection geometry of video SAR on a circular path.

Figure 7 illustrates the antenna geometry of the proposed MIMO ViSAR system. We suppose that there are M Tx antennas and N Rx antennas. We assume in this paper that, in general, the virtual array formed by the MIMO antenna is a uniform linear array in the azimuth. Therefore, a large co-array can be realized using the MIMO concept. Also, it is advantageous to increase the number of phase centers if M and N are greater than two, since M Tx antennas and N Rx antennas give MN antenna elements in the virtual array.

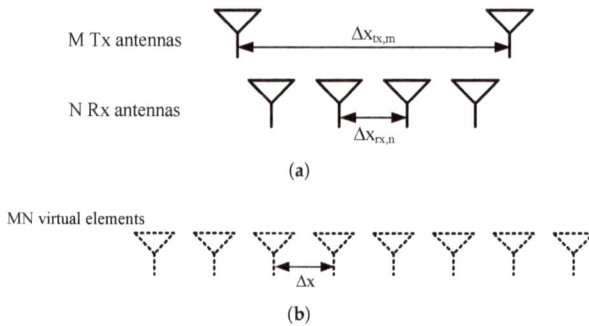

Figure 7. Antenna geometry: (**a**) The actual array; (**b**) The equivalent virtual array. Tx: transmitting; Rx: receiving.

3.2. Signal Model

Two types of signal models are formulated in two different domains: the fast time–slow time domain and the fast time–azimuth angle domain. First, we derive the signal model in the fast time–slow time domain. We then translate this model into the fast time-azimuth angle domain. As the fast time–azimuth angle domain is based on the MIMO radar signal model, it can be used more conveniently in subsequent signal processing steps.

Figure 8 shows the generic MIMO video SAR model [25]:

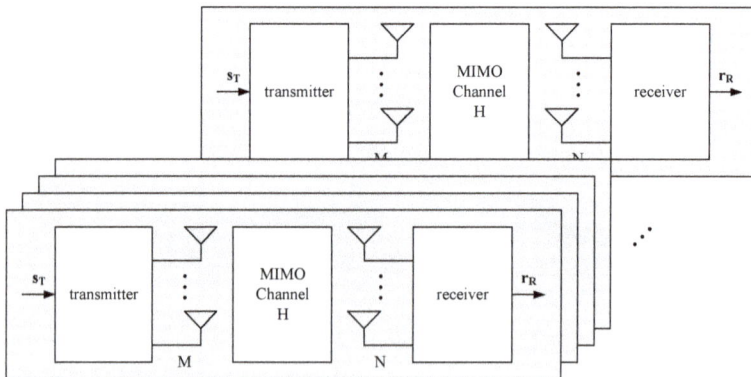

Figure 8. Signal model of the generic multiple-input multiple-output (MIMO) video SAR.

In the system model, a broadside, uniform spatial sampling condition is assumed. Therefore, the PRF is selected such that the following relationship is satisfied [6]:

$$f_p = \frac{v}{MN \cdot \Delta x} \qquad (8)$$

where f_p is the sweep repetition frequency, and Δx is the distance between antenna elements of the virtual array.

- Signal model in the fast time–slow time domain.

In this sub-section, we define the transmitted waveform of the proposed MIMO ViSAR, and derive the received signal, the reference signal for dechirping in fast time, and the dechirped signal in the fast time–slow time domain.

The transmitted waveform vector of the proposed MIMO ViSAR system can be expressed in vector form as [3]:

$$
\mathbf{s}_T\left(t, t_a\right) \equiv
\begin{bmatrix}
ss_0\left(t, t_a\right) \\
\vdots \\
ss_m\left(t, t_a\right) \\
\vdots \\
ss_{M-1}\left(t, t_a\right)
\end{bmatrix}
\tag{9}
$$

where the transmitted signal of the m-th Tx antenna, $ss_m\left(t, t_a\right)$, is the BFD FMCW waveform [28] and can be written in complex form as follows:

$$
ss_m\left(t, t_a\right) = \mathrm{rect}\left(\frac{t - t_a}{T_d}\right) \exp\left[j2\pi\left(f_c + m\Delta f_b\right)\left(t - t_a\right) + j\pi k_r\left(t - t_a\right)^2\right]
\tag{10}
$$

where $t_a = n_s T_d$ is slow time, n_s is the sweep number, t is the time variable, B_r and T_d are the bandwidth and sweep duration of the BFD FMCW waveform, respectively, and $k_r = \frac{B_r}{T_d}$, which is the chirp rate of the waveform.

The beat frequency division offset, which means the frequency offset between the transmitted signals, should be decided such that the orthogonality within the range swath is maintained as follows:

$$
\Delta f_b > \frac{B_r}{(MN - 1)T_d} \frac{2R_{sw}}{c}
\tag{11}
$$

where R_{sw} is the range swath and c is the speed of light.

In the proposed MIMO ViSAR system, the signal from an ideal point scatterer, received at the th Rx antenna, is a delayed version of the transmitted signal and can be derived using convolution as [25],

$$
\begin{aligned}
rr_n\left(t, t_a\right) &= \sum_{m=0}^{M-1} ss_m\left(t, t_a\right) * h_{nm}\left(t, t_a\right) = \sum_{m=0}^{M-1} a_{nm} ss_m\left(t - \tau_{nm}\left(x_t, y_t, z_t\right), t_a\right) \\
&= \sum_{m=0}^{M-1} a_{nm} \mathrm{rect}\left(\frac{t - t_a - \tau_{nm}\left(x_t, y_t, z_t\right)}{T_d}\right) \\
&\quad \cdot \exp\left[j2\pi\left(f_c + m\Delta f_b\right)\left(t - \tau_{nm}\left(x_t, y_t, z_t\right)\right) + j\pi k_r\left(t - t_a - \tau_{nm}\left(x_t, y_t, z_t\right)\right)^2\right]
\end{aligned}
\tag{12}
$$

where, a_{nm} denotes the complex coefficient representing the scattering and the path loss, $h_{nm}\left(t, t_a\right)$ is the ideal channel impulse response between the n-th Rx antenna and the m-th Tx antenna, and $\tau_{nm}\left(x_t, y_t, z_t\right)$ is the round trip time delay from the m-th Tx antenna to the point target at position $\left(x_t, y_t, z_t\right)$ and from the point target to the n-th Rx antenna.

The reference signal for dechirping in fast time, which is a replica of the transmitted waveform delayed by the time to the center of the swath, can be written as,

$$
\begin{aligned}
ss_{ref}&\left(t, t_a\right) \\
&= s\, s_m\left(t - \tau_{ref}, t_a\right)\Big|_{m=0} \\
&= \mathrm{rect}\left(\frac{t - t_a - \tau_{ref}}{T_{d,ref}}\right) \exp\left[j2\pi f_c\left(t - t_a - \tau_{ref}\right) + j\pi k_r\left(t - t_a - \tau_{ref}\right)^2\right]
\end{aligned}
\tag{13}
$$

where, τ_{ref} is the reference delay time to the center of the swath, and $T_{d,ref}$ is the sweep duration of the reference signal.

The dechirp-on-receive technique is widely used with FMCW systems. The dechirped signal that results from mixing the received signal in Equation (12) with the reference signal in Equation (13) is:

$$
\begin{aligned}
rr_{n,dc}(t, t_a) &= rr_n(t, t_a)\, ss^*_{ref}(t, t_a) \\
&= \sum_{m=0}^{M-1} a_{nm} \mathrm{rect}\left(\frac{t - t_a - \tau_{nm}}{T_d}\right) \cdot \mathrm{rect}\left(\frac{t - t_a - \tau_{ref}}{T_{d,ref}}\right) \\
&\quad \cdot \exp\left[j2\pi\left(m\Delta f_b - k_r\left(\tau_{nm} - \tau_{ref}\right)\right)(t - t_a - \tau_{nm})\right] \cdot \exp\left[-j\pi k_r\left(\tau_{nm} - \tau_{ref}\right)^2\right] \\
&\quad \cdot \exp\left[-j2\pi f_c\left(\tau_{nm} - \tau_{ref}\right)\right]
\end{aligned}
\tag{14}
$$

The round-trip range from the *m*-th Tx antenna to a point scatterer and to the *n*-th Rx antenna, can be expanded and approximated using Taylor series expansion as,

$$
\begin{aligned}
r_{n,m}(t; r_0) &= c\frac{\tau_{nm}}{2} = r_0\sqrt{1 + \left(\frac{vt - \Delta x_{tx,m}}{r_0}\right)^2} + r_0\sqrt{1 + \left(\frac{vt - \Delta x_{rx,n}}{r_0}\right)^2} \\
&\approx r_0\sqrt{1 + \left(\frac{vt_a - \Delta x_{tx,m}}{r_0}\right)^2} + r_0\sqrt{1 + \left(\frac{vt_a - \Delta x_{rx,n}}{r_0}\right)^2} \\
&\approx r_0\left(1 + \frac{1}{2}\left(\frac{vt_a - \Delta x_{tx,m}}{r_0}\right)^2\right) + r_0\left(1 + \frac{1}{2}\left(\frac{vt_a - \Delta x_{rx,n}}{r_0}\right)^2\right) \\
&= 2r_0 + \frac{v^2\left(t_a - \frac{\Delta x_{tx,m} + \Delta x_{rx,n}}{2v}\right)^2}{r_0} + \frac{\left(\Delta x_{tx,m} - \Delta x_{rx,n}\right)^2}{4r_0}
\end{aligned}
\tag{15}
$$

where, r_0 is the closest distance between the antenna and the scatterer, $\Delta x_{tx,m}$ is the distance between the reference Tx antenna and the *m*-th Tx antenna, and $\Delta x_{rx,n}$ is the distance between the reference Rx antenna and the *n*-th Rx antenna. The range can be interpreted as a phase given by:

$$
\begin{aligned}
&\frac{2\pi}{\lambda} r_{n,m}(t; r_0) \\
&= \frac{2\pi}{\lambda}\left[2r_0 + \frac{v^2\left(t_a - \frac{\Delta x_{tx,m} + \Delta x_{rx,n}}{2v}\right)^2}{r_0} + \frac{\left(\Delta x_{tx,m} - \Delta x_{rx,n}\right)^2}{4r_0}\right] \\
&= \frac{4\pi}{\lambda} r_0 + \frac{2\pi}{\lambda}\frac{v^2\left(t_a - \frac{\Delta x_{tx,m} + \Delta x_{rx,n}}{2v}\right)^2}{r_0} + \frac{\pi\left(\Delta x_{tx,m} - \Delta x_{rx,n}\right)^2}{2\lambda r_0}
\end{aligned}
\tag{16}
$$

where λ is the wave length.

After substituting Equation (15) or (16) into Equation (14), the dechirped signal at the *n*-th Rx channel can be rewritten as,

$$
\begin{aligned}
&rr_{n,dc}(t_r, t_a) \\
&\approx \sum_{m=0}^{M-1} a_{nm} \mathrm{rect}\left(\frac{t_r - \tau_{nm}}{T_d}\right) \cdot \mathrm{rect}\left(\frac{t_r - \tau_{ref}}{T_{d,ref}}\right) \\
&\quad \cdot \exp\left[j2\pi m\Delta f_b (t_r - \tau_0)\right] \\
&\quad \cdot \exp\left[-j2\pi k_r (t_r - \tau_0)\left(\tau_0 - \tau_{ref}\right) - j\pi k_r\left(\tau_0 - \tau_{ref}\right)^2\right] \\
&\quad \cdot \exp\left[-j\frac{4\pi}{\lambda} r_0 - j\frac{2\pi}{\lambda}\frac{v^2\left(t_a - \frac{\Delta x_{tx,m} + \Delta x_{rx,n}}{2v}\right)^2}{r_0} - j\frac{\pi\left(\Delta x_{tx,m} - \Delta x_{rx,n}\right)^2}{2\lambda r_0}\right] \\
&\quad \cdot \exp\left(j2\pi f_d t_r\right) \cdot \exp\left(j2\pi f_c \tau_{ref}\right)
\end{aligned}
\tag{17}
$$

where, $t_r = t - t_a$ is fast time, τ_0 is the time delay corresponding to r_0, and f_d is the Doppler frequency shift within the transmission of one sweep. f_d results from the continuous antenna motion of the platform, since the traditional stop-and-go approximation is not valid for FMCW SAR [11]. In Equation (17), the first exponential term is the frequency and phase shift due to the BFD offset. The second exponential term represents the range signal including the RVP component, which is the second term in the square brackets. The third exponential term is the azimuth phase history of the proposed MIMO ViSAR. The fourth exponential term represents the Doppler shift due to continuous antenna motion within one sweep. The fifth exponential term is the phase delay corresponding to the time delay to the scene center.

Therefore, the vector of a dechirped received signal in the proposed MIMO ViSAR system can be expressed as:

$$
\mathbf{r}_R (t_r, t_a) \equiv
\begin{bmatrix}
rr_{0,dc} (t_r, t_a) \\
\vdots \\
rr_{n,dc} (t_r, t_a) \\
\vdots \\
rr_{N-1,dc} (t_r, t_a)
\end{bmatrix}
\tag{18}
$$

- Signal model in the fast time–azimuth angle domain.

This model is convenient for understanding the subsequent signal processing steps of the MCRA, as it is implemented in the fast time–azimuth frequency domain. The model has a close relationship with the digital beamforming technique mentioned in [7]. The received signal can be expressed using a steering vector in the azimuth angle domain. This signal model explicitly shows the relationship between the digital beamforming techniques used for Tx and Rx in the proposed MIMO ViSAR system.

The relationship between the azimuth frequency domain, f_a, and the azimuth angle domain, θ, is as follows [7]:

$$
f_a = \frac{2v}{\lambda} \sin \theta
\tag{19}
$$

The azimuth time delay term in Equation (17) can be expressed in the azimuth frequency domain, using the Fourier transform property as:

$$
b_n (f_a) a_m (f_a) = 2\pi f_a \frac{\Delta x_{tx,m} + \Delta x_{rx,n}}{2v}
\tag{20}
$$

which can be written in the azimuth angle domain using Equation (19) as:

$$
\begin{aligned}
& b_n (\theta) a_m (\theta) \\
&= 2\pi f_a \frac{\Delta x_{tx,m} + \Delta x_{rx,n}}{2v} \\
&= 2\pi \left(\frac{2v}{\lambda} \sin \theta \right) \frac{\Delta x_{tx,m} + \Delta x_{rx,n}}{2v} \\
&= \frac{2\pi (\Delta x_{tx,m} + \Delta x_{rx,n})}{\lambda} \sin \theta
\end{aligned}
\tag{21}
$$

where $a_m (\theta)$ are individual elements of the Tx steering vector, $\mathbf{a} (\theta)$, and are given by:

$$
a_m (\theta) = \frac{2\pi \Delta x_{tx,m}}{\lambda} \sin \theta
\tag{22}
$$

and $b_n (\theta)$ are individual elements of the Rx steering vector, $\mathbf{b} (\theta)$, and are given by:

$$
b_n (\theta) = \frac{2\pi \Delta x_{rx,n}}{\lambda} \sin \theta
\tag{23}
$$

The Tx steering vector, $\mathbf{a}\,(\theta)$, can be expressed as an $M \times 1$ vector [3]:

$$\mathbf{a}\,(\theta) \equiv \begin{bmatrix} \frac{2\pi}{\lambda}\Delta x_{tx,0}\sin\theta \\ \vdots \\ \frac{2\pi}{\lambda}\Delta x_{tx,m}\sin\theta \\ \vdots \\ \frac{2\pi}{\lambda}\Delta x_{tx,M-1}\sin\theta \end{bmatrix} \tag{24}$$

and the Rx steering vector, $\mathbf{b}\,(\theta)$, can be expressed as an $N \times 1$ vector:

$$\mathbf{b}\,(\theta) \equiv \begin{bmatrix} \frac{2\pi}{\lambda}\Delta x_{rx,0}\sin\theta \\ \vdots \\ \frac{2\pi}{\lambda}\Delta x_{rx,n}\sin\theta \\ \vdots \\ \frac{2\pi}{\lambda}\Delta x_{rx,N-1}\sin\theta \end{bmatrix} \tag{25}$$

Therefore, after applying the Fourier transform in the azimuth frequency domain to Equation (17), the dechirped signal can be written as:

$$
\begin{aligned}
&rR_{n,dc}\,(t_r, f_a) \\
&= F_a\left[rr_{n,dc}\,(t_r, t_a)\right] \\
&= b_n\,(f_a)\sum_{m=0}^{M-1} a_{nm}\mathrm{rect}\left(\frac{t_r - \tau_{nm}}{T_d}\right) \cdot \mathrm{rect}\left(\frac{t_r - \tau_{ref}}{T_{d,ref}}\right) \\
&\quad \cdot \exp\left[j2\pi m\Delta f_b\,(t_r - \tau_0)\right] \\
&\quad \cdot \exp\left[-j2\pi k_r\,(t_r - \tau_0)\left(\tau_0 - \tau_{ref}\right) - j\pi k_r\left(\tau_0 - \tau_{ref}\right)^2\right] \\
&\quad \cdot \exp\left[-j\frac{4\pi}{\lambda}r_0 - j\frac{\pi\,(\Delta x_{tx,m} - \Delta x_{rx,n})^2}{2\lambda r_0}\right] \\
&\quad \cdot \exp\left(j2\pi f_d t_r\right) \cdot \exp\left(j2\pi f_c \tau_{ref}\right) \\
&\quad \cdot a_m\,(f_a) \cdot R_{az}\,(f_a)
\end{aligned}
\tag{26}
$$

where $F_a\,[\cdot]$ is the Fourier transform operator with respect to slow time, and $R_{az}\,(f_a)$ is calculated as follows:

$$R_{az}\,(f_a) = T_a\mathrm{sinc}\left(\pi T_a\,(f_a + k_a\Delta t_a)\right) \tag{27}$$

which is the Fourier transform of the following azimuth time function, which is given by:

$$r_{az}\,(t_a) = \exp\left[-j2\pi k_a\Delta t_a t_a\right] \tag{28}$$

where $k_a = \frac{2v^2}{\lambda r_0}$ is the chirp rate in azimuth and Δt_a is the differential azimuth time between the scatterer and the scene center. Note that $R_{az}\,(f_a)$ represents the result of the azimuth dechirping, which is implemented by geometrical steering on the circular path.

After substituting Equations (19)–(21) into Equation (26) and changing the notation, we can obtain the dechirped signal at the n-th Rx channel in the fast time–azimuth angle domain as follows:

$$rR_{n,dc}\left(t_r,\theta\right)$$

$$\approx b_n\left(\theta\right)\sum_{m=0}^{M-1}a_{nm}\mathrm{rect}\left(\tfrac{t_r-\tau_{nm}}{T_d}\right)\cdot\mathrm{rect}\left(\frac{t_r-\tau_{ref}}{T_{d,ref}}\right)$$

$$\cdot\exp\left[j2\pi m\Delta f_b\left(t_r-\tau_0\right)\right]$$

$$\cdot\exp\left[-j2\pi k_r\left(t_r-\tau_0\right)\left(\tau_0-\tau_{ref}\right)-j\pi k_r\left(\tau_0-\tau_{ref}\right)^2\right]\tag{29}$$

$$\cdot\exp\left[-j\frac{4\pi}{\lambda}r_0\right]\cdot\exp\left(j2\pi f_d t_r\right)\cdot\exp\left(j2\pi f_c\tau_{ref}\right)$$

$$\cdot a_m\left(\theta\right)\cdot R_{az}\left(\theta\right)$$

$$\equiv b_n\left(\theta\right)\sum_{m=0}^{M-1}rR_{m,dc}\left(t_r,\theta\right)\cdot a_m\left(\theta\right)$$

where $rR_{m,dc}\left(t_r,\theta\right)$ is defined as a function of fast time and azimuth angle at the m-th Tx channel.

Therefore, in the fast time–azimuth angle domain, the dechirped signal in the N Rx channels of the proposed MIMO ViSAR system, described by Equation (29), can be compactly expressed as an $N\times1$ vector using Equations (24) and (25) as follows [3]:

$$\mathbf{rR}_{R,dc}\left(t_r,\theta\right)=\mathbf{b}\left(\theta\right)\mathbf{a}\left(\theta\right)^T\mathbf{rR}_{T,dc}\left(t_r,\theta\right)\tag{30}$$

where the dechirped signal from the M Tx channels, described in Equation (30), can be defined in the fast time–azimuth angle domain as an $M\times1$ vector:

$$\mathbf{rR}_{T,dc}\left(t_r,\theta\right)=\begin{bmatrix}rR_{0,dc}\left(t_r,\theta\right)\\\vdots\\rR_{M-1,dc}\left(t_r,\theta\right)\end{bmatrix}=\begin{bmatrix}r_0\left(t_r\right)\\\vdots\\r_{M-1}\left(t_r\right)\end{bmatrix}R_{az}\left(\theta\right)\tag{31}$$

where $r_m\left(t_r\right)$ is defined as a function of fast time as follows:

$$r_m\left(t_r\right)$$

$$=a_{nm}\mathrm{rect}\left(\frac{t_r-\tau_{nm}}{T_d}\right)\cdot\mathrm{rect}\left(\frac{t_r-\tau_{ref}}{T_{d,ref}}\right)$$

$$\cdot\exp\left[j2\pi m\Delta f_b\left(t_r-\tau_0\right)\right]\tag{32}$$

$$\cdot\exp\left[-j2\pi k_r\left(t_r-\tau_0\right)\left(\tau_0-\tau_{ref}\right)-j\pi k_r\left(\tau_0-\tau_{ref}\right)^2\right]$$

$$\cdot\exp\left[-j\tfrac{4\pi}{\lambda}r_0\right]\cdot\exp\left(j2\pi f_d t_r\right)\cdot\exp\left(j2\pi f_c\tau_{ref}\right)$$

and $R_{az}\left(\theta\right)$ is calculated in the azimuth angle domain from $R_{az}\left(f_a\right)$ using Equation (27). Note that the $N\times M$ MIMO channel matrix is defined as follows [3]:

$$\mathbf{A}\left(\theta\right)=\mathbf{b}\left(\theta\right)\mathbf{a}\left(\theta\right)^T\tag{33}$$

In the case of orthogonal waveforms, the MIMO steering vector for angle θ is $\mathbf{g}\left(\theta\right)\equiv\mathrm{Vec}\left\{\mathbf{\Lambda}\left(\theta\right)\right\}$, where Vec [ ] denotes the vectorization operator.

3.3. Signal Processing Procedure

Signal processing for the proposed MIMO ViSAR system proceeds in the following parts: BFD FMCW demodulation, including Doppler shift compensation, azimuth ambiguity suppression using the MCRA, and video frame formation, with image rotation to the CDU using the PFA.

The RVP terms in the dechirped signal described in Equation (32) can be ignored or compensated using the RVP correction method ("range deskew") described in [2] as follows:

$$
\begin{aligned}
& r_{m,DRVP}\left(t_r\right) \\
& = a_{nm} \mathrm{rect}\left(\tfrac{t_r - \tau_{nm}}{T_d}\right) \cdot \mathrm{rect}\left(\tfrac{t_r - \tau_{ref}}{T_{d,ref}}\right) \\
& \quad \cdot \exp\left[j2\pi m\Delta f_b\left(t_r - \tau_0\right)\right] \\
& \quad \cdot \exp\left[-j2\pi k_r\left(t_r - \tau_0\right)\left(\tau_0 - \tau_{ref}\right)\right] \\
& \quad \cdot \exp\left[-j\tfrac{4\pi}{\lambda}r_0\right] \cdot \exp\left(j2\pi f_d t_r\right) \cdot \exp\left(j2\pi f_c \tau_{ref}\right)
\end{aligned}
\tag{34}
$$

where the fourth exponential term represents the Doppler shift due to continuous antenna motion within one sweep, as mentioned earlier.

The continuous antenna motion at the n-th Rx channel is compensated by multiplying Equation (34) with the Doppler frequency correction factor giving the following expression:

$$
\begin{aligned}
& r_{m,DDOP}\left(t_r\right) \\
& = r_{m,DRVP}\left(t_r\right) \cdot \exp\left(-j2\pi f_d t_r\right) \\
& = a_{nm} \mathrm{rect}\left(\tfrac{t_r - \tau_{nm}}{T_d}\right) \cdot \mathrm{rect}\left(\tfrac{t_r - \tau_{ref}}{T_{d,ref}}\right) \\
& \quad \cdot \exp\left[j2\pi m\Delta f_b\left(t_r - \tau_0\right)\right] \\
& \quad \cdot \exp\left[-j2\pi k_r\left(t_r - \tau_0\right)\left(\tau_0 - \tau_{ref}\right)\right] \\
& \quad \cdot \exp\left[-j\tfrac{4\pi}{\lambda}r_0\right] \cdot \exp\left(j2\pi f_c \tau_{ref}\right)
\end{aligned}
\tag{35}
$$

where the first exponential term represents the frequency and phase offset due to BFD FMCW modulation.

Next, BFD FMCW demodulation is carried out at the n-th Rx channel in two processing steps: frequency shift and phase shift compensation. The frequency shift, which results from the beat frequency division offset, is compensated by multiplying Equation (35) with the correction factor, giving the following expression:

$$
\begin{aligned}
& r_{m,BFD1}\left(t_r\right) \\
& = \exp\left(-j2\pi m\Delta f_b t_r\right) \cdot r_{m,DDOP}\left(t_r\right) \\
& = \exp\left(-j2\pi m\Delta f_b t_r\right) \exp\left[j2\pi\left\{m\Delta f_b - k_r\left(\tau_0 - \tau_{ref}\right)\right\}\left(t_r - \tau_0\right)\right] \\
& \quad \cdot \exp\left[-j\tfrac{4\pi}{\lambda}r_0\right] \cdot \exp\left(j2\pi f_c \tau_{ref}\right) \\
& = \exp\left[-j2\pi k_r\left(\tau_0 - \tau_{ref}\right)\left(t_r - \tau_0\right)\right] \cdot \exp\left(-j2\pi m\Delta f_b \tau_0\right) \\
& \quad \cdot \exp\left[-j\tfrac{4\pi}{\lambda}r_0\right] \cdot \exp\left(j2\pi f_c \tau_{ref}\right)
\end{aligned}
\tag{36}
$$

where the second exponential term represents the phase offset due to BFD FMCW modulation.

The phase shift is then compensated by multiplying Equation (36) with the correction factor, giving the following:

$$
\begin{aligned}
& r_{m,BFD2}\left(t_r\right) \\
& = r_{m,BFD1}\left(t_r\right) \cdot \exp\left(j2\pi m\Delta f_b \tau_0\right) \\
& = \exp\left[-j2\pi k_r\left(\tau_0 - \tau_{ref}\right)\left(t_r - \tau_0\right)\right] \cdot \exp\left[-j\tfrac{4\pi}{\lambda}r_0\right] \cdot \exp\left(j2\pi f_c \tau_{ref}\right)
\end{aligned}
\tag{37}
$$

Therefore, using Equation (37), after BFD FMCW waveform demodulation, the signal from Equation (30) can be expressed as:

$$
\mathbf{rR}_{BFD}\left(t_r, \theta\right) = \mathbf{b}\left(\theta\right) \mathbf{a}\left(\theta\right)^T \mathbf{M}_{BFD}\left(t_r, \theta\right)
\tag{38}
$$

where the BFD FMCW waveform demodulation matrix can be defined using Equations (27) and (37) as the following $M \times M$ diagonal matrix:

$$
\begin{aligned}
&\mathbf{M}_{BFD}\left(t_r, \theta\right) \\
&\equiv \begin{bmatrix} rR_{0,BFD2}\left(t_r, \theta\right) & 0 & 0 \\ 0 & \ddots & 0 \\ 0 & 0 & rR_{M-1,BFD2}\left(t_r, \theta\right) \end{bmatrix} \\
&= \begin{bmatrix} r_{0,BFD2}\left(t_r\right) & 0 & 0 \\ 0 & \ddots & 0 \\ 0 & 0 & r_{M-1,BFD2}\left(t_r\right) \end{bmatrix} R_{az}\left(\theta\right)
\end{aligned} \tag{39}
$$

Note that each element of the matrix, $\mathbf{rR}_{BFD}\left(t_r, \theta\right)$, represents the signal after BFD FMCW demodulation of the corresponding elements of the virtual array.

Following a range Fourier transform of Equation (38), the $N \times M$ range compression output matrix can be obtained as follows:

$$
\mathbf{RR}_{RC}\left(f_r, \theta\right) = \int_{-\infty}^{\infty} \mathbf{rR}_{BFD}\left(t_r, \theta\right) e^{-j2\pi f_r t_r} dt_r \tag{40}
$$

where the nm-th element of this matrix is the Fourier transform of the dechirped transmitted signal vector, with respect to fast time, as shown in Equation (38) as follows:

$$
\begin{aligned}
& RR_{RC,nm}\left(f_r, \theta\right) \\
&= F_r\left[rR_{m,BFD2}\left(t_r, \theta\right) b_n\left(\theta\right) a_m\left(\theta\right)\right] \\
&= F_r\left[\exp\left[-j2\pi k_r\left(\tau_0 - \tau_{ref}\right)\left(t_r - \tau_0\right)\right]\right] \cdot \exp\left[-j\frac{4\pi}{\lambda} r_0\right] \\
&\quad \cdot \exp\left(j2\pi f_0 \tau_{ref}\right) \cdot R_{az}\left(\theta\right) \cdot b_n\left(\theta\right) a_m\left(\theta\right) \\
&= T_d \text{sinc}\left(\pi T_d\left(f_r + k_r\left(\tau_0 - \tau_{ref}\right)\right)\right) \cdot \exp\left(-j2\pi \tau_0 f_r\right) \\
&\quad \cdot \exp\left[-j\frac{4\pi}{\lambda} r_0\right] \cdot \exp\left(j2\pi f_c \tau_{ref}\right) \cdot R_{az}\left(\theta\right) \cdot b_n\left(\theta\right) a_m\left(\theta\right)
\end{aligned} \tag{41}
$$

where $F_r\left[\cdot\right]$ is the Fourier transform operator with respect to fast time. Using the fast time relationship, $t_r = \frac{T_d}{B_r} f_r$, and the range relationship, $r = \frac{cT_d}{2B_r} f_r$, we can define the point spread function in range as follows:

$$
\delta_r\left(r\right) \equiv T_d \text{sinc}\left(\pi T_d f_r\right)\big|_{f_r = \frac{2B_r}{cT_d} r} \tag{42}
$$

It is also assumed that all targets exist only within the range swath. The $MN \times 1$ range compression vector, which is output from the virtual array, is given from Equation (40) as [3],

$$
\begin{aligned}
&\mathbf{w}\left(f_r, \theta\right) \\
&\equiv \text{Vec}\left\{\mathbf{RR}_{RC}\left(f_r, \theta\right)\right\} \\
&= \begin{bmatrix} w_0\left(f_r, \theta\right) \\ \vdots \\ w_l\left(f_r, \theta\right) \\ \vdots \\ w_{MN-1}\left(f_r, \theta\right) \end{bmatrix}
\end{aligned} \tag{43}
$$

Using Equations (19), (41) and (42), this equation is represented in the range-azimuth frequency domain, (r, f_a), as:

$$\begin{aligned} \mathbf{w}\,(r, f_a) \\ = \delta_r\,(r - (r_0 - R_a)) \cdot \exp\left(-j2\pi \tfrac{4B_r r_0}{c^2 T_d} r\right) \\ \cdot \exp\left[-j\tfrac{4\pi}{\lambda}\,(r_0 - R_a)\right] \cdot R_{az}\,(f_a) \cdot g\,(f_a) \end{aligned} \tag{44}$$

This equation represents MN channel data of the virtual array formed by BFD FMCW demodulation described above, without taking into account the azimuth ambiguities. When the azimuth ambiguities are considered, the signal from the MN virtual elements can be expressed as an $MN \times 1$ vector, using Equation (44), as follows [9]:

$$\begin{aligned} \mathbf{c}\,(r, f_a) = \Big\{ \textstyle\sum_{l=0}^{MN-1} \delta_r\,(r - (r_0 - R_a)) \cdot \exp\left(-j2\pi \tfrac{4B_r r_0}{c^2 T_d} r\right) \cdot \exp\left[-j\tfrac{4\pi}{\lambda}\,(r_0 - R_a)\right] \\ \cdot R_{az}\,(f_a + l f_p) \cdot \Phi\,(f_a + l f_p) \Big\} \cdot \mathbf{1}_{MN} \end{aligned} \tag{45}$$

where $l = 0, \cdots, MN - 1$ is the ambiguity number, and $\mathbf{1}_{MN} = \begin{bmatrix} 1 & 1 & \cdots & 1 \end{bmatrix}^T$ is the vector of MN ones. The $MN \times MN$ channel phase delay matrix is a diagonal matrix whose elements can be defined as, $(\Phi\,(f_a))_{k,k} = g_k\,(f_a)$, where $g_k\,(f_a)$ is k-th element of the MIMO steering vector, $\mathbf{g}\,(f_a)$. In Equation (45), we assume ideal and identical patterns for all antennas. This equation is included here for completeness of the signal processing procedure. The measured $MN \times 1$ multi-channel signal vector is equal to:

$$\mathbf{z}\,(r, f_a) = \mathbf{c}\,(r, f_a) + \mathbf{n}\,(r, f_a) \tag{46}$$

where $\mathbf{n}\,(r, f_a)$ is the noise vector. As mentioned in the introduction, there are many reconstruction algorithms. In the following, the matrix inversion method [9] is described due to the simplicity of its implementation. The reconstruction filters can be calculated through minimization as [9]:

$$F_M = \min_{\mu_p(f_a)} \|z_{org}\,(r, t_a) - z_{rec}\,(r, t_a)\| \tag{47}$$

where $\|\cdot\|$ denotes the L$_2$ norm and $z_{org}\,(r, t_a)$ is the original signal, sampled at a frequency of $MN \cdot f_p$. The reconstructed signal is obtained as:

$$z_{rec}\,(r, f_a + l f_p) = \boldsymbol{\mu}_p\,(f_a)\,\mathbf{z}\,(r, f_a) \tag{48}$$

where $\mathbf{z}\,(r, f_a)$ is the multi-channel data given by:

$$\mathbf{z}\,(r, f_a) = \begin{bmatrix} z_0\,(r, f_a) & z_1\,(r, f_a) & \cdots & z_{MN-1}\,(r, f_a) \end{bmatrix}^T \tag{49}$$

and the reconstruction filter can be written as:

$$\boldsymbol{\mu}_l\,(f_a) = \mathbf{H}_{inv}^H \cdot \mathbf{1}_{MN}^l \tag{50}$$

where the superscript, H, denotes the Hermitian transpose, and $\mathbf{1}_{MN}^l = \begin{bmatrix} 0 & 0 & \cdots & 1 & \cdots & 0 \end{bmatrix}^T$ is an $MN \times 1$ vector of zeros except for the $l + 1$th position which contains a one, and,

$$\mathbf{H}_{inv}\,(f_a) = \mathbf{H}\,(f_a)^{-1} \tag{51}$$

$$\mathbf{H}\,(f_a) = \begin{bmatrix} \mathbf{g}_0\,(f_a) & \mathbf{g}_1\,(f_a) & \cdots & \mathbf{g}_{MN-1}\,(f_a) \end{bmatrix} \tag{52}$$

$$\mathbf{g}_l\,(f_a) = \text{Vec}\,\{\mathbf{A}\,(f_a + l f_p)\} \tag{53}$$

Note that, as mentioned in [7], in the case of a single-platform system or collocated antennas, the columns of the channel matrix, $\mathbf{H}\left(f_a\right)$, consist of the steering vectors, $\mathbf{g}_l\left(f_a\right)$.

After reconstruction, SAR video frames are formed by the PFA, with $z_{rec}\left(r, f_a + l f_p\right)$ as input data. In the circular path case, image rotation is implemented in the frequency domain using Equation (54), so that the image is in a fixed orientation. This orientation is called the cardinal direction. Image rotation is performed in the frequency domain to reduce distortion. Rotation proceeds as follows:

$$\mathbf{k}_{CDU} = \mathbf{R} \cdot \mathbf{k}_{FRU} \tag{54}$$

where $\mathbf{k}_{CDU}$ is the spatial frequency vector in the cardinal direction up (CDU) coordinate system (global coordinate), $\mathbf{k}_{FRU}$ is the spatial frequency vector in the far range up (FRU) coordinate system (local coordinate), and the rotation matrix is defined as follows:

$$\mathbf{R} = \begin{bmatrix} \cos\theta_{as} & \sin\theta_{as} \\ -\sin\theta_{as} & \cos\theta_{as} \end{bmatrix} \tag{55}$$

where θ_{as} is the aspect angle with respect to the cardinal direction, as shown in Figure 2.

A block diagram of signal processing procedures in the proposed ViSAR system is shown in Figure 9. The first part of signal processing is BFD waveform demodulation, which includes Doppler compensation. In the second part, the MCRA [6–9] is implemented. In the third part, the PFA is used for SAR video frame formation, including image rotation in the 2D spatial frequency domain.

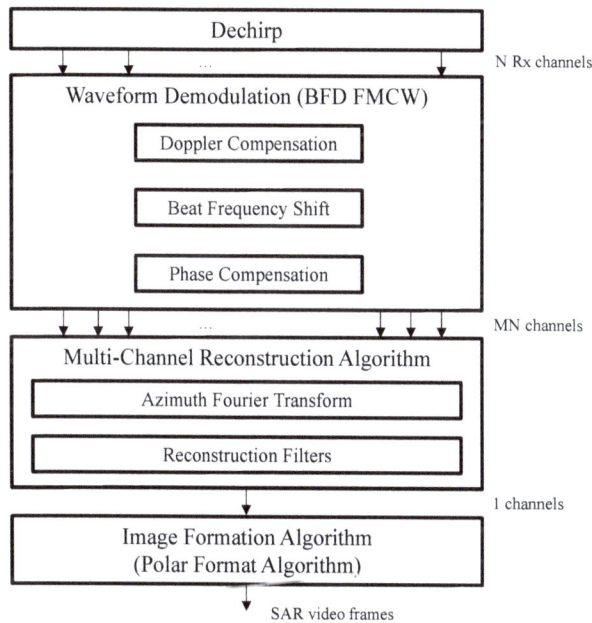

Figure 9. Signal processing block diagram for MIMO video SAR with a beat frequency division FMCW waveform. FMCW: frequency-modulated continuous wave; BFD: beat frequency division.

4. Simulation Results

In this section, we design an exemplary MIMO video SAR system based on BFD FMCW. The performance of the proposed system is evaluated in terms of peak sidelobe ratio (PSLR), resolution, image size, and frame rate, using numerical simulations. As well as antenna parameters, other system and geometric parameters are chosen according to traditional rules, such as the radar equation, or to mirror parameters of existing systems.

4.1. System Parameters

The transmitted frequency is selected to get a higher frame rate as dictated by Equation (4). When considering commercial availability of RF components, 94 GHz is an acceptable frequency for small UAVs. A one-millisecond sweep duration corresponds to a PRF of 1 kHz. According to the Nyquist theorem, the PRF of conventional FMCW SAR systems should be greater than the Doppler bandwidth. However, the PRF of the proposed system can be chosen to be 4000/4 Hz, using four virtual array channels consisting of two Tx and two Rx channels ($M = 2$, $N = 2$). To ensure a slant range resolution of 0.15 m, a chirp bandwidth of 1 GHz is necessary. The beat frequency offset is selected to be 2 MHz for an 80-m range swath width or scene size, based on calculations using Equation (11). The sampling frequency for analog-to-digital conversion (ADC) of the proposed system and a reference single channel ViSAR system are, 4 MHz, and 2 MHz, respectively.

To achieve the required resolution in the azimuth, the integration angle of the synthetic aperture should be greater than 1.14°. The integration angle in this simulation is 1.17°. To observe the required scene size, the transmitting and receiving antenna beamwidth can be chosen to 4°. The distance between Tx antennas is 0.04 m. The distance between Rx antennas is 0.02 m. System parameters are selected such that uniform spatial sampling is applied, as defined by Equation (8). The distance between the phase centers in the virtual array is 0.02 m. In the single channel ViSAR system, the receiving antenna is identical to the transmit antenna. In the MIMO ViSAR system, the transmitting and receiving antennas are placed in the azimuth. Table 1 lists technical and geometric parameters of both systems investigated. The operation velocities of 20 m/s and 40 m/s are selected considering those of commercial or military UAVs such as NEO S-300 by Swiss UAV [37] and RQ-7 Shadow 200 by AAI [38]. As mentioned in Section 2.2, two times higher operation velocity of the proposed system is applied to increase the frame rate due to its proportionality to the platform velocity.

Table 1. System parameters. ViSAR: video synthetic aperture radar.

Parameters	MIMO ViSAR	Single Channel ViSAR
Transmitted frequency	94 GHz	94 GHz
Bandwidth	1 GHz	1 GHz
Slant range to scene center	1000 m	1000 m
Operation velocity	40 m/s	20 m/s
Scene size	80 m	40 m
Resolution	0.15 × 0.08 m	0.15 × 0.08 m
Radar losses	10 dB	10 dB
Number of transmit apertures	2	1
Number of receive apertures	2	1
Transmit antenna gain	15 dB	15 dB
Receive antenna gain	15 dB	15 dB
Sweep duration	1 msec	1 msec
Sampling frequency	4 MHz	2 MHz

Figure 10 shows an illustration of the block diagram for the proposed MIMO video SAR system, with two Tx and two Rx channels. As described in Section 3, there are four channels in the equivalent virtual array in this case.

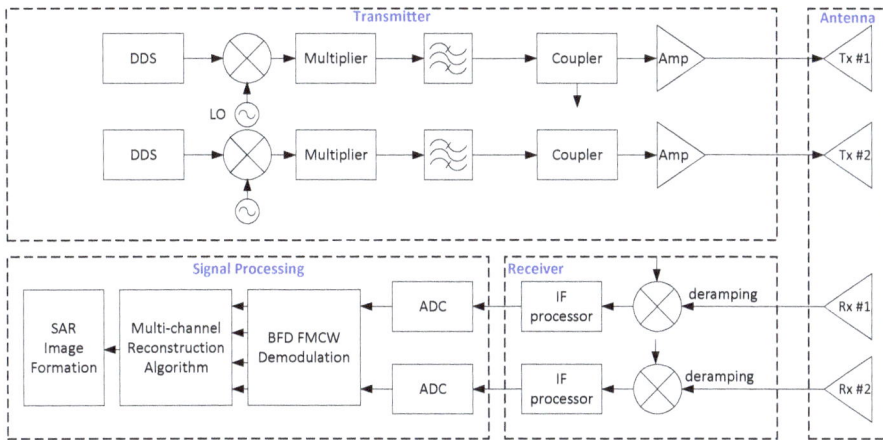

Figure 10. System block diagram for MIMO video SAR with beat frequency division FMCW waveform in the case of two Tx channels and two Rx channels ($M = 2$, $N = 2$). DDS: direct digital synthesis; LO: local oscillator; IF: intermediate frequency (IF); Tx: transmit; ADC: analog-to-digital conversion.

4.2. Point Target Simulation

In this sub-section, a single point target simulation, developed in MATLAB with the system parameters listed in Table 1, is used to verify the performance of the proposed MIMO ViSAR system under ideal sensor motion conditions. Figure 11 illustrates the collection geometry, including the target location, for a circular path simulation. Ideal patterns are assumed for all antennas.

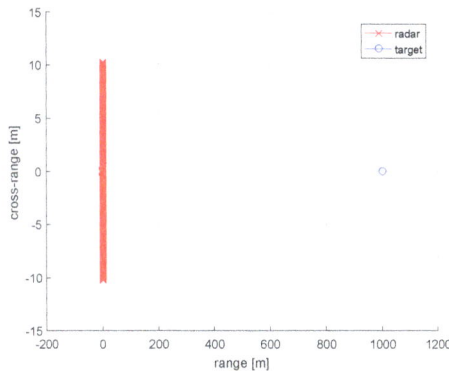

Figure 11. Collection geometry.

The transmitted waveforms are generated using Equation (9) and (10), with $\Delta f_b = 2$ MHz, $B_r = 1$ GHz, and $T_d = 1$ ms. The reference signal for dechirping in fast time was simulated as the delayed version of the transmitted signal at Tx channel 1 using Equation (13). τ_{ref} was set to 6.67 μs in this simulation, which corresponds to the time delay to the scene center. The signal processing procedures presented in Section 3.3 are simulated using the scene geometry and the parameters of the MIMO ViSAR system listed in Table 1.

RVP terms can be ignored for the parameters used in this simulation. The continuous antenna motion effect for FMCW is compensated using Equation (35).

BFD FMCW demodulation is shown in Figure 12. Figure 12a,b shows the signal in range-frequency domain before and after BFD FMCW demodulation, respectively. Note that in Figure 12a, the responses to different transmitters are separated by the beat frequency division offset, 2 MHz. Due to this spectral separation, transmitted waveforms from each Tx antenna can be distinguished. BFD FMCW demodulation is implemented using the frequency shift compensation detailed by Equation (36), and the phase compensation detailed by Equation (37), as described in Section 3.3.

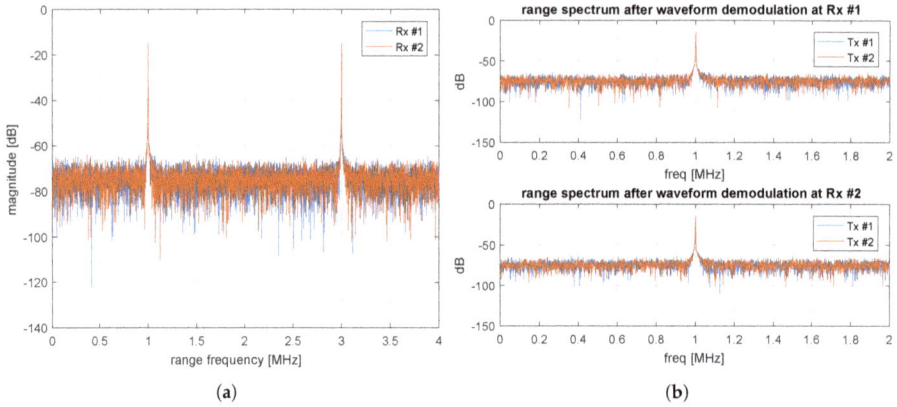

(a)   (b)

Figure 12. Results of BFD FMCW waveform demodulation: (a) before demodulation; (b) after demodulation.

Since two Tx and two Rx channels are used, there are four channels in the virtual array formed by MIMO technology after demodulation, as shown in Figure 12. The distance between the phase centers in the virtual array is 0.02 m. The distance between Tx antennas is 0.04 m, and the distance between Rx antennas is 0.02 m. In this case, uniform spatial sampling is applied according to Equation (8).

Azimuth spectrum reconstruction is shown in Figures 13 and 14. Figure 13a,b show the azimuth aliasing spectrum of the first channel ($l = 0$) in the proposed MIMO ViSAR system before the reconstruction filters.

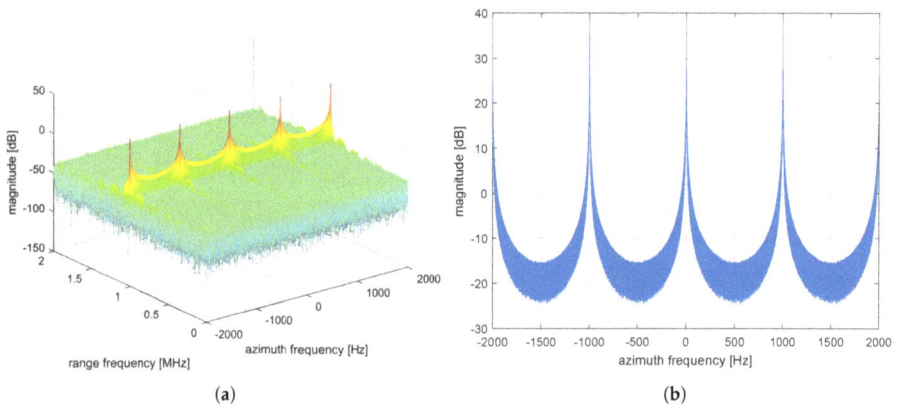

(a)   (b)

Figure 13. Azimuth spectrum before MCRA: (a) three-dimensional plot; (b) two-dimensional plot at target. MCRA: multi-channel reconstruction algorithm.

The azimuth spectrum is completely reconstructed, after applying the reconstruction filters of the MCRA, $\mu_0(f_a)$, $\mu_1(f_a)$, $\mu_2(f_a)$ and $\mu_3(f_a)$, as shown in Equation (50), to the data received by the four MIMO channels, respectively, and stitching all four output signals together, as shown in Figure 14. Note that the azimuth frequency axis is expanded to four times its magnitude before applying the MCRA. The MCRA coherently combines the four channel signals in the azimuth and reconstructs a 4 kHz Doppler spectrum, which is four times wider than the single channel signal.

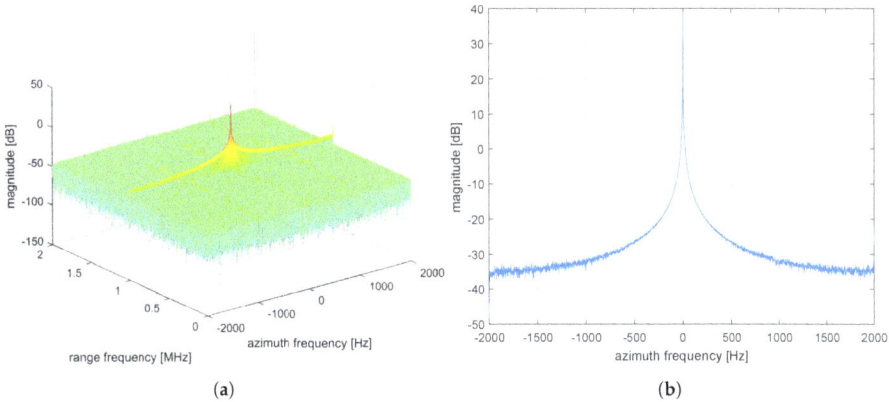

(a) (b)

Figure 14. Azimuth spectrum after MCRA: (**a**) three-dimensional plot; (**b**) two-dimensional plot at target.

Figure 15 plots the simulated 2D SAR image (video frame) of the impulse response function of a point target. Figure 15b shows the contour plot at −3, −6, −9, −30 dB from the peak level.

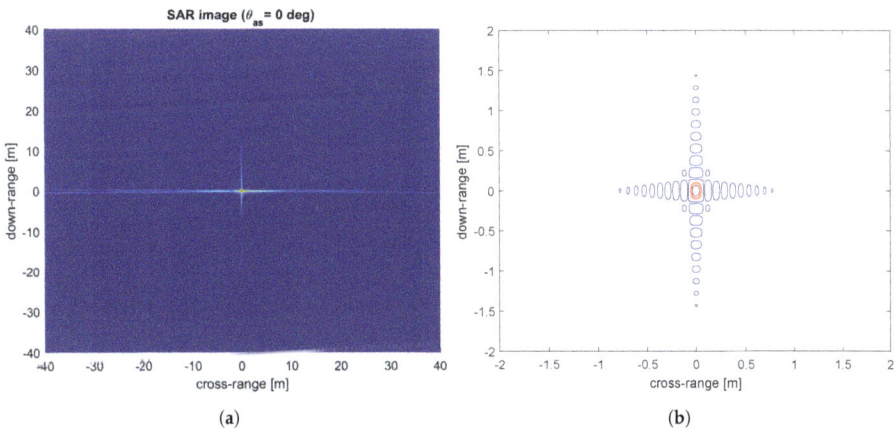

(a) (b)

Figure 15. SAR image: (**a**) two-dimensional plot; (**b**) contour plot (zoomed).

The results of impulse response analysis in the down-range and cross-range are shown in Figure 16a,b respectively. The resolutions at −3.9 dB from the peak level are 0.149 m in the down-range, and 0.081 m in the cross-range. The PSLRs are −13.42 dB in the down-range, and

−13.41 dB in the cross-range. No windows are applied for sidelobe reduction, which means $K_a = 1$ at −3.9 dB.

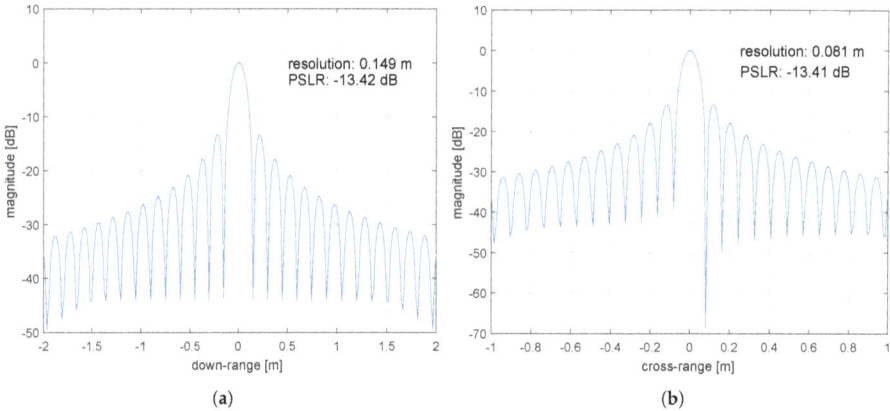

Figure 16. Impulse response analysis (**a**) in down-range; (**b**) in cross-range. PSLR: peak sidelobe ratio

A simulation with multiple synthetic point targets is used to verify the improvements in the frame rate and the image size achieved by the proposed MIMO ViSAR system, by suppressing azimuth ambiguities using MCRA. The five targets are located within a maximum of 30 m from the scene center, as shown in Figure 17.

As shown in Figure 5, the maximum scene size of the PFA, which, according to Equation (6), is limited by wave front curvature, is 126.7 m, when $R_a = 1000$ m and $f_c = 94$ GHz. As shown in Figure 4, the Doppler bandwidth is proportional to the image size. The Doppler bandwidths for 2°-wide azimuth beamwidths are 437 Hz, and 1750 Hz for $v = 20$ m/s, and 80 m/s, respectively. The Doppler bandwidths for 4°-wide azimuth beamwidths are 874 Hz, and 1750 Hz for $v = 20$ m/s, and 40 m/s, respectively. Therefore, a wider image size requires a wider Doppler bandwidth, which results in the need for a higher PRF in the FMCW SAR system. According to Equation (5), when the frame rate and the image size are increased by a factor of two, the Doppler bandwidth is four times wider. Thus, a four-times-higher PRF is required, which is 4 kHz for the system under investigation. In this case, increases to both the frame rate and image size can be achieved through the suppression of Doppler ambiguities in the azimuth spectrum, with the application of the MCRA, while maintaining the same PRF, which is 1 kHz.

The dechirped azimuth signal, gotten after BFD FMCW demodulation is performed in this multiple point target simulation, is shown in Figure 18. The SAR signal on the circular path is azimuth dechirped as described in Section 3.2. Scatterers farther from the scene center have a higher Doppler frequency, which is proportional to the scene size, according to Equation (27). We observe three azimuth ambiguities for each target in Figure 18 ($l = 1, 2, 3$). Due to these azimuth ambiguities, the ambiguous azimuth frequency of Scatterer E is measured as 248.3 Hz, as shown in Figure 18c. As a result of these errors, the image size of the single channel ViSAR system is limited to 40 m.

After applying the MCRA, azimuth ambiguities are suppressed, as shown in Figure 19. The unambiguous azimuth frequency of Scatterer E, which is measured as 751.3 Hz, is reconstructed correctly by the MCRA, as shown in Figure 19c. The other peaks in Figure 19c include the sidelobe of scatterer A, the sidelobe of scatterer C, and residual azimuth ambiguities due to mismatch of the azimuth reconstruction model. No windows are applied for sidelobe reduction for clear visualization the sidelobes.

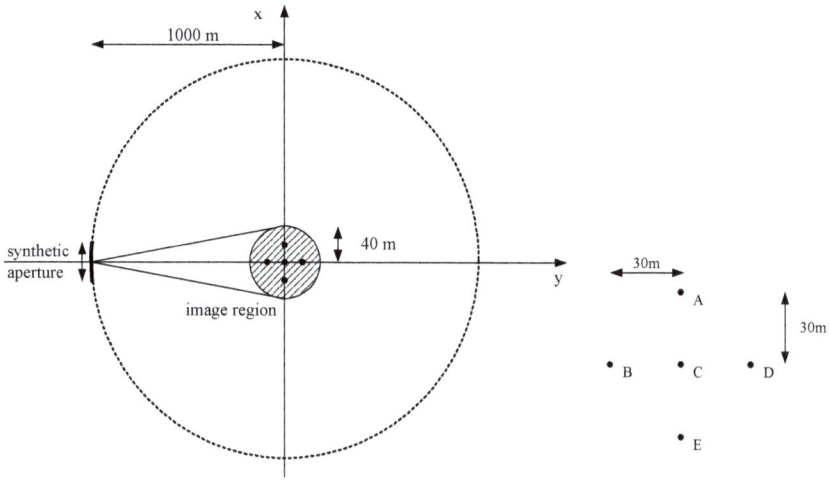

Figure 17. Collection geometry and scene geometry.

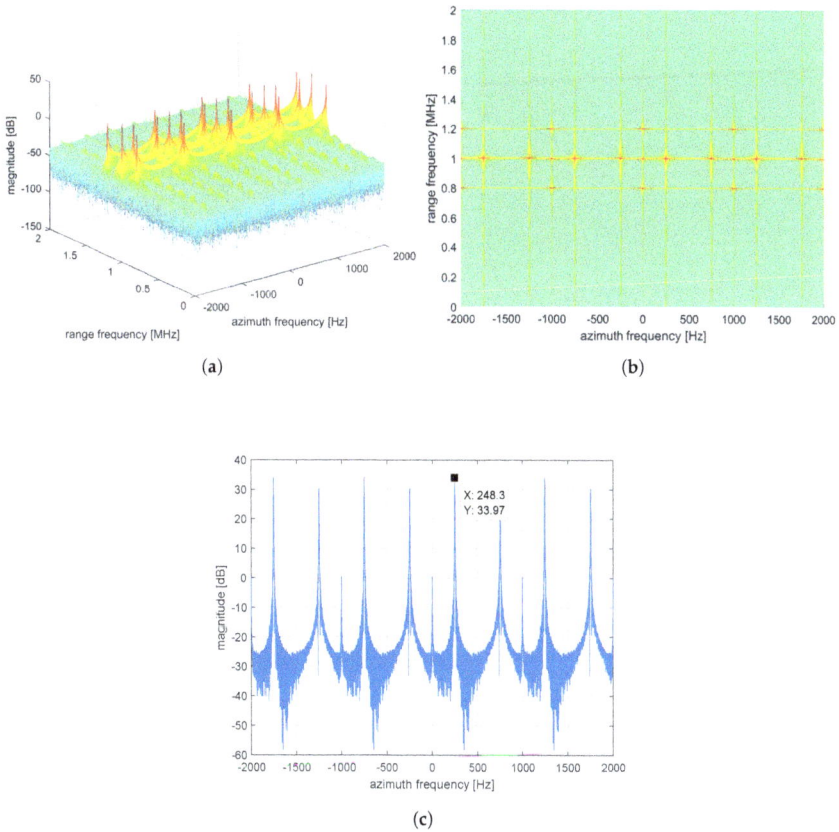

(a)

(b)

(c)

Figure 18. Simulation results before MCRA: (**a**) three-dimensional plot; (**b**) two-dimensional plot; (**c**) azimuth frequency cut at Scatterer E.

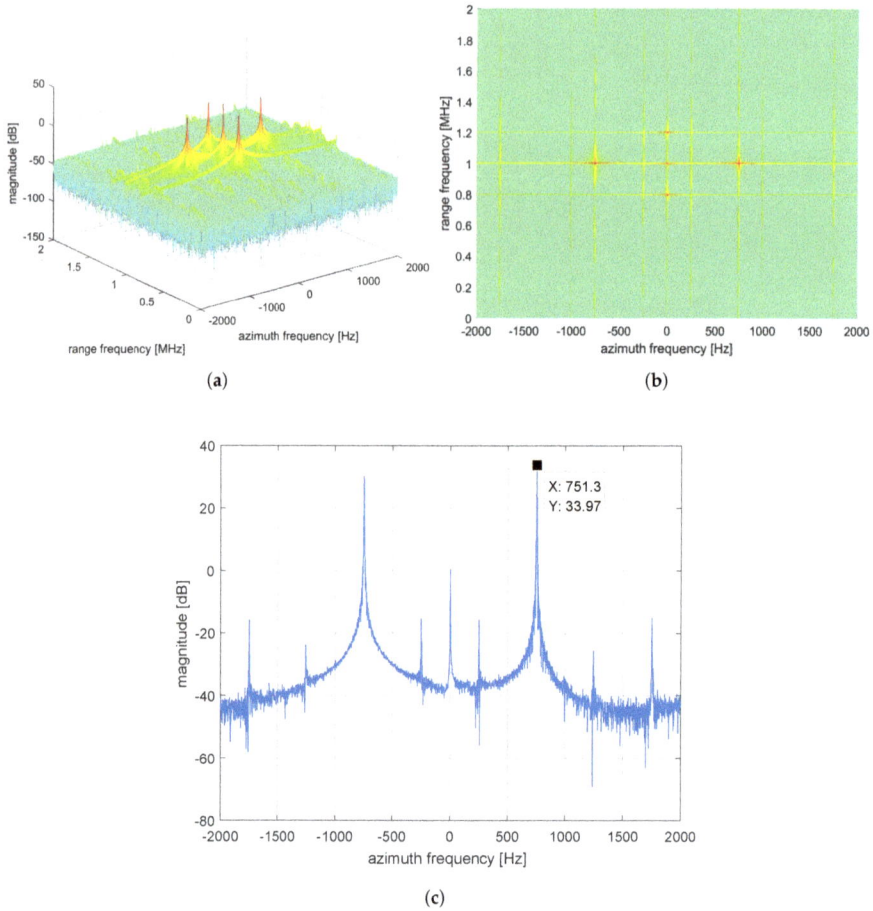

Figure 19. Simulation results after MCRA: (**a**) three-dimensional plot; (**b**) two-dimensional plot; (**c**) azimuth frequency cut at Scatterer E.

The SAR video frame is shown in Figure 20. The video frame is 80 m in the down-range and cross-range, which is two times wider than the single channel system. As both the frame rate and image size are doubled, the azimuth frequency bandwidth is four times bigger than in the single channel case, as suggested by Equation (5).

Figures 21 and 22 show the SAR video frames before and after image rotation from 20° to 70°, with a 10° step in aspect angle, in order to recognize the difference between the SAR video frames. Only the values above −60 dB from the peak level are plotted. As shown in Figure 21, stationary targets appear to rotate, since the aspect angle changes continuously while the radar moves on a circular path.

Image rotation is implemented using Equation (55). As shown in Figure 22, when viewed successively, stationary targets appear fixed after image rotation to CDU, even though the aspect angle changes continuously.

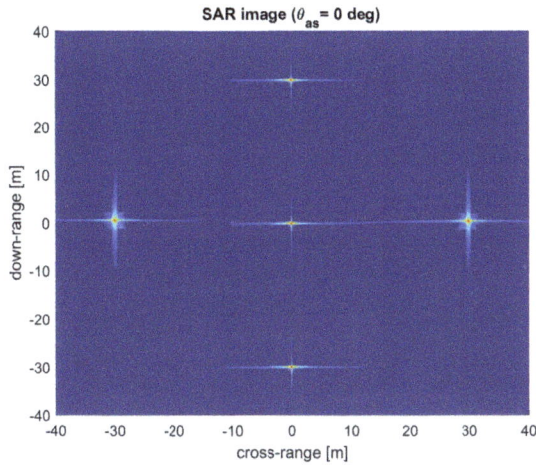

Figure 20. SAR video frame at aspect angle 0°.

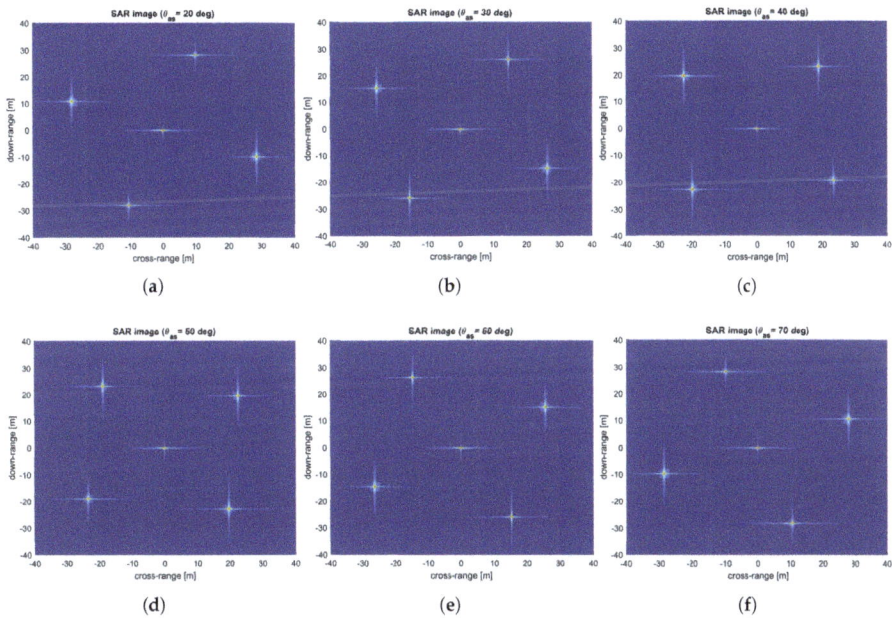

(a)

(b)

(c)

(d)

(e)

(f)

Figure 21. SAR video frames before image rotation at aspect angle: (**a**) 20°; (**b**) 30°; (**c**) 40°; (**d**) 50°; (**e**) 60°; (**f**) 70°.

In this case, the frame rate of the SAR video is calculated from Equation (4) as 2.005 Hz when $v = 40$ m/s, $\rho_a = 0.08$ m, $R_a = 1000$ m, and $f_c = 94$ GHz. This frame rate is two times higher than the single channel system, with $v = 20$ m/s. Note that the platform velocity in the single channel system is limited to 20 m/s, due to the maximum Doppler frequency limitation dictated by Equation (5). In summary, the simulations show improved performance of the proposed MIMO ViSAR system measured by the frame rate and image size of the SAR video, which are two times higher than observed with a single channel system.

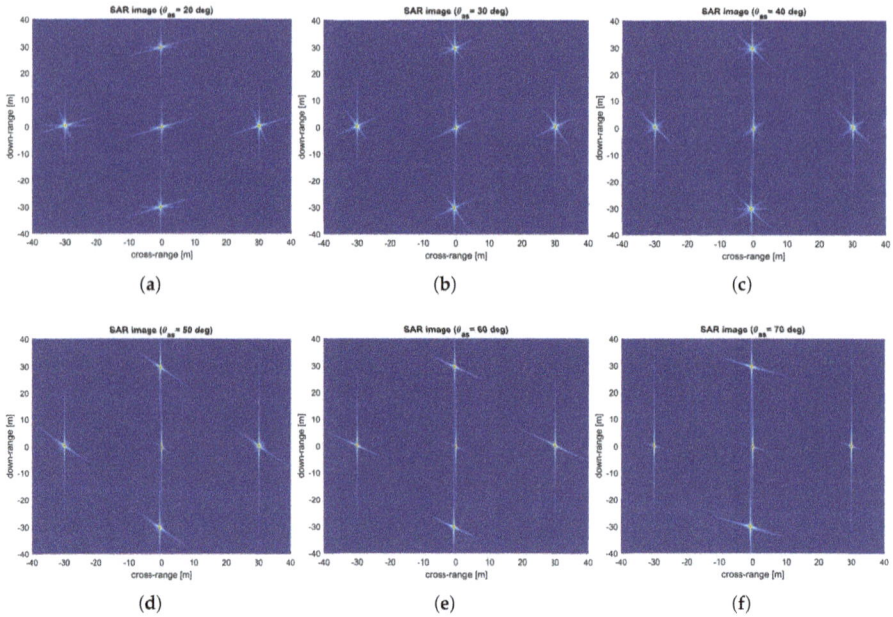

Figure 22. SAR video frames after image rotation at aspect angle: (a) 20°; (b) 30°; (c) 40°; (d) 50°; (e) 60°; (f) 70°.

5. Conclusions

A novel MIMO ViSAR system and signal processing method were presented. This paper described theoretical aspects of ViSAR in terms of two important parameters: frame rate and image size. A MIMO signal model was selected for the proposed system, assuming broadside antenna beam orientation, and a collocated antenna configuration. The signal processing procedures for generating SAR video were then proposed.

The proposed system was able to overcome frame rate and image size limitations caused by azimuth ambiguities, using two notable advanced techniques: the MCRA, which is a multi-channel azimuth processing technique, and MIMO technology. The MCRA is used to increase the Doppler bandwidth by suppressing the azimuth ambiguities in multiple azimuth channels. MIMO technology was used to increase the equivalent number of receiving channels, by forming a virtual array using BFD FMCW as the orthogonal waveform.

The signal model and the signal processing method took into consideration Doppler compensation, required for mitigating effects caused by continuous antenna motion in FMCW-based SAR systems. Since the model used is based on a MIMO signal, it is easy to understand the relationship between beamsteering techniques for Tx and Rx, and multi-channel azimuth processing techniques.

Simulation results showed that the proposed MIMO ViSAR system and signal processing method improved performance, as measured by the frame rate and image size of the SAR video. Further research can be easily extended to multi-mode operation of ViSAR systems with MIMO technology, for example, InSAR and PolSAR.

Acknowledgments: This research was supported by the MSIP (Ministry of Science, ICT and Future Planning), Korea, under the "ICT Consilience Creative Program" (IITP-2017-2017-0-01015) supervised by the IITP (Institute for Information & Communications Technology Promotion), and the Civil Military Technology Cooperation Program. This work was supported by ICT R&D program of MSIP/IITP [2017-0-00286].

Remote Sens. **2017**, *9*, 491

Author Contributions: Seok Kim is responsible for the theoretical work, simulation and the writing of the manuscript. Jiwoong Yu and Se-Yeon Jeon assisted the theoretical study. Aulia Dewantari assisted to write the manuscript. Min-Ho Ka supervised the research.

Conflicts of Interest: The authors declare no conflict of interest.

References

1. Curlander, J.C.; McDonough, R.N. *Synthetic Aperture Radar: Systems and Signal Processing*; John Wiley & Sons: New York, NY, USA, 1991, ISBN:978-0-471-85770-9.
2. Carrara, W.G.; Goodman, R.S.; Majewski, R.M. *Spotlight Synthetic Aperture Radar: Signal Processing Algorithms*; Artech House: Norwood, MA, USA, 1995, ISBN:0890067287.
3. Melvin, W.; Scheer, J. *Principles of Modern Radar Vol. II: Advanced Techniques.* *Edison*; Scitech Publishing: Mendham, NJ, USA, 2013, ISBN:978-1-891121-53-1.
4. Willey, C. Synthetic aperture radars—A paradigm for technology evolution. *IEEE Trans. Aerosp. Electron. Syst.* **1985**, *21*, 440–443.
5. Currie, A.; Brown, M.A. Wide-swath SAR. *IEE Proc. F Radar Signal Process.* **1992**, *139*, 122–135, doi:10.1049/ip-f-2.1992.0016.
6. Krieger, G.; Gebert, N.; Moreira, A. Unambiguous SAR signal reconstruction from nonuniform displaced phase center sampling. *IEEE Geosci. Remote Sens. Lett.* **2004**, *1*, 260–264, doi:10.1109/LGRS.2004.832700.
7. Gebert, N.; Krieger, G.; Moreira, A. Digital beamforming on receive: Techniques and optimization strategies for high-resolution wide-swath SAR imaging. *IEEE Trans. Aerosp. Electron. Syst.* **2009**, *45*, doi:10.1109/TAES.2009.5089542.
8. Gebert, N. *Multi-Channel Azimuth Processing for High-Resolution Wide-Swath SAR Imaging*; DLR (German Aerospace Center): Köln, Germany, 2009.
9. Cerutti-Maori, D.; Sikaneta, I.; Klare, J.; Gierull, C.H. MIMO SAR processing for multichannel high-resolution wide-swath radars. *IEEE Trans. Geosci. Remote Sens.* **2014**, *52*, 5034–5055, doi:10.1109/TGRS.2013.2286520.
10. Liu, B.; He, Y. Improved DBF algorithm for multichannel high-resolution wide-swath SAR. *IEEE Trans. Geosci. Remote Sens.* **2016**, *54*, 1209–1225, doi:10.1109/TGRS.2015.2476496.
11. Meta, A.; Hoogeboom, P.; Ligthart, L.P. Signal processing for FMCW SAR. *IEEE Trans. Geosci. Remote Sens.* **2007**, *45*, 3519–3532, doi:10.1109/TGRS.2007.906140.
12. Kim, J.H.; Younis, M.; Moreira, A.; Wiesbeck, W. Spaceborne MIMO synthetic aperture radar for multimodal operation. *IEEE Trans. Geosci. Remote Sens.* **2015**, *53*, 2453–2466, doi:10.1109/TGRS.2014.2360148.
13. Damini, A.; Balaji, B.; Parry, C.; Mantle, V. A videoSAR mode for the X-band wideband experimental airborne radar. *Proc. SPIE Def. Secur. Sens. Int. Soc. Opt. Photonics* **2010**, *7699*, 76990E, doi:10.1117/12.855376.
14. Wallace, H.B. Development of a video SAR for FMV through clouds. *Proc. SPIE Def. Secur. Sens. Int. Soc. Opt. Photonics* **2015**, *9749*, 94790L, doi:10.1117/12.2181420.
15. Miller, J.; Bishop, E.; Doerry, A. An application of backprojection for video SAR image formation exploiting a subaperture circular shift register. *Proc. SPIE Def. Secur. Sens. Int. Soc. Opt. Photonics* **2013**, *8746*, 874609, doi:10.1117/12.2016417.
16. Wallace, H.; Gorman, J.; Maloney, P. *Video Synthetic Aperture Radar (ViSAR)*; Defense Advanced Research Projects Agency: Arlington, VA, USA, 2012.
17. Baumgartner, S.V.; Krieger, G. Simultaneous high-resolution wide-swath SAR imaging and ground moving target indication: Processing approaches and system concepts. *IEEE J. Sel. Top. App. Earth Obs. Remote Sens.* **2015**, *8*, 5015–5029, doi:10.1109/JSTARS.2015.2450019.
18. De Wit, J.; Meta, A.; Hoogeboom, P. Modified range-Doppler processing for FM-CW synthetic aperture radar. *IEEE Geosci. Remote Sens. Lett.* **2006**, *3*, 83–87, doi:10.1109/LGRS.2005.856700.
19. Jiang, Z.H.; Huang-Fu, K. Squint LFMCW SAR data processing using Doppler-centroid-dependent frequency scaling algorithm. *IEEE Trans. Geosci. Remote Sens.* **2008**, *46*, 3535–3543, doi:10.1109/TGRS.2008.2000755.
20. Xin, Q.; Jiang, Z.; Cheng, P.; He, M. Signal processing for digital beamforming FMCW SAR. *Math. Probl. Eng.* **2014**, *2014*, 859890, doi:10.1155/2014/859890.
21. Paulraj, A.J.; Kailath, T. Increasing Capacity in Wireless Broadcast Systems Using Distributed Transmission/directional Reception (DTDR). U.S. Patent 5,345,599, 6 September 1994.

22. Fishler, E.; Haimovich, A.; Blum, R.; Chizhik, D.; Cimini, L.; Valenzuela, R. MIMO radar: An idea whose time has come. In Proceedings of the 2004 IEEE Radar Conference, Philadelphia, PA, USA, 26–29 April 2004; pp. 71–78, doi:10.1109/NRC.2004.1316398.

23. Ponce, O.; Rommel, T.; Younis, M.; Prats, P.; Moreira, A. Multiple-input multiple-output circular SAR. In Proceedings of the 2014 IEEE 15th International Radar Symposium (IRS), Gdańsk, Poland, 16–18 June 2014; pp. 1–5, doi:10.1109/IRS.2014.6869262.

24. Kim, J.H.; Younis, M.; Moreira, A.; Wiesbeck, W. A novel OFDM chirp waveform scheme for use of multiple transmitters in SAR. *IEEE Geosci. Remote Sens. Lett.* **2013**, *10*, 568–572, doi:10.1109/LGRS.2012.2213577.

25. Kim, J.H. Multipe-Input Multiple-Output Synthetic Aperture Radar for Multimodal Operation. Ph.D. Thesis, Karlsruher Institut für Technologie (KIT), Karlsruhe, Germany, 2011.

26. Krieger, G. MIMO-SAR: Opportunities and pitfalls. *IEEE Trans. Geosci. Remote Sens.* **2014**, *52*, 2628–2645, doi:10.1109/TGRS.2013.2263934.

27. Wang, W.Q. *Multi-Antenna Synthetic Aperture Radar*; CRC Press: Nottingham, UK, 2013, ISBN:978-1-4665-1051-7.

28. De Wit, J.; Van Rossum, W.; De Jong, A. Orthogonal waveforms for FMCW MIMO radar. In Proceedings of the 2011 IEEE Radar Conference (RADAR), Kansas City, MO, USA, 23–27 May 2011; pp. 686–691, doi:10.1109/RADAR.2011.5960625.

29. Cheng, P.; Wang, Z.; Xin, Q.; He, M. Imaging of FMCW MIMO radar with interleaved OFDM waveform. In Proceedings of the 2014 IEEE 12th International Conference on Signal Processing (ICSP), Hangzhou, China, 19–23 October 2014; pp. 1944–1948, doi:10.1109/ICOSP.2014.7015332.

30. Sikaneta, I.; Gierull, C.H.; Cerutti-Maori, D. Optimum signal processing for multichannel SAR: With application to high-resolution wide-swath imaging. *IEEE Trans. Geosci. Remote Sens.* **2014**, *52*, 6095–6109, doi:10.1109/TGRS.2013.2294940.

31. Li, X.; Xing, M.; Xia, X.G.; Sun, G.C.; Liang, Y.; Bao, Z. Simultaneous stationary scene imaging and ground moving target indication for high-resolution wide-swath SAR system. *IEEE Trans. Geosci. Remote Sens.* **2016**, *54*, 4224–4239, doi:10.1109/TGRS.2016.2538564.

32. Walker, J.L. Range-Doppler imaging of rotating objects. *IEEE Trans. Aerosp. Electron. Syst.* **1980**, *16*, 23–52, doi:10.1109/TAES.1980.308875.

33. Doren, N.; Jakowatz, C.; Wahl, D.E.; Thompson, P.A. General formulation for wavefront curvature correction in polar-formatted spotlight-mode SAR images using space-variant post-filtering. In Proceedings of the IEEE International Conference on Image Processing, Santa Barbara, CA, USA, 26–29 October 1997; Volume 1, pp. 861–864, doi:10.1109/ICIP.1997.648102.

34. Jakowatz, C.V., Jr.; Wahl, D.E.; Thompson, P.A.; Doren, N.E. Space-variant filtering for correction of wavefront curvature effects in spotlight-mode SAR imagery formed via polar formatting. In Proceedings of the International Society for Optics and Photonics (AeroSense'97), Orlando, FL, USA, 21 July 1997; pp. 33–42, doi:10.1117/12.281576.

35. Doren, N.E. *Space-Variant Post-Filtering for Wavefront Curvature Correction in Polar-Formatted Spotlight-Mode SAR Imagery*; Technical Report; Sandia National Labs.: Albuquerque, NM, USA; Livermore, CA, USA, 1999.

36. Jakowatz, C.V.; Wahl, D.E.; Eichel, P.H.; Ghiglia, D.C.; Thompson, P.A. *Spotlight-Mode Synthetic Aperture Radar: A Signal Processing Approach: A Signal Processing Approach*; Springer Science & Business Media: Berlin, Germany, 2012, ISBN:978-0-7923-9677-2.

37. Johannes, W.; Essen, H.; Stanko, S.; Sommer, R.; Wahlen, A.; Wilcke, J.; Wagner, C.; Schlechtweg, M.; Tessmann, A. Miniaturized high resolution Synthetic Aperture Radar at 94 GHz for microlite aircraft or UAV. In Proceedings of the 2011 IEEE Sensors, Limerick, Ireland, 28–31 October 2011; pp. 2022–2025, doi:10.1109/ICSENS.2011.6127301.

38. Cheng, S.W. Rapid deployment UAV. In Proceedings of the 2008 IEEE Aerospace Conference, Big Sky, MT, USA, 1–8 March 2008; pp. 1–8, doi:10.1109/AERO.2008.4526564.

remote sensing

MDPI

Article

Fast and Efficient Correction of Ground Moving Targets in a Synthetic Aperture Radar, Single-Look Complex Image

Jeong-Won Park [1], Jae Hun Kim [2] and Joong-Sun Won [2,*]

[1] Nansen Environmental and Remote Sensing Center, 5006 Bergen, Norway; jeong-won.park@nersc.no
[2] Department of Earth System Sciences, Yonsei University, Seoul 03722, South Korea; jhkim90@yonsei.ac.kr
* Correspondence: jswon@yonsei.ac.kr; Tel.: +82-2-2123-2673

Received: 1 August 2017; Accepted: 28 August 2017; Published: 6 September 2017

Abstract: Ground moving targets distort normally-focused synthetic aperture radar (SAR) images. Since most high-resolution SAR data providers only offer single-look complex (SLC) data rather than raw signals to general users, they need to apply a simple and efficient residual SAR focusing to SLC data containing moving targets. This paper presents an efficient and effective SAR residual focusing method that is practically applicable to SLC data. The residual Doppler spectrum of the moving target is derived from a general SAR configuration and normal SAR focusing. The processing steps are simple and straightforward, with a limited size of the processing window, e.g., 64×64. Application results using simulation data and actual TerraSAR-X SLC data with a speed-controlled vehicle demonstrate the effectiveness of the method, which particularly improves the -3 dB width, integrated sidelobe ratio, and symmetry of the reconstructed signals. In particular, the azimuthal symmetry becomes seriously distorted when the target speed is higher than 8 m/s (or 28.8 km/h), and the symmetry is well recovered by the proposed method.

Keywords: SAR; ground moving target; single-look complex data; Doppler spectrum; residual focusing

1. Introduction

The imaging characteristics of ground moving targets using synthetic aperture radar (SAR) have been well known since the early stages of SAR development [1,2]. Ground moving objects in SAR single-look complex (SLC) images are typically characterized by three features: target displacement in the azimuth dimension and range walking according to the range component of the ground target velocity, azimuth image blurring (mainly due to the azimuth component of velocity and the range component of acceleration), and residual Doppler centroid. While these features distort SAR images, they have been exploited as ground moving target indicators (GMTIs) to retrieve the target's velocity [3]. Thus, the main concerns related to ground moving objects are twofold: the detection of a moving target within an SAR image, and the estimation of physical parameters such as velocity or original location. Numerous algorithms have been proposed for GMTI, and most of them are based on sensing the difference in Doppler parameters between the moving object [4–8] and the fixed clutter or on detection by focusing [9–16]. For more efficient detection of ground moving targets, the theory and systems for the along-track interferometry (ATI) also have been extensively researched [13,17–25]. Bistatic ATI SAR systems have recently gained growing popularity [26–28].

While the SAR imaging characteristics are exploited for the GMTI, precise focusing remains an important issue, especially as the resolution of SAR images becomes ever higher. Various focusing methods for SAR have previously been developed, but most of them are based on raw signal processing [9,12], and recently, many have involved the keystone transformation [29–31]. Although several SAR focusing algorithms have been developed for ground moving targets, raw signals rather than single-look complex (SLC) data are required in most cases. However, it is not practical for general users to process raw signals because the raw signal data acquired by most current high-resolution SAR systems are not provided to users, mainly because of their complexity. Thus, it is necessary for general users to apply a residual focusing to SLC data rather than the raw signals. This study proposes a simple and straightforward method of residual focusing for ground moving targets that is practically applicable to SLC data by general users. Generally, there are two types of ground moving targets: targets moving in groups such as ocean currents and waves, and isolated small but fast-moving targets such as moving ships or cars. The proposed method is particularly applicable for the latter type, based on the point-target spectrum in the 2D frequency domain. The residual Doppler phase in the normally-focused SLC data is to be elaborately formulated and discussed. This paper presents formulae related to the residual Doppler spectrum caused by ground moving targets after azimuth and range compression, and proposes a simple and straightforward method for residual focusing of SLC images based on the derived formulae. The derived residual Doppler spectrum accounts for target distortion by the asymmetry of the compressed signals as well as image blurring. For evaluation and demonstration of the performance, the algorithm is applied to simulated data and TerraSAR-X SLC data in which a speed-controlled vehicle is imaged.

The advantages of the proposed method are twofold. First, the residual focusing is based on the derived formulae of the Doppler spectrum after image formation, which implies that targets can be precisely reconstructed in terms of main-to-sidelobe ratio and symmetry. Second, the method is simple and practical because only SLC data of high-resolution SAR systems (rather than raw signal data) are normally provided to general users. A processing window for each ground moving target is relatively small (few tens of pixels) because the image is already focused. This maximizes the computational efficiency and minimizes the distortion of neighboring stationary objects. In this paper, Sections 2 and 3 describe the derivation of the residual Doppler spectrum after azimuth and range compression and the processing tactics. Section 4 presents the application results that demonstrate the efficiency and effectiveness of the residual focusing in terms of the −3 dB width, integrated sidelobe ratio, and symmetry of the focused signals. Finally, the discussion and conclusions follow in Sections 5 and 6, respectively.

2. Correction Formula for SAR SLC data

2.1. Phase Effects of a Moving Target in 2D Frequency Domain

The received signal from a ground point target in a monostatic SAR configuration after demodulation is as follows [32]:

$$s_r(t, \tau) = rect\left(\frac{\tau}{T_a}\right) \cdot s_t\left(t - \frac{2R(\tau)}{c}\right) \cdot \exp\left\{-i2\pi f_0 \frac{2R(\tau)}{c}\right\} \tag{1}$$

where $s_t(\ )$ is the transmitted signal, $R(\ )$ is the slant range distance, f_0 is the carrier frequency, c is the speed of light, T_a is the length of full aperture time, and t and τ are the range time (or fast-time) and azimuth time (or slow-time), respectively. In general the amplitude modulation is described by two-way antenna pattern; however, since our interest here is the phase component of the returned signal, the amplitude component in Equation (1) is neglected without losing generality. The point-target spectrum of a stationary ground object in the 2D frequency domain is as follows [32,33]:

$$S_{r,ST}(f, f_\tau) \approx S_t(f) \cdot rect\left(\frac{f_\tau}{B_a}\right) \cdot \exp\left\{-i2\pi \frac{2R_0}{c}\sqrt{(f + f_0)^2 - \frac{c^2}{4V^2}f_\tau^2}\right\} \tag{2}$$

where $S_t(\ )$ is the Fourier transform of $s_t(\ )$, R_0 is the range distance at the closest approach, V is the effective antenna velocity along the azimuth direction, B_a is the full aperture bandwidth, and f and f_τ are the range frequency and azimuth (or Doppler) frequency, respectively.

Let us consider a ground moving object now. Figure 1 shows an imaging geometry of SAR to a ground object with a Cartesian coordinate (x, y, z). x, y, and z axes are with the azimuth direction, the zero-Doppler ground range direction, and the normal to the earth at the antenna position, respectively. For a space-borne imaging scenario, the use of earth ellipsoid model is more appropriate. However for describing the time-varying distance of a moving object during short observation time (less than 1 second for the stripmap mode), the real imaging geometry can be approximated by plane-earth geometry. The time-varying distance, $R(\tau)$, from a ground target located in $(0, y_0, 0)$ at $\tau = 0$ to the antenna, changing with a velocity of $v_0 = (v_x, v_y, 0)$ and an acceleration of $a_0 = (a_x, a_y, 0)$, is given as follows:

$$
\begin{aligned}
R(\tau; R_0) &= \sqrt{H^2 + \left(V\tau - v_x\tau - \tfrac{a_x}{2}\tau^2\right)^2 + \left(y_0 + v_y\tau + \tfrac{a_y}{2}\tau^2\right)^2} \\
&\approx \sqrt{R_0^2 + \left(1 - 2\tfrac{v_x}{V} + \tfrac{a_y}{2}\tfrac{y_0}{V^2}\right)V^2\tau^2 + v_y\tfrac{y_0}{R_0}\tau} \\
&\equiv R_m(t, \tau) - \tfrac{\lambda}{2}\alpha\tau
\end{aligned}
\tag{3}
$$

where $R_m(\tau; R_0) = \sqrt{R_0^2 + V_m^2\tau^2}$, $V_m^2 = V^2\left(1 - 2\tfrac{v_x}{V} + \tfrac{a_y y_0}{2V^2}\right)$, $\alpha = -2\tfrac{v_y}{\lambda}\tfrac{y_0}{R_0}$, λ is the wavelength, H is the altitude of antenna, and $R_0 = \sqrt{H^2 + y_0^2}$ in the given imaging geometry. Thus, the returned signal from a ground moving target in signal space is

$$
s_{r,MT}(t, \tau) \approx rect\left(\frac{\tau}{T_a}\right) \cdot s_t\left(t - \left(\frac{2R_m(\tau)}{c} - \frac{\alpha}{f_0}\tau\right)\right) \cdot \exp\left\{-2\pi f_0 \frac{2R_m(\tau)}{c}\right\} \cdot \exp\{+i2\pi\alpha\tau\}
\tag{4}
$$

The point target spectrum in the range frequency-azimuth time domain is given by

$$
S_{r,MT}(f, \tau) \approx S_t(f) \cdot rect\left(\frac{\tau}{T_a}\right) \cdot \exp\left\{-i2\pi\frac{2R_m(\tau)}{c}(f_0 + f)\right\} \cdot \exp\left\{+i2\pi\frac{\alpha}{f_0}(f + f_0)\tau\right\}
\tag{5}
$$

The range walk due to the moving target is expressed by the last phase component in Equation (5) and a modification of the relative antenna velocity, $V \to V_m$, in $R_m(\tau)$ of the first phase component, which in turn affects the Doppler slope. Then, the moving target in the 2D frequency domain can be obtained from Equation (2) by replacing the Doppler frequency, f_τ, with $[f_\tau - \alpha(1 + f/f_0)]$ and the effective velocity, V, with V_m:

$$
S_{r,MT}(f, f_\tau) \approx S_t(f) \cdot rect\left(\frac{f_\tau - \alpha(1 + f/f_0)}{B_a}\right) \cdot \exp\{-i2\pi\Phi(f, f_\tau)\}
\tag{6}
$$

where

$$
\begin{aligned}
\Phi(f, f_\tau) &= \tfrac{2R_0}{c}\sqrt{(f + f_0)^2 - \tfrac{c^2}{4V_m^2}\left[f_\tau - \alpha\left(1 + \tfrac{f}{f_0}\right)\right]^2} \\
&= \tfrac{2R_0}{\lambda}\sqrt{\left(1 + \tfrac{f}{f_0}\right)^2 - \tfrac{\lambda^2}{4V_m^2}\left[f_\tau - \alpha\left(1 + \tfrac{f}{f_0}\right)\right]^2}
\end{aligned}
\tag{7}
$$

Applying Taylor expansion of $\Phi(f, f_\tau)$ by $\tfrac{f}{f_0}$ up to second order results in a simplified form as follows:

$$
\begin{aligned}
\Phi(f, f_\tau) &\approx \tfrac{2R_0}{\lambda}\left\{a_m(f_\tau; \alpha) + \tfrac{1}{a_m(f_\tau; \alpha)}\tfrac{f}{f_0} - \tfrac{1}{2}\tfrac{1}{a_m^3(f_\tau; \alpha)}\left(\tfrac{\lambda^2}{4V_m^2}\right)f_\tau^2\left(\tfrac{f}{f_0}\right)^2\right\} \\
&+ \tfrac{2R_0}{\lambda}\left\{\tfrac{\alpha\cdot(f_\tau - \alpha)}{a_m(f_\tau; \alpha)}\left(\tfrac{\lambda^2}{4V_m^2}\right)\tfrac{f}{f_0}\right\}
\end{aligned}
\tag{8}
$$

where $a_m(f_\tau; \alpha) = \sqrt{1 - \tfrac{\lambda^2}{4V_m^2}(f_\tau - \alpha)^2}$. Compared with the stationary target case in Equation (2), the terms with α and V_m that are involved distort the point target spectrum of a ground moving target. It is necessary to compensate these terms when fine-tuning the SAR image of each ground moving target.

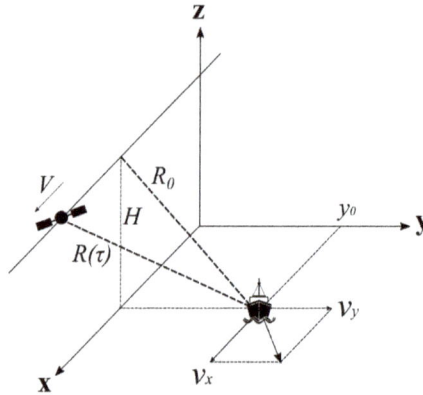

Figure 1. A stripmap synthetic aperture radar (SAR) observation geometry for a ground moving object.

2.2. The Residual Phase Removal in the SAR SLC Image

Since the SAR SLC image, rather than raw signals, is usually provided to general users, it is practical to consider phase compensation in SLC data. The SAR SLC data is formed by a series of processing steps, including range-curvature-migration correction (RCMC), azimuth compression, and the secondary range compression. These three processing steps for a point target can be modeled as multiplication by the complex conjugate of the following function [33]:

$$H(f, f_\tau) = \exp\left\{ -i2\pi \frac{2R_0}{\lambda} \left(a(f_\tau) + \frac{1}{a(f_\tau)} \left(\frac{f}{f_0}\right) - \frac{1}{2} \frac{1}{a^3(f_\tau)} \left(\frac{\lambda^2}{4V^2}\right) f_\tau^2 \left(\frac{f}{f_0}\right)^2 \right) \right\} \qquad (9)$$

where $a(f_\tau) = \sqrt{1 - \frac{\lambda^2}{4V^2} f_\tau^2}$. The first term represents the transfer function of the azimuth chirp, the second that of the RCMC, and the last that of the secondary range compression [33,34]. Then, the compressed point target spectrum after compensating Equation (9) from Equation (6) in the 2D point target spectrum is given by

$$
\begin{aligned}
C_1(f, f_\tau) &= S_{r,MT}(f, f_\tau) \cdot H^*(f, f_\tau) \\
&= rect\left(\frac{f}{B_r}\right) \cdot rect\left(\frac{f_\tau - a(1+f/f_0)}{B_a}\right) \cdot \exp\{-i2\pi\Phi_{res}(f, f_\tau)\}
\end{aligned}
\qquad (10)
$$

where

$$\Phi_{res}(f, f_\tau) = \frac{2R_0}{\lambda} \left(\frac{\lambda^2}{4V_m^2}\right) \alpha \cdot \left(f_\tau - \alpha\left(1 + \frac{f}{f_0}\right)\right) + \Phi_1(f_\tau) + \Phi_2(f, f_\tau) + \Phi_3(f, f_\tau) \qquad (11)$$

when the high-order terms are neglected. The first residual phase, $\Phi_1(f_\tau)$, stands for the residual azimuth chirp,

$$
\begin{aligned}
\Phi_1(f_\tau) &= -\frac{2R_0}{\lambda}\left(\frac{v_x}{V_m} - \frac{a_y}{4}\frac{y_0}{V_m^2}\right) f_\tau^2 \\
&\approx -\frac{2R_0}{\lambda}\left(\frac{v_x}{V} - \frac{a_y}{4}\frac{y_0}{V^2}\right) f_\tau^2
\end{aligned}
\qquad (12)
$$

The second residual phase, $\Phi_2(f, f_\tau)$, is for the range shift and a coupling between the two frequencies, f_τ and f,

$$\Phi_2(f, f_\tau) = \frac{2R_0}{\lambda}\left\{ \left(\frac{1}{a_m(f_\tau; \alpha)} - \frac{1}{a(f_\tau)}\right) + \frac{1}{a_m(f_\tau; \alpha)}\left(\frac{\lambda^2}{4V_m^2}\right)\alpha f_\tau \right\} \cdot \left(\frac{f}{f_0}\right) \qquad (13)$$

Finally, the third residual phase, $\Phi_3(f, f_\tau)$, is for the residual range compression that is usually very small unless the target's velocity is very high:

$$\Phi_3(f, f_\tau) = -\frac{2R_0}{\lambda}\frac{1}{2}\left\{\frac{1}{a_m^3(f_\tau; \alpha)}\left(\frac{\lambda^2}{4V_m^2}\right)\alpha^2\right\} \cdot \left(\frac{f}{f_0}\right)^2 \tag{14}$$

Among the three residual phases in Equations (12)–(14), the second term, $\Phi_2(f, f_\tau)$, is the most complicated and has not been well reviewed while the first term, $\Phi_1(f_\tau)$, has the biggest effect. $\Phi_2(f, f_\tau)$ in Equation (13) is composed of two terms: a slight range-time shift due to V_m as in Equation (3) and a coupling term between Doppler frequency, f_τ, and range frequency, f.

The effect of the latter is particularly significant. The coupling between the two frequencies in Equation (13) projects the azimuth-compressed signals on a slanted line rather than a horizontal line in the range frequency-azimuth time domain, as shown in Figure 2.

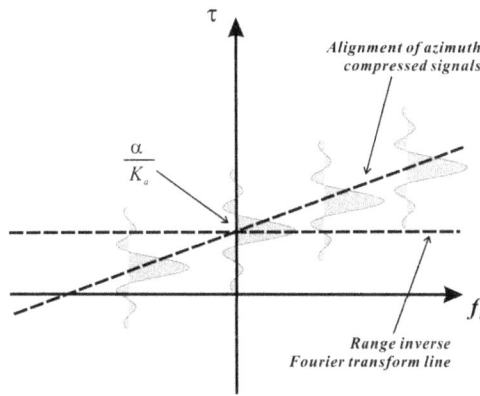

Figure 2. Schematic of the effect of residual phase $\Phi_2(f, f_\tau)$ in the range frequency-azimuth time domain. The coupling between the range and azimuth frequencies projects the azimuth-compressed signals on a slanted line rather than a horizontal line in the range frequency-azimuth time domain. The slope largely depends on α (or v_y); consequently, the ground object with large range speed suffers a significant distortion.

The inverse Fourier transform along the horizontal range frequency results in asymmetric compressed signals in both the azimuth and range dimensions, as well as dispersion of the signal power to some extent. The steeper the slope is, the more serious the distortion of the compressed signal is. The slope largely depends on α (or v_y); consequently, the ground object with large range speed suffers a significant distortion. After removal of the three residual phases, the refocused 2D point target spectrum in Equation (10) becomes as follows:

$$\begin{aligned}
C_2(f, f_\tau) &= C_1(f, f_\tau) \cdot \exp\{+i2\pi\Phi_{res}(f, f_\tau)\} \\
&= rect\left[\frac{f}{B_r}\right] \cdot rect\left[\frac{f_\tau - \alpha(1+f/f_0)}{B_a}\right] \cdot \exp\left\{-i2\pi\frac{\alpha}{K_a}\left[f - \alpha\left(1+\frac{f}{f_0}\right)\right]\right\}
\end{aligned} \tag{15}$$

where $K_a = \frac{2R_0}{\lambda}\left(\frac{\lambda^2}{4V_m^2}\right) \approx \frac{\lambda R_0}{2V^2}$, which is called the Doppler slope. The point target spectrum in the range frequency-azimuth time domain after inverse Fourier transformation of the Doppler frequency is given by

$$C_2(f, \tau) = rect\left[\frac{f}{B_r}\right] \cdot \sin c\left[B_a \cdot \left(\tau - \frac{\alpha}{K_a}\right)\right] \cdot \exp\left(+i2\pi\alpha\left(1+\frac{f}{f_0}\right) \cdot \tau\right) \tag{16}$$

It is necessary to remove the last term in the range frequency-azimuth time domain by multiplying with the complex conjugate of following term:

$$K(f, \tau) = \left\{ \exp\left[+i2\pi\alpha \cdot \left(\frac{f}{f_0} \right) \cdot \tau \right] \right\} \tag{17}$$

This process can also be achieved by applying the Keystone transform [35–37]. This transform is very effective if applied to raw or range-compressed signals [29–31,38], but this process is not a computationally efficient method. Instead of raw signals, here we consider the SLC data with which both azimuth and range compression are already performed, and unlike the rectangular function in Equation (1), the extension of the sinc function in the azimuth dimension in Equation (16) is very limited. Thus, the simple multiplication of Equation (17) would be sufficient to remove the coupling of the azimuth and range frequencies in the SLC data. After inverse-Fourier transformation (F^{-1}) of the range frequency, the fine-tuned SLC image finally becomes as follows:

$$\begin{aligned} c_3(t, \tau) &= F^{-1}(C_2(f, \tau) \cdot K^*(f, \tau)) \\ &= \sin c[B_r t] \cdot \sin c\left[B_a \cdot \left(\tau - \frac{\alpha}{K_a} \right) \right] \cdot \exp\{+i2\pi\alpha\tau\} \end{aligned} \tag{18}$$

Now, the resulting effect of a ground moving target is an azimuthal shift by α/K_a and a linear phase of α in the fine-tuned SLC image. Recall that $\alpha = -2\frac{v_y}{\lambda}\frac{y_0}{R_0}$ is a function of the range velocity of the target, the incidence angle, and the wavelength.

3. Residual Focusing of SAR SLC Data

A fine-tuning of the not fully focused ground moving targets in an SLC image can be achieved by removing the three phases in Equations (12)–(14) in the 2D frequency domain, followed by applying Equation (17) in the range frequency-azimuth time domain. Since the detection of moving targets is beyond the scope of this paper, detection tactics are not discussed herein. To apply the residual focusing, it is necessary to estimate the velocity and acceleration components of an individual target. To retrieve velocity and acceleration components for isolated fast moving targets in SLC data, it is necessary to estimate Doppler parameters from a single range bin, or at most, a few range bins. There are already a wide variety of detection and Doppler parameter estimation techniques, and therefore a short review of prevalent techniques is presented.

Classical approaches assume that the Doppler shift of a moving target is directly observable in the returned SAR signals [1]. Space-borne SAR configuration needs to consider additional factors including Earth curvature and rotation [39]. For SAR data from the single-channel system, Doppler filtering must be used first to reduce contributions from clutter [40]. A Doppler-filtering method that requires a pulse-repetition frequency (PRF) four times larger than the clutter bandwidth was proposed [10]. The amplitude and phase modulations of the returned signal in the Fourier domain can be utilized to detect and resolve multiple moving targets, and the skew of the received signal in the 2D frequency domain was also discussed to resolve an aliased range velocity component [5]. Exploiting the image-blurring effect caused by the along-track velocity component, a Doppler rate detector was proposed and its performance was compared with that of the two-channel ATI and the DCPA method in [7]. Among the various approaches, the joint time-frequency analysis (JTFA) has been popularly applied to measure object motion directly from a chirp signal and was successful in retrieving the velocity by estimating the Doppler frequency rate of a moving object in the time-frequency domain [40–42]. The JTFA demonstrated the potential to extract a time sequence of motion parameters [43,44]. It is possible to measure the velocity of a ground moving vehicle with an error of less than 5% for a velocity higher than 3 m/s [45]. When the target contains prominent high backscatters, an iterative approach for searching maximum image contrast can be used [46,47].

In addition to the problem of a low signal-to-clutter ratio, the fundamental limitation of single-channel SAR is that the Doppler shift must be greater than the clutter Doppler spectrum width, which can be achieved using a high pulse-repetition frequency (PRF) [11]. To overcome the shortcomings of single-channel SAR systems, multi-channel SAR systems using two or more antenna displaced in the along-track direction have become popular. In multi-channel approaches, moving target detection and motion parameter estimation can be achieved by adopting so-called clutter cancellation techniques. The main classes include the displaced phase center antenna (DPCA) [48–50], the ATI method [48,49,51], and raw data-based methods such as space-time adaptive processing (STAP) [50,52,53]. The DPCA algorithm directly subtracts the complex signals received at two different phase centers, while the ATI computes the phase difference of the two channels by exploiting interferometric techniques. In both approaches, clutter signals are cancelled out to remain signals contributed by moving objects. In STAP, instead of subtracting two signals on each range-azimuth pixel dimension, statistical approaches are used in order to suppress the clutter and noise. In general, STAP is the superior scheme when raw data are available, since it has additional processing gain (maximizing SNR) over optimized SAR processing. The acceleration component can also cause significant bias on the estimation of along-track velocity [54], including acceleration as an additional unknown parameter, leading to insufficient degrees of freedom to solve for the other parameters in a two-channel SAR system. However, its influence in the case of a space-borne SAR geometry is usually very small when compared with the case of an airborne SAR geometry, such that the target's motion is assumed to be constant in most cases of the SAR-GMTI problem.

In summary, there is a wide variety of approaches for Doppler parameter estimation. Since the efficiency of each method largely depends on the system parameters and configuration, one should carefully examine the properties of a given system and data. Once the Doppler parameters and velocity are retrieved from a given SAR SLC data, the next step is to apply the residual focusing. As described in the previous section, the proposed algorithm consists of two phase-multiplications in two different domains. Figure 3 shows the entire operations in the proposed correction scheme. Note that the residual phase, $\Phi_{res}(f, f_\tau)$, and the range-azimuth frequency coupling, $K(f, \tau)$, in Figure 3 correspond to the Equations (11) and (17), respectively. Since the data has already been azimuth and range compressed, and the signal bandwidth of the moving target remains unchanged after taking a subset in the time domain, it would be sufficient to use a limited window size for the processing. From various tests, a sub-window of 64×64 is large enough for high-resolution, X-band SLC data from space-borne SAR such as TerraSAR-X and COSMO-SkyMed.

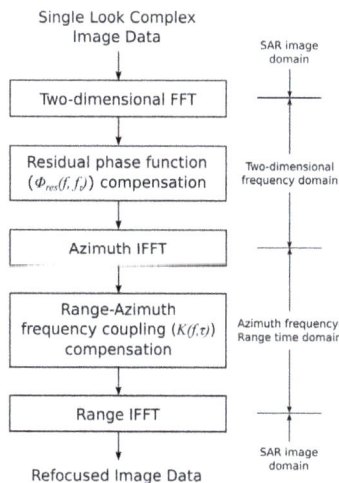

Figure 3. The operations in the proposed correction algorithm, which consists of two phase-multiplications in two different domains.

4. Simulation and Application Results

4.1. Simulation Results

Simulation tests have been carried out as follows using the system parameters of the TerraSAR-X stripmap mode (Table 1). First, SAR raw signals were simulated with a point target which moved at 45° oblique to the azimuth and range directions with various velocities ranging from −30 m/s to +30 m/s with 1 m/s interval. The motion of the point target was assumed to be constant, i.e., without acceleration for simplicity, as the effect of acceleration is often insignificant in the case of the space-born SAR stripmap mode, which has a relatively short azimuth integration time (for instance, less than 0.6 s for TerraSAR-X). The simulated raw signals were then focused into standard SLC images using a chirp-scaling algorithm [32]. After detection and Doppler parameter estimation of each moving target, a sub window of 64 × 64 centered on the moving target was extracted and then corrected by applying the proposed method. To evaluate the performance of the method, we substituted the exact motion parameters into the fine-tuning filter, which means that there was no error in the detection and Doppler parameter estimation scheme. Finally, various quality parameters were measured from both the original and fine-tuned SLC images in order to evaluate the improvement achieved by the proposed refocusing method. Three parameters are typically used for SAR point target quality assessment: −3 dB width, integrated sidelobe ratio (ISLR), and symmetry. The −3 dB width is commonly used for spatial resolution estimation. Since the reconstructed signal from a moving point target is asymmetrical and the side lobe is difficult to determine, ISLR is used instead of the peak sidelobe ratio (PSLR). Figure 4a,b show −3 dB widths of both directions with varying moving speeds for the point target. Before applying the fine-tuning, the spatial resolution in the azimuth direction degrades rapidly as the target speed increases, while that in the range direction shows no significant changes. After applying the fine-tuning process, the spatial resolution in the azimuth direction is significantly improved. There is a small amount of residual broadening in the azimuth direction, which is proportional to the target speed. The ISLR in Figure 4c,d display the ISLR in both the azimuth and range directions. The azimuth ISLR increases steeply as the target speed increases up to −3 dB, while the range ISLR is almost unaffected by the target's motion. The improvement in the azimuth direction is significant particularly up to about 8 m/s (or 28.8 km/h).

Table 1. Summary of sensor model parameters used for simulation

Parameter	Value
Chirp length	47.17 μm
Range sampling rate	109.88 MHz
Chirp bandwidth	100 MHz
Carrier frequency	9.65 GHz
Antenna length	4.8 m
Effective velocity	7371.1 m/s
Pulse repetition frequency	3815.49 Hz
Slant range to scene center	650.79 km
Indicence angle at scene center	39.24°
Doppler centroid	0 Hz
Sensor height	513.08 km

Symmetry is a measure of the energy balance of the compressed signal. In order to measure this quantity, we decomposed the power of the compressed signal $P(x) = c_3(x) \cdot c_3^*(x)$ into symmetric and antisymmetric parts as follows:

$$\begin{aligned} P(x) &= \frac{1}{2}[P(x) + P(-x)] + \frac{1}{2}[P(x) - P(-x)] \\ &= P_+(x) + P_-(x) \end{aligned} \tag{19}$$

where x is a cell number along range or azimuth, axis centered at 0, which ranges between $-N/2$ and $N/2$ where N is the size of fine-tuning filter. Note that $P_+(x)$ and $P_-(x)$ are essentially symmetric and anti-symmetric, respectively. Then, the symmetry can be measured from vector norms ($\| \ \|$) as follows:

$$\psi = \frac{\|P_+(x)\|}{\|P_+(x)\| + \|P_-(x)\|} \tag{20}$$

which ranges from 0 (fully anti-symmetric) to 1 (fully symmetric). Figure 5 displays the changes of symmetry in both azimuth and range. As expected, the symmetry also becomes worse as the target's speed increases. Unlike the former two measures, the symmetry relies on range velocity. The simulation with varying speed only in the azimuth component showed no changes in symmetry. This is based on the fact that the range migration of a scatterer moving in the range direction is asymmetrical to the closest range distance, while that of a scatterer moving in the azimuth direction is fully symmetric in a zero-Doppler geometry. It may be of interest to note that all three parameters change abruptly when the target moves at around 8 m/s. This is nearly coincident with the moment when the peak power of a focused range cell migrates into the next range cell. This range walk effect is abrupt and discrete. Since we set the range center of the fine-tuning window to where we observe the maximum energy of the blurred target in units of integer number, the change in range cell number leads to inaccurate selection of the range distance, which is required for generating a proper filter. Consequently, the apparent performance of the algorithm, based on the quality parameters, seems to be gradually reduced; however, the overall achievement by the fine-tuning filter is sufficiently high to retrieve the target's true nature. The maximum error of post-correction in our simulation in the 30 m/s case was comparable to that of pre-correction at 3 m/s or less.

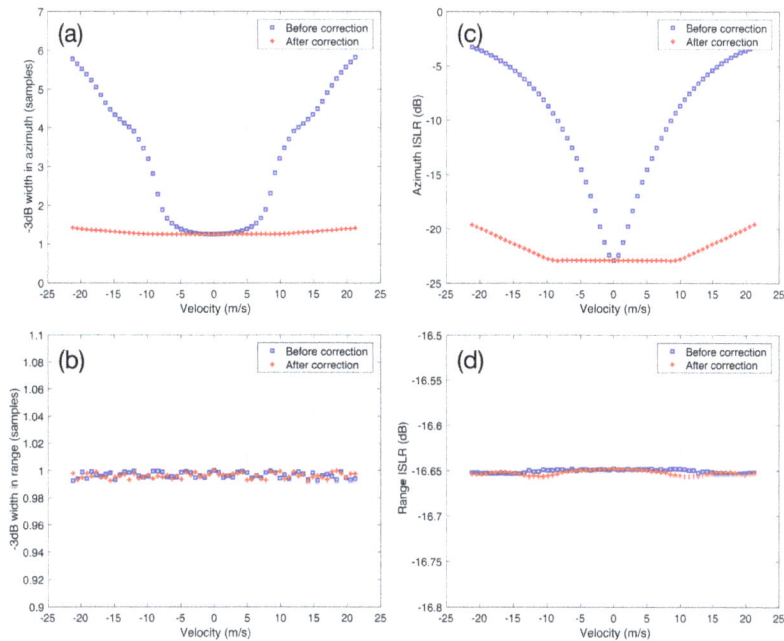

Figure 4. Simulation results of the -3 dB width and the integrated sidelobe ratio (ISLR) in the (**a,c**) azimuth and (**b,d**) range directions, respectively. The simulation was carried out using system parameters of the TerraSAR-X stripmap mode (see Table 1). The improvement in the azimuth direction is significant particularly up to about 8 m/s (or 28.8 km/h).

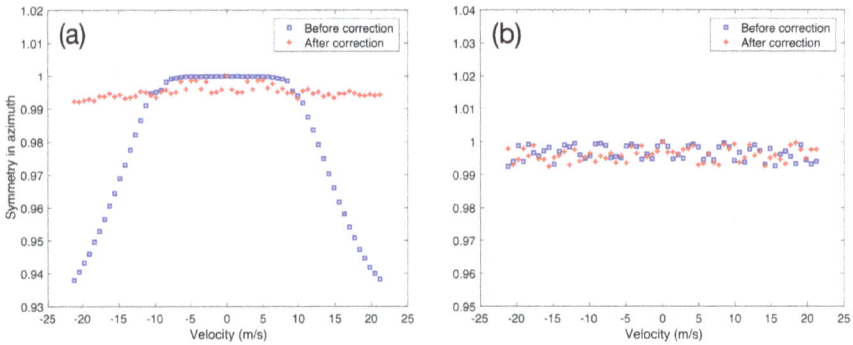

Figure 5. Simulation results of (**a**) azimuth and (**b**) range symmetry of the reconstructed signals. While the range symmetry was not seriously affected by the target's moving speed, the azimuthal symmetry deteriorated when the moving speed was higher than 8 m/s. The azimuthal symmetry is well recovered by the proposed method as in (**a**).

Although the quantitative performance evaluation was carried out using a point target simulation, a visual interpretation of the correction result for an extended target helps to convey how the image is restored. Figure 6 shows a simulated aircraft taxiing on the ground. The upper panels show the normally focused images, and the lower panels show the corresponding fine-tuned images. The correction results are not only well focused but also shifted a bit in the range direction because of the correction for the range walk.

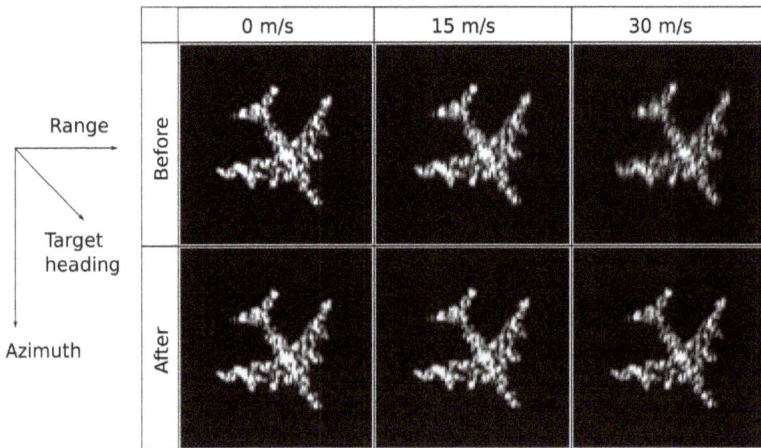

Figure 6. Simulation results of an extended target with varying velocities. The upper panels are for the single-look complex (SLC) sub-images processed by a standard synthetic aperture radar (SAR) focusing, and the lower panels for the fine-tuned images processed by the proposed algorithm. The fine-tuned images are not only well focused but also shifted a bit in the range direction because of the correction for the range walk.

4.2. Example of Application to TerraSAR-X Data

Test data were obtained by TerraSAR-X from a speed-controlled vehicle moving on the road with velocities of −6.6 and −13.8 m/s (or −23.8 and −49.6 km/h, respectively) in the azimuth and range directions, respectively. The speed of the vehicle was precisely measured by GPS as well as speedometer, and was retrieved from the TerrSAR-X data itself by Doppler frequency analysis. The details of the data used in the test and velocity retrieval process are given in [45]. The application result is shown in Figures 7 and 8, which demonstrate the effectiveness of the proposed fine-tuning tactics for high-resolution SAR SLC data.

As seen in Figures 7 and 8, the target movement caused significant energy dispersion (or image blurring) in the azimuth and range directions, losing the symmetry of the typical sinc function, and causing a slight shift of peak locations along both the azimuth and range. Image blurring in the azimuth (see Figures 7a and 9a) has been previously well known, and the improvement of the peak sidelobe ratio is about 4 dB in this example. In addition to image blurring, asymmetry of the compressed signal in the azimuth direction is significant, as in Figure 8a. The symmetry value of 0.92 in the azimuth of the original SLC data is improved to 0.94 after fine-tuning. Also, this resulted in distortion removal of the target shape. Slight shifts in the peak locations by about 1.1 samples (or 2.13 m) and 0.3 samples (or 0.65 m) in the azimuth and ground range directions are also noted. The residual range compression by Equation (14) is, however, not significant in this example because the value of $\Phi_3(f, f_\tau)$ is relatively small compared with those of $\Phi_1(f_\tau)$ and $\Phi_2(f, f_\tau)$.

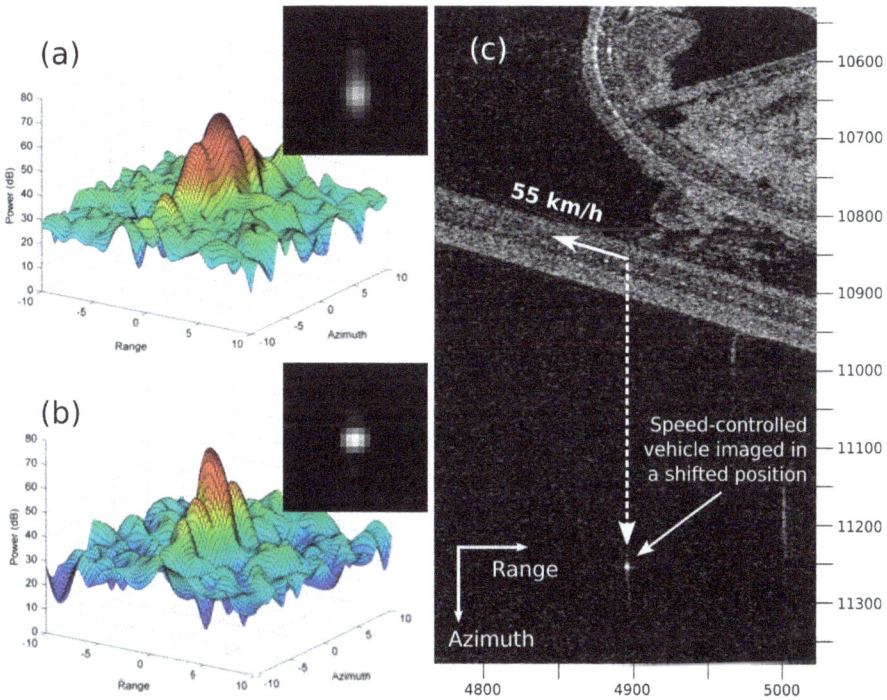

Figure 7. Sub-window image and power distribution of (**a**) the original SLC data and (**b**) the data after removing the residual phase. (**c**) The test was carried out using a speed-controlled vehicle moved with velocities of −6.6 and −13.8 m/s (or −23.8 and −49.6 km/h, respectively) in the azimuth and range directions, respectively [45]. Note the improvement of symmetry around the target as well as improvement of the compression ratio and peak sidelobe ratio by about 4 dB.

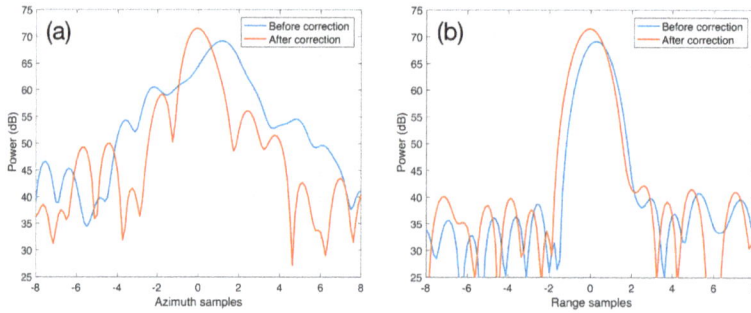

Figure 8. Power profiles crossing the center of the ground moving vehicle along the (**a**) azimuth and (**b**) range directions in Figure 7. Note the improvement in the peak sidelobe ratio by about 4 dB and that in the symmetry from 0.92 to 0.94 in the azimuth dimension, as in (**a**). In addition, the refocused object shifted by about 1.1 and 0.3 samples in azimuth and range direction, respectively.

Besides the speed-controlled target, there were several other moving objects in the same test image. As an example of the extended target, Figure 9 shows a large ship moving along an oblique to the azimuth and range direction. Although an in-situ measurement for the movement was not available, the joint time-frequency analysis [45] was used for the velocity estimation. The estimated velocity was −5.6 and −5.1 m/s (or −20.1 and 18.4 km/h) in the azimuth and range directions, respectively. Although the simulation results showed in the previous section indicates that the distortion would not be serious compared to the case with velocities higher than 8 m/s, the original image and the corresponding fine-tuned image in Figure 9a,b have notable differences.

Unlike the small object in Figure 7, this target is extended to few tens of pixels, and the radar reflectivities along the ship orientation look not much changing in the scaled image in Figure 9c. However, the fine-tuned image in Figure 9d indicates that the ship body actually has a structure with countable prominent points.

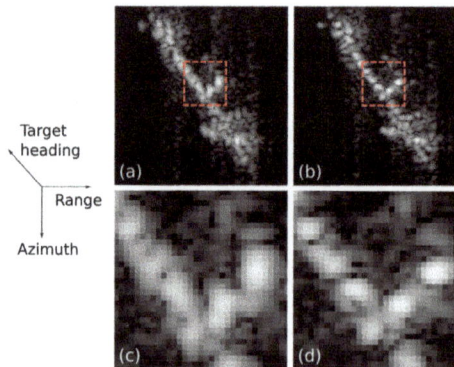

Figure 9. A moving ship observed in (**a**) the original SLC data and (**b**) the data after removing the residual phase. The original image size here is 64 by 64 pixels, however, the image was interpolated by a factor of two in order to show the details. The estimated velocity by using the joint time-frequency analysis [45] was −5.6 and −5.1 m/s (or −20.1 and −18.4 km/h, respectively) in the azimuth and range directions, respectively. Although the target speed is low considering the image distortion rapidly increases with speed higher than 8 m/s as noted in the simulation test, the scaled images, (**c**,**d**) show clear improvement in image focusing.

5. Discussion

Both simulation and TerraSAR-X results demonstrated the capability of the proposed method to improve the signal compression of ground moving objects. The improvement is a function of the target speed and monotonically is increased with the target speed up to 20 m/s. When the target speed was 7 m/s, the simulated signal in the azimuth direction was improved by 134% and 196% in −3 dB width and ISLR, respectively. The application to the TerraSAR-X also showed similar results with an improvement of 193% in −3 dB width for the vehicle whose speed was controlled at an azimuth velocity of 6.6 m/s. It should also be noted that the improvement of the compression ratio and peak sidelobe ratio was by about 4 dB. The refocused object was also shifted to the true position by about 1.1 and 0.3 samples in the azimuth and range direction, respectively. Symmetry of the compressed signal is also important to characterize the target, and the proposed method improved the symmetry by up to 0.94. Various SAR focusing methods have derived from raw radar signals [9,12], but they are neither computationally efficient nor practically applicable because of the data distribution policy of most high resolution SAR data. Among recent publications, the methods in [46,47] deal with a similar topic of ground moving target imaging. The method in [46] adopted the Stolt interpolation in the 2D frequency domain. Although the compression ratio of the method was very competitive, the asymmetric sidelobes were not fully suppressed [47]. A symmetry of 0.94 achieved from the TerraSAR-X results of this study in Figure 8 showed the superior performance of the proposed method. An improved method was also proposed by [47], which exploits the parameter sparse representation method and iteration. Image entropy and symmetry of compressed signals were significantly improved by the method in [47]. As far as the performance of refocusing is concerned, the method in [47] is superior to [46] at the cost of computational efficiency. The method in [47] reached the state of convergence after about one hundred iterations which significantly increased computation time.

As discussed before, the main advantages of the proposed method are twofold. First, the proposed residual focusing is based on the derived formulae of the Doppler spectrum after conventional image formation. Second, the method is simple and practically applicable to SLC data of high-resolution SAR systems. Although the proposed method demonstrated the competitive ability of residual focusing, it does have some limitations. First, the correction formulae in Equations (12)–(14) are approximated to the second-order terms, and consequently the effects of high-order terms remain to be further accounted. Second, the final quality of the residual focused SAR image depends on the accuracy of the Doppler parameters used for processing. The Doppler parameters can be obtained by parameter estimation from SLC data, which is beyond the scope of this paper. A number of approaches for Doppler parameter estimation have been developed and typical examples are referred to in [43–45]. One should carefully examine the properties of a given system and data before Doppler parameter estimation because the efficiency of each method largely depends on the system parameters and configuration. Third, the coupling of the azimuth and range frequencies in the SLC data described in Equation (16) and Figure 2 is sensitive to the slope α which is a function of the range velocity of the target, the incidence angle, and the wavelength. A more sophisticated approach than the simple phase compensation by Equation (17) might be necessary if a further refinement in range dimension is required. For this purpose, raw SAR signals rather than SLC data should be involved in the processing.

6. Conclusions

Theoretical formulae were derived for the fine-tuning of ground moving targets in SAR SLC data and processing tactics were proposed. The proposed fine-tuning significantly improved the image quality in three aspects: residual compression, the symmetry of the compressed signal in both the azimuth and range directions, and the slight shift of the peak positions. While various focusing methods for SAR raw signals have been developed previously, a general approach for SAR SLC data has not been well investigated. The proposed method is practical for post-processing of SAR SLC data by general users. The importance of the method lies in the fact that most SAR data provided to general users are SLC data rather than raw signals. The residual focusing is simple and straightforward, based on an elaborately derived residual Doppler spectrum with a relatively small processing window. Simulation results support that the target's motion deteriorates image quality particularly in azimuth in terms of the −3 dB width, ISLR, and asymmetry, and the proposed fine-tuning efficiently restores the image quality. It may be of interest to note that all three parameters (−3 dB width, ISLR and asymmetry) change abruptly when the target moves faster than 8 m/s. The application results using TerraSAR-X and a speed-controlled ground moving vehicle demonstrate the effectiveness of the method without a heavy computational burden. The residual focusing ideally reconstructs ground moving targets with particular improvements of the compression ratio and symmetry.

Acknowledgments: This work was supported by the National Space Lab program through the Korea Science and Engineering Foundation funded by the Ministry of Science, ICT and Future Planning (2013M1A3A3A02042314. The TerraSAR-X data were provided to J.-S. Won as a part of TerraSAR-X Science Team Project (PI No. COA0047).

Author Contributions: J.-S. Won conceived the main idea; J.-W. Park and J.H. Kim performed the experiments; J.-W. Park and J.-S. Won analyzed the data and wrote the paper.

Conflicts of Interest: The authors declare no conflict of interest.

References

1. Raney, R.K. Synthetic aperture imaging radar and moving targets. *IEEE Trans. Aerosp. Electron. Syst.* **1971**, *7*, 499–505. [CrossRef]
2. Werness, S.; Carrara, W.; Joyce, L.; Franczak, D. Moving target imaging algorithm for SAR data. *IEEE Trans. Aerosp. Electron. Syst.* **1990**, *26*, 57–67. [CrossRef]
3. Kirscht, M. Detection and imaging or arbitrarily moving targets with single-channel SAR. In Proceedings of the 2002 International Radar Conference, Edinburgh, UK, 15–17 October 2002. [CrossRef]
4. Linnehan, R.; Perlovsky, L.; Mutz, I.C.; Rangaswamy, M.; Schindler, J. Detecting multiple slow-moving targets in SAR images. In Proceedings of the Sensor Array and Multichannel Signal Processing Workshop, Barcelona, Spain, 18–21 July 2004; pp. 643–647.
5. Marques, P.A.C.; Dias, J.M.B. Velocity estimation of fast moving targets using a single SAR sensor. *IEEE Trans. Aerosp. Electron. Syst.* **2005**, *41*, 75–89. [CrossRef]
6. Weihing, D.; Hinz, S.; Meyer, F.; Laika, A.; Bamler, R. Detection of along-track ground moving targets in high resolution spaceborn SAR images. In Proceedings of the ISPRS Commission VII Symposium, Enschede, The Netherlands, 8–11 May 2006; pp. 81–85.
7. Meyer, F.; Hinz, S.; Laika, A.; Suchandt, S.; Bamler, R. Performance analysis of space-borne SAR vehicle detection and velocity estimation. In Proceedings of the ISPRS Commission III Symposium, Born, Germany, 20–22 September 2006.
8. Baumgartner, S.V.; Krieger, G. Fast GMTI algorithm for traffic monitoring based on a priori knowledge. *IEEE Trans. Geosci. Remote Sens.* **2012**, *50*, 4626–4641. [CrossRef]
9. Chen, C.C.; Andrews, H.C. Target motion induced radar imaging. *IEEE Trans. Aerosp. Electron. Syst.* **1980**, *16*, 2–14. [CrossRef]
10. Freeman, A.; Currie, A. Synthetic aperture radar (SAR) images of moving targets. *GEC J. Res.* **1987**, *5*, 106–115.
11. Moreira, J.R.; Keydel, W. A new MTI-SAR approach using the reflectivity displacement method. *IEEE Trans. Geosci. Remote Sens.* **1995**, *33*, 1238–1244. [CrossRef]
12. Fienup, J.P. Detecting moving targets in SAR imagery by focusing. *IEEE Trans. Aerosp. Electron. Syst.* **2001**, *37*, 749–809.
13. Pettersson, M. Extraction of moving ground targets by a bistatic ultra-wideband SAR. *IEE Proc. Radar Sonar Navig.* **2001**, *148*, 35–49. [CrossRef]
14. Dias, J.M.B.; Marques, P.A.C. Multiple moving target detection and trajectory estimation using a single SAR sensor. *IEEE Trans. Aerosp. Electron. Syst.* **2003**, *39*, 604–624. [CrossRef]
15. Sparr, T. Moving target motion estimation and focusing in SAR images. In Proceedings of the IEEE Radar Conference, Arlington, VA, USA, 9–12 May 2005; pp. 290–294.
16. Chapman, R.D.; Hawes, C.M.; Nord, M.E. Target motion ambiguities in single-aperture synthetic aperture radar. *IEEE Trans. Aerosp. Electron. Syst.* **2010**, *46*, 459–468. [CrossRef]

17. Goldstein, R.; Zebker, H. Interferometric radar measurements of ocean surface currents. *Nature* **1987**, *328*, 707–709. [CrossRef]
18. Rodriguez, E.; Martin, J.M. Theory and design of interferometric synthetic aperture radars. *IEE Proc. F Radar Signal Process.* **1992**, *139*, 147–159. [CrossRef]
19. Ainsworth, T.; Chubb, S.; Fusina, R.; Goldstein, R.; Jansen, R.; Lee, J.; Valenzuela, G. InSAR imagery of surface currents, wave fields, and fronts. *IEEE Trans. Geosci. Remote Sens.* **1995**, *33*, 1117–1123. [CrossRef]
20. Soumekh, M. Moving target detection in foliage using along track monopulse synthetic aperture radar imaging. *IEEE Trans. Image Process.* **1997**, *6*, 1148–1163. [CrossRef] [PubMed]
21. Romeiser, R.; Thompson, D.R. Numerical study on the along-track interferometric radar imaging mechanism of oceanic surface currents. *IEEE Trans. Geosci. Remote Sens.* **2000**, *38*, 446–458. [CrossRef]
22. Rosen, P.; Hensley, S.; Joughin, I.; Li, F.; Madsen, S.; Rodriguez, E.; Goldstein, R. Synthetic aperture radar interferometry. *Proc. IEEE* **2000**, *88*, 333–382. [CrossRef]
23. Moccia, A.; Rufino, G. Spaceborne along-track SAR interferometry: performance analysis and mission scenarios. *IEEE Trans. Aerosp. Electron. Syst.* **2001**, *37*, 199–213. [CrossRef]
24. Chen, C.W. Performance assessment of along-track interferometry for detecting ground moving targets. In Proceedings of the IEEE Radar Conference, Philadelphia, PA, USA, 26–29 April 2004; pp. 99–104.
25. Chiu, S.; Livingstone, C.E. A comparison of displaced phase centre antenna and along-track interferometry techniques for RADARSAT-2 ground moving target indication. *Can. J. Remote Sens.* **2005**, *31*, 37–51. [CrossRef]
26. Gierull, C.H.; Cerutti-Maori, D.; Ender, J.H.G. Ground moving target indication with tandem satellite constellations. *IEEE Geosci. Remote Sens. Lett.* **2008**, *5*, 710–714. [CrossRef]
27. Yang, L.; Wang, T.; Bao, Z. Ground moving target indication using an InSAR system with a hybrid baseline. *IEEE Geosci. Remote Sens. Lett.* **2008**, *5*, 373–377. [CrossRef]
28. Cerutti-Maori, D.; Gierull, C.H.; Ender, J.H.G. Experimental verification of SAR-GMTI improvement through antenna switching. *IEEE Trans. Geosci. Remote Sens.* **2010**, *48*, 2066–2075. [CrossRef]
29. Kirkland, D. Imaging moving targets using the second-order keystone transform. *IET Radar Sonar Navig.* **2011**, *5*, 902–910. [CrossRef]
30. Sun, G.; Xing, M.; Xia, X.-G.; Wu, Y.; Bao, Z. Robust ground moving-target imaging using deramp-keystone processing. *IEEE Trans. Geosci. Remote Sens.* **2013**, *51*, 966–982. [CrossRef]
31. Yang, J.; Liu, C.; Wang, Y. Imaging and parameter estimation of fast-moving targets with single-antenna SAR. *IEEE Geosci. Remote Sens. Lett.* **2014**, *11*, 529–533. [CrossRef]
32. Raney, R.K.; Runge, H.; Bamler, R.; Cumming, I.G.; Wong, F.H. Precision SAR processing using chirp scaling. *IEEE Trans. Geosci. Remote Sens.* **1994**, *32*, 786–799. [CrossRef]
33. Bamler, R. A comparison of range-Doppler and wavenumber domain SAR focusing algorithms. *IEEE Trans. Geosci. Remote Sens.* **1992**, *30*, 706–713. [CrossRef]
34. Moreira, A.; Mittermayer, J.; Scheiber, R. Extended chirp scaling algorithm for air- and spaceborne SAR data processing in stripmap and ScanSAR imaging modes. *IEEE Trans. Geosci. Remote Sens.* **1996**, *34*, 1123–1136. [CrossRef]
35. Perry, P.R.; DiPietro, R.C.; Fante, R. SAR imaging of moving targets. *IEEE Trans. Aerosp. Electron. Syst.* **1999**, *35*, 188–200. [CrossRef]
36. Zhou, F.; Wu, R.; Xing, M.; Bao, Z. Approach for single channel SAR ground moving target imaging and motion parameter estimation. *IET Radar Sonar Navig.* **2007**, *1*, 59–66. [CrossRef]
37. Kirkland, D. Using the keystone transform for detection of moving targets. In Proceedings of the EUSAR, Aachen, Germany, 7–10 June 2010; pp. 305–308.
38. Kirkland, D.M. An alternative range migration correction algorithm for focusing moving targets. *Prog. Electromagn. Res.* **2012**, *131*, 227–241. [CrossRef]
39. Raney, R.K. Considerations for SAR image quantification unique to orbital systems. *IEEE Trans. Geosci. Remote Sens.* **1991**, *29*, 754–760. [CrossRef]
40. Barbarossa, S.; Farina, A. Space-time-frequency processing of synthetic aperture radar signals. *IEEE Trans. Aerosp. Electron. Syst.* **1994**, *30*, 341–358. [CrossRef]
41. Barbarossa, S. Detection and imaging of moving objects with synthetic aperture radar. Part 1. Optimal detection and parameter estimation theory. *IEE Proc. F Radar Signal Process.* **1992**, *139*, 79–88. [CrossRef]

42. Barbarossa, S.; Farina, A. Detection and imaging of moving objects with synthetic aperture radar. Part 2: Joint time-frequency analysis by Wigner-Ville distribution. *IEE Proc. F Radar Signal Process.* **1992**, *139*, 89–97. [CrossRef]

43. Kersten, P.R.; Janse, R.W.; Luc, K.; Ainsworth, T.L. Motion analysis in SAR images of unfocused objects using time-frequency methods. *IEEE Trans. Geosci. Remote Sens. Lett.* **2007**, *4*, 527–531. [CrossRef]

44. Kersten, P.R.; Topokov, J.V.; Ainsworth, T.L.; Sletten, M.A.; Jansen, R.W. Estimating surface water speeds with a single-phase center SAR versus an along-track interferometric SAR. *IEEE Trans. Geosci. Remote Sens.* **2010**, *48*, 3638–3646. [CrossRef]

45. Park, J.-W.; Won, J.-S. An efficient method of Doppler parameter estimation in the time-frequency domain for a moving object from TerraSAR-X data. *IEEE Trans. Geosci. Remote Sens.* **2011**, *49*, 4771–4787. [CrossRef]

46. Zhang, Y.; Sun, J.; Lei, P.; Li, G.; Hong, W. High-resolution SAR-based ground moving target imaging with defocused ROI data. *IEEE Trans. Geosci. Remote Sens.* **2016**, *54*, 1062–1073. [CrossRef]

47. Chen, Y.; Li, G.; Zhang, Q.; Sun, J. Refocusing of Moving Targets in SAR Images via Parametric Sparse Representation. *Remote Sens.* **2017**, *9*, 795. [CrossRef]

48. Livingstone, C.; Sikaneta, I.; Gierull, C.H.; Chiu, S.; Beaudoin, A.; Campbell, J.; Beaudoin, J.; Gong, S.; Knight, T. An airborne SAR experiment to support RADARSAT-2 GMTI. *Can. J. Remote Sens.* **2002**, *28*, 1–20. [CrossRef]

49. Gierull, C.H. Ground moving target parameter estimation for two-channel SAR. *IEE Proc. Radar Sonar Navig.* **2006**, *153*, 224–233. [CrossRef]

50. Cerutti-Maori, D.; Sikaneta, I. Generalization of DPCA processing for multichannel SAR/GMTI radars. *IEEE Trans. Geosci. Remote Sens.* **2013**, *51*, 560–572. [CrossRef]

51. Suchandt, S.; Runge, H.; Breit, H.; Steinbrecher, U.; Kotenkov, A.; Balss, U. Automatic extraction of traffic flows using TerraSAR-X along-track interferometry. *IEEE Trans. Geosci. Remote Sens.* **2010**, *48*, 807–819. [CrossRef]

52. Ender, J.H.G. Space-time processing for multichannel synthetic aperture radar. *Electron. Commun. Eng. J.* **1999**, *11*, 29–38. [CrossRef]

53. Melvin, W.L. A STAP overview. *IEEE Aerosp. Electron. Syst. Mag.* **2004**, *19*, 19–35. [CrossRef]

54. Sharma, J.; Gierull, C.H.; Collins, M.J. The influence of target acceleration on velocity estimation in dual-channel SAR-GMTI. *IEEE Trans. Geosci. Remote Sens.* **2006**, *44*, 134–147. [CrossRef]

remote sensing

MDPI

Article

A Unified Algorithm for Channel Imbalance and Antenna Phase Center Position Calibration of a Single-Pass Multi-Baseline TomoSAR System

Yuncheng Bu [1,2], Xingdong Liang [1,*], Yu Wang [1,*], Fubo Zhang [1] and Yanlei Li [1]

[1] National Key Laboratory of Science and Technology on Microwave Imaging, Institute of Electronics, Chinese Academy of Sciences, Beijing 100190, China; buyuncheng13@mails.ucas.ac.cn (Y.B.); zhangfubo8866@126.com (F.Z.); radar_sonar@163.com (Y.L.)

[2] University of Chinese Academy of Sciences, Beijing 100049, China

* Correspondence: xdliang@mail.ie.ac.cn (X.L.); wangyu@mail.ie.ac.cn (Y.W.); Tel.: +86-10-5888-7101 (X.L.); +86-10-5888-7524 (Y.W.)

Received: 3 January 2018 ; Accepted: 12 March 2018; Published: 14 March 2018

Abstract: The multi-baseline synthetic aperture radar (SAR) tomography (TomoSAR) system is employed in such applications as disaster remote sensing, urban 3-D reconstruction, and forest carbon storage estimation. This is because of its 3-D imaging capability in a single-pass platform. However, a high 3-D resolution of TomoSAR is based on the premise that the channel imbalance and antenna phase center (APC) position are precisely known. If this is not the case, the 3-D resolution performance will be seriously degraded. In this paper, a unified algorithm for channel imbalance and APC position calibration of a single-pass multi-baseline TomoSAR system is proposed. Based on the maximum likelihood method, as well as the least squares and the damped Newton method, we can calibrate the channel imbalance and APC position. The algorithm is suitable for near-field conditions, and no phase unwrapping operation is required. The effectiveness of the proposed algorithm has been verified by simulation and experimental results.

Keywords: TomoSAR; multi-baseline SAR; unified algorithm; channel imbalance; APC position

1. Introduction

In recent years, synthetic aperture radar (SAR) tomography (TomoSAR) has become a popular research topic due to its 3-D imaging capability [1–6]. TomoSAR has been successfully applied in many application contexts, such as forestry [7,8], 3D urban reconstruction [9,10], and glaciers [11]. The single-pass multi-baseline TomoSAR system has 3D resolution including the height resolving ability in a single-pass platform because there are multiple channels in the cross-track direction. However, the Rayleigh resolution in the height direction is very limited due to the limitation of the baseline length in a single-pass platform. As shown in Equation (25), the Rayleigh resolution of a TomoSAR system is about 35 m. If super-resolution performance is desired in the height direction, at a resolution of, say, 5 m, then a super-resolution algorithm must be introduced in that direction so as to distinguish multiple targets in small intervals within the Rayleigh resolution. Tebaldini [11] concludes that even a subwavelength accuracy of the antenna phase center (APC) position will hinder the focusing result in the height direction. The authors of [12] conclude that, when one wants to distinguish multiple point-like targets with different heights within a slant range-azimuth resolution cell, the requirements for phase stability or phase calibration accuracy are higher than those for traditional InSAR. Therefore, if we hope to obtain a super-resolution performance in the height direction, high requirements for the channel imbalance (also known as amplitude and phase inconsistency) and APC position calibration are required in a single-pass multi-baseline TomoSAR system.

Several calibration algorithms have been proposed in recent years. Pardini [13] calibrated the phase error due to APC position by an algorithm based on a minimum entropy criterion. In [14], Gocho compensates for the phase screen caused by APC position errors using eigenvalue decomposition and phase interpolation/extrapolation. However, these two methods cannot obtain the exact APC position, and they fail to take channel imbalance into account. A phase center double localization algorithm is proposed by Tebaldini [11,15] to obtain the APC position. However, this method loses its effectiveness when there is channel imbalance. A different method has been proposed in [16], which estimates the APC position based on raw data. Such processing seems infeasible for users with single-look complex (SLC) images only. Regarding the correction of amplitude and phase errors in multi-channel array systems, Kuoye Han [17] proposes a calibrator-based approach. On the basis of [17], Xiaolin Yang takes into account both channel imbalance and APC position in [18], but they failed to decouple the phase errors caused by the channel itself and the APC position. In general, these calibration algorithms above either do not consider both the channel imbalance and the APC position or are unable to decouple channel imbalance and the APC position error. In addition, a plane wave model is used in most of these algorithms. However, even in the far-field condition, the plane wave model will bring some non-negligible phase error when the ratio of the baseline length to line-of-sight distance is not sufficiently close to zero [19]. As an example, Figure 6 will show that the plane wave model brings a non-negligible phase error even in the far-field condition, where b = 0.6 m and r = 1625 m. Therefore, we use the Fresnel approximation, which assumes that spherical waves can be approximated by quadratic waves in our signal model.

In this paper, we propose a unified algorithm for the channel imbalance and APC position calibration of a single-pass multi-baseline TomoSAR system, which can not only calibrate channel imbalance but can also calibrate the APC position. Features of this algorithm are as follows: (1) the channel imbalance and APC position can be calibrated individually rather than confusing the phase error caused by the APC position error with the phase error of the channel itself; (2) the Fresnel approximation is used in the calibration signal model, which heightens the accuracy of the calibration signal model; and (3) there is no need for phase unwrapping.

This paper is structured as follows. Section 2 is devoted to establishing a signal model. The proposed calibration algorithm is described in Section 3. The effectiveness of the calibration algorithm is validated in Section 4 with a real data set acquired by the array InSAR system [20,21] developed by the Institute of Electronics, Chinese Academy of Sciences (IECAS). The discussion and conclusions follow in Sections 5 and 6, respectively.

2. A Signal Model

Supposing that the TomoSAR system is working in side-looking mode, the TomoSAR acquisition geometry is depicted in Figure 1. Assuming that the number of the APC is N, axes x, y, and z are the cross-track, azimuth, and height directions, respectively. APC is supposed in the zero Doppler plane (see Figure 2), where s is the cross-range direction. Let (x_n, z_n) denote the position of the nth APC; without loss of generality, we assume that APC1 is at the origin of the coordinate system, that is, APC1 is the reference APC and $(x_1, z_1) = (0,0)$. Applying a classical imaging algorithm to the raw SAR data collected in each channel, we obtain N 2-D SAR images, usually referred to as SLC images. After some sub-pixel accuracy coregistration to the reference channel, and under the Born weak-scattering approximation, the focused complex value of an azimuth-range pixel (y_0, r_0) of the nth channel is [1]:

$$\widehat{\gamma}_n(y_0, r_0) = \iint dy dr\, f(y_0 - y, r_0 - r) \int ds\, \gamma(y, r, s) \exp\left[-j\frac{4\pi}{\lambda} R_n(r, s)\right] \qquad (1)$$

where λ is the wavelength, $f(y_0 - y, r_0 - r)$ is the 2-D point spread function (PSF), $\gamma(y, r, s)$ is the function that models the 3-D scene scattering properties, and $R_n(r, s)$ represents the slant range between point targets located at r and s coordinates, and the nth APC. Under the Fresnel approximation, $R_n(r,s)$ can be written as

$$R_n(r,s) = \sqrt{(r - b_{//n})^2 + (b_{\perp n} - s)^2} \approx |r - b_{//n}| + \frac{(b_{\perp n} - s)^2}{2|r - b_{//n}|} \tag{2}$$

where $b_{//n}$ is the nth horizontal baseline, $b_{\perp n}$ is the nth orthogonal baseline, the relation between the baseline and APC position is shown as follows:

$$\begin{cases} b_{\perp n} = x_n \cos\theta + z_n \sin\theta \\ b_{//n} = x_n \sin\theta - z_n \cos\theta \end{cases} \tag{3}$$

where θ is the off-nadir angle.

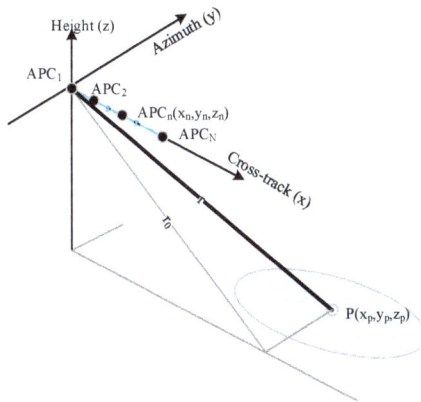

Figure 1. TomoSAR acquisition geometry.

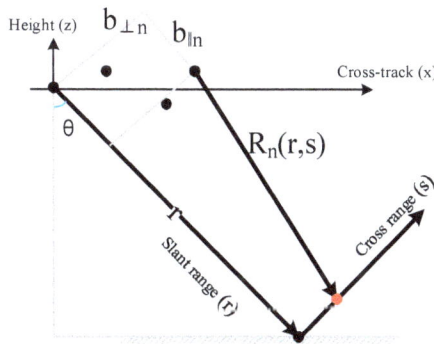

Figure 2. Geometry in the zero Doppler plane.

Based on the assumption of the point target, $f(y_0 - y, r_0 - r)$ can be regarded as the 2-D Dirac function. Following this, Equation (1) can be written as

$$\widehat{\gamma}_n(y_0, r_0) = \int ds\, \gamma(y_0, r_0, s) \exp\left[-j\frac{4\pi}{\lambda} R_n(r_0, s)\right]. \tag{4}$$

In the calibration processing, the ground control point (GCP) is usually placed in the non-layover area (e.g., the bare ground). That is, there is only one target in the s direction within a slant range-azimuth cell corresponding to the GCP. Therefore, Equation (4) can be rewritten as

$$\widehat{\gamma}_n(y_0, r_0) = \exp\left[-j\frac{4\pi}{\lambda}R_n(r_0)\right]\gamma(y_0, r_0) \tag{5}$$

where

$$R_n(r_0) = |r_0 - b_{//n}| + \frac{b_{\perp n}{}^2}{2|r_0 - b_{//n}|}. \tag{6}$$

For notational convenience, the target position coordinates r_0 and y_0 in each symbol are omitted, and the focused complex value acquired by the nth channel is denoted as g_n. Considering that $r_0 \gg b_{//n}$, we have $R_n = r_0 - b_{//n} + \frac{b_{\perp n}{}^2}{2r_0}$; therefore, Equation (5) can be rewritten as

$$g_n = \exp\left(-j\frac{4\pi}{\lambda}R_n\right)\gamma. \tag{7}$$

By combining the exponential term of the reference channel into the backscatter coefficient, and by taking into account channel imbalance, the calibration signal model of the multi-baseline TomoSAR system would be of the form:

$$\mathbf{g} = \mathbf{C}\boldsymbol{\alpha}\gamma' + \mathbf{E} \tag{8}$$

where $\mathbf{g} = [g_1, g_2, ..., g_N]^T$ is the $N \times 1$ observation vector, the calibration matrix $\mathbf{C} = diag\{\rho_1 e^{j\varphi_1}, ..., \rho_n e^{j\varphi_n}, ..., \rho_N e^{j\varphi_N}\}$, where ρ_n and φ_n are the amplitude and phase of the nth channel (note that, if all channels are exactly the same, then $\mathbf{C} = \mathbf{I}$), represents the channel imbalance, $\boldsymbol{\alpha} = \left[1, e^{-j\frac{4\pi}{\lambda}(R_2-R_1)}, ..., e^{-j\frac{4\pi}{\lambda}(R_N-R_1)}\right]^T$, and $\gamma' = \gamma \exp\left(-j\frac{4\pi}{\lambda}R_1\right)$, which is related to the backscatter coefficient of GCP.

3. The Calibration Algorithm

Let θ_m ($m = 1, 2, ..., M$) be the off-nadir angle of the mth GCP, and r_m be the slant range. Then,

$$\mathbf{g} = \mathbf{C}\boldsymbol{\alpha}(\psi, \theta_m)\gamma_m + \mathbf{n} \tag{9}$$

where $\mathbf{g} = [g_1, g_2, ..., g_N]^T$ is an observation vector of the mth GCP, γ_m is the backscatter coefficient, $\boldsymbol{\alpha}(\psi, \theta_m)$ is an array manifold, and $\psi = [x_1, x_2, ..., x_N, z_1, z_2, ..., z_N]^T$ is a vector of the unknown APC position. By executing an eigenvalue decomposition of the covariance matrix $E[\mathbf{g}\mathbf{g}^T]$, and normalizing the first element of the eigenvector corresponding to the largest eigenvalue to 1, we obtain the array manifold estimation $\boldsymbol{\alpha}_{mea}(\theta_m)$, and

$$\boldsymbol{\alpha}_{mea}(\theta_m) = \mathbf{C}\boldsymbol{\alpha}(\psi, \theta_m) + \mathbf{n}_m \tag{10}$$

where $\mathbf{n}_m \in \mathbb{C}^{N \times 1}$ is a random vector of additive noise. We shall assume that $\mathbf{n}_m$ has a zero-mean Gaussian distribution with a covariance ζ^2.

Considering all M GCPs, the probability density function of the set of estimated array manifold is

$$p\left(\boldsymbol{\alpha}_{mea}(\theta_1), \boldsymbol{\alpha}_{mea}(\theta_2), ..., \boldsymbol{\alpha}_{mea}(\theta_M)\big|\mathbf{C}, \psi, \zeta^2\right)$$
$$= (\pi\zeta^2)^{-MN}\exp\left\{-\frac{1}{\zeta^2}\sum_{m=1}^{M}\|\boldsymbol{\alpha}_{mea}(\theta_m) - \mathbf{C}\boldsymbol{\alpha}(\psi, \theta_m)\|^2\right\}. \tag{11}$$

It can be easily shown that the maximum likelihood estimates of **C** and ψ are the corresponding values that minimize the following cost function:

$$\min_{\mathbf{C}, \psi} \|\mathbf{C}\mathbf{A}(\psi) - \mathbf{A}_m\|_F^2 \tag{12}$$

where $\|\cdot\|_F$ denotes the Frobenius norm, $\mathbf{A}_m \triangleq [\alpha_{mea}(\theta_1), \alpha_{mea}(\theta_2), ..., \alpha_{mea}(\theta_M)]$, and $\mathbf{A}(\psi) \triangleq [\alpha(\psi, \theta_1), \alpha(\psi, \theta_2), ..., \alpha(\psi, \theta_M)]$. Since the calibration matrix **C** and the APC position contain $2N^2$ and $2N$ real unknowns, respectively, and since the M sources can provide $2MN$ independent measurements, $M \geq N+1$ is a necessary condition. This condition can also be found in [22]. According to this condition, at least 9 GCPs are needed to meet the requirement for our 8-channel array InSAR system.

Assuming that $\mathbf{A}(\psi)$ is of full rank N, the cost equation, Equation (12), is a separable nonlinear least-squares optimization problem. When we keep ψ fixed, the least-square estimation of the calibration matrix **C** is

$$\mathbf{C} = \mathbf{A}_m \mathbf{A}^H(\psi) \left(\mathbf{A}(\psi)\mathbf{A}^H(\psi) \right)^{-1}. \tag{13}$$

Substituting Equation (13) into Equation (12), we obtain the maximum likelihood estimation for the APC position:

$$\hat{\psi}_{ML} = \operatorname{argmin} Tr\left(P^{\perp}_{\mathbf{A}^H(\psi)} \mathbf{A}_m^H \mathbf{A}_m \right) = \operatorname{argmin} f(\psi) \tag{14}$$

where $P^{\perp}_{\mathbf{A}^H(\psi)} = I - \mathbf{A}^H(\psi)\left(\mathbf{A}(\psi)\mathbf{A}^H(\psi)\right)^{-1}\mathbf{A}(\psi)$, and the minimization objective function $f(\psi) \triangleq Tr\left(P^{\perp}_{\mathbf{A}^H(\psi)} \mathbf{A}_m^H \mathbf{A}_m \right)$.

The estimation problem is now decoupled into two steps. The APC position estimation $\hat{\psi}_{ML}$ is obtained at first by solving the optimization problem Equation (14) before the estimation of the calibration matrix **C** is derived by substituting $\hat{\psi}_{ML}$ estimated in the first step of Equation (13). We can then obtain the channel imbalance estimation by extracting the diagonal elements of the calibration matrix **C**.

3.1. Step 1: APC Position Calibration

In this step, the APC position estimation $\hat{\psi}_{ML}$ is obtained by solving the optimization problem (Equation (14)). The minimization objective function $f(\psi)$ may have many local minimums. However, when the nominal values of the APC position ψ_0 are close enough to the true values, we can rewrite the minimization objective function as

$$f(\psi_0 + \mathbf{p}) \approx f(\psi_0) + \mathbf{g}^t(\psi_0)\mathbf{p} + \frac{1}{2}\mathbf{p}^t\mathbf{H}(\psi_0)\mathbf{p} \tag{15}$$

where **g** and **H** are the gradient and Hessian of $f(\psi)$, respectively. **p** is the search direction that minimizes the right hand of Equation (15).

In order to solve this optimization problem, the damped Newton method can be applied. The nominal values were set as the initial APC position ψ_0, and the search direction $\mathbf{p}_{k+1} = \hat{\psi}_{k+1} - \hat{\psi}_k = -\mu_k \mathbf{H}_k^{-1} \mathbf{g}_k$, where $\mathbf{H}_k \triangleq \mathbf{H}|_{\psi = \hat{\psi}_k}$ and $\mathbf{g}_k \triangleq \mathbf{g}|_{\psi = \hat{\psi}_k}$. The step length $\mu_k = (0.5)^l$, where l is the smallest nonnegative integer that satisfies $f\left(\hat{\psi}_{k+1}\right) < f\left(\hat{\psi}_k\right)$. Then the estimation of APC position is obtained when the damped Newton method converges or reaches the maximum number of iterations.

The solution is re-derived in the Appendix A, and the required **g** and **H** are shown directly to be

$$\mathbf{g} = -2\text{Re}\left\{ \begin{array}{c} vecd\left(\mathbf{A_x} P^{\perp}_{\mathbf{A}^H(\psi)} \mathbf{A}_m^H \mathbf{A}_m \mathbf{A}^H(\psi)\left(\mathbf{A}(\psi)\mathbf{A}^H(\psi)\right)^{-1}\right) \\ vecd\left(\mathbf{A_z} P^{\perp}_{\mathbf{A}^H(\psi)} \mathbf{A}_m^H \mathbf{A}_m \mathbf{A}^H(\psi)\left(\mathbf{A}(\psi)\mathbf{A}^H(\psi)\right)^{-1}\right) \end{array} \right\} \tag{16}$$

$$\mathbf{H} = 2\text{Re}\left\{\mathbf{DP}_{\mathbf{A}^{H}(\mathbf{\psi})}^{\perp}\mathbf{D}^{H} \odot \left(\mathbf{1}_{2\times2} \otimes \mathbf{E}\right)^{T}\right\} \tag{17}$$

where

$$\mathbf{A_x} = j\frac{4\pi}{\lambda}\mathbf{A}(\mathbf{\psi})\Lambda_{\sin} - j\frac{4\pi}{\lambda}\mathbf{A}(\mathbf{\psi}) \odot \begin{bmatrix} x_1, z_1 \\ x_2, z_2 \\ \vdots & \vdots \\ x_N, z_N \end{bmatrix} \begin{bmatrix} \frac{\cos^2\theta_1}{r_1}, & \frac{\cos^2\theta_2}{r_2}, & \dots, & \frac{\cos^2\theta_M}{r_M} \\ \frac{\sin\theta_1\cos\theta_1}{r_1}, & \frac{\sin\theta_2\cos\theta_2}{r_2}, & \dots, & \frac{\sin\theta_M\cos\theta_M}{r_M} \end{bmatrix} \tag{18}$$

$$\mathbf{A_z} = -j\frac{4\pi}{\lambda}\mathbf{A}(\mathbf{\psi})\Lambda_{\cos} - j\frac{4\pi}{\lambda}\mathbf{A}(\mathbf{\psi}) \odot \begin{bmatrix} x_1, z_1 \\ x_2, z_2 \\ \vdots & \vdots \\ x_N, z_N \end{bmatrix} \begin{bmatrix} \frac{\sin\theta_1\cos\theta_1}{r_1}, & \frac{\sin\theta_2\cos\theta_2}{r_2}, & \dots, & \frac{\sin\theta_M\cos\theta_M}{r_M} \\ \frac{\sin^2\theta_1}{r_1}, & \frac{\sin^2\theta_2}{r_2}, & \dots, & \frac{\sin^2\theta_M}{r_M} \end{bmatrix} \tag{19}$$

$$\Lambda_{\sin} = diag\{[\sin\theta_1, \sin\theta_2, \dots, \sin\theta_M]\} \tag{20}$$

$$\Lambda_{\cos} = diag\{[\cos\theta_1, \cos\theta_2, \dots, \cos\theta_M]\} \tag{21}$$

$$\mathbf{D} = \left[\mathbf{A_x^T}, \mathbf{A_z^T}\right]^{T} \tag{22}$$

$$\mathbf{E} = \left(\mathbf{A}(\mathbf{\psi})\mathbf{A}^{H}(\mathbf{\psi})\right)^{-1}\mathbf{A}^{H}(\mathbf{\psi})\mathbf{A}_{m}^{H}\mathbf{A}_{m}\mathbf{A}^{H}(\mathbf{\psi})\left(\mathbf{A}(\mathbf{\psi})\mathbf{A}^{H}(\mathbf{\psi})\right)^{-1} \tag{23}$$

where $vecd(\mathbf{V})$ represents a vector formed from the diagonal elements of the matrix $\mathbf{V}$, the symbol $\odot$ represents the matrix multiplication of the elements, $\otimes$ represents the Kronecker product, and $\mathbf{1}_{p\times q}$ represents the $p \times q$ matrix with all entries equal to one.

3.2. Step 2: Channel Imbalance Calibration

After solving the APC position calibration problem, Equation (13) is solved in order to retrieve the channel imbalance. The diagonal element of the calibration matrix is the estimation of the channel imbalance. Non-diagonal elements are actually mutual coupling factors. In general, the mutual coupling factor is relatively small if the channels are spaced out far enough. In particular, the correction matrix will have some characteristics when the multi-channels array has some regular geometry. For example, the calibration matrix of the equidistant line array is the Toeplitz matrix.

3.3. Validation with Simulation Data

The simulation data is necessary since it makes it possible to directly compare the true channel imbalance and the APC position with those yielded by the calibration algorithm. This test would be quite hard to implement using real data, for which the true channel imbalance and APC position are, in general, not known with sufficient precision. The simulation data set consists of 8-channel 2-D focused SAR SLC images. Eight APCs corresponding to the eight-channel were distributed in the cross-track direction, and the longest baseline is 0.6 m. In this section, we will begin by showing a special case of the simulation data set in order to illustrate the effectiveness of the proposed algorithm. A Monte Carlo simulation is then carried out to assess the performance of the proposed algorithm in a statistical framework.

In the special case, eight APC trajectories—equally spaced out in the cross-track direction—are shown in Figure 3. Amplitude and phase inconsistency are set as AmpErr = [1, 1, 1, 1, 1, 1, 1, 1] and PhaseErr = [0, 0.3, 0.1, −0.2, 0.3, 0.1, 1, 0.4] rad. Other relevant system parameters are summarized in Table 1.

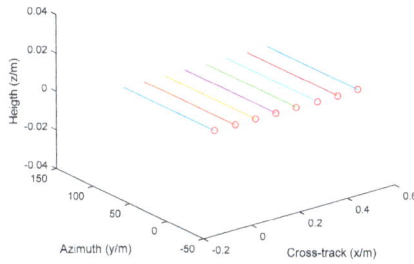

Figure 3. Antenna phase center (APC) trajectories.

Table 1. System parameters used in the simulation data set.

Simulation Data Parameters	
Frequency	15 GHz (Ku-Band)
Bandwidth	500 MHz
PRF	2000 Hz
Airplane altitude	1000 m
Airplane Velocity	60 m/s
Azimuth beam width	2°
Depression angle	25°–41°

The observation scene includes 3 × 11 non-layover GCPs, and four pairs of layover GCPs spaced out with super-resolution factors of 0.5, 1, 2, and 8, respectively. The super-resolution factor [4] is defined as the ratio of the Rayleigh resolution and layover scatters' interval. The observation scene's 2-D SAR image of the simulation data set is depicted in Figure 4. When taking a look at the 2-D SAR image in Figure 4, one can see that all non-layover GCPs are included in the imaging area and are well focused, and that four pairs of layover GCPs in red circles are appropriate superpositions, as expected.

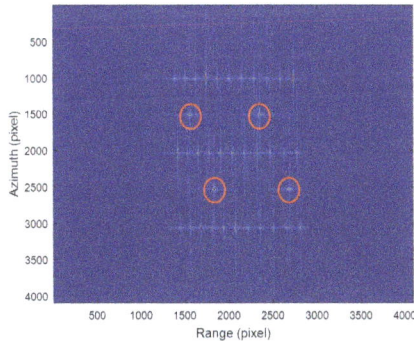

Figure 4. Single-look complex (SLC) image of Channel 1, four pairs of layover ground control points (GCPs) are shown in the red circles.

After being coregistered to Channel 1 (the reference channel) with sub-pixel accuracy, the calibration algorithm proposed in this paper was carried out by utilizing all 33 non-layover GCPs. After calibration, we obtained the calibration values of the channel imbalance and APC position (see Figure 5). The true values of the channel imbalance and APC position are compared with the calibration values in order to verify the validity of the calibration algorithm. In Figure 5, the true values and calibration values are drawn in red and blue lines, respectively. Graphs in the left panel show the true and calibration values in contrast, while the differences (estimation errors) between them are plotted in the right panel. One can immediately note that the APC position estimation results are very accurate. The maximum and standard deviation are 0.16 mm and 0.105 mm, respectively. The maximum error of amplitude is only -30 dB. The maximum and standard deviation of the phase error are 0.12 rad and 0.06 rad, respectively. This seems a little disappointing because the maximum phase error is not as highly accurate as expected. An additional step, which can be helpful for improving the phase calibration accuracy, involves checking the same GCP in all channels or averaging the results of multiple GCPs, because we have obtained the exact APC position after the previous calibration.

In order to examine the effectiveness of this calibration values, a comparison experiment was performed on four pairs of layover GCPs. The true values and the estimation results of the target number and height are listed in Table 2.

Table 2. Height resolution experiment on four pairs of layover GCPs.

Layover GCPs	Number of Targets			Height of Targets (m)		
	True Value	Before Calibration	After Calibration	True Value	Before Calibration	After Calibration
1st pair	2	3	2	0, 56.9	9.7, 138.2, 166.9	0, 56.9
2nd pair	2	3	2	0, 32.7	0, 34, 143	0, 32.7
3rd pair	2	3	2	0, 12.4	8.4, 34.9, 78.1	0, 12.4
4th pair	2	4	2	0, 4.5	36, 126, 173.5, 176	0, 5.0

As shown in Table 2, each pair of layover GCPs is correctly identified as being two targets within the slant range-azimuth resolution cell after calibration. After calibration, the height of each target is also accurately estimated. Prior to calibration, however, the situation is disappointing, since not only the height estimation but even the estimate of the target number is wrong.

A precise performance assessment of the proposed algorithm has been carried out by means of Monte Carlo simulations. The systems parameters are the same as those in Table 1. The true APC positions, amplitude inconsistency, and phase inconsistency are modeled as follows:

$$\text{Amplitude inconsistency (dB)} \sim N(0,1)$$
$$\text{Phase inconsistency (rad)} \sim U(-0.5, 0.5)$$
$$\text{APC position (m)}: \mathbf{x} = [0,1,2,3,4,5,6,7] \times 0.6/7 + \Delta\mathbf{x}, \quad (24)$$
$$\text{where } \Delta\mathbf{x} = [\Delta x_1, \Delta x_2, ..., \Delta x_n, ..., \Delta x_8] \text{ and } \Delta x_n \sim N(0, 5 \times 10^{-3})$$
$$\mathbf{z} = [z_1, z_2, ..., z_n, ..., z_8], \text{ where } z_n \sim N(0, 10 \times 10^{-3})$$

One hundred trials were carried out. For each trial, the mean and standard deviation of the calibration errors (amplitude and phase) are calculated. Then, results of the different trials are averaged. The mean and standard deviation of calibration errors are shown in Table 3. The calibration error of amplitude is below -30 dB in nearly all trials. As for the calibration error of the phase, the standard deviation is less than 0.06 rad. Regarding the calibration error of APC position, it is computed via averaging the RMSE of all trials, where $RMSE_i = \sqrt{\left[\sum_{n=1}^{8} (\Delta x_{n,i})^2 + (\Delta z_{n,i})^2 \right]/8}$, and $\Delta x_{n,i}$ and $\Delta z_{n,i}$ represent the calibration errors of the APC position in the cross-track and height directions, respectively, in the ith trial. The simulation result shows that the calibration error of the APC position is less than 0.127 mm. Based on these simulation results, it can be concluded that the proposed algorithm performs well.

Table 3. Calibration errors.

	Mean μ	Standard Deviation σ
Amplitude (dB)	-35.10	4.15
Phase (rad)	-0.0054	0.0577

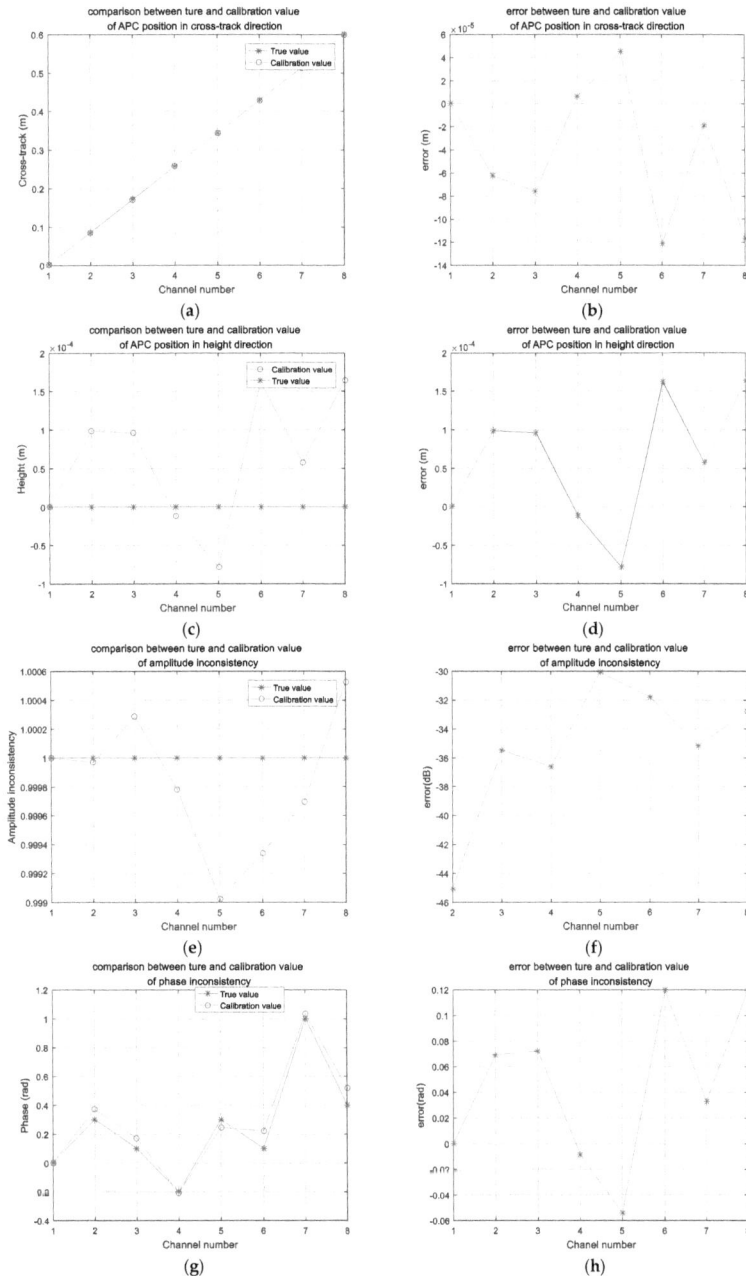

Figure 5. Calibration results of the simulation SAR data set. (**a**) The true and calibration values of axis x (cross-track direction); (**b**) The error between the true and calibration values of axis x; (**c**) The true and calibration values of axis z (height direction); (**d**) The error between the true and calibration values of axis z; (**e**) The true and calibration values of the amplitude; (**f**) The error between the true and calibration values of the amplitude; (**g**) The true and calibration values of the phase; (**h**) The error between the true and calibration values of the phase.

4. Experimental Results

4.1. Data Acquisition

In this section, we will validate the effectiveness of the calibration algorithm by using a real data set acquired by the array InSAR system. The array InSAR system was developed by IECAS in 2014, usually installed onboard a Y-12 aircraft. The radar system operates at 15 GHz (Ku-band), and has eight channels in the cross-track direction. Other system parameters are the same as those in Table 1. As a multi-baseline system, the array InSAR system can obtain 3-D images of the observed scene in a single-pass platform. The Rayleigh resolution [5] in the height direction of the array InSAR system can be derived approximately as

$$\rho_h = \frac{\lambda R \sin(\theta)}{B_n} \simeq 35\,\text{m} \tag{25}$$

where λ represents the wavelength, the slant range R is 1625 m, and the off-nadir angle θ is 38°. Super-resolution techniques must be used since many targets are under 35 m in height; if not, TomoSAR would be meaningless, especially in urban monitoring and mountain mapping tasks. The utilization of a super-resolution technique can achieve high resolution in the height direction, but it is based on the premise that the channel imbalance and APC position are precisely known. Therefore, calibration is particularly important.

It is clear in Figure 6 that, under the system parameters described above, the quadratic wave model is more accurate than the plane wave model, and that is why we use the Fresnel approximation in our signal model.

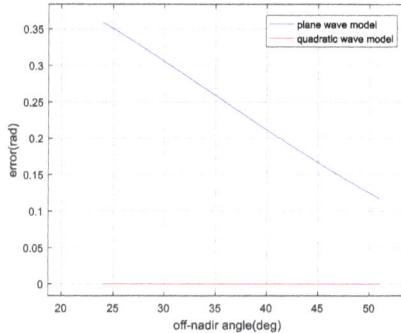

Figure 6. Phase errors of the plane wave model and quadratic wave model.

The real SAR data set used in this section was acquired by the array InSAR system in April 2015. The calibration site is located in Yuncheng county in the province of Shanxi; its SLC image of Channel 1 is shown in Figure 7. In this single-pass airborne TomoSAR campaign, images of nine GCPs and layover scenes, such as urban buildings, have been acquired. This provides valuable data for our subsequent calibration and validation experiments.

Figure 7. SLC image of Channel 1; this image, zoomed in on in Figure 8, contains GCPs and a flat ground.

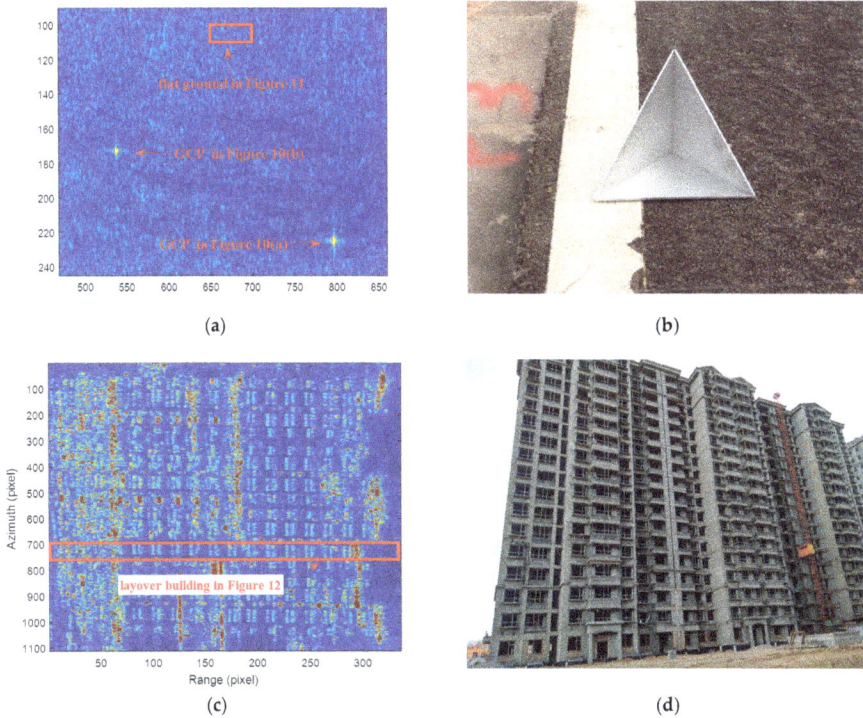

(a)

(b)

(c)

(d)

Figure 8. The SLC images and corresponding optical images. (**a**) SLC image of GCPs and the flat ground; (**b**) Optical image of one GCP; (**c**) SLC image of the layover building (the red block is the area corresponding to Figure 12); (**d**) Optical image of the building in (**c**).

4.2. Validation Data Description

Three special scenes, including a flat ground, non-layover GCP, and layover building, were selected to verify the effectiveness of the unified calibration algorithm. The chosen flat ground is located at the calibration site with an altitude of about 550 m. The 3-D image of this flat ground should be like a horizontal plane. The selected GCPs are trihedral corner reflectors with a leg length of 20 cm, which are placed at a non-layover area; there should be only one target in the height slice for each non-layover GCP. As for the chosen building, serious layover phenomena appear due to the SAR side view principle. Note that there are a maximum of three layover points in a slant range-azimuth resolution cell of the chosen building. Figure 8 shows SLC images of the chosen scenes and corresponding optical images.

4.3. Calibration Algorithm Validation

The calibration processing of the channel imbalance and APC position was carried out as discussed in Section 3. Nine GCPs were used in the calibration site. The results shown in Figure 9 are the channel imbalance and APC position calibration results (blue line) of the array InSAR system campaign from April 2015. The nominal values of the APC position are depicted by a red line. Figure 9a,b show the nominal values and the calibration values of the APC position. The difference between the nominal values and calibration values may be due to the fact that the nominal values are obtained by rough measurement, which is not guaranteed to perfectly correspond to the electromagnetic phase center of the antenna. Figure 9c shows the amplitude difference between different channels. Channel 2 appears slightly smaller than the other channels. We found that the estimate amplitude consistency of each channel is similar to the amplitude calibration results in Figure 9c when the power of the same GCP is extracted in different channels. As the true phase inconsistency and APC position cannot be exactly known, special scenes were chosen to verify the effectiveness of the calibration results in Figure 9. As we shall see later, the 3D imaging results in Figures 10–12 show the correctness of the calibration results.

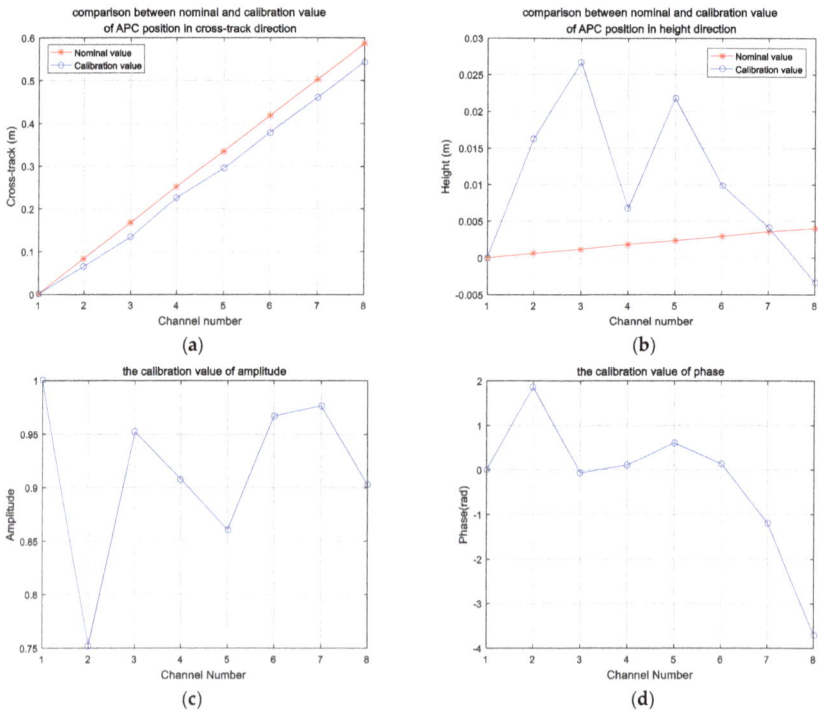

Figure 9. Calibration results of the real SAR data set. (**a**) The nominal and calibration values of axis *x* (cross-track direction); (**b**) The nominal and calibration values of axis *z* (height direction); (**c**) The calibration values of the amplitude; (**d**) The calibration values of the phase.

In order to verify whether the calibration values are correct, TomoSAR focusing experiments were carried out for two non-layover GCPs, for flat ground and for a layover building; they are described above. In these experiments, a compressive sensing ()-based TomoSAR spectral estimation algorithm—SL1MMER [4]—was used in the imaging of the height direction. Note that SL1MMER can only achieve the best super-resolution power if the parameters are precisely known. Therefore, the more accurate the parameter calibration is, the better the imaging and resolving power in the height direction will be.

- GCP Results

Figure 10 shows the TomoSAR focusing results in the height slice of GCPs. The APC calibration only means that we only compute the estimation values of the APC position by means of a phase center double localization (PCDL) algorithm [11]. The channel imbalance calibration only means that we only compute the estimation results of the channel imbalance by means of the algorithm proposed in [18]. It is apparent that the TomoSAR focusing results of the unified calibration are more accurate in the target number and position estimation and have less disturbance points than the results of the APC calibration only or the results of the channel imbalance calibration only. The heights of the two chosen GCPs are 550.93 m and 552.79 m, respectively. The reconstructed heights obtained by utilizing the unified algorithm proposed in this paper are 551.06 m and 552.75 m, which is very close to the true heights. However, the results of the other methods are not satisfactory. This is not difficult to understand, as the phase error caused by the channel imbalance and APC position error would result in disturbance points and the target position offset.

- Flat Ground Validation

Figure 11 shows the 3D imaging results of the selected flat ground. The 3D imaging results of the unified calibration algorithm (see Figure 11c,f) are almost clustered on a plane, but the results of the APC calibration only and channel imbalance only are scattered, which is inconsistent with the actual flat terrain. It can therefore be said that the 3D imaging quality is improved by utilizing the system parameters calibrated by the proposed unified algorithm.

- Layover Building Validation

In this partition, the 3D imaging results of the layover building are shown in Figure 12. Figure 12a shows the building observation geometry. In this situation, Points A, B, and C with the slant range r1 will be superimposed within the same slant range-azimuth resolution cell in the 2D SAR image. The same is true for Points D and E. If the channel imbalance and APC position are calibrated precisely, these layover points will be perfectly reconstructed in the right height position utilizing a super-resolution algorithm. On the contrary, these layover points will not be properly reconstructed, such as the occurrence of disturbance points, location shifts, or even defocusing.

When the channel imbalance and APC position have been calibrated by the proposed algorithm, the 3D imaging result is shown in Figure 12b. In this figure, we see that Points A, B, and C are resolved and reconstructed in the corresponding position. This is similar to the other points (i.e., Points D and E) of this layover building. For comparison, the 3D imaging results of the APC calibration only and the channel imbalance calibration only are shown in Figure 12c,d, respectively. Compared with Figure 12c,d, there are fewer disturbance points in Figure 12b, especially in the area near the bottom of the building. This is because the imaging results of these areas are more sensitive to system parameters due to the very small difference in height between the building and the ground. It is therefore safe to say that the imaging results of Figure 12b is better and shows the validity of the unified calibration algorithm proposed in this paper.

To qualitatively exhibit the validity of the calibration algorithm we propose in this paper, the 3D focusing result of a residential area after channel imbalance and APC position calibration is shown below. The SAR 3D focusing result can be approached as follows. Applying a classical imaging algorithm to the raw SAR data collected by each channel, we obtain N 2D SLC images. After some sub-pixel accuracy coregistration to the reference channel and a unified calibration for the channel imbalance and APC position, a super-resolution algorithm, such as compressed sensing [3] or the direction of arrivals [12], is applied to each slant range-azimuth resolution cell. Following this, the 3D imaging result is obtained. The Google Earth image, SLC image, and 3D image of the residential area are shown in Figure 13. As shown in the SLC image, in the 2D SAR image, we cannot even correctly estimate the number of buildings owing to the serious layover phenomenon. When 3D imaging was carried out after precise calibration of the array parameters, the building footprints and the texture of the roof and façade were all clearly visible.

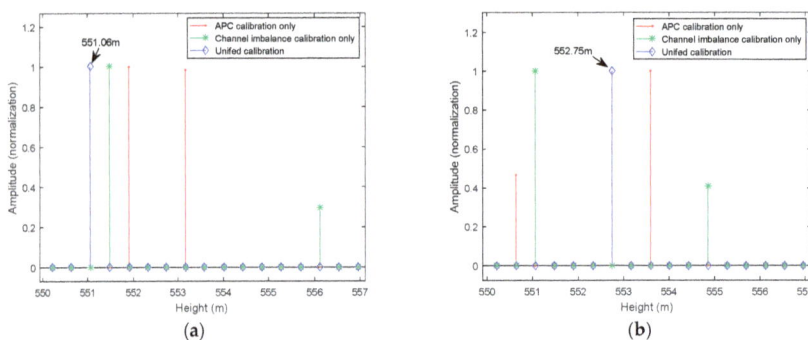

Figure 10. TomoSAR height slices of GCPs. (a,b) TomoSAR height slices of GCP1 and GCP2, respectively.

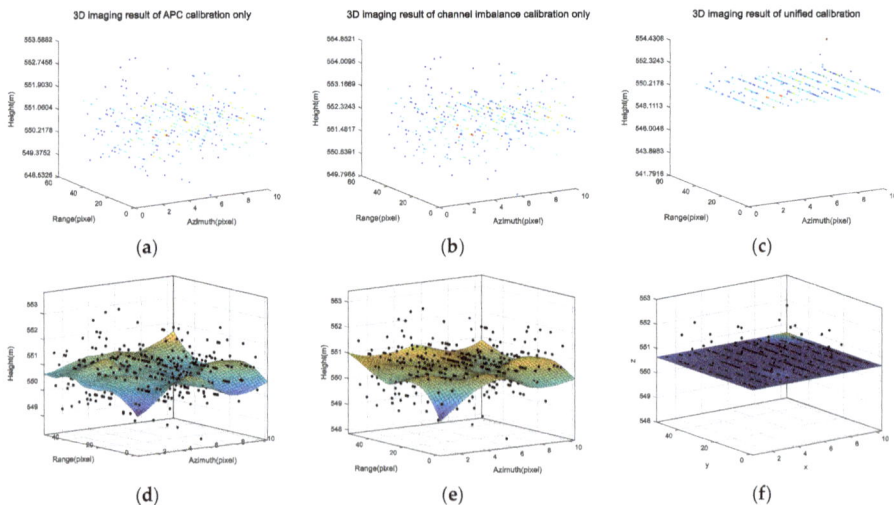

Figure 11. 3D imaging results of the selected flat ground. (a–c) 3D imaging results of the flat ground obtained by utilizing APC calibration only, channel imbalance calibration only, and a unified calibration algorithm, respectively; (d–f) The surface fitting results of (a–c), respectively.

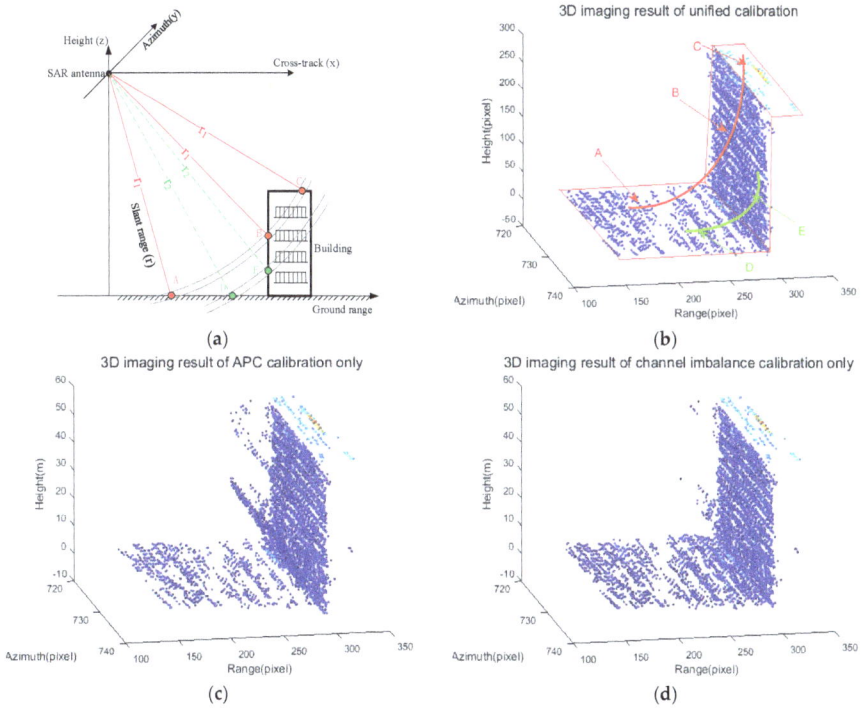

(a)

(b)

(c)

(d)

Figure 12. 3D imaging results of the layover building. (**a**) The building observation geometry; (**b**–**d**) 3D imaging results of the layover building utilizing the proposed unified calibration algorithm, APC calibration only, and channel imbalance calibration only, respectively.

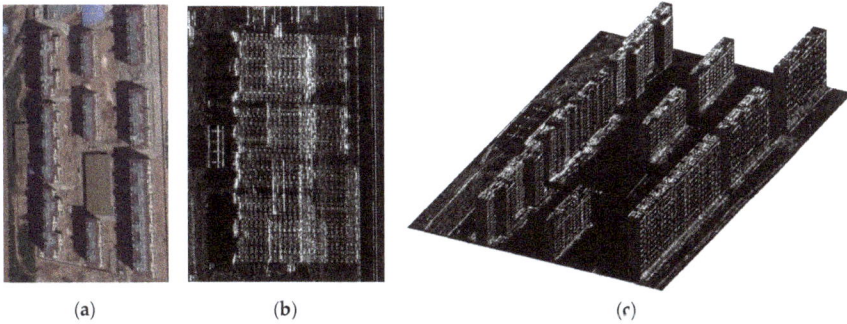

(a)   (b)   (c)

Figure 13. 3D imaging results of a residential area. (**a**,**b**) Google Earth and SLC images of the building, respectively; (**c**) The reconstruction result of the building after calibration and focusing processing.

5. Discussion

Both the simulation and array InSAR system results demonstrated the effectiveness of the proposed unified calibration algorithm. The simulation results showed that the calibration accuracy of the phase inconsistency and APC position went up to 0.06 rad and 0.105 mm, respectively. The application to the array InSAR system, a single-pass multi-baseline TomoSAR system, demonstrated the effectiveness of the unified calibration algorithm. Both the point-like targets (e.g., GCPs) and distributed target (e.g., flat ground) presented more reliable 3D imaging results after calibration. In the simulation, the calibration accuracy mentioned above provided the super-resolution ability to distinguish two targets with a 4.5 m height interval (e.g., the 4th pair layover GCPs in Table 2) despite the system's Rayleigh resolution of only 35 m. The resolution results in the height direction of the array InSAR system showed a similar super-resolution performance with a better separation of the building façade and ground (e.g., Figure 12b).

As discussed before, our proposed algorithm calibrated not only the channel imbalance but also the APC position. This is of great importance since both the channel imbalance (especially the phase inconsistency) and the APC position are the key parameters of the super-resolution algorithm. This proposed algorithm makes it possible to obtain a super-resolution performance in the height direction where the Rayleigh resolution is usually about 10–50 times below that in range or azimuth. In addition, the signal model uses the Fresnel approximation, which increases the accuracy of the calibration signal model. Although the proposed algorithm is of great accuracy, other tasks lie ahead: First, the question of how to choose GCPs for calibration, including the choice of size and position, needs to be studied, since the quality of GCPs will directly affect calibration results. In addition, the derivation of Cramer-Rao lower bound (CRLB) [23,24] and the quantitative assessment method require further study.

6. Conclusions

The multi-baseline TomoSAR system-like array InSAR system, whose antennas are placed in the cross-track direction, can obtain 3-D reconstruction images of the observation area in a single-pass flight campaign. In particular, when using the super-resolution algorithm in height focusing, we can obtain a high height resolution close to the azimuth and ground range resolutions. However, the high resolution in the height direction is based on the condition that the channel imbalance and APC position are precisely known. This is almost impossible, even though high-precision laboratory measurements have been carried out, since there are many factors that may lead to parameters changing in a flight campaign, such as installation error and the difference between the mechanical position and electromagnetic phase center. When a system error occurs, the 3-D resolution performance will be seriously degraded. In order to ensure the 3-D resolution performance of the TomoSAR system, especially the super-resolution performance in the height direction, the calibration processing of the channel imbalance and APC position is therefore necessary.

In our signal model, the Fresnel approximation is used. It is of great significance since the quadratic wave model is more accurate than the plane wave model, and most small aircraft platforms and UAVs work in relatively low flight height conditions. The importance of the proposed algorithm also lies in the fact that both the channel imbalance and APC position are accurately calibrated, which provides the basis for super-resolution in the height direction. This conclusion is supported by experiments based on a simulation data set and real data set acquired by the array InSAR system.

Acknowledgments: This research work has been partly supported by NSFC (the National Natural Science Foundation of China, No.: 61771453) and the Equipment Development Department pre-research fund No. 6140416010202. The optical image of the building was provided by Hang Li. The authors thank Ruichang Cheng and Song Wang in IECAS for plotting support and Longyong Chen for insightful suggestions. The authors would also like to thank the anonymous reviewers for their meticulous work in enhancing the quality of this paper.

Author Contributions: Xingdong Liang was the project manager and played a leading role in the preparation of this article. Yuncheng Bu established the signal model and developed the unified calibration algorithm. Yu Wang contributed the simulation analysis. The acquisition of real SAR data was organized by Fubo Zhang. Yanlei Li and Fubo Zhang provided support in SAR 3-D focusing and experimental validation. Yu Wang and Yanlei Li contributed to the structure and revision of this article and provided insightful comments and suggestions.

Conflicts of Interest: The authors declare no conflict of interest.

Appendix A

The required **g** and **H** are given in [22] under the plane wave model. Due to the use of the quadratic wave model in this article, the required **g** and **H** should be re-derived.

Under the quadratic wave model, the slant range between the point targets located at r and s coordinates and the nth APC is given in Equation (6), considering that the GCP is in the reference plane where $s = 0$; therefore, Equation (6) can be rewritten as

$$R_n(r_m) = r_m - b_{\|n} + \frac{b_{\perp n}^2}{2r_m}. \tag{A1}$$

Meanwhile, the slant range under the plane wave model could be written in the following form:

$$R_n(r_m) \approx \left| r_m - b_{\|n} \right| \tag{A2}$$

where r_m represents the distance between the reference channel and the mth GCP.

The slant range difference between Equations (A1) and (A2) is the root cause that leads to the re-derivation. This difference would result in a different form of $\mathbf{A}(\psi)$, $P_{\mathbf{A}^H(\psi)}^\perp$, $\mathbf{A_x}$, and $\mathbf{A_z}$. $\mathbf{A}(\psi)$ and $P_{\mathbf{A}^H(\psi)}^\perp$ have been given in Section 3; we will deduce $\mathbf{A_x}$ and $\mathbf{A_z}$ below.

As with [22], the n-m th element of matrix $\mathbf{A_x}$ is defined as

$$a_{nm} = \frac{\partial \mathbf{A}(\psi)_{nm}}{\partial x_n} = \frac{\partial e^{-j\frac{4\pi}{\lambda}(R_n(r_m) - R_1(r_m))}}{\partial x_n}. \tag{A3}$$

Substituting Equation (A1) into Equation (A3), we obtain

$$\begin{aligned} R_n(r_m) - R_1(r_m) &= -b_{\|n} + \frac{b_{\perp n}^2}{2r_m} \\ &= -(x_n \sin\theta_m - y_n \cos\theta_m) + \frac{(x_n \cos\theta_m + y_n \sin\theta_m)^2}{2r_m} \end{aligned} \tag{A4}$$

Therefore,

$$a_{nm} = e^{-j\frac{4\pi}{\lambda}(R_n(r_m) - R_1(r_m))} j\frac{4\pi}{\lambda} \left(\sin\theta_m - \frac{(x_n \cos\theta_m + z_n \sin\theta_m)\cos\theta_m}{r_m} \right). \tag{A5}$$

Then, $\mathbf{A_x}$ can be easily written as in Equation (18). The derivation process of $\mathbf{A_z}$ is similar to that of $\mathbf{A_x}$ and is not described here.

References

1. Fornaro, G.; Lombardini, F.; Serafino, F. Three-dimensional multipass SAR focusing: Experiments with long-term spaceborne data. *IEEE Trans. Geosci. Remote Sens.* **2005**, *43*, 702–714. [CrossRef]
2. Reigber, A.; Moreira, A. First demonstration of airborne SAR tomography using multibaseline L-band data. *IEEE Trans. Geosci. Remote Sens.* **2000**, *38*, 2142–2152. [CrossRef]
3. Schmitt, M.; Stilla, U. Compressive sensing based layover separation in airborne single-pass multi-baseline InSAR data. *IEEE Geosci. Remote Sens. Lett.* **2013**, *10*, 313–317. [CrossRef]
4. Zhu, X.X.; Bamler, R. Super-Resolution Power and Robustness of Compressive Sensing for Spectral Estimation With Application to Spaceborne Tomographic SAR. *IEEE Trans. Geosci. Remote Sens.* **2012**, *50*, 247–258. [CrossRef]
5. Schmitt, M.; Stilla, U. Maximum-likelihood-based approach for single-pass synthetic aperture radar tomography over urban areas. *IET Radar Sonar Navig.* **2014**, *8*, 1145–1153. [CrossRef]
6. Urasawa, F.; Yamada, H.; Yamaguchi, Y.; Sato, R. Fundamental study on multi-baseline SAR tomography by Pi-SAR-L2. In Proceedings of the URSI Asia-Pacific Radio Science Conference (URSI AP-RASC), Seoul, Korea, 21–25 August 2016; IEEE: Piscataway, NJ, USA, 2016; pp. 514–515.

7. Frey, O.; Morsdorf, F.; Meier, E. Tomographic imaging of a forested area by airborne multi-baseline P-band SAR. *Sensors* **2008**, *8*, 5884–5896. [CrossRef] [PubMed]

8. Schmitt, M.; Shahzad, M.; Zhu, X.X. Reconstruction of individual trees from multi-aspect TomoSAR data. *Remote Sens. Environ.* **2015**, *165*, 175–185. [CrossRef]

9. Zhu, X.X.; Bamler, R. Superresolving SAR tomography for multidimensional imaging of urban areas: Compressive sensing-based TomoSAR inversion. *IEEE Signal Process. Mag.* **2014**, *31*, 51–58. [CrossRef]

10. Zhu, X.X.; Bamler, R. Demonstration of Super-Resolution for Tomographic SAR Imaging in Urban Environment. *IEEE Trans. Geosci. Remote Sens.* **2012**, *50*, 3150–3157. [CrossRef]

11. Tebaldini, S.; Rocca, F.; D'Alessandro, M.M.; Ferro-Famil, L. Phase Calibration of Airborne Tomographic SAR Data via Phase Center Double Localization. *IEEE Trans. Geosci. Remote Sens.* **2016**, *54*, 1775–1792. [CrossRef]

12. Tebaldini, S.; Guarnieri, A.M. On the Role of Phase Stability in SAR Multibaseline Applications. *IEEE Trans. Geosci. Remote Sens.* **2010**, *48*, 2953–2966. [CrossRef]

13. Pardini, M.; Papathanassiou, K.; Bianco, V.; Iodice, A. Phase calibration of multibaseline SAR data based on a minimum entropy criterion. In Proceedings of the Geoscience and Remote Sensing Symposium, Munich, Germany, 22–27 July 2012; pp. 1–4.

14. Gocho, M.; Yamada, H.; Arii, M.; Sato, R.; Yamaguchi, Y.; Kojima, S. Verification of simple calibration method for multi-baseline SAR tomography. In Proceedings of the 2016 International Symposium on Antennas and Propagation (ISAP), Okinawa, Japan, 24–28 October 2016; IEEE: Piscataway, NJ, USA, 2016; pp. 312–313.

15. Tebaldini, S.; Rocca, F.; D'Alessandro, M.M.; Ferro-Famil, L. Point-target free phase calibration of InSAR data stacks. In Proceedings of the 2016 IEEE International Geoscience and Remote Sensing Symposium, Beijing, China, 10–15 July 2016; pp. 1440–1443.

16. Zhu, H. Study on Phase Center Calibration Methods of Linear Array Downward-Looking 3D-SAR. Master's Thesis, University of Chinese Academy of Sciences, Beijing, China, 2012.

17. Han, K. Study on Multi-Channel Amplitude-Phase Errors Calibration and Imaging Methods of Downward-Looking 3D-SAR Based on Array Antennas. Master's Thesis, University of Chinese Academy of Sciences, Beijing, China, 2011.

18. Yang, X.L.; Tan, W.X.; Qi, Y.L.; Wang, Y.P.; Hong, W. Amplitude and Phase Errors Correction for Array 3D SAR System Based on Single Prominent Point Like Target Echo Data. *J. Radars* **2014**, *3*, 409–418.

19. Wang, Y. Studies on Calibration Model and Algorighm for Airborne Interferoemtric SAR. Ph.D. Thesis, University of Chinese Academy of Sciences, Beijing, China, 2003.

20. Zhang, F. Research on Signal Processing of 3-D reconstruction in Linear Array Synthetic Aperture Radar Interferometry. Ph.D. Thesis, University of Chinese Academy of Sciences, Beijing, China, 2015.

21. Li, H.; Ding, C.; Zhang, F.; Liang, X.; Wu, Y. A novel 3-D reconstruction approach based on group-sparsity of array InSAR. *Sci. Sinica Inform.* **2017**. [CrossRef]

22. Ng, B.C.; See, C.M.S. Sensor-array calibration using a maximum-likelihood approach. *IEEE Trans. Antennas Propag.* **1996**, *44*, 827–835.

23. Rockah, Y.; Schultheiss, P.M. Array shape calibration using sources in unknown locations—Part I: Far-field sources. *IEEE Trans. Acoust. Speech Signal Process.* **1987**, *35*, 286–299. [CrossRef]

24. Rockah, Y.; Schultheiss, P.M. Array shape calibration using sources in unknown locations—Part II: Near-field sources and estimator implementation. *IEEE Trans. Acoust. Speech Signal Process.* **1987**, *35*, 724–735. [CrossRef]

remote sensing

MDPI

Review

InSAR Deformation Analysis with Distributed Scatterers: A Review Complemented by New Advances

Markus Even *,† and **Karsten Schulz**

Fraunhofer IOSB, Gutleuthausstr. 1, 76275 Ettlingen, Germany; karsten.schulz@iosb.fraunhofer.de
* Correspondence: markus_even@web.de
† Worked at IOSB until June 2017.

Received: 31 March 2018; Accepted: 9 May 2018; Published: 11 May 2018

Abstract: Interferometric Synthetic Aperture Radar (InSAR) is a powerful remote sensing technique able to measure deformation of the earth's surface over large areas. InSAR deformation analysis uses two main categories of backscatter: Persistent Scatterers (PS) and Distributed Scatterers (DS). While PS are characterized by a high signal-to-noise ratio and predominantly occur as single pixels, DS possess a medium or low signal-to-noise ratio and can only be exploited if they form homogeneous groups of pixels that are large enough to allow for statistical analysis. Although DS have been used by InSAR since its beginnings for different purposes, new methods developed during the last decade have advanced the field significantly. Preprocessing of DS with spatio-temporal filtering allows today the use of DS in PS algorithms as if they were PS, thereby enlarging spatial coverage and stabilizing algorithms. This review explores the relations between different lines of research and discusses open questions regarding DS preprocessing for deformation analysis. The review is complemented with an experiment that demonstrates that significantly improved results can be achieved for preprocessed DS during parameter estimation if their statistical properties are used.

Keywords: InSAR; Persistent Scatterer; Distributed Scatterer; preprocessing; adaptive neighborhood; covariance; coherence; deformation

1. Introduction

The subject of this review will be multitemporal deformation analysis with spaceborne (Synthetic Aperture Radar) SAR interferometry. More precisely, the methods that have been developed pertaining to preprocessing of Distributed Scatterers (DS) for use in Persistent Scatterers (PS) algorithms will be discussed with a focus on progress in the last decade.

Interferometric Synthetic Aperture Radar (InSAR) is a technique that has its origin in the late 1970s, when spaceborne imaging radars began to play an important role in remote sensing [1–4]. It became popular when, after the launch of the European Space Agency (ESA) satellite ERS-1 in 1991, an enormous amount of suitable SAR data became available. Since that time, its importance has increased steadily and today about one and a half dozen SAR satellites are orbiting the earth that are continuously acquiring data for scientific, governmental, and commercial purposes (e.g., Sentinel 1, TerraSAR-X, TanDEM-X, CosmoSkymed, RADARSAT-2, ALOS II, SAOCOM, PAZ). Data are used to gather information over land, ice, and sea. They allow mapping and change detection for a multitude of purposes. Applications comprise, for example, land cover classification, mapping of ocean currents, intelligence, or situational awareness in case of natural catastrophes, e.g., mapping of flooded or destroyed areas. However, the unique capability of spaceborne SAR is the acquisition of large area interferometric data. Devoted missions (SRTM, TanDEM-X) have provided Digital Elevation Models (DEMs) of the whole surface of the earth, allowing for glaciologists to study extent, flow, and mass balance of glaciers and ice sheets, for climatologists to estimate the biomass of the world's forests, and for geologists to use SAR data to study phenomena like earthquakes, volcanoes, and tectonic processes. For other geoscientists and for governmental and economic stakeholders, it is of high importance to monitor movements of the earth's surface that go along with tunneling, mining, gas, water, and oil withdrawal.

Deformation analysis with InSAR is based on the following idea that will be given for the moment largely simplified, ignoring varying positions of the sensor, atmospheric delay, phase ambiguity, presence of complex scattering mechanisms, physical changes in the illuminated scene, and other complications [1–4]: The SAR data processing gives a complex valued image, where the amplitude of a pixel is conceived as the magnitude of the signal scattered back from a resolution cell on the ground and where the argument is interpreted as the phase shift between emitted and received signal (instead of phase shifts, one simply speaks of phases). If a movement of the earth's surface occurs between acquisitions, the signal travels a different distance and the phases change accordingly. By integrating spatially and temporally these changes of phases, the deformation is obtained. The changes of phase are found in the name giving interferograms, which are formed by multiplying the pixel values of the one acquisition with the complex conjugated pixel values of the other acquisition. However, processing of real data cannot ignore the mentioned complications and requires solutions. The basis for developing corresponding algorithms is usually a decomposition of the interferogram phase as in the following formula (observe that phases are only known modulo 2π):

$$\varphi_m - \varphi_s = \varphi_{synth} - \frac{4\pi}{\lambda}\left(\frac{B_\perp}{r_m}\cdot\frac{\Delta h}{tan(\theta)} + \Delta r\right) + \alpha_m - \alpha_s + \nu \tag{1}$$

where φ_m and φ_s are the phases of acquisitions m and s. φ_{synth} is the synthetic phase corresponding to geometric path lengths calculated based on orbit information and a DEM, λ is the wavelength, $B_\perp$ is the perpendicular baseline, r_m is the distance corresponding to the pixel center, θ is the looking angle, Δh is the DEM error, Δr is the displacement in line of sight of the sensor, α_m and α_s account for atmospheric delay and other spatially correlated errors (e.g., caused by imprecise orbits)—in the sequel named atmospheric phase screen (APS)—and ν is everything else, usually called noise. The relevant contribution for deformation analysis is the displacement, and the question is under what circumstances it can be extracted. In general, the phase model tells us that the deformation signal can be accurately determined if the other terms can either be compensated or are insignificant (e.g., DEM and orbit data are precise or atmosphere over an arid region might be stable). Of particular interest here is the miscellaneous term ν. It accounts for sensor noise and processing errors, which can be assumed to be small. However, it comprises also decorrelation effects and changes in reflectivity that might make estimation infeasible. Deformation analysis is feasible mainly for two categories of scattering mechanisms. The first are Persistent Scatterers (PS), the case where ν is small. This corresponds most often to one dominant scatterer in the resolution cell, e.g., a trihedral manmade structure, a pole, or a single rock. There have been several papers that investigated the physical origin of PS, e.g., [5], where six main types of PS are described. The second are Distributed Scatterers (DS), which is the case where a sufficiently large group of adjacent pixels shares the same scattering mechanism and ν can be mitigated by statistical methods. Usually, these are pixels with many small scatterers of similar size. If the resolution is some 10 meters, this is true for most natural scatterers (forest, agricultural fields, bare soil, rock surfaces). If the resolution is some meters, DS are mostly found in arid areas with low vegetation and debris, but even rough asphalt or plaster can constitute a DS. There are also exploitable pixels, where a small number (e.g., two or three) of pointlike scatterers are contained. The corresponding field of research is SAR tomography (cp. e.g., [6,7]) and will be left aside as DS are the focus of this review.

The history of InSAR deformation analysis exploiting DS commenced with Differential InSAR (DInSAR), for the first time described in [8] for L-band data from Seasat (a comprehensive overview on DInSAR is given in [2]). The initial approach consisted of using three images that were used to form interferograms and a DEM. The DEM served to calculate and remove the synthetic phase. Furthermore, phase unwrapping was applied to generate the DEM and to obtain the deformation field. An algorithm for phase unwrapping was developed somewhat earlier for DEM generation from InSAR data in [9]. During the following years, techniques for most of the basic challenges of InSAR were developed. Early work on phase statistics and the phenomenon of decorrelation can be found in [10–15]. Enhancement of signal quality by filtering was considered, e.g., in [16,17]. Investigations, where stacking of interferograms is used to mitigate atmospheric delay, are discussed in [18–20]. Theory and algorithms were based on DS until the existence of PS was observed in the late 1990s by Ferretti while working on DEM reconstruction from a stack of SAR images [21]. In the sequel, they developed the first Persistent Scatterer Interferometry (PSI) algorithm [22,23], which extended the applicability of InSAR to scenes where enough PS are found but large parts are strongly decorrelated and hence unwrapping on the full interferograms cannot succeed. For several years from this time on, DInSAR and PSI developed in parallel. The next big step for DInSAR was the small-baseline subset (SBAS) technique [24]. By considering a redundant graph of small baseline interferograms, the effects of decorrelation could be mitigated and the redundancy enhanced robustness of estimation. In this first version, only DS were considered (using boxcar multilooking), but the next step [25] was to include processing of PS: coherent targets in the full resolution interferograms were recognized as having small residuals relative to the spatially filtered interferograms. In the same year, a new approach [26] for processing PS was proposed that later became the Stanford Method for PS (StaMPS; [27,28]). It aims at exploiting low amplitude PS on volcanoes and other natural terrains and likewise detects these PS as pixels that have small phase differences to the filtered interferograms. A peculiarity of StaMPS is the application of an extensive iterative spatiotemporal filtering. This might be seen as an example from a third line of development beside PS and SBAS techniques, where SAR image filters progress from boxcar filtering to ever more sophisticated approaches. Later, the ideas from SBAS were included in StaMPS, allowing joint processing of PS and DS [29]. At that time, several research groups worked towards integrated processing of DS and PS with the goal of increasing the spatial coverage with measurements. Also, at Milano progress was made. An important step was the first estimator making use of the full covariance matrix for estimating the parameters of a deterministic phase model (linear deformation rate and height error) of a DS relative to a reference PS [30,31]. The effect is that, during estimation, phases are weighted in an optimal way; under assumption of Gaussianity, it is the maximum likelihood estimator (MLE). To prevent APS from deteriorating results, it is necessary that DS are added to the result of a PS analysis. This allows for extension of the APS estimated for the PS to DS positions and removal it before the MLE is applied. This way, DS are not used to bridge gaps in the PS net, which would make results more robust. DS are added in a postprocessing step. Transforming DS in a preprocessing step in such a way that they can be used like PS in any PS algorithm was the next stage of development. De Zan [31] describes an experiment where he observes that the phases of the eigenvector of the covariance matrix to the largest eigenvalue correspond to deformation, DEM error, and APS averaged over the DS pixel. [32,33] derived an MLE (likewise under assumption of Gaussianity) for the phase history of DS that approximates the phases of the complex covariance matrix by triangular phases, assuming that all pixels in the neighborhood corresponding to a DS are affected by the same deformation, DEM error, and APS. The original phases of the DS are then replaced by the estimated phase history. At Fringe 2009, the power of this idea was demonstrated when SqueeSAR was presented, a framework for the preprocessing of DS [34]. In [35], this approach was explained in more detail, adding suggestions for adaptive neighborhood (AN) forming (DeSpecKS) and DS quality assessment. Shortly afterwards, AN forming and phase triangulation were integrated in the SBAS framework in [36]. For SBAS, this opened the applicability of spatiotemporal unwrapping [37–40], which is not possible directly from independently spatially unwrapped interferograms because phases are not triangular (cp. [41] for phase triangulation without consideration of statistics).

Here, this historical survey ends. It has been intended to show on a very coarse scale how the processing of DS and PS evolved over time and how they fused a decade ago with spatiotemporal filtering techniques together in approaches that can be described as preprocessing of DS for the use in PSI algorithms. Preprocessing of DS for the use in PSI algorithms is the main subject of this review. It has seen tremendous further progress since that time, which is visible, e.g., in a number of excellent doctoral theses related to the subject [42–46]. Although in each of them the state of the art has been discussed, not all aspects are covered, partly because of new developments that resulted since their publication, partly because they necessarily concentrated on a certain issue. The present review is intended to give a broad view of the subject, with a focus on giving a survey on methods and ideas and presenting phase triangulation as a unifying concept that allows extraction of DS signals in a general manner from weighted filtered interferograms. Furthermore, a complement was included in the review. Although preprocessing of DS is often simplistically depicted as transforming DS into PS, preprocessed DS are statistically not equivalent to PS. An experiment demonstrates that parameter estimation from preprocessed DS gives significantly better results if statistical information is considered.

In Section 2, statistical modeling of DS is surveyed. Section 3 is the core of the review. It commences with an estimation of DS signals because of the pivotal role we assign to this element. Then filtering of interferograms and coherence estimation are treated. Points to be addressed are nonstationary phases, grouping of statistically homogeneous neighborhoods resp. of adaptive neighborhoods, nonlocal methods, bias correction and regularization, and quality numbers for DS. In Section 4, phase model parameter estimation for preprocessed DS is discussed and the announced experiment presented. In Section 5, a short discussion of what has been achieved is given and interesting possibilities for future research are indicated. Conclusions and a brief synopsis in Section 6 complete the review.

2. Statistics for Distributed Scatterers

A DS pixel is supposed to originate from many small scatterers of comparable size in a resolution cell. If the SAR image contains a larger area with such a scattering mechanism, a so-called speckle pattern is visible that can be stochastically modeled. This does not contradict the fact that the scattering process is deterministic, and if the acquisition is repeated from precisely the same position and with no changes having affected the terrain, the same pattern would result again. One should rather think of a repeated random experiment, where a random number of scatterers is randomly distributed in each resolution cell and the range positions are uniformly distributed. In case the range extension of the pixel is much larger than the wavelength, the latter has the consequence that phases can be described with good precision by a uniform distribution. This concept allows for successfully dealing with a situation where the detailed information is missing that would be necessary for a deterministic treatment. To derive a specific statistical model, traditionally several assumptions are made [47,48]:

1. the backscatter from a resolution cell is the superposition of the backscatter of stochastically independent elementary scatterers;
2. their number is large;
3. amplitude and phase are independent random variables;
4. the phase is uniformly distributed;
5. no individual scatterer dominates the resolution cell;
6. the resolution cell is large compared to the single scatterer.

From the generalized central limit theorem, it can be concluded that the real and imaginary parts of backscatter are approximately α-stable distributed ($0 < \alpha \leq 2$). The α-stable distributions form a four-parameter family: location, scale, stability, and skewness parameter (note: skewness requiring the third central moment is not defined). The particular case $\alpha = 2$ occurs if standard deviations of the elementary random variables are bounded and the central limit theorem can be applied. The limiting distribution is then consequentially normal. While Goodman [49] assumed bounded standard deviations and obtained a complex circular (i.e., $z \sim e^{i\delta} \cdot z$ independent of angle δ) normal distribution in the limit, other authors favor the more general framework of symmetric α-stable distributions [47,50] to be able to account for impulsive behavior of the signal, e.g., found in high-resolution SAR images of urban areas (or of the sea surface).

Symmetric α-stable random vectors belong to the larger class of complex elliptically symmetric (CES) distributed random vectors [51–53]. CES distributions comprise, e.g., complex normal, complex t-, complex K-, generalized Gaussian, and inverse Gaussian distributions that are used to model radar clutter. They provide alternative statistic models for DS in cases where the assumption of complex normal distribution does not hold, e.g., because of high-resolution SAR data or, more importantly, because deviating scattering mechanisms are wrongly included in the DS neighborhood. A comprehensive theory of robust estimation has been developed for CES distributions that will be discussed at the end of the section. A survey on statistical modeling of SAR images was given by [54].

Usually for DS, Goodmann's model is adopted, i.e., that they can modeled as circular complex normally distributed random vectors, and it will also be the basis for most of the work presented here. A circular complex normally distributed random vector $y \sim CN(0, C, 0)$, $C = E[yy^H]$ is complex normally distributed with mean and relation matrix equal to zero [55]. For the entries of the covariance matrix C, let $c_{mn} = |c_{mn}| \cdot exp(i \cdot \phi_{mn})$, and $\sigma_m = \sqrt{c_{mm}}$ is the square root of the backscatter coefficient in acquisition m. Then, the complex correlation or coherence is

$$\gamma_{mn} = |\gamma_{mn}| \cdot e^{i\phi_{mn}} = \frac{c_{mn}}{\sigma_m \cdot \sigma_n}. \tag{2}$$

Because of its importance for InSAR, this correlation has been investigated by many authors. Zebker and Villasenor [12] studied the causes for loss of correlation between two images in basic situations:

1. presence of thermal noise (thermal decorrelation);
2. effect of different viewing geometry (spatial baseline and rotation decorrelation);
3. small random movements of the scatterers (temporal decorrelation).

They derived a formula presenting the total correlation as the product of the basic correlations. In [56] the formula for the total correlation of [12] is modified by thresholding with a bias term dependent on the number of independent looks and replacing the critical baseline by an effective baseline that is intended to account for volumetric effects. In [57], the authors investigate the development of a temporal correlation for sensors in L-, C- and X-band and different revisit times over drained peat soils in the Netherlands. To this end, a model for correlation is formulated that contains a long-term coherence and its parameters are estimated (e.g., about 10 days for C-band in summer). The finding is that, "it is the combination of longer wavelengths, shorter repeat interval, and higher spatial resolution that increases the likelihood to obtain a coherent signal" [57]. Because of the large decorrelation rate, it is difficult to perform deformation estimations on this terrain. Afterwards, they succeeded estimating deformation via a multisatellite approach presented in [58]. Models considering a periodic factor are given in [46] and later in Section 4.2.

An observation that is of importance for the stochastic model for DS that will be introduced next is that $\gamma_{mn} \in \mathbb{R}_{\geq 0}$ holds for the complex correlation coefficients in the formula of [12]. Likewise, this is the case for temporal correlation as modeled in [59] or [31]. If a common phase history $\phi = (\phi_1, \cdots, \phi_N)^t$ is superposed that accounts for deformation, atmospheric delay and large area DEM errors $c_{mn} = |c_{mn}| \cdot exp(i \cdot (\phi_m - \phi_n))$ are obtained. This is equivalent to saying that phase triangularity is given, i.e., $\phi_{mn} = \phi_{ml} - \phi_{ln}$, for all l, m, and n. In many situations, this is a plausible model for the shape of the covariance matrix of a DS, but not always. For certain scattering phenomena connected with soil moisture changes or thawing permafrost, it is known that phase triangularity might be corrupted [60–62].

In [33], the following stochastic model for a DS is given that consists of a neighborhood Ω of pixels: The complex vectors of N image values are realizations of random vectors $y_k = (y_{k1}, \cdots, y_{kN})^t \sim CN(0, C, 0)$ $(k \in \Omega)$ that result from independent identically circular complex normal distributed random vectors. (For the sake of simple notation, it is not distinguished between the random vectors and their realizations in the following.) $C \in \mathbb{C}^{N \times N}$ denotes the covariance matrix. It is assumed that the shape of the covariance matrix is as discussed above. All pixels in Ω have a common phase history $\phi = (\phi_1, \cdots, \phi_N)^t$ that accounts for deformation, atmospheric delay, large area DEM errors, and other contributions that do not vary spatially. In the hypothetical case that there were no such contributions and ϕ were equal zero, $c_{mn} \in \mathbb{R}_{\geq 0}$ for all entries of C can be assumed. The consequence is the assumption that, in general, $c_{mn} = |c_{mn}| \cdot exp(i \cdot (\phi_m - \phi_n))$ holds. All y_k are collected in one random vector with covariance matrix $C_\Omega = C_\Omega(\phi) \in \mathbb{C}^{KN \times KN}$, where $K = \#\Omega$:

$$
y = \begin{pmatrix} y_1 \\ \vdots \\ y_K \end{pmatrix} \sim CN(0, C_\Omega, 0). \tag{3}
$$

Then the following holds

$$
(C_\Omega)_{kmln} = E\left[y_{km} y_{ln}{}^H\right] = \delta_{kl} \cdot |c_{mn}| \cdot exp(i \cdot (\phi_m - \phi_n)). \tag{4}
$$

Furthermore, the pdf (probability density function) for a given ϕ is

$$
p(y|\phi) = const. \cdot exp\left(-y^H C_\Omega{}^{-1} y\right). \tag{5}
$$

Note that the constant is not dependent on ϕ. A short calculation leads to

$$
y^H C_\Omega{}^{-1} y = \cdots = K \cdot \zeta^H \cdot \left(|C|^{-1} \circ \hat{C}\right) \cdot \zeta \tag{6}
$$

where $\hat{C}$ is the sample covariance matrix (SaCM). $\hat{C}$ is the MLE for the covariance matrix of circular complex normally distributed random vectors and its probability density function is the complex Wishart distribution [63]. This last equation is the basis for the MLE for the phase history of a DS discussed later. From $\hat{C} = (\hat{c}_{mn})$, the coherence matrix is obtained:

$$
\hat{\Gamma} = \begin{pmatrix} \sqrt{\hat{c}_{11}} & 0 \\ & \ddots & \\ 0 & \sqrt{\hat{c}_{NN}} \end{pmatrix}^{-1} \cdot \hat{C} \cdot \begin{pmatrix} \sqrt{\hat{c}_{11}} & 0 \\ & \ddots & \\ 0 & \sqrt{\hat{c}_{NN}} \end{pmatrix}^{-1}. \tag{7}
$$

Its entries are the sample complex coherences for each interferogram:

$$
\hat{\gamma}_{mn} = |\hat{\gamma}_{mn}| \cdot e^{i\hat{\phi}_{mn}} = \frac{\sum_{k \in \Omega} y_{km} y_{kn}^*}{\sqrt{\sum_{k \in \Omega} |y_{km}|^2 \cdot \sum_{k \in \Omega} |y_{kn}|^2}} \tag{8}
$$

where $|\hat{\gamma}_{mn}|$ is a measure of the variation of phase inside Ω and the MLE for coherence magnitude [64] (p. 581). $\hat{\phi}_{mn}$ is the MLE for the joint interferogram phase under circular complex normal distribution (stochastic model and proof in [14], already stated in [11]). Note that in [14], the MLE of the coherence magnitude was derived under the assumption that the variance in both acquisitions is the same. Its expectation is always smaller than that of the magnitude of the sample coherence [14]. $|\hat{\gamma}_{mn}|$ is known [65–67] to be biased towards larger values but is asymptotically unbiased (with growing number of looks). The bias for a given number of looks is worst for a small magnitude of coherence. In [68], a refined speckle noise model was given and used to derive a bias corrected estimator for coherence magnitude. Formulas for pdf, mean, and moments of $|\hat{\gamma}_{mn}|$ have been given [66,67]. To use $|\hat{\gamma}_{mn}|$ as a reliable indicator of phase quality, no phase ramp must be present [69], as quality is otherwise underestimated. The pdf for $\hat{\phi}_{mn}$ can be found, e.g., in [65,66,70]. Its standard deviation drops with increasing coherence and with an increasing number of looks. For a more detailed discussion of errors in coherence estimation, see [44]. As a reliable estimation of $\hat{\gamma}_{mn}$ is paramount, the following crucial issues will be discussed in Section 3.2 in more detail:

1. removal of residual fringes;
2. grouping of a statistically homogeneous neighborhood Ω;
3. bias reduction.

Still under the assumption that the statistics of a DS in two repeat-pass SAR images can be described as a complex circular normal random vector, formulas for several related random variables were derived: joint pdf of magnitude and phase of the interferogram [70,71], pdf of interferometric phase [15,70], pdf of interferogram magnitude [70], pdf, and expectation and standard deviation of the multilooked interferometric phase [64]. Inspection of the joint pdf of magnitude and phase shows that samples with a phase close to the mean phase more likely have a high amplitude, while larger phase deviations more often correspond to small amplitudes. Although the simplified exposition in the introduction might have given the impression that only phases matter, amplitudes are also relevant as they reflect the quality of the phase, and it often makes sense to use the complex signal for processing. A trivial example is the use for estimation of the SaCM. Further examples can be found in Equations (26) and (27) of the subsection on estimation of model parameters.

For the case of symmetric α-stable distributions, a modified estimator for coherence based on fractional lower order statistics was given in [72]. Their examples of coherence estimation with the proposed estimator show less artifacts near strong scatterers. DS are supposed to be statistically homogeneous, so assuming a distribution made to account for strong heterogeneous scattering would improve DS-processing means that pixels that do not belong to the DS may be contained in the neighborhood and hence, at least for high resolution data, grouping was suboptimal. Jiang [44] reports that neighborhoods generated with his adaptive neighborhood (AN) selection algorithm are approximately Gaussian distributed and therefore no advantage can be expected from an estimator modified for symmetric α-stable distributions. Nevertheless, if there is reason to think that neighborhoods are less homogeneous than necessary, it can make sense to invest the additional computational effort and use a robust M-estimator of scatter [52]. Scatter means the scatter matrix, one of the defining parameters of a CES distribution. It is a positive constant times the covariance matrix and hence provides the same useful information as the covariance matrix. Compared to amplitude-based outlier rejection, M-estimators of scatter have the advantage of being sensitive versus phase when weighting down outlying pixels. As they use the Mahalanobis length and therefore weight down all values of a pixel, it still makes sense to detect outliers beforehand and discard them before estimating the scatter matrix. They are not recommended for small neighborhoods as they involve inversion of the estimated covariance matrix. In this case, regularized M-estimators perform better [53]. Robust M-estimators of scatter matrix are robust in the sense that they have a bounded influence function. This means that small contaminations may not have an arbitrarily large effect on the estimation result, e.g., the SaCM is an M-estimator of scatter but not robust. Robust examples are the Huber estimator, the MLE for the complex t-distribution, or the S-estimator with Rocke's weight function according (for implementing S-estimators, see [73]). The latter M-estimators of scatter lend themselves for MLE of phase history, as has been derived in [74]. There is a trade-off between robustness and precision of estimation that can be measured via the asymptotic relative efficiency [52,75]. While the MLE might be sensitive to outliers, a very robust estimator might have a too strongly varying asymptotic distribution, and a better solution is found in the middle between those extremes. Finally, under reasonable conditions, M-estimators of scatter for CES distributions are asymptotically normal and the limiting covariance matrix can be calculated based on the parameters of the underlying CES distribution [52]. This could be a starting point to develop new quality numbers for the scatter matrix.

3. Estimation of Distributed Scatterer Signals for Preprocessing of Multitemporal InSAR Data

In this section, the estimation of DS signals from the wrapped interferogram phases is discussed. The estimated DS signal constitutes a filtered version of the original data and can be used afterwards for any InSAR application which might benefit from a filtered input. Good DS can be used like PS in any PSI algorithm. This latter conception is from our point of view a key idea of SqueeSAR [35]. In [35] and other approaches, it is realized via the following steps for estimation of DS signals:

1. Grouping of a neighborhood Ω;
2. Estimation of the covariance matrix;
3. Phase triangulation or more generally estimation of the DS signal;
4. Calculation of a quality number for the DS.

For grouping of a neighborhood for a pixel, a search window is centered on it. In case of DespecKS, a method suggested in [35], the amplitudes of all other pixels in the search window are compared with those of the center pixel via the KS two-sample test. Those pixels accepted to have the same distribution of amplitudes form the neighborhood. Often, the connectedness of the neighborhood is enforced with the argument that pixels then are more likely to belong to the same physical structure. For the pixels in the neighborhood, the SaCM is calculated. A phase history is estimated that optimally fits to the phases of the SaCM. As a quality number for goodness of fit, the phase triangulation coherence is calculated. These steps and also the scheme itself can be modified in various ways. An important further example is the SBAS approach. It has been demonstrated that it gives improved results if boxcar multilooking is replaced with more refined techniques, and if due to triangular phases, 3D-unwrapping algorithms are applicable [36,40,76]. A difference here is that not all possible interferograms are calculated but only those with small baselines. There are also other algorithms that do not exactly fit this scheme. This section is devoted to discussing the different solutions found in the literature. As the unifying ingredient common to all preprocessing schemes discussed in this work is phase triangulation, the exposition does not follow the succession of the above steps but starts with explaining estimators of DS signal. This facilitates the discussion in the sequel.

3.1. Estimators of Distributed Scatterer Signal

In this section, estimators of the DS signal are presented. In some cases, amplitudes are neglected and only the phase history of the DS is provided. They allow to preprocess the data stack and to replace the noisy original signal with the estimated signal. If the estimation is successful, then these pixels can be used like PS. Some of these estimators can also be used to determine the parameters of a phase model. This will be the subject of Section 4.

The first estimator of phase history for multitemporal InSAR was introduced and investigated in [32,33]. It is the maximum likelihood estimator (ML), which is asymptotically optimal and close to the Cramér–Rao lower bound:

$$\hat{\phi} = \arg\max_{\phi} exp\left(-y^H C_\Omega^{-1} y\right) = \arg\min_{\phi} \xi^H \cdot \left(|C|^{-1} \circ \hat{C}\right) \cdot \xi \tag{9}$$

where C is the covariance matrix, $\hat{C}$ is the sample covariance matrix, and

$$\xi = \begin{pmatrix} e^{i\phi_1} \\ \vdots \\ e^{i\phi_N} \end{pmatrix} \tag{10}$$

contains the sought phase history $\phi_1, \dots, \phi_N$. Please observe that in [33], it was assumed that all variances are one, and therefore, the coherence matrix replaces the covariance matrix in their formulas. An historical side note: this estimator has an early predecessor that was developed at the end of the

1990s to retrieve heights from data of an airborne three-antenna SAR system, cp. e.g., [77]. To make the result unique, the master phase ϕ_m is assumed to be zero. The ML estimator is not available for real data, as it requires the unknown covariance matrix. In [33], a test case is presented, where no deformation was expected and a replacement for C was calculated from acquisition geometry and a SRTM DEM under the assumption that only spatial decorrelation matters. If the covariance matrix is replaced by an estimation $\hat{C}$ of the covariance matrix and if $|\hat{C}|^{-1}$ exists, an applicable estimator is obtained, which is different from the ML estimator and that could be named a pseudo ML estimator:

$$\hat{\phi} = \arg\min_{\phi} \xi^H \cdot \left(|\hat{C}|^{-1} \circ \hat{C} \right) \cdot \xi = \arg\min_{\phi} \sum_{m,n} \hat{\zeta}_{mn} \cdot |\hat{c}_{mn}| \cdot exp(i \cdot (\hat{\phi}_{mn} - (\phi_m - \phi_n))). \tag{11}$$

Here, $|\hat{C}|^{-1} = (\hat{\zeta}_{mn})$, $\hat{C} = (\hat{c}_{mn})$, and $\hat{c}_{mn} = |\hat{c}_{mn}| \cdot exp(i \cdot \hat{\phi}_{mn})$. In cases where $|\hat{C}|^{-1}$ does not exist, some regularization has to be applied or the pseudoinverse can be taken. If a PSI algorithm is applied that is able to benefit from a DS signal comprising phases and amplitudes, a natural choice for amplitudes would be the square roots of the diagonal entries of $\hat{C}$. We will refer to this type of estimator, which consists of estimation of covariance or coherence matrix plus execution of the phase linking algorithm also as phase linking (PhL), although the authors of [33] introduced the notion of phase linking for the iterative determination of the minimum with the following formula:

$$\phi_p^{(k)} = \angle \left\{ -\sum_{n(\neq p)} \hat{\zeta}_{pn} \cdot \hat{c}_{pn} \cdot exp(i \cdot \phi_p^{(k-1)}) \right\}. \tag{12}$$

The minimization can also be solved by more advanced algorithms, e.g., the Broyden–Fletcher–Goldfarb–Shanno algorithm [78], but probably less effectively. As $|\hat{C}|^{-1} \circ \hat{C} = |\hat{\Gamma}|^{-1} \circ \hat{\Gamma}$ holds, PhL can also be stated using the coherence matrix. However, for other estimators of DS signal, the choice between $\hat{C}$ and $\hat{\Gamma}$ might result in different estimators. [32,33] provide also the hybrid Cramér–Rao bound for PhL. We give a slightly modified formulation. Let $\phi(\vartheta) = \Theta \cdot \vartheta + \tilde{I} \cdot \omega$, $\Theta = \tilde{I} \cdot \tilde{\Theta}$ with $\tilde{\Theta} \in \mathbb{R}^{(N-1) \times p}$, so that $\tilde{I} \in \mathbb{R}^{N \times (N-1)}$ is obtained by the identity matrix by removing the master column, ϑ contains the sought model parameters (PhL corresponds to the case where $\tilde{\Theta}$ is the identity matrix), and ω denotes interferogram atmosphere. Furthermore, assume that atmosphere α can be modelled as a Gaussian iid signal with standard deviation σ_a. $\omega_n = \alpha_n - \alpha_{master}$ has then covariance matrix $V \in \mathbb{R}^{(N-1) \times (N-1)}$ with entries $v_{mn} = \sigma_a^2 \cdot (1 + \delta_{mn})$. The Fisher information matrix is $X = 2L \cdot (|\Gamma|^{-1} \circ |\Gamma| - I)$, where L is the number of looks. From a theorem of Fiedler, it follows that it is positive semidefinite [79]. Define $\tilde{X} := \tilde{I}^t \cdot X \cdot \tilde{I}$. Assume $\Theta^t \cdot X \cdot \Theta$ is invertible. Then, the following inequality is obtained:

$$E_{y,\omega} \left[(\hat{\vartheta} - \vartheta)(\hat{\vartheta} - \vartheta)^t \right] \geq \left(\tilde{\Theta}^t \cdot \left(\tilde{X} - \tilde{X} V^{\frac{1}{2}} \left(V^{\frac{1}{2}} \tilde{X} V^{\frac{1}{2}} + I \right)^{-1} V^{\frac{1}{2}} \tilde{X} \right) \cdot \tilde{\Theta} \right)^{-1} \tag{13}$$

and the inverse matrices on the right-hand side exist (here A $\geq$ B means A-B is positive semidefinite). Although this formulation looks on first sight more complicated than the one given in [32,33], it has the advantage of avoiding a limit process and allows for setting $V = 0$ in case ω is negligible without further thinking. In case $V = 0$, the equation simplifies to the standard Cramér–Rao bound. Furthermore, it is still easily verified that the matrix is symmetric.

SAR polarimetry inspired a second way of estimating DS signals [31,80,81]. The method can be applied either to $\hat{C}$ or $\hat{\Gamma}$ and is derived from the dyadic decomposition

$$\hat{\Gamma} = \sum_{n=1}^{N} \lambda_n \cdot u_n u_n^H \tag{14}$$

with eigenvalues λ_n and orthonormal eigenvectors u_n. Here, the eigenvector to the largest eigenvalue is taken as an estimator of DS signal (abbreviation for the method: EVG). If there is more than one significant eigenvalue in analogy to polarimetry, this often is interpreted as the superposition of several scattering mechanisms. In such a case, this approach is supposed to capture the dominant scattering mechanism, while the other estimators of phase history give degraded results. The presence of more than one scattering mechanism can be detected with the help of entropy.

A third possibility to estimate DS signals from $\hat{C}$ or from $\hat{\Gamma} = (\hat{\gamma}_{mn})$ is phase triangulation coherence maximization (PTCM), as described in [82]:

$$\hat{\phi} = \arg\max_{\phi} \Sigma_{m,n} |\hat{\gamma}_{mn}|^{\alpha} \cdot exp(i \cdot (\hat{\phi}_{mn} - (\phi_m - \phi_n))).\tag{15}$$

Here, α is a positive real number, e.g., 1 or 2. Analogous to PhL, the maximum can be found iteratively with the help of the following formula:

$$\phi_p^{(k)} = \measuredangle \left\{ \sum_{n(\neq p)} |\hat{\gamma}_{pn}|^{\alpha-1} \cdot \hat{\gamma}_{pn} \cdot exp(i \cdot \phi_p^{(k-1)}) \right\}.\tag{16}$$

A related approach can be found in [83]. Although they do estimate the parameters of a model with linear deformation and DEM error and not the phase history, they also perform PTCM. An interesting difference is that they consider a more general situation, where the summation does not necessarily take over the full set of all possible interferograms, but over graphs that are for each target individually optimized. They state that, "the links of a complete graph are not necessarily all informative" and argue that different decorrelation mechanisms require different graphs. For example, in the same scene, one DS might be mostly sensitive to perpendicular baselines (debris), while another is afflicted strongly by temporal decorrelation (sparse vegetation). A third might display a seasonal dependence (changes in vegetation or occasional snow cover). As a rule, to construct such a graph, they suggest commencing with a spanning tree with edges of maximal coherence and to complement it with all edges having coherence larger than a threshold. A similar idea was presented by [40], who, under the designation improved EMCF-SBAS processing, also applied PTCM over an optimized graph to estimate phase history. Starting from a reduced Delaunay triangulation in the baseline plane, they optimized their triangulation with the help of a simulated annealing approach. Other than suggested by [83], the same SBAS graph was taken for all points. A very noteworthy observation of [40] is that results achieved with this optimized graph were significantly improved compared to the use of the full covariance matrix. For algorithms that apply PTCM with the full covariance or coherence matrix, these ideas can easily be adopted by simply setting the coherences to zero for interferograms that are not used. For PhL or EVG, there is no obvious way of doing this. Finally, it is an advantage that EVG and PTCM are still valid in case the coherence matrix approaches the coherence matrix of an ideal quasi-PS: $\Gamma \longrightarrow \zeta\zeta^H$ (see Section 4.1 for more on this). On the other hand, PhL is prone to diverge in this transition.

As a fourth method, an estimator using a weighted integer least squares (ILS) approach has been introduced [46,84] that solves for the integer ambiguities to unwrap the phase. It searches a solution for the following problem:

$$E[\hat{\phi}_{mn}] = \begin{cases} \phi_m & n = master \\ -\phi_n & m = master \\ \phi_m - \phi_n + 2\pi \cdot a_{mn} & otherwise \end{cases}\tag{17}$$

where $a_{mn} \in \{-1, 0, 1\}$ is an integer. This can be reformulated for suitable arrangement of $\hat{\phi} = (\hat{\phi}_{mn})$, $a = (a_{mn})$ and $\phi = (\phi_n)$ and with appropriate matrices A and B as

$$[\hat{a}, \hat{\phi}] = \arg \min_{a, \phi} ||\hat{\phi} - 2\pi \cdot A \cdot a - B \cdot \phi||_W^2 \tag{18}$$

where the constraints $\phi_n \in [-\pi, \pi)$, $\phi_{master} = 0$ and $a_{mn} \in \{-1, 0, 1\}$ have to be obeyed. W is a weight matrix. In case of normally distributed data, the inverse of the covariance matrix would be a natural choice for W. However, as phases are far from being normally distributed, other options might provide better estimators. Nevertheless, Samiei-Esfahany [46] derives an approximation to the covariance matrix Q_ϕ of interferometric phases of a DS pixel:

$$(Q_\phi)_{ij,kl} = cov[\varphi_{ij}, \varphi_{kl}] \approx \frac{|\gamma_{ik}||\gamma_{jl}| - |\gamma_{ij}||\gamma_{kl}|}{2L|\gamma_{ij}||\gamma_{kl}|}. \tag{19}$$

For this formula, he demonstrates, with the help of Monte Carlo simulation, that it provides a good approximation if the number of looks L is >50 and a better approximation than a formula derived earlier based on simpler assumptions [59,60]. Beside the inverse of the approximated covariance matrix, he considers for W the diagonal matrices with coherences $\hat{\gamma}_{mn}$ respectively with the Fisher information index $2L\hat{\gamma}_{mn}^2 \cdot (1 - \hat{\gamma}_{mn}^2)^{-1}$ as entries. His experiments with simulated data for an exponential decay and a seasonal decay scenario show best results for the Fisher information index. For these two scenarios, he also performs comparisons between PhL, EVG, PTCM, and ILS. Best results were achieved for PTCM and ILS. PhL performed distinctly worse than the other estimators. This is due to the small 5×5 search window, which leads to an imprecise estimation of $|\hat{\Gamma}|$ and corresponding problems with its inversion. Furthermore, for both scenarios, experiments with ILS plus Fisher info are performed with true and estimated coherences as well as the complete graph and a small baseline graph. For the complete graph, standard deviations double for estimated compared to true coherences, while those for the small baseline graph are very similar. In the exponential decay scenario, the results for the small baseline graph are distinctly better than for the complete graph, and for the seasonal scenario, it is vice versa. In an experiment with real data, ILS outperforms StaMPS. ILS performs very well but has the drawback of high computation time. Finally, a big advantage of ILS is that it provides quality control via the covariance matrix for the estimated phase history:

$$Q_{\hat{\phi}} = (B^t W B)^{-1} B^t W Q_\phi W B (B^t W B)^{-1}. \tag{20}$$

In [85], the authors introduce their concept of Joint-Scatterer InSAR (JSInSAR). They estimate a covariance matrix from blocks of pixels, that is of dimension $PN \times PN$, where P is the size of a patch in the spatial domain. By requiring that the signal and the noise space obtained from the covariance matrix are orthogonal, they derive an expression that must be a minimized analog of that occurring during PhL to find the phase history.

An independent approach with the name Multi-Link SAR has been developed in [86]. The idea is to improve multilooked interferograms. For two acquisitions in the interferogram graph, the paths connecting them are integrated and weighted. The result of the integration is an estimation of the phase for the interferogram between these two acquisitions obtained by adding up the multilooked phases of the consecutive interferograms. It is demonstrated that in case all these phases are reliable, e.g., because they have a short baseline, this wrapped sum of phases is for problematic interferograms a significantly better estimate than the original multilooked phase. The weighted sum of these integrals serves to replace the original wrapped phase. The weights reflect the reliability of the integrated phases and are obtained based on the quality criterion colinearity introduced by the authors. Results from simulated data demonstrate that colinearity measures phase errors significantly more reliably than coherence [87] for a 3×3 estimation window on multilooked data.

As already noted by [46], the estimators for DS signal PhL, EVG, PTCM, and JSInSAR explained and discussed in this subsection can be interpreted as special cases of the following general estimation approach:

$$\hat{\phi} = \arg \max_{\phi} \Sigma_{m,n} w_{mn} \cdot exp(i \cdot (\hat{\phi}_{mn} - (\phi_m - \phi_n))). \tag{21}$$

$W \in \mathbb{R}^{N \times N}$ is a symmetric weight matrix (depending on the DS). If phases $\hat{\phi}_{mn}$ are only available for certain interferograms, as in SBAS approaches, the corresponding weights are set to zero. Indications that this can be advantageous have been reported. The estimators named before weights will be nonnegative, with the exemption of PhL, where negative weights might occur. The merit of this formulation, and this is likewise true for ILS, is that it is obvious that anyhow filtered wrapped interferogram phases can be triangulated, while weights are a steering quality. Notwithstanding this very general formulation, in all the cases discussed here, weights can be calculated from the scatter matrix $\hat{C}$. Consequentially, the next section will review (phase) filtering of interferograms and coherence estimation with regard to the purpose of preprocessing DS.

3.2. Filtering of Interferograms and Coherence Estimation

In the preceding subsection, different possibilities for estimating a DS signal for preprocessing of InSAR data stacks were presented. The required input to all these estimators consisted in interferogram phases and weights, where phases were filtered and weights were derived from coherence, or more generally, from the scatter matrix. In the current subsection, it will be studied how these can be obtained from techniques that either are applied separately to each interferogram or work on the stack. Some basic facts were already addressed in the section on DS statistics: estimators of scatter matrix, sample coherence, and the MLE of [14]. Furthermore, it was reported on intrinsic biases of estimators, the consequences of heterogeneous data and biases caused by nonstationary phases. Now, methods will be discussed that have been developed to deal with these issues and to get the best out of the data.

3.2.1. Nonstationary Phases

In this section, approaches will be addressed for dealing with the presence of nonstationary phases during preprocessing of an InSAR data stack. We assume that the synthetic phase has already been removed [8]. There are approaches that implicitly handle nonstationarity and such that estimate the interferogram phase explicitly for correcting the bias in coherence estimation. Examples for implicit approaches will be given in the section on nonlocal methods (e.g., InSAR-BM3D). The explicit approach occurs in several variants. Either it is applied separately for each DS or it is realized on the interferogram or stack level and passes through the following steps:

1. denoising of the phases and correction of interferograms;
2. estimation of covariance or coherence from the corrected interferograms;
3. adding back denoised phases to covariances;
4. DS signal estimation.

This approach is compatible with most of the methods developed for interferogram filtering by the InSAR community during the last 20 years. Examples are [16] (several suggestions, e.g., MUSIC; applied in [88]), Goldstein, et al. [17] that works in the frequency domain, Davidson, et al. [89] an adaptive multiresolution defringe algorithmus (e.g., applied in [90]), a modification of the filter of Goldstein and Werner that reduces overfiltering by adapting the parameters to coherence [91], [62] was mentioned before, a combination of the filter of Goldstein and Werner with a narrow low-pass filter iteratively applied in StaMPS [28], or [92] that is devised for frequency estimation on adaptive neighborhoods (cp. IDAN in the section on grouping of statistically homogeneous neighborhoods).

An approach for InSAR stacks that works DS-wise and is based on a model describing the totality of local phase ramps at the DS position in all interferograms caused by DEM errors was given in [93]. Slopes in range and azimuth are estimated from a sum of periodograms over the interferograms.

Each periodogram is calculated on the pixels of the adaptive neighborhood corresponding to the DS. This approach was extended to gradients in the deformation field in [94] monitoring. They point to the importance of including periodograms calculated for interferograms with large baselines for the precision of this approach.

In [44], likewise, local phase ramps at the DS position are estimated, but in each interferogram separately. The fringe frequency is obtained as the position of the peak after FFT with optimal window size. The optimal window size is defined to result in minimal mean phase standard deviation.

3.2.2. Grouping of Statistically Homogeneous Neighborhoods

Heterogeneous data are the rule. The use of all pixels in a rectangular window entails the dilemma of either using a small window and hoping that homogeneity is thus achieved or taking a larger window, which would lead to precise estimation if the statistical assumptions remained valid but often spoils the result by including unsuitable pixels. Therefore, it is an important question how to build up effectively so-called adaptive neighborhoods (AN) that have variable shape but are statistically homogeneous. ANs seem to have been used for the first time for speckle filtering of multitemporal InSAR imagery in [95]. The authors report to be inspired by the use of ANs in other fields of image exploitation [96]. It is also noteworthy that they already sought for a proper 3D-neighborhood, by which is meant that although a pixel is included in the neighborhood, some of its values corresponding to certain channels (polarimetry) or points in time (multitemporal InSAR) may be excluded. From Lee's sigma filter [97], they borrowed the idea of checking if the amplitudes of neighbors of the pixel to be processed have less than two standard deviations difference from the processed pixels amplitude. As the speckle effect in SAR imagery behaves like multiplicative noise, some modifications to this approach developed for additive noise have been introduced. In particular, a region growing in two steps proved to be seminal. The idea is to apply first a stricter criterion (confidence interval for amplitudes at level 50%) in order that the region does not grow into statistically unsuitable areas. Furthermore, this neighborhood provides a larger sample that allows for a more precise re-estimation of mean and standard deviation, which are used to define the confidence interval used during the second step. Here, pixels in gaps and at the rim of the region of the first step are added to the region if their amplitudes fulfill a weaker criterion (amplitudes are contained in a larger confidence interval at level 95%). This approach was adopted also from other researchers. For the intensity-driven AN (IDAN) technique, [98] also let the region grow in two steps. However, they do not compare the value of a pixel corresponding to a channel or to a point in time with another but compare the vectors assigned to the two pixels. This might have been an inspiration for the authors of [35], where the application of two-sample tests for the amplitudes of the pixels is advocated. In particular, they introduce DeSpecKS, where the Kolmogorov–Smirnov two-sample test (KS) is used. As a second example, they name the Anderson–Darling two sample test (AD). They do not build up a region in two steps but test every pixel in a search window versus the center pixel. The accepted pixels form the AN. Finally, pixels not belonging to the connected component of the center pixel are discarded in order "to increase the probability that nearby pixels belong to the same radar target and share the same geophysical parameters". In [99], four two-sample tests are compared: generalized likelihood ratio test (GLRT) for the scale parameter of the Rayleigh distribution, AD, KS, and Kullback–Leibler. The best detection rates in different simulation scenarios were achieved for GLRT and AD. In particular, GLRT performed best when Rayleigh-distributed amplitudes or different scale parameters for the K-distribution were simulated but was third when the shape parameter of the K-distribution was varied. KS was somewhat inferior to AD. Kullback–Leibler performed always worst. The subjective impression from results of filtering real data with GLRT and AD is that GLRT has a more confetti like appearance. These results have convinced several authors [42,58,90] that AD is the better test to be used for forming an AN. Its superiority over KS is explained with the higher sensitivity towards big amplitudes. However, findings of [100] show that this advantage is lost in case an outlier removal was performed before the tests. Furthermore, as outlier removal is highly recommended and KS is faster, there is a little advantage for KS. Another small advantage of KS is that critical values can be precisely calculated by a simple recursion that also works for samples of different sizes [101] (Section 6.3), while for AD, usually an approximation described in [102] is used. On the other hand, critical values for KS come in discrete steps, which for small sample size and high significance level restricts the possible choices. [103] again grew a region requiring the relaxed criterion and then applied k-means clustering to separate the homogeneous neighborhood of the center pixel from unsuitable pixels. In [104], a new approach was taken for preventing running into unsuitable areas. The idea is to replace the noisy stack of SAR amplitudes by a denoised extract of its information. To this purpose, a new image is generated. The vector of amplitudes is projected to the main principal component of the covariance matrix for amplitudes calculated by averaging over all pixels. The result is an image that gets denoised in a further step. The denoised image is now the basis for determining the neighborhood of a center pixel by thresholding on the square of the difference of image value of the center pixel and the other pixels inside a search window. The advantage is faster processing.

In [105], the authors introduced a criterion for similarity that also makes use of phase information. In a small neighborhood of the center pixel, the covariance matrix gets estimated with the MLT. This allows for checking the other pixels in the search window. A pixel is accepted if a certain threshold on the probability density corresponding to the estimated covariance matrix is exceeded. All these approaches continue from here the same way. The four- or eight-connected component of the neighborhood containing the center pixel is taken to estimate the SaCM. This step is carried out in order to enhance the probability that all pixels of the adaptively chosen neighborhood actually belong to a homogeneous area. An analysis of results in [105] demonstrates that the probabilistic method performs best for small stacks up to 16 images when compared with boxcar multilooking, DeSpecKS, or PCA-TV (the method of [104]). DeSpecKS proves even inferior to boxcar multilooking in this study. If applied to a single interferogram, its results are comparable to the NL-InSAR filter of Deledalle [106], discussed in the section on nonlocal methods.

In [44], the author proposes two different algorithms for forming an AN, introducing important new ideas. A third is suggested in [107], which aims at fast processing. The first proposed algorithm starts with a classification of pixels based on their amplitudes. A boxplot approach is used to detect and remove outliers and afterwards determine the skewness and tailweight of the pixels. These characteristics are decisive for an adaptive two-sample test (ADT). They serve to select the appropriate test that decides over the statistical similarity of the two pixels compared. The pixels statistically similar to the center pixel and in its connected component form the AN. The ADT scheme has been developed starting from a set of candidate tests with the help of simulated data in order to compile an optimal configuration. Regarding the power of the test, it is demonstrated that the ADT significantly outperforms nonadaptive tests (KS, AD, Wilcoxon–Mann–Whitney) for several scenarios. The second proposed algorithm provides a solution for the problem of low test power for small data stacks. To this purpose, the number of available samples is enlarged by considering all amplitudes of all pixels in a little neighborhood of each of the two pixels to be compared. The little neighborhood is chosen among 8 directed windows containing 15 pixels each (as suggested in [108,109]) to be the one with the smallest coefficient of variance of amplitudes. For the chosen directed window, amplitudes that lie outside a relaxed confidence interval are discarded. The remaining samples are compared with a differently set up ADT adapted to more strongly varying sample sizes. The third proposed algorithm makes use of the observation that the mean of amplitudes of a pixel (in a multilook image) over time is approximately normally distributed according to the central limit theorem for sufficiently big stacks (e.g., $N \geq 10$). For the mean amplitude image, an AN is grown with the help of a two-step procedure like the one described above. The definition of confidence intervals uses an estimation of the equivalent number of looks from the data cleaned from outliers with help of the adjusted boxplot [110].

In this section, different possibilities of forming neighborhoods have been discussed that serve for phase estimation (usually via the argument of the complex coherence). Most approaches proceed in two steps: first, a conservative estimation in order to prepare a more precise second one. The shape of the neighborhood developed from boxcar to ANs to 3D-ANs. Grouping criteria were applied to values adjacent, either in space or time, to pixels or to blocks of pixels. They were based only on amplitudes or also considering phase. Region growing was applied or all blocks inside a search window were included that are similar to the center block. Then, several years ago, the next level of generality has been entered. Approaches were introduced, where patches (in case of a single interferogram) or blocks (in case of stacks of interferograms) are not used for grouping but rather for weighting. The use for InSAR of the so-called nonlocal methods will be the subject of the next section.

3.2.3. Nonlocal (NL) Methods

The origin of nonlocal (NL) methods for image denoising is the NL means algorithm for optical data introduced in [111]. The name-giving basic idea is to obtain the denoised pixel value as a weighted sum over all pixel values in the image (or in a not too small search window). The weights are computed from the distance between the vectors of the pixel values in a small patch around the pixel to be denoised and the vector of pixel values of the patch shifted to the pixel to be weighted. [111] argues that under the assumption of additive white Gaussian noise, the weighted Euclidean distance has desirable statistical properties. Their approach already comprises the three basic steps characteristic of the NL methods discussed in this section:

1. for each pixel to be estimated, a patch is shifted around and a similarity measure (based on the statistical characteristics of the data) is calculated for every position; for multichannel data, it can be a 3D block instead of a patch;
2. weights are computed from the calculated similarity measure;
3. a weighted mean or a weighted MLE provides the result.

The weighted mean is generalized to a weighted maximum likelihood approach in [112], where weights are defined via probability of patch similarity given a noise model (probabilistic patch-based (PPB) filter). In particular, they derive weights applicable for speckle noise in SAR images based on the Nakagami–Rayleigh distribution. Furthermore, an iterative application of PPB is suggested using the result of the previous iteration as a prior. The same authors extend their approach in [106] to InSAR data (named the NL-InSAR estimator), obtaining estimations of reflectivity, phase, and coherence. Weights are now defined under assumption of zero-mean circular Gaussian distribution, with patch similarity making use of amplitudes as well as phases. Comparisons of simulated data with the boxcar, the refined Lee [108,109], the IDAN, and the noniterative NL-InSAR estimator demonstrate a better bias-variance trade-off and better signal-to-noise ratio of the iterative NL-InSAR estimator. Likewise, the subjective impression from comparisons on simulated and on real data of the same estimators indicates a superior performance of the iterative NL-InSAR estimator. Similarity criteria for patches were studied systematically for different types of noise in different types of imagery, including InSAR data, but also X-ray, in [113]. The finding was that the generalized likelihood ratio test is the best basis for defining patch similarity criteria among the numerous investigated alternatives. Building on this, a survey is given in [114] on patch-based nonlocal filtering of SAR imagery (speckle filtering, InSAR, PolSAR, PolInSAR), e.g., estimation of covariance matrices for multitemporal InSAR is discussed. Finally, a framework for nonlocal filtering of SAR imagery (NL-SAR) is presented in [115] that displays several new features. It is adaptive to scale and contrast of local structures by trying multiple parameter settings and automatically choosing locally the best suited one. With the help of the empirical cumulative distribution function of the dissimilarities determined on a homogeneous region selected by the user, the weights are defined in a way such that they are independent of choice of patch size, scale of averaging, number of looks, and number of channels. Following the strategy of the local linear minimum mean square estimator (LLMMSE), the weighted mean of SaCM and the NL estimate of the covariance is calculated to debias the covariance matrix. The criterion for automatic selection of parameters is the maximum equivalent number of looks calculated in a way that respects the debiasing step. Comparisons among IDAN, refined Lee filter, and NL-InSAR for several data sets demonstrate the superiority of NL-SAR. Examples prove that each of the newly introduced improvements is necessary to achieve this success. In particular, the occurrence of the "rare patch" effect can be avoided by adaptive parameter settings. It consists of large local variation, where, for a unique structure, only few similar partner patches are found. If the patch size is optimized to find many partners, it is chosen in such a way that the unique structure is not contained if possible. Thus, the surroundings of the unique structure are smoothed and show no artefacts. Also, for filtering of speckle and PolSAR data, NL-SAR obtains better results than the techniques used for comparison. Open source code for NL-SAR is available (see [115]).

The potential of NL filtering for SBAS processing was investigated in [76]. To limit the computational effort, the algorithm was kept simple. Amplitudes were despeckled. For these three variants were tested: not despeckled, boxcar, and SAR-Block Matching 3D (SAR-BM3D, cp. [116]). The similarity measure was calculated for pairs of pixels based on their vectors of despeckled amplitudes and the filtered stack was obtained as a weighted mean. Among the studied similarity measures were KS and a probabilistic distance based on the assumption of multiplicative noise. The latter, unlike KS, depends on the succession of values over time. The clear winner of the comparison on synthetic and real data was the combination SAR-BM3D plus probabilistic distance.

InSAR-BM3D is introduced in [117] (remark: block is here synonymous to patch). Processing runs through two passes. In the first pass, a basic estimate is obtained that serves to steer the filtering during the second pass. Both passes consist of three steps: During grouping, similar patches are collected to a stack. This stack is filtered considering intra- and inter-patch dependencies (collaborative filtering). Each pixel in the image is now contained in multiple filtered patches from different stacks. During the aggregation step, the final value for the pixel is calculated as the weighted average over all these patches. During the first pass, collaborative filtering involves a hard threshold that during the second pass is replaced by Wiener filtering based on the statistics of the result from the first pass. As adaptations for InSAR data, the real and imaginary part of the interferogram are transformed to decorrelate their noise. The transforms are filtered and the result is transformed back. Together with the phase, an estimate of coherence is obtained. The coherence is calculated such that identical phase patterns in the reference and the partner patch cancel out, thereby preventing bias caused by phase gradients. Comparisons among boxcar, some version of Lee filter, Goldstein–Werner, NL-InSAR, and NL-SAR are performed on several simulated and real data sets. InSAR-BM3D proves superior on simulated data and shows good results on real data. On real data, the method noise seems almost white, while for NL-InSAR and NL-SAR, artifacts are visible. Subjectively, the Goldstein–Werner filter gives the best results on real data but was inferior on simulated data at higher noise levels to InSAR-BM3D. NL-InSAR and NL-SAR have problems in recovering the simulated phase fields, while Goldstein–Werner and InSAR-BM3D perform this task much better. The executable code and simulated data are available (see [117]).

An interesting new option is proposed in [118] under the name multichannel logarithm with Gaussian denoising (MuLoG). It transforms the field of sample covariance matrices of a stack of multichannel SAR data in such a way that denoising algorithms for additive white noise are applicable. After the transform of the denoised data backwards, filtered covariance matrices for the SAR data are available. Comparisons of this approach with two transforms, different Gaussian denoisers, and NL-SAR demonstrate that NL-SAR better preserves details and contrast but is a bit less smooth in homogeneous areas. The Gaussian denoiser TV distinctly displays artefacts. The combinations of MuLoG or homomorphic and DDID or BM3D give results of similar quality, while the homomorphic approach tends to oversmooth bright targets and MuLoG gives slightly better values of SSIM. DDID and BM3D show small oscillatory artefacts. The open source code for MuLoG is available (see [118]).

The NL methods discussed in this section count as state of the art in image filtering. Its success is often explained by the use of more intelligent prediction. The assumption of "local" methods was that similar pixels belong to the same radar target and therefore are found nearby. Often, they enforced the connectedness of the DS neighborhood to make this sure. NL methods do explicitly check for similarity and invest in its reliability by using a patch. They can search on a larger area for suitable partners and do not require connectedness, which in a richly structured area may prevent growing sufficiently large neighborhoods even though suitable pixels are available. Furthermore, weighting appears to be more efficient than deciding over membership in a neighborhood (this is a feature shared with robust estimators of scatter). Hence, more pixels contribute to the result and make it more reliable. Nevertheless, it still seems miraculous that enough similar patches are found and that averaging with them improves results even if they cannot be ascribed to the same physical phenomenon. However, the success of these approaches indicates that this requirement is often fulfilled.

3.2.4. Bias Correction and Regularization

As mentioned in the section on DS statistics, $|\hat{\gamma}_{mn}|$ is a biased but asymptotically unbiased estimator for coherence magnitude. Also, as coherence approaches one, the bias tends versus zero. For small coherences and a small number of looks, values are overestimated. Correcting for this bias is an important task because the quality of many InSAR applications depends on precise values of coherence. For estimation of DS signal, it has for all DS with a small to medium number of pixels an adverse effect as soon as large baselines occur in the stack.

In [67], several methods of coherence estimation with bias correction were investigated. The first step is estimation of the complex coherence $|\hat{\gamma}_{mn}|$, e.g., as sample coherence or as mean over a sample from a coherence map estimated for a certain number of looks (e.g., L = 20). The latter is a nearly unbiased estimator. The second step makes use of the analytic expression for the expectation value of $|\hat{\gamma}_{mn}|$ in dependence of the number of looks and true coherence. The unbiased estimation is that value of true coherence which has the expectation value $|\hat{\gamma}_{mn}|$. Unfortunately, the standard deviation of this estimator is significant for small number of looks, a situation where debiasing is most needed. [31] (p. 42) comments on the difficulty of obtaining a positive definite covariance matrix from this approach. In [119], several methods for bias correction were compared. For simulated Gaussian data, bias corrections with log-sample coherence (cp. [120]) and double bootstrapping were able to mitigate bias, while double bootstrapping was more effective. For simulated contaminated Gaussian data with true coherences in the range 0.5–1.0, bias corrections with double bootstrapping were very effective, although the bias was now towards lower values. The bias correction of the second method of [67], as explained before, decreases coherence, further making things even worse. Furthermore, an experiment was performed with ASAR and TSX data sets of a scene where large homogeneous areas of different types were contained, having a different parameter α. Again, double bootstrapping mitigated bias more effectively than the method of [120]. Moreover, the performance of double bootstrapping proved less dependent on α. In conclusion, double bootstrapping proved the most accurate among the investigated estimators. Unfortunately, it is computationally quite expensive. Because of that, the jackknife was investigated as an alternative [121]. It proved to be approximately 30 times faster. An experiment with simulated data and true coherence values 0.2 and 0.6 demonstrated almost perfect debiasing for sample sizes bigger than 20. Furthermore, ADT plus jackknife lead to a distinctly better signal-to-noise ratio than ADT alone or DeSpecKS. A coherence image from real data generated with DeSpecKS seems blurred compared to ADT plus jackknife.

Another strategy in case of a small sample size is not to debias each $|\hat{\gamma}_{mn}|$ separately but to improve on the estimated coherence matrix. In [108], the local linear minimum mean square estimator was given for multiplicative noise. For each pixel, coefficients for a convex combination of mean signal and signal of the pixel are determined that minimize the mean square error of estimation of the noise free signal. These coefficients are also used to obtain the estimation of the covariance matrix as a convex combination of SaCM and a dyad of the pixel signal, thus constituting a shrinkage estimator. This idea is used until today, e.g., in IDAN, SAR-BM3D, NL-SAR, and in a wide sense, also in InSAR-BM3D. Similar to the approach of [108], the same starting point was taken by [122], where the well-known Ledoit–Wolf estimator has been introduced. No specific assumption on the probability distribution is required, only that fourth order moments are finite. They also give several interpretations of the minimum mean square approach, e.g., as a trade-off between bias and variance. An extension of shrinkage estimators of the SaCM (also termed general linear estimation estimators) were developed in [53] as an alternative for regularized M-estimators of the scatter matrix in situations with insufficient sample support. That is where the inverse of the SaCM cannot be computed or is poorly conditioned and hence robust estimators of the scatter matrix cannot be applied. Regularized M-estimators of scatter share with M-estimators of scatter the disadvantage of a computationally expensive iterative calculation involving the repeated inversion of the scatter matrix.

3.3. Quality Numbers for Distributed Scatterers for Preprocessing

A prerequisite for the successful use of preprocessed DS is to be able to assess the quality of the estimated signal. Remember that the phase standard deviation is a function of the coherence magnitude and the number of looks [1]. In [35], phase triangulation coherence was introduced as a measure of successful phase triangulation:

$$\gamma_{PTA} = \frac{1}{N(N-1)} \sum_m \sum_{n(\neq m)} e^{i \cdot (\phi_{mn} - (\hat{\phi}_m - \hat{\phi}_n))}. \tag{22}$$

Although this is a measure of goodness of fit, it is rather improbable that a very high γ_{PTA} corresponds to a meaningless signal. It should be used in combination with other criteria, e.g., requiring a minimum number of samples. In [46], this approach was taken. He required in one experiment $\gamma_{PTA} \geq 0.7$ and a number of samples ≥ 50, a and in a second $\gamma_{PTA} \geq 0.4$ and a number of samples ≥ 25 for DS candidates. γ_{PTA} can be sharpened by weighting the phasors with the coherence magnitudes (cp. Equation (20) in [33]):

$$\gamma_{PTAw} = \frac{\sum_m \sum_{n(\neq m)} |\hat{\gamma}_{mn}| \cdot e^{i \cdot (\phi_{mn} - (\hat{\phi}_m - \hat{\phi}_n))}}{\sum_m \sum_{n(\neq m)} |\hat{\gamma}_{mn}|}. \tag{23}$$

In [123], those signals are accepted as DS that have a mean coherence magnitude larger than 0.25 (4×20 looks). This measure is also used in [76].

In [124], those are accepted that have coherence magnitude larger than 0.15 in at least 60% of the interferograms (64 looks).

In [42,125], a minimum average coherence and minimum number of samples were used (e.g., 0.3 or 0.4 and 20).

In the context of multitemporal polarimetric InSAR, [126] suggest establishing a common quality criterion for DS and PS measuring phase standard deviation. In both cases, it can be approximately calculated: In the case of PS for small values, the phase standard deviation is approximately equal to the amplitude dispersion [23]. For DS, they use an approximation depending on coherence magnitude and the number of looks. Coherence magnitude is replaced by the average coherence magnitude and number of looks is calculated as the number of DS pixels divided by the oversampling factors in range and azimuth.

In [58], a low coherence situation is given. Therefore, the authors calculate from the formulas for expectation and standard deviation of coherence magnitude the corresponding values for coherence magnitude zero. The sum serves as threshold for DS selection.

One should be aware that thresholds suited for an SBAS framework might have to be adopted if all interferograms are used.

3.4. Algorithmic Approaches to Reduce Run Time

Preprocessing of DS is computationally very expensive. Therefore, it is necessary to optimize algorithms for better utilization of computing resources. Besides basic improvements like parallelization, there is also the possibility to modify the formulation of the task. The crucial property of a DS that can be used to achieve some time savings is its spatial extension. Given a coarse mask, either derived from the data themselves or from GIS, areas where no DS can be expected are annotated and do not have to be processed (water, forest, shadow, layover, etc.). While a PS consisting of a single pixel can hide in the forest, a DS necessarily consists of a larger group of pixels and cannot. Likewise, not every pixel belonging to the neighborhood determined for a DS must be processed on its own. An approach that uses a raster, where each cell at most contains one DS, was given in [93]. Later, this algorithm was completed by fitting a smooth deformation field to estimations [94]. Also, NL methods could be adapted for the use of DS processing. If the reference patch is recognized as inhomogeneous, it needs not be processed.

Sentinel-1 and the future missions NISAR and Tandem-L with wide swaths and short revisit times will provide huge data volumes. In addition, near real-time monitoring has been defined as a future objective, e.g., for use in early warning systems. To answer to this challenge, the Sequential Estimator [127] has been developed. Long time series are subdivided in ministacks that are sequentially processed. A compression method allows for representation of the information of each ministack needed for further processing in artificial interferograms. This results in an impressive reduction of computing operations without significant loss of quality and even displays a more balanced performance than conventional estimators in two scenarios (fast exponentially decaying and long-term coherence) with simulated data.

4. Phase Model Parameter Estimation for Distributed Scatterers

This section is devoted to an experiment that proves that parameter estimation from preprocessed DS provides significantly improved results if statistical information available for the DS is used. The modeled phase accounts for linear deformation rates and DEM errors.

4.1. Estimators of Model Parameters

A big advantage of estimation of DS signals is that a start net can be built up containing DS as well as PS. This allows bridging gaps between PS by DS. Phase histories of DS can be used as if DS has been transformed to PS. Nevertheless, DS are not PS and have other statistical properties that still matter after preprocessing is finished. The experiments with simulated data described in this section show that using the additional information (covariances, amplitudes) available for DS allows to obtain better estimates of model parameters, here, linear deformation rates and DEM errors, for DS–PS pairs and DS–DS pairs. To formulate the new approach, some notation is needed. It is assumed that the signal of the PS can be written as $p = c \cdot \xi$, $c \in \mathbb{R}_{>0}$. By abuse of terminology, we write in the case of a PS $\hat{C} = p \cdot p^H$ and $\hat{\Gamma} = \xi \cdot \xi^H$ to achieve a uniform notation for PS and DS. This is close to what [31] (p. 52) named quasi-PS, only that the noise is omitted. The trick here is that a zero mean Gaussian random vector with nonzero variance and covariance matrix of rank 1 is the same as a one-dimensional zero mean Gaussian random variable times the (nonrandom) eigenvector of the covariance matrix with an eigenvalue greater than zero. With M as the model matrix, ϑ as the parameter vector, and $\phi(\vartheta) = M \cdot \vartheta \in \mathbb{R}^N$:

$$\eta = \begin{pmatrix} e^{i\phi_1(\vartheta)} \\ \vdots \\ e^{i\phi_N(\vartheta)} \end{pmatrix}. \tag{24}$$

Given a pixel pair with matrices $\hat{\Gamma}_1$ and $\hat{\Gamma}_2$, the model increments can now be estimated by

$$\hat{\vartheta} = \max_{\vartheta} \eta^H \cdot \left(\hat{\Gamma}_1 \circ \overline{\hat{\Gamma}_2} \right) \cdot \eta. \tag{25}$$

In the case of two PS, the estimate is the same as with the periodogram. This estimator will be denoted as pair-PTCM (pPTCM). Another estimator of model parameters that is of interest is

$$\hat{\vartheta} = \min_{\vartheta}(z_1 \circ \overline{z_2} \circ \eta)^H \cdot \left| \hat{C}_1 \circ \overline{\hat{C}_2} \right|^{-1} \cdot (z_1 \circ \overline{z_2} \circ \eta) \tag{26}$$

where z_1 and z_2 are the complex signals of the two pixels of the pair as estimated during preprocessing. It can be considered a $|\hat{C}|^{-1}$-weighted periodogram (wPdg). In the case of DS, it is supposed that the signal z has been estimated during preprocessing with some estimator of the DS signal, e.g., the eigenvector $\hat{z}$ to the largest eigenvalue of the covariance matrix $\hat{C}$. For a PS–DS pair, this corresponds to the estimator introduced in [30] for a single pixel, only that the authors did use the original signal from the center pixel of the DS and not an estimated signal. In case the true covariance matrix is used, the latter is the ML estimator:

$$\hat{\vartheta} = \min_{\vartheta}(\overline{z} \circ \eta)^H \cdot |C|^{-1} \cdot (\overline{z} \circ \eta). \tag{27}$$

The use of the original pixel phase by [30] is a crucial difference to the other estimators explained here. In [31] (p. 77), it was remarked that this prevents compromising the resolution. An opposed view is that all pixels of the DS neighborhood share the same phase history (plus re-added fringes if necessary). Any adverse effects caused by wrongly grouped pixels or because of imprecise estimation of fringes are estimation errors but do not pertain to resolution. A further development that retained the use of the original pixel phase is the RIO estimator of [45,128]. It has the interesting feature of providing a robust estimation of $|C|$ also for nonstationary data without needing a prior estimation and subtraction of residual fringes.

4.2. Results of Investigations on Simulated Data for Parameter Estimation from Pixel Pairs

In this section some tests with simulated data are described that were run with the goal to compare performance of some of the estimators introduced before, in particular regarding estimation of parameters from pixel pairs. First, the simulated data are described. Afterwards, tests and their results are presented and discussed.

The data are simulated based on acquisition parameters of a stack of 26 TSX high-resolution spotlight-mode images from the town of Lüneburg in Germany that is available to the scientific community via ISPRS. The basic model used for the coherence matrix is the following (cp. [12,56]):

$$c_{kl} = \gamma_0 \cdot exp\left(-\frac{|t_k - t_l|}{\tau} \right) \cdot max\left\{ 0,\ 1 - \frac{|B_k - B_l|}{B_{crit}} \right\} \tag{28}$$

where γ_0 accounts for noise and processing artefacts, t_k are the acquisition times, τ is a parameter describing temporal deccorelation, B_k are the perpendicular baselines, and B_{crit} is the critical baseline. In some of the simulations, the covariance matrix was modified by the introduction of one or two snow days, i.e., for the corresponding acquisition dates, all nondiagonal coherences were multiplied by 0.25. Furthermore, we defined a seasonal model to complement the basic model:

$$c_{kl} = \gamma_{kl}^{season} \cdot exp\left(-\frac{|t_k - t_l|}{\tau} \right) \cdot max\left\{ 0,\ 1 - \frac{|B_k - B_l|}{B_{crit}} \right\} \tag{29}$$

where for given $\gamma_0 = (A + B)^2$ and $\gamma_{min} = (A - B)^2$

$$\gamma_{kl}^{season} = \left(A + B \cos\left(\frac{2\pi t_k}{365} \right) \right) \cdot \left(A + B \cos\left(\frac{2\pi t_l}{365} \right) \right). \tag{30}$$

Note that the coherence matrix remains positive definite after introduction of γ_{kl}^{season}. The seasonal model is intended to capture a situation where a good DS is periodically deteriorated by correlation, e.g., debris or enduring parts of low vegetation partly covered by grass or leaves in the growth phase.

In all simulations, complex circular normally distributed data with a given covariance matrix were generated and superposed with a phase history corresponding to some linear deformation and some DEM error. Additionally, in several cases, the data were contaminated by replacing a certain percentage of values by complex circular normally independently distributed numbers of twice the standard deviation as the original data. A list of the simulation settings used for the generation of the test data can be found in Table 1. For each setting, 1000 DS were simulated.

Table 1. Settings for simulations.

γ_0	τ (Days)	Modifications
0.9	30, 45, 60, 90, 720, 1440	-
0.9	90	One snow date
0.9	60	Two snow dates
0.95	720	Seasonal model $\gamma_{min} = 0.05$
0.9	60, 720	Contaminated with 10% or 20% outliers

For tests of the PS–DS pairs, the PS signal was assumed to be constant over time. As DS in these pairs, all simulated data sets described in Table 1 were considered. For DS–DS pairs, 19 representative combinations between data sets described in Table 1 were investigated.

The first comparison that will be discussed is between two types of estimation strategies for PS–DS pairs. The older one was introduced by [30] and uses an estimate of the covariance matrix $\hat{C}$ for a $|\hat{C}|^{-1}$-weighted periodogram estimation. What is characteristic for this strategy is that it takes the unmodified signal of the DS center pixel as the input to the estimator. A more refined version of this strategy that is not included in the present comparison is the RIO estimator of [128]. The newer strategy originates in the SqueeSAR paper of 2011 [35]. Its characteristic is that during preprocessing, the signal of the DS is estimated by one of the estimators introduced earlier and replaces the original signal of the center pixel henceforth, in particular for model parameter estimation. The finding is that the first strategy as suggested by the De Zan performed distinctly worse in all tests than the second. As illustration Figure 1 displays, the histograms of error of deformation velocity estimation for three estimator combinations and for different search window sizes obtained for the data simulated for the basic covariance matrix model with $\gamma_0 = 0.9$ and $\gamma_0 = 60$ days. The second strategy is represented by the result of PhL combined with the periodogram (Pdg). As a benchmark, the combination of the two ML estimators is added.

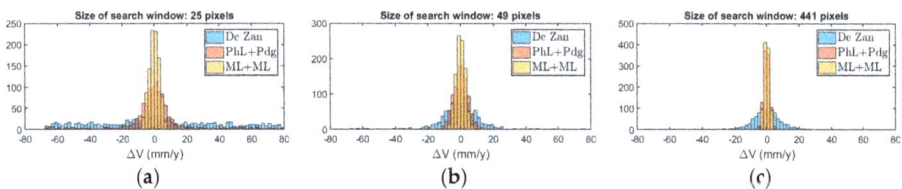

Figure 1. Histograms of error of deformation velocity estimation for three estimator combinations and for different search window sizes obtained for the data simulated for the basic covariance matrix model with $\gamma_0 = 0.9$ and $\tau = 60$ days. The search window sizes are (**a**) 25 pixels, (**b**) 49 pixels and, (**c**) 441 pixels.

The second comparison is between estimators following the second strategy. Figure 2 displays results for the given search window size for all datasets of PS–DS and DS–DS pairs as described above. The combination of marker and color identifies the combination of estimators. For each test case, the marker is plotted at the position corresponding to the medians of absolute values of estimation errors for the parameters velocity and height. Best results are achieved with the benchmark ML + ML. From the estimators applicable for real data, pPTCM performs best, followed by EV + Pdg and PTCM

+ Pdg. For a larger search window size, PhL + Pdg and PhL + $|\hat{C}|^{-1}$-weighted Pdg estimation are of comparable quality, but they fail for small window sizes. Using the 90% percentiles instead of the median confirms this assessment.

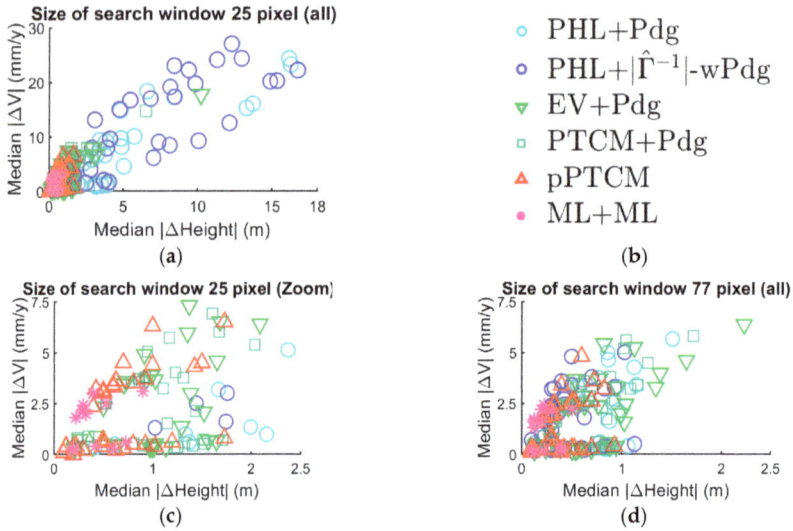

Figure 2. Comparison between estimators for all data sets of PS–DS and DS–DS pairs for two search window sizes. (a) Search window size 25 pixels, (b) legend, (c) search window size 25 pixels (zoom) and, (d) search window size 77 pixels. The combination of marker and color identifies the combination of estimators. For each test case, the marker is plotted at the position corresponding to the medians of absolute values of estimation errors for the parameters velocity and height.

The observation that estimators making use of an inverse of the covariance or coherence matrix give for small search window sizes worse results is easily brought into connection with their bad condition. However, plotting the condition number or its logarithm versus the absolute estimation error does not clearly confirm this expectation. What happens seems to be more indirectly caused by the indeed bad condition of the covariance matrices. For PhL, the coherence matrix is weighted, allowing negative numbers, with $|\hat{\Gamma}|^{-1}$. The bad condition entails that sometimes these weights are very adversely distributed. To capture this in a number, the ratio of the sum of the absolute values of the entries of $|\hat{\Gamma}|^{-1}$ in diagonals of higher order divided by the sum of the absolute values of all entries has been calculated. In Figure 3, evidence for this hypothesis is given by showing the plots of the absolute values of errors in height estimation versus these weight ratios for two examples (the main diagonal and the first secondary diagonal were spared).

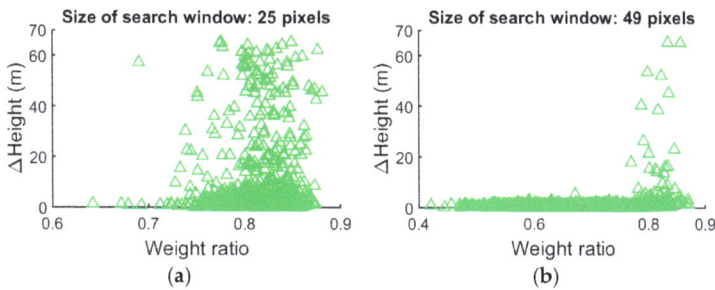

Figure 3. Absolute error in height estimation plotted versus weight ratio for different search window sizes obtained for the data simulated for the basic covariance matrix model with $\gamma_0 = 0.9$ and $\tau = 60$ days. The search window sizes are (**a**) 25 pixels and, (**b**) 49 pixels.

5. Discussion

In this review, preprocessing of DS for use in PS algorithms was explored. The extraction of DS signals from weighted interferograms, which then can be used like PS in further processing, was identified as a key concept. Because of this concept's general nature, elements from a large variety of different approaches can be combined to realize a preprocessing algorithm. Naturally, this poses the question: what would a preprocessing algorithm look like that provides optimal quality of results? For the moment, a concluding answer seems out of scope given the large number of techniques and the relative low number of comparative studies. Nevertheless, some very promising approaches have been suggested that give indications of what should be part of the solution. For the choice of the estimator of DS signal, one could make the answer dependent on circumstances:

1. large #Ω, entropy close to 0: PhL;
2. small #Ω, entropy close to 0: PTCM;
3. entropy not close to 0: EVG.

If time is not critical, ILS could be used, having the advantage of providing quality control. To estimate the DS signal, it is necessary to determine the coherence matrix or more generally phases and weights. A feasible way would be to follow [44]: use the ADT to find a 3D AN; defringe; estimate the SaCM; and debias with double bootstrapping or jackknife. However, there are many other options, e.g., for removal of fringes, there are algorithms with more evidence for good performance. InSAR-BM3D has been run on a representative selection of test cases with superior results in phase restoration. This could be the basis for an alternative. However, this approach has just been published and nothing is known about its use for deformation analysis. Moreover, although the concept of DS preprocessing via interferogram filtering plus phase triangulation allows many possible combinations of algorithms, to our knowledge, there are only a few publications concerning such an approach (cp. [36,40,76,83]). In all of these examples, presumably better results could be achieved with advanced filtering taking into account:

1. use of a proper 3D neighborhood in the sense that, although a pixel is included in the neighborhood, some of its values corresponding to certain points in time may be excluded; alternatively, a NL analog of this might be taken;
2. robust and effective treatment of fringes;
3. some bias correction or regularization.

Furthermore, NL-SAR provides coherence matrices ready to use with phase triangulation. It would be interesting to see comparisons of deformation maps generated with all these different approaches. The techniques presented in this review use various methods and it is not obvious which are to be preferred. This demands systematic comparisons with the goal of identifying best practices. A suggestion would be to define a set of standardized test cases for interferogram filtering available to everyone that allows for the assessment and comparison of the performance of algorithms in the most relevant typical situations. Steps in this direction have already been taken by the authors of [115,117,118] by providing downloadable code for their algorithms and by [117] by also providing part of their test data.

Another aspect is that for most algorithms, no theoretical framework is known that would provide quality indicators like standard deviation or covariance matrices together with the estimations. Exceptions are, e.g., PhL and ILS (see Equation (13) or (20)). It would be advantageous to have this available at least for the basic estimators.

Finally, an issue that we ignored so far in this discussion is that today's best performing algorithms with respect to precision are often not applicable to very large datasets because of enormous computing times or costly investment in computing facilities. Of course, further progress also in this direction is required.

6. Conclusions

During the last decade, different lines of development in InSAR research have converged. Today, it is common that algorithms have some capability of jointly processing DS and PS, that advanced filtering algorithms are applied, and estimators of DS signal provide triangular phases. Jointly processing DS and PS allows for more stable algorithms and increases coverage with the desired information. Triangular phases enable 3D unwrapping, which is superior to 2D unwrapping. Consequentially, one main focus of this review has been the preprocessing of DS, which enables their use in PS software without the need of further adaptation of the algorithms. In this regard, relevant work on estimation of DS signals has been discussed. It has been pointed out that this is the key that makes available the whole variety of InSAR filtering algorithms for DS preprocessing. Referring to this matter, recently developed new techniques for filtering of interferograms and coherence estimation have been presented and been put into context. Interesting possibilities for future research have been highlighted (cp. Discussion).

As second leg of this work, this review on DS preprocessing has been complemented by preliminary experiments demonstrating that statistical information on DS is still valuable for post-preprocessing. A heuristically motivated method was described for parameter estimation for DS–PS and DS–DS pairs that makes use of the coherence matrices extracted for DS during preprocessing. It was demonstrated that significantly more precise results for transformed DS can be obtained this way than are achieved when treating them as PS. A solid theoretical underpinning is lacking for the moment, but its possibilities are sure worth to be further explored.

Finally, it can be stated that the progress and success of InSAR is an ongoing story. There are many important applications in geoscience, the economy, and governance that are reflected in the investments in today's and scheduled future systems, with their tight orbit tubes, short repeat cycles, high resolution, and large swaths ensuring good conditions for exploitation of DS. Research and improvement of algorithms to make optimal use of data is of high importance also in the future.

Author Contributions: M.E. and K.S. jointly wrote the review. The experiment was conceived, performed and analyzed by M.E.

Funding: This research received no external funding.

Conflicts of Interest: The authors declare no conflict of interest.

References

1. Bamler, R.; Hartl, P. Synthetic Aperture Radar Interferometry. *Inverse Probl.* **1998**, *14*, 1–54. [CrossRef]
2. Massonnet, D.; Feigl, K.L. Radar interferometry and its application to changes in the Earth's surface. *Rev. Geophys.* **1998**, *36*, 441–500. [CrossRef]

3. Rosen, P.A.; Hensley, S.; Joughin, I.R.; Li, F.; Madsen, S.; Rodrìguez, E.; Goldstein, R.M. Synthetic Aperture Radar Interferometry. *Proc. IEEE* **2000**, *83*, 333–382. [CrossRef]
4. Hanssen, R. *Radar Interferometry: Data Interpretation and Error Analysis*, 1st ed.; Kluwer Academic Publishers: Dordrecht, The Netherlands, 2001; ISBN 978-0-7923-6945-5.
5. Perissin, D.; Ferretti, A. Urban-Target Recognition by Means of Repeated Spaceborne SAR Images. *IEEE Trans. Geosci. Remote Sens.* **2007**, *45*, 4043–4058. [CrossRef]
6. Fornaro, G.; Lombardini, F.; Pauciullo, A.; Reale, D.; Viviani, F. Tomographic Processing of Interferometric SAR data: Developments, applications and future research perspectives. *IEEE Signal Process. Mag.* **2014**, *31*, 41–50. [CrossRef]
7. Zhu, X.; Montazeri, S.; Gisinger, C.; Hanssen, R.F.; Bamler, R. Geodetic SAR Tomography. *IEEE Trans. Geosci. Remote Sens.* **2016**, *54*, 18–35. [CrossRef]
8. Gabriel, A.K.; Goldstein, R.M.; Zebker, H.A. Mapping Small Elevation Changes over Large Areas: Differential Radar Interferometry. *J. Geophys. Res.* **1989**, *94*, 9183–9191. [CrossRef]
9. Goldstein, R.M.; Zebker, H.A.; Werner, C.L. Satellite Radar Interferometry: Two-dimensional phase unwrapping. *Radio Sci.* **1988**, *23*, 713–720. [CrossRef]
10. Li, F.; Goldstein, R.M. Studies of multi-baseline spaceborne interferometric synthetic aperture radars. *IEEE Trans. Geosci. Remote Sens.* **1990**, *28*, 88–97. [CrossRef]
11. Rodrìguez, E.; Martin, J.M. Theory and design of interferometric synthetic aperture radars. *IEE Proc. F* **1992**, *139*, 147–159. [CrossRef]
12. Zebker, H.; Villasenor, J. Decorrelation in interferometric radar echoes. *IEEE Trans. Geosci. Remote Sens.* **1992**, *30*, 950–959. [CrossRef]
13. Prati, C.; Rocca, F. Improving slant-range resolution with multiple SAR surveys. *IEEE Trans. Aerosp. Electr. Syst.* **1993**, *29*, 135–143. [CrossRef]
14. Seymour, M.S.; Cumming, I.G. Maximum likelihood estimator for SAR interferometry. In Proceedings of the IGARSS, Pasadena, CA, USA, 8–12 August 1994; pp. 2272–2275. [CrossRef]
15. Just, D.; Bamler, R. Phase Statistics of Interferograms with Applications to Synthetic Aperture Radar. *Appl. Opt.* **1994**, *33*, 4361–4368. [CrossRef] [PubMed]
16. Trouvé, E.; Caramma, M.; Maître, H. Fringe detection in noisy complex interferograms. *Appl. Opt.* **1996**, *35*, 3799–3806. [CrossRef] [PubMed]
17. Goldstein, R.M.; Werner, C.L. Radar interferogram filtering for geophysical applications. *Radio Sci.* **1998**, *25*, 4035–4038. [CrossRef]
18. Zebker, H.A.; Rosen, P.A.; Hensley, S. Atmospheric effects in interferometric synthetic aperture radar surface deformation and topographic maps. *J. Geophys. Res.* **1997**, *102*, 7547–7563. [CrossRef]
19. Sandwell, D.T.; Price, E.J. Phase gradient approach to stacking interferograms. *J. Geophys. Res.* **1998**, *103*, 30183–30204. [CrossRef]
20. Wright, T.; Parsons, B.; Fielding, E. Measurement of interseismic strain accumulation across the North Anatolian Fault by satellite radar interferometry. *Geophys. Res. Lett.* **2001**, *28*, 2117–2120. [CrossRef]
21. Ferretti, A.; Monti Guarnieri, A.; Prati, C.; Rocca, F. Multi-image DEM reconstruction. In Proceedings of the IGARSS, Seattle, WA, USA, 6–10 July 1998; pp. 1367–1369. [CrossRef]
22. Ferretti, A.; Prati, C.; Rocca, F. Nonlinear subsidence rate estimation using permanent scatterers in differential SAR Interferometry. *IEEE Trans. Geosci. Remote Sens.* **2000**, *38*, 2202–2212. [CrossRef]
23. Ferretti, A.; Prati, C.; Rocca, F. Permanent scatterers in SAR Interferometry. *IEEE Trans. Geosci. Remote Sens.* **2001**, *39*, 8–20. [CrossRef]
24. Berardino, P.; Fornaro, G.; Lanari, R.; Sansosti, E. A new algorithm for surface deformation monitoring based on small baseline differential SAR interferograms. *IEEE Trans. Geosci. Remote Sens.* **2002**, *40*, 2375–2383. [CrossRef]
25. Lanari, R.; Mora, O.; Manunta, M.; Mallorqui, J.J.; Berardino, P.; Sansosti, E. A small baseline approach for investigating deformations on full-resolution differential SAR interferograms. *IEEE Trans. Geosci. Remote Sens.* **2004**, *42*, 1377–1386. [CrossRef]
26. Hooper, A.; Zebker, H.; Segall, P.; Kampes, B. A new method for measuring deformation on volcanoes and other natural terrains using InSAR persistent scatterers. *Geophys. Res. Lett.* **2004**, *31*. [CrossRef]
27. Hooper, A.J. Persistent Scatterer Radar Interferometry for Crustal Deformation Studies and Modeling of Volcanic Deformation. Ph.D. Thesis, Stanford University, Stanford, CA, USA, May 2006.

28. Hooper, A.; Segall, P.; Zebker, H. Persistent scatterer interferometric synthetic aperture radar for crustal deformation analysis, with application to Volcán Alcedo, Galápagos. *J. Geophys. Res.* **2007**, *112*. [CrossRef]

29. Hooper, A. A multi-temporal InSAR method incorporating both persistent scatterer and small baseline approaches. *Geophys. Res. Lett.* **2008**, *35*. [CrossRef]

30. De Zan, F.; Rocca, F. Coherent Processing of Long Series of SAR Images. In Proceedings of the IGARSS, Seoul, Korea, 25–29 July 2005; pp. 1987–1990.

31. De Zan, F. Optimizing SAR Interferometry for Decorrelating Scatterers. Ph.D. Thesis, Politecnico di Milano, Milano, Italy, 2008.

32. Monti Guarnieri, A.; Tebaldini, S. Hybrid Cramér-Rao Bounds for Crustal Displacement Field Estimators in SAR Interferometry. *IEEE Signal Proc. Lett.* **2007**, *14*, 1012–1015. [CrossRef]

33. Monti Guarnieri, A.; Tebaldini, S. On the Exploitation of Target Statistics for SAR Interferometry Applications. *IEEE Trans. Geosci. Remote Sens.* **2008**, *46*, 3436–3443. [CrossRef]

34. Ferretti, A.; Fumagalli, A.; Novali, F.; Prati, C.; Rocca, F.; Rucci, A. The second generation PSInSAR approach: SqueeSAR. In Proceedings of the International Fringe Workshop, Frascati, Italy, 30 November–4 December 2009.

35. Ferretti, A.; Fumagalli, A.; Novali, F.; Prati, C.; Rocca, F.; Rucci, A. A New Algorithm for Processing Interferometric Data-Stacks: SqueeSAR. *IEEE Trans. Geosci. Remote Sens.* **2011**, *49*, 3460–3470. [CrossRef]

36. Fornaro, G.; Reale, D.; Verde, S. Adaptive spatial multilooking and temporal multilinking in SBAS interferometry. In Proceedings of the IGARSS, Munich, Germany, 22–27 July 2012; p. 4. [CrossRef]

37. Piyush Shanker, A.; Zebker, H. Edgelist phase unwrapping algorithm for time series InSAR analysis. *J. Opt. Soc. Am. A* **2010**, *27*, 605–612. [CrossRef] [PubMed]

38. Fornaro, G.; Pauciullo, A.; Reale, D. A Null-Space Method for the Phase Unwrapping of Multitemporal SAR Interferometric Stacks. *IEEE Trans. Geosci. Remote Sens.* **2011**, *49*, 2323–2334. [CrossRef]

39. Costantini, M.; Malvarosa, F.; Minati, F. A General Formulation for Redundant Integration of Finite Differences and Phase Unwrapping on a Sparse Multidimensional Domain. *IEEE Trans. Geosci. Remote Sens.* **2012**, *50*, 758–768. [CrossRef]

40. Pepe, A.; Yang, Y.; Manzo, M.; Lanari, R. Improved EMCF-SBAS processing chain based on advanced techniques for the noise-filtering and selection of small baseline multilook DInSAR interferograms. *IEEE Trans. Geosci. Remote Sens.* **2015**, *53*, 4394–4417. [CrossRef]

41. Pepe, A.; Lanari, R. On the Extension of the Minimum Cost Flow Algorithm for Phase Unwrapping of Multitemporal Differential SAR Interferograms. *IEEE Trans. Geosci. Remote Sens.* **2006**, *44*, 2374–2383. [CrossRef]

42. Goel, K. Advanced Stacking Techniques and Applications in High Resolution SAR Interferometry. Ph.D. Thesis, Technische Universität München, München, Germany, 27 January 2014.

43. Schmitt, M. Reconstruction of Urban Surface Models from Multi-Aspect and Multi-Baseline Interferometric SAR. Ph.D. Thesis, Technische Universität München, München, Germany, 2014.

44. Jiang, M. InSAR Coherence Estimation and Applications to Earth Observation. Ph.D. Thesis, Hong Kong Polytechnic University, Hong Kong, China, 2014.

45. Wang, Y. Advances in Meter-resolution Multipass Synthetic Aperture Radar Interferometry. Ph.D. Thesis, Technische Universität München, München, Germany, 19 October 2015.

46. Samiei-Esfahany, S. Exploitation of Distributed Scatterers in Synthetic Aperture Radar Interferometry. Ph.D. Thesis, Technische Universiteit Delft, Delft, The Netherlands, 31 May 2017.

47. Kuruoğlu, E.E.; Zerubia, J. Modeling SAR Images with a Generalization of the Rayleigh Distribution. *IEEE Trans. Image Process.* **2004**, *13*, 527–533. [CrossRef] [PubMed]

48. Ulaby, F.T.; Dobson, M.C. *Handbook of Radar Statistics for Terrain*; Artech House: Dedham, MA, USA, 1989; ISBN 978-0890063361.

49. Goodman, J.W. Some fundamental properties of speckle. *J. Opt. Soc. Am.* **1976**, *66*, 1145–1150. [CrossRef]

50. Nikias, C.L.; Shao, M. *Signal Processing with Alpha-Stable Distributions and Applications*; Wiley: New York, NY, USA, 1995; ISBN 978-0471106470.

51. Ollila, E.; Eriksson, J.; Koivunen, V.; Poor, H.V. Complex Elliptically Symmetric Random Variables—Generation, Characterization, and Circularity Tests. *IEEE Trans. Signal Process.* **2011**, *59*, 58–69. [CrossRef]

52. Ollila, E.; Tyler, D.E.; Koivunen, V. Complex Elliptically Symmetric Distributions: Survey, New Results and Applications. *IEEE Trans. Signal Process.* **2012**, *60*, 5597–5625. [CrossRef]

53. Ollila, E.; Tyler, D.E. Regularized M-Estimators of Scatter Matrix. *IEEE Trans. Signal Process.* **2014**, *62*, 6059–6070. [CrossRef]

54. Gao, G. Statistical Modeling of SAR Imagery: A Survey. *Sensors* **2000**, *10*, 775–795. [CrossRef] [PubMed]

55. Picinbono, B. Second-Order Complex Random Vectors and Normal Distributions. *IEEE Trans. Signal Process.* **1996**, *44*, 2637–2640. [CrossRef]

56. Ferretti, A.; Colesanti, C.; Perrisin, D.; Prati, C.; Rocca, F. Evaluating the Effect of the Observation Time on the Distribution of SAR Permanent Scatterers. In Proceedings of the Fringe Workshop, Frascati, Italy, 1–5 December 2003; p. 6.

57. Morishita, Y.; Hanssen, R.F. Temporal Decorrelation in L-, C-, and X-band Satellite Radar Interferometry for Pasture on Drained Peat Soils. *IEEE Trans. Geosci. Remote Sens.* **2015**, *53*, 1096–1104. [CrossRef]

58. Morishita, Y.; Hanssen, R.F. Deformation Parameter Estimation in Low Coherence Areas Using a Multisatellite InSAR Approach. *IEEE Trans. Geosci. Remote Sens.* **2015**, *53*, 4275–4283. [CrossRef]

59. Rocca, F. Modeling interferogram stacks. *IEEE Trans. Geosci. Remote Sens.* **2007**, *45*, 3289–3299. [CrossRef]

60. De Zan, F.; Zonno, M.; López-Dekker, P. Phase Inconsistencies and Multiple Scattering in SAR Interferometry. *IEEE Trans. Geosci. Remote Sens.* **2015**, *53*, 6608–6616. [CrossRef]

61. De Zan, F.; Parizzi, A.; Prats-Iraola, P.; López-Dekker, P. A SAR interferometric Model for Soil Moisture. *IEEE Trans. Geosci. Remote Sens.* **2014**, *52*, 418–425. [CrossRef]

62. Zwieback, S.; Liu, X.; Antonova, S.; Heim, B.; Bartsch, A.; Boike, J.; Hajnsek, I. A Statistical Test of Phase Closure to Detect Influences on DInSAR Deformation Estimates besides Displacements and Decorrelation Noise: Two Case Studies in High-Latitude Regions. *IEEE Trans. Geosci. Remote Sens.* **2016**, *54*, 5588–5601. [CrossRef]

63. Goodman, N.R. Statistical analysis based on a certain multivariate complex Gaussian distribution (an introduction). *Ann. Math. Stat.* **1963**, *34*, 152–177. [CrossRef]

64. Tough, R.J.A.; Blacknell, D.; Quegan, S. A statistical Description of Polarimetric and Interferometric Synthetic Aperture Radar. *Proc. R. Soc.* **1995**, *449*, 567–589. [CrossRef]

65. Joughin, I.R.; Winebrenner, D.P.; Percival, D.B. Probability density functions for multilook polarimetric signatures. *IEEE Trans. Geosci. Remote Sens.* **1994**, *32*, 562–574. [CrossRef]

66. Touzi, R.; Lopes, A. Statistics of the Stokes parameters and of the complex coherence parameters in one-look and multilook speckle fields. *IEEE Trans. Geosci. Remote Sens.* **1996**, *34*, 519–531. [CrossRef]

67. Touzi, R.; Lopes, A.; Bruniquel, J.; Vachon, P.W. Coherence estimation for SAR imagery. *IEEE Trans. Geosci. Remote Sens.* **1999**, *37*, 135–149. [CrossRef]

68. López-Martínez, C.; Pottier, E. Coherence estimation in synthetic aperture radar data based on speckle noise modeling. *Appl. Opt.* **2007**, *46*, 544–558. [CrossRef] [PubMed]

69. Zebker, H.A.; Chen, K. Accurate estimation of correlation in InSAR observations. *IEEE Geosci. Remote Sens. Lett.* **2005**, *2*, 124–127. [CrossRef]

70. Lee, J.S.; Hoppel, K.W.; Mango, S.A.; Miller, A.R. Intensity and phase statistics of multilook polarimetric and interferometric SAR imagery. *IEEE Trans. Geosci. Remote Sens.* **1994**, *32*, 1017–1028. [CrossRef]

71. Sarabandi, K. Derivation of phase statistics from the Mueller matrix. *Radio Sci.* **1992**, *27*, 553–560. [CrossRef]

72. Bian, Y.; Mercer, B. Interferometric SAR Extended Coherence Calculation Based on Fractional Lower Order Statistics. *IEEE Geosci. Remote Sens. Lett.* **2010**, *7*, 841–845. [CrossRef]

73. Maronna, M.A.; Martin, R.D.; Yohai, V.J. *Robust Statistics—Theory and Methods*, 1st ed.; John Wiley & Sons: New York, NY, USA, 2006; ISBN 978-0-470-01092-1.

74. Even, M. Advanced InSAR Processing in the Footsteps of SqueeSAR. In Proceedings of the Fringe Workshop, Frascati, Italy, 23–27 March 2015; p. 6.

75. Shevlyakov, G.L.; Oja, H. *Robust Correlation—Theory and Applications*, 1st ed.; John Wiley & Sons: Hoboken, NJ, USA, 2016; ISBN 978-1-11849345-8.

76. Sica, F.; Reale, D.; Poggi, G.; Verdoliva, L.; Fornaro, G. Nonlocal Adaptive Multilooking in SAR Multipass Differential Interferometry. *IEEE J. Sel. Top. Appl. Earth Obs. Remote Sens.* **2015**, *8*, 1727–1742. [CrossRef]

77. Lombardini, F. Optimum absolute phase retrieval in three-element SAR interferometer. *IET Electr. Lett.* **1998**, *34*, 1522–1524. [CrossRef]

78. Press, W.H.; Teukolsky, S.A.; Vetterling, W.T.; Flannery, B.P. *Numerical Recipes in C: The Art of Scientific Computing*, 2nd ed.; Cambridge University Press: Cambridge, UK, 1992; pp. 394–455. ISBN 0-521-43108-5.

79. Bapat, R.B.; Kwong, M.K. A generalization of $A^{-1} \circ A \geq I$. *Linear Algebra Its Appl.* **1987**, *93*, 107–112. [CrossRef]

80. Rocca, F.; Rucci, A.; Ferretti, A.; Bohane, A. Advanced InSAR interferometry for reservoir monitoring. *First Break* **2013**, *31*, 77–85.

81. Fornaro, G.; Verde, S.; Reale, D.; Pauciullo, A. CAESAR: An Approach Based on Covariance Matrix Decomposition to Improve Multibaseline–Multitemporal Interferometric SAR Processing. *IEEE Trans. Geosci. Remote Sens.* **2015**, *53*, 2050–2065. [CrossRef]

82. Ferretti, A.; Fumagalli, A.; Novali, F.; De Zan, F.; Rucci, A.; Tebaldini, S. Process for Filtering Interferograms Obtained from SAR Images Acquired on the Same Area. CA Patent 2,767,144, 13 January 2011.

83. Peressin, D.; Wang, T. Repeat-Pass SAR Interferometry with Partially Coherent Targets. *IEEE Trans. Geosci. Remote Sens.* **2012**, *50*, 271–280. [CrossRef]

84. Samiei-Esfahany, S.; Esteves Martins, J.; van Leijen, F.; Hanssen, R.F. Phase Estimation for Distributed Scatterers in InSAR Stacks Using Integer Least Squares Estimation. *IEEE Trans. Geosci. Remote Sens.* **2016**, *54*, 5671–5687. [CrossRef]

85. Lv, X.; Birsen, Y.; Zeghal, M.; Bennett, V.; Abdoun, T. Joint-Scatterer Processing for Time-Series InSAR. *IEEE Trans. Geosci. Remote Sens.* **2014**, *52*, 7205–7221. [CrossRef]

86. Pinel-Puysségur, B.; Rémi, M.; Avouac, J.-P. Multi-Link InSAR Time Series: Enhancement of a Wrapped Interferometric Database. *IEEE JSTARS* **2012**, *5*, 784–794. [CrossRef]

87. Pinel-Puysségur, B.; Karnoukian, M.; Granin, R.; Lasserre, C.; Doin, M.-P. Wrapped Interferograms Enhanced By MuLSAR Method: Applications And Comparison To Other Methods. In Proceedings of the ESA Living Planet Symposium, Edinburgh, UK, 9–13 September 2013.

88. Trouvé, E.; Nicolas, J.-M.; Maître, H. Improving Phase Unwrapping Techniques by the Use of Local Frequency Estimates. *IEEE Trans. Geosci. Remote Sens.* **1998**, *36*, 1963–1972. [CrossRef]

89. Davidson, G.W.; Bamler, R. Multiresolution Phase Unwrapping for SAR Interferometry. *IEEE Trans. Geosci. Remote Sens.* **1999**, *37*, 163–174. [CrossRef]

90. Wang, Y.; Zhu, X.; Bamler, R. Retrieval of phase history parameters from distributed scatterers in urban areas using very high resolution SAR data. *ISPRS J. Photogr. Remote Sens.* **2012**, *73*, 89–99. [CrossRef]

91. Baran, I.; Stewart, M.P.; Kampes, B.M.; Perski, Z.; Lilly, P. A Modification to the Goldstein Radar Interferogram Filter. *IEEE Trans. Geosci. Remote Sens.* **2003**, *41*, 2114–2118. [CrossRef]

92. Vasile, G.; Trouvé, E.; Petillot, I.; Bolon, P.; Nicolas, J.-M.; Gay, M.; Chanussot, J.; Landes, T.; Grussenmeyer, P.; Buzuloiu, V.; et al. High-Resolution SAR Interferometry: Estimation of Local Frequencies in the Context of Alpine Glaciers. *IEEE Trans. Geosci. Remote Sens.* **2008**, *46*, 1079–1090. [CrossRef]

93. Goel, K.; Adam, N. An advanced algorithm for deformation estimation in non-urban areas. *ISPRS J. Photogr. Remote Sens.* **2012**, *73*, 100–110. [CrossRef]

94. Goel, K.; Adam, N.A. Distributed Scatterer Interferometry Approach for Precision Monitoring of Known Surface Deformation Phenomena. *IEEE Trans. Geosci. Remote Sens.* **2014**, *52*, 5454–5468. [CrossRef]

95. Ciuc, M.; Bolon, P.; Trouvé, E.; Buzuloiu, V.; Rudant, J.-P. Adaptive-neighborhood speckle removal in multitemporal synthetic aperture radar images. *Appl. Opt.* **2001**, *40*, 5954–5966. [CrossRef] [PubMed]

96. Gordon, R.; Rangayyan, R.M. Feature enhancement of film mammograms using fixed and adaptive neighborhoods. *Appl. Opt.* **1984**, *23*, 560–564. [CrossRef] [PubMed]

97. Lee, J.-S. Digital image smoothing and the sigma filter. *Comput. Vis. Graph. Image Process.* **1983**, *24*, 255–269. [CrossRef]

98. Vasile, G.; Trouvé, E.; Lee, J.-S.; Buzuloiu, V. Intensity-Driven Adaptive-Neighborhood Technique for Polarimetric and Interferometric SAR Parameters Estimation. *IEEE Geosci. Remote Sens. Lett.* **2006**, *44*, 1609–1621. [CrossRef]

99. Parizzi, A.; Brcic, R. Adaptive InSAR Stack Multilooking Exploiting Amplitude Statistics: A Comparison Between Different Techniques and Practical Results. *IEEE Trans. Geosci. Remote Sens.* **2011**, *8*, 441–445. [CrossRef]

100. Even, M. Advanced InSAR Processing for Distributed Scatterers. In Proceedings of the Fringe Workshop, Helsinki, Finland, 5–9 June 2017.

101. Gibbons, J.; Chakraborti, S. *Nonparametric Statistical Inference*, 4th ed.; Marcel Dekker, Inc.: New York, NY, USA; Basel, Switzerland, 2003; ISBN 0-8247-4052-1.

102. Scholz, F.W.; Stephens, M.A. K-Sample Anderson-Darling Tests. *J. Am. Stat. Assoc.* **1987**, *82*, 918–924. [CrossRef]

103. Schmitt, M.; Stilla, U. Adaptive multilooking of airborne Ka-band multi-baseline InSAR data of urban areas. In Proceedings of the IGARSS, Munich, Germany, 22–27 July 2012; p. 4. [CrossRef]

104. Schmitt, M.; Stilla, U. Adaptive Multilooking of Airborne Single-Pass Multi-Baseline InSAR Stacks. *IEEE Trans. Geosci. Remote Sens.* **2014**, *52*, 305–312. [CrossRef]

105. Schmitt, M.; Schönberger; Stilla, U. Adaptive Covariance Matrix Estimation for Multi-Baseline InSAR Data Stacks. *IEEE Trans. Geosci. Remote Sens.* **2014**, *52*, 6807–6817. [CrossRef]

106. Deledalle, C.-A.; Loic, D.; Tupin, F. NL-InSAR: Nonlocal Interferogram Estimation. *IEEE Trans. Geosci. Remote Sens.* **2011**, *49*, 2661–2672. [CrossRef]

107. Jiang, M.; Ding, X.; Hanssen, R.F.; Malhotra, R.; Chang, L. Fast Statistically Homogeneous Pixel Selection for Covariance Matrix Estimation for Multitemporal InSAR. *IEEE Trans. Geosci. Remote Sens.* **2015**, *53*, 1213–1224. [CrossRef]

108. Lee, J.-S.; Grunes, M.R.; de Grandi, G. Polarimetric SAR speckle filtering and its implication for classification. *IEEE Trans. Geosci. Remote Sens.* **1999**, *37*, 2363–2373. [CrossRef]

109. Lee, J.-S.; Cloude, S.; Papathanassiou, C.; Grunes, M.; Woodhouse, I. Speckle filtering and coherence estimation of polarimetric SAR interferometry data for forest applications. *IEEE Trans. Geosci. Remote Sens.* **2003**, *41*, 2254–2263. [CrossRef]

110. Hubert, M.; Vandervieren, E. An adjusted boxplot for skewed distributions. *Comput. Stat. Data Anal.* **2008**, *52*, 5186–5201. [CrossRef]

111. Buades, A.; Coll, B.; Morel, J.-M. A non-local algorithm for image denoising. In Proceedings of the IEEE Computer Society Conference on Computer Vision and Pattern Recognition, San Diego, CA, USA, 20–25 June 2005.

112. Deledalle, C.-A.; Loic, D.; Tupin, F. Iterative Weighted Maximum Likelihood Denoising With Probabilistic Patch-Based Weights. *IEEE Trans. Image Process.* **2009**, *18*, 2661–2672. [CrossRef] [PubMed]

113. Deledalle, C.-A.; Loic, D.; Tupin, F. How to compare noisy patches? Patch similarity beyond Gaussian noise. *Int. J. Comput. Vis.* **2012**, *99*, 86–102. [CrossRef]

114. Deledalle, C.-A.; Loic, D.; Poggi, G.; Tupin, F.; Verdoliva, L. Exploiting patch similarity for SAR image processing: The nonlocal paradigm. *IEEE Signal Process. Mag.* **2014**, *31*, 69–78. [CrossRef]

115. Deledalle, C.-A.; Loic, D.; Tupin, F.; Reigber, A.; Jäger, M. NL-SAR: A Unified Nonlocal Framework for Resolution-Preserving (Pol)(In)SAR Denoising. *IEEE Trans. Geosci. Remote Sens.* **2015**, *53*, 2021–2038. [CrossRef]

116. Parrilli, S.; Poderico, M.; Angelino, C.V.; Verdoliva, L. A Nonlocal SAR Image Denoising Algorithm Base don LLMMSE Wavelet Shrinkage. *IEEE Trans. Geosci. Remote Sens.* **2012**, *50*, 606–616. [CrossRef]

117. Sica, F.; Cozzolino, D.; Zhu, X.; Verdoliva, L.; Poggi, G. InSAR-BM3D: A Nonlocal Filter for SAR Interferometric Phase Restoration. *IEEE Trans. Geosci. Remote Sens.* **2018**, *56*, 1–12. [CrossRef]

118. Deledalle, C.-A.; Loic, D.; Tabti, S.; Tupin, F. MuLoG, or How to apply Gaussian denoisers to multi-channel SAR speckle reduction? *IEEE Trans. Image Process.* **2017**, *26*, 4389–4403. [CrossRef] [PubMed]

119. Jiang, M.; Ding, X.; Li, Z. Hybrid Approach for Unbiased Coherence Estimation for Multitemporal InSAR. *IEEE Trans. Geosci. Remote Sens.* **2014**, *52*, 2459–2473. [CrossRef]

120. Abdelfattah, R.; Nicolas, J.-M. Interferometric SAR coherence magnitude estimation using second kind statistics. *IEEE Trans. Geosci. Remote Sens.* **2006**, *44*, 1942–1953. [CrossRef]

121. Jiang, M.; Ding, X.; Li, Z.; Tian, X.; Wang, C.; Zhu, W. InSAR Coherence Estimation for Small Data Sets and Its Impact on Temporal Decorrelation Extraction. *IEEE Trans. Geosci. Remote Sens.* **2014**, *52*, 6584–6596. [CrossRef]

122. Ledoit, O.; Wolf, M. A well-conditioned estimator for large-dimensional covariance matrices. *J. Multivar. Anal.* **2004**, *88*, 365–411. [CrossRef]

123. González, P.J.; Fernández, J. Error estimation in multitemporal InSAR deformation time series, with application to Lanzarote, Canary Islands. *J. Geophys. Res.* **2011**, *116*, 1–17. [CrossRef]

124. Lauknes, T.R.; Zebker, H.A.; Larsen, Y. InSAR Deformation Time Series Using an L_1-Norm Small Baseline Approach. *IEEE Trans. Geosci. Remote Sens.* **2011**, *49*, 536–546. [CrossRef]

Remote Sens. **2018**, *10*, 744

125. Goel, K.; Adam, N. High Resolution Deformation Time Series Estimation for Distributed Scatterers Using TerraSAR-X Data. *ISPRS Ann. Photogr. Remote Sens. Spat. Inf. Sci.* **2012**, *I-7*, 29–34. [CrossRef]

126. Navarro-Sanchez, V.D.; Lopez-Sanchez, J.M. Spatial Adaptive Speckle Filtering Driven by Temporal Polarimetric Statistics and Its Application to PSI. *IEEE Trans. Geosci. Remote Sens.* **2014**, *52*, 4548–4557. [CrossRef]

127. Ansari, H.; De Zan, F.; Bamler, R. Sequential Estimator: Toward Efficient InSAR Time Series Analysis. *IEEE Trans. Geosci. Remote Sens.* **2017**, *55*, 5637–5652. [CrossRef]

128. Wang, Y.; Zhu, X. Robust Estimators for Multipass SAR Interferometry. *IEEE Trans. Geosci. Remote Sens.* **2016**, *54*, 968–980. [CrossRef]

remote sensing

MDPI

Article

Modeling Orbital Error in InSAR Interferogram Using Frequency and Spatial Domain Based Methods

Xin Tian [1,*], Rakesh Malhotra [2], Bing Xu [3], Haoping Qi [1] and Yuxiao Ma [1]

[1] Department of Surveying and Mapping Engineering, School of Transportation, Southeast University, Nanjing 211189, China; qhp@seu.edu.cn (H.Q.); maxiaoxiao0928@163.com (Y.M.)
[2] Department of Environmental, Earth and Geospatial Sciences, North Carolina Central University, Durham, NC 27707, USA; rmalhotra@nccu.edu
[3] School of Geosciences and Info-Physics, Central South University, Changsha 410083, China; xubing@csu.edu.cn
* Correspondence: tianxin@seu.edu.cn; Tel.: +86-25-83795386

Received: 27 January 2018; Accepted: 19 March 2018; Published: 23 March 2018

Abstract: Synthetic Aperture Radar Interferometry (SAR, InSAR) is increasingly being used for deformation monitoring. Uncertainty in satellite state vectors is considered to be one of the main sources of errors in applications such as this. In this paper, we present frequency and spatial domain based algorithms to model orbital errors in InSAR interferograms. The main advantage of this method, when applied to the spatial domain, is that the order of the polynomial coefficient is automatically determined according to the features of the orbital errors, using K-cross validation. In the frequency domain, a maximum likelihood fringe rate estimate is deployed to resolve linear orbital patterns in strong noise interferograms, where spatial-domain-based algorithms are unworkable. Both methods were tested and compared with synthetic data and applied to historical Environmental Satellite Advanced Synthetic Aperture Radar (ENVISAT ASAR) sensor and modern instruments such as Gaofen-3 (GF-3) and Sentinel-1. The validation from the simulation demonstrated that an accuracy of ~1mm can be obtained under optimal conditions. Using an independent GPS measurement that is discontinuous from the InSAR measurement over the Tohoku-Oki area, we found a 31.45% and 73.22% reduction in uncertainty after applying our method for ASAR tracks 347 and 74, respectively.

Keywords: InSAR; orbital error; deformation monitoring; fringe rate estimation; K-cross validation

1. Introduction

The precise orbiting position information of space-borne Synthetic Aperture Radar (SAR) systems is of great significance in many Interferometric Synthetic Aperture Radar (InSAR) applications, especially in the case of ground motion monitoring [1–3]. The InSAR technique uses two sensors, carried on satellites, with slightly different incidence angles, to measure a displacement along the radar line-of-sight (LOS) between two SAR acquisitions. The theoretical accuracy can be as high as a centimeter to as low as a millimeter. This technique has been widely used to identify many geophysical processes that usually cause long-wavelength crustal deformation, including strain accumulation along locked continental faults, coseismic deformation caused by the occurrence of faulting in the lithosphere, and postseismic deformation caused by afterslip and viscoelastic relaxation [4–6].

The radar, the ground, and the intervening medium create a compound observing system; within this, there are several potential sources of errors that have a detrimental impact on the accuracy of InSAR measurements. In particular, satellite orbital errors [7], temporal and spatial decorrelation effects [8,9], atmospheric screens [10], and high-deformation gradients are the main limitations [11]. As far as orbital errors are concerned, they cause long-wavelength phase contributions to interferogram and are often referred to as 'the phase ramp'. The pattern of the phase ramp usually depends on the satellite–state–vector error, especially on the radial and cross-track components of the orbital error [7]. The influence of orbital errors on the final accuracy of deformation products is largely contingent on the SAR instruments, i.e., the trajectory of the satellite orbit, the radar frequency, and the degree of overlap between the phase ramp and the deformation signals.

Several efforts have been made during the last two decades to mitigate the orbital error, and most of them start with spatial domain analyses of InSAR interferograms. A simple but very effective solution is de-ramping from the InSAR interferogram using a linear or quadratic surface fitting [3,12]. The phase ramp is either subtracted from the original phase or used to refine the spatial baseline to infer a revised interferogram [13,14]. Given a wrapped-phase pattern, the unwrapping operation is always required before fitting. The accuracy of the estimated coefficients in the polynomial model relies on the quality of the observations to be fitted. Error propagation in the unwrapping procedure over fast decorrelation areas with low coherence may distort the signal and, therefore, contribute to surface artifacts. To solve this problem, pixel offsets between bi-temporal-SAR data can be estimated using the cross-correlation algorithm; the baseline is then re-estimated to compensate for the residual phase in the interferogram [15]. Despite many efforts, the accuracy of the estimated offsets, from meter to decimeter, restricts the correction of the baseline components. Kohlhase et al. (2003) suggest counting the fringes caused by orbital errors according to phase gradients in differential interferograms and to adjust the trajectories of satellite orbits [16]. This does not work for coseismic scenes, where a stronger deformation causes dense fringes [7].

Advances related to orbital error correction refer to the time-series-SAR dataset. One such suggestion is to independently estimate the phase ramp on each interferogram using least squares scheme. To ensure consistency in the interferometric network, the polynomial coefficients are refined in a network sense by time-series inversions [17,18]. Considering the inseparability of phase ramp and long-wavelength deformation signals in the spatial dimension, the alternative is to estimate both deformation and orbital errors by exploiting spatio-temporal characteristics of both quantities [17,19]. However, resolving large, linear systems of algebraic equations, error propagation, and underdetermined problems have become the main challenges associated with these methods.

One promising alternative to separating long-wavelength displacement signals and orbit errors is to employ external data, for example, GPS measurement [20–22]. GPS displacements located in the SAR scene are projected onto the line-of-sight direction according to the incidence angle and the heading of the SAR geometry. The phase ramp is then estimated by minimizing the residuals between the InSAR interferogram and the phase inverted by the GPS. The final accuracy of the InSAR deformation product depends on the accuracy of the GPS measurements, the number of GPS stations and their spatial distribution.

In summary, the current problems of orbital error removal are two-fold. The first problem originates from the fact that all long-wavelength signals are treated as orbital errors during the correction without ancillary data. Although a few studies try to use wavelet-multiresolution analysis to classify the different components in the unwrapped interferogram automatically [23], the method is still empirical, as the number of levels of wavelet decomposition must be specified manually according to the features of the interferogram. The second problem involves robust regression. As stated in Reference [24], the model selection is subjective and relies on the distribution and the density of the targets showing high coherence, while errors in unwrapping are likely to mislead surface fitting. In fact, the method of estimating the phase ramp in the frequency domain may avoid the unwrapping error. However, few studies discuss using this method for orbital error correction. In the following section, we focus our analyses on the individual interferogram, as it is essential for time-series inversion and many geophysical applications.

The remainder of the paper is organized as follows: In Section 2, we introduce the frequency and spatial domain based methods, respectively, from the image-processing viewpoint. In Section 3, the results and quantitative comparisons are presented using synthetic and real data. In Section 4, the feasibility of these methods, including their merits and faults, are investigated. In Section 5, conclusions are given.

2. Methods

2.1. Modeling Orbital Error in Frequency Domain

The fringe rate of long-wavelength signals in interferogram can be estimated by means of maximum likelihood (ML) frequency algorithms [25–27]. The idea is that the orbital error has a dominant term in its Fourier expansion. Assuming original interferogram I can be expressed as:

$$I = e^{i(\phi_{def}+\phi_{topo}+\phi_{orbit}+\phi_{other})} = D \cdot e^{i\phi_{topo}} \tag{1}$$

where ϕ_{def}, ϕ_{topo}, ϕ_{orbit}, ϕ_{other} denote deformation phase, topographic phase, orbital ramp, and other phase components related to the atmosphere and decorrelation, respectively. D is the differential interferogram. After atmospheric correction and/or phase filtering, we rewrite D as a sinusoidal model:

$$D = e^{i\phi_{orbit}} \cdot e^{i\phi_{def}} = e^{i2\pi f_x x} \cdot e^{i2\pi f_y y} \cdot e^{i\rho} \cdot e^{i\phi_{def}} \tag{2}$$

where x and y denote the pixel index in radar coordinate, f_x and f_y are true fringe frequencies along range and azimuth direction, respectively, and ρ is the residual phase.

$$F = fft(D) \tag{3}$$

The Fourier filter can be used to obtain the ML estimate of the average stripe rate in Equation (2). The peak location obtained by maximizing the 2D discrete Fourier transform (DFT) corresponds to the estimated frequencies $\hat{f}_x$ and $\hat{f}_y$, respectively, and the phase $\hat{\rho}$ at the DFT peak is approximate to ρ. To better identify the exact frequencies, we increase the DFT frequency sampling by padding the signal in the window with enough zeros. The size of the padded zeros in both directions, n_x and n_y, can be determined by the Cramer-Rao bound of the variance of estimated frequency [26]:

$$\begin{cases} \left(\frac{1}{N_x+n_x}\right)^2 \le \frac{6}{SNR \cdot N_x \cdot N_y \cdot (N_x^2 - 1)} \\ \left(\frac{1}{N_y+n_y}\right)^2 \le \frac{6}{SNR \cdot N_x \cdot N_y \cdot (N_y^2 - 1)} \end{cases} \tag{4}$$

where SNR represents the signal-to-noise ratio, N_x and N_y are equivalent to each of the image sizes. The de-ramped interferogram can be obtained from the cross multiplication of the differential interferogram and the estimated phase ramp under the assumption $\hat{f}_x \approx f_x$, $\hat{f}_y \approx f_y$ and $\hat{\rho} \approx \rho$.

The significant advantage of this method is its simple computation without the need for phase unwrapping and higher reliability over low-coherence areas (see Section 3.1). However, this method can only estimate the linear terms of orbital error. For non-linear fringe patterns, which are usually present in long-strip images, polynomial fitting in the spatial domain is recommended.

2.2. Modeling Orbital Error in Spatial Domain

The most popular method used in orbital-error correction is surface fitting with linear or quadratic models, using unwrapped phase observations. Its general form can be defined as [23]:

$$z = P_1 x + P_2 y + P_3 \tag{5}$$

where z is the polynomial model, x and y indicate the pixel index in radar coordinates. To solve the equation for the unknown coefficients P_1, P_2 and P_3, the least squares scheme is used. The solution can be written as:

$$P = (A^T A)^{-1} A^T Z \tag{6}$$

where $P^T = [\hat{P}_1, \hat{P}_2, \hat{P}_3]$ is the estimated vector in which the elements correspond to the coefficients in Equation (5). Z is the observations with size N, and A is the $N \times 3$ matrix containing the coordinates x and y. However, there are several drawbacks to using this method. Firstly, Equation (6) is not robust to abnormal values, such as unwrapping errors. In addition, the observations contribute to the equivalent weight without the consideration of the interferogram qualities. Secondly, the local deformation will contribute to the polynomial coefficients with the increase of the dimension in P^T. Finally, the determination of the order of the polynomial model is not clear yet and is usually empirical. A more sophisticated method is therefore needed.

2.2.1. Preprocess: Multi-Looking and Manually Masking

An increase in *looks* can enhance the quality of long-wavelength signals and simultaneously mitigate the effect of the short-wavelength signals during the regression. On the other hand, the decreased dimension can reduce the computational burden for unwrapping and the iterative least squares scheme. For those regional deformation signals, manual masking should be undertaken. Although this step is simpler than wavelet decomposition [23], it is still effective without expert knowledge and computational complexity. In addition, masking is important for areas with a high phase gradient, where phase aliasing causes unexpected unwrapping errors [11].

2.2.2. Polynomial Model

For long-strip InSAR interferograms, it is difficult to use the linear or quadric models to incorporate orbit errors. A more general form is needed:

$$z = XMY^T \tag{7}$$

where $X = [1 \; x \; x^2 \; x^3 \ldots]_{1 \times (n+1)}$ and $Y^T = [1 \; y \; y^2 \; y^3 \ldots]_{1 \times (m+1)}$ is the vector with the length $n + 1$ and $m + 1$ respectively, M is $(n + 1) \times (m + 1)$ coefficient matrix.

$$M = \begin{bmatrix} P_{1,1} & P_{1,2} & \cdots & P_{1,m} & P_{1,m+1} \\ P_{2,1} & P_{2,2} & \cdots & P_{2,m} & \\ \vdots & \vdots & & & \\ P_{n,1} & P_{n,2} & & & \\ P_{n+1,1} & & & & 0 \end{bmatrix}$$

The parameters n and m should be carefully determined to prevent overfitting in terms of the spatial distribution of observations.

2.2.3. Iteratively Reweighted Least Squares Fitting

The main disadvantage of ordinary least squares is that there is a constant deviation in the errors. For observations with different coherence magnitudes, the method of weighted least squares should be used. Moreover, iterative behavior can be employed to reduce the influence of outliers on regression. We designed an iterative version for the least-squares solution according to the phase standard deviation (STD) and robust Bi-square weight. The modification of the parameter estimate P in Equation (6) can be rewritten as:

$$P = (A^T W A)^{-1} A^T W z \tag{8}$$

In this paper, we define weight W in two steps. The prior weight $V = \{v_1, v_2, \ldots, v_N\}$, defined by interferometric phase STD, can be simply written as [27]:

$$v_i = \frac{1}{\sigma_i} = \frac{\sqrt{2L}\gamma_i}{\sqrt{1-\gamma_i^2}} L \geq 4 \qquad (9)$$

where L is the *looks* and γ_i is the coherence observations. Compared with the coherence, phase STD σ_i is a more appropriate weight, as it takes *looks* into account. The function model Equation (9) is a close approximation to the complicated stochastic model when *looks* are greater than four [27]. After initial regression, the robust Bi-square weight $B = \{b_1, b_2, \ldots, b_N\}$, which minimizes a weighted sum of squares, is used during the iteration [28]:

$$u_i = \frac{r_i}{cs\sqrt{1-h_i}}$$
$$b_i = \begin{cases} (1-u_i^2)^2 & |u_i| < 1 \\ 0 & |u_i| \geq 1 \end{cases} \qquad (10)$$

where r_i is least-squares residual for each observation from the previous iteration, c is a tuning constant (4.685), and s is the robust variance given by $MAD/0.6745$, where MAD is the absolute deviation of the residuals from their median; h_i is leverage that adjust the residuals. The total weight function W, therefore, is:

$$W = \begin{cases} V & l=0 \\ V \cdot B^{(l)} & l > 0 \end{cases} \qquad (11)$$

where l denotes the number of iterations. The iterations will stop if the fit converges (minimum change in coefficients $<10^{-5}$), or the maximum number of iterations allowed for fitting is reached (400).

2.2.4. Model Selection

The presence of the unwanted terms in Equation (7) and the overfitting lead to an unexpected bias or variance in interferograms. A trade-off should be made by taking advantage of K-cross validation, which has been widely used in machine learning [29]. Specifically, we obtained some test data from the same distribution and picked appropriate m and n in Equation (7) to minimize the same sum-of-squares difference that we used for fitting to the training data. This can be achieved by making several different splits in data Z. Each subsample is used once for testing, and the rest is used for training. The arrangement of this method is described as follows.

1. Split data Z into K subsamples with equivalent size N.
2. For $k = 1, 2, \ldots, K$, set validation data Z_{test} to be the kth subsample, and training data Z_{train} to be the other $K-1$ subsamples.
3. Fit each model to Z_{train} and evaluate its performance on Z_{test} through weighted root-mean-square error (WRMSE).
4. Pick m and n that leads to minimum WRMSE by averaging K results.

Uniform sampling without replacement is used to separate data Z, and $K = 10$ is set throughout this paper. Considering the different qualities of the validation data, we define WRMSE as:

$$WRMSE = \sqrt{\frac{\sum\limits_{i=1}^{N} v_i (z_i - Z_{test,i})^2}{\sum\limits_{i=1}^{N} v_i}} \qquad (12)$$

3. Results

3.1. Synthetic Data

In this section, the performance of the orbital-error corrections are evaluated and compared using a simulation. For notional convenience, we refer to the method in Equations (3) and (4) as DFT, and the method developed in Section 2.2 as adaptive ADP-Poly.

Three phase components, including deformation, random noise, and orbital errors were simulated according to the ENVISAT ASAR geometry. The surface deformation caused by a finite rectangular source was first simulated by the Okada 85 model [30] and then inverted to an interferometric phase (Figure 1b). To add phase noise, the coherence map was simulated under three principle sources of error, i.e., the Doppler, spatial, and temporal decorrelations [25,31], in which the temporal decorrelation was created using an isotropic-2D-fractal surface with a power law spectrum [32]. On the basis of the simulated coherence map under different *looks* (Figure 1a), the phase-STD map was obtained using the stochastic model (4.2.24) and (4.2.26) in [31]. The phase-noise map was finally generated using pointwise multiplication of the phase-STD map and the standard normal distributed random numbers. The linear and non-linear phase ramps were added to the orbital errors (Figure 1c,d). The combination of all components in synthetic interferograms is given in Figure 2.

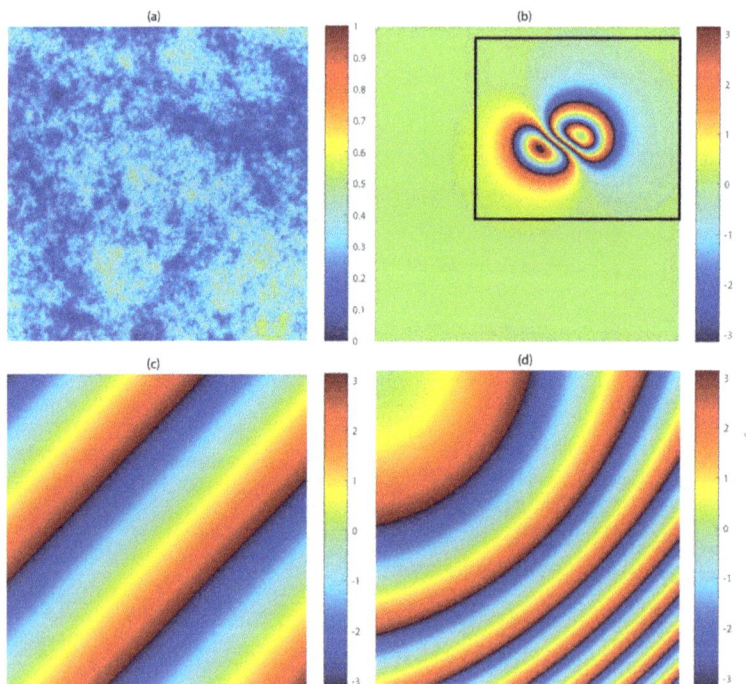

Figure 1. Simulated InSAR parameters (**a**) coherence map; (**b**) noise-free differential interferogram. The black frame denotes the masked area for surface fitting in spatial domain; (**c**) linear orbital error and (**d**) nonlinear orbital error with non-uniform phase gradient.

To test the methods over fast decorrelation areas, we adjusted the coherence magnitude to an average value of $\gamma = 0.2$ under single *look*, and then added the phase-noise to the interferogram in Figure 2a. Likewise, we set $\gamma = 0.4$ under *2-looks* to test the capability of both methods on moderate noise scenes, as shown in Figure 2f. For the frequency–domain-based method, the DFT is applied to noisy interferograms directly, while adaptive Goldstein filtering driven by phase STD [33] and minimum cost–flow unwrapping [34] were used before ADP-Poly surface fitting.

Visually, ADP-Poly could not detect the phase ramp in very noisy scenes and left phase residuals in Figure 2c. By contrast, DFT accurately captured the linear phase ramp. The residual between the truth and estimate confirms the potential of using DFT without phase unwrapping (Figure 2d). When the quality of interferogram became better (Figure 2f), ADP-Poly showed its superior performance at detecting nonlinear features (Figure 2h), while DFT could not completely remove the error and left nonlinear residuals in the interferogram (Figure 2g).

To make the results more statistically significant, we performed 500 simulations for each case and evaluated the phase residuals using root-mean-square error (RMSE). It can be seen from Figure 3 that DFT works well in linear orbital error correction over low coherence areas, and the averaged RMSE is up to 0.16 rad (4.99 rad for ADP-Poly). For nonlinear orbital features, ADP-Poly obtained RMSE of 0.10 rad (2.77 rad for DFT), which is equivalent to ~1 mm in deformation monitoring for C-band InSAR measurements.

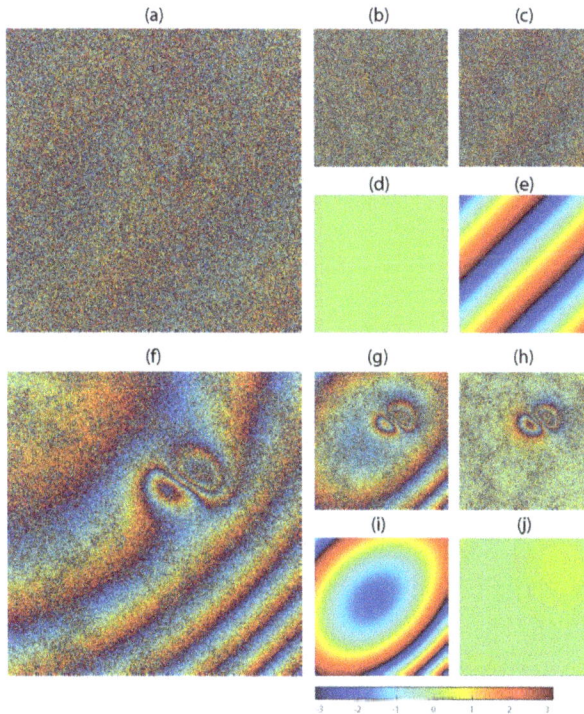

Figure 2. Simulated linear and nonlinear orbit errors in InSAR interferograms. (**a**) noise added interferogram with averaged coherence $\gamma = 0.2$ and single *look*; (**b**) orbit-corrected interferogram using DFT; (**c**) orbit-corrected interferogram using ADP-Poly; the difference between truth and phase ramp estimated from DFT (**d**) and ADP-Poly (**e**); (**f**) noise added interferogram with averaged coherence $\gamma = 0.4$ and 2-*looks*; (**g**) orbit-corrected interferogram using DFT; (**h**) orbit-corrected interferogram using ADP-Poly; the difference between truth and phase ramp estimated from DFT (**i**) and ADP-Poly (**j**).

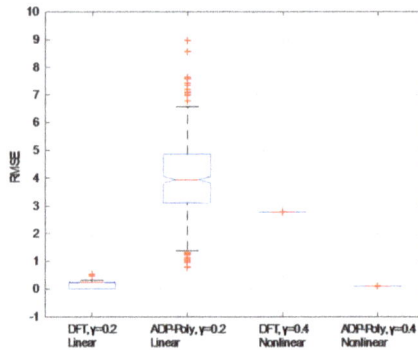

Figure 3. Accuracy assessment for DFT and ADP-Poly under different noise degree and phase ramp patterns.

3.2. Real Data

Four SAR datasets, covering the Datong and Tohoku-Oki areas, were used to validate the methods of orbital error correction, respectively. The information for the interferogram is summarized in Table 1. The pre-processing includes the co-registration to subpixel accuracy with cross correlation algorithm, multi-*looking* (3 × 15 *looks* for ASAR data set along range and azimuth, 4 × 4 *looks* for GF-3 with stripmap model, 9 × 3 *looks* for Sentinel-1 with TOPSAR mode), the topography–component removal using the 90-m Shuttle Radar Topography Mission v. 4.1 digital elevation model, and the adaptive Goldstein filtering. The atmospheric delay is not considered any further due to the lack of the external data. For this paper, we assume that this component, together with orbital error, comprises the linear/nonlinear phase ramps.

Table 1. ENVISAT ASAR acquisitions over Datong and Tohoku-Oki areas.*

Location	Sensor	Track	Master (yyyy-mm-dd)	Slave (yyyy-mm-dd)	$B_\perp$ (m)	Pass	Char. Def.
Datong	GF-3	-	2017-04-01	2017-06-27	536	D	local
Datong	Sentinel-1	40	2015-10-15	2015-10-27	87	A	local
Tohoku-Oki	ASAR	347	2011-02-19	2011-03-21	163	D	global
Tohoku-Oki	ASAR	74	2011-03-02	2011-04-01	−121	D	global

* Columns show location, satellite platform, track number, acquisition dates, perpendicular baseline, satellite direction (ascending or descending) and characteristics of deformations. "-" denotes unknown track number (Orbit number: 003387/004641).

3.2.1. Datong Area

The deformation feature in the GF-3 interferogram is locally distributed because of the dynamics of land subsidence caused by underground mining activities over the Datong area, China. Errors in the GF-3 satellite state vectors can be observed in the differential interferogram (Figure 4c) and cause a linear phase ramp along the azimuth direction. Because of the relatively low coherence over the area (Figure 4b), we employed DFT to remove the long-wavelength signals (Figure 4d) without the phase unwrapping. The results in the amplified sub-regions (1–3) show that mining-related surface deformation could be monitored after removal of orbital errors. The multi-looking GF-3 SAR image is shown in Figure 4a.

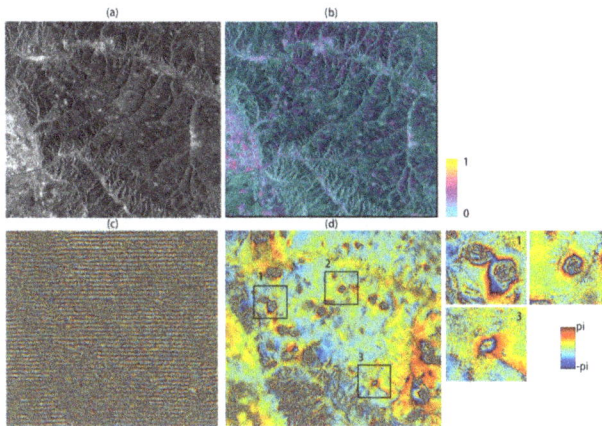

Figure 4. Orbital error correction for GF-3 sensor over Datong area: (**a**) SAR intensity acquired on 1 April 2017; (**b**) corresponding coherence map; (**c**) original interferogram with linear phase ramp; (**d**) de-ramping interferogram estimated from DFT. The sub-image 1–3 corresponds to frames shown in Figure 4d.

In contrast to GF-3, the better orbital qualities of Sentinel-1 instruments mitigate the phase ramp in the interferogram (Figure 5c). After burst and sub-swath merging, a nonlinear phase ramp with a relatively large spatial coverage can be observed in the upper right part of image. To remove the phase ramp, ADP-Poly is motivated by the high coherence magnitude (Figure 5b). A ten-fold cross validation was performed to select the optimal order of polynomial model. It can be seen from Figure 6a that the overfitting causes a larger bias for regression with the increase in order in polynomial, while the polynomial coefficient with $n = 3$ and $m = 3$ reaches the minimum WRMSE (2.15 rad, corresponding to ~9 mm). The final de-ramping interferogram is shown in Figure 5d, where the spatially distributed subsidence bowls can be clearly seen. The multi-looking sentinel-1 SAR image is shown in Figure 5a.

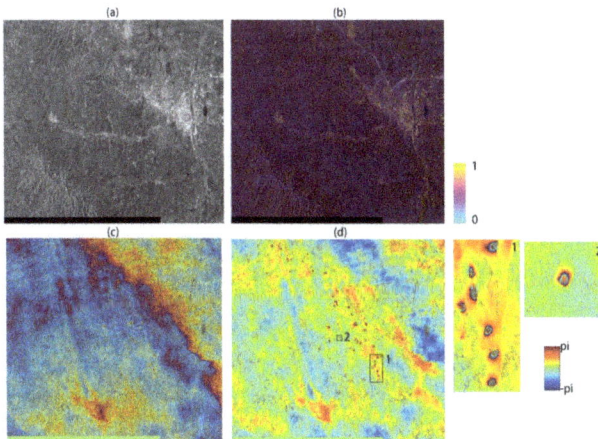

Figure 5. Orbital error correction for Sentinel-1 sensor over Datong area: (**a**) SAR intensity acquired on 15 October 2015; (**b**) corresponding coherence map; (**c**) original interferogram with nonlinear phase ramp; (**d**) de-ramping interferogram estimated from ADP-Poly. The sub-image 1–2 corresponds to frames shown in Figure 5d.

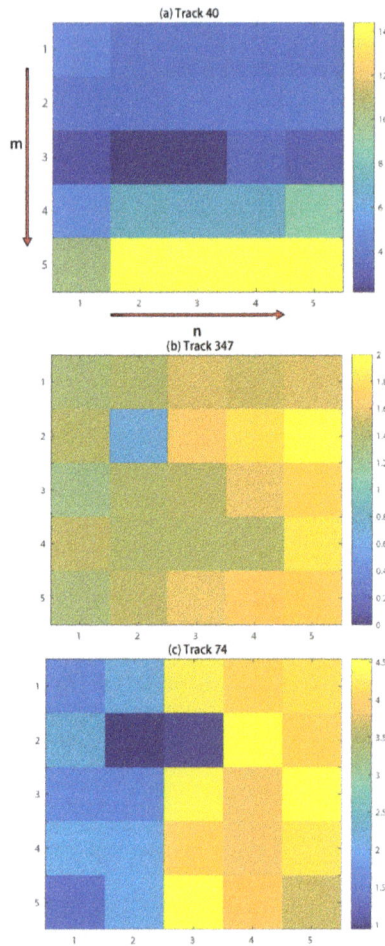

Figure 6. Adaptive model selection of ADP-Poly for (**a**) track 40 (Sentinel-1); (**b**) track 347 (ASAR) and (**c**) track 74 (ASAR).

3.2.2. Tohoku-Oki Area

For long-wavelength displacement signals covering the whole image, external GPS measurements should be used to correct the displacement before phase-ramp estimation. To this end, GPS observations, including total 120 stations, were collected from the Advanced Rapid Imaging and Analysis (ARIA) team at Jet Propulsion Laboratory (JPL) and Caltech [20]. During the process, the GPS observations were first projected into the LOS direction using the central unit vector for east, north and vertical directions in descending ASAR (Table 2). Then, all GPS locations were transformed into radar coordinates. The nearest SAR neighbors, with respect to each GPS measurement, were averaged using 3×3 boxcar windows, and the weights were determined by means of averaged phase STD with a factor of $1/3$. When the phase ramp fitted using ADP-Poly, 90% of the GPS observations located in each SAR track were used to correct the deformation in unwrapped InSAR interferogram, and the remaining 10% were used to evaluate the discrepancies between the GPS observations and the orbit-corrected InSAR coseismic measurements.

Table 2. Deformation validation and RMSE before and after orbital error correction over Tohoku-Oki area.

Sensor	Track	Unit Vector of LOS [East North Up]	Number of GPS Station	RMSE before Correction (cm)	RMSE after Correction (cm)	RMSE Reduction (%)
ASAR	347	[0.64 0.11 0.75]	97	35.55	9.52	73.22
ASAR	74	[0.65 0.11 0.75]	23	12.24	8.39	31.45

We present the results of before and after ADP-Poly orbital error correction in Figures 6 and 7 and Table 2. It can be seen that ADP-Poly suggests a quadratic model to fit the ramps for both tracks (Figure 6b,c). The RMSE values, before and after the orbital error correction, are 35.55 cm and 9.52 cm for track 347, and 12.24 cm and 8.39 cm for track 74, respectively, demonstrating the usefulness of the ADP-poly method developed by the authors. After removal of phase ramp, the displacement caused by earthquakes is rewrapped to 11.8 cm for each fringe, as shown in Figure 7. As stated in previous studies [17,20,35], the discontinuity between GPS measurements and InSAR displacement can be attributed to unwrapping errors over very low coherence area and the aftershock and postseismic deformation that were not covered by the GPS observation period.

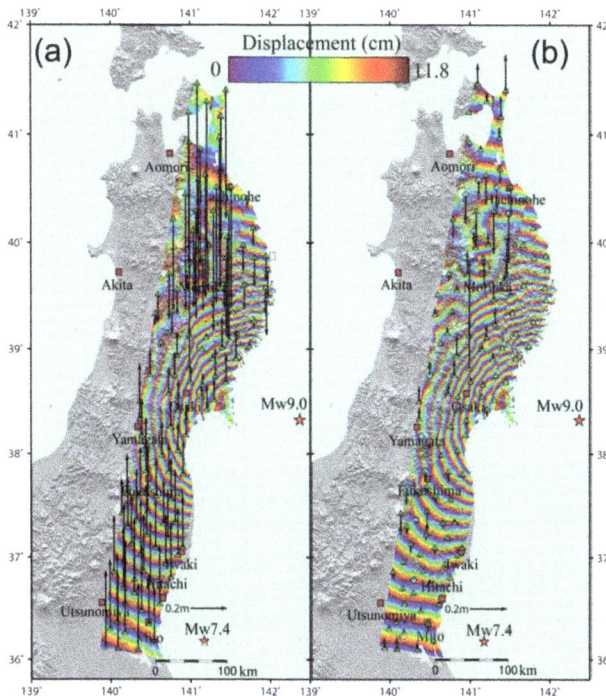

Figure 7. Coseismic deformation interferograms generated using data from ASAR descending tracks 347 and 74, before (**a**) and after (**b**) orbital error correction. The locations of the cities (khaki square) and GPS stations (green triangle) that are used to remove the orbital errors. The arrows indicate the biases in the InSAR results as calculated from the GPS observations. Each color cycle represents 11.8 cm of LOS displacement toward or away from the satellite. The epicenters of the magnitude 9.0 main shock and a magnitude 7.4 aftershock are marked as red stars.

4. Discussion

Two methods of modeling orbital error in InSAR interferograms that deploy DFT and the polynomial model with adaptive order determination were presented. DFT is a robust estimator of linear components of the long-wavelength signals, even in strong noise environment. It transfers spatial phase ramps into the frequency spectrum and finds the fringe rate at the peak location of the Fourier transformation. The interpolation of the DFT of the interferogram can better identify the exact fringe rate. A significant advantage of this method is that phase wrapping is not required, and, therefore, the errors caused by the poor quality of the interferograms are avoided. For nonlinear fringe patterns, the polynomial model (ADP-Poly) can compensate for nonlinear components that cannot be estimated with DFT. The development of adaptive order determination for the polynomial model using K-cross validation ensures the model can follow the features of orbital errors more accurately. This step is missed in previous studies, which usually determine the order empirically.

Two problems in orbital error correction should be further pointed out. Firstly, the approach has no capacity to distinguish between orbital errors and long-wavelength atmospheric delays. To our knowledge, this is a common problem in most related studies [12–14,20,23]. We therefore made the assumption that the long-wavelength signal only represents the orbital error, implying that part of the tropospheric delay will also be removed from the interferogram. A good alternative is to remove the atmospheric delays prior to orbital error correction, for example, by using the Generic Atmospheric Correction Online Service, applying the Iterative Tropospheric Decomposition model [34].

Secondly, manual masking is used to remove local deformation signals, leading to gaps in the local region, and so, no observation will yield contributions during de-ramping. To solve this problem, a deformation free interferogram can be inverted using a source geophysical model, e.g., Mogi's formulation for an infinitesimal spherical source in an elastic half-space, by finding the global minimum in the misfit space using a simulated annealing algorithm [36]. To do this, General Inversion of Phase Technique, which estimates parameters in a quantitative model directly from the wrapped-phase data, can be employed [36].

5. Conclusions

This paper has explored orbital error correction in InSAR interferogram using frequency (DFT) and spatial domain (ADP-Poly) based methods. The results reveal that the modeling of orbital error should follow the features of fringe patterns and the quality of the interferogram. To detect linear features of orbital error, the method based on maximum likelihood fringe rate estimation should be deployed. The main advantage of this technique is that it is workable over fast decorrelation areas without the need of unwrapping. However, only linear components can be estimated using this method. For nonlinear features, ADP-Poly, which determines the order of polynomial automatically using K-cross validation, should be used. We evaluated both methods using historical and modern sensors, including ENVISAT ASAR, GF-3 and Sentinel-1. The experiments show that both historical and modern systems can benefit from our method to varying degrees, leading to clearer ground deformation signals, both locally and globally.

Acknowledgments: The authors wish to thank Guangcai Feng, from Central South University, for helping to collect GPS data and drawing Figure 7. This work was supported by the National Natural Science Foundation of China under Grant 41471373, the ESA-MOST Dragon 4 project under Grant 32248_2, the Fundamental Research Funds for the Central Universities, Surveying and Mapping Research Fund of Jiangsu Province JSCHKY201601. The GPS displacement data (Version 0.3) was provided by the ARIA team at JPL based on GPS data from the Geospatial Information Authority of Japan. The ENVISAT ASAR raw data were supplied through the GEO Geohazards Supersite.

Author Contributions: Xin Tian designed and performed the experiments, and also wrote the article. Rakesh Malhotra revised the manuscript. Bing Xu performed GF-3 and Sentinel-1 data processing. Haoping Qi and Yuxiao Ma searched information about the data and contributed to the editing of the manuscript.

Conflicts of Interest: The authors declare no conflicts of interest.

Remote Sens. **2018**, *10*, 508

References

1. Berardino, P.; Fornaro, G.; Lanari, R.; Sansosti, E. A new algorithm for surface deformation monitoring based on small baseline differential sar interferograms. *IEEE Trans. Geosci. Remote Sens.* **2002**, *40*, 2375–2383. [CrossRef]

2. Ferretti, A.; Prati, C.; Rocca, F. Permanent scatterers in sar interferometry. *IEEE Trans. Geosci. Remote Sens.* **2001**, *39*, 8–20. [CrossRef]

3. Massonnet, D.; Feigl, K.L. Radar interferometry and its application to changes in the earth's surface. *Rev. Geophys.* **1998**, *36*, 441–500. [CrossRef]

4. Diao, F.; Xiong, X.; Wang, R.; Zheng, Y.; Walter, T.R.; Weng, H.; Li, J. Overlapping post-seismic deformation processes: Afterslip and viscoelastic relaxation following the 2011 M_w9.0 Tohoku (Japan) earthquake. *Geophys. J. Int.* **2013**, *196*, 218–229. [CrossRef]

5. Copley, A.; Hollingsworth, J.; Bergman, E. Constraints on fault and lithosphere rheology from the coseismic slip and postseismic afterslip of the 2006 M_w7.0 Mozambique earthquake. *J. Geophys. Res. Solid Earth* **2012**, *117*. [CrossRef]

6. Liu, Y.; Xu, C.; Li, Z.; Wen, Y.; Chen, J.; Li, Z. Time-dependent afterslip of the 2009 M_w6.3 Dachaidan earthquake (China) and viscosity beneath the qaidam basin inferred from postseismic deformation observations. *Remote Sens.* **2016**, *8*, 649. [CrossRef]

7. Fattahi, H.; Amelung, F. Insar uncertainty due to orbital errors. *Geophys. J. Int.* **2014**, *199*, 549–560. [CrossRef]

8. Thiel, C.; Schmullius, C. Impact of tree species on magnitude of palsar interferometric coherence over siberian forest at frozen and unfrozen conditions. *Remote Sens.* **2014**, *6*, 1124–1136. [CrossRef]

9. Zebker, H.A.; Villasenor, J. Decorrelation in interferometric radar echoes. *IEEE Trans. Geosci. Remote Sens.* **1992**, *30*, 950–959. [CrossRef]

10. Ding, X.-L.; Li, Z.-W.; Zhu, J.-J.; Feng, G.-C.; Long, J.-P. Atmospheric effects on insar measurements and their mitigation. *Sensors* **2008**, *8*, 5426–5448. [CrossRef] [PubMed]

11. Jiang, M.; Li, Z.; Ding, X.; Zhu, J.-J.; Feng, G. Modeling minimum and maximum detectable deformation gradients of interferometric sar measurements. *Int. J. Appl. Earth Obs. Geoinf.* **2011**, *13*, 766–777. [CrossRef]

12. Xu, B.; Li, Z.-W.; Wang, Q.-J.; Jiang, M.; Zhu, J.-J.; Ding, X.-L. A refined strategy for removing composite errors of sar interferogram. *IEEE Geosci. Remote Sens. Lett.* **2014**, *11*, 143–147. [CrossRef]

13. Rosen, P.A.; Hensley, S.; Peltzer, G.; Simons, M. Updated repeat orbit interferometry package released. *Eos Trans. Am. Geophys. Union* **2004**, *85*, 47. [CrossRef]

14. Knedlik, S.; Loffeld, O.; Hein, A.; Arndt, C. A novel approach to accurate baseline estimation. In Proceedings of the IEEE 1999 International Geoscience and Remote Sensing Symposium (IGARSS'99), Hamburg, Germany, 28 June–2 July 1999; pp. 254–256.

15. Pepe, A.; Berardino, P.; Bonano, M.; Euillades, L.D.; Lanari, R.; Sansosti, E. Sbas-based satellite orbit correction for the generation of dinsar time-series: Application to radarsat-1 data. *IEEE Trans. Geosci. Remote Sens.* **2011**, *49*, 5150–5165. [CrossRef]

16. Kohlhase, A.; Feigl, K.; Massonnet, D. Applying differential inSAR to orbital dynamics: A new approach for estimating ERS trajectories. *J. Geod.* **2003**, *77*, 493–502. [CrossRef]

17. Biggs, J.; Wright, T.; Lu, Z.; Parsons, B. Multi-interferogram method for measuring interseismic deformation: Denali fault, Alaska. *Geophys. J. Int.* **2007**, *170*, 1165–1179. [CrossRef]

18. Bähr, H.; Hanssen, R.F. Reliable estimation of orbit errors in spaceborne SAR interferometry. *J. Geod.* **2012**, *86*, 1147–1164. [CrossRef]

19. Wang, H.; Wright, T.; Biggs, J. Interseismic slip rate of the northwestern Xianshuihe fault from inSAR data. *Geophys. Res. Lett.* **2009**, *36*. [CrossRef]

20. Feng, G.; Ding, X.; Li, Z.; Jiang, M.; Zhang, L.; Omura, M. Calibration of an insar-derived coseimic deformation map associated with the 2011 M_w-9.0 Tohoku-oki earthquake. *IEEE Geosci. Remote Sens. Lett.* **2012**, *9*, 302–306. [CrossRef]

21. Béjar-Pizarro, M.; Socquet, A.; Armijo, R.; Carrizo, D.; Genrich, J.; Simons, M. Andean structural control on interseismic coupling in the north Chile subduction zone. *Nat. Geosci.* **2013**, *6*, 462–467. [CrossRef]

22. Gourmelen, N.; Amelung, F.; Lanari, R. Interferometric synthetic aperture radar–GPS integration: Interseismic strain accumulation across the Hunter Mountain fault in the eastern California shear zone. *J. Geophys. Res. Solid Earth* **2010**, *115*. [CrossRef]

23. Shirzaei, M.; Walter, T.R. Estimating the effect of satellite orbital error using wavelet-based robust regression applied to inSAR deformation data. *IEEE Trans. Geosci. Remote Sens.* **2011**, *49*, 4600–4605. [CrossRef]

24. Bähr, H. *Orbital Effects in Spaceborne Synthetic Aperture Radar Interferometry*; KIT Scientific Publishing: Karlsruhe, Germany, 2013.

25. Zebker, H.A.; Chen, K. Accurate estimation of correlation in inSAR observations. *IEEE Geosci. Remote Sens. Lett.* **2005**, *2*, 124–127. [CrossRef]

26. Spagnolini, U. 2-D phase unwrapping and instantaneous frequency estimation. *IEEE Trans. Geosci. Remote Sens.* **1995**, *33*, 579–589. [CrossRef]

27. Jiang, M.; Ding, X.; Li, Z. Hybrid approach for unbiased coherence estimation for multitemporal inSAR. *IEEE Trans. Geosci. Remote Sens.* **2014**, *52*, 2459–2473. [CrossRef]

28. Holland, P.W.; Welsch, R.E. Robust regression using iteratively reweighted least-squares. *Commun. Stat. Theory Methods* **1977**, *6*, 813–827. [CrossRef]

29. Bishop, C.M. *Pattern Recognition and Machine Learning*; Springer: Berlin, Germany, 2006.

30. Okada, Y. Surface deformation due to shear and tensile faults in a half-space. *Bull. Seismol. Soc. Am.* **1985**, *75*, 1135–1154.

31. Jiang, M.; Ding, X.; Li, Z.; Tian, X.; Zhu, W.; Wang, C.; Xu, B. The improvement for baran phase filter derived from unbiased inSAR coherence. *IEEE J. Sel. Top. Appl. Earth Obs. Remote Sens.* **2014**, *7*, 3002–3010. [CrossRef]

32. Hanssen, R.F. *Radar Interferometry: Data Interpretation and Error Analysis*; Springer: New York, NY, USA, 2001; Volume 2.

33. Jiang, M.; Yong, B.; Tian, X.; Malhotra, R.; Hu, R.; Li, Z.; Yu, Z.; Zhang, X. The potential of more accurate insar covariance matrix estimation for land cover mapping. *ISPRS J. Photogramm. Remote Sens.* **2017**, *126*, 120–128. [CrossRef]

34. Chen, C.W.; Zebker, H.A. Phase unwrapping for large SAR interferograms: Statistical segmentation and generalized network models. *IEEE Trans. Geosci. Remote Sens.* **2002**, *40*, 1709–1719. [CrossRef]

35. Jónsson, S.; Zebker, H.; Segall, P.; Amelung, F. Fault slip distribution of the 1999 M_w7.1 Hector mine, California, earthquake, estimated from satellite radar and GPS measurements. *Bull. Seismol. Soc. Am.* **2002**, *92*, 1377–1389. [CrossRef]

36. Feigl, K.L.; Thurber, C.H. A method for modelling radar interferograms without phase unwrapping: Application to the M 5 Fawnskin, California earthquake of 1992 December 4. *Geophys. J. Int.* **2009**, *176*, 491–504. [CrossRef]

remote sensing

MDPI

Article

Ionospheric Reconstructions Using Faraday Rotation in Spaceborne Polarimetric SAR Data

Cheng Wang [1,*], Liang Chen [1], Haisheng Zhao [2], Zheng Lu [3], Mingming Bian [3], Running Zhang [3] and Jian Feng [2]

[1] Qian Xuesen Laboratory of Space Technology, China Academy of Space Technology, Beijing 100094, China; chenliang@qxslab.cn

[2] National Key Laboratory of Electromagnetic Environment, China Research Institute of Radiowave Propagation, Qingdao 266107, China; zhaohaisheng213@163.com (H.Z.); fengjian428@163.com (J.F.)

[3] Beijing Institute of Spacecraft System Engineering, China Academy of Space Technology, Beijing 100094, China; lvzheng_cast@163.com (Z.L.); bianmingming2008@163.com (M.B.); 13661051645@139.com (R.Z.)

* Correspondence: solskjaer2006@126.com; Tel.: +86-186-1815-4639

Received: 3 August 2017; Accepted: 13 November 2017; Published: 14 November 2017

Abstract: It is well known that the Faraday rotation (FR) is obviously embedded in spaceborne polarimetric synthetic aperture radar (PolSAR) data at L-band and lower frequencies. By model inversion, some widely used FR angle estimators have been proposed for compensation and provide a new field in high-resolution ionospheric soundings. However, as an integrated product of electron density and the parallel component of the magnetic field, FR angle measurements/observations demonstrate the ability to characterize horizontal ionosphere. In order to make a general study of ionospheric structure, this paper reconstructs the electron density distribution based on a modified two-dimensional computerized ionospheric tomography (CIT) technique, where the FR angles, rather than the total electron content (TEC), are regarded as the input. By using the full-pol (full polarimetric) data of Phase Array L-band Synthetic Aperture Radar (PALSAR) on board Advanced Land Observing Satellite (ALOS), International Reference Ionosphere (IRI) and International Geomagnetic Reference Field (IGRF) models, numerical simulations corresponding to different FR estimators and SAR scenes are made to validate the proposed technique. In simulations, the imaging of kilometer-scale ionospheric disturbances, a spatial scale that is rarely detectable by CIT using GPS, is presented. In addition, the ionospheric reconstruction using SAR polarimetric information does not require strong point targets within a SAR scene, which is necessary for CIT using SAR imaging information. Finally, the effects of system errors including noise, channel imbalance and crosstalk on the reconstruction results are also analyzed to show the applicability of CIT based on spaceborne full-pol SAR data.

Keywords: Faraday rotation; polarimetric synthetic aperture radar; Phase Array L-band Synthetic Aperture Radar; computerized ionospheric tomography; International Reference Ionosphere; International Geomagnetic Reference Field

1. Introduction

Due to the dispersive nature of ionosphere and the existence of Earth's magnetic field, the polarization rotation of a linearly polarized wave will occur after traveling through the ionosphere. This phenomenon is known as Faraday rotation (FR) and depends on the frequency, the electron density, the Earth's magnetic field, and the geometry of observation [1]. For spaceborne polarimetric synthetic aperture radar (PolSAR) systems at L-band and lower frequencies, FR will distort the scattering matrix (i.e., complex backscattering coefficients in the four channels of PolSAR) and become a significant error source [2]. Thus for a space-borne PolSAR system, some mitigation techniques are required.

It is known that the key in mitigation techniques is retrieval of the accurate FR angle by measuring the polluted scattering matrix. On the other hand, the FR retrieval from space using PolSAR is also a new capability of high-resolution ionospheric sounding, and various FR estimators have been proposed. After the transformation from Cartesian linear polarization to circular polarization, Bickel and Bates [3] propose a widely used FR estimator. The scattering matrix can be converted to the covariance matrix. By measuring the covariance matrix, one FR estimator is proposed by Freeman [2]. Chen and Quegan [4] propose six further FR estimators based on the off-diagonal terms of covariance matrix. It is important to note that none of above estimators is insensitive to all system errors (i.e., system noise, channel phase/amplitude imbalance, and crosstalk) and scattering types in SAR scenes. For example, the Chen and Quegan's third estimator is the preferred one to channel amplitude imbalance but worse than Bickel and Bates estimator when the channel phase imbalance is the dominant error [4]. As discussed by Rogers and Quegan [5], the performance of Chen and Quegan's third estimator is scattering dependent. Thus, in order to obtain the accurate FR angle, the choice of FR estimator should depend on the domain error and scattering types.

Assume the magnetic field along the path is approximately equal to a median value; the vertical TEC (i.e., the integration of electron density) distribution with kilometer-scale in terms of latitude and longitude can further be obtained, which is clearly beneficial to the studies of small-scale ionospheric features [6–11]. However, the information of TEC distribution is still limited to the detection of ionospheric horizontal structure. Compared with TEC, the spatial distribution of electron density can give a better study of ionospheric inhomogeneity or irregularity caused by the magnetic storms, earthquakes, etc. [12–15]. Thus, the electron density reconstruction based on the computerized ionospheric tomography (CIT) technique is required. By setting a series of ground-based GPS (Global Positioning System) receivers, the CIT technique was proposed to reconstruct the electron density distribution [16,17]. The TEC values for different look angles can be retrieved from GPS signals and regarded as the input of CIT. However, it is only suitable for hundred kilometers scale electron density monitoring [13–15]. In order to improve the resolution, previous studies have considered the CIT based on the information of spaceborne SAR imaging [18–20]. After the signal has passed twice through the ionosphere, its linear frequency modulated (FM) rate will be changed [21]. An autofocus algorithm is applied here to iteratively search the change of FM rate, which can further be used to derive the TEC value [22]. Although it can provide a high resolution reconstruction, the autofocus algorithm is insensitivity to TEC because of the limitation of small bandwidth for current low-frequency spaceborne SAR systems, e.g., the ALOS Phase Array L-band Synthetic Aperture Radar (PALSAR) [23,24]. In addition, it also requires strong point targets with high signal-to-clutter (SCR) ratio in a SAR scene [22,25].

In contrast to the TEC retrieval based on SAR imaging information, the TEC derived from FR using polarimetric information is independent of above limitations [10,26,27]. Recently, we have reconstructed the ionosphere by using the TEC values derived from FR [28]. However, as discussed above, the TEC derived from FR will introduce the error that the magnetic field must be approximated by a fixed value. Thus, in order to avoid it, the CIT reconstruction based on spaceborne PolSAR will be a promising direction where the FR angles are directly regarded as the input. FR can be defined as the integration of electron density weighted by the magnetic field along the ray path. Since in most cases, the magnetic field distribution is known with high precision from the International Geomagnetic Reference Field (IGRF) model, this information can be used to realize the final reconstruction of electron density after modifying the traditional CIT technique. In addition, this paper also focuses on the systems errors on the proposed CIT reconstruction. We start with a brief review of the main FR estimators from the full-pol data in Section 2. By using the PALSAR full-pol data sets, International Reference Ionosphere (IRI) and IGRF models, a modified two-dimensional CIT technique based on the spaceborne PolSAR system is analyzed in Section 3. The effects of system errors on the reconstructions are analyzed in Section 4. In addition, the results based on different FR estimators are also compared. Last, our conclusions are presented in Section 5.

2. Review of FR Estimators Based on the Spaceborne PolSAR Data

Due to the existence of Earth's magnetic field in the ionosphere, a linearly polarized wave will split into ordinary and extraordinary waves with different phase velocities. The linearly polarized wave is therefore rotated by an angle called FR after traveling through the ionosphere, and can be derived by the half integration of the phase difference along the ray path [1,29]

$$\Omega = \frac{2.365 \times 10^4 \int_{path} Ne(s)|B(s)|\cos\theta_B(s)ds}{f_0^2} \tag{1}$$

where θ_B is the angle between signal propagation direction and magnetic field, $|B|$ is the magnitude of magnetic field, Ne is the electron density (unit is electrons·m^{-3}), and f_0 is the frequency. We can see that Ω is inversely proportional to the square of the frequency and depends on the θ_B, $|B|$ and Ne along the path. For a full-pol SAR system at L-band or lower, all the linearly polarized waves in each channel will encounter the ionospheric effects. The measured scattering matrix can then be written as Rogers and Quegan [5]:

$$\begin{bmatrix} M_{hh} & M_{vh} \\ M_{hv} & M_{vv} \end{bmatrix} = \begin{bmatrix} 1 & \delta_2 \\ \delta_1 & f_1 \end{bmatrix} \times \begin{bmatrix} \cos\Omega & \sin\Omega \\ -\sin\Omega & \cos\Omega \end{bmatrix} \times \begin{bmatrix} S_{hh} & S_{vh} \\ S_{hv} & S_{vv} \end{bmatrix} \times \begin{bmatrix} \cos\Omega & \sin\Omega \\ -\sin\Omega & \cos\Omega \end{bmatrix} \times \begin{bmatrix} 1 & \delta_3 \\ \delta_4 & f_2 \end{bmatrix} + \begin{bmatrix} N_{hh} & N_{vh} \\ N_{hv} & N_{vv} \end{bmatrix} \tag{2}$$

Here, N_{hh}, N_{vh}, N_{hv} and N_{vv} are the independent complex Gaussian noise in each measurement, S_{hh}, S_{vh}, S_{hv} and S_{vv} are the true scattering matrix, f_1 and f_2 denote the channel imbalance on receive and transmit, respectively δ_1 and δ_2 are the crosstalk on receive, and δ_3 and δ_4 are the crosstalk on transmit. In an ideal system, the noise and crosstalk are zero, and the channel imbalance is equal to 1, the covariance matrix can then be derived as follow as Freeman [2]

$$\begin{bmatrix} C_{11} & C_{12} & C_{13} & C_{14} \\ C_{21} & C_{22} & C_{23} & C_{24} \\ C_{31} & C_{32} & C_{33} & C_{34} \\ C_{41} & C_{42} & C_{43} & C_{44} \end{bmatrix} = \begin{bmatrix} \overline{M_{hh}M_{hh}^*} & \overline{M_{hh}M_{vh}^*} & \overline{M_{hh}M_{hv}^*} & \overline{M_{hh}M_{vv}^*} \\ \overline{M_{vh}M_{hh}^*} & \overline{M_{vh}M_{vh}^*} & \overline{M_{vh}M_{hv}^*} & \overline{M_{vh}M_{vv}^*} \\ \overline{M_{hv}M_{hh}^*} & \overline{M_{hv}M_{vh}^*} & \overline{M_{hv}M_{hv}^*} & \overline{M_{hv}M_{vv}^*} \\ \overline{M_{vv}M_{hh}^*} & \overline{M_{vv}M_{vh}^*} & \overline{M_{vv}M_{hv}^*} & \overline{M_{vv}M_{vv}^*} \end{bmatrix} \tag{3}$$

where $\overline{\bullet}$ and $(\ )^*$ represent averaging and conjugate, respectively. By assuming reflection symmetry, Freeman [2] has proposed one FR estimator formulated as

$$\Omega_F = \pm\frac{1}{2}\tan^{-1}\left(\sqrt{\frac{C_{22} + C_{33} - 2\Re(C_{23})}{C_{11} + C_{44} + 2\Re(C_{14})}}\right) \tag{4}$$

where

$$\frac{C_{22} + C_{33} - 2\Re(C_{23})}{C_{11} + C_{44} + 2\Re(C_{14})} = \frac{\sin^2 2\Omega}{\cos^2 2\Omega} \tag{5}$$

and $\Re(\bullet)$ denotes the real part. According to the off-diagonal terms of Equation (3), Chen and Quegan [4] proposed six further FR estimators, where the third one performs the best. This estimator can be written as

$$\Omega_C = \frac{1}{2}\arg\left(\Im(C_{14}) + i\Im\left(\frac{C_{13} + C_{34} - C_{12} - C_{24}}{2}\right)\right) \tag{6}$$

where

$$\begin{aligned} \Im(C_{14}) &= \overline{\Im(S_{hh}S_{vv}^*)}\cos 2\Omega \\ \Im(C_{13} + C_{34} - C_{12} - C_{24}) &= 2\overline{\Im(S_{hh}S_{vv}^*)}\sin 2\Omega \end{aligned} \tag{7}$$

and $\Im(\bullet)$ denotes the imaginary part. Bickel and Bates [3] note that, in the absence of system errors and assuming reciprocity, the scattering of Equation (2) can be transformed into a circular polarization basis, that is,

$$
\begin{bmatrix} Z_{11} & Z_{12} \\ Z_{21} & Z_{22} \end{bmatrix} = \begin{bmatrix} 1 & i \\ i & 1 \end{bmatrix} \times \begin{bmatrix} M_{hh} & M_{vh} \\ M_{hv} & M_{vv} \end{bmatrix} \times \begin{bmatrix} 1 & i \\ i & 1 \end{bmatrix} = \begin{bmatrix} S_{hh} - S_{vv} + 2iS_{hv} & (S_{hh} + S_{vv})\exp\left(i\frac{\pi}{2} - i2\Omega\right) \\ (S_{hh} + S_{vv})\exp\left(i\frac{\pi}{2} + i2\Omega\right) & S_{vv} - S_{hh} + 2iS_{hv} \end{bmatrix} \tag{8}
$$

Thus, the FR estimated from Equation (8) can be written as

$$
\Omega_{B\&B} = -\frac{1}{4}\arg\left(\overline{Z_{12} \times Z_{21}^*}\right) \tag{9}
$$

3. Principle of the Proposed CIT Using FR Angles

For the traditional CIT reconstruction, the TEC value is first retrieved and regarded as the input parameter. The well-known multiplicative algebraic reconstruction technique (MART) using iterative scheme is then applied to reconstruct the true electron density distribution [16,17,30]. In the iterative scheme of MART, the electron density distribution obtained from IRI model is used as the initial value. However, for the proposed CIT based on the spaceborne PolSAR data, the FR angle, rather than TEC value, is first retrieved and regarded as the input. Correspondingly, the spatial distribution of the product of electron density and the magnitude of magnetic field are used as the initial value in iteration.

The map of the proposed two-dimensional CIT technique is shown in Figure 1, where the annotations in red denote the main differences from traditional CIT due to the consideration of magnetic field [28]. The magnetic field varies in altitude and azimuth directions. The whole ionosphere region of interest is subdivided into H grids, where the product of electron density and magnetic field, rather than only electron density for traditional CIT, are constant in each grid. It is assumed that the whole synthetic aperture of PolSAR is divided into L sub-apertures and corresponding sampling position is located at the center of each sub-aperture. For traditional CIT simulations based on GPS [14–17], a set of ground-based receiver stations are required. At each sampling position of GPS, different TEC values corresponding to different receiver stations can then be obtained. From Figure 1, the explored ground scene is divided into K subimages, which is similar to the ground-based receiver. At one sub-aperture, by applying averaging within each subimage, the distorted scattering matrix can be measured and corresponding K FR angles values, which are assumed as the integration of the product of electron density and magnetic field from the sampling position to the center of each subimage, can be retrieved. That means there are $K \cdot L$ FR angle values in one CIT simulation. According to Equation (1) and Figure 1, each FR value can then be estimated with a discrete sum, i.e.,

$$
\begin{cases}
\frac{f_0^2\Omega_1}{2.365\cdot10^4} = a_{11}\cos\theta_{B11}Ne_1|B|_1 + \cdots + a_{1q}\cos\theta_{B1q}Ne_q|B|_q + \cdots a_{1H}\cos\theta_{B1H}Ne_H|B|_H \\
\vdots \\
\frac{f_0^2\Omega_p}{2.365\cdot10^4} = a_{p1}\cos\theta_{Bp1}Ne_1|B|_1 + \cdots + a_{pq}\cos\theta_{Bpq}Ne_q|B|_q + \cdots a_{pH}\cos\theta_{BpH}Ne_H|B|_H \\
\vdots \\
\frac{f_0^2\Omega_{K\cdot L}}{2.365\cdot10^4} = a_{K\cdot L\cdot1}\cos\theta_{BK\cdot L\cdot1}Ne_1|B|_1 + \cdots + a_{K\cdot L\cdot q}\cos\theta_{BK\cdot L\cdot q}Ne_q|B|_q + \cdots a_{K\cdot L\cdot H}\cos\theta_{BK\cdot L\cdot H}Ne_L|B|_H
\end{cases} \tag{10}
$$

here, Ω_p denotes the pth FR angles, Ne_q and $|B|_q$ are the electron density and magnitude of magnetic field in grid point q, respectively. a_{pq} and θ_{Bpq} are the projection length and angle between the pth ray path and magnetic field in grid point q, respectively. It should be noted that θ_{Bpq} is not considered in traditional CIT. Figure 1 shows that both Ne_q and $|B|_q$ are constants in each grid point, and both a_{pq}

and θ_{Bpq} are relative to the geometry of ray path. Thus, Ne_q and $|B|_q$ as well as a_{pq} and θ_{Bpq} can be regarded as a whole. The iterative equation is then written as

$$(Ne|B|)_q^{(l+1)} = (Ne|B|)_q^{(l)} \left(\frac{f_0^2 \Omega_p}{2.365 \times 10^4 \left\langle (a\cos\theta_B)_p^T, (Ne|B|)^{(l)} \right\rangle} \right)^{\lambda_k (a\cos\theta_B)_{pq} / \|(a\cos\theta_B)_p\|} \tag{11}$$

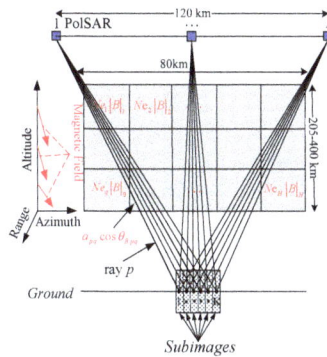

Figure 1. Map of two-dimensional CIT using FR angles.

Equation (11) means the $(l + 1)$th iterative result of $Ne|B|$ in grid q. $\langle \bullet \rangle$, $\|\bullet\|$, and $(\bullet)^T$ denotes the inner product, norm, and transposition, respectively. λ_k is the relaxation factor and is set to 0.5, and $f_0 = 1.27$ GHz. The initial distribution of electron density and the magnetic field are derived from IRI and IGRF models, respectively. When the values of all grids satisfy the terminating threshold after several iterations, i.e., the root-mean-square (RMS) of the difference of two adjacent iterations is smaller than a specified value $\Delta\zeta = 1 \times 10^8$ electrons $\cdot$ m^{-3}, the final spatial distribution $(Ne|B|)_{q\_final}$ will be reconstructed. It should be noted that the true magnetic field can be accurately obtained from the IGRF model. That means during the process of iteration, $|B|_q$ is always unchanged and equal to the initial values. Thus, the final spatial distribution of Ne_{q_final} can further be obtained by removing the $|B|_q$ in $(Ne|B|)_{q\_final}$. Above processes of the proposed CIT technique are also shown in the flowchart in Figure 2.

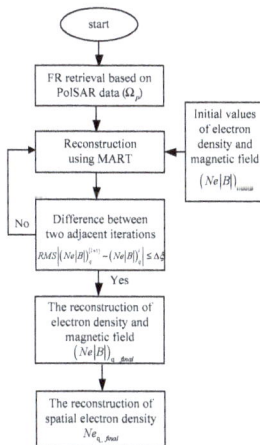

Figure 2. The flowchart of the proposed CIT technique.

4. Error Analysis of the Proposed CIT Technique

In order to analyze the effects of system errors (system noise, channel phase/amplitude imbalance, and crosstalk) on the proposed CIT individually, the semi-physical simulations using synthetic data of calibrated PALSAR full-pol data, IRI and IGRF models are required. In numerical simulations, two calibrated PALSAR full-pol data sets with different scattering types are used, namely, one from an area near Changbai Mountain (42.17°N, 128.0°E) acquired on 3 December 2007 and one from a seaside area of Qingdao (35.83°N, 120.75°E) acquired on 29 March 2011. Figure 3 shows the corresponding Pauli false-color images [31], and both data sets are composed of 1200 × 8000 pixels in range (*x*-axis) and azimuth directions (*y*-axis). *N* denotes geographical North. The resolutions are about 9.369 m and 3.557 m in range and azimuth directions, respectively.

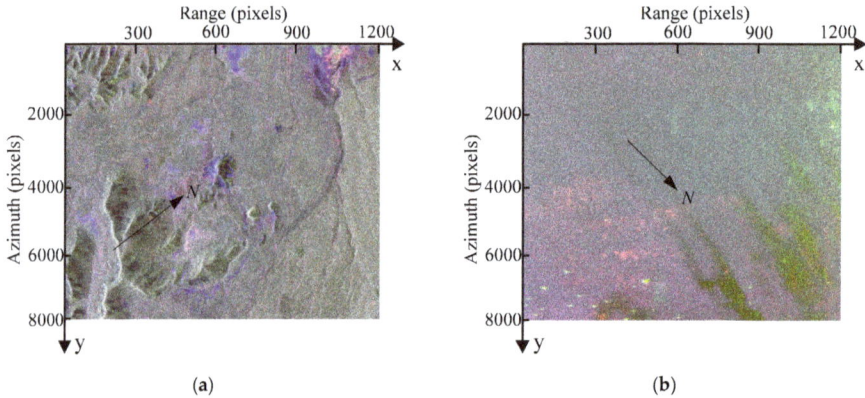

Figure 3. The PALSAR polarimetric images in two areas. (**a**) Changbai Mountain (42.17°N, 128.0°E) acquired on 3 December 2007 and (**b**) Qingdao (35.83°N, 120.75°E) acquired on 29 March 2011.

According to the operations in Figure 1, we assume that during one CIT simulation, the whole orbiting length of PolSAR along the azimuth direction is about 120 km and has 75 sampling positions (*N* = 75). The reconstructed area of the ionosphere is set to 80 km long and 204–400 km along azimuth and altitude directions, respectively. Correspondingly, each grid spacing is set to 5 km and 2.5 km along altitude and azimuth directions, respectively. For the distribution of the electron density, two small-scale artificial disturbances are then embedded. This new distribution Ne_{q_true} with ionospheric disturbances is regarded as the "true distribution" to be reconstructed and used for comparison, as shown in Figure 4. On the ground, the imaging scene will be divided into 16 parts along azimuth direction, where each subimage is composed of 1200 × 500 pixels. Thus, according to the position of azimuth sampling, vectors of ray path and magnetic field, the $a_{pq} \cos \theta_{Bpq}$ can be calculated. By combining $a_{pq} \cos \theta_{Bpq}$, Ne_{q_true} and $|B|_q$ into Equation (10), the true Ω_{p_true} corresponding to each ray path can further be determined. The scattering matrix measured for each ray path is then corrupted by the system errors and Ω_{p_true}.

Based on the proposed CIT method without system errors, Figures 5a and 6a show the two-dimensional CIT reconstructions in Changbai and Qingdao, respectively. Figures 5b and 6b show the corresponding two-dimensional absolute deviations between the true (i.e., Figure 4a,b) and reconstructed distributions (i.e., Figures 5a and 6a), and the RMS over the whole images are 1.96×10^9 electrons·m^{-3} and 2.21×10^9 electrons·m^{-3}, respectively. Similar, Figures 5c and 6c are the two-dimensional CIT reconstructions based on the traditional CIT method [28], where TEC is regarded as the input and first derived from FR when the magnetic field is approximated by a fix value of 300 km. The corresponding absolute deviations are shown in Figures 5d and 6d. The RMS of Figures 5d and 6d are 2.92×10^9 electrons·m^{-3} and 4.29×10^9 electrons·m^{-3}, respectively. We can see that due to the errors caused by the approximation of magnetic field, obvious errors occur for traditional CIT. Thus, compared with traditional CIT, the proposed CIT method can avoid the approximate error of magnetic field.

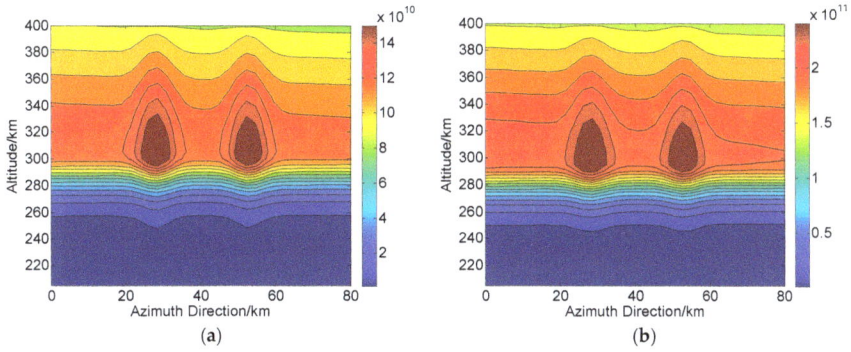

Figure 4. The true distributions of electron density in areas of (**a**) Changbai and (**b**) Qingdao. The unit of the electron density is electrons $\cdot$ m^{-3}.

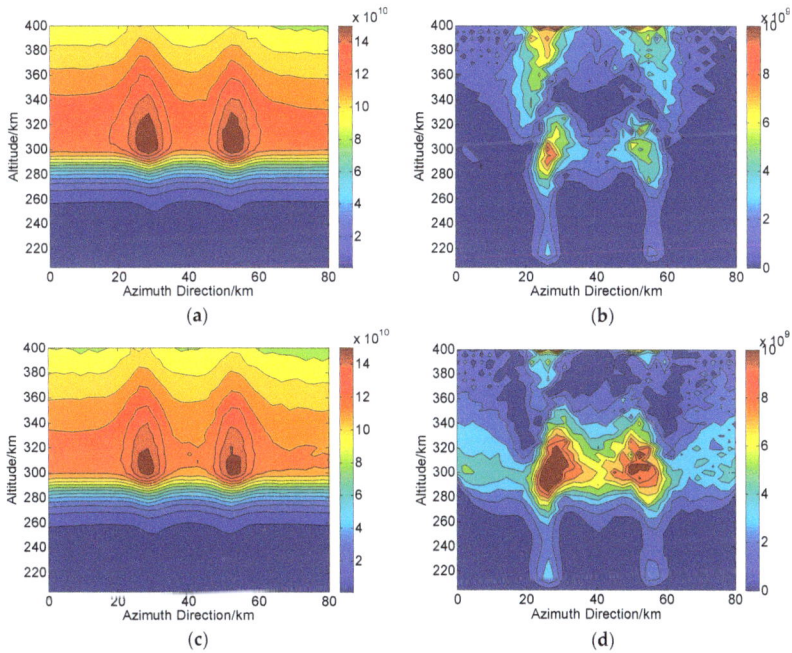

Figure 5. Two-dimensional CIT reconstructions without system errors in Changbai. (**a,c**) are the reconstructed results based on proposed and traditional CIT methods, respectively. (**b,d**) are the corresponding absolute deviations between the true and reconstructed results. The unit of the electron density is electrons $\cdot$ m^{-3}.

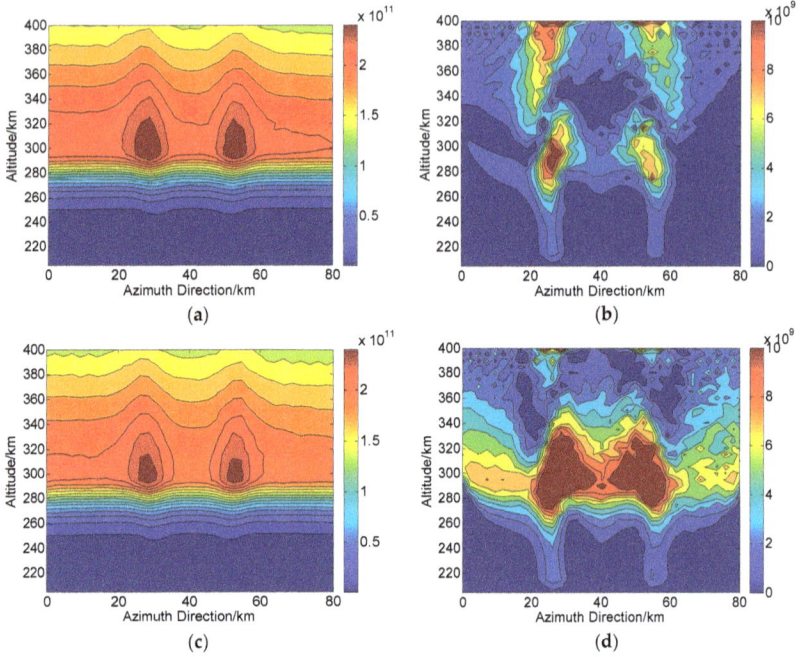

Figure 6. Two-dimensional CIT reconstructions without system errors in Qingdao. (**a,c**) are the reconstructed results based on proposed and traditional CIT methods, respectively. (**b,d**) are the corresponding absolute deviations between the true and reconstructed results. The unit of the electron density is electrons $\cdot$ m^{-3}.

4.1. CIT Reconstructions Under the Condition of System Noise

Assume the system noise and FR are the only errors, and the noise power in each channel is the same (i.e., $\overline{|N_{hh}|^2} = \overline{|N_{hv}|^2} = \overline{|N_{vh}|^2} = \overline{|N_{vv}|^2}$), the scattering matrix of Equation (2) can be written as (see Appendix A)

$$
\begin{aligned}
M_{hh\_n} &= S_{hh}\cos^2\Omega - S_{vv}\sin^2\Omega + N_{hh} \\
M_{vh\_n} &= S_{vh} + (S_{hh} + S_{vv})\sin\Omega\cos\Omega + N_{vh} \\
M_{hv\_n} &= S_{hv} - (S_{hh} + S_{vv})\sin\Omega\cos\Omega + N_{hv} \\
M_{vv\_n} &= S_{vv}\cos^2\Omega - S_{hh}\sin^2\Omega + N_{vv}
\end{aligned}
\tag{12}
$$

According to the process of CIT discussed above, we first evaluate the performances of the FR estimators, as shown in Tables 1–3. The RMS error, bias and standard deviation (SD) are defined as

$$
\begin{aligned}
\sigma_{RMS} &= \sqrt{\frac{100}{K\times L}\sum_{p=1}^{K\cdot L}\left(\frac{\Omega_{p\_re}-\Omega_{p\_true}}{\Omega_{p\_true}}\right)^2} \\
\sigma_{bias} &= \frac{100}{K\times L}\sum_{p=1}^{K\cdot L}\left(\frac{\Omega_{p\_re}-\Omega_{p\_true}}{\Omega_{p\_true}}\right) \\
\sigma_{sd} &= \sqrt{\frac{1}{K\times L}\sum_{p=1}^{K\cdot L}\left(100\cdot\left(\frac{\Omega_{p\_re}-\Omega_{p\_true}}{\Omega_{p\_true}}\right)-\sigma_{bias}\right)^2}
\end{aligned}
\tag{13}
$$

where Ω_{p_re} (Ω_{BB}, Ω_F or Ω_C) is the retrieved FR value corresponding to each ray path. The units of σ_{RMS}, σ_{bias} and σ_{sd} are %. In addition, the signal-to-noise ratio (SNR) can be defined by Chen and Quegan [4]

$$SNR = \frac{\overline{|S_{hh}|^2 + |S_{hv}|^2 + |S_{vh}|^2 + |S_{vv}|^2}}{4\overline{|N_{hh}|^2}} \tag{14}$$

Table 1. RMS error (σ_{RMS}) under the condition of system noise.

SNR (dB)	RMS Error (%)					
	Changbai			Qingdao		
	Ω_{BB}	Ω_F	Ω_C	Ω_{BB}	Ω_F	Ω_C
5	19.1023	866.7750	3.1956	10.6599	457.1695	14.1860
10	9.2369	475.1484	1.4204	4.0307	237.6931	7.8032
15	3.4335	239.2308	0.8279	1.3728	109.9127	4.0608
20	1.5564	109.0368	0.3182	0.5284	44.7311	1.9436
25	0.9041	44.1177	0.2153	0.2032	16.1787	1.4127
30	0.3327	15.9300	0.0991	0.1384	5.4124	0.5717

Table 2. Bias (σ_{bias}) under the condition of system noise.

SNR (dB)	Bias (%)					
	Changbai			Qingdao		
	Ω_{BB}	Ω_F	Ω_C	Ω_{BB}	Ω_F	Ω_C
5	18.6803	−858.3574	1.1071	7.8405	−452.2203	7.8584
10	8.0336	−469.8182	−0.9973	3.2375	−234.7492	3.3109
15	3.1830	−235.7411	0.2154	1.2264	−108.1120	1.3629
20	1.1940	−107.1453	−0.1585	0.5594	−43.7774	0.6791
25	0.3286	−43.1128	−0.0424	0.0342	−15.7346	0.2223
30	0.1523	−15.4490	0.0385	−0.0017	−5.2271	0.0380

Table 3. SD (σ_{SD}) under the condition of system noise.

SNR (dB)	SD (%)					
	Changbai			Qingdao		
	Ω_{BB}	Ω_F	Ω_C	Ω_{BB}	Ω_F	Ω_C
5	8.0833	121.1693	2.9652	1.2226	65.3801	17.1186
10	5.3539	70.6254	1.5441	0.8802	37.3999	7.5438
15	2.1559	39.2181	0.7358	0.4353	19.7724	4.6861
20	1.0907	20.4720	0.3469	0.2412	9.2965	1.7415
25	0.5715	9.5487	0.1984	0.1454	3.7366	1.1294
30	0.2959	3.8175	0.1311	0.1013	1.3163	0.5045

From Tables 1–3, we can see that Ω_C performs the best in Changbai while Ω_{BB} is the preferred one in Qingdao. Thus, under the condition of noise, the choice of FR estimators in CIT depends on the scattering type. For Ω_F, large errors are occurred in both SAR scenes. This is because the mean FR values Ω_{p_true} in areas of Changbai and Qingdao are about 0.8° and 1.45°, respectively, and the Freeman's estimator is sensitivity to noise near $\Omega_{p\_true} = 0°$ [5].

Based on the results of Tables 1–3, the final CIT reconstructions in Changbai (Ω_C is applied) and Qingdao (Ω_{BB} is applied) are shown in Figures 7 and 8, respectively. Figures 7a and 8a are the reconstructed results when $SNR = 5$ dB. Figures 7b and 8b show the corresponding two-dimensional absolute deviations between the true and reconstructed distributions, and the RMS over the whole images are

8.46×10^9 electrons $\cdot$ m^{-3} and 14.66×10^9 electrons $\cdot$ m^{-3}, respectively. It can be seen that severe distortions are yielded when SNR is as low as 5 dB, and the two small-scale disturbances are barely identified. Figures 7c and 8c are the reconstructed results when $SNR = 20$ dB, a typical condition for radar systems [4], and corresponding absolute deviations are shown in Figures 7d and 8d. We can see that the performance of reconstructions is significantly improved and the two small-scale disturbances are clear. The RMS of Figures 7d and 8d are 2.05×10^9 electrons $\cdot$ m^{-3} and 2.43×10^9 electrons $\cdot$ m^{-3}, respectively.

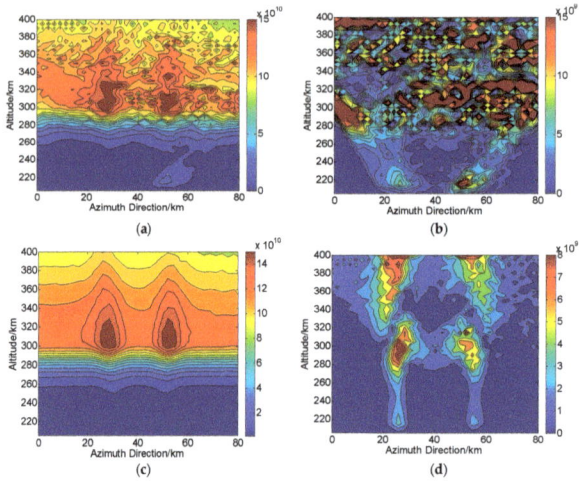

Figure 7. Two-dimensional CIT reconstructions under the condition of system noise in Changbai (Ω_C is used). (**a,c**) are the reconstructed results with $SNR = 5$ dB and $SNR = 20$ dB, respectively. (**b,d**) are the corresponding absolute deviations between the true and reconstructed results. The unit of the electron density is electrons $\cdot$ m^{-3}.

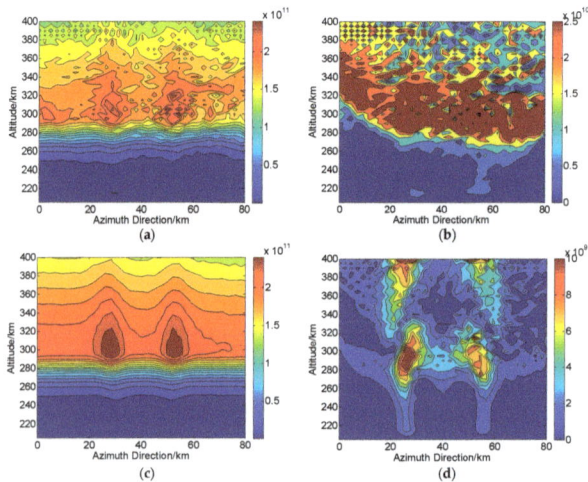

Figure 8. Two-dimensional CIT reconstructions under the condition of system noise in Qingdao (Ω_{BB} is used). (**a,c**) are the reconstructed results with $SNR = 5$ dB and $SNR = 20$ dB, respectively. (**b,d**) are the corresponding absolute deviations between the true and reconstructed results. The unit of the electron density is electrons $\cdot$ m^{-3}.

4.2. CIT Reconstructions Under the Condition of Channel Imbalance

When the FR and channel phase/amplitude imbalance are considered, the scattering matrix can be written as (see Appendix B)

$$
\begin{aligned}
M_{hh\_f} &= S_{hh} \cos^2 \Omega - S_{vv} \sin^2 \Omega \\
M_{vh\_f} &= f(S_{hv} - S_{hh} \sin \Omega \cos \Omega - S_{vv} \sin \Omega \cos \Omega) \\
M_{hv\_f} &= f(S_{hv} + S_{hh} \sin \Omega \cos \Omega + S_{vv} \sin \Omega \cos \Omega) \\
M_{vv\_f} &= f^2(S_{vv} \cos^2 \Omega - S_{hh} \sin^2 \Omega)
\end{aligned}
\tag{15}
$$

where the channel imbalances in the receiver and transmitter are assumed to be identical (i.e., $f_1 = f_2 = f$) to simplify the analysis [2,4,32]. The effects of channel phase imbalance, i.e., $f = 1 \times \exp(i\phi)$, on the FR estimators are first evaluated under typical values [2,5]. By using the full-pol data sets of Changbai and Qingdao, the results are shown in Tables 4–6. We can see that when the phase imbalance is the dominant error, Ω_{BB} can perform the smallest RMS error and SD while Ω_F has the smallest biases. Thus, in order to make a comparison, Ω_{BB} and Ω_F are respectively regarded as the input of CIT simulations.

Table 4. RMS error (σ_{RMS}) under the condition of channel phase imbalance.

| Channel Phase Imbalance (°) | RMS Error (%) | | | | | |
| | Changbai | | | Qingdao | | |
	Ω_{BB}	Ω_F	Ω_C	Ω_{BB}	Ω_F	Ω_C
5	1.5075	1.5863	5.0152	0.4340	0.4644	5.7890
10	3.6991	3.9594	9.2409	1.5537	1.6645	7.4793
15	6.6352	7.2038	13.9008	3.3994	3.6423	9.8274
20	10.4286	11.4409	19.5022	6.0230	6.4697	13.0879

Table 5. Bias (σ_{bias}) under the condition of channel phase imbalance.

| Channel Phase Imbalance (°) | Bias (%) | | | | | |
| | Changbai | | | Qingdao | | |
	Ω_{BB}	Ω_F	Ω_C	Ω_{BB}	Ω_F	Ω_C
5	−1.4987	−1.5859	−4.8996	−0.4337	−0.4643	−5.7350
10	−3.6691	−3.9582	−9.0977	−1.5528	−1.6643	−7.4330
15	−6.6716	−7.2012	−13.7565	−3.3974	−3.6420	−9.7901
20	−10.3174	−11.4362	−19.3646	−6.0199	−6.4692	−13.0582

Table 6. SD (σ_{SD}) under the condition of channel phase imbalance.

| Channel Phase Imbalance (°) | SD (%) | | | | | |
| | Changbai | | | Qingdao | | |
	Ω_{BB}	Ω_F	Ω_C	Ω_{BB}	Ω_F	Ω_C
5	0.1620	0.0371	1.0709	0.0180	0.0096	0.7895
10	0.4708	0.0983	1.6208	0.0523	0.0251	0.8311
15	0.9164	0.1910	1.9986	0.1175	0.0482	0.8558
20	1.5199	0.3261	2.3139	0.1954	0.0809	0.8814

When $\phi = 5°$, Figures 9 and 10 are the final CIT reconstructions in Changbai and Qingdao, respectively. In Changbai, the RMS of corresponding absolute deviations based on Ω_F and Ω_{BB} (i.e., Figure 9b,d are 2.24×10^9 electrons $\cdot$ m^{-3} and 2.22×10^9 electrons $\cdot$ m^{-3}, respectively. Similar, the RMS based on Ω_F and Ω_{BB} in Qingdao (i.e., Figure 10b,d) are 2.56×10^9 electrons $\cdot$ m^{-3} and 2.54×10^9 electrons $\cdot$ m^{-3}, respectively. We can see that both two FR estimators can maintain good performance and the two disturbances are clearly visible. However, the CIT errors become obvious when the phase imbalance is as large as 20°, as shown in Figures 11 and 12. Here, the RMS based on Ω_F

and Ω_{BB} in Changbai (i.e., Figure 11b,d) are 6.78×10^9 electrons $\cdot$ m^{-3} and 6.25×10^9 electrons $\cdot$ m^{-3}, respectively. Similar, the RMS in Figure 12b,d are 9.09×10^9 electrons $\cdot$ m^{-3} and 8.56×10^9 electrons $\cdot$ m^{-3}, respectively. Thus, it can be see that for all conditions, the results based on Ω_{BB} are better than that based on Ω_F, especially for large phase imbalance.

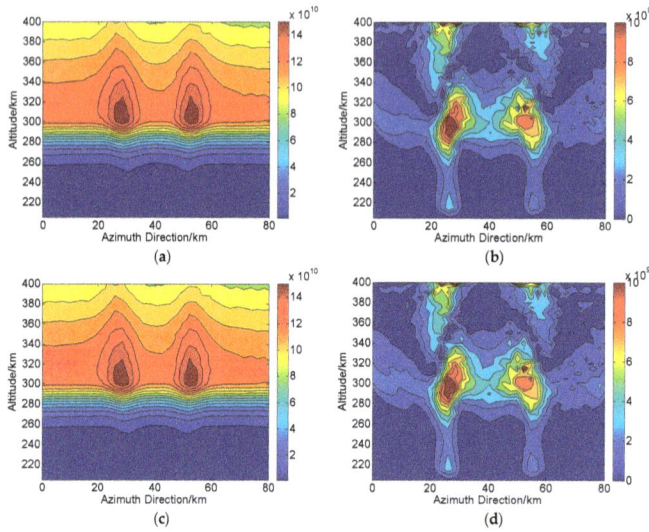

Figure 9. Two-dimensional CIT reconstructions under the condition of channel phase imbalance (5°) in Changbai. (**a,c**) are the reconstructed results based on Ω_F and Ω_{BB}, respectively. (**b,d**) are the corresponding absolute deviations between the true and reconstructed results. The unit of the electron density is electrons $\cdot$ m^{-3}.

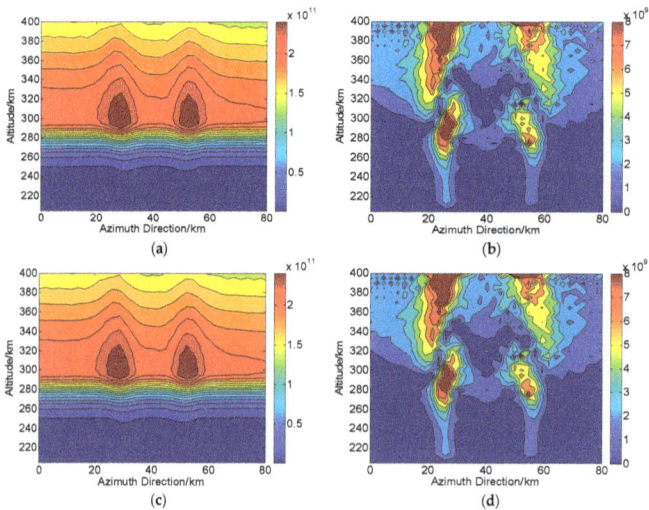

Figure 10. Two-dimensional CIT reconstructions under the condition of channel phase imbalance (5°) in Qingdao. (**a,c**) are the reconstructed results based on Ω_F and Ω_{BB}, respectively. (**b,d**) are the corresponding absolute deviations between the true and reconstructed results. The unit of the electron density is electrons $\cdot$ m^{-3}.

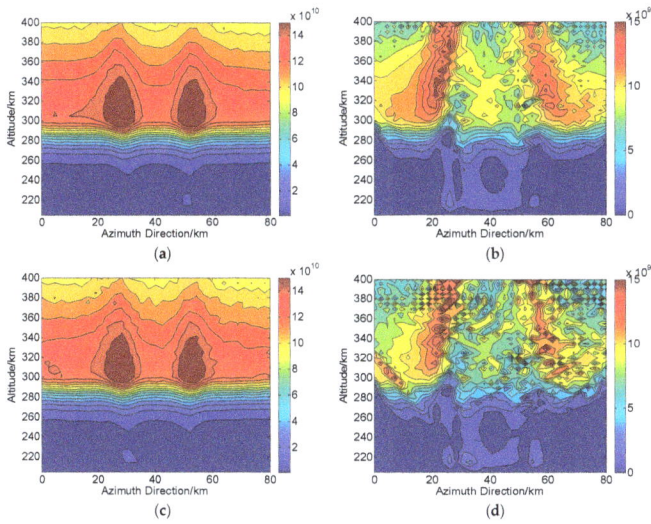

Figure 11. Two-dimensional CIT reconstructions under the condition of channel phase imbalance ($20°$) in Changbai. (**a**,**c**) are the reconstructed results based on Ω_F and Ω_{BB}, respectively. (**b**,**d**) are the corresponding absolute deviations between the true and reconstructed results. The unit of the electron density is electrons $\cdot$ m^{-3}.

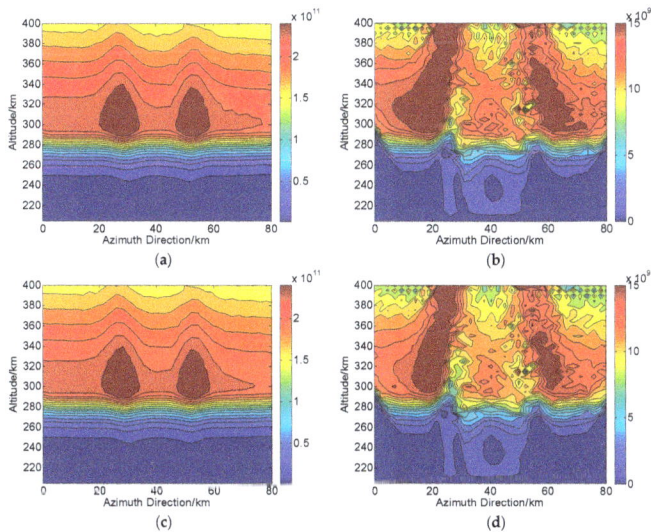

Figure 12. Two-dimensional CIT reconstructions under the condition of channel phase imbalance ($20°$) in Qingdao. (**a**,**c**) are the reconstructed results based on Ω_F and Ω_{BB}, respectively. (**b**,**d**) are the corresponding absolute deviations between the true and reconstructed results. The unit of the electron density is electrons $\cdot$ m^{-3}.

Similarly, the effects of amplitude imbalance on the performances of FR estimators are shown in Tables 7–9, where the magnitude $|f|$ is less than 0.5 dB. We can see that under the conditions of amplitude imbalance, Ω_C is the preferred estimator in CIT reconstructions. In addition, the statistical characters of Ω_C are almost unchanged both in Changbai and Qingdao. This is because according to Equation (6) and Equation (15), Ω_C becomes (see Appendix C)

$$\Omega_C = \frac{1}{2}\arg\left(\Im(S_{hh}S_{vv}^*)\left(|f|^2\cos 2\Omega + i0.5\left(|f| + |f|^3\right)\sin 2\Omega\right)\right) \tag{16}$$

thus, Ω_C only depends on $|f|$.

Table 7. RMS error (σ_{RMS}) under the condition of channel amplitude imbalance.

| Channel Amplitude Imbalance (dB) | RMS Error (%) | | | | | |
| | Changbai | | | Qingdao | | |
	Ω_{BB}	Ω_F	Ω_C	Ω_{BB}	Ω_F	Ω_C
0.1	0.1525	0.2570	0.0066	0.0688	0.0659	0.0066
0.2	0.2958	0.5045	0.0266	0.1233	0.1187	0.0266
0.3	0.4270	0.7358	0.0594	0.1632	0.1570	0.0594
0.4	0.5509	0.9570	0.1060	0.1890	0.1820	0.1060
0.5	0.6630	1.1639	0.1654	0.1982	0.1933	0.1652

Table 8. Bias (σ_{bias}) under the condition of channel amplitude imbalance.

| Channel Amplitude Imbalance (dB) | Bias (%) | | | | | |
| | Changbai | | | Qingdao | | |
	Ω_{BB}	Ω_F	Ω_C	Ω_{BB}	Ω_F	Ω_C
0.1	−0.1198	−0.2539	−0.0066	−0.0684	−0.0655	−0.0066
0.2	−0.2281	−0.4980	−0.0266	−0.1225	−0.1178	−0.0266
0.3	−0.3215	−0.7257	−0.0594	−0.1617	−0.1555	−0.0594
0.4	−0.4041	−0.9430	−0.1060	−0.1868	−0.1796	−0.1058
0.5	−0.4738	−1.1459	−0.1654	−0.1949	0.1897	−0.1652

Table 9. SD (σ_{SD}) under the condition of channel amplitude imbalance.

| Channel Amplitude Imbalance (dB) | SD (%) | | | | | |
| | Changbai | | | Qingdao | | |
	Ω_{BB}	Ω_F	Ω_C	Ω_{BB}	Ω_F	Ω_C
0.1	0.0945	0.0401	0.00001	0.0072	0.0074	0.00001
0.2	0.1884	0.0811	0.00001	0.0142	0.0149	0.00001
0.3	0.2811	0.1216	0.00002	0.0222	0.0222	0.00002
0.4	0.3745	0.1629	0.00002	0.0281	0.0296	0.00003
0.5	0.4639	0.2042	0.00002	0.0356	0.0370	0.00005

Figures 13 and 14 show the final reconstructions in Changbai and Qingdao, respectively. In simulations of Figures 13a and 14a, the $|f|$ is set to 0.1 dB. The RMS of corresponding absolute deviations in Figures 13b and 14b are 2.07×10^9 electrons $\cdot$ m^{-3} and 2.42×10^9 electrons $\cdot$ m^{-3}, respectively. For Figures 13c and 14c, the $|f|$ is set to 0.5 dB and the RMS in Figures 13d and 14d are 2.11×10^9 electrons $\cdot$ m^{-3} and 2.45×10^9 electrons $\cdot$ m^{-3}, respectively. From the results, we can see that the CIT reconstructions still performs well even if the $|f|$ is as high as 0.5 dB. Thus, it can be concluded that compared with noise and phase imbalance, amplitude imbalance is not a problem for CIT reconstruction.

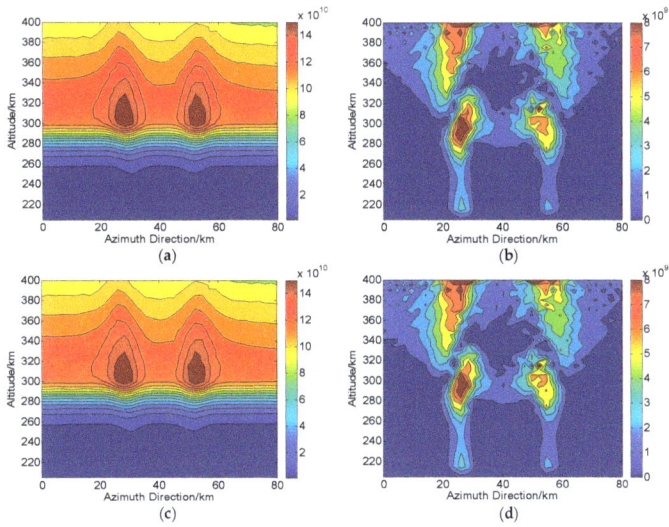

Figure 13. Two-dimensional CIT reconstructions under the condition of channel amplitude imbalance in Changbai (Ω_C is used). (**a,c**) are the reconstructed results with 0.1 dB and 0.5 dB, respectively. (**b,d**) are the corresponding absolute deviations between the true and reconstructed results. The unit of the electron density is electrons $\cdot$ m^{-3}.

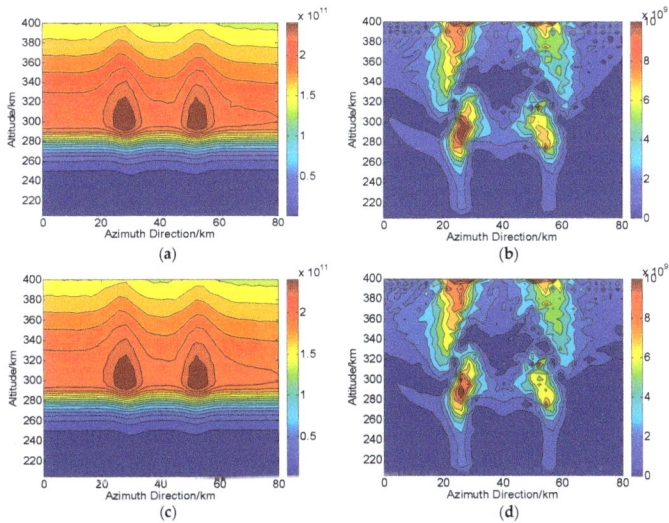

Figure 14. Two-dimensional CIT reconstructions under the condition of channel amplitude imbalance in Qingdao (Ω_C is used). (**a,c**) are the reconstructed results with 0.1 dB and 0.5 dB, respectively. (**b,d**) are the corresponding absolute deviations between the true and reconstructed results. The unit of the electron density is electrons $\cdot$ m^{-3}.

4.3. CIT Reconstructions Under the Condition of Crosstalk

The last system error that should be considered is the crosstalk. Assuming $\delta_1 = \delta_2 = \delta_3 = \delta_4 = \delta$ for convenience [2,4,32], the scattering matrix of Equation (2) can be written as (see Appendix D)

$$
\begin{aligned}
M_{hh\_\delta} &= S_{hh}\left(\cos^2\Omega - \delta^2\sin^2\Omega\right) + 2\delta S_{hv} + S_{vv}\left(\delta^2\cos^2\Omega - \sin^2\Omega\right) \\
M_{vh\_\delta} &= (S_{hh} + S_{vv})\left[\delta\cos 2\Omega + 0.5\left(\delta^2 - 1\right)\sin 2\Omega\right] + S_{hv}\left(1 + \delta^2\right) \\
M_{hv\_n} &= (S_{hh} + S_{vv})\left[\delta\cos 2\Omega + 0.5\left(1 - \delta^2\right)\sin 2\Omega\right] + S_{hv}\left(1 + \delta^2\right) \\
M_{vv\_n} &= S_{hh}\left(\delta^2\cos^2\Omega - \sin^2\Omega\right) + 2\delta S_{hv} + S_{vv}\left(\cos^2\Omega - \delta^2\sin^2\Omega\right)
\end{aligned}
\tag{17}
$$

Tables 10–12 show the performances of FR estimators under the typical values of crosstalk ranging from −15 dB to −35 dB [2,5]. We can see that in area of Changbai, Ω_F performs the best and can be applied in CIT reconstructions. However, by using the full-pol data of Qingdao, the error of Ω_C is smaller than that of Ω_F when the crosstalk is higher than −20 dB. Thus, if the crosstalk is the domain error, the choice of FR estimator in CIT depends on the scattering type and magnitude of crosstalk. The Ω_C should be applied in our CIT simulations of Qingdao when the crosstalk is equal to −15 dB and −20 dB, otherwise Ω_F is the preferred estimator.

Table 10. RMS error (σ_{RMS}) under the condition of crosstalk.

Crosstalk (dB)	RMS Error (%)					
	Changbai			Qingdao		
	Ω_{BB}	Ω_F	Ω_C	Ω_{BB}	Ω_F	Ω_C
−15	8.8421	7.6382	7.7358	6.2014	6.0407	2.5106
−20	3.7897	2.8994	2.9776	1.9464	1.8950	1.1415
−25	1.7467	1.1683	1.2207	0.5906	0.5700	0.9784
−30	0.8633	0.5115	0.5437	0.1698	0.1623	0.6808
−35	0.4496	0.2430	0.2621	0.0437	0.0413	0.4264

Table 11. Bias (σ_{bias}) under the condition of crosstalk.

Crosstalk (dB)	Bias (%)					
	Changbai			Qingdao		
	Ω_{BB}	Ω_F	Ω_C	Ω_{BB}	Ω_F	Ω_C
−15	8.6232	7.5837	7.6724	5.4012	5.4406	−1.8124
−20	3.5849	2.8467	2.9162	1.6463	1.6949	−1.5488
−25	1.5885	1.1245	1.1700	0.5925	0.5669	−0.8016
−30	0.7551	0.4788	0.5064	0.1684	0.1612	−0.6044
−35	0.3811	0.2207	0.2369	0.0426	0.0416	−0.3885

Table 12. SD (σ_{SD}) under the condition of crosstalk.

Crosstalk (dB)	SD (%)					
	Changbai			Qingdao		
	Ω_{BB}	Ω_F	Ω_C	Ω_{BB}	Ω_F	Ω_C
−15	1.9561	0.9114	0.9888	1.1464	1.1382	1.0981
−20	1.2295	0.5506	0.6017	0.8232	0.8218	0.7813
−25	0.7268	0.3172	0.3481	0.1390	0.1224	0.5612
−30	0.4186	0.1800	0.1980	0.0832	0.0681	0.3135
−35	0.2387	0.1017	0.1120	0.0393	0.0385	0.1757

When the crosstalk is set to −15 dB, Figures 15a and 16a show the CIT reconstructions in Changbai and Qingdao, respectively. The RMS of corresponding absolute deviations in Figures 15b and 16b are 5.96×10^9 electrons $\cdot$ m^{-3} and 4.98×10^9 electrons $\cdot$ m^{-3}, respectively. It can be seen that the small-scale distributions are difficult to be recognized in both areas when the crosstalk is as high as −15 dB, an extreme value. It should be noted that from the evaluations in Table 4, the RMS error rapidly decreases with the decrease of crosstalk, which further improves the accurate of CIT reconstructions. Thus, the effect of crosstalk on the CIT is not serious in most conditions. Figures 15c and 16c show the CIT reconstructions when the crosstalk is set to -35 dB. We can see that significant improvements are demonstrated to clearly show the small-scale distributions. The deviations in Figures 15d and 16d are 2.06×10^9 electrons $\cdot$ m^{-3} and 2.39×10^9 electrons $\cdot$ m^{-3}, respectively.

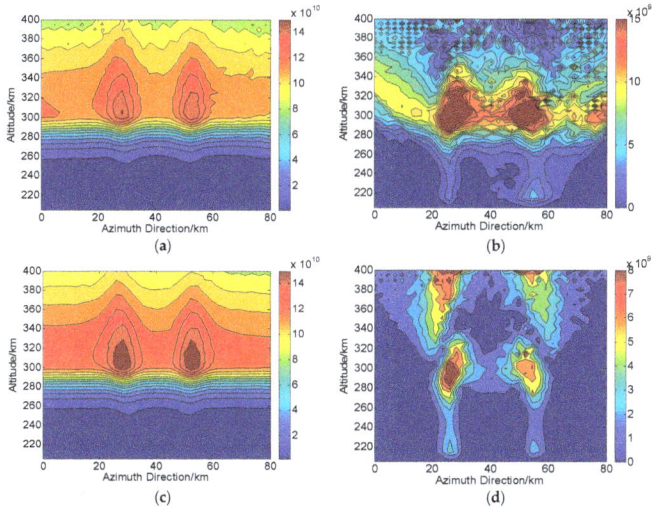

Figure 15. Two-dimensional CIT reconstructions under the condition of crosstalk in Changbai (Ω_F is used). (**a**,**c**) are the reconstructed results with -15 dB and -35 dB, respectively. (**b**,**d**) are the corresponding absolute deviations between the true and reconstructed results. The unit of the electron density is electrons $\cdot$ m^{-3}.

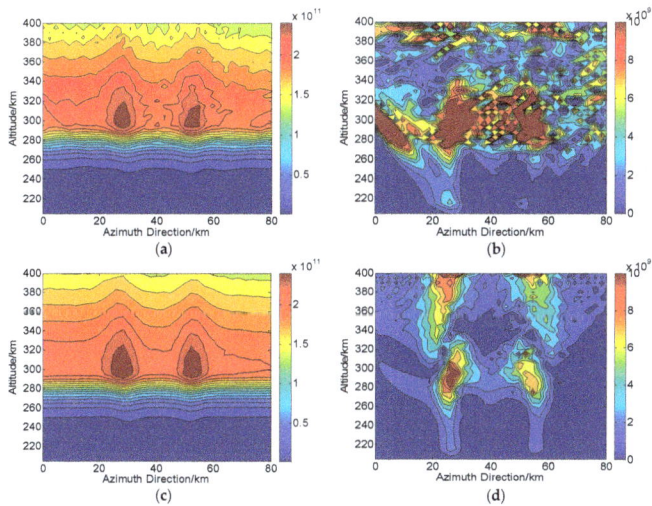

Figure 16. Two-dimensional CIT reconstructions under the condition of crosstalk in Qingdao. (**a**) (Ω_C is used) and (**c**) (Ω_F is used) are the reconstructed results with -15 dB and -35 dB, respectively. (**b**,**d**) are the corresponding absolute deviations between the true and reconstructed results. The unit of the electron density is electrons $\cdot$ m^{-3}.

4.4. CIT Reconstructions Under the Combination of System Errors

For a realistic situations of the PALSAR system, these calibration errors will appear together. Thus, in order to make a practical value for our CIT technique, the system errors (i.e., noise, channel imbalance and crosstalk) are all considered in this subsection. According to the calibration accuracy of the PALSAR system [32], in simulations, we assume $SNR = 15$ dB, channel phase imbalance ϕ is $2°$, channel amplitude imbalance $|f|$ is 0.5 dB, and crosstalk δ is -35 dB. Table 13 shows the performances of FR estimators under the condition of joint errors, we can see that for the area of Changbai, Ω_C will be the preferred estimator while Ω_{BB} performs the best for the area of Qingdao.

Table 13. The performances of FR estimators under the condition of joint errors.

Performances	Changbai			Qingdao		
	Ω_{BB}	Ω_F	Ω_C	Ω_{BB}	Ω_F	Ω_C
RMS error (%)	3.6572	262.1338	1.8965	1.4348	120.5995	6.6420
Bias (%)	3.5491	−258.7012	−1.7362	1.3959	−118.7056	−5.5683
SD (%)	2.9604	42.3003	0.8634	0.5181	21.2925	4.9092

Figure 17a,c shows the CIT reconstructions in Changbai (Ω_C is used) and Qingdao (Ω_{BB} is used), respectively. We can see that when the all system errors are considered, the true ionospheric distribution can still be accurately reconstructed based on the proposed CIT technique, and the two small-scale disturbances are clearly visible. The RMS of corresponding absolute deviations shown in Figure 17b,d are 2.32×10^9 electrons $\cdot$ m^{-3} and 2.69×10^9 electrons $\cdot$ m^{-3}, respectively. Thus, we can see that after calibration of the PALSAR systems, the proposed CIT technique can give an accurate ionospheric reconstruction in consideration of the residual errors.

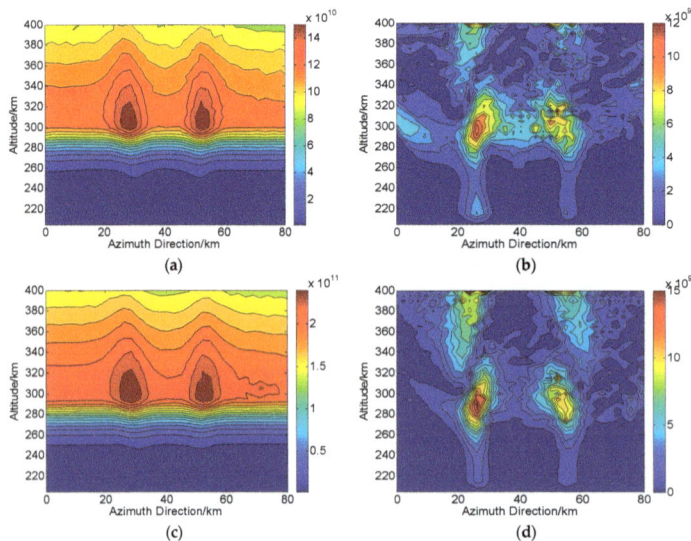

Figure 17. Two-dimensional CIT reconstructions under the condition of joint errors. (**a**) (Ω_C is used) and (**c**) (Ω_{BB} is used) are the reconstructed results in areas of Changbai and Qingdao, respectively. (**b,d**) are the corresponding absolute deviations between the true and reconstructed results. The unit of the electron density is electrons $\cdot$ m^{-3}.

5. Conclusions

The distribution of ionospheric electron density is an important part of solar-terrestrial space environment, which can demonstrate the solar and earth activities (magnetic storms, plasma bubbles, midlatitude troughs, ionospheric anomalies caused by earthquakes [8,15], etc.). Monitoring the ionospheric behaviors, especially small-scale ionospheric anomaly, based on the CIT technique is therefore beneficial to these studies. In order to obtain the electron density distribution with high resolution, a modified CIT technique based on the spaceborne PolSAR data is proposed in this paper, where the FR, rather than TEC, is regarded as the input. From the results in Section 4, small-scale distributions can be reconstructed by this proposed CIT technique due to the high spatial resolution of spaceborne SAR, which is inaccessible by conventionally used data source. However, the accuracy of FR retrieval will affect the final CIT reconstructions. The evaluations of three typical FR estimators considering different system errors and scattering types were made. Some conclusions are as follow:

(1) The effect of system noise on FR retrieval depends on both the scattering types and SNR. According to the evaluations, Ω_C and Ω_{BB} are the optimal estimators of CIT in areas of Changbai and Qingdao, respectively. The performances of CIT in both areas are improved with the increase of SNR. The small-scale distributions are visible in reconstructions when $SNR = -20$ dB, a typical configuration for the PALSAR sensors.
(2) For considering the effects of channel phase imbalance, Ω_{BB} can give the smallest error both in areas of Changbai and Qingdao. From the simulation results, it can be seen that the CIT errors are sensitive to phase imbalance. For amplitude imbalance, Ω_C should be applied. However, in contrast to phase imbalance and noise, the effects of amplitude imbalance on CIT is small.
(3) The choice of FR estimator considering the crosstalk depends on both scattering types and magnitude of the crosstalk. When the crosstalk is as high as -15 dB, serious CIT reconstructions are shown. However, with the decrease of the crosstalk, the error of FR retrieval is sharply decreased. The effects of crosstalk on the CIT are limited when the crosstalk is equal to -35 dB.

In general, the main system effects on CIT are the noise and channel phase imbalance while channel amplitude imbalance and crosstalk can be ignored for most cases. According to the calibration accuracy of the PALSAR system, we evaluate the reconstructions when all the system errors are considered. Accurate results can still be obtained by the proposed CIT technique. It should be noted that we have analyzed other full-pol SAR scenes, and the same results are obtained. In addition, there are other factors that can affect the CIT reconstructions. For example, the lack of horizontal ray paths and the choice of initial distribution in iteration also degrade the accuracy of CIT [14,15]. Combining of the PolSAR and occultation or ground-based ionosonde data can reduce these problems and will be done in our future work.

Acknowledgments: This work was supported by the National Natural Science Foundation of China (NSFC) under Grants 41604157, 41601483 and 61401022, and by the National Key Laboratory of Electromagnetic Environment.

Author Contributions: Cheng Wang, Liang Chen, Haisheng Zhao, Mingming Bian and Zheng Lu initiated the research. Under supervision of Running Zhang, Jian Feng, and Cheng Wang performed the analysis and wrote the manuscript. Liang Chen, Haisheng Zhao and Mingming Bian revised the manuscript. All authors read and approved the final version of the manuscript.

Conflicts of Interest: The authors declare no conflict of interest.

Appendix A

When the system noise and FR are the only errors, the scattering matrix of Equation (2) can be written as

$$
\begin{bmatrix} M_{hh} & M_{vh} \\ M_{hv} & M_{vv} \end{bmatrix} = \begin{bmatrix} \cos\Omega & \sin\Omega \\ -\sin\Omega & \cos\Omega \end{bmatrix} \times \begin{bmatrix} S_{hh} & S_{vh} \\ S_{hv} & S_{vv} \end{bmatrix} \times \begin{bmatrix} \cos\Omega & \sin\Omega \\ -\sin\Omega & \cos\Omega \end{bmatrix} + \begin{bmatrix} N_{hh} & N_{vh} \\ N_{hv} & N_{vv} \end{bmatrix} \tag{A1}
$$

Thus, each factor of M matrix can be written as

$$
\begin{aligned}
M_{hh\_n} &= S_{hh}\cos^2\Omega - S_{vv}\sin^2\Omega + N_{hh} \\
M_{vh\_n} &= S_{vh} + (S_{hh}+S_{vv})\sin\Omega\cos\Omega + N_{vh} \\
M_{hv\_n} &= S_{hv} - (S_{hh}+S_{vv})\sin\Omega\cos\Omega + N_{hv} \\
M_{vv\_n} &= S_{vv}\cos^2\Omega - S_{hh}\sin^2\Omega + N_{vv}
\end{aligned}
\tag{A2}
$$

Appendix B

Similarly, if the FR and channel phase/amplitude imbalance are considered, and the channel imbalances in the receiver and transmitter are assumed to be identical (i.e., $f_1 = f_2 = f$), the scattering matrix of Equation (2) can be written as

$$
\begin{bmatrix} M_{hh} & M_{vh} \\ M_{hv} & M_{vv} \end{bmatrix}
= \begin{bmatrix} 1 & 0 \\ 0 & f \end{bmatrix}
\times \begin{bmatrix} \cos\Omega & \sin\Omega \\ -\sin\Omega & \cos\Omega \end{bmatrix}
\times \begin{bmatrix} S_{hh} & S_{vh} \\ S_{hv} & S_{vv} \end{bmatrix}
\times \begin{bmatrix} \cos\Omega & \sin\Omega \\ -\sin\Omega & \cos\Omega \end{bmatrix}
\times \begin{bmatrix} 1 & 0 \\ 0 & f \end{bmatrix}
\tag{A3}
$$

Thus, each factor of M matrix is derived as

$$
\begin{aligned}
M_{hh\_f} &= S_{hh}\cos^2\Omega - S_{vv}\sin^2\Omega \\
M_{vh\_f} &= f(S_{hv} - S_{hh}\sin\Omega\cos\Omega - S_{vv}\sin\Omega\cos\Omega) \\
M_{hv\_f} &= f(S_{hv} + S_{hh}\sin\Omega\cos\Omega + S_{vv}\sin\Omega\cos\Omega) \\
M_{vv\_f} &= f^2(S_{vv}\cos^2\Omega - S_{hh}\sin^2\Omega)
\end{aligned}
\tag{A4}
$$

Appendix C

When we only consider the amplitude imbalance ($|f|$) in (A4), each factor of Ω_C (Equations (6) and (7)) is

$$
\begin{aligned}
\Im(C_{14}) &= \Im\overline{\left(M_{hh\_f}\times M^*_{vv\_f}\right)} \\
&= |f|^2\Im\left((S_{hh}\cos^2\Omega - S_{vv}\sin^2\Omega)\times(S^*_{vv}\cos^2\Omega - S^*_{hh}\sin^2\Omega)\right) \\
&= |f|^2\Im(S_{hh}S^*_{vv})\cos(2\Omega)
\end{aligned}
\tag{A5}
$$

$$
\begin{aligned}
\Im(C_{34}-C_{24}) &= \Im\overline{\left(M^*_{vv\_f}\left(M_{hv\_f}-M_{vh\_f}\right)\right)} \\
&= \Im\left(|f|^3(S^*_{vv}\cos^2\Omega - S^*_{hh}\sin^2\Omega)\times(S_{hh}+S_{vv})\sin(2\Omega)\right) \\
&= |f|^3\Im(S_{hh}S^*_{vv})\sin(2\Omega)
\end{aligned}
\tag{A6}
$$

$$
\begin{aligned}
\Im(C_{13}-C_{12}) &= \Im\overline{\left(M_{hh\_f}\left(M^*_{hv\_f}-M^*_{vh\_f}\right)\right)} \\
&= \Im\left((S_{hh}\cos^2\Omega - S_{vv}\sin^2\Omega)\times|f|(S^*_{hh}+S^*_{vv})\sin(2\Omega)\right) \\
&= |f|\Im(S_{hh}S^*_{vv})\sin(2\Omega)
\end{aligned}
\tag{A7}
$$

Then,

$$
\begin{aligned}
\Omega_C &= \tfrac{1}{2}\arg\left(\Im(C_{14})+i\Im\left(\tfrac{C_{13}+C_{34}-C_{12}-C_{24}}{2}\right)\right) \\
&= \tfrac{1}{2}\arg\left(\Im(S_{hh}S^*_{vv})\left(|f|^2\cos 2\Omega + i0.5\left(|f|+|f|^3\right)\sin 2\Omega\right)\right)
\end{aligned}
\tag{A8}
$$

Appendix D

If the crosstalk is considered and $\delta_1 = \delta_2 = \delta_3 = \delta_4 = \delta$ is assumed, the scattering matrix of Equation (2) becomes

$$
\begin{bmatrix} M_{hh} & M_{vh} \\ M_{hv} & M_{vv} \end{bmatrix}
= \begin{bmatrix} 1 & \delta \\ \delta & 1 \end{bmatrix}
\times \begin{bmatrix} \cos\Omega & \sin\Omega \\ -\sin\Omega & \cos\Omega \end{bmatrix}
\times \begin{bmatrix} S_{hh} & S_{vh} \\ S_{hv} & S_{vv} \end{bmatrix}
\times \begin{bmatrix} \cos\Omega & \sin\Omega \\ -\sin\Omega & \cos\Omega \end{bmatrix}
\times \begin{bmatrix} 1 & \delta \\ \delta & 1 \end{bmatrix}
\tag{A9}
$$

Thus, each factor of the M matrix is

$$
\begin{aligned}
M_{hh\_\delta} &= S_{hh}\left(\cos^2\Omega - \delta^2\sin^2\Omega\right) + 2\delta S_{hv} + S_{vv}\left(\delta^2\cos^2\Omega - \sin^2\Omega\right) \\
M_{vh\_\delta} &= \left(S_{hh} + S_{vv}\right)\left[\delta\cos 2\Omega + 0.5\left(\delta^2 - 1\right)\sin 2\Omega\right] + S_{hv}\left(1 + \delta^2\right) \\
M_{hv\_n} &= \left(S_{hh} + S_{vv}\right)\left[\delta\cos 2\Omega + 0.5\left(1 - \delta^2\right)\sin 2\Omega\right] + S_{hv}\left(1 + \delta^2\right) \\
M_{vv\_n} &= S_{hh}\left(\delta^2\cos^2\Omega - \sin^2\Omega\right) + 2\delta S_{hv} + S_{vv}\left(\cos^2\Omega - \delta^2\sin^2\Omega\right)
\end{aligned}
\tag{A10}
$$

References

1. Lawrence, R.S.; Little, C.G.; Chivers, H.A. A survey of ionospheric effects upon earth-space radio propagation. *Proc. IEEE* **1964**, *52*, 4–27. [CrossRef]
2. Freeman, A. Calibration of linearly polarized polarimetric SAR data subject to Faraday rotation. *IEEE Trans. Geosci. Remote Sens.* **2004**, *42*, 1617–1624. [CrossRef]
3. Bickel, S.H.; Bates, R.H.T. Effects of magneto-ionic propagation on the polarization scattering matrix. *Proc. IEEE* **1964**, *53*, 1089–1091. [CrossRef]
4. Chen, J.; Quegan, S. Improved estimators of Faraday rotation in spaceborne polarimetric SAR data. *IEEE Geosci. Remote Sens. Lett.* **2010**, *7*, 846–850. [CrossRef]
5. Rogers, N.C.; Quegan, S. The accuracy of Faraday rotation estimation in satellite synthetic aperture radar images. *IEEE Trans. Geosci. Remote Sens.* **2014**, *52*, 4799–4807. [CrossRef]
6. Meyer, F.; Bamler, R.; Jakowski, N.; Fritz, T. The potential of low-frequency SAR systems for mapping ionospheric TEC distributions. *IEEE Geosci. Remote Sens. Lett.* **2006**, *3*, 560–564. [CrossRef]
7. Jehle, M.; Ruegg, M.; Zuberbuhler, L.; Small, D.; Meier, E. Measurement of ionospheric faraday rotation in simulated and real spaceborne SAR Data. *IEEE Trans. Geosci. Remote Sens.* **2009**, *47*, 1512–1523. [CrossRef]
8. Pi, X.Q.; Freeman, A.; Chapman, B.; Rosen, P.; Li, Z.H. Imaging ionospheric inhomogeneities using spaceborne synthetic aperture radar. *J. Geophys. Res.* **2011**, *116*, 1451–1453. [CrossRef]
9. Pi, X.Q. Ionospheric effects on spacebrone synthetic aperture radar and a new capability of imaging the ionosphere from space. *Space Weather* **2015**, *13*, 737–741. [CrossRef]
10. Kim, J.S.; Papathanassiou, K.P.; Scheiber, R.; Quegan, S. Correcting Distortion of Polarimetric SAR Data Induced by Ionospheric Scintillation. *IEEE Trans. Geosci. Remote Sens.* **2015**, *53*, 6319–6335.
11. Wang, C.; Liu, L.; Chen, L.; Feng, J.; Zhao, H.S. Improved TEC retrieval based on spaceborne PolSAR data. *Radio Sci.* **2017**, *52*, 288–304. [CrossRef]
12. Yizengaw, E.; Dyson, P.L.; Essex, E.A. Ionosphere dynamics over the southern hemisphere during the 31 March 2001 severe magnetic storm using multi-instrument measurement data. *Ann. Geophys.* **2005**, *23*, 707–721. [CrossRef]
13. Wen, D.B.; Yuan, Y.B.; Ou, J.K.; Zhang, K.F. Ionospheric response to the geomagnetic storm on August 21, 2003 over China using GNSS-Based tomographic technique. *IEEE Trans. Geosci. Remote Sens.* **2010**, *48*, 3212–3217.
14. Zhao, H.S.; Xu, Z.W.; Wu, J.; Quegan, S. Ionospheric tomography of small-scale disturbances with a triband beacon: A numerical study. *Radio Sci.* **2010**, *45*. [CrossRef]
15. Zhao, H.S.; Xu, Z.W.; Wu, J.; Wang, Z.G. Ionospheric tomography by combining vertical and oblique ionograms with TEC retrieved from a tri-band beacon. *J. Geophys. Res.* **2010**, *115*, 788–802. [CrossRef]
16. Austen, J.R.; Franke, S.J.; Liu, C.H. Ionospheric imaging using computerized tomography. *Radio Sci.* **1988**, *23*, 299–307. [CrossRef]
17. Pryse, S.E. Radio tomography: A new experimental technique. *Surv. Geophys.* **2003**, *24*, 1–38. [CrossRef]
18. Li, L.L.; Li, F. Ionosphere tomography based on spaceborne SAR. *Adv. Space Res.* **2008**, *42*, 1187–1193. [CrossRef]
19. Wang, C.; Zhang, M.; Xu, Z.W.; Zhao, H.S. TEC retrieval from spacebrone SAR data and its applications. *J. Geophys. Res.* **2014**, *119*, 8648–8659. [CrossRef]
20. Hu, C.; Tian, Y.; Dong, X.; Wang, R.; Long, T. Computerized ionospheric tomography based on geosynchronous SAR. *J. Geophys. Res.* **2017**, *122*, 2686–2705. [CrossRef]
21. Hu, C.; Li, Y.; Dong, X.; Cui, C.; Long, T. Impacts of temporal-spatial variant background ionosphere on repeat-track GEO D-InSAR system. *Remote Sens.* **2016**, *8*, 916. [CrossRef]

22. Jehle, M.; Frey, O.; Small, D.; Meier, E. Measurement of ionospheric TEC in spaceborne SAR data. *IEEE Geosci. Remote Sens. Lett.* **2010**, *48*, 2460–2468. [CrossRef]
23. Lizuka, K.; Tateishi, R. Estimation of CO_2 sequestration by the forests in Japan discriminating precise tree age category using remote sensing techniques. *Remote Sens.* **2015**, *7*, 15082–15113.
24. Xiong, S.; Peter, J.; Li, G. The application of ALOS/PALSAR InSAR to measure subsurface penetration depths in deserts. *Remote Sens.* **2017**, *9*, 638. [CrossRef]
25. Meyer, F. Performance Requirements for Ionospheric Correction of Low-Frequency SAR Data. *IEEE Trans. Geosci. Remote Sens.* **2011**, *49*, 3694–3702. [CrossRef]
26. Yuel, S.H. Estimates of Faraday rotation with passive microwave polarimetry for microwave remote sensing of Earth surfaces. *IEEE Trans. Geosci. Remote Sens.* **2000**, *38*, 2434–2438.
27. Cushley, A.C.; Noel, J.M. Ionospheric tomography using ADS-B signals. *Radio Sci.* **2014**, *49*, 549–563. [CrossRef]
28. Wang, C.; Chen, L.; Liu, L.; Yang, J.; Lu, Z.; Feng, J.; Zhao, H.S. Robust computerized ionospheric tomography based on spaceborne polarimetric SAR data. *IEEE J. Sel. Top. Appl. Earth Obs. Remote Sens.* **2017**, *10*, 4022–4031. [CrossRef]
29. Budden, K.G. *Radio Waves in the Ionosphere: The Mathematical Theory of the Reflection of Radio Waves from Stratified Ionized Layers*; Cambridge University Press: Cambridge, UK, 1961.
30. Bust, G.S.; Mitchell, C.N. History, current state, and future directions of ionospheric imaging. *Rev. Geophys.* **2008**, *46*, 394–426. [CrossRef]
31. Lee, J.S.; Pottier, E. *Polarimetric Radar Imaging: From Basics to Applications*; CRC Press: Florida, FL, USA, 2009.
32. Meyer, F.; Nicoll, J.B. Prediction, detection, and correction of Faraday rotation in full-polarimetic L-band SAR data. *IEEE Trans. Geosci. Remote Sens.* **2008**, *46*, 3076–3086. [CrossRef]

remote sensing

MDPI

Article

An Efficient Maximum Likelihood Estimation Approach of Multi-Baseline SAR Interferometry for Refined Topographic Mapping in Mountainous Areas

Yuting Dong [1], Houjun Jiang [2], Lu Zhang [1,3] and Mingsheng Liao [1,3,*]

[1] State Key Laboratory of Information Engineering in Surveying, Mapping and Remote Sensing,
    Wuhan University, 129 Luoyu Road, Wuhan 430079, China; yuting_dong@whu.edu.cn (Y.D.);
    luzhang@whu.edu.cn (L.Z.)
[2] College of Geographic and Biologic Information, Nanjing University of Posts and Telecommunications,
    Nanjing 210023, China; jianghouj@njupt.edu.cn
[3] Collaborative Innovation Center of Geospatial Technology, Wuhan University, 129 Luoyu Road,
    Wuhan 430079, China
* Correspondence: liao@whu.edu.cn; Tel.: +86-27-6877-8070

Received: 4 January 2018; Accepted: 5 March 2018; Published: 14 March 2018

Abstract: For InSAR topographic mapping, multi-baseline InSAR height estimation is known to be an effective way to facilitate phase unwrapping by significantly increasing the ambiguity intervals and maintaining good height measurement sensitivity, especially in mountainous areas. In this paper, an efficient multi-baseline SAR interferometry approach based on maximum likelihood estimation is developed for refined topographic mapping in mountainous areas. In the algorithm, maximum likelihood (ML) height estimation is used to measure the topographic details and avoid the complicated phase unwrapping process. In order to be well-adapted to the mountainous terrain conditions, the prior height probability is re-defined to take the local terrain conditions and neighboring height constraint into consideration in the algorithm. In addition, three strategies are used to optimize the maximum likelihood height estimation process to obtain higher computational efficiency, so that this method is more suitable for spaceborne InSAR data. The strategies include substituting a rational function model into the complicated conversion process from candidate height to interferometric phase, discretizing the continuous height likelihood probability, and searching for the maximum likelihood height with a flexible step length. The experiment with simulated data is designed to verify the improvement of the ML height estimation accuracy with the re-defined prior height distribution. Then the optimized processing procedure is tested with the multi-baseline L-band ALOS/PALSAR data covering the Mount Tai area in China. The height accuracy of the generated multi-baseline InSAR DEM can meet both standards of American DTED-2 and Chinese national 1:50,000 DEM (mountain) Level 2.

Keywords: multi-baseline InSAR; maximum likelihood (ML); DEM; L-band; ALOS/PALSAR

1. Introduction

SAR interferometry (InSAR) is an effective tool for large-area topographic mapping due to its all-weather imaging and high sensitivity to terrain relief [1,2]. The InSAR height measurement accuracy is greatly influenced by the phase unwrapping accuracy and the length of normal baselines [3,4]. Longer normal baselines allow more accurate height estimation but also generate higher frequency of the interferometric fringes, which increases the complexity of phase unwrapping. On the other hand, shorter normal baselines reduce the complexity of phase unwrapping but suffer from poorer phase-to-height sensitivity [5,6]. Therefore, the contradiction between the sensitivity of height measurement and the reliability of phase unwrapping caused by the length of normal baseline over rough terrain is inevitable for single-baseline InSAR.

To combine the advantages of large and short baselines in topographic mapping, a multi-baseline InSAR principle has been proposed to estimate terrain height by joint analysis of multiple interferometric pairs with diverse normal baselines [7,8]. Researchers have proposed a variety of methods for multi-baseline InSAR processing. The standard method to increase the accuracy of interferometric DEMs is to stack multiple geocoded layers, weighted with the individual error estimates [9–11]. However, all DEM stacking approaches assume correctly unwrapped interferograms. Moreover, it is difficult to co-register different geocoded DEM layers accurately and the horizontal displacements can introduce height errors. Other multi-baseline estimation methods have been published to facilitate the phase unwrapping process by taking advantage of baseline diversity, such as the Least Square estimation method [5], the iterative multi-baseline method [12], the Chinese Remainder Theorem (CRT) method [13,14], and so on. These multi-baseline phase unwrapping methods can significantly increase the ambiguity intervals of interferometric phases and keep the topographic details as well; however, these methods still have to solve the phase unwrapping problem correctly.

In order to avoid the phase unwrapping process and determine the target height directly, the statistical method using the criterion of maximum likelihood [9,15–18] is exploited to combine the multi-baseline information for target height estimation. However, in actual data processing, atmospheric effects, orbital errors, and decorrelation will introduce phase noise that cannot be ignored. Therefore, the maximum likelihood estimation method is not robust enough to search for the target height within all elevation values. In order to realize more robust and reliable height estimation, the prior height information from the reference DEM is incorporated into the ML estimation to restrict the height searching range [18], but the problem is that the local terrain conditions and the neighboring height constraint are not considered in the prior height distribution. The maximum a posteriori (MAP) estimation tries to introduce the neighboring height constraints by using Markov random fields to model the prior distribution of the unknown images [19,20]. This method allows recovering topographic profiles affected by strong height discontinuities and performing efficient noise rejections. However, the MAP methods also have limits concerning the computational time and the optimization step because there is no guarantee of finding the global optimum.

In this article, we apply the maximum likelihood estimation for multi-baseline InSAR DEM generation. The prior height probability is re-defined to take the local terrain conditions and neighboring height constraint into consideration. An experiment with simulated data is designed to verify the improvement of the ML height estimation accuracy with the well-defined prior height distribution. Furthermore, to make the maximum likelihood estimation method well adapted to spaceborne SAR data, the processing flow is optimized for higher computational efficiency with the following innovative points: (1) Replacing the rigorous height-to-phase conversion with the rational function model (RFM); (2) substituting the complicated height likelihood probability function with two-dimension lookup table; (3) searching for the maximum likelihood height with flexible search step length instead of fixed search step length. This processing flow is testified with the L-band ALOS/PALSAR data, which can be less influenced by the temporal and volume decorrelation and have a longer critical baseline than InSAR data with a shorter wavelength (such as X band or C band) [21]. Since the ALOS/PALSAR data were acquired in the repeat-pass mode, the above processing flow integrates atmospheric effect correction to improve the reliability of multi-baseline estimation. The rest of this article is structured as follows: Section 2 introduces the principle of ML estimation with prior DEM; Section 3 presents the improved proposed processing flow for spaceborne datasets; detailed descriptions of the experiments and results are given in Section 4; Section 5 is a discussion of the experimental results; finally, the conclusions are drawn.

2. Maximum Likelihood Height Estimation Assisted by Prior DEM

2.1. Basic Principle

The maximum likelihood (ML) height estimation method of multi-baseline InSAR considers the target height h as a parameter of the probability distribution of the interferometric phase ϕ, denoted as $pdf(\phi|h)$, and combines the probability distributions of all the interferometric phase observations to estimate the target height with the maximum likelihood criterion [22]. Aiming at reducing the undesired variation in the maximum likelihood height estimation caused by non-negligible phase noise, the prior height distribution from a reference DEM is integrated into the maximum likelihood estimation, which not only greatly improves the estimation reliability but also narrows the search range and increases the computational efficiency. Therefore, the maximum likelihood height estimation assisted by the reference DEM is shown in Equation (1):

$$\hat{H}_{ML} = \arg\max_{h \in \mathbb{F}} \left\{ \left[\prod_{i=1}^{K} pdf(\phi_i|h) \right] pdf(h) \right\},$$
(1)

where $pdf(h)$ is the a priori distribution function of height provided by the reference DEM. $pdf(\phi_i|h)$ is the height likelihood function for the ith interferogram. According to [23], $pdf(\phi|h)$ can be estimated as the edge probability density function (PDF) of the interferometric phase ϕ, as shown in Equation (2):

$$
pdf(\phi|h) = pdf(\phi|\phi_0 = F_{h2\phi}(h)) = \frac{(1-|\rho|^2)^L}{2\pi} \left\{ \frac{(2L-2)!}{[(L-1)!]^2 2^{2(L-1)}} \right.
$$
$$
\times \left[\frac{(2L-1)\beta}{(1-\beta^2)^{L+1/2}} \left(\frac{\pi}{2} + \arcsin\beta \right) + \frac{1}{(1-\beta^2)^L} \right]
$$
$$
\left. + \frac{1}{2(L-1)} \sum_{r=0}^{L-2} \frac{\Gamma(L-1/2)}{\Gamma(L-1/2-r)} \frac{\Gamma(L-1-r)}{\Gamma(L-1)} \frac{1+(2r+1)\beta^2}{(1-\beta^2)^{r+2}} \right\}
$$
(2)

where $\beta = |\rho|\cos(\phi - \phi_0)$; ϕ_0 is the mathematical expectation of interferometric phase; ρ is the complex coherence coefficient; and L is the effective number of looks (ENL). ϕ_0 can be represented by the target height h through height-to-phase conversion function $F_{h2\phi}(h)$. When ϕ_0 is set as zero, $pdf(\phi|0, \rho, L)$ describes the probability distribution of the interferometric phase noise. Hence the standard deviation of phase noise σ_ϕ can be derived by Equation (3). Given L, σ_ϕ is determined by the coherence coefficient ρ:

$$\sigma_\phi = sqrt\left\{ \int_{-\pi}^{+\pi} f^2 pdf(f|0, \rho, L) df \right\}.$$
(3)

The standard deviation of the height errors σ_h can be approximated as:

$$\sigma_h = -\frac{\lambda R \sin\theta}{4\pi B_\perp} \sigma_\phi,$$
(4)

where λ is the wavelength; R is the slant range; θ is the incidence angle; and $B_\perp$ is the normal baseline.

2.2. Definition of the Prior Height Probability

Suppose the height acquired from the prior DEM corresponding to a resolution unit of the interferogram is h_{prior}. Generally, we think that the system error of the prior DEM has been corrected. Then the height error is the accidental error and obeys Gaussian distribution. The standard deviation of the height errors for each cell in the interferogram is σ_h. Hence, the prior height probability distribution for each cell can be defined as in Equation (5) [18]:

$$pdf(h) = \frac{1}{\sqrt{2\pi\sigma_h^2}} \exp\left\{ \frac{-(h - h_{prior})^2}{2\sigma_h^2} \right\}.$$
(5)

In this article, the SRTM DEM is used as the reference DEM with standard deviation of the height errors σ_{SRTM}. The σ_{SRTM} of the SRTM DEM in the Eurasian continent is 3.8 m [24]. For each cell of the interferogram, if we suppose that $\sigma_h = \sigma_{SRTM}$ directly, Equation (5) defines the prior height probability distribution over all the geographic coverage of SRTM DEM, which can be much larger than the SAR image coverage, causing the following problems: (1) the local terrain conditions such as plains or mountains, are not taken into account, while σ_h under different terrain conditions is not consistent; (2) the neighboring height constraints are not considered that the height probability distribution defined by Equation (5) is only related to the height of the resolution unit itself. Aimed at these two problems, Equation (5) is modified to take the local terrain conditions in the neighborhood into consideration. The prior height probability distribution is re-defined as in Equation (6):

$$pdf(h) = \frac{1}{\sqrt{2\pi\sigma_h^2}} \exp\left\{ \frac{1}{T} \sum_{i\in\mathcal{N}} \frac{-(h-h_{\mathrm{DEM},i})^2}{2\sigma_h^2} \right\}$$

$$\sigma_h = \begin{cases} \sigma_{local}, & \sigma_{local} > \sigma_{DEM} \\ \sigma_{DEM}, & \sigma_{local} \le \sigma_{DEM} \end{cases}$$

(6)

where $\mathcal{N}$ is composed of the resolution units and its adjacent units, which are usually four-neighbor, eight-neighbor, or 24-neighbor. $h_{SRTM,i}$, $i\in\mathcal{N}$ represent the heights of the resolution units, T represents the number of the pixels in the neighborhood, and σ_{local} is the standard deviation of height errors in the neighborhood. Both h_{SRTM} and σ_{local} are obtained from the SRTM DEM. It can be seen from Equation (6) that when σ_{local} is larger than σ_{SRTM}, σ_h is set to σ_{local}. Under this condition, σ_h is no longer a constant. In the undulating terrain areas, σ_h depends on the local terrain conditions, while in the flat areas it is still conservatively set as σ_{SRTM}.

The size of neighborhood used to define the prior height distribution probability is determined based on the spatial resolution of the prior DEM and the coherence level of the interferogram. When the spatial resolution of the prior DEM is much lower than that of the interferogram and the coherence level of the interferogram is high, a smaller neighborhood such as four-neighbor is preferred to reduce the influence of the prior height and neighboring heights in the height probability distribution so that the resulting DEM will not be too smooth. Otherwise, when the spatial resolution of the prior DEM is close to that of the interferogram or the coherence level of the interferogram is not high, a larger neighborhood such as eight-neighbor or even 24-neighbor is selected to enhance the constraints of the prior height and neighboring heights on the height probability distribution so that the impact of the phase noise is suppressed and the height estimate is more robust.

In this article, SRTM DEM with a cell size of 30 m is used as the prior DEM and the interferogram has a comparable cell size, about 22 m after 3×7 multi-look processing. The mean coherence of the interferogram is about 0.5, which is a moderate coherence level. Therefore, eight-neighbor is chosen in the experiment.

The correspondence between SRTM DEM and the resolution cell of the interferogram needs to be established by radar coding. The height error introduced by the radar coding procedure will increase the value of σ_{SRTM}, hence, in practical calculations, σ_{SRTM} will be adjusted empirically to make the height distribution curve more reasonable.

3. Optimized Processing Flow for Spaceborne Multi-Baseline InSAR Datasets

When applying the multi-baseline InSAR height estimation with maximum likelihood criterion in spaceborne InSAR datasets, the processing flow can be divided into three major stages. First is the interferometric processing including interferometric pairs combination and differential interferogram generation, as in Section 3.1. Second is the maximum likelihood height estimation process based on the principle introduced in Section 2. In order to make the ML estimation method well adapted to the spaceborne data, the processing flow is optimized for higher computational efficiency with the following new points: (1) Replacing the rigorous height-to-phase conversion with the rational function model (RFM), as in Section 3.2.1; (2) substituting the complicated height likelihood probability function with two-dimension lookup table, as in Section 3.2.2; (3) searching for the maximum likelihood height with flexible search step length instead of the fixed search step length, as in Section 3.2.3. Thirdly, the estimated height map in SAR image coordinate can be geocoded into the geographic coordinate system or the universal transverse mercator (UTM) system, as in Section 3.3. The flowchart of this approach is outlined in Figure 1. A brief description and rationale for each step are given as follows.

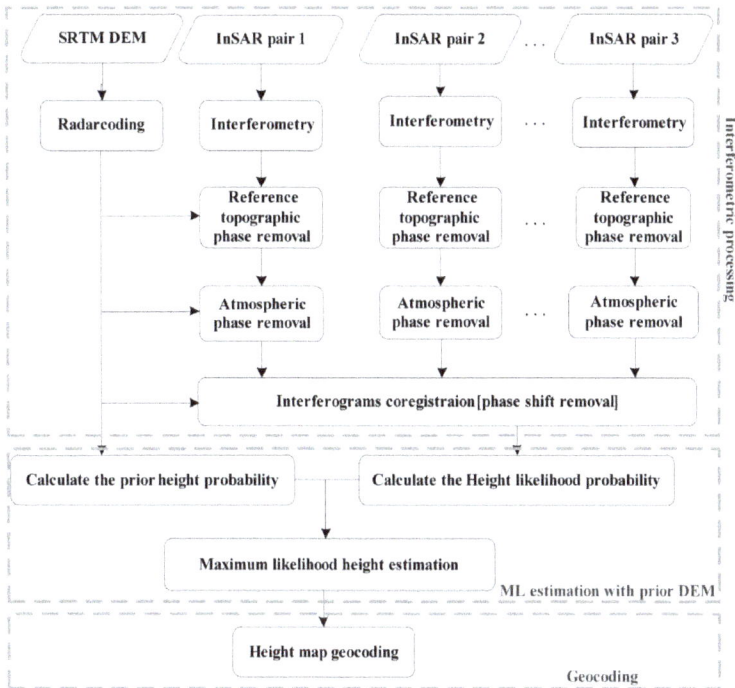

Figure 1. Flowchart for spaceborne multi-baseline InSAR DEM generation.

3.1. Interferometric Processing

For the repeat-pass interferometry, we first need to select the suitable SAR images in the given dataset to constitute interferometric pairs and then choose the master interferogram for other interferograms to register with, while for single-pass interferometry we merely need to choose the master interferogram. To keep good coherence, the interferometric pairs should have temporal baselines as short as possible under the premise that the normal baseline is shorter than the critical normal baseline. In order to select the proper master interferogram, the principle is to select a master interferogram at around the center of the time axis among all the interferograms to make it easier for the other interferograms to register to it.

Next is the interferometric processing step for each interferometric pair, which is the basic processing unit for the maximum likelihood height estimation, including complex image co-registration, interferogram generation, flattening, and filtering. The SRTM DEM is projected into the azimuth and slant-range coordinate system of the master SAR image in the interferometric pair. Then the interferometric phase is flattened by the radar-coded DEM, i.e., major 2π phase jumps due to topography are removed. For repeat-pass interferometry, the atmospheric phase screen (APS) introduces non-negligible height error into the InSAR height [25]. The APS consists of a vertically stratified component and a turbulent mixing one [6]. Based on spatial pattern analyses of these two

APS components, a SRTM elevation-to-phase regression model and a low-pass plus adaptive combined filter are employed to estimate and remove them from the differential interferogram sequentially [25].

After interferometric processing, the differential interferograms are registered to the image space of the chosen master interferogram. The phase shift between the interferograms is a constant. In the repeat-pass interferometry mode, the phase shift can be removed through the combined filter used for the APS correction [25], while in the single-pass interferometry mode it can be obtained through statistical average value of the differential phase maps.

3.2. Maximum Likelihood Height Estimation with the Prior DEM

3.2.1. Rational Function Model (RFM) for Height-to-Phase Conversion

The candidate height must be converted to the interferometric phase to calculate the corresponding height likelihood probability as in Equation (2). This height-to-phase conversion starts from a pixel in the master image with the image coordinates (i_M, j_M), and its slant range R_M can be determined with the orbital information. The candidate height for pixel (i_M, j_M) is represented by h_{DEM}. The Range-Doppler (RD) model can be iteratively solved to calculate the geographic coordinates for pixel (i_M, j_M), which is the direct positioning process. With the calculated geographic coordinates and orbital information of the slave image, the RD model can be iteratively solved again to calculate its image coordinates (i_S, j_S) in the slave image, which is the indirect positioning process. The slant range for the pixel (i_S, j_S) can be determined for the slave image as R_S. The corresponding interferometric phase for the chosen pixel with the candidate height h_{DEM} can be determined through Equation (7), where λ is the wavelength. For repeat-pass interferometry, $k = 2$ and for single-pass interferometry, $k = 1$.

$$\phi = -\frac{2\pi k}{\lambda}(R_M - R_S) \tag{7}$$

As we can see, this height-to-phase conversion process requires iteratively solving the RD model twice for every candidate height in the height search range; this process has to be performed for every pixel in the SAR image, which is extremely time-consuming.

In order to improve the efficiency of height-to-phase conversion, we try to no longer care about the specific analytical form of height-to-phase conversion functions, but rather write them directly as a rational function of the image coordinates (L_a, P_r), the height h, as shown in Equation (8):

$$\phi = F_{h2\phi}(h) = \frac{Num_h(L\_a, P\_r, h)}{Den_h(L\_a, P\_r, h)} = \frac{\sum\limits_{i=0}^{3}\sum\limits_{j=0}^{3-i}\sum\limits_{n=0}^{3-i-j} a_{i,j,n} \cdot L\_a^i \cdot P\_r^j \cdot h^n}{\sum\limits_{i=0}^{3}\sum\limits_{j=0}^{3-i}\sum\limits_{n=0}^{3-i-j} b_{i,j,n} \cdot L\_a^i \cdot P\_r^j \cdot h^n}. \tag{8}$$

They are called the rational function model of the height-to-phase conversion function $F_{h2\phi}(h)$. $a_{i,j,n}$, $b_{i,j,n}$ are the unknown parameters to be solved. The value of $b_{0,0,0}$ is set to 1. The way to solve the unknown parameters is the same as in [26].

The rational function model for the height-to-phase conversion has been established in this article with the spaceborne InSAR data and tested to see whether it can replace the rigorous method. The number of the control points is $50 \times 50 \times 10$, that is, there are 50×50 regular grid points in the plane and 10 layers in the height range. The height interval is between -500 m and 10,000 m. We randomly generate 10,000 checkpoints at which the rational function model and the rigorous method are used for the height-to-phase conversion, respectively. The conversion error of the rational function model is calculated using the results of the rigorous method as a reference. Table 1 lists the conversion error for different spaceborne SAR data, indicating that the conversion error for the rational function model is completely negligible in practical applications. Therefore, the rational function model can replace the rigorous method for height-to-phase conversion at each resolution unit.

Table 1. Height-to-phase conversion errors of rational function model.

Spaceborne InSAR Data	Height Ambiguity	Height-To-Phase	
		Max. Error	RMSE
ALOS/PALSAR	82 m	$1.97 \times 10^{-3}°$	$2.14 \times 10^{-4}°$
COSMO-SkyMed	164 m	$-4.08 \times 10^{-4}°$	$6.78 \times 10^{-5}°$
TerraSAR-X	59 m	$-1.81 \times 10^{-3}°$	$2.15 \times 10^{-4}°$

3.2.2. Height Likelihood Probability Lookup Table

Equation (2) is used to calculate the height likelihood probability, which is a very complicated expression. In order to improve the computational efficiency, we propose using the two-dimensional look-up table as a substitution of Equation (2) to calculate the height likelihood probability. It is a regular sampling of the continuous likelihood probability that is to replace the continuous function surface with a discrete numerical table. The height likelihood probability is related to the phase, coherence, and effective number of looks. For a multi-baseline InSAR dataset, the effective number of looks of each interferogram is the same and hence the look-up table to be established is indexed by phase and coherence. The value range of the phase is $[-\pi, \pi]$, and the sampling interval is $\pi/180$ rad. The value range of the coherence coefficient is $[0,1]$, and the sampling interval is $1/100$. Figure 2a shows a three-dimensional view of the look-up table with an effective number of looks (16) and Figure 2b shows the profile perpendicular to the coherence axis.

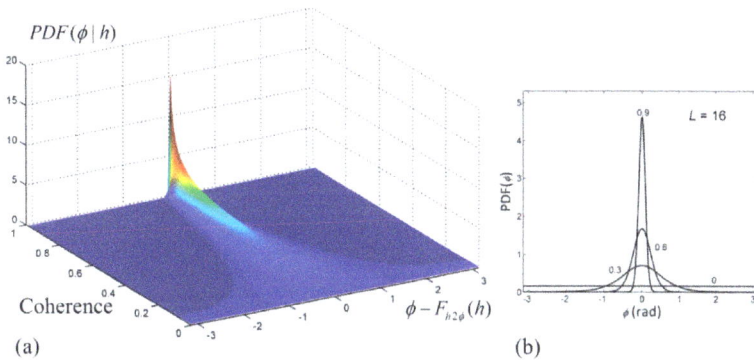

Figure 2. (**a**) The fitted surface by the look-up table with ENL $L = 16$ and (**b**) the profile perpendicular to the coherence axis with $\rho = 0.3, 0.6, 0.9$.

3.2.3. ML Height Estimation with Flexible Search Step Length

From Section 3.1, the ML height satisfying Equation (1) can be estimated by searching the candidate height range. The search step length determines the accuracy of the ML height and the amount of computation time. Instead of the fixed step length, we propose using a flexible search step length to ensure the efficiency and accuracy of the ML height searching.

First, larger steps are used to narrow the candidate height range. Then a smaller step is applied to search for the ML height. Repeat the process until the target height changes less than the given threshold. The specific implementation flow is as follows:

(1) Obtain the initial height value h_0 from the prior DEM, which is radar-coded to the SAR image coordinates;

(2) Set the height search range to $[h_0 - \Delta H_i, h_0 + \Delta H_i]$ and the search step to Δh_i. ΔH_i can be set to an integral multiple of σ_h. The optimal height obtained by the maximum likelihood estimation is h_i;

(3) The height search range becomes $[h_i - \Delta H_{i+1}, h_i + \Delta H_{i+1}]$, $\Delta H_{i+1} = \Delta H_i / 2$ and the search step is $\Delta h_{i+1} = \Delta h_i / 2$. The optimal height obtained by the maximum likelihood estimation is h_{i+1};

(4) Test whether $(h_{i+1} - h_i)$ is less than the given threshold. If yes, then stop the search and return h_{i+1} as the optimal height. If no, then repeat Step (3).

It should be noted that the step length Δh must be less than half of the minimum height ambiguity to satisfy the Nyquist sampling law and ensure that the correct position of the height likelihood probability peak can be detected.

3.3. Geocoding

The geographic coordinates of each cell with estimated height are calculated through the Range Doppler (RD) model with the World Geodetic System 1984 (WGS84) as the reference spheroid. Afterward, the multi-baseline InSAR DEM could be geo-referenced and gridded in the geographic coordinate system (latitude and longitude) or in the projection coordinates such as the universal transverse mercator (UTM) system with regular spacing in the East and North dimensions [11].

4. Experiments and Results

The purpose of the simulated experiment in Section 4.1 is to testify the improvement of the ML height accuracy with the re-defined prior height probability distribution by comparing the height accuracy of ML height with or without the prior DEM. Then, the proposed processing flow was applied in the L-band ALOS/PALSAR data covering the Mount Tai area of China, as in Section 4.2.

4.1. Simulated Experiment

4.1.1. Simulation of SAR Interferograms

An American NED DEM of 10 m resolution is used to simulate three SAR interferograms with different normal baselines and evaluate height accuracy of the generated DEMs. The wavelength of the simulated radar system is 0.031 m (X band). The orbit height is about 600 km. The incidence angle is 35°. The effective number of looks (ENL) of the interferogram is 16. For the three simulated interferograms, their respective normal baseline, height ambiguity (H_{amb}), coherence coefficient, and standard deviation (Std.) of phase noise are shown in Table 2.

Table 2. The simulation parameters of the interferograms.

	Interferogram I	Interferogram II	Interferogram III
Normal baseline	47 m	83 m	178 m
Height ambiguity	139.54 m	79.02 m	36.84 m
Coherence coefficient	0.60	0.57	0.51
Std. of phase noise	0.254 rad	0.277 rad	0.333 rad

The geometric decorrelation caused by terrain changes is not considered and the temporal decorrelation for the three interferograms is assumed to be the same. Therefore, the coherence differences among the three interferograms are induced only by the normal baseline differences. The standard deviation of the phase noise induced by decorrelation is calculated by Equation (3). Figure 3a shows the hillshade of NED DEM and Figure 3b is obtained by 15 × 15 smoothing filtering of the NED DEM, which is used for calculating the prior height probabilities and lots of detailed topographic information is lost, as in Figure 3c. Figure 3d–f represents the simulated interferograms I, II, and II, which are calculated from $W\{2\pi h/H_{amb}\}$, ($W\{\cdot\}$ is the wrapping operator) and h is the height value acquired from the NED DEM (Figure 3a). Figure 3g–i shows atmospheric turbulence phase generated by statistical simulation. Figure 3j–l is the simulated interferogram I, II, and III superimposed by the atmospheric phase (g–i), respectively.

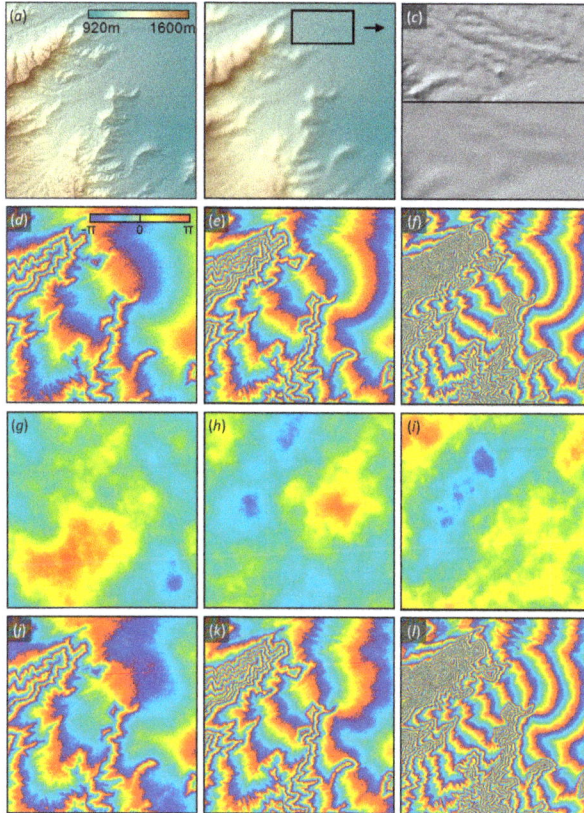

Figure 3. (a) Hillshade of 10 m NED DEM; (b) hillshade of 15 × 15 smoothing filtered NED DEM; (c) NED DEM (upper) and smoothing filtered NED DEM (lower); (d–f) are the simulated interferograms I, II, and III, respectively, and the phase noise introduced by decorrelation is superimposed; (g–i) are the simulated atmospheric phase for interferograms I, II, and III, respectively; (j–l) are the simulated interferograms I, II, and III superimposed by the atmospheric phase (g–i), respectively.

4.1.2. Test of the Impact of the Prior Height on ML Estimation

The simulated interferograms superimposed by the phase noise (Figure 3d–f) are used to generate a multi-baseline InSAR DEM with ML estimation method with or without the prior DEM. The height search range is set to between 500 m and 2000 m and the search step is 1 m. For ML estimation without the prior DEM, the prior height probability is evenly distributed within this range, while for ML estimation with the prior DEM it is calculated through Equation (6). The standard deviation of the prior DEM height error σ_{DEM} is calculated from the height error map generated by differentiating the NED DEM (Figure 3a) and the smoothed NED DEM (Figure 3b). σ_{DEM} is 4.7 m by calculation and empirically enlarged to 6 m.

Figure 4 shows the generated DEMs and their error maps for ML estimation with the prior DEM, (Figure 4a,b) and without the prior DEM (Figure 4c,d). We can see the topographic information is almost all covered by the noise value in Figure 4a and the height error map in Figure 4b ranges between −960 m and 960 m. Compare Figure 4b to Figure 3d–f, it can be seen that the spatial distribution of the height errors is still related to the wrapped terrain phase, and hence the ML estimation without the prior DEM does not solve the height ambiguity problem and is very sensitive to phase noise. The DEM obtained with ML estimation with the prior DEM depicts the terrain condition well in Figure 4c. The height error map ranges between −8 m and 8 m and is almost randomly spatially distributed except that it is slightly larger in the fluctuating area. From all the above comparisons, it is obvious that the height estimation accuracy and anti-noise ability of ML estimation with the prior DEM are much better than those of ML estimation without the prior DEM.

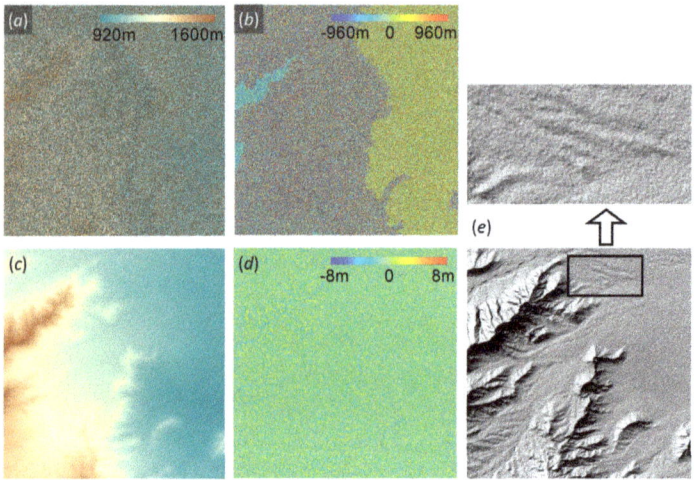

Figure 4. Multi-baseline InSAR DEMs (without atmospheric effects) and height error maps, without the prior DEM (**a,b**) and with the prior DEM (**c–e**) is the hillshade of (**c**) with the enlarged topographic details of a small area.

The statistical values of the height errors in the simulation experiment are calculated as shown in Table 3. The theoretical error of the single-baseline interferogram is caused by the decorrelation phase noise without the phase unwrapping error. From Table 3, the Std. of the height errors are up to 408.8 m for ML estimated height without the prior DEM while shrink to 1.6 m for ML estimated height with the prior DEM. The height accuracy of ML estimation with the prior DEM is better than that of the single-baseline InSAR DEMs in terms of the standard deviation of height errors.

Table 3. Statistical values of height errors without atmospheric effects.

	Mean	Std.
Prior DEM	0.007 m	4.7 m
Interferogram I (Figure 3d)	0.002 m	5.6 m
Interferogram II (Figure 3e)	−0.010 m	3.5 m
Interferogram III (Figure 3f)	−0.001 m	2.0 m
ML without prior DEM	70.072 m	408.8 m
ML with prior DEM	**−0.003 m**	**1.6 m**

In Figure 3b, a small area in the black rectangle is selected and enlarged as shown in Figure 3c. The upper part of Figure 3c is the hillshade of the 10 m NED DEM and the lower part of Figure 3c is the hillshade of the 15 × 15 smoothing filtered NED DEM. Most of the terrain details are lost in the 15 × 15 smoothing filtered NED DEM. The same area of the ML-estimated InSAR DEM is also selected and enlarged in Figure 4e. By comparative analysis of Figures 3c and 4e, it can be seen that although there is high-frequency random noise (caused by decorrelation noise) in the ML-estimated DEM in Figure 4e, the ML-estimated DEM well reconstructs the topographic details lost in the prior DEM as in the lower part of Figure 3c and obviously improves the spatial resolution of the prior DEM. The Std. of height error of ML estimation with a prior DEM (1.6 m) is less than that of prior DEM (4.7 m) in Table 3.

4.1.3. Test of the Impact of the Atmospheric Effects on ML Estimation

Figure 5 shows the multi-baseline InSAR DEM generated from interferograms with the atmospheric effects shown in Figure 3j–l and the corresponding height error map. Comparing the height error map Figure 5b to Figure 4d, it can be seen that there is a trend towards height error in Figure 5b, caused by the atmospheric effects.

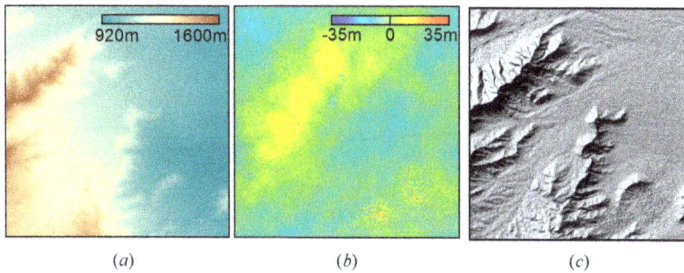

(a)   (b)   (c)

Figure 5. (a) Multi-baseline InSAR DEM with prior DEM (with atmospheric effects); (b) height error map; (c) is the hillshade map of (a).

Furthermore, Table 4 shows the statistical values of height errors of the single and multi-baseline InSAR DEMs generated from interferograms with the atmospheric effects in Figure 3j–l. For both single-baseline interferometry and multi-baseline InSAR height estimation, the atmospheric effects can evidently increase the height errors according to Table 4. The multi-baseline InSAR DEM has better height accuracy than the single-baseline InSAR DEMs, indicating that the ML estimation with prior DEM can effectively suppress the atmospheric effects to some extent. However, comparing Table 3 with Table 4, it can be seen the standard height error of the ML estimation with prior DEM has increased from 1.6 m to 4.1 m, revealing that the atmospheric effects can cause non-negligible height errors and should be removed from each interferogram.

Table 4. Statistical values of height errors with atmospheric effects.

	Mean	Std.
Interferogram I (Figure 3j)	−2.5 m	16.2 m
Interferogram II (Figure 3k)	2.5 m	8.7 m
Interferogram III (Figure 3l)	−0.6 m	4.6 m
ML with prior DEM	**−0.006 m**	**4.1 m**

4.2. ALOS/PALSAR Data Experiment

4.2.1. Experimental Area

The experimental area covers the central and southern parts of Mount Tai (including its main peak, Yuhuangding), and the the central plains and hills (including Taian city at the southern foot of Mount Tai) of Shandong Province. A large portion of this area is covered by vegetation. From the optical image acquired from Google Earth in Figure 6a, we can see that the northern part of the experimental area, including Mount Tai and hills, and Mount Culai in the southeast corner, are basically covered by woods. Moreover, there is a large amount of farmland in the central and southern parts of the experimental area, which is covered by crops. In this vegetation-covered area, the L-band ALOS/PALSAR data can maintain better temporal coherence and penetrate the vegetation cover to some extent. The terrain types of the experimental area are diverse, including plains, hills, and mountains. The height varies greatly such that the height of the plains is about 100 m above the sea level while the height of Mount Tai peak is about 1533 m. Therefore, we can say it is a suitable experimental area for multi-baseline InSAR DEM generation experiment.

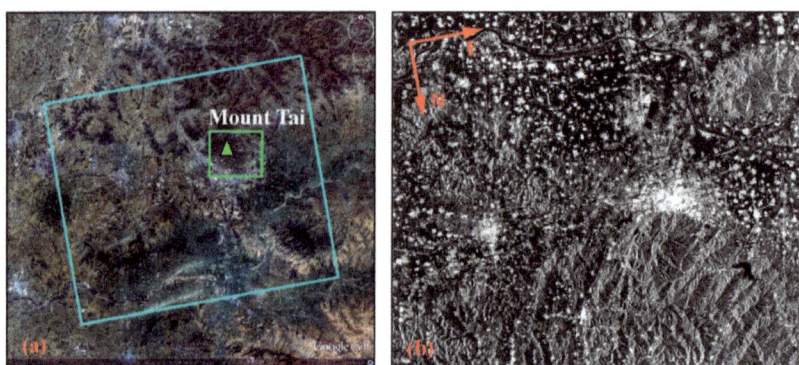

Figure 6. (a) The coverage of ALOS/PALSAR images marked by the blue rectangle and 1:25,000 DEM marked by the green rectangle shown in Google Earth; (b) the amplitude image of ALOS/PALSAR data (acquisition time: 6 February 2008).

4.2.2. ALOS/PALSAR Data

The SAR interferometric data used in this experiment are ALOS/PALSAR data obtained in the FBS mode and have a total of six images. Table 5 lists the detailed image parameters. The amplitude image shown in Figure 6b is acquired on 6 February 2008 and has been through multi-look processing with a ratio of 7:3 (azimuth:range). Since the data are acquired from ascending orbital direction with right-looking imaging, the amplitude image in Figure 6b is flipped over vertically compared with the image coverage in Figure 6a.

Table 5. Image parameters for ALOS/PLASAR data.

Acquisition Time	22 December 2007/6 February 2008/23 March 2008/ 27 December 2009/11 February 2010/29 March 2010
Orbit direction	Ascending
Imaging mode	Stripmap
Polarization	HH
Central incidence angle	38.7°
Sampling space of azimuth/range direction	3.18 m/4.68 m
Band width of azimuth/range direction	1522 Hz/28 MHz

The ALOS/PALSAR multi-baseline dataset used in the experiment has six images and Figure 7 shows the spatial–temporal baseline distribution of the interferometric dataset. We selected four interferometric pairs based on the criteria given in Section 3.1, connected by the blue lines as shown in Figure 7. It can be seen that the temporal baselines of interferometric pairs are all about 46 d. For interferometric pairs 1–2 and 2–3, image 2 is chosen as the master image and for interferometric pairs 4–5 and 5–6, image 5 is chosen as the master image. Then we register interferograms 4–5 and 5–6 to the image space of image 2, as shown by the red line in Figure 7.

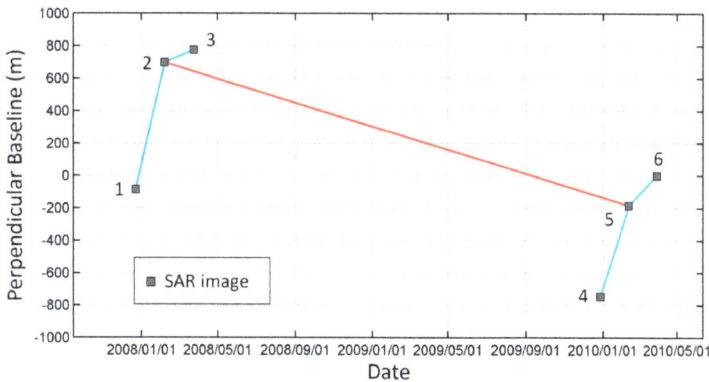

Figure 7. The temporal and spatial baseline distribution of ALOS/PALSAR interferometric pairs. The images connected by the blue line constitute interferometric pairs and the interferograms connected by the red line are co-registered to the same image space.

Table 6 shows the parameters of the four interferograms; the distribution of the normal baselines ranges from −784 m to 185 m and the corresponding height ambiguity ranges from 82 m to 833 m. These are the suitable normal baseline combinations for multi-baseline InSAR processing. In order to improve the signal-to-noise ratio, we perform 7×3 multi-look processing of the SAR images in the azimuth and range direction, and then the spatial resolution of the interferogram is about 22 m $\times$ 22 m.

Table 6. Parameters for ALOS/PALSAR interferometric pairs.

	Interferogram I	Interferogram II	Interferogram III	Interferogram IV
Acquisition time of the Master image	6 February 2008	6 February 2008	11 February 2010	11 February 2010
Acquisition time of the Slave image	22 December 2007	23 March 2008	27 December 2009	29 March 2010
Temporal baseline	46 days	46 days	46 days	46 days
Normal baseline $B_\perp$	−784 m	77 m	−561 m	185 m
Height ambiguity $h_{2\pi}$	82 m	833 m	115 m	347 m
Central Doppler frequency	74/75 Hz	74/80 Hz	68/57 Hz	68/46 Hz
Mean coherence coefficient	0.52	0.53	0.58	0.50

4.2.3. Elevation Datasets

In the experiment, the SRTM DEM with a spatial resolution of 90 m $\times$ 90 m is used as the prior DEM and the standard deviation of the global height error of SRTM DEM is about 5 m [21]. In order to validate the accuracy of the single/multi-baseline InSAR DEMs, the 1:25,000 aerial photogrammetric DEM provided by the Shandong Provincial Land and Surveying and Mapping Institute is used as the reference DEM with the root mean square error (RMSE) less than 4 m [27]. The spatial resolution of 1:25,000 DEM is 10 m $\times$ 10 m. The area of the 1:25,000 DEM coverage is about 12 km $\times$ 14 km marked as the green rectangle in Figure 6a, including the southern part of Mount Tai and Tai'an City.

4.2.4. Experimental Results

Figure 8 shows the flattened interferograms of the ALOS/PALSAR data. The interferometric fringes are sparse in the interferogram with short baseline in Figure 8b,d, and become very dense in the interferogram with long baseline in Figure 8a,c. The interferograms are clearly visible even in the mountainous and vegetation-covered areas, indicating that the L-band SAR data have good capability of keeping geometric and temporal coherence.

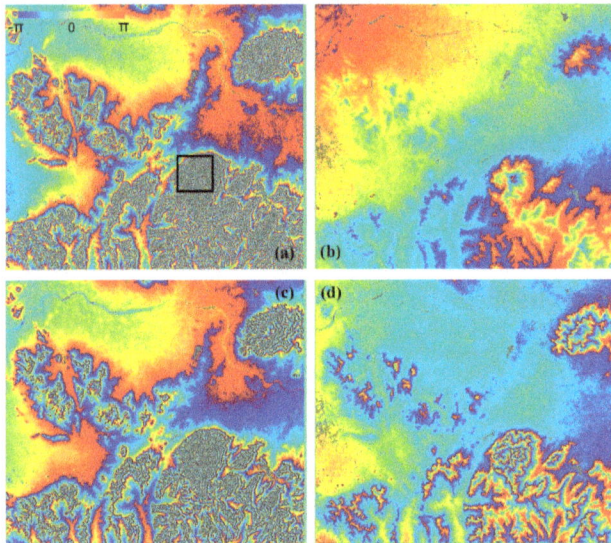

Figure 8. Flattened interferograms of the ALOS/PALSAR data. (**a**) Interferogram I, $B_\perp = -784$ m; (**b**) Interferogram II, $B_\perp = 77$ m; (**c**) Interferogram III, $B_\perp = -561$ m; (**d**) Interferogram IV, $B_\perp = 185$ m.

In the calculation of the prior height probability, σ_{SRTM} is empirically set to 20 m (considering the spatial resolution difference and the interpolation error of the radar coding) with the neighborhood system $\mathcal{N} = 8$. When calculating the height likelihood probability, the initial height value is obtained from the radar-coded SRTM DEM and the initial search height range and the step size are set at ±300 m and 20 m, respectively.

Figure 9 shows the hillshades of the single-/multi-baseline DEMs and SRTM DEM corresponding to the black rectangle in Figure 8a. Besides the multi-baseline InSAR DEM (Figure 8a), the radar-coded SRTM DEM (Figure 8b) and single-baseline InSAR DEMs (Figure 8c–f) of the same area are also presented as comparisons.

As to quantitatively evaluate the height accuracy with the reference 1:25,000 DEM, all the InSAR DEMs and are geocoded into the geographic coordinate system and then projected into the UTM coordinate system with spatial resolution of 20 m. Since the spatial resolution of 1:25,000 DEM is about 10 m, it needs to be down-resampled to 20 m when calculating the height error map. Moreover, in order to verify the height accuracy improvement of the multi-baseline estimation, the SRTM DEM and single-baseline InSAR DEMs are also evaluated as comparison. Figure 10 shows the hillshade of the generated multi-baseline InSAR DEM with height ranging between 20 m and 1535 m, which clearly shows the topographic conditions of Mount Tai, Mount Culai (located at the southeast corner) and a large number of hills.

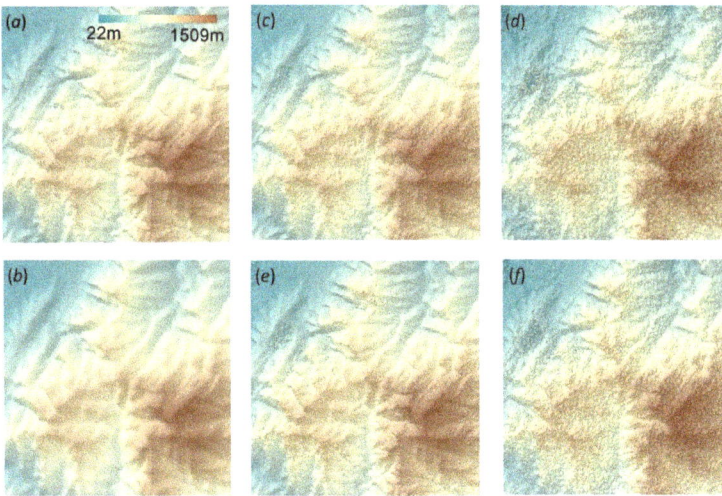

Figure 9. Hillshades of single/multi-baseline DEMs and SRTM DEM corresponding to the black rectangle in Figure 8a. (**a**) Multi-baseline InSAR DEM; (**b**) radar-coded SRTM DEM; (**c**) Interferogram I DEM, $h_{2\pi} = 82$ m; (**d**) Interferogram II DEM, $h_{2\pi} = 833$ m; (**e**) Interferogram III DEM, $h_{2\pi} = 115$ m; (**f**) Interferogram IV DEM, $h_{2\pi} = 347$ m.

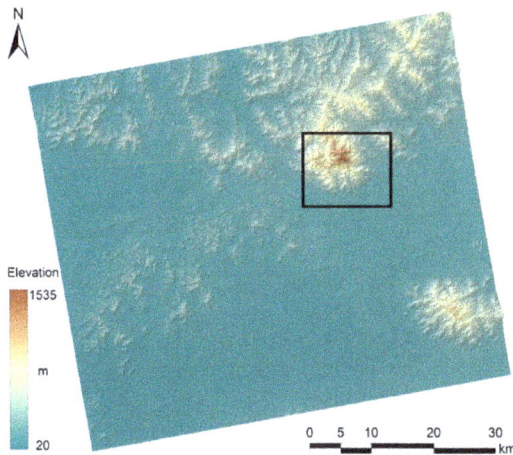

Figure 10. Hillshade of multi-baseline InSAR DEM. The black rectangle marks the coverage of the 1:25,000 DEM used for accuracy validation.

Figure 11 shows the height error maps of single/multi-baseline InSAR DEMs and SRTM DEM, which directly reflect the distribution and amount of the height error for different DEMs. Table 7 shows the statistical values of the height error maps of the corresponding InSAR DEMs and SRTM DEM shown in Figure 10, which can clearly reflect the statistical characteristics of the height error distribution.

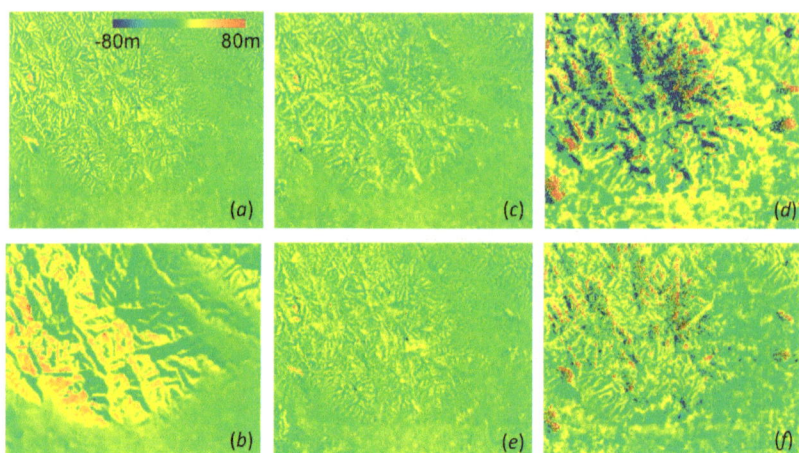

Figure 11. The height error map of InSAR DEMs, (**a**) the DEM generated by ML estimation with prior DEM; (**b**) SRTM DEM; (**c–f**) are the DEMs generated by single-baseline interferograms I, II, III, and IV.

Table 7. Statistical values of height errors of single/multi-baseline InSAR DEMs and SRTM DEM.

	Mean	Std.	Absolute Value ≤ 10 m
SRTM DEM	4.9 m	15.4 m	58.9%
Interferogram I DEM	1.9 m	11.3 m	81.4%
Interferogram II DEM	−4.4 m	43.0 m	32.8%
Interferogram III DEM	2.3 m	10.6 m	83.0%
Interferogram IV DEM	−0.3 m	27.7 m	51.8%
multi-baseline DEM	**1.7 m**	**8.6 m**	**86.3%**

5. Discussion

5.1. Comparative Analysis of the Single- and Multi-Baseline InSAR DEMs

For the single-baseline InSAR, the interferograms in Figure 7 with longer normal baselines (Figure 7a,c) have denser fringes than the interferograms with shorter normal baselines (Figure 7b,d). After the phase unwrapping and phase-to-height conversion, the single-baseline InSAR DEMs with longer normal baselines (Figure 8c,e) have more topographic detail than InSAR DEMs with shorter normal baselines (Figure 8d,f). Quantitatively, the height error of single-baseline InSAR DEMs with longer normal baselines (Figure 10c,e) is evidently smaller than the InSAR DEMs with shorter normal baselines (Figure 10d,f), which can also be reflected in Table 6—the interferograms I, III with longer normal baselines a have smaller standard deviation of height errors and a larger percentage of the absolute height error less than 10 m than the interferograms II and IV with shorter normal baselines. Then, can it be concluded that the longer the normal baseline, the better height measurement accuracy for single-baseline InSAR? The answer is obviously no. Interferogram I (Figure 7a) has a longer normal baseline than interferogram III (Figure 7c); however, the height accuracy of the InSAR DEM generated by interferogram II, as in Figures 8e and 10e, is better than that of the InSAR DEM generated by interferogram I, as in Figures 8c and 10c. According to Equation (4), the height accuracy of single-baseline InSAR is determined by both the length of the normal baseline and the phase noise, while too long a normal baseline can also introduce non-negligible phase noise during the delicate phase unwrapping.

In this article, this contradiction is solved by a multi-baseline InSAR with maximum likelihood estimation criterion. In the ALOS/PALSAR data experiment, the multi-baseline InSAR DEM (Figure 9a)

have more topographic detail than the single-baseline InSAR DEMs (Figure 9c–f). The height error of
the single-baseline InSAR DEMs (Figure 10c–f) is larger than that of the multi-baseline InSAR DEM
(Figure 10a). In Table 6, the standard deviation of height error of the multi-baseline InSAR DEM is the
smallest and the percentage of the absolute height error less than 10 m of the multi-baseline InSAR
DEM is the largest among all the DEMs. Therefore, the height accuracy of the multi-baseline DEM
is better than that of the single-baseline DEMs on the whole, indicating that the proposed method
is effective.

5.2. Comparative Analysis of the Prior Height's Impact on ML Estimation

From the simulated data experiment in Section 4.1.2, the ML estimated DEM with the prior height
probability distribution in Figure 4c,d has higher height accuracy than the ML estimated DEM without
the prior height probability distribution in Figure 4a,b. Here, we analyze the influence of the prior
height on ML estimation. A resolution unit in the simulated interferogram with true height of 1320.4 m
is used as an example. Figure 12 shows the height probability distribution of the chosen resolution unit.
Figure 12a shows the joint height probability distribution of three simulated interferograms without
phase noise; when h is 1320 m, the joint probability density function (PDF) reaches its peak. Figure 12b
shows the joint PDF with the decorrelation noise and, although there is a local peak at $h = 1320$ m,
the highest peak is located at $h = 1469$ m with height error up to 149 m. Figure 12c shows the prior
height probability distribution, with the highest peak located at $h = 1332$ m. Figure 12d shows the
joint height PDF by multiplying the height PDF in Figure 12b and the prior height distribution in
Figure 12c. There is only one peak in Figure 12d, located at $h = 1320$ m, very close to the true height of
$h = 1320.4$ m. Therefore, we can say that the prior DEM forms an external height constraint on the
multi-baseline InSAR observations and can effectively remove the erroneous estimate and solve the
height ambiguity problem.

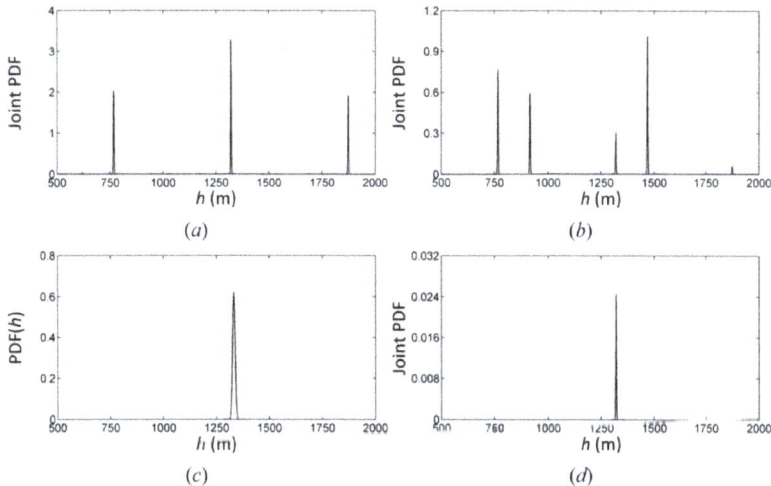

Figure 12. The height probability distribution of a cell in the simulated interferogram with a true
height of $h = 1320.4$ m. (**a,b**) The joint height probability distribution with and without decorrelation
phase noise, respectively; (**c**) the prior height probability distribution acquired from the prior DEM;
(**d**) the joint height probability distribution with the prior height probability distribution, obtained by
multiplying the two probability distributions in (**b,c**).

5.3. Comparative Analysis of the Multi-Baseline InSAR DEM and SRTM DEM

The height accuracy of the multi-baseline DEM and SRTM DEM is comparatively evaluated. The multi-baseline InSAR DEM (Figure 8a) has more topographic details than the radar-coded SRTM DEM (Figure 8b). In Figure 10, the height error of multi-baseline InSAR DEM (Figure 10a) is obviously smaller than that of SRTM DEM (Figure 10b), especially in the mountainous areas. From the statistical values of height errors in Table 6, it can also be seen that the multi-baseline has better height accuracy than SRTM DEM. The spatial resolution of the generated multi-baseline InSAR DEM is 20 m, while the SRTM DEM used for the prior DEM in the experiment is at 90 m resolution. The generation of the multi-baseline DEM with the SRTM DEM as the prior DEM can be viewed as a topographical information update of the SRTM DEM in terms of height accuracy and resolution.

According to the Digital Terrain Elevation Data (DTED) Standard defined by the American National Geospatial Intelligence Agency (NGA) [28], the multi-baseline InSAR DEM generated in this article meets the DTED-2 standard for spatial resolution and height accuracy based on the height error statistical values in Table 6. Similarly, with reference to the Chinese National 1:50,000 DEM Standard released by the National Administration of Surveying, Mapping and Geoinformation of China, the height accuracy of the multi-baseline InSAR DEM in mountainous areas reaches 1:50,000 DEM (mountain) Level 2.

Since high-resolution spaceborne InSAR data acquired by TerraSAR-X/TanDEM-X, COSMO-SkeMed, ALOS2/PALSAR-2 and so on are increasingly available, how can the optimized multi-baseline InSAR DEM generation with maximum likelihood estimation be applicable with these data? Two factors have to be considered. First, the cell size of the reference DEM, e.g., SRTM DEM, is much larger than that of the interferogram. The reference DEM should be oversampled and then radar-coded into the SAR image space. The height error introduced by the oversampling and radar coding processing will increase the value of σ_{SRTM} to make it even larger than in the condition when the prior DEM has a cell size comparable to that of the interferogram. Hence, in the actual calculation, σ_{SRTM} should be adjusted accordingly. Second, the size of the neighborhood should also be adjusted. Since the high spatial resolution spaceborne images have a much smaller cell size than the SRTM DEM and, moreover, these InSAR images usually have high coherence, a smaller neighborhood such as four-neighbor is preferred.

6. Conclusions

Multi-baseline InSAR height estimation can combine the advantages of both short and long normal baselines and generate DEMs with higher height accuracy than single-baseline InSAR DEMs. In this article, a multi-baseline InSAR with maximum likelihood criterion is used to generate DEM in mountainous areas. The prior height probability distribution is incorporated to suppress the phase noise and re-defined to take the local terrain conditions and neighboring height constraints into consideration. Furthermore, the processing flow is optimized for better computational efficiency. Simulation data and ALOS/PALSAR data experiments are performed to test the effectiveness of the proposed method. Our major findings in this article are as follows:

(1) The height accuracy of the ML estimation with re-defined prior height probability distribution is much better than that of the ML estimation without prior height probability, indicating that well-defined height probability can suppress phase noise and help solve the height ambiguity problem.

(2) The processing strategy proposed in this article, including (1) replacing the rigorous height-to-phase conversion with the rational function model (RFM); (2) substituting the complicated height likelihood probability function with a two-dimensional lookup table; (3) searching for the maximum likelihood height with flexible search step length instead of the fixed search step length, is effective, making the proposed processing flow applicable to spaceborne datasets.

(3)   Compared with SRTM DEM, the multi-baseline InSAR DEM has obvious advantages in terms of resolution and precision. Hence the multi-baseline InSAR estimation can be viewed as a topographical information update of the historical low-resolution DEMs.

(4)   The multi-baseline InSAR DEM generated from ALOS/PALSAR datasets meets the American DTED-2 standard and Chinese 1:50,000 DEM (mountain) Level 2 in the case of spatial resolution and height accuracy.

In the future, the proposed multi-baseline InSAR topographic mapping flow can be tested and improved further with more spaceborne datasets. The maximum likelihood height estimation method with the prior DEM provides a promising solution for DEM mass production in mountainous areas.

Acknowledgments: This work was financially supported by the National Natural Science Foundation of China (Grant Nos. 61331016, 41774006, 41501497, and 41271457). The authors would like to thank the Japan Aerospace Exploration Agency (JAXA) for providing the ALOS/PALSAR datasets via the ALOS RA project (PI: 1247, 1440 and 3248) and the Shandong Provincial Land and Surveying and Mapping Institute for providing the 1:25,000 photogrammetric DEM for validation.

Author Contributions: Houjun Jiang and Lu Zhang designed and performed the experiments; Mingsheng Liao analyzed the data; Yuting Dong validated the experiments and wrote the paper.

Conflicts of Interest: The authors declare no conflict of interest.

References

1.   Zebker, H.A.; Goldstein, R.M. Topographic mapping from interferometric synthetic aperture radar observations. *J. Geophys. Res. Solid Earth* **1986**, *91*, 4993–4999. [CrossRef]

2.   Zebker, H.A.; Werner, C.L.; Rosen, P.A.; Hensley, S. Accuracy of topographic maps derived from ERS-1 interferometric radar. *IEEE Trans. Geosci. Remote Sens.* **1994**, *32*, 823–836. [CrossRef]

3.   Li, W.; Zhang, L.; Xu, L.; Xu, M.; Lou, L.; Xu, H.; Liao, M.; Chen, J.; Yu, W. Spaceborne D-InSAR system: Conceptual overview. In Proceedings of the 2015 IEEE 5th Asia-Pacific Conference on Synthetic Aperture Radar (APSAR), Singapore, 1–4 September 2015; pp. 99–102.

4.   Gao, X.; Liu, Y.; Li, T.; Wu, D. High Precision DEM Generation Algorithm Based on InSAR Multi-Look Iteration. *Remote Sens.* **2017**, *9*, 741. [CrossRef]

5.   Ghiglia, D.C.; Wahl, D.E. Interferometric synthetic aperture radar terrain elevation mapping from multiple observations. In Proceedings of the IEEE 6th Digital Signal Processing Workshop, Yosemite National Park, CA, USA, 2–5 October 1994; pp. 33–36.

6.   Hanssen, R.F. *Radar Interferometry: Data Interpretation and Error Analysis*; Springer Science & Business Media: Berlin, Germany, 2001.

7.   Gini, F.; Lombardini, F.; Montanari, M. Layover solution in multi-baseline SAR interferometry. *IEEE Trans. Aerosp. Electron. Syst.* **2002**, *38*, 1344–1356. [CrossRef]

8.   Fornaro, G.; Pauciullo, A.; Sansosti, E. Phase difference-based multichannel phase unwrapping. *IEEE Trans. Image Process.* **2005**, *14*, 960–972. [CrossRef] [PubMed]

9.   Ferretti, A.; Prati, C.; Rocca, F.; Monti Guarnieri, A. Multi-baseline SAR interferometry for automatic DEM reconstruction (DEM). In Proceedings of the Third ERS Symposium on Space at the service of our Environment, Florence, Italy, 14–21 March 1997.

10. Ferretti, A.; Prati, C.; Rocca, F. Multi-baseline InSAR DEM reconstruction: The wavelet approach. *IEEE Trans. Geosci. Remote Sens.* **1999**, *37*, 705–715. [CrossRef]

11. Jiang, H.J.; Zhang, L.; Wang, Y.; Liao, M.S. Fusion of high-resolution DEMs derived from COSMO-SkyMed and TerraSAR-X InSAR datasets. *J. Geodesy* **2014**, *88*, 587–599. [CrossRef]

12. Thompson, D.G.; Robertson, A.E.; Arnold, D.V.; Long, D.G. Multi-baseline interferometric SAR for iterative height estimation. In Proceedings of the 1999 IEEE International Geoscience and Remote Sensing Symposium, Hamburg, Germany, 28 June–2 July 1999; pp. 251–253.

13. Xia, X.-G.; Wang, G. Phase unwrapping and a robust Chinese remainder theorem. *IEEE Signal Process. Lett.* **2007**, *14*, 247–250. [CrossRef]

14. Yuan, Z.H.; Deng, Y.K.; Li, F.; Wang, R.; Liu, G.; Han, X.L. Multichannel InSAR DEM reconstruction through improved closed-form robust Chinese Remainder Theorem. *IEEE Geosci. Remote Sens. Lett.* **2013**, *10*, 1314–1318. [CrossRef]

15. Lombardo, P.; Lombardini, F. Multi-baseline SAR interferometry for terrain slope adaptivity. In Proceedings of the IEEE National Radar Conference, Syracuse, NY, USA, 13–15 May 1997; pp. 196–201.

16. Pascazio, V.; Schirinzi, G. Multifrequency InSAR height reconstruction through maximum likelihood estimation of local planes parameters. *IEEE Trans. Image Process.* **2002**, *11*, 1478–1489. [CrossRef] [PubMed]

17. Fornaro, G.; Guarnieri, A.M.; Pauciullo, A.; De-Zan, F. Maximum likelihood multi-baseline SAR interferometry. *IEEE Proc. Radar Sonar Navig.* **2006**, *153*, 279–288. [CrossRef]

18. Eineder, M.; Adam, N. A maximum-likelihood estimator to simultaneously unwrap, geocode, and fuse SAR interferograms from different viewing geometries into one digital elevation model. *IEEE Trans. Geosci. Remote Sens.* **2005**, *43*, 24–36. [CrossRef]

19. Ferraiuolo, G.; Pascazio, V.; Schirinzi, G. Maximum a posteriori estimation of height profiles in InSAR imaging. *IEEE Geosci. Remote Sens. Lett.* **2004**, *1*, 66–70. [CrossRef]

20. Ferraiuolo, G.; Meglio, F.; Pascazio, V.; Schirinzi, G. DEM Reconstruction Accuracy in Multichannel SAR Interferometry. *IEEE Trans. Geosci. Remote Sens.* **2009**, *47*, 191–201. [CrossRef]

21. Xiong, S.; Muller, J.-P.; Li, G. The Application of ALOS/PALSAR InSAR to Measure Subsurface Penetration Depths in Deserts. *Remote Sens.* **2017**, *9*, 638. [CrossRef]

22. Pascazio, V.; Schirinzi, G. Estimation of terrain elevation by multifrequency interferometric wide band SAR data. *IEEE Signal Process. Lett.* **2001**, *8*, 7–9. [CrossRef]

23. Tough, R.; Blacknell, D.; Quegan, S. *A Statistical Description of Polarimetric and Interferometric Synthetic Aperture Radar Data*; Royal Society: London, UK, 1995; pp. 567–589.

24. Rodriguez, E.; Morris, C.S.; Belz, J.E. A global assessment of the SRTM performance. *Photogramm. Eng. Remote Sens.* **2006**, *72*, 249–260. [CrossRef]

25. Liao, M.; Jiang, H.; Wang, Y.; Wang, T.; Zhang, L. Improved topographic mapping through high-resolution SAR interferometry with atmospheric effect removal. *ISPRS J. Photogramm. Remote Sens.* **2013**, *80*, 72–79. [CrossRef]

26. Zhang, L.; He, X.Y.; Balz, T.; Wei, X.H.; Liao, M.S. Rational function modeling for spaceborne SAR datasets. *ISPRS J. Photogramm. Remote Sens.* **2011**, *66*, 133–145. [CrossRef]

27. National Administration of Surveying, Mapping and Geoinformation of China. *National Standards for 1:5000, 1:10,000, 1:25,000, 1:50,000, 1:100,000 Digital Elevation Model*; Surveying and Mapping Publishing House: Bejing, China, 2010.

28. National Geospatial-Intelligence Agency. Geospatial Standards and Specifications, 2003. Available online: https://www.nga.mil/ProductsServices/TopographicalTerrestrial/Pages/DigitalTerrainElevationData.aspx (accessed on 10 March 2018).

remote sensing

MDPI

Technical Note

Elevation Extraction and Deformation Monitoring by Multitemporal InSAR of Lupu Bridge in Shanghai

Jingwen Zhao [1,2], Jicang Wu [1,*], Xiaoli Ding [2] and Mingzhou Wang [3]

[1]   College of Surveying and Geo-Informatics, Tongji University, Shanghai 200092, China; mnls0226@gmail.com
[2]   Department of Land Surveying and Geo-Informatics, Faculty of Construction and Environment,
      The Hong Kong Polytechnic University, Hong Kong, China; xl.ding@polyu.edu.hk
[3]   Shenzhen Water Science and Technology Development Company, Shenzhen 518000, China;
      wangmingzhou@whu.edu.cn
*   Correspondence: jcwu@tongji.edu.cn; Tel.: +86-21-6598-2709

Academic Editors: Timo Balz, Uwe Soergel, Mattia Crespi, Batuhan Osmanoglu, Zhong Lu and Prasad S. Thenkabail
Received: 11 July 2017; Accepted: 28 August 2017; Published: 30 August 2017

Abstract: Monitoring, assessing, and understanding the structural health of large infrastructures, such as buildings, bridges, dams, tunnels, and highways, is important for urban development and management, as the gradual deterioration of such structures may result in catastrophic structural failure leading to high personal and economic losses. With a higher spatial resolution and a shorter revisit period, interferometric synthetic aperture radar (InSAR) plays an increasing role in the deformation monitoring and height extraction of structures. As a focal point of the InSAR data processing chain, phase unwrapping has a direct impact on the accuracy of the results. In complex urban areas, large elevation differences between the top and bottom parts of a large structure combined with a long interferometric baseline can result in a serious phase-wrapping problem. Here, with no accurate digital surface model (DSM) available, we handle the large phase gradients of arcs in multitemporal InSAR processing using a long–short baseline iteration method. Specifically, groups of interferometric pairs with short baselines are processed to obtain the rough initial elevation estimations of the persistent scatterers (PSs). The baseline threshold is then loosened in subsequent iterations to improve the accuracy of the elevation estimates step by step. The LLL lattice reduction algorithm (by Lenstra, Lenstra, and Lovász) is applied in the InSAR phase unwrapping process to rapidly reduce the search radius, compress the search space, and improve the success rate in resolving the phase ambiguities. Once the elevations of the selected PSs are determined, they are used in the following two-dimensional phase regression involving both elevations and deformations. A case study of Lupu Bridge in Shanghai is carried out for the algorithm's verification. The estimated PS elevations agree well (within 1 m) with the official Lupu Bridge model data, while the PS deformation time series confirms that the bridge exhibits some symmetric progressive deformation, at 4–7 mm per year on both arches and 4–9 mm per year on the bridge deck during the SAR image acquisition period.

Keywords: deformation monitoring; elevation extraction; InSAR; LLL lattice reduction; long–short baseline iteration; Lupu Bridge

---

1. Introduction

Space borne interferometric synthetic aperture radar (InSAR) technology makes use of interferometric image pairs of the same ground area obtained from repeating satellite orbits. Interferometric phases from the image pairs can be used to determine elevation and deformation in the radar line of sight (LOS) direction [1,2]. With a high spatial resolution and a short revisit period, InSAR can be used for large-scale deformation monitoring and elevation extraction and has great application potential in areas such as health monitoring of large structures [3–9]. As interferograms are produced by the complex multiplication of coherent synthetic aperture radar (SAR) images, phase unwrapping is required to determine the number of whole phase cycles for arcs of interferometric phase observables in multitemporal InSAR data processing. Phase unwrapping is a core of InSAR technology. Many methods have been developed for phase unwrapping, including, e.g., the two-dimensional branch-cut method [10], the quality map guidance algorithm [11], the region-growing algorithm [12], the network flow method [13,14], three-dimensional phase unwrapping [15], and some others [16–18]. Each of the methods has its advantages and limitations.

In mathematics, phase unwrapping can be seen as the closest lattice vector problem; the lattice reduction algorithm is designed to find the shortest vector in a two-dimensional grid. In 1982, Lenstra, Lenstra, and Lovász proposed the LLL lattice reduction algorithm [19] to extend the search space to an n-dimensional space. Since the lattice reduction method can rapidly reduce the search radius, compress the search space, and improve the successful rate of ambiguity resolution, it has been widely used in integer programming [20], cryptography [21,22], number theory [23], and other fields. For example, it was applied by Liu to resolve Global Navigation Satellite System (GNSS) phase ambiguity [24]. We propose in this paper to use the LLL lattice reduction algorithm for InSAR phase unwrapping.

The classical permanent scatterer interferometry (PSI) model estimates linear deformation rate and elevation error simultaneously [1]. However, because the interferometric fringes are relatively dense in the case of a long baseline, a non-continuous steep slope phase corresponding to a large elevation gap may bring various challenges to a permanent scatterer (PS) arc's solution. For example, the mean/sigma ratio of an arc may exceed the threshold value, so that some PSs may be eliminated. The obtained ambiguity may not be accurate, or the phase may be no longer continuous. Long-baseline interferometric pairs correspond to a smaller elevation ambiguity (namely, more accurate elevation); however, a long baseline also increases the difficulty of phase unwrapping. Conversely, short-baseline interferometric pairs correspond to a large elevation ambiguity (namely, less accurate elevation), but the interferometric fringes are relatively smooth and much easier to unwrap. In view of this, this paper makes use of the long–short baseline iteration method [25–28] for multitemporal InSAR data processing, which first selects short baseline interferometric pairs for a one-dimensional elevation solution, and then gradually enlarges the spatial baseline threshold and reduces the phase gradient with the elevation components calculated from the previous iteration. Once the elevations of the selected PSs are obtained with suitable accuracy, they are used in the following two-dimensional phase regression involving both the elevations and deformations. Finally, the linear and seasonal deformations are extracted from the multitemporal InSAR time series.

The paper demonstrates a method suitable for the high-phase-gradient phase unwrapping problem with no digital surface model (DSM) available in multitemporal InSAR processing. The long–short baseline iteration method is adopted to deal with the problem of the large phase gradients of the arcs, while the LLL lattice reduction algorithm is applied to rapidly resolve phase ambiguity. A case study of Lupu Bridge validates the usefulness of the proposed method.

2. Research Area, Data and Methods

2.1. Research Area and Data

The research area is the Lupu Bridge (Figure 1) in Shanghai, China. This bridge has been in operation since 2003. The bridge is about 750 m long and 100 m tall. As the first arch bridge on the Huangpu River and the world's second longest span all-steel arch bridge at that time, the Lupu Bridge soon became a famous scenic spot in Shanghai. The bridge's axis is almost perpendicular to the radar line of sight (LOS) direction. Since the bridge structure is rather large, and the PS points of the bridge are sparse, it is quite difficult to form a connected PS network and resolve the arcs by the traditional PSI method.

Thirty-five (35) ascending X-band Cosmo-SkyMed SAR images are used. The key parameters of the images are shown in Table 1.

(a)   (b)

Figure 1. Lupu Bridge: (**a**) side view; (**b**) top view (red circle surrounding the bridge) (Cr. Baidu).

Table 1. Parameters of the ascending Cosmo-SkyMed images.

Time Range	Number of Scenes	Azimuth Lines	Range Columns	Incident Angle (°)	Heading (°)	Azimuth Resolution (m)	Range Resolution (m)
10 December 2008–6 November 2010	35	400	250	40	−10.34	2.25	1.25

2.2. Data Processing Chain

2.2.1. Long–Short Baseline Iteration PSInSAR Method

Figure 2 shows the workflow of the proposed method. As there is no accurate DSM data available, it is difficult to perform traditional two-dimensional phase unwrapping as the initial elevation value contributes considerably to the convergence of the algorithm. Therefore, the long–short baseline iteration PSInSAR method is applied for one-dimensional, accurate elevation extraction. Only interferometric pairs with a short temporal baseline are selected assuming no deformation exists. The thresholds of the interferometric perpendicular baseline in each iteration are loosened gradually as shown in Table 2 to improve the elevation accuracy. Note that the elevation ambiguity Δh is calculated from the perpendicular baseline,

$$\Delta h = \frac{\lambda \times \gamma \times \sin \theta}{2b} \tag{1}$$

where Δh is the elevation ambiguity, i.e., the elevation change when the phase varies by 2π; λ is the radar wavelength; γ is the range between the satellite and the illuminating scene; θ is the incident angle; and b is the perpendicular baseline.

Table 2. Interferometric perpendicular baseline thresholds in a long–short baseline permanent scatterer interferometric synthetic aperture radar (PSInSAR) iteration.

Iteration Round	Temporal Baseline (Day)	Perpendicular Baseline (m)	Number of Interferometric Pairs	Number of Arcs Used in the Net	Elevation Ambiguity (m)
1	<65	<50	8	1320	150.0
2	<65	<200	38	1271	37.6
3	<65	<360	58	1395	20.9
4	<65	<600	77	1233	12.5
5	<65	<1000	119	1237	7.5

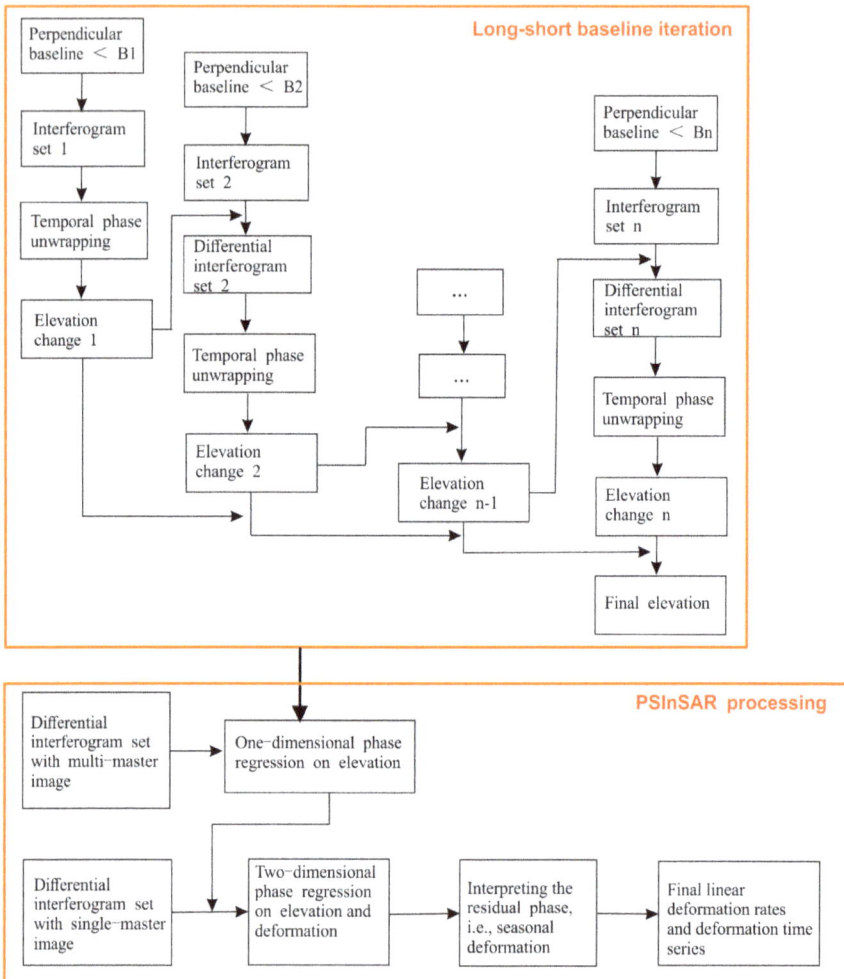

Figure 2. Workflow of the proposed InSAR processing method.

The selection of PSs is conducted in GAMMA software using criteria such as the mean/standard deviation ratio, the minimum intensity, and coherence. In general, the PSs on the bridge are distributed along both the arches and the pavement of the bridge deck. Based on the PSs, an initial Delaunay triangular network is formed with 1700 arcs. As the large elevation differences between the arches and the deck are likely to result in the transmission and accumulation of elevation errors in the network adjustment, long arcs or those with poor coherence are screened out. Subsequently, phase unwrapping and network adjustment are carried out. In each iteration, the elevation of the PS on the deck with minimum azimuth is set as the reference. This value is 53.6 m, based on the bridge model offered by the official Lupu Bridge maintenance company. After five iterations, with a perpendicular baseline threshold condition of 1000 m, the PSs elevation corrections become very small; therefore, the elevation values obtained in the fifth iteration are considered the final solution.

2.2.2. The LLL Lattice Reduction Algorithm

InSAR phase unwrapping can be considered as a mixed integer least squares problem. The LLL lattice reduction algorithm separates the unknowns into an integer part and a real part, solving the integer part first and then the real part. In this way, the LLL algorithm provides a fast and numerically reliable routine to the mixed integer least squares problem.

$$y = Ax + Bz + \delta \tag{2}$$

where x is a real unknown vector with k elements, $x \in R^{k*1}$; z is an integer unknown vector with n elements, $z \in Z^{n*1}$; $A \in R^{m*k}$ and $B \in R^{m*n}$ are known coefficient matrices with full column rank; m is the number of observations; and $y \in R^{m*1}$ is the vector of observations. Z represents the set of integers; R represents the set of real numbers; δ is the noise vector. The aim is to solve for the unknowns x and z based on the known matrices A, B, and observations y. The solutions should minimize the 2-norm of vector $y - Ax - Bz$:

$$\min_{x \in R^k, z \in Z^n} \|y - Ax - Bz\|_2^2 \tag{3}$$

If matrix A has QR factorization (A decomposition of a matrix A into a product A = QR of an orthogonal matrix Q and an upper triangular matrix R),

$$A = \begin{bmatrix} Q_A & \overline{Q}_A \end{bmatrix} \begin{bmatrix} R_A \\ 0 \end{bmatrix} \tag{4}$$

where $\begin{bmatrix} Q_A & \overline{Q}_A \end{bmatrix} \in R^{m*m}$ is orthogonal, and $R_A \in R^{k*k}$ is a nonsingular upper triangular matrix. Then

$$\|y - Ax - Bz\|_2^2 = \left\| \begin{bmatrix} Q_A^T \\ \overline{Q}_A^T \end{bmatrix} y - \begin{bmatrix} R_A \\ 0 \end{bmatrix} x - \begin{bmatrix} Q_A^T B \\ \overline{Q}_A^T B \end{bmatrix} z \right\|_2^2$$

$$= \|Q_A^T y - R_A x - Q_A^T Bz\|_2^2 + \|\overline{Q}_A^T y - \overline{Q}_A^T Bz\|_2^2 \tag{5}$$

If z is fixed, there must be an appropriate $x \in R^{k*1}$ that ensures that the first term ($\|Q_A^T y - R_A x - Q_A^T Bz\|_2^2 \geq 0$) in Equation (5) is 0 so as to satisfy the minimization requirement of Equation (3). Therefore, the problem can be decomposed into the following two problems,

1. An ordinary integer least squares problem to calculate $\hat{z}$

$$\min_{z \in Z^n} \|\overline{Q}_A^T y - \overline{Q}_A^T Bz\|_2^2 \tag{6}$$

Specifically, a reduction algorithm and a search algorithm are presented to obtain the integer z which satisfies Equation (6).

2. With z known, Equation (3) becomes a least squares problem. With $\hat{z}$ brought back into Equation (5) and setting the first term into 0, $\hat{x}$ can be obtained from

$$R_A x = Q_A^T y - Q_A^T B\hat{z} \tag{7}$$

For simplicity, the above problem 1 is noted as:

$$\min_{z \in Z^n} \|y - Bz\|_2^2 \tag{8}$$

where y is a known vector; z is the least squares solution required; and Bz is a vector in the grid. Thus, seeking a solution for Equation (8) can be interpreted as searching for the grid vector that is nearest to y. This is a closest vector problem (CVP), which has been proven to be an NP-hard (non-deterministic polynomial hard) problem. To make the search process simple and efficient, many reduction methods have been proposed. In this study, we use the LLL method, which has two steps,

- Reduction

First, using a minimum main-element method, the QR decomposition of matrix B is carried out to transform it into an upper triangular matrix R and an orthogonal matrix Q. Second, the non-diagonal elements in R are reduced using an integer Gaussian transform to remove any correlation and enable efficient searching. Third, the columns are rearranged using the minimum-column pivoting strategy to meet the LLL reduction criterion.

- Search

After reduction, we need to search for the optimal integer solution $z \in Z^n$ to satisfy $\min_{\bar{z} \in Z^n} \|\bar{y} - R\bar{z}\|_2^2$. Given a threshold β, we assume that the optimal integer solution z satisfies

$$f(z) \triangleq \|y - Rz\|_2^2 < \beta \tag{9}$$

This corresponds to searching for the optimal solution within an ellipsoid.

R is then decomposed into the first $(n-1)$-order submatrix and the last line, and y is decomposed into the $(n-1)$-dimensional sub-vector and the last element. Thus,

$$\|y - Rz\|_2^2 = \left\| \begin{pmatrix} y_1 \\ y_n \end{pmatrix} - \begin{pmatrix} R_1 & r_{1:n-1,n} \\ 0 & r_{nn} \end{pmatrix} \begin{pmatrix} z_1 \\ z_n \end{pmatrix} \right\|_2^2$$
$$= \|(y_1 - z_n r_{1:n-1,n}) - R_1 z_1\|_2^2 + (y_n - r_{nn} z_n)^2 \tag{10}$$

To satisfy Equation (9), the following conditions need to be met,

$$(y_n - r_{nn} z_n)^2 < \beta \tag{11}$$

and

$$\|(y_1 - z_n r_{1:n-1,n}) - R_1 z_1\|_2^2 < \beta - (y_n - r_{nn} z_n)^2 \tag{12}$$

Equation (12) is an $(n-1)$-dimensional integer least squares problem, and the corresponding search radius is $\rho = \sqrt{(\beta - (y_n - r_{nn} z_n)^2)}$. The integer solution to Equation (11) falls within $[(y_n - \beta)/r_{nn}, (y_n + \beta)/r_{nn}]$. Using this algorithm recursively, we can solve the upper triangular integer least squares problem.

Once the p optimal integer solutions $\hat{z}$ are obtained, we can use the following upper triangular matrix to solve for the corresponding p real solutions: $R_A \hat{x} = Q_A^T (ye^T - B\hat{z})$, where $e = [1, \cdots, 1]^T \epsilon R^p$.

2.2.3. LLL Lattice Reduction Algorithm Used for PSInSAR

When applying the above LLL lattice reduction algorithm to PSInSAR data processing, by contrast, the phase ambiguities correspond to the integer unknowns, while the elevation error and the linear deformation rates correspond to the real unknowns, and the interferometric phases correspond to the observations in Section 2.2.2.

Assume that there are m + 1 SAR images of the same area, obtained at time $t_1, \ldots, t_{m+1}$, respectively. One of the images is chosen as the master image and the other m images are the slave images, to form m interferograms. The unwrapped phase between pixel i and pixel j in interferogram p is expressed as

$$\Delta\varnothing_{ij}^p = \alpha^p \times v_{ij} + \beta^p \times h_{ij} + 2\pi \times z + \delta \tag{13}$$

where v_{ij} and h_{ij} are the relative displacement rate and relative elevation error between the two pixels, respectively. β^p changes with the perpendicular baseline, and α^p changes with the temporal baseline. z is the unknown number of whole phase cycles, and δ is the noise resulting possibly from decorrelation error, nonlinear deformation, thermal noise, and so on. Note that the atmospheric phase is considered to be correlated in space and can be significantly reduced by differencing interferometric phases between adjacent PSs to form an arc observation.

As an arc in an interferogram contains one-integer ambiguity, together with real unknowns v_{ij} and h_{ij}, there are $m + 2$ unknowns in the m interferograms corresponding to the arc. As there are only m observations, the observation equations formed according to Equation (13) have rank defects. To solve this problem, the initial values of two unknown parameters are assumed to be equal to 0, i.e.,

$$\begin{cases} h_{ij} = 0 \\ v_{ij} = 0 \end{cases} \tag{14}$$

and will be updated iteratively. The new observation equations can be expressed as follows:

$$y = \begin{pmatrix} A1 \\ A2 \end{pmatrix} x + \begin{pmatrix} B1 \\ B2 \end{pmatrix} z + \delta \tag{15}$$

where A1 has m rows, and its two columns are $\left[\alpha^1, \ldots, \alpha^m\right]^T$ and $\left[\beta^1, \ldots, \beta^m\right]^T$, respectively. A2 is a 2×2 identity matrix. The real unknowns x include v_{ij} and h_{ij}. B1 is an $m \times m$ identity matrix times 2π. B2 is a $2 \times m$ zero matrix. If the signal-to-noise ratio of the observations is high, the solution can be found with a few iterations.

3. Results and Discussion

3.1. Bridge Elevation Extraction

At first, only interferometric pairs with a short temporal baseline are selected assuming no deformation exists, and only the relative elevation errors are considered as the real unknowns in the one-dimensional elevation extraction step with LLL. The thresholds of the interferometric perpendicular baseline length in each iteration are loosened gradually, as shown in Table 2, to improve the elevation estimation accuracy obtained.

Figure 3 shows the side views of the PS elevations on the bridge obtained in iterations 1 ($B_\perp < 50$ m), 3 ($B_\perp < 360$ m), and 5 ($B_\perp < 1000$ m), respectively. In iteration 1, it is obvious that the elevation variations are relatively larger and the elevations on the arch are discontinuous, even with some obvious errors. However, with the loosening of the spatial baseline threshold and the increase of iterations, the elevations become smoother. The elevations obtained in iteration 5 are accepted as the final solution.

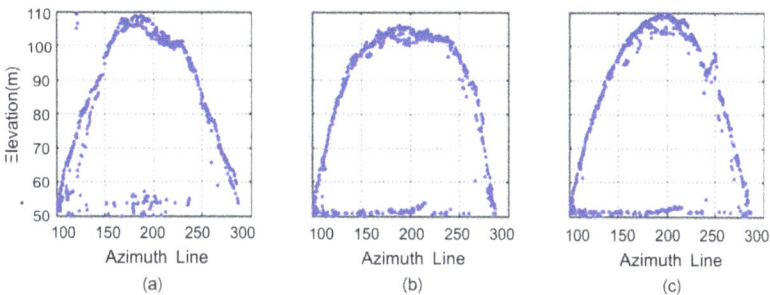

Figure 3. Estimated permanent scatterer (PS) elevations on the bridge obtained in iterations 1 ($B_\perp < 50$ m) (**a**); 3 ($B_\perp < 360$ m) (**b**); and 5 ($B_\perp < 1000$ m) (**c**), respectively (Azimuth-Elevation plane).

Figure 4 shows the elevations of PSs on the bridge obtained in iteration 5 after geocoding (three-dimensional (3D) view). The results clearly show the bridge arches and the deck. In order to further evaluate the accuracy of the obtained elevations, two external bridge datasets are used for comparison. One is the official bridge model obtained from the Shanghai Lupu Bridge Investment Development Co., Ltd. (No. 449, Yaohua Road, Shanghai, China). Three key parameters of the bridge structure (namely, maximum arch elevation, minimum deck elevation, and maximum deck elevation) obtained in iterations 1, 3, and 5 are compared with the official bridge model dataset (Table 3). The accuracy of the estimated elevations improves with the number of iterations; the final estimated elevations from iteration 5 are in good agreement with the bridge model. Note that Table 3 only presents a rough comparison, as it is difficult to match the PSs with their exact location in the bridge model, and therefore the elevation differences between the InSAR solution and the model do not necessarily represent the accuracy of the proposed method. The second external bridge dataset is downloaded from Google 3D Warehouse. The 3D PSs are transformed to the model coordinates for visualization in the Meshlab software. As shown in Figure 5, the arch shape and the location of the PSs are in excellent agreement.

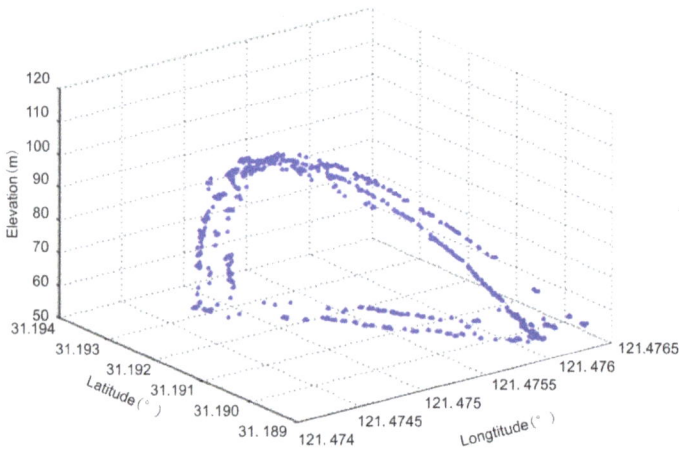

Figure 4. Elevations of PSs on the bridge from iteration 5 (three-dimensional (3D) view).

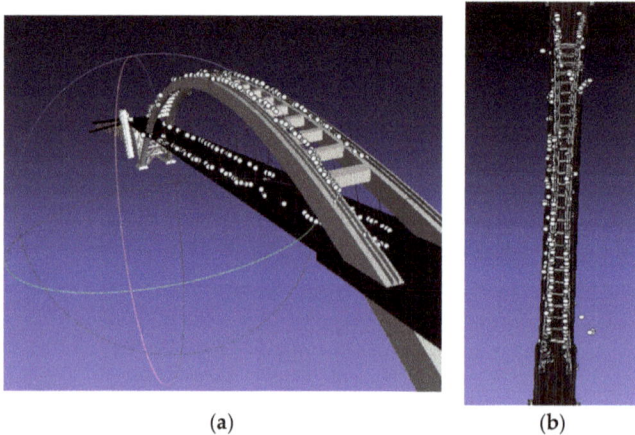

| (a) | (b) |

Figure 5. PSs (white dots) superimposed on the bridge model in Meshlab: (**a**) side view; (**b**) top view.

Table 3. Key parameters of the estimated bridge structure from iterations 1, 3, and 5 compared with the bridge model data (only direct measurements are listed).

	Iteration 1	Iteration 3	Iteration 5	Official Model
Maximum arch elevation (m)	109.1	106.0	109.6	109.35
Minimum deck elevation (m)	41.1	47.3	50.1	
Maximum deck elevation (m)	59.1	54.2	53.2	
Mean deck elevation (m)				53.60

3.2. Bridge Deformation Extraction

As shown in [29], the Lupu Bridge is an all-welded steel arch bridge connected by a set of components with a misalignment error of less than 1 mm. Moreover, even with nearly 20 arch ribs and a span of more than 500 m connecting Puxi and Pudong, the axial deviation of the central arch joints is less than 5 mm.

In fact, it is challenging to interpret InSAR-derived deformation results of man-made structures, particularly bridges, because it can be difficult to separate the major components of the InSAR phase, such as the linear deformation rates, seasonal deformation, elevation of structures, and atmosphere phase screen (APS). For example, the elevation-related atmospheric phase and the temperature-related deformation tend to have the same pattern. The elevation errors leak easily to deformation solutions. A few InSAR time series studies have investigated the thermal expansion of bridges and other structures [30–35]; however, many technical details are yet to be resolved.

In the Lupu Bridge case, once the elevations of the PSs are resolved with suitable accuracy, they are then used in a two-dimensional phase regression involving both elevation and linear deformation, and to obtain the linear deformation map as in Figure 6. By the way, PS No. 175 on the riverside is chosen as the reference point. As the research area is relatively small, the APS is neglected.

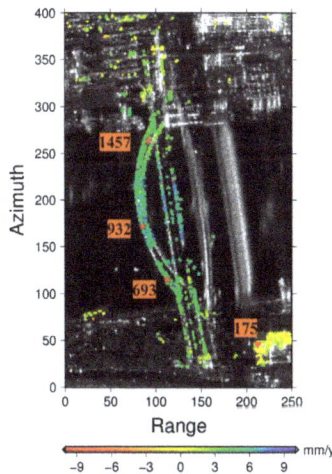

Figure 6. Linear deformation rates (in the line of sight (LOS) direction) of PSs. The labeled numbers are the IDs of chosen PSs. Red stars and the inverted triangle indicate the monitoring PSs (No. 693, 932, and 1457) and the reference PS (No. 175), respectively.

As shown in Figure 6, the linear deformation rates of the main part of the bridge are uniform, indicating that the bridge is stable as a whole. The LOS deformation rates compared to the reference point No. 175 vary from 4 to 7 mm per year during the SAR image acquisition period. Note that a positive value represents motion away from the satellite along the LOS, while a negative value indicates motion towards the satellite along the LOS. Progressive deformation appears on the two bridge arches and bridge deck. The largest deformation rates occur at the central part of the arches (7 mm per year) and the bridge deck (9 mm per year).

The thermal expansion of metallic or reinforced concrete structures can significantly affect the interferometric phase signature [36]. Typically, thermal dilation provides progressive patterns due to its accumulation over the structure's length. This is in agreement with our result. To further study thermal expansion effects on the deformation results, we collected temperature records of Shanghai during the SAR image acquisition period. Unfortunately, only some scattered monthly averaged temperature records can be found on the internet for the period, i.e., from December 2008 to November 2010. The SAR sensor passed over Shanghai at about 06:00 Beijing time. However, if we calculate the average temperature on the date of the data acquisition two, three, and four years later (T2010–2012, T2011–2013, T2012–2014 in Figure 7a) and extrapolate the monthly temperature during 2008 to 2010 (T2008–2010 in Figure 7a), the trends of variation of the temperature are almost the same. Thus, the interpolated temperature (T2008–2010) is used for the seasonal deformation analysis.

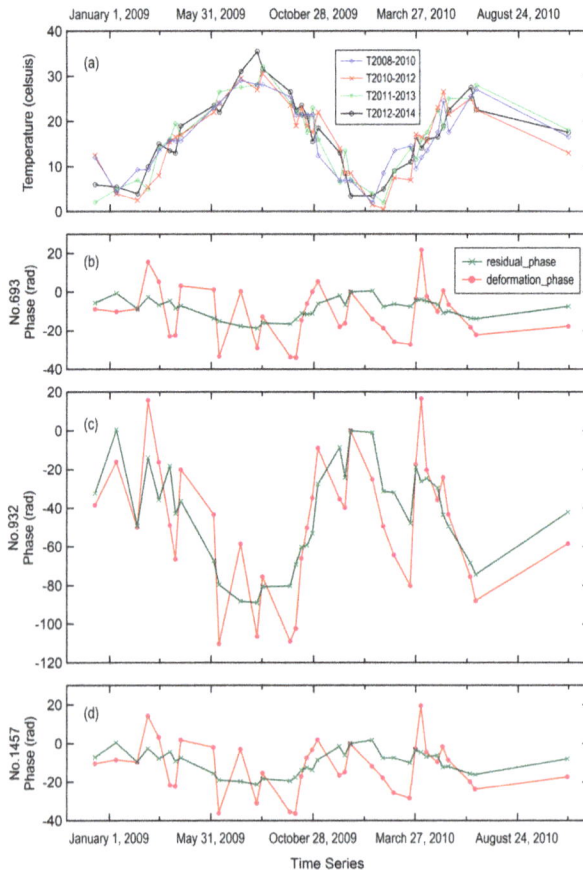

Figure 7. (**a**) Temporal variations of temperature (T2008–2010: interpolated averaged daily temperature records on SAR data acquisition dates. T2010–2012, T2011–2013, T2012–2014: averaged daily temperature records on acquisition dates two, three, and four years later); (**b–d**): Unwrapped deformation phase time series (red dots) and residual time series (green crosses) of PS points 693, 932, and 1457 whose locations are shown in Figure 6.

Three PSs (No. 693, 932, and 1457), whose locations are shown in Figure 6, are chosen for a more detailed seasonal deformation analysis. In the two-dimensional phase regression as shown in the last line of the bottom box in Figure 2, the unwrapped deformation phase time series (red dots) and the residual time series (green crosses) of these three points are shown in Figure 7b–d. Note that the unwrapped deformation phase time series are calculated by removing the elevation phase from the unwrapped interferometric phase, while the residual unwrapped phase is obtained by removing both the linear deformation and elevation phase from the unwrapped phase.

Both the unwrapped deformation phase time series and residual time series of the three points are used to calculate the correlation coefficients with the interpolated temperature (T2008–2010) and the results are listed in Table 4. In Table 4, the residual unwrapped phase time series corresponding to the three PSs have a strong negative correlation with the temperature. The correlation coefficients of all the three target points are larger than 0.9. When the phases of the linear deformation and the minor elevation correction are added back, the unwrapped phase time series have a much weaker negative correlation with the temperature, with the correlation coefficients ranging from 0.2526 to 0.646, with 0.646 corresponding to the PS at the center of the bridge arches, indicating that this PS was affected much more by the temperature than the other two PSs.

As expected, the central parts of the bridge arches and the bridge deck were experiencing the largest deformation.

Table 4. Coefficients between the residual unwrapped phase/unwrapped deformation phase and temperature variations.

PS Point Number	Location	Correlation Coefficient (Residual Unwrapped Phase vs. Temperature)	Correlation Coefficient (Unwrapped Deformation Phase vs. Temperature)
693	Southern end of arch	−0.9225	−0.2526
932	Center	−0.9163	−0.6460
1457	Northern end of arch	−0.9240	−0.3421

4. Conclusions and Outlook

We use a long–short baseline iteration method for elevation extraction in PSInSAR data processing to overcome the high-phase-gradient problem in a case where no DSM is available, so as to improve the accuracy of the estimated deformation rates. The LLL lattice reduction algorithm is used to rapidly reduce the search radius, compress the search space, and improve the success rate of resolving the ambiguities in phase unwrapping. To validate the method, elevations of 577 PSs on the Lupu Bridge have been obtained and compared with elevation data of the bridge model. The results are in excellent agreement. Besides, the linear deformation rates and seasonal deformation of the PSs have been extracted from InSAR deformation time series, which indicates that the bridge is stable in general, although symmetric progressive deformation has been found on the bridge arches and the bridge deck. The results agree with the Lupu Bridge design, where the arch joints would absorb most of the thermal deformation to mitigate the thermal dilation of the bridge as much as possible. Compared to the traditional PSInSAR approach, our method obtained more accurate elevation estimations. Consequently, the deformation estimation results are also more reliable.

As a whole, multitemporal InSAR is a useful tool for elevation reconstruction and the health monitoring of large infrastructures, such as bridges, dams, and high-rise buildings. Future work should be focused on interpreting the deformation; for example, linking individual PSs with the local structural elements and evaluating the results. It should also be interesting to consider to model the temperature-related deformation in the InSAR observation equation and to carry out close comparison of the results with in-situ measurements.

Acknowledgments: This research was supported by the State Key Development Program for Basic Research of China under Grant 2013CB733304 and Research Grants Council (RGC) of Hong Kong Special Administrative Region (PolyU 152043/14E). The authors would like to thank the Italian Space Agency for providing the COSMO-SkyMed images and thank the GAMMA Remote Sensing and Consulting AG for the software to process SAR data.

Author Contributions: Jingwen Zhao conceived and designed the experiments, performed the experiments, analyzed the results, and wrote the paper. Jicang Wu and Xiaoli Ding helped to conceive and design the experiments and analyze the results, and contributed ideas that helped improve the final paper. Jicang Wu also obtained the external bridge data. Mingzhou Wang helped to process the SAR data for bridge deformation extraction.

Conflicts of Interest: The authors declare no conflict of interest.

References

1. Ferretti, A.; Prati, C.; Rocca, F. Permanent scatterers in SAR interferometry. *IEEE Trans. Geosci. Remote Sens.* **2001**, *39*, 8–20. [CrossRef]
2. Kampes, B.M. *Radar Interferometry: Persistent Scatterer Technique*; Springer: Dordrecht, The Netherlands, 2006.
3. Gernhardt, S.; Bamler, R. Deformation monitoring of single buildings using meter-resolution SAR data in PSI. *ISPRS J. Photogramm. Remote Sens.* **2012**, *73*, 68–79. [CrossRef]
4. Zhu, X.X. Very High Resolution Tomographic SAR Inversion for Urban Infrastructure Monitoring—A Sparse and Nonlinear Tour. Ph.D. Thesis, University of München, München, Germany, 2011.
5. Adam, N.; Gonzalez, F.R.; Parizzi, A.; Brcic, R. Wide area persistent scatterer interferometry: Current developments, algorithms and examples. In Proceedings of the 2013 IEEE International Geoscience and Remote Sensing Symposium (IGRASS), Melbourne, Australia, 21–26 July 2013; pp. 1857–1860.
6. Gernhardt, S. High Precision 3D Localization and Motion Analysis of Persistent Scatterers Using Meter-Resolution Radar Satellite Data. Ph.D. Thesis, University of München, München, Germany, 2012.
7. Zhu, X.X.; Shahzad, M. Facade reconstruction using multiview spaceborne TomoSAR point clouds. *IEEE Trans. Geosci. Remote Sens.* **2014**, *52*, 3541–3552. [CrossRef]
8. Bakon, M.; Perissin, D.; Lazecky, M.; Papco, J. Infrastructure nonlinear deformation monitoring via satellite radar interferometry. In Proceedings of the Conference on Enterprise Information Systems (ICEIS), Lisbon, Portugal, 27–30 April 2014; pp. 294–300.
9. Lazecky, M.; Perissin, D.; Bakon, M.; Sousa, J.J.M.; Hlavacova, I.; Real, N. Potential of satellite InSAR techniques for monitoring of bridge deformations. Presented at the Joint Urban Remote Sensing Event, Lausanne, Switzerland, 1 April 2015.
10. Goldstein, R.M.; Zebker, H.A.; Werner, C.L. Satellite radar interferometry: Two-dimensional phase unwrapping. *Radio Sci.* **1988**, *23*, 713–720. [CrossRef]
11. Flynn, T.J. Consistent 2-D phase unwrapping guided by a quality map. In Proceedings of the 1996 IEEE International Geoscience and Remote Sensing Symposium (IGRASS), Piscataway, NE, USA, 31 May 1996; pp. 2057–2059.
12. Xu, W.; Cumming, I. A region-growing algorithm for InSAR phase unwrapping. *IEEE Trans. Geosci. Remote Sens.* **1999**, *37*, 124–134. [CrossRef]
13. Carballo, G.F.; Fieguth, P.W. Probabilistic cost functions for network flow phase unwrapping. *IEEE Trans. Geosci. Remote Sens.* **2000**, *38*, 2192–2201. [CrossRef]
14. Chen, C.W. Statistical-Cost Network-Flow Approaches to Two-Dimensional Phase Unwrapping for Radar Interferometry. Ph.D. Thesis, Stanford University, Stanford, CA, USA, 2001.
15. Hooper, A.; Zebker, H.A. Phase unwrapping in three dimensions with application to InSAR time series. *J. Opt. Soc. Am. A* **2007**, *24*, 2737–2747. [CrossRef]
16. Herráez, M.A.; Burton, D.R.; Lalor, M.J.; Gdeisat, M.A. Fast two-dimensional phase-unwrapping algorithm based on sorting by reliability following a noncontinuous path. *Appl. Opt.* **2002**, *41*, 7437–7444. [CrossRef] [PubMed]
17. Bioucas-Dias, J.M.; Valadao, G. Phase unwrapping via graph cuts. *IEEE Trans. Image Process.* **2007**, *16*, 698–709. [CrossRef] [PubMed]
18. Li, C.; Zhu, D.Y. A residue-pairing algorithm for InSAR phase unwrapping. *Prog. Electromagn. Res.* **2009**, *95*, 341–354. [CrossRef]

19. Lenstra, A.K.; Lenstra, H.W.; Lovász, L. Factoring polynomials with rational coefficients. *Math. Ann.* **1982**, *261*, 515–534. [CrossRef]

20. Fincke, U.; Pohst, M. On reduction algorithms in non-linear integer mathematical programming. In Proceedings of the Operation Research, Mannheim, Germany, 21–23 September 1983; pp. 289–295.

21. Lagarias, J.C. *Knapsack Public Key Cryptosystems and Diophantine Approximation*; Springer US: New York, NY, USA, 1984.

22. Coppersmith, D. Small solutions to polynomial equations and low exponent vulnerabilities. *J. Cryptol.* **1997**, *10*, 223–260. [CrossRef]

23. Kaltofen, E.; Yui, N. *Explicit Construction of the Hilbert Class Fields of Imaginary Quadratic Fields by Integer Lattice Reduction*; Springer US: New York, NY, USA, 1991.

24. Liu, J.N.; Yu, X.W.; Zhang, X.H. GNSS ambiguity resolution using the lattice theory. *Acta Geod. Cartogr. Sin.* **2012**, *41*, 636–645.

25. Zhou, L.F. Monitoring Ground Deformation in Urban Major Project Area with High-Resolution Persistent Scatterer SAR Interferometry. Ph.D. Thesis, Zhejiang University, Hangzhou, China, 2014.

26. Robertson, A.E. Multi-Baseline Interferometric SAR for Iterative Height Estimation. Master's Thesis, Brigham Young University, Provo, UT, USA, 1998.

27. Thompson, D.G.; Robertson, A.E.; Arnold, D.V.; Long, D.G. Multi-baseline interferometric SAR for iterative height estimation. In Proceedings of the 1999 IEEE International Geoscience and Remote Sensing Symposium (IGRASS), Hamburg, Germany, 28 June–2 July 1999; pp. 251–253.

28. Pieraccini, M.; Luzi, G.; Atzeni, C. Terrain mapping by ground-based interferometric radar. *IEEE Geosci. Remote Sens. Lett.* **2001**, *39*, 2176–2181. [CrossRef]

29. Academician Lin Yuanpei Tells the History of Shanghai Bridges. Available online: http://tieba.baidu.com/p/1108985575 (accessed on 14 June 2011).

30. Reale, D.; Fornaro, G.; Pauciullo, A. Extension of 4-D SAR imaging to the monitoring of thermally dilating scatterers. *IEEE Trans. Geosci. Remote Sens.* **2013**, *51*, 5296–5306. [CrossRef]

31. Fornaro, G.; Reale, D.; Verde, S. Bridge thermal dilation monitoring with millimeter sensitivity via multidimensional SAR imaging. *IEEE Geosci. Remote Sens. Lett.* **2013**, *10*, 677–681. [CrossRef]

32. Monserrat, O.; Crosetto, M.; Cuevas, M.; Crippa, B. The thermal expansion component of persistent scatterer interferometry observations. *IEEE Geosci. Remote Sens. Lett.* **2011**, *8*, 864–868. [CrossRef]

33. Goel, K.; Gonzalez, F.R.; Adam, N.; Duro, J.; Gaset, M. Thermal dilation monitoring of complex urban infrastructure using high resolution SAR data. In Proceedings of the 2014 IEEE International Geoscience and Remote Sensing Symposium (IGRASS), Quebec, QC, Canada, 13–18 July 2014; pp. 954–957.

34. Cuevas, M.; Monserrat, O.; Crosetto, M.; Crippa, B. A new product from persistent scatterer interferometry: The thermal dilation maps. In Proceedings of the 2011 Joint Urban Remote Sensing Event (JURSE), Munich, Germany, 11–13 April 2011; pp. 285–288.

35. Montazeri, S.; Zhu, X.X.; Eineder, M.; Bamler, R. Three-Dimensional Deformation Monitoring of Urban Infrastructure by Tomographic SAR Using Multitrack TerraSAR-X Data Stacks. *IEEE Trans. Geosci. Remote Sens.* **2016**, *54*, 6868–6878. [CrossRef]

36. Lazecky, M.; Hlavacova, I.; Bakon, M.; Sousa, J.J.; Perissin, D.; Patricio, G. Bridge Displacements Monitoring Using Space-Borne X-Band SAR Interferometry. *IEEE J. Sel. Top. Appl. Earth Obs. Remote Sens.* **2017**, *10*, 205–210. [CrossRef]

remote sensing

MDPI

Article

Ground Deformations around the Toktogul Reservoir, Kyrgyzstan, from Envisat ASAR and Sentinel-1 Data—A Case Study about the Impact of Atmospheric Corrections on InSAR Time Series

Julia Neelmeijer [1,2,*], **Tilo Schöne** [1], **Robert Dill** [1], **Volker Klemann** [1] and **Mahdi Motagh** [1,2]

[1]   GFZ German Research Centre for Geosciences, Telegrafenberg, 14473 Potsdam, Germany;
     tschoene@gfz-potsdam.de (T.S.); dill@gfz-potsdam.de (R.D.); volkerk@gfz-potsdam.de (V.K.);
     motagh@gfz-potsdam.de (M.M.)
[2]   Institute of Photogrammetry and GeoInformation, Leibniz University Hannover, 30167 Hannover, Germany
*   Correspondence: neelmeijer@gfz-potsdam.de; Tel.: +49-331-2882-8621

Received: 18 January 2018; Accepted: 12 March 2018; Published: 15 March 2018

Abstract: We present ground deformations in response to water level variations at the Toktogul Reservoir, located in Kyrgyzstan, Central Asia. Ground deformations were measured by Envisat Advanced Synthetic Aperture Radar (ASAR) and Sentinel-1 Differential Interferometric Synthetic Aperture Radar (DInSAR) imagery covering the time periods 2004–2009 and 2014–2016, respectively. The net reservoir water level, as measured by satellite radar altimetry, decreased approximately 60 m ($\sim$13.5 km^3) from 2004–2009, whereas, for 2014–2016, the net water level increased by approximately 51 m ($\sim$11.2 km^3). The individual Small BAseline Subset (SBAS) interferograms were heavily influenced by atmospheric effects that needed to be minimized prior to the time-series analysis. We tested several approaches including corrections based on global numerical weather model data, such as the European Centre for Medium-Range Weather Forecasts (ECMWF) operational forecast data, the ERA-5 reanalysis, and the ERA-Interim reanalysis, as well as phase-based methods, such as calculating a simple linear dependency on the elevation or the more sophisticated power-law approach. Our findings suggest that, for the high-mountain Toktogul area, the power-law correction performs the best. Envisat descending time series for the period of water recession reveal mean line-of-sight (LOS) uplift rates of 7.8 mm/yr on the northern shore of the Toktogul Reservoir close to the Toktogul city area. For the same area, Sentinel-1 ascending and descending time series consistently show a subsidence behaviour due to the replenishing of the water reservoir, which includes intra-annual LOS variations on the order of 30 mm. A decomposition of the LOS deformation rates of both Sentinel-1 orbits revealed mean vertical subsidence rates of 25 mm/yr for the common time period of March 2015–November 2016, which is in very good agreement with the results derived from elastic modelling based on the TEA12 Earth model.

Keywords: DInSAR; SBAS; ground deformation; atmosphere correction; elastic modelling; reservoir monitoring; Toktogul Reservoir

---

1. Introduction

The water levels of large artificial water reservoirs constructed for hydroelectric power generation and irrigation are prone to significant changes over the course of a year. This periodic loading of the crust causes ground deformations of the surrounding area, alters pore pressure and changes stress on underlying faults and fractures, which may ultimately induce seismicity [1]. The amount of ground deformations of a reservoir's surrounding can either be measured on individual points with levelling [2] or Global Navigation Satellite System (GNSS) measurements [3,4] or measured in a spatially continuous manner by means of Differential Synthetic Aperture Radar Interferometry (DInSAR). Thus far, in studies that are based on SAR data, either ERS-1/2 [5,6], Envisat Advanced Synthetic Aperture Radar (ASAR) [7] or a combination of those sensors [8,9] were used to quantify the regional deformation around a lake. Recently, ground deformations due to the water level changes in the Tehri Reservoir in the Himalaya region was investigated with ALOS PALSAR data [10].

Our aim in this study is to measure ground deformations induced by water level changes in the Toktogul Reservoir, which is located at N 41.8° E 72.9° in the northwest of Kyrgyzstan, Central Asia (Figure 1). This reservoir is fed by the Naryn River, which originates from glacial melt water of the Tien Shan mountain range. The lake is located at an elevation of approximately 870 m, and surrounding mountains reach elevations of 4300 m. It has existed since 1975, when the construction of the 214 m high and 293 m wide Toktogul Dam was completed [11,12]. At high water (Figure 2c), the reservoir has a length of 65 km, a width of 12 km, a surface area of 284 km^2, and a maximum depth of 200 m [12]. As the water level decreases, the eastern elongated part, where the Naryn River enters the lake, goes dry (Figure 2b). Toktogul is the largest artificial water reservoir in the Syr Darya Basin, with a maximum capacity of 19.5 km^3 [13]. Its main purposes are power generation for the Kyrgyz population in winter time and irrigation of agricultural areas located downstream in Uzbekistan and Kazakhstan in summer time [13,14]. These activities lead to a trans-boundary water policy conflict, which resulted in an exaggerated use of water in some years that could not be compensated by the incoming amount of water until the beginning of the following winter season (Figure 2).

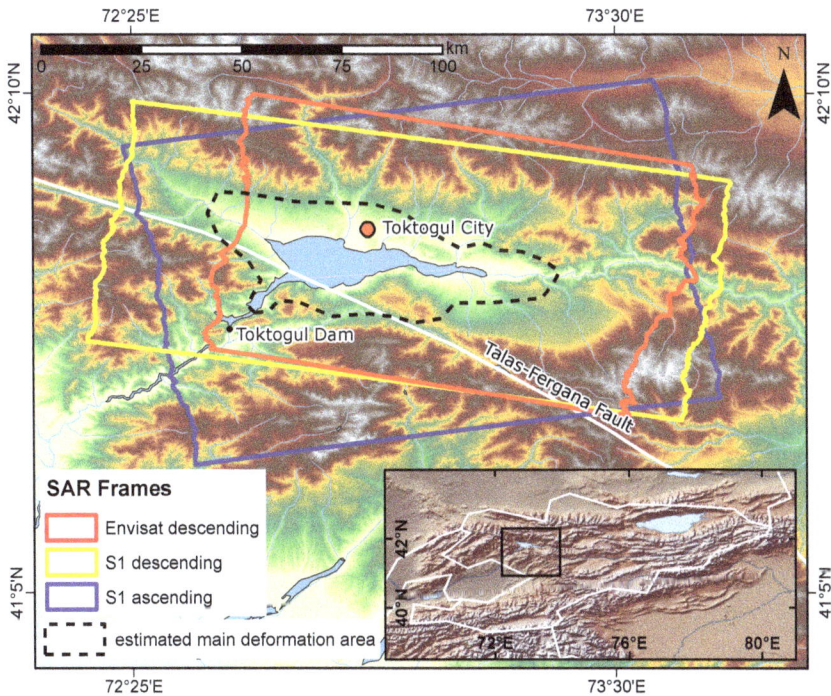

Figure 1. Location of the Toktogul Reservoir, city and dam with the outlines of the cut SAR data frames that are used for the final analysis. The alignment of the Talas-Fergana Fault is based on vector data from the Kyrgyzstan Disaster Risk Data Platform [15]. The dashed outline denotes the estimated area of the main deformation. The inset shows the location of the area within Kyrgyzstan.

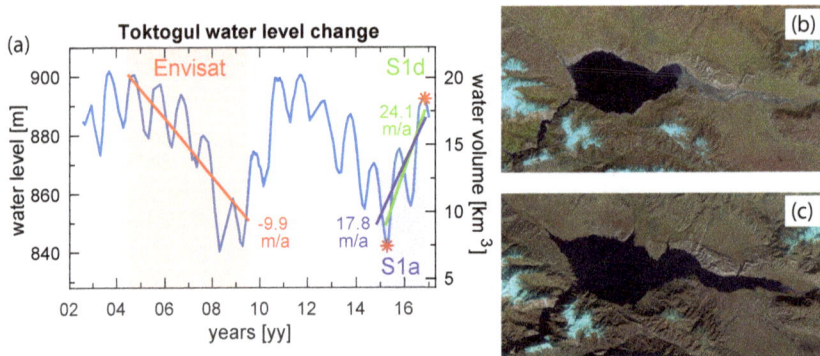

Figure 2. (**a**) Toktogul water level change between 2002 and 2016 obtained from satellite radar altimetry. Red, purple and green highlighted periods correspond to Envisat and Sentinel-1 ascending (S1a) and Sentinel-1 descending (S1d) acquisition times, respectively. The corresponding regression lines denote the average water increase per year for each of the three SAR time series (note that differences between S1a and S1d are due to different covered time periods). The red asterisks correspond to the reservoir extents at low and high water levels, which are shown by Landsat-8 images from (**b**) 11.04.2015 and (**c**) 07.11.2016, respectively.

The southwestern edge of the reservoir coincides at a length of 20 km with the Talas-Fergana Fault, an area with moderate-to-high seismicity. Larger earthquakes of magnitude M 7.6 have been reported for the Chatkal Range in 1946, 65 km west of the Toktogul Reservoir [11], and of magnitude M_S 7.3 for the Suusamyr Valley in 1992, 70 km northeast of the Toktogul Reservoir [16]. No major events have been recorded in the direct lake area since the construction of the dam. Seismic activity is still constantly monitored at the power station with seismometers [17], but no ground-based geodetic observations are available to assess the deformation of the surrounding area.

Consequently, we measure the ground deformations of this particular region by interferometrically analysing a time series of Envisat ASAR data for the time period 2004–2009, in which the net water level decreased by approximately 60 m ($\sim$13.5 km^3), and a time series of Sentinel-1 data for the time period 2014–2016, in which the net water level increased by approximately 51 m ($\sim$11.2 km^3) (Figure 2). We expect that these large load changes on the ground lead to an uplift of the surrounding area in the case of water recession and to a subsidence response in the case of water replenishing.

Sentinel-1 is the latest generation of the European Space Agency's (ESA) SAR missions and consists of two satellites, Sentinel-1A and Sentinel-1B, that were launched in April 2014 and April 2016, respectively. Together, these C-band-based SAR satellites are able to cover most regions of the world with the interferometric wide (IW) swath mode (swath width: 250 km; spatial resolution: 5×20 m in range and azimuth, respectively) from the same relative orbit every twelve days, whereas Europe and some selected areas are even monitored with a temporal resolution of six days. Compared to the Envisat ASAR C-band sensor, which only acquired data every 35 days with a swath width of 56–100 km (image mode single-look complex (IMS); spatial resolution: 8×4 m in range and azimuth, respectively), this mission is predestined for monitoring not only persistent linear deformations, but also intra-annual deformation changes on a large spatial scale.

Atmospheric effects in the SAR data caused by vertical stratification and turbulent water vapour variations play an important role in time-series investigations [6,18–20]. We therefore apply various correction approaches based on either global numerical weather models (in particular, the European Centre for Medium-Range Weather Forecasts (ECMWF) operational forecast analysis, the ERA-5 reanalysis and the ERA-Interim (ERA-I) reanalysis) or empirical models that rely on the dependency of the phase on the elevation of the terrain (in particular, the linear dependency and the power-law approach of Bekaert et al. [21]). The time series with the best working atmospheric correction approach is then used for a comparison to the reservoir's water level variations that are extracted from satellite altimetry data. The measured ground deformation rates are further compared to the results obtained from elastic modelling of the surface deformations.

2. Materials and Methods

2.1. Lake Altimetry

At present, radar altimetry (RA) is widely used not only for monitoring global sea level changes, but increasingly also for measuring the water levels of rivers and lakes for hydrology applications [22–24]. Since the early 1990s, a series of RA missions has provided continuous measurements of water surface heights with 10- and 35-day repeat intervals. Novel processing technologies, such as retracking, allow the extraction of the water levels of smaller inland water bodies and reservoirs [23]. The accuracies of the derived water levels are slightly worse compared to open ocean applications but can still reach 5 cm. For hydrological applications, the water levels can be converted into volume changes using supplementary information such as hypsometry or lake extents extracted from remote sensing data.

The water level of the Toktogul Reservoir has been measured with RA since 1995—in particular, mostly every 35 days by the European ERS-2, Envisat and later by the Indo-French AltiKa missions. Some data are also available from the US-French Jason-1 and Jason-2 and the European CryoSat-2 missions. Using all available RA data, applying up-to-date environmental correction models and cross-checking for and applying inter-mission biases, a homogeneous time series of reservoir heights and reservoir volumes is constructed (Figure 2a). The internal accuracy is estimated from all high-rate measurements of one reservoir crossing (e.g., ~50 measurements for AltiKa) and is mostly within the expected 5 cm root mean square error (RMSE) range with slightly higher values for the earlier missions.

In the case of the Toktogul Reservoir, sparse historical monthly volume information of the total volume is available for the full range of water levels from CA WATER Info [25] between 1984 and 2000, and some more recent information is made available by the reservoir operator [26]. This allows the construction of a polynomial transfer function (R^2 = 0.9998), which can be used for converting all water height levels into reservoir volumes. These data can then be used to verify the accuracy of the RA-derived water heights, which is approximately ±0.3 m [24].

2.2. DInSAR Processing of Envisat ASAR and Sentinel-1 Data

We use Envisat ASAR IMS and Sentinel-1 IW SAR data to monitor deformations around the Toktogul Reservoir area. The Envisat data, only available in the descending orbit, were acquired between December 2003 and July 2009 and cover a time of decreasing annual water level, whereas Sentinel-1 ascending (S1a) and descending (S1d) acquisitions analysed for the time period of October 2014 until December 2016 correspond to an increasing annual water level. In the Envisat acquisition period, the highest water level measured with RA (referenced to the EIGEN-6C3 static gravity field [27]) was 900.5 m on 10 May 2004, and the lowest was 840.4 m on 22 April 2008. In the Sentinel-1 acquisition period, the lowest measured water level was 842.2 m on 24 April 2015, and the highest was 892.9 m on 29 October 2016 (Figure 2).

Data preprocessing is performed with the GAMMA software [28] as follows. First, for both sensors, single-look complex images are imported taking into account precise orbit ephemerides; in the case of Sentinel-1 data only, bursts covering the area of interest are concatenated. Second, images of each SAR time series are individually coregistered and cropped to the desired area of interest. Because Sentinel-1 data require a precise coregistration accuracy of a few thousands of a pixel in azimuth to prevent contamination with phase variations due to along-track differences in the Doppler centroids [29,30], we rely on the spectral diversity method for coregistration [31,32]. The outlines of the cropped SAR data frames are shown in Figure 1.

The single-look, coregistered SAR images are used to construct a network of interferograms by applying the Small BAseline Subset (SBAS) technique [33] implemented in StaMPS/MTI, the Stanford Method for Persistent Scatterers and Multi-Temporal InSAR [34]. The main objective of this approach is to generate interferograms from pixels that decorrelate only slightly over short time intervals. To identify such pixels, interferograms are built from SAR acquisitions that match the criteria of having small perpendicular, temporal and Doppler baselines, but with the restriction that all selected interferograms should be connected; thus, no isolated cluster is allowed in the network [35]. We constrain the Envisat network by a maximum spatial baseline of 500 m and a maximum temporal baseline of 2000 days, whereas, at the same time, the overall coherence between two interferograms should be at least 0.4. In the Sentinel-1 case, we use constraint values of 200 m, 365 days and 0.5. In the following, the selected interferograms are treated with topography removal and geocoding, for which we rely on the 1-arc resolution Shuttle Radar Topography Mission (SRTM) digital elevation model (DEM). The results are visually inspected, and decorrelated interferograms are discarded from the network. Unwrapping of the remaining interferograms is achieved by using a three-dimensional phase unwrapping approach [36]. Displacement values in line-of-sight (LOS) are subsequently retrieved by least-squares inversion of the unwrapped interferograms with respect to a reference area selected outside of the main deformation region (N 41.6930° E 73.1660°, radius: 2 km).

By carefully investigating the residuals of the unwrapped phase of the SBAS interferograms and the inverted interferograms, we neglect scenes that introduce errors to the time series. The main error source is thus the snow coverage in winter time. After some problematic scenes are removed, we iteratively repeat the process of unwrapping, inverting and discarding until all remaining interferograms could be reliably unwrapped. A summary of the amounts of the used scenes and interferograms along with the corresponding SAR sensor specifications is presented in Table 1. The final network for all three SAR time series is shown in Figure 3.

Table 1. SAR data specifications and summary of the amount of images used in the final networks. The covered time period that could be reliably unwrapped is as follows for the individual time series: Envisat: 24.10.2004–05.07.2009, Sentinel-1, descending (desc.): 23.03.2015–12.11.2016, ascending (asc.): 24.10.2014–18.11.2016.

Satellite	Orbit	Path	Acquisition Time (UTC)	Mean Angle of Incidence	Heading Angle	Amount of Scenes	Amount of Interferograms
Envisat	desc.	277	05:23	23.4°	−167.8°	22	53
Sentinel 1	desc.	5	01:13	39.7°	−170.1°	20	49
Sentinel 1	asc.	100	13:06	43.3°	9.4°	28	96

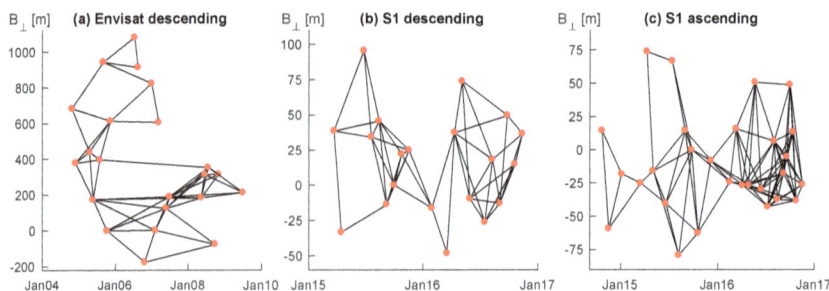

Figure 3. Small BAseline Subset (SBAS) networks after offending interferograms are removed for (**a**) Envisat descending; (**b**) Sentinel-1 descending and (**c**) Sentinel-1 ascending time series. Red dots denote the time of the image acquisitions, and black lines show the interferograms.

All interferograms remaining in the network are further treated by removing (1) individual phase ramps; (2) the overall topography error; (3) estimated atmospheric influences; and (4) the phase oscillator drift (Envisat data only) [37]. From all these aspects, the most influential and simultaneously most challenging error aspect to remove is the effect of the atmosphere. Multiple approaches dealing with this issue are discussed in the following section.

2.3. Atmospheric Correction

Because the Toktogul Reservoir is located in a high-mountain area, atmospheric disturbances in the data are inevitable and must be corrected to avoid a misinterpretation as a loading signal. The success of atmospheric correction methods is highly dependent on the characteristics of the area of interest in terms of topography and its dominance either in stratified tropospheric delay or dynamical local weather and turbulence [38]. To reduce the impact of atmospheric artifacts in the Toktogul SAR data, we apply a range of tropospheric correction methods implemented in the MATLAB-based Toolbox for Reducing Atmospheric InSAR Noise (TRAIN, version 2beta) from Bekaert et al. [38].

The applied techniques can be divided into two main categories: (1) global numerical weather-model-based and (2) phase-based correction methods. In theory, weather-model-based approaches should be more effective because they should be able to compensate not only for vertical stratification, but also for turbulent water vapour variations in the lower troposphere. However, previous studies have shown that the success of the exact representation of stratification and turbulence is highly dependent on the area of interest [39] and that global models, such as ERA-I, also suffer from coarse temporal and spatial resolutions [40]. This disadvantage is now compensated for by newer available weather model data such as the ERA-5 reanalysis that are shipped with increased spatial and temporal resolutions. However, note that this type of data is not always easy to access and that it is currently only available for a limited range of time.

For comparison, we further apply phase-based corrections. These corrections have the advantage that the required external data are readily available but the disadvantage that they can only be used to treat vertical stratification, not turbulence mixtures. However, as Bekaert et al. [38] noted in his study, in regions where tropospheric delay is mainly correlated to topography, phase-based methods potentially outperform weather-based approaches.

We neglect spectrometer-related correction methods based on the Medium Resolution Imaging Spectrometer (MERIS) or the Moderate Resolution Imaging Spectroradiometer (MODIS) for the following reasons: MERIS data are available for Envisat data only and MODIS data are acquired between approximately 5:00 a.m. and 7:30 a.m. UTC. This differs by more than an hour from Sentinel-1 acquisition times (Table 1); thus, changes in atmospheric water vapour conditions may introduce more errors rather than correcting for turbulence. Furthermore, correction with spectrometer data requires daytime acquisitions under cloud-free conditions [41], which is often not the case in high-mountain areas.

Finally, we determine the best atmospheric correction by evaluating the RMSE values of deformation in time. To avoid an influence of loading-induced deformation on the analysis, we excluded for the RMSE calculation the area of main deformation (cf. Figure 1) from the overall atmosphere-corrected interferograms. For comparison, we also calculate the RMSE for unwrapped results that are only corrected for the DEM error and orbital plane but not for atmosphere. The derived RMSE values are an indicator of the best-performing atmosphere removal algorithm, but note that the absolute values of different SAR frames cannot be compared because the extents of the frames differ. The Sentinel-1 ascending image, for example, covers much more high-mountain areas compared to the Sentinel-1 descending images; thus, higher error values can be expected. Furthermore, the amount and location of the SBAS-derived points within the SAR frames also vary.

2.3.1. Tropospheric Delays from Numerical Weather Models

Temperature, relative humidity and pressure information from numerical weather models can be used to compute the hydrostatic and wet tropospheric delay [38,39,42]. For the Toktogul case, we apply three global models based on ECMWF data: first, the ECMWF operational forecast analysis (opECMWF); second, the ERA-5 global atmospheric reanalysis; and third, the ERA-I global atmospheric reanalysis. From these models, only the ERA-I solution is currently freely accessible at a reduced spatial resolution.

The opECMWF data are expected to be the most accurate of the three models because these data are used for routine short-term predictions. These data are distributed with a temporal resolution of 6 h, a spatial resolution of 0.1° and 25° pressure levels. Since July 2017, the new ERA-5 reanalysis has been available for the time period 2010–2016. Similar to opECMWF, ERA-5 comes with a high spatial resolution of 0.1° but has an increased temporal resolution of 1 h. Upper-air information is delivered at 37 pressure levels. The available ERA-5 data currently do not cover the Envisat acquisition time period. We therefore also consider the former ERA-I reanalysis that is available for the time period from 1979 to present. This reanalysis is also delivered with a temporal resolution of 6 h and contains 37 pressure levels, but it has a coarse spatial resolution of 0.75° [43].

To compare the influence of the temporal resolution, we apply two versions of the ERA-5 data: the hourly reanalysis and an artificially reduced version with a 6 h temporal resolution, similar to the ERA-I data. An overview of the model specifications is presented in Table 2.

Table 2. Parameters of the applied numerical weather-model-based atmosphere corrections.

Model	Spatial Resolution	Temporal Resolution	Pressure Levels
opECMWF	0.1°	6 h	25
ERA5 1 h	0.1°	1 h	37
ERA5 6 h	0.1°	6 h	37
ERA-I	0.75°	6 h	37

2.3.2. Phase-Based Tropospheric Delays

The Toktogul Reservoir is surrounded by high-mountain ranges, which influence the moisture content of the troposphere, which consequently has an impact on the phase delay of the radar signal. It is therefore straightforward to apply correction methods such as the power-law and linear tropospheric approaches that use the correlation of the phase signal with the topography.

The linear tropospheric correction assumes a uniform troposphere that is directly correlated to the elevation of the terrain. In principle, a linear relationship between phase delay and terrain height is estimated and subtracted from the entire interferogram. To prevent real tectonic signals from being taken into account during the linear dependency analysis, we exclude the estimated deformation area around the Toktogul Reservoir from the calculation (cf. Figure 1).

The power-law correction technique [21] is more sophisticated than the linear approach, as it considers a spatially varying troposphere within an interferogram. The method assumes a non-varying delay at the relative top of the troposphere and then applies a power-law function on the phase delay variations depending on elevation. It thus considers phase delays mainly due to hydrostatic and wet components of the refractivity. Delays due to the liquid component and the influence of the ionosphere on C-band SAR data are neglected because their influence is assumed to be small [21].

First approximations of the tropospheric delays that are used as coefficients for the power-law method can be calculated from balloon sounding data as distributed by the University of Wyoming [21]. We extract data for the Envisat and Sentinel-1 acquisition periods from the Taraz station (station no. 38341 at N 42.85° E 71.38°), which is located approximately 170 km northwest of Toktogul. We constrain the upper troposphere height to 10 km and extract a corresponding mean power-law decay coefficient of 1.51 ± 0.01 from the sounding data, which we use for all three InSAR time series.

The estimation of the spatially varying relation between topography and tropospheric phase is based on the assumption that the tropospheric signal is present in all wavelength scales. Tropospheric effects should thus be removed from the interferogram by band-filtering the signal, choosing a band for filtering that is insensitive to other signals such as turbulent troposphere, orbital errors and deformation. Furthermore, spatial variability of the phase delay is provided by dividing the area into multiple smaller windows, in which the coefficient describing the relation between topography and tropospheric phase is calculated locally [21,38].

In the Toktogul case, we have a pronounced topography around the reservoir, allowing us to set the window size to be comparatively small. Empirical tests show that, for the SAR time series, the following window sizes work best (window overlap: 50%): Sentinel-1, descending: 77 × 59 km; Sentinel-1, ascending: 68 × 78 km; and Envisat: 60 × 62 km. The relationship between topography and phase delay is computed for the following filtering band ranges: 2–4, 2–8, 2–16, 4–8, 4–16, 4–32, 8–16, 8–32, 8–64, 16–32, 16–64, 32–64, 32–128, and 64–128 km.

2.4. Deformation Decomposition of Sentinel-1 Data

Because Sentinel-1 data are available from two different orbits, it is possible to decompose the deformation into a vertical part and a horizontal part. However, because there are only two observations available, we cannot directly compute the 3D vector components. We thus neglect potential displacements in the north–south direction, for which LOS measurements are the least sensitive in any case due to the near polar orbit of the spacecraft.

The average S1d and S1a LOS displacement points are interpolated to 200 × 200 m grids, which are used as input for the deformation decomposition. Furthermore, we ensure that only results covering the same time period (March 2015–November 2016) are considered for the decomposition. The mean LOS displacements of the ascending (d_a) and descending (d_d) orbits are then used to discriminate between vertical d_v and east-west d_e displacements by solving the following equation [44,45]:

$$\begin{pmatrix} d_a \\ d_d \end{pmatrix} = \begin{pmatrix} \cos \theta_a & -\cos \alpha_a \sin \theta_a \\ \cos \theta_d & -\cos \alpha_d \sin \theta_d \end{pmatrix} \begin{pmatrix} d_v \\ d_e \end{pmatrix}, \tag{1}$$

where θ_a and θ_d represent the incidence angles and α_a and α_d are the heading angles of Sentinel-1's ascending and descending orbits, respectively.

2.5. Modelling of Elastic Surface Deformations

Considering surface deformations induced by short periodic mass variations, such as intra-annual water level changes, the purely elastic, instantaneous response of the Earth is an adequate approximation. In a spherical harmonic representation, the coefficients of the vertical and horizontal deformations and the geoid changes can be related linearly to the surface mass load through degree-dependent load Love numbers. Farrell [46] outlines the calculation of properly weighted sums of the load Love numbers for a given Earth model to form Green's functions that provide the distance-dependent elastic response of the Earth model due to a unit point mass. Assigning the point mass response to any extended mass distribution by means of a convolution integral over the loaded region leads to the global displacement field. Because the convolution occurs in the spatial domain, the Green's function approach is particularly useful if the spherical harmonic representation of the surface mass load is dominated by high-degree coefficients, such as in our case of the highly heterogeneous distribution of non-loaded and loaded regions around the Toktogul lake.

Rather than using one globally defined Green's function for a customary idealization of the Earth by a model composed of spherically symmetric layers, we calculated geographically dependent local Green's functions [47] that are valid especially for the crustal structure beneath the Toktogul region. For small-scale heterogeneous mass loads, the geological structure of the shallow crust becomes the most important; thus, we replaced the outermost 71 km of the 1D PREM Earth model [48] by the lateral variability given in the crustal model TEA12 provided by Tesauro et al. [49]. The deformation response depends mainly on the lithology of the upper and lower crystalline crustal layers (granite, mafic granite, diabase, diorite, and olivine), their thicknesses, and the varying thickness of the overlaying sediments. The changes in the crustal properties from PREM to TEA12 affect the near-field values of the local Green's functions for distances to the load lower than 100 km.

To simulate elastic Earth surface deformations, the local Green's function model was applied to the water storage variations composed from the Toktogul lake level changes as observed by satellite RA (Figure 2a), in combination with the changes of the lake surface area given by Landsat-8 images (Figure 2b,c). Finally, the trend in the modelled surface deformation was calculated for the same period as for the Sentinel-1 descending acquisition time S1d.

3. Results

A first general result of our study is that Sentinel-1 products are superior to Envisat results in the following aspects: (1) the wide swath of the Sentinel-1 sensor allows more freedom in cutting the scenes to the desired area of interest; hence, we capture the western region of the reservoir better with the Sentinel-1 time series than with the Envisat ones; (2) due to the higher temporal acquisition sampling, Sentinel-1 interferograms are affected less by decorrelation, which leads to a higher point density than in the Envisat time series. This again results in a better area coverage that can be taken into account for the deformation analysis; (3) since the orbital tube of Sentinel-1 is very narrow, the length of the spatial baseline between two images is a no critical rejection criterion, which ultimately leads again to a denser network of interferograms. In the following, we will provide more details regarding the improvement of the results due to different atmospheric corrections and then regarding the ground deformation correlated to water level changes. Furthermore, we provide a comparison of Sentinel-1-derived vertical ground deformation to elastic modelling results.

3.1. Atmospheric Corrections

First, we analyse the RMSE values of the individual power-law results, where we had applied different filter bands. It appears that, in the Toktogul case, small- to medium-scale bands generally perform better than longer ones (Table 3), but the results are not consistent among the three different SAR time series, which may again be explained by the different SAR frame extents. In the case of the Envisat descending time series, the lowest RMSE value of 8.0 mm is achieved with the 8–64 km filter band; in the case of S1d, the lowest RMSE value is 7.0 mm, derived by applying filter bands for a range of 4–32 km and 8–64 km; and, in the case of S1a, the lowest RMSE of 10.6 mm is found for filter bands of 2–8 km and 4-8 km range. In the following sections, whenever the results of the power-law correction method are mentioned, we are referring to the results from the best-performing filter band.

Table 3. Root mean square errors for different power-law filtering bands. The best results are highlighted in bold. The area of main deformation (cf. Figure 1) is excluded from this estimation.

Band [km]:	2–4	2–8	2–16	4–8	4–16	4–32	8–16	8–32	8–64	16–32	16–64	32–64	32–128	64–128
Envisat RMSE [mm]:	8.5	8.2	8.1	8.3	8.1	8.2	8.2	8.1	**8.0**	9.0	8.1	8.2	8.2	9.3
S1d RMSE [mm]:	7.1	7.3	7.3	7.2	7.2	**7.0**	7.5	7.1	**7.0**	7.5	7.3	7.6	7.2	7.6
S1a RMSE [mm]:	10.7	**10.6**	10.8	**10.6**	10.8	10.8	10.9	10.9	11.1	11.1	11.2	11.4	11.2	11.6

We now compare the RMSE values of the different correction techniques to determine the best atmospheric correction solution. The results show that, although phase-based methods are not able to represent turbulence mixtures, their correction performance is superior to the weather-model-based methods for all three SAR time series, which in all cases yield even higher RMSE values compared to the non-atmosphere-corrected time series (Table 4). This is also true for the ERA-5 1h solution with the highest temporal and spatial resolutions, which surprisingly does not lead to any improvement compared to the other numerical weather-based approaches.

Table 4. Root mean square errors for different atmospheric correction techniques. ERA-5 data cover only the Sentinel-1 acquisition period and thus cannot be used for improving the Envisat time series. The best results are highlighted in bold. The area of main deformation (cf. Figure 1) is excluded from this estimation.

Atmospheric Correction:	None	Best Power-Law	Linear	opECMWF	ERA-I	ERA-5 1 h	ERA-5 6 h
Envisat RMSE [mm]:	8.8	**8.0**	8.0	13.1	11.5	-	-
S1d RMSE [mm]:	9.9	**7.0**	7.5	10.1	10.7	12.0	11.3
S1a RMSE [mm]:	12.6	**10.6**	11.0	12.9	13.4	13.8	14.0

(1) Envisat descending

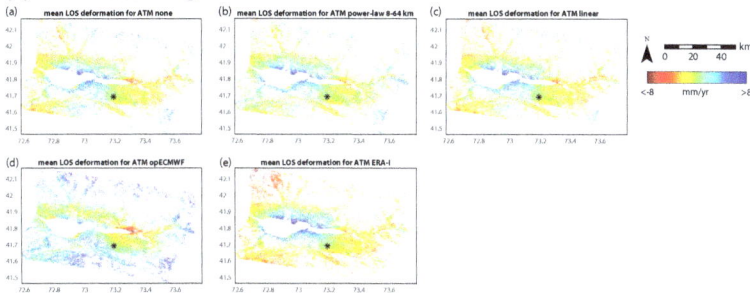

(2) Sentinel-1 descending

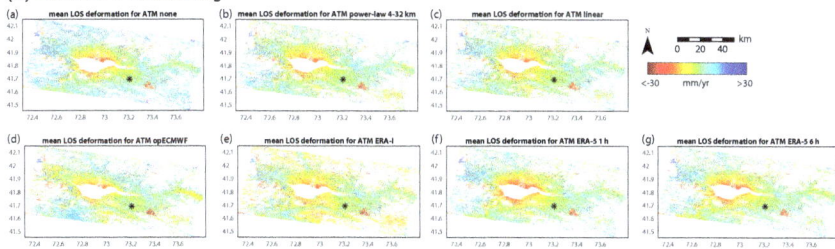

(3) Sentinel-1 ascending

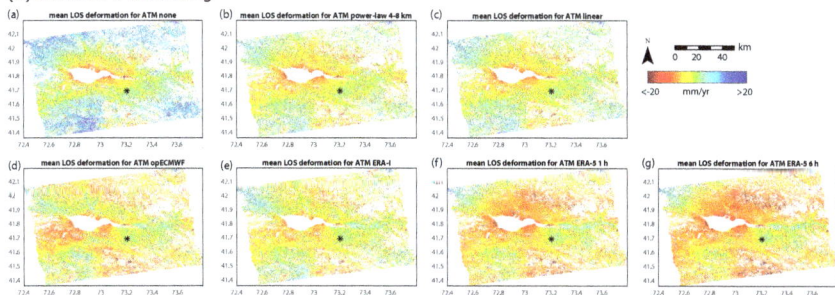

Figure 4. Mean line-of-sight (LOS) deformation values for various atmospheric corrections of the (**1**) Envisat descending; (**2**) Sentinel-1 descending and (**3**) Sentinel-1 ascending time series. (**a**) no atmospheric correction applied; (**b**) best results of the power-law technique; (**c**) linear dependency on the topography; (**d**) correction with the operational weather model (opECMWF) analysis; (**e**) ERA-I weather model correction; (**f**) the ERA-5 1 h temporal resolution solution and (**g**) the ERA-5 6 h temporal resolution solution. The black asterisks show the location of the reference point.

Among the phase-based corrections, the power-law method performs better than the linear one, although differences are small or (in the case of Envisat) even non-existent (Figure 4). This behaviour is expected because the power-law method is able to adapt to the variation of the vertical stratification and calculates individual corrections for windows smaller than the entire interferogram, whereas linear correction works on the entire image only. At the end, the best filter bands of the power-law technique improve the RMSE of Envisat by 9% from 8.8 mm to 8.0 mm, of S1d by 29% from 9.9 mm to 7.0 mm and of S1a by 16% from 12.6 mm to 10.6 mm.

Analysing the performance of different atmospheric correction methods in time (Figure 5) reveals that most variances appear in summer time and that winter acquisitions are much less affected. Furthermore, in the S1a time series, ground deformation variations due to different applied atmospheric correction approaches are significantly higher than in S1d or Envisat time series.

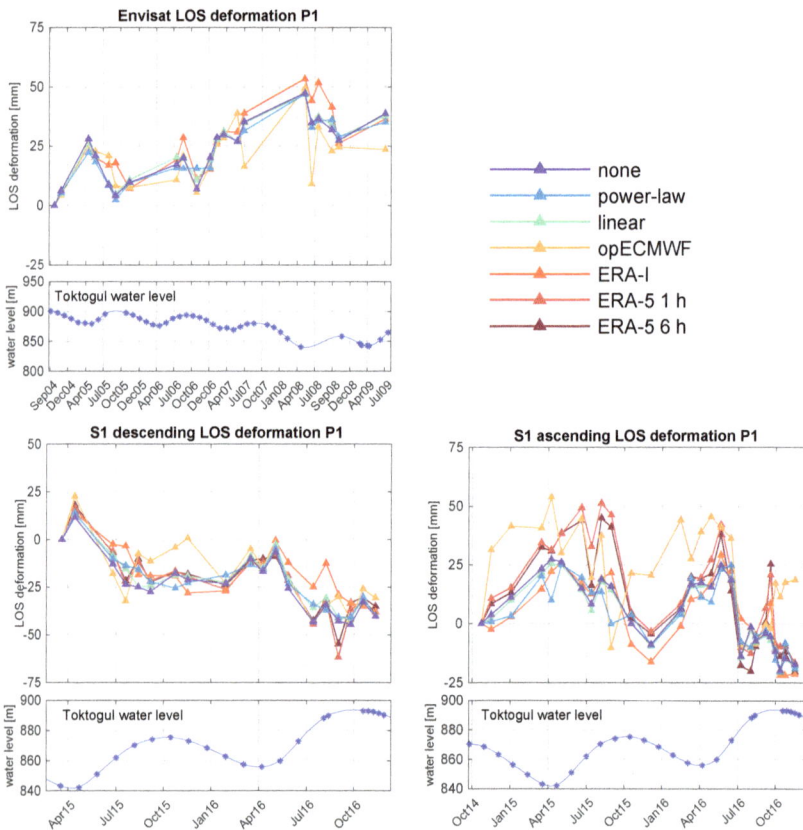

Figure 5. Variances of the LOS deformation in time for all applied atmospheric corrections. The location of the exemplary point P1 is shown in Figure 6. For better visualization, deformation markers have been connected with lines, although this does not imply that we expect linear deformation in between.

3.2. Ground Deformation

The ground deformation pattern is discussed on the basis of the average Envisat, S1d and S1a power-law-corrected deformation maps (Figure 6—note that the red triangle area at the southeastern corner of the S1d time series shows an unwrapping artifact that we did not correct for because it is located far from the reservoir area and thus did not hamper our analysis). We additionally show the correlation of the water level data to the corresponding deformation data for four selected points (Figure 7). For these examples, the water level amplitudes are fitted to the power-law-corrected SAR time series to highlight the amount of their correlation. Note that we do not correct for any potential time lag between water level change and deformation during this fit and that differences between the overall ascending and descending Sentinel-1 LOS deformations are mainly due to the different time period coverages.

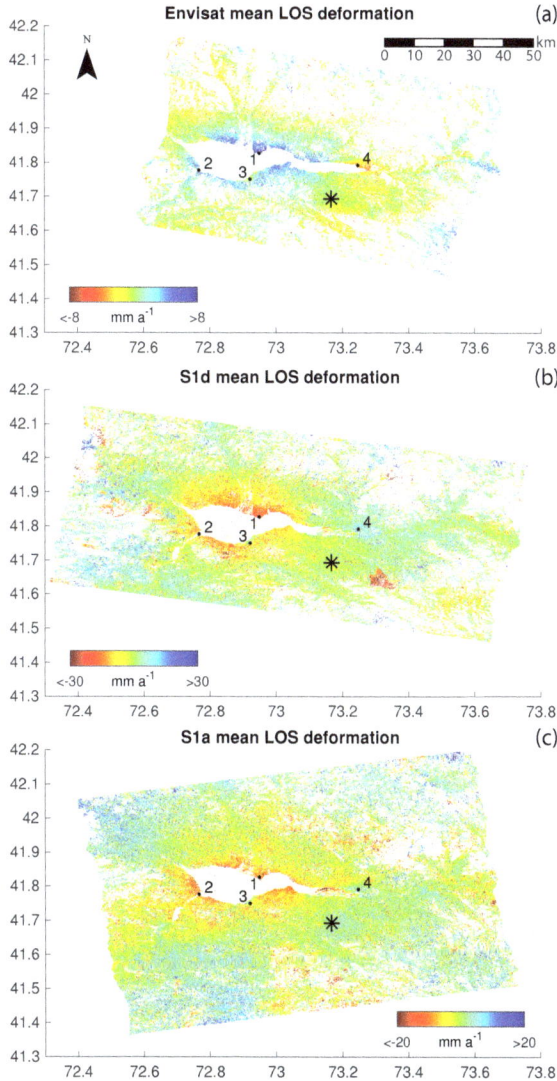

Figure 6. Power-law-corrected mean LOS deformation maps for (**a**) Envisat; (**b**) S1d and (**c**) S1a SAR time series. Points 1-4 show the locations of the deformation time series shown in Figure 7, and the black asterisk denotes the reference point used in the time series. The red triangle area in the southeast corner of the S1d time series shows an unwrapping artifact that we did not correct for.

From the spatial perspective, all three mean deformation maps reveal a significant LOS deformation of the bedrock areas around the entire Toktogul Reservoir, whereas the strongest deformations are found north of the lake close to the Toktogul city area (point P1). In the time of water recession (Envisat time series), we observe for P1 LOS uplift rates of 7.8 mm/yr, which converts to 0.78 mm per one metre of water level loss. The reverse behaviour of subsidence is monitored for the time of water filling (Sentinel-1 time series). Here, the mean LOS values are on the order of −19.8 mm/yr (−0.82 mm per 1 m water level increase) and −11 mm/yr (−0.62 mm per 1 m water level increase) in the case of S1d and S1a, respectively. The intra-annual time series (Figure 7) for point P1 shows that the correlation between LOS deformation and water fit are higher for the Envisat ($R^2 = 0.85$) and S1d ($R^2 = 0.88$) time series compared to the S1a ($R^2 = 0.62$) time series. From the Sentinel-1 data, it is clear that intra-annual LOS deformation changes appear simultaneously with the water level changes, which indicates an elastic response of the surface. At location P1, these intra-annual LOS deformation variations are on the order of 30 mm and more.

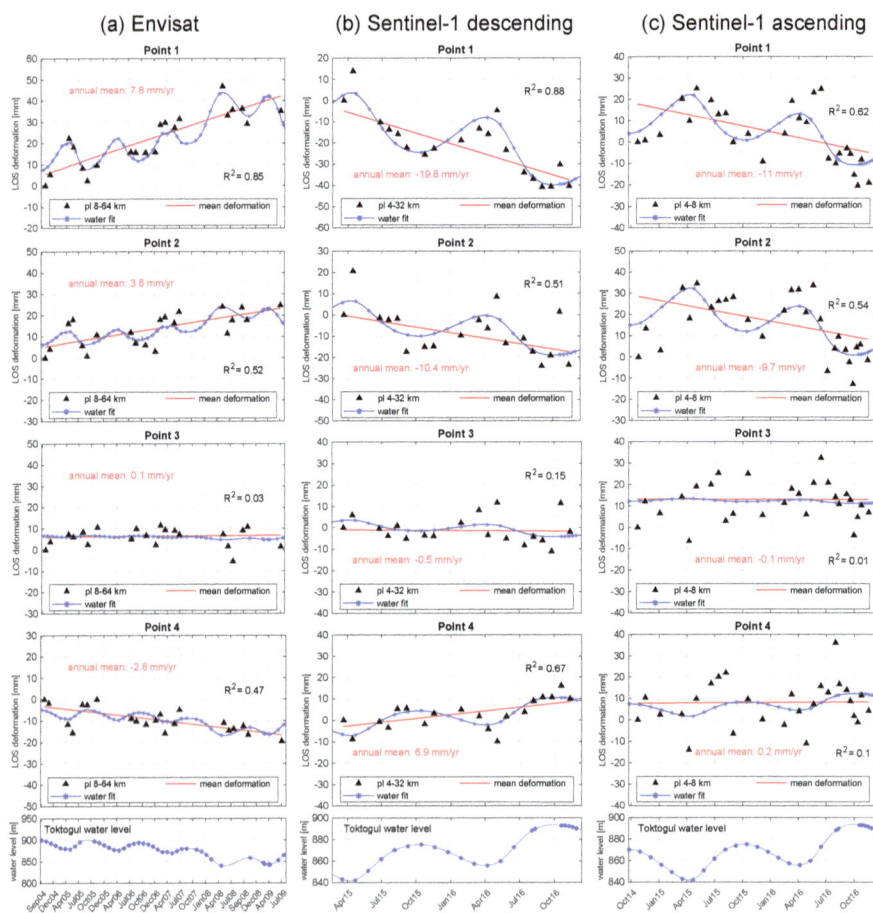

Figure 7. (**a**) Envisat; (**b**) S1d and (**c**) S1a power-law-corrected LOS deformation in time (black triangles) for points 1-4 as in Figure 6. The red lines indicate the annual mean LOS deformation for the investigated time period. The blue lines in the bottom diagrams show the true Toktogul water level change, whereas the blue lines in the point figures represent the best fit of this water level to the shown deformation.

The same deformation process but with decreased rates and lower correlation rates are found west of the Talas-Fergana Fault (point P2). LOS values (and water level fit correlation rates) are on the order of 3.8 mm/yr ($R^2 = 0.52$), -10.4 mm/yr ($R^2 = 0.51$), and -9.7 mm/yr ($R^2 = 0.54$) for Envisat, S1d and S1a, respectively, which convert to 0.38 mm, -0.43 mm and -0.54 mm per 1 m water level change.

Very intriguing are the results of a small area close to the reservoir (point P3) that do not show a distinct correlation to any water level changes. In all three SAR data time series, we find no significant deformation rates (LOS values are on the order of 0.1 mm/yr, -0.5 mm/yr and -0.1 mm/yr for Envisat, S1d and S1a, respectively), and intra-annual water level fit correlation rates (R^2) are consistently below 0.2.

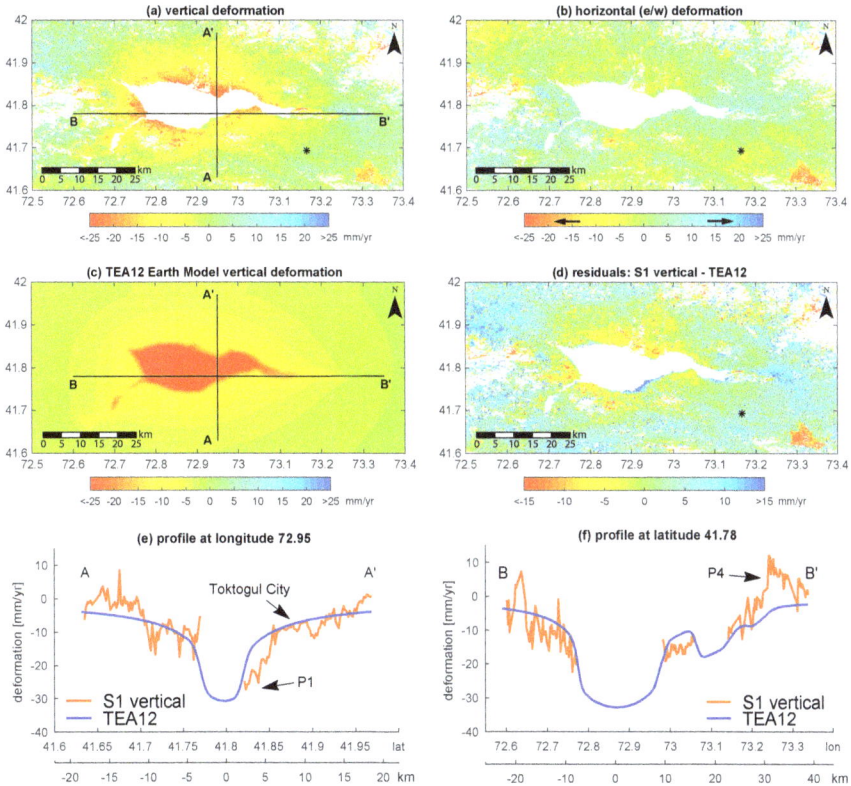

Figure 8. (**a**) vertical and (**b**) horizontal (east/west) components of the decomposed Sentinel-1 ascending and descending LOS deformations for the time period March 2015–November 2016. Input data for the decomposition are the power-law-corrected mean LOS deformation maps; (**c**) vertical deformation from elastic modelling with the TEA12 Earth model; and (**d**) residuals from Sentinel-1 minus modelled vertical deformation. Negative values in the vertical case refer to subsidence, and blue values refer to uplift. In the horizontal case, positive values denote a motion towards the east, and negative values denote a motion towards the west. The black asterisk shows the location of the reference point. Measured and modelled vertical deformation rates for profiles $\overline{AA'}$ and $\overline{BB'}$ are given in (**e**, **f**), respectively.

In the eastern region, at the entrance of the Naryn River (point P4), we observe contradictory but clearly intra-annual correlated deformation rates compared to points P1 and P2. Here, we find mean LOS subsidence rates of -2.8 mm/yr (-0.28 mm per 1 m water level change and $R^2 = 0.47$) during the water recession phase captured by Envisat data and mean LOS uplift rates of 6.9 mm/yr (0.29 mm per

1 m water level change and $R^2 = 0.67$) in the S1d time series that cover a water replenishing phase. Compared to S1d, S1a LOS deformation is less explicit and also only yields a low water level fit correlation value of $R^2 = 0.1$.

The availability of ascending and descending data in the Sentinel-1 case allows a decomposition of the mean deformation data in the vertical and east–west directions (Figure 8) that can be compared to vertical deformation rates obtained from elastic modelling. For the purpose of decomposition, we truncated the ascending SAR time series to fit the time period of the descending SAR time series. Overall, the long-wave spatial pattern of the Sentinel-1-derived vertical deformation around the reservoir correlates very well with the trend in the vertical displacement for the same time interval as calculated by the elastic response of the Earth model due to water load changes (Figure 8a,c,d). In the case of the horizontal deformation, S1 measurements yield only noise but no definite deformation in one direction or the other (Figure 8b).

Two perpendicular profiles further illustrate how well the measured and modelled data fit together and up to which distance the area is deforming. The 2 km closest to the reservoir are affected the most by subsidence rates of approximately -25 mm/yr (-1.07 mm per 1 m water level increase) (Figure 8e). The Toktogul city area is affected less by approximately -10 mm/yr (-0.41 mm per 1 m per 1 m water level increase). Non-affected areas are only found at a distance of approximately 15 km away from the reservoir shoreline. The comparison to modelled data also clearly outlines the anomaly of the measured uplift signal at point P4 (Figure 8f).

4. Discussion

Our study shows that DInSAR remote-sensing-derived displacements that were properly corrected for atmospheric effects can explain loading-induced ground deformations at the Toktogul Reservoir. The very good agreement of Sentinel-1 decomposed results to predictions of an elastic surface deformation model based on TEA12 proves that the derived LOS deformations can be mainly attributed to vertical displacements. In the following, we will discuss the artifacts remaining after atmospheric corrections and provide reasons for the observed variances of the ground deformations.

4.1. Atmospheric Corrections

It is often argued that the occurrence of atmospheric turbulences is random and thus cancel each other out when calculating the average LOS deformation [18–20]. However, we observe that artifacts remaining after the atmospheric correction that we mainly attribute to non-corrected turbulence do have an influence on the averaged LOS deformations in our case (Figure 4). Doin et al. [39] argue that the sign and amplitude of stratified tropospheric delay in high-mountain areas do not appear randomized. It is thus likely that non-corrected vertical stratification leaks into the results, leading to the observed differences and also to significant alterations in the LOS time series (Figure 5). These alterations appear mainly in the summer months, which can be related to increased water vapour content in the atmosphere compared to winter time. Furthermore, in summer, high evaporation rates originating from the lake surface also contribute to SAR signal delays.

The low performance of the weather-model-driven corrections may be explained by several reasons. First, the Toktogul Reservoir is located in a valley that is surrounded by high mountains, which leads to micro-climate artifacts that are not well captured by global numerical models. Second, within the region of Central Asia, in situ weather stations used to constrain the models are sparse; thus, the accuracy of the model predictions might not be comparable to a region such as Europe, where significantly more in situ data are available [50]. Third, especially in high-mountain areas, weather conditions are prone to rapid changes; hence, even a 1-h temporal resolution of a weather model might not be sufficient to represent the atmospheric conditions at the SAR acquisition time. Fourth, an increase in RMSE values after weather-model-based corrections was also found by Bekaert et al. [38], who attributed this to the incorrect estimation of the location of turbulence in the weather models.

4.2. Ground Deformation

The area with the highest deformation rates is north of the lake at P1, which is reasonable because it is located closest to the lake centre and its shoreline is rather flat (Figure 9a). Here, the two short time series of Sentinel-1 show a clear intra-annual correlation to the water level changes. Although the R^2 is also high in the Envisat case, intra-annual variations are less obvious due to the much lower temporal sampling. The area in the west (P2), close to the Talas-Fergana Fault, is also prone to the same intra-annual deformation, but at lower rates. Here, the slopes of the high mountains are much steeper than at P1 and even reach the shoreline (Figure 9b).

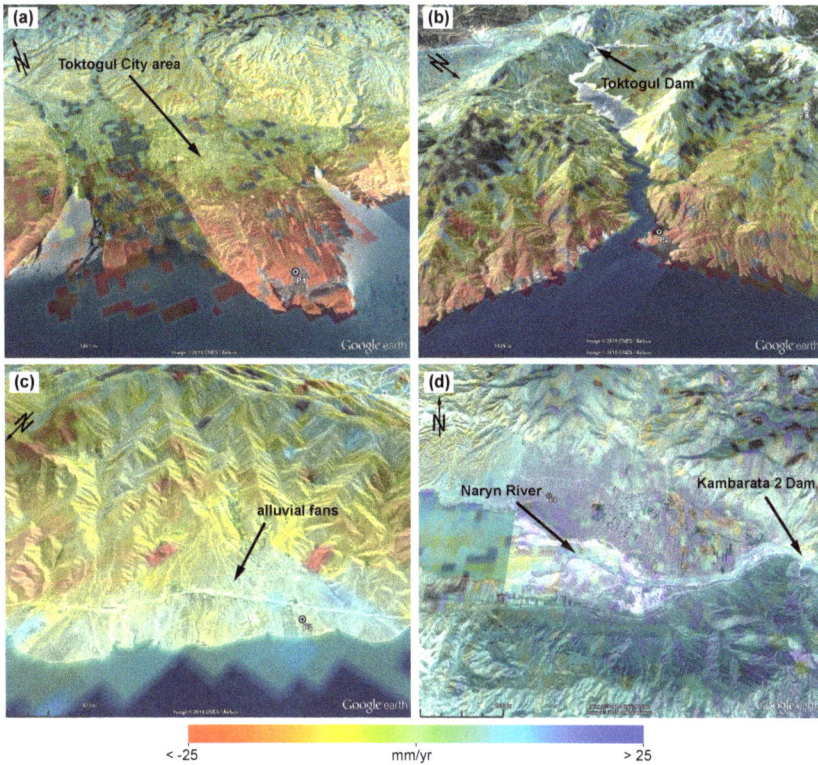

Figure 9. Vertical deformation extracted from Sentinel-1 decomposition around (**a**) point P1 and the Toktogul city area; (**b**) point P2; (**c**) alluvial fans at point P3 and (**d**) the Naryn River entrance area at point P4. For the locations of the points (refer to Figure 6). Background imagery provided by Google ® Earth.

Anomalies in the overall deformation pattern are found at P3 and P4. As these are stringent in all time series, it is straightforward to assume that they are induced by local characteristics of the ground and not due to remaining atmospheric effects, artificial changes of landcover such as construction work or the influence of the Kambarata-2 hydropower plant, which is located east of the Naryn River entrance (Figure 9d) and only started operating in 2010 [51].

For P3, where no deformation could be measured, an overlay on DEM and optical remote sensing data reveals that this area is characterized by alluvial fans that are rather flat and consist of fluvial sediments (Figure 9c). The inverse deformation at P4 is also located in a rather flat and sedimentary area at the entrance of the Naryn River (Figure 9d). We believe that, in both areas, the height change in ground water level is the main source for the observed deformation pattern. At times of high reservoir water level, the ground water level in the sedimentary areas at P3 and P4 will also be higher compared to times at low reservoir water level. The increased amount of ground water and soil moisture prevents the SAR signal from penetrating as deep into the ground as in the case of dry material. This effect leads to a measured "uplift" signal at P4 in the Sentinel-1 time series, although the area is rather subsiding (cf. the model results in Figure 8c). However, because the area at P4 is very shallow and thus affected by only a small amount of load change, the reduced signal penetration into the ground is dominating. Conversely, we observe a "subsidence" signal in the Envisat time series because the ground water level is becoming lower here. In the case of P3, however, the area is located much more closely to the lake centre; thus, it is likely here that both effects, loading/unloading and radar penetration changes, are compensating each other, leading to the observed "non-deformation". The hypothesis that the same process is affecting the deformation measurements at both locations is further underlined by the fact that residuals between measured and modelled deformations are on the same order at P3 and P4 (Figure 8d).

Another process that has the same contradictory effect is the compaction of the sediments at times of lower ground water levels and soil swelling at times of higher ground water levels. The related deformation might also contribute to the observed signals, but presumably at a much lower rate.

Compared to Envisat and S1d, a significantly lower correlation between LOS deformation and water level fit is found for the S1a time series. This result indicates that the atmospheric correction was rather poor for the S1a data. We relate this to increased lake evaporation rates that locally increase the moisture content in the low-level atmosphere since S1a data were acquired during local evening hours (UTC+6 h). In comparison, Envisat and S1d images were captured at noon and during morning hours, respectively. The relation to evaporation is further indicated by the fact that LOS deformation rates with the highest offsets from the water level fit can be found in summer time (Figure 7c). We view this mismatch of atmospheric correction also as the main driver for why we do not observe a clear uplift signal at P4 in the S1a time series. In principle, this can be overcome by an extension of the time series, although here it is not likely that the water level will increase much in the upcoming year because the reservoir is already at full capacity.

5. Conclusions

We have investigated the suitability of Sentinel-1 and Envisat SAR data for capturing surface loading effects due to water level changes in the Toktogul Reservoir. Since the study area is situated in a high-mountain terrain, the SAR acquisitions are severely affected by atmospheric noise that had to be removed prior to the deformation analysis. In our study, we have tested a global numerical weather model and phase-based removal approaches and found that, for the Toktogul area, the power-law method of Bekaert et al. [21] performed the best. However, the S1a evening acquisitions in summer time were significantly influenced by remaining atmospheric effects that we relate to non-corrected lake evaporation. We thus recommend for similar studies to either focus on morning or noon acquisitions or to make an effort to reduce the impact of such remaining effects.

Although only two years of Sentinel-1 data (2014–2016) were used, we found a strong correlation between the increase in the water level and the subsidence of the surrounding region. Areas north of the lake and within a range of 2 km to the shore were affected the most (25 mm/yr, which corresponds to a deformation rate of −1.07 mm per 1 m water level change), but measurable deformations occurred as far as 15 km away from the shore. Due to the dense acquisition interval, we were also able to retrieve intra-annual LOS deformation variations on the order of 30 mm. The analysis of intra-annual time series and water level change further revealed that ground deformations occurred simultaneously with water level changes, which indicated an elastic deformation response. We therefore estimated ground deformation rates with an elastic forward model that was based on the TEA12 Earth model. The results showed that the modelled ground deformation rates fit very well with our Sentinel-1 measured vertical deformations.

The SAR time series from Envisat was considerably longer (2004–2009). This time period was characterized by an overall release of the Toktogul Reservoir water; hence, we retrieved a corresponding inter-annual uplift deformation of the area. However, clearly correlated intra-annual changes were not extracted, which was related to the less temporal density of SAR acquisitions.

In terms of the Toktogul case study, our estimations of the dimension and spatial extent of deformation can contribute to seismic hazard prediction maps as presented by Abdrakhmatov et al. [52] or Bindi et al. [53]. The Talas-Fergana Fault to the southwest of the reservoir is a potential trigger for earthquakes. Such an event is especially dangerous if it induces a landslide as in the M_S 7.3 Suusamyr earthquake in 1992. If such a landslide collapses into the reservoir, it may induce a tsunami, which poses a severe threat to the dam [12].

We see several avenues for further research in the study area that would benefit from continuous intra-annual monitoring with Sentinel-1 data: (1) for areas such as the city of Toktogul, it is important to assess the potential consequences of the identified deformations and their implications on the stability of the buildings; (2) In combination with the observed Toktogul water level changes, the effects of varying water levels at the Kambarata 2 dam should be investigated. Focus should therefore be on the connections of the corresponding ground deformations to those of the Toktogul Reservoir; (3) Continuous SAR monitoring of the mountain slopes facing towards the lake can help in the early detection of instabilities such as slope failures that may eventually collapse into the lake [12]; and (4) via the elastic modelling, the deformations estimated from remote sensing in combination with the directly observed water mass loads allow constraining the specific Earth structure in the Toktogul region.

Acknowledgments: The Sentinel-1 and Envisat data used for this study are provided by the European Space Agency/Copernicus. The SRTM DEM data were derived from the Consortium for Spatial Information of the Consultative Group for International Agricultural Research (CGIAR–CSI). The Landsat data were obtained from the U.S. Geological Survey. We thank Deutscher Wetterdienst, Offenbach, Germany, and the European Centre for Medium Range Weather Forecast, Reading, U.K., for providing data from ECMWF's operational forecast model, the ERA-5 reanalysis and the ERA-I reanalysis. This work was funded by the Initiative and Networking Fund of the Helmholtz Association in the framework of the Helmholtz Alliance "Remote Sensing and Earth System Dynamics" (EDA) and by the CAWa project (Contract No. AA7090002), funded by the German Federal Foreign Office. The study contributes to the "Advanced Earth System Modelling Capacity" (ESM) project of the Helmholtz Association of German Research Centres. We also appreciate the valuable comments and constructive suggestions of four anonymous reviewers, which greatly helped us to improve our manuscript.

Author Contributions: Mahdi Motagh and Julia Neelmeijer designed the study together. Julia Neelmeijer accomplished the InSAR and atmosphere correction processing, analysis and wrote the manuscript. Tilo Schöne processed the lake altimetry data and wrote the corresponding section of the manuscript. Robert Dill and Volker Klemann performed the modelling and provided the content for the related section. Mahdi Motagh contributed to the analysis, discussion and writing.

Conflicts of Interest: The authors declare no conflict of interest.

References

1. Simpson, D.W.; Leith, W.; Scholz, C. Two types of reservoir-induced seismicity. *Bull. Seismol. Soc. Am.* **1988**, *78*, 2025–2040.

2. Kaufmann, G., Amelung, F. Reservoir-induced deformation and continental rheology in vicinity of Lake Mead, Nevada. *J. Geophys. Res.-Solid Earth* **2000**, *105*, 16341–16358, doi:10.1029/2000JB900079.

3. Bevis, M.; Kendrick, E.; Cser, A.; Smalley, R. Geodetic measurement of the local elastic response to the changing mass of water in Lago Laja, Chile. *Phys. Earth Planet. Inter.* **2004**, *141*, 71–78, doi:10.1016/j.pepi.2003.05.001.

4. Wahr, J.; Khan, S.A.; van Dam, T.; Liu, L.; van Angelen, J.H.; van den Broeke, M.R.; Meertens, C.M. The use of GPS horizontals for loading studies, with applications to northern California and southeast Greenland. *J. Geophys. Res.-Solid Earth* **2013**, *118*, 1795–1806, doi:10.1002/jgrb.50104.

5. Cavalié, O.; Doin, M.P.; Lasserre, C.; Briole, P. Ground motion measurement in the Lake Mead area, Nevada, by differential synthetic aperture radar interferometry time series analysis: Probing the lithosphere rheological structure. *J. Geophys. Res.* **2007**, *112*, B03403, doi:10.1029/2006JB004344.
6. Nof, R.N.; Ziv, A.; Doin, M.P.; Baer, G.; Fialko, Y.; Wdowinski, S.; Eyal, Y.; Bock, Y. Rising of the lowest place on Earth due to Dead Sea water-level drop: Evidence from SAR interferometry and GPS. *J. Geophys. Res.-Solid Earth* **2012**, *117*, B05412, doi:10.1029/2011JB008961.
7. Zhao, W.; Amelung, F.; Doin, M.P.; Dixon, T.H.; Wdowinski, S.; Lin, G. InSAR observations of lake loading at Yangzhuoyong Lake, Tibet: Constraints on crustal elasticity. *Earth Planet. Sci. Lett.* **2016**, *449*, 240–245, doi:10.1016/j.epsl.2016.05.044.
8. Furuya, M.; Wahr, J.M. Water level changes at an ice-dammed lake in west Greenland inferred from InSAR data. *Geophys. Res. Lett.* **2005**, *32*, L14501, doi:10.1029/2005GL023458.
9. Doin, M.P.; Twardzik, C.; Ducret, G.; Lasserre, C.; Guillaso, S.; Jianbao, S. InSAR measurement of the deformation around Siling Co Lake: Inferences on the lower crust viscosity in central Tibet. *J. Geophys. Res.-Solid Earth* **2015**, *120*, 5290–5310, doi:10.1002/2014JB011768.
10. Gahalaut, V.; Yadav, R.K.; Sreejith, K.M.; Gahalaut, K.; Bürgmann, R.; Agrawal, R.; Sati, S.; Kumar, A. InSAR and GPS measurements of crustal deformation due to seasonal loading of Tehri reservoir in Garhwal Himalaya, India. *Geophys. J. Int.* **2017**, *209*, 425–433, doi:10.1093/gji/ggx015.
11. Simpson, D.W.; Hamburger, M.W.; Pavlov, V.D.; Nersesov, I.L. Tectonics and seismicity of the Toktogul Reservoir Region, Kirgizia, USSR. *J. Geophys. Res.* **1981**, *86*, 345–358, doi:10.1029/JB086iB01p00345.
12. Tibaldi, A.; Corazzato, C.; Rust, D.; Bonali, F.; Pasquarè Mariotto, F.; Korzhenkov, A.; Oppizzi, P.; Bonzanigo, L. Tectonic and gravity-induced deformation along the active Talas–Fergana Fault, Tien Shan, Kyrgyzstan. *Tectonophysics* **2015**, *657*, 38–62, doi:10.1016/j.tecto.2015.06.020.
13. Savoskul, O.; Chevnina, E.; Perziger, F.; Vasilina, L.; Baburin, V.; Danshin A.I.; Matyakubov, B.; Murakaev, R. Water, climate, food, and environment in the Syr Darya Basin. In *Contribution to the Project ADAPT*; Savoskul, O.S., Ed.; The Pennsylvania State University: State College, PA, USA, 2003.
14. Keith, J.E.; McKinney, D.C. Options Analysis of the Operation of the Toktogul Reservoir. Available online: http://www.ce.utexas.edu/prof/mckinney/papers/aral/Issue7.html (accessed on 15 December 2017).
15. Kyrgyzstan Disaster Risk Data Platform. Available online: http://geonode.mes.kg (accessed on 10 March 2017).
16. Ghose, S.; Mellors, R.J.; Korjenkov, A.M.; Hamburger, M.W.; Pavlis, T.L.; Pavlis, G.L.; Omuraliev, M.; Mamyrov, E.; Muraliev, A.R. The M_S = 7.3 1992 Suusamyr, Kyrgyzstan, Earthquake in the Tien Shan: 2. Aftershock Focal Mechanisms and Surface Deformation. *Bull. Seismol. Soc. Am.* **1997**, *87*, 23–38.
17. Dovgan, V. Seismometric Monitoring of Toktogul Hydroelectric Power Station. In Proceedings of the IV International Conference "Problems of Cybernetics and Informatics" (PCI'2012), Baku, Azerbaijan, 12–14 September 2012; pp. 81–84.
18. Zebker, H.A.; Rosen, P.A.; Hensley, S. Atmospheric effects in interferometric synthetic aperture radar surface deformation and topographic maps. *J. Geophys. Res.-Solid Earth* **1997**, *102*, 7547–7563, doi:10.1029/96JB03804.
19. Fialko, Y. Interseismic strain accumulation and the earthquake potential on the southern San Andreas fault system. *Nature* **2006**, *441*, 968–971, doi:10.1038/nature04797.
20. Puysségur, B.; Michel, R.; Avouac, J.P. Tropospheric phase delay in interferometric synthetic aperture radar estimated from meteorological model and multispectral imagery. *J. Geophys. Res.* **2007**, *112*, B05419, doi:10.1029/2006JB004352.
21. Bekaert, D.P.S.; Hooper, A.; Wright, T.J. A spatially variable power law tropospheric correction technique for InSAR data. *J. Geophys. Res.-Solid Earth* **2015**, *120*, 1345–1356, doi:10.1002/2014JB011558.
22. Birkett, C.M. Radar altimetry: A new concept in monitoring lake level changes. *Eos Trans. Am. Geophys. Union* **1994**, *75*, 273–275, doi:10.1029/94EO00944.
23. Crétaux, J.F.; Abarca-del Río, R.; Bergé-Nguyen, M.; Arsen, A.; Drolon, V.; Clos, G.; Maisongrande, P. Lake Volume Monitoring from Space. *Surv. Geophys.* **2016**, *37*, 269–305.
24. Schöne, T.; Dusik, E.; Illigner, J.; Klein, I. Water in Central Asia: Reservoir Monitoring with Radar Altimetry Along the Naryn and Syr Darya Rivers. In Proceedings of the International Symposium on Earth and Environmental Sciences for Future Generations, Prague, Czech Republic, 22 June–2 July 2015; Springer: Cham, Switzerland, 2017; Volume 147, pp. 349–357.
25. CA WATER Info. Available online: www.cawater-info.net (accessed on 31 January 2017).
26. JSC "Electric Stations". Available online: www.energo-es.kg (accessed on 31 January 2017).

27. Förste, C.; Bruinsma, S.; Abrykosov, O.; Flechtner, F.; Dahle, C.; Neumayer, K.H.; Barthelmes, F.; König, R.; Marty, J.-C.; Lemoine, J.M.; et al. EIGEN-6C3—The newest high resolution global combined gravity field model based on the 4th release of the GOCE Direct Approach. In Proceedings of the 2013 IAG Scientific Assembly, 150th Anniversary of the IAG, Potsdam, Germany, 1–6 September 2013.

28. Werner, C.L.; Wegmüller, U.; Strozzi, T.; Wiesmann, A. GAMMA SAR and Interferometric Processing Software. In Proceedings of the ERS-ENVISAT Symposium, Gothenburg, Sweden, 16–20 October 2000.

29. Prats, P.; Marotti, L.; Wollstadt, S.; Scheiber, R. Investigations on TOPS interferometry with TerraSAR-X. In Proceedings of the 2010 IEEE International Geoscience and Remote Sensing Symposium (IGARSS), Honolulu, Hawaii, 25–30 July 2010; pp. 2629–2632, doi:10.1109/IGARSS.2010.5650037.

30. Prats-Iraola, P.; Scheiber, R.; Marotti, L.; Wollstadt, S.; Reigber, A. TOPS Interferometry With TerraSAR-X. *IEEE Trans. Geosci. Remote Sens.* **2012**, *50*, 3179–3188, doi:10.1109/TGRS.2011.2178247.

31. Scheiber, R.; Moreira, A. Coregistration of interferometric SAR images using spectral diversity. *IEEE Trans. Geosci. Remote Sens.* **2000**, *38*, 2179–2191, doi:10.1109/36.868876.

32. Wegmüller, U.; Werner, C.; Strozzi, T.; Wiesmann, A.; Frey, O.; Santoro, M. Sentinel-1 IWS mode support in the GAMMA software. *Procedia Comput. Sci.* **2016**, *100*, 431–436, doi:10.1109/APSAR.2015.7306242.

33. Berardino, P.; Fornaro, G.; Lanari, R.; Sansosti, E. A new algorithm for surface deformation monitoring based on small baseline differential SAR interferograms. *IEEE T. Geosci. Remote* **2002**, *40*, 2375–2383, doi:10.1109/TGRS.2002.803792.

34. Hooper, A.J.; Bekaert, D.; Spaans, K.; Arıkan, M. Recent advances in SAR interferometry time series analysis for measuring crustal deformation. *Tectonophysics* **2012**, *514–517*, 1–13, doi:10.1016/j.tecto.2011.10.013.

35. Hooper, A. A multi-temporal InSAR method incorporating both persistent scatterer and small baseline approaches. *Geophys. Res. Lett.* **2008**, *35*, L16302, doi:10.1029/2008GL034654.

36. Hooper, A.J.; Zebker, H.A. Phase unwrapping in three dimensions with application to InSAR time series. *J. Opt. Soc. Am.* **2007**, *24*, 2737–2747, doi:10.1364/JOSAA.24.002737.

37. Marinković, P.; Larsen, Y. On Resolving the Local Oscillator Drift Induced Phase Ramps in ASAR and ERS1/2 Interferometric Data—The Final Solution. In Proceedings of the Fringe 2015 workshop (ESA SP-731), Frascati, Italy, 23–27 March 2015.

38. Bekaert, D.; Walters, R.; Wright, T.; Hooper, A.; Parker, D. Statistical comparison of InSAR tropospheric correction techniques. *Remote Sens. Environ.* **2015**, *170*, 40–47, doi:10.1016/j.rse.2015.08.035.

39. Doin, M.P.; Lasserre, C.; Peltzer, G.; Cavalié, O.; Doubre, C. Corrections of stratified tropospheric delays in SAR interferometry: Validation with global atmospheric models. *J. Appl. Geophys.* **2009**, *69*, 35–50, doi:10.1016/j.jappgeo.2009.03.010.

40. Jolivet, R.; Agram, P.S.; Lin, N.Y.; Simons, M.; Doin, M.P.; Peltzer, G.; Li, Z. Improving InSAR geodesy using Global Atmospheric Models. *J. Geophys. Res.-Solid Earth* **2014**, *119*, 2324–2341, doi:10.1002/2013JB010588.

41. Barnhart, W.D.; Lohman, R.B. Characterizing and estimating noise in InSAR and InSAR time series with MODIS. *Geochem. Geophys.* **2013**, *14*, 4121–4132, doi:10.1002/ggge.20258.

42. Jolivet, R.; Grandin, R.; Lasserre, C.; Doin, M.P.; Peltzer, G. Systematic InSAR tropospheric phase delay corrections from global meteorological reanalysis data. *Geophys. Res. Lett.* **2011**, *38*, L17311, doi:10.1029/2011GL048757.

43. Dee, D.P.; Uppala, S.M.; Simmons, A.J.; Berrisford, P.; Poli, P.; Kobayashi, S.; Andrae, U.; Balmaseda, M.A.; Balsamo, G.; Bauer, P.; et al. The ERA-Interim reanalysis: Configuration and performance of the data assimilation system. *Q. J. R. Meteorol. Soc.* **2011**, *137*, 553–597, doi:10.1002/qj.828.

44. Fialko, Y.; Simons, M.; Agnew, D. The complete (3-D) surface displacement field in the epicentral area of the 1999 M W 7.1 Hector Mine Earthquake, California, from space geodetic observations. *Geophys. Res. Lett.* **2001**, *28*, 3063–3066, doi:10.1029/2001GL013174.

45. Motagh, M.; Shamshiri, R.; Haghshenas Haghighi, M.; Wetzel, H.U.; Akbari, B.; Nahavandchi, H.; Roessner, S.; Arabi, S. Quantifying groundwater exploitation induced subsidence in the Rafsanjan plain, southeastern Iran, using InSAR time-series and in situ measurements. *Eng. Geol.* **2017**, *218*, 134–151, doi:10.1016/j.enggeo.2017.01.011.

46. Farrell, W.E. Deformation of the Earth by surface loads. *Rev. Geophys.* **1972**, *10*, 761–797, doi:10.1029/RG010i003p00761.

47. Dill, R.; Klemann, V.; Martinec, Z.; Tesauro, M. Applying local Green's functions to study the influence of the crustal structure on hydrological loading displacements. *J. Geodyn.* **2015**, *88*, 14–22, doi:10.1016/j.jog.2015.04.005.

48. Dziewonski, A.M.; Anderson, D.L. Preliminary reference Earth model. *Phys. Earth Planet. Inter.* **1981**, *25*, 297–356, doi:10.1016/0031-9201(81)90046-7.

49. Tesauro, M.; Audet, P.; Kaban, M.K.; Bürgmann, R.; Cloetingh, S. The effective elastic thickness of the continental lithosphere: Comparison between rheological and inverse approaches. *Geochem. Geophys. Geosys.* **2012**, *13*, Q09001, doi:10.1029/2012GC004162.

50. Haiden, T.; Janousek, M.; Bauer, P.; Bidlot, J.; Dahoui, M.; Ferranti, L.; Prates, F.; Richardson, D.; Vitart, F. Evaluation of ECMWF forecasts, including 2014–2015 upgrades. In *ECMWF Technical Memoranda*; European Centre for Medium-Range Weather Forecasts: Reading, UK, 2015; Volume 765, pp. 1–51.

51. Havenith, H.B.; Torgoev, I.; Torgoev, A.; Strom, A.; Xu, Y.; Fernandez-Steeger, T. The Kambarata 2 blast-fill dam, Kyrgyz Republic: Blast event, geophysical monitoring and dam structure modelling. *Geoenviron. Disasters* **2015**, *2*, 11, doi:10.1186/s40677-015-0021-x.

52. Abdrakhmatov, K.; Havenith, H.B.; Delvaux, D.; Jongmans, D.; Trefois, P. Probabilistic PGA and Arias Intensity maps of Kyrgyzstan (Central Asia). *J. Seismol.* **2003**, *7*, 203–220, doi:10.1023/A:1023559932255.

53. Bindi, D.; Abdrakhmatov, K.; Parolai, S.; Mucciarelli, M.; Grünthal, G.; Ischuk, A.; Mikhailova, N.; Zschau, J. Seismic hazard assessment in Central Asia: Outcomes from a site approach. *Soil Dyn. Earthq. Eng.* **2012**, *37*, 84–91, doi:10.1016/j.soildyn.2012.01.016.

remote sensing

MDPI

Article

Time Series Analysis of Very Slow Landslides in the Three Gorges Region through Small Baseline SAR Offset Tracking

Luyi Sun [1], Jan-Peter Muller [2,*] and Jinsong Chen [1,*]

[1]   Shenzhen Institutes of Advanced Technology, Chinese Academy of Sciences, 1068 Xueyuan Avenue, Shenzhen University Town, Shenzhen 518055, China; ly.sun@siat.ac.cn
[2]   Mullard Space Science Laboratory, University College London, Holmbury St Mary, Surrey, RH5 6NT, UK
*   Correspondence: j.muller@ucl.ac.uk (J.-P.M.); js.chen@siat.ac.cn (J.C.)

Received: 19 October 2017; Accepted: 11 December 2017; Published: 14 December 2017

Abstract: Sub-pixel offset tracking has been used in various applications, including measurements of glacier movement, earthquakes, landslides, etc., as a complementary method to time series InSAR. In this work, we explore the use of a small baseline subset (SBAS) Offset Tracking approach to monitor very slow landslides with centimetre-level annual displacement rate, and in challenging areas characterized by high humidity, dense vegetation cover, and steep slopes. This approach, herein referred to as SBAS Offset Tracking, is used to minimize temporal and spatial de-correlation in offset pairs, in order to achieve high density of reliable measurements. This approach is applied to a case study of the Tanjiahe landslide in the Three Gorges Region. Using the TerraSAR-X Staring Spotlight (TSX-ST) data, with sufficient density of observations, we estimate the precision of the SBAS offset tracking approach to be 2–3 cm on average. The results demonstrated accord well with corresponding GPS measurements.

Keywords: sub-pixel offset tracking; small baseline subset (SBAS); TerraSAR-X Staring Spotlight (TSX-ST); very slow landslide; Three Gorges Region (TGR)

---

1. Introduction

As a major natural hazard, landslides cause enormous direct and indirect damage worldwide every year. Remote sensing has become the most convenient and feasible tool widely applied in deformation mapping, including in the monitoring of landslides. In the study area, due to the often limited access to Global Positioning System (GPS) measurements, and the high costs of skilled labour and instrumentation, it is difficult to collect sufficient geodetic measurements. Due to the high humidity caused by the monsoon climate of this region, optical sensors are often limited in obtaining an effective time series of measurements. Thus, microwave remote sensing using Synthetic Aperture Radar (SAR) imagery has been recognized as an effective tool for landslide monitoring. It is able to work both day and night during all weather conditions, and repeatedly acquires time series of images over large areas.

DInSAR techniques have been conventionally used for mapping of landslide activities. However, several difficulties arise when attempting to apply DInSAR in areas with steep slopes and rugged topography, high humidity, and dense vegetation cover. In addition to these difficulties, in previous studies [1–3], it is shown that the maximum detectable displacement gradient (DDG) of DInSAR can be exceeded in some case of very slow landslides (16 mm·year^{-1}–1.6 m·year^{-1}, as defined in Cruden and Varnes [4], Hungr et al. [5]) even when using high resolution SAR imagery.

As an alternative method, Offset Tracking (sometimes also referred to as intensity tracking) can be used to address some of the technical limitations of DInSAR, particularly the limitation of maximum detectable displacement gradient (DDG) and low coherence due to vegetation changes [6–9]. Offset

Tracking allows the measurement of two-dimensional (2D) ground surface displacement with sub-pixel accuracy, by analysing the 2D offsets of the master and slave images based on cross-correlation of SAR intensity and amplitude.

Sub-pixel correlation of optically sensed imagery from spaceborne or airborne platforms has been proven as a very useful technique for investigation of landslides [10–15]. A more recent study proposed a multiple pairwise image correlation (MPIC) technique based on a sub-pixel correlation analysis of optical data [16]. This method was tested with time series Pléiades monoscopic and stereoscopic images to investigate a landslide-prone landscape in the South French Alps. It demonstrated the capability of this method to improve detection accuracy, benefiting from averaging redundant measurements from multiple pair combinations. However, in some areas, such as the Three Gorges Region where our case study is located, due to the frequent cloud cover throughout the year, it is difficult to obtain multi-temporal optical satellite images for time series analysis.

For time series offset tracking of SAR imagery, the commonly adopted approach is to use a single master image, usually the first acquisition. This simple strategy is suitable when no significant dependence is found between the number of reliable measurements and the temporal or spatial baseline. Under such a scenario, connecting offset pairs by a small baseline network has limited benefits and leads to much higher time consumption.

However, in many cases, due to a larger dynamic range of spatial baseline or temporal de-correlation effects, the number of reliable measurements decreases significantly with the increase in temporal or spatial baseline. As indicated in Yonezawa and Takeuchi [17], Offset Tracking requires similar speckle patterns between the master and slave images to obtain a sharp correlation peak. Long baseline distances will result in significant speckle geometrical de-correlation. The correlation coefficient between offset pairs decreases with the increase of spatial baseline, which leads to a higher standard deviation (STD) error in cross-correlation [18]. In addition, in densely vegetated areas, the temporal de-correlation effects are significant. Higher accuracy is required to measure very slow landslides, which again leads to lower density of final measurements. In particular, in rural areas with dense vegetation cover, there are few high-contrast surface features (e.g., artificial corner reflectors, houses, bare rocks, etc.), but a number of natural scatterers can maintain a medium correlation within a certain time period rather than over the whole time series. Thus, constructing a small baseline network based on proper thresholds of temporal and spatial baseline can help to minimize temporal and speckle geometric de-correlation effects, and take advantage of the scatterers with temporary medium correlation, so as to increase the density of measurements.

Small baseline approaches have previously been combined with offset tracking to measure large deformation magnitudes. Casu et al. proposed a PO-SBAS (pixel-offset small baseline subset) approach applied to medium resolution ENVISAT SAR data to measure large displacements (several metres) occurring in the inner part of the Sierra Negra caldera due to the October 2005 eruption [19]. The measured deformation reached one to several metres in both azimuth and range directions. Manconi et al. produced post-event deformation maps for emergency evaluation of a large, rapidly-moving (10–20 m) landslide [20]. The PO-SBAS approach was applied to ascending and descending pairs of COSMO-SkyMed images to retrieve three-dimensional (3D) deformation of the Montescaglioso landslide (Italy), of which the main movement occurred in 15–20 min at an average velocity of 0.5–1 m per minute.

For measurements of large displacement, the topographic component of offsets is not significant with regard to the deformation magnitude. Topographic distortions are usually modeled using a reference DEM and orbital data, and removed from offset results [19–22]. In the case study presented in Raucoules et al. [23], considering the selected small baselines (ranging from 1 to 200 m) with regard to the large deformation magnitude, the topographic component was neglected. In addition, co-registration errors (about 1/10 pixel size) are not significant either, in the case of large deformation.

However, in the case of much smaller displacement rates (several to dozens of centimetres per year), the residual offsets in both range and azimuth directions due to co-registration errors and orbit inaccuracies in topographic distortion removal can even obscure the real displacement, and thus are non-negligible.

This research, to the best of our knowledge, is the first to explore the use of SBAS offset tracking technique to monitor very slow landslides, in which scenario the removal of residual offsets becomes a crucial step to derive correct displacement rates even using sub-metre resolution SAR data. A step of 2D polynomial fitting is applied to both range and azimuth offset measurements to estimate and remove the residual offsets before the inversion step. Furthermore, the study area, the Tanjiahe landslide in the Three Gorges Region, China, is characterized by high humidity and dense vegetation cover on steep slopes, posing more difficulties on the application of time series InSAR and offset tracking. In our preliminary study, time series InSAR analysis of the landslide did not obtain satisfactory results. Rapid loss of phase coherence combined with topographical phase residuals lead to very low redundancy of connections (less than three per acquisition) in the SBAS InSAR network. The dense vegetation cover in this area lead to a very low density of Persistent Scatter (PS) candidates. For this reason, time series InSAR cannot provide reliable measurements. Similarly, for offset tracking, there are very few high-contrast surface features (e.g., artificial corner reflectors, houses, bare rocks, etc.) in the study area. This means this area lacks strong scatterers with constantly high correlation coefficient throughout the time series. An experiment using conventional offset tracking approach using a single master image yielded rather sparse coverage, because the number of reliable measurements decreases significantly with the temporal baseline, due to the lack of strong scatterers. Hence, SBAS offset tracking is applied to make use of scatterers showing medium correlation within a certain time period, to increase the density of reliable measurements. An assessment is then made on the potential and limitations of SBAS offset tracking in the challenging conditions.

This paper is organized as follows: Section 2 describes the study area, employed data, and proposed method; Section 3 presents the application results of this method to the Tanjiahe landslide area, followed by discussions in Section 4. Finally, some concluding remarks are reported in Section 5.

2. Materials and Methods

2.1. Study Area

The case study is carried out in the Tanjiahe landslide area in the Three Gorges Region of China. The Three Gorges Region, situated on the middle Yangtze River from Chongqing to Yichang, covers an area of 58,000 km^2 [24]. The terrain is composed of a succession of limestone gorges and ridges, and inter-gorge valleys. Frequent and wide distributed landslides in the Three Gorges Region have caused a lot of wasted resources, damage to properties and public facilities, and even loss of human lives. They also pose great threats to the normal operation of the Three Gorges Dam. The land cover within this region is dominated by cultivated land and mixed deciduous forest. The terrain is featured by steep slopes and dense vegetation cover [25].

The Tanjiahe landslide area is representative of the hillsides of the Three Gorges Region, sparsely populated by small villages filled with single-story buildings amongst dense orange trees. The Tanjiahe landslide area is an ancient landslide, located on the southern bank of Yangtze River with centre coordinates of 31.030°N, 110.509°E, about 56 km upstream from the Three Gorges Dam. The landslide body is underlain by mudstone, sandstone, and siltstone. The trailing edge is 432 m high. The front edge extends into the Yangtze River at an altitude of 135 m. The landslide body is about 400 m wide and 1000 m long, with a slope ranging from 10° to 25° and a volume of 9×10^7 m^3. The sliding direction is 340° clockwise counting from the North, predominantly towards the Yangtze River. As shown in Figure 1, the boundary of the Tanjiahe landslide looks like a boot [26,27].

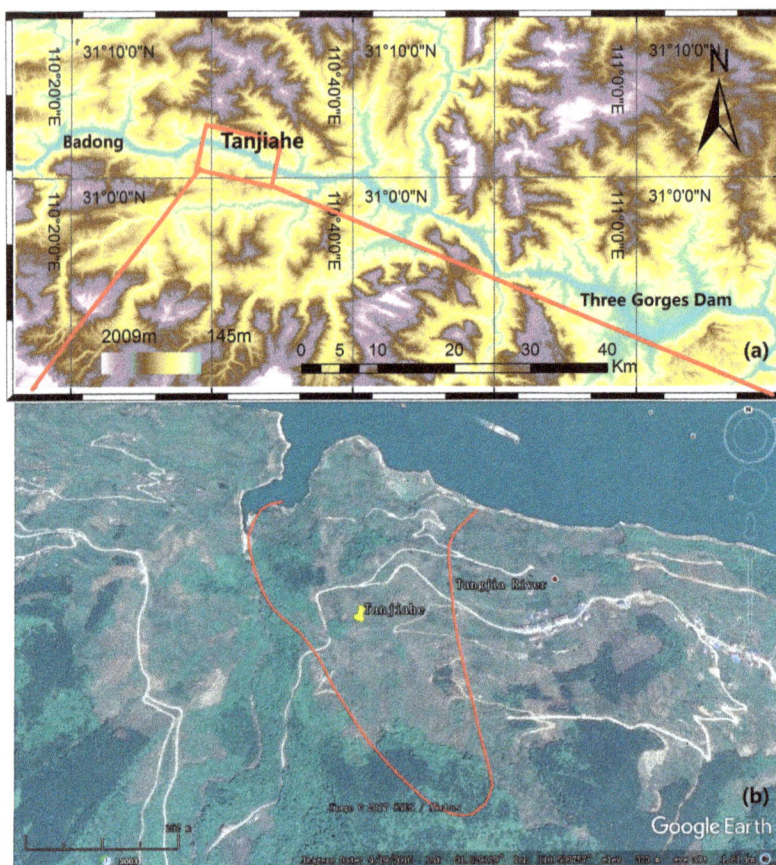

Figure 1. (a) Location of Tanjiahe landslide area shown in SRTM DEM. (b) Tanjiahe landslide area shown in Google Earth with landslide body highlighted in red. Map data: Google Earth, Image@ 2017 CNES/Airbus.

The Tanjiahe landslide has not been well studied especially in the English-language literature, but some historical measurements from GPS monitoring stations can be found in a few Chinese articles. The monitoring was started in October 2006. Notable deformation development of Tanjiahe landslide was observed in 2007 [28]. By December 2009, the accumulated deformation measured from one of the GPS points (ZG289) reached 757.9 mm [26], predominantly towards the Yangtze River. By December 2015, the accumulative displacements rise up to 1800–1900 mm [29]. From 2006 to 2015, seasonal accelerations can be observed from the deformation time series plot, which is suspectedly linked to hydrological factors, such as the local rainfall and water level changes of Three Gorges Reservoir.

2.2. Data

A stack of TSX Staring Spotlight (TSX-ST) images is employed in this research supplied under data grant GEO2630 of the German Aerospace Centre (DLR), acquired in a right-looking orientation on a descending orbit over the Tanjiahe landslide area mostly at 11-day intervals. The data stack spans a time period of one year, from February 2015 to February 2016. The metadata of this annual time series of TSX-ST data is listed in Table 1. The estimated perpendicular baselines of all subsequent images with regard to the first acquisition are listed in Table 2.

Table 1. Metadata of the data stack of TSX Staring Spotlight (TSX-ST) data using the parameters from the first image. These values remain very close for all subsequent acquisitions.

TerraSAR-X Staring Spotlight Data	
First acquisition	8 February 2015
Last acquisition	28 February 2016
Satellite orbit heading (°)	189.555
Wavelength (m)	0.031
Incidence angle (°)	44.303
Polarization	HH
Range pixel spacing (m)	0.455
Azimuth pixel spacing (m)	0.169
Range resolution (m)	0.84
Azimuth resolution (m)	0.23
Maximum DDG	0.0059 (Range looks = 2)

Table 2. Perpendicular baseline of each slave image with regard to the first acquisition.

Common Master	Slave			
	Acquisition Date	Perpendicular Baseline (m)	Acquisition Date	Perpendicular Baseline (m)
	19 February 2015	391.3	27 September 2015	46.2
	2 March 2015	80.8	8 October 2015	119.4
	4 April 2015	46.8	19 October 2015	199.0
	15 April 2015	192.4	30 October 2015	12.2
	18 May 2015	42.0	10 November 2015	37.7
	20 June 2015	60.2	21 November 2015	128.9
8 February 2015	1 July 2015	19.4	2 December 2015	53.4
	12 July 2015	28.6	24 December 2015	28.6
	23 July 2015	147.7	4 January 2016	49.5
	3 August 2015	123.1	15 January 2016	140.0
	25 August 2015	3.5	26 January 2016	121.3
	5 September 2015	77.0	17 February 2016	252.8
	16 September 2015	122.0	28 February 2016	11.8

2.3. Method: SBAS Offset Tracking

We briefly summarize the SBAS offset tracking algorithm based on the work reported by Berardino et al. [30] and Casu et al. [19]. Similar to the SBAS InSAR approach proposed by Berardino et al. [30], we here consider the scenario in the amplitude domain. We assume that there is a stack of co-registered full resolution SAR data consisting of $N + 1$ images $\phi_0, \phi_1, \phi_2, \cdots, \phi_N$. Based on selected thresholds of spatial and temporal baseline, small baseline subsets are formed by M data pair connections.

The 2D offset measurements of the M connected pairs are represented by $\delta\phi^T = [\delta\phi_1, \delta\phi_2, \cdots, \delta\phi_M]$, where $\delta\phi_j = \phi_{IS_j} - \phi_{IE_j}$ $\forall j = 1, 2, \cdots, M$, ϕ_{IS_j} is the slave image and ϕ_{IE_j} is the master image of a generic offset pair. Assuming that:

$$\phi_{IS_j} - \phi_{IE_j} = \sum_{k=IE_j+1}^{IS_j} (t_k - t_{k-1}) \frac{\phi_k - \phi_{k-1}}{t_k - t_{k-1}} = \sum_{k=IE_j+1}^{IS_j} (t_k - t_{k-1}) v_k \tag{1}$$

where v_k is the mean azimuth or range displacement velocity between time-adjacent acquisitions of a connected pair. Thus, a vector v^T consist of a time series displacement velocity and can be expressed as:

$$v^T = \left[v_1 = \frac{\phi_1 - \phi_0}{t_1 - t_0}, v_2 = \frac{\phi_2 - \phi_1}{t_2 - t_1}, \cdots, v_N = \frac{\phi_N - \phi_{N-1}}{t_N - t_{N-1}} \right]. \tag{2}$$

Set B as a matrix recording all connections in the small baseline network,

$$B(j,k) = \begin{cases} t_k - t_{k-1}, & IE_j + 1 \le k \le IS_j, \forall j = 1, 2, \cdots, M \\ 0, & elsewhere \end{cases}. \tag{3}$$

So, the following relationship holds

$$Bv = \delta\phi. \tag{4}$$

In the vast majority of cases that B exhibits a rank deficiency, the velocity vector v can be retrieved by solving the over-determined equations in Equation (4) using Singular Value Decomposition (SVD), as:

$$[U, S, V] = SVD(B). \tag{5}$$

Then the displacement velocity vector can be derived by

$$v = VS^+U^T\delta\phi \tag{6}$$

where

$$S = \mathrm{diag}(\sigma_1, \sigma_2, \cdots, \sigma_{N-L+1}, 0, \cdots, 0)$$
$$S^+ = \mathrm{diag}(1/\sigma_1, 1/\sigma_2, \cdots, 1/\sigma_{N-L+1}, 0, \cdots, 0) \tag{7}$$

σ_i represent the singular values; L is the number of different subsets; N is the number of images in the data stack; the rank of matrix B is $N - L + 1$.

The step-by-step processing strategy is described as follows:

(1) Co-registered images are cropped to cover the landslide body and the surrounding stable area. Each data pair is processed by sub-pixel offset tracking.

(a) Topographic distortions are modeled using a reference DEM (SRTM 1 arc-second global DEM) with orbital parameters and subtracted.

(b) The azimuth and range offsets are derived using cross-correlation. As described in Sun and Muller [3] and recalled herein, the Normalized Cross Correlation (NCC) is applied to the amplitudes of the master and slave images, to derive two-dimensional (2D) offsets. The offsets of a point in any dimension are determined by its different positions in the master and slave images. The corresponding position is determined by a measure of similarity calculated between the point-centred window in the master image and a sliding window of same pixel size in the slave image. The similarity, which is defined as the correlation coefficient, is computed as follows:

$$NCC = \frac{\sum\limits_{m=1}^{N_x} \sum\limits_{n=1}^{N_y} [(i_1(m,n) - \overline{i_1}) \cdot (i_2(m,n) - \overline{i_2})]}{\sqrt{\sum\limits_{m=1}^{N_x} \sum\limits_{n=1}^{N_y} (i_1(m,n) - \overline{i_1})^2} \sqrt{\sum\limits_{m=1}^{N_x} \sum\limits_{n=1}^{N_y} (i_2(m,n) - \overline{i_2})^2}} \tag{8}$$

where i_1 and i_2 denote pre-event and post-event images with a two-dimensional offset (a, b), which can be described as $i_2(x,y) = i_1(x - a, y - b)$. $N_x \times N_y$ is the correlation window size which can be modified by the application requirements. $\overline{i_1}$ and $\overline{i_2}$ are the mathematical expectation values of the cross-event image pair:

$$\overline{i_1} = \frac{1}{N_x \times N_y} \sum\limits_{m=1}^{N_x} \sum\limits_{n=1}^{N_y} i_1(m,n) \tag{9}$$

$$\overline{i_2} = \frac{1}{N_x \times N_y} \sum_{m=1}^{N_x} \sum_{n=1}^{N_y} i_2(m,n). \tag{10}$$

The NCC method searches for maximum correlation (i.e., maximum similarity) between window pairs formed by the master and slave images. Those window pairs for which a maximum correlation detected is considered as corresponding pairs. After locating the corresponding pixels in the master and slave images, the 2D offsets of the slave image with regard to the master image can be obtained. To achieve a sub-pixel accuracy of correlation, image amplitudes are oversampled prior to cross-correlation. Positive values of range displacement correspond to an increase of sensor to target distance. Positive values of azimuth offsets refer to an increase of along-track displacement.

(c)   Residuals offsets due to orbit inaccuracies and co-registration errors are estimated by 2D polynomial fitting of selected reference points in the stable area, and reconstructed for the whole subset, including both the landslide body and stable area.

(2)   After correction of residual offsets, Singular Value Decomposition (SVD) is applied to invert the range and azimuth offset measurements of all connected offset pairs, to derive displacements at each acquisition time.

(3)   To discard unreliable measurements, a mask is built based on the root mean square error (RMSE) of the time series range and azimuth offset measurements, calculated pixel by pixel. For the stable area, RMSE is calculated against zero offset measurements. For the landslide area, a polynomial function is used to fit the displacement time series. RMSE is estimated between the offset measurements and the fitted polynomial. The degree of the polynomial function is selected by multiple fitting tests to obtain best goodness of fit. For the case study of Tanjiahe landslide, a third order polynomial function is used.

(4)   Time series azimuth and range offset maps can be produced to reflect the temporal evolution and spatial distribution of the landslide; time series analysis is carried out on displacement rates of selected pixels in the landslide area.

3. Results

3.1. Small Baseline Network Construction

To determine the spatial and temporal baseline thresholds to create a small baseline network, we made an experiment using the conventional offset tracking method, i.e., computing the 2D offsets of 26 offset pairs, using the first acquisition on 8 February 2015 as the common master image and neglecting any constraint of spatial baseline. All images in the data stack are cropped to the sub-area covering the landslide body and the surrounding stable ground, as shown in Figure 2. Orbital data and a reference DEM (~30 m resolution SRTM DEM) are used to model and correct the topographic components of offsets. A correlation window of 32×128 pixels is exploited, corresponding to a $27\ m \times 29\ m$ resolution grid on the ground. The images are oversampled by a factor of 16 before the cross-correlation.

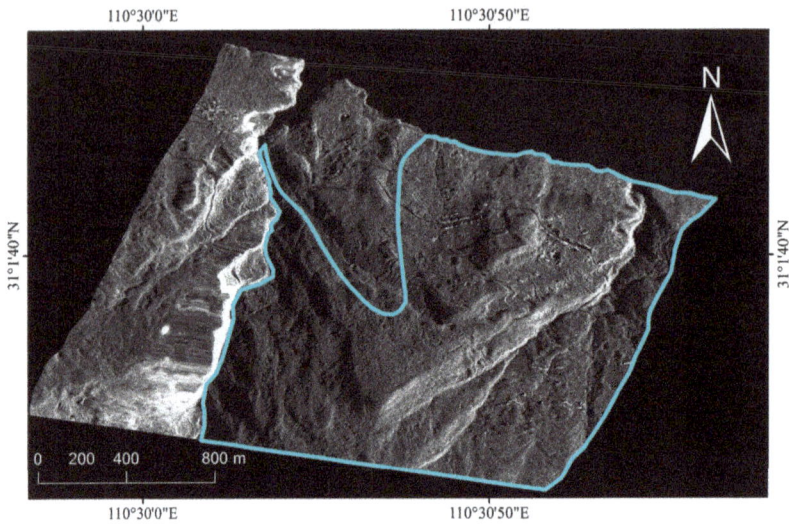

Figure 2. Boundary of the stable area surrounding Tanjiahe landslide, marked in blue polygon on the geocoded Synthetic Aperture Radar (SAR) amplitude. Data source: TerraSAR-X Staring Spotlight © DLR <2015>. All "stable area" mentioned and used in this study refer to this area inside the blue boundary. Apart from the Tanjiahe landslide body, the rest of area adjacent to the blue boundary belongs to another landslide active zone, therefore is excluded from the stable area.

The reliability of offset measurements is assessed by their noise level on the stable ground (Figure 2), which in this case is the local variances of azimuth and range offset measurements in the stable area. This is calculated in the spatial domain using a window of 33 × 33 pixels centred by each pixel for each acquisition, each corresponding to a temporal and spatial baseline value. A proper threshold of the variances is used to select pixels of reliable measurements. Constraints of the temporal–spatial baseline are then determined based on the relationship observed between the number of reliable measurements (in pixels) and the perpendicular baseline and temporal baseline, as displayed in Figure 3.

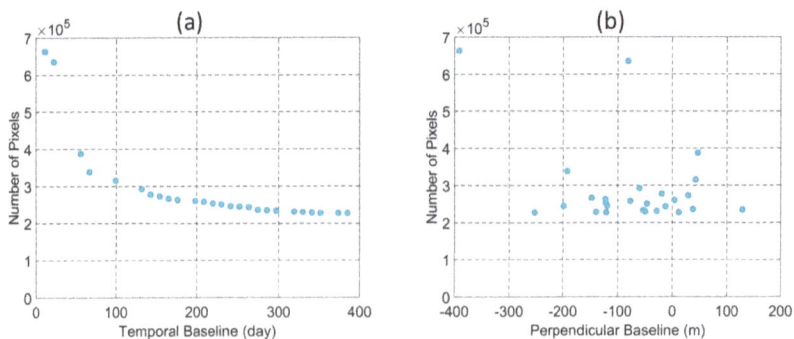

Figure 3. (a) Number of reliable measurements in the stable area plotted in relation to the temporal baseline; (b) Number of reliable measurements in the stable area plotted in relation to the perpendicular baseline.

In Figure 3a, we can see that the number of valid pixels decreases with the temporal baseline, following an approximate exponential trend. In Figure 3b, the same perpendicular baseline corresponds to varied number of reliable measurements, and no significant dependence of the number of pixels is found upon the perpendicular baseline. This suggests that the key factor affecting the number of reliable measurements is the time interval of offset pairs.

A temporal baseline of 99 days and a perpendicular baseline of 400 m are selected as constraints to construct a small baseline network. In total, 157 offset pairs are connected, with a mean connection redundancy of 5.6 per acquisition. The resulting small baseline network is shown in Figure 4, with the relative position (perpendicular baseline with respect to the first acquisition on 8 February 2015) plotted versus the temporal baseline.

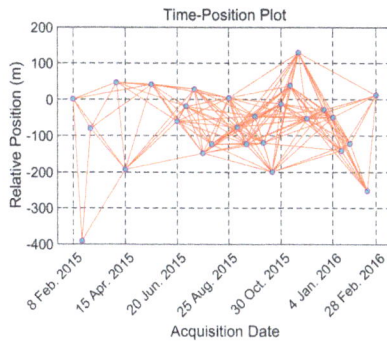

Figure 4. Time–Position plot of the small baseline connection network.

3.2. Removal of Residual Offsets

Sub-pixel offset tracking is applied to each offset pair connected by the small baseline network, using the same procedure and parameters described in Section 3.1. Removal of residual offsets is carried out before inversion of the 157 offset pairs. The standard deviations over time are respectively estimated for the azimuth and range offsets measured from the stable area. By imposing a proper threshold to the standard deviations of 2D measurements, a number of pixels on the stable ground are selected as reference points for correction of residual offsets. A 2D first order polynomial function is then fitted to the range and azimuth offsets measured from these points. Using the fitted parameters, the overall residual offsets are reconstructed for the whole area, including both the landslide body and the stable area. After correction, the mean of the azimuth and range offset measurements derived from the reference points on the stable ground are estimated in the spatial domain for each offset pair, as displayed in Figure 5.

In Figure 5, we can see the mean offset measurements on the stable ground are extremely close to zero. This suggests the correction is successful.

3.3. Two Dimensional Displacement Measured by SBAS Offset Tracking

After correction of residual offsets, all 157 pairs of offset measurements are inverted by SVD decomposition, to derive the azimuth and range displacement at each acquisition time. Noise-dominant pixels are discarded by a RMSE mask using the method described in Section 2.3. Displacement maps are produced to show the temporal evolution and spatial distribution of the landslide, as displayed in Figures 6 and 7.

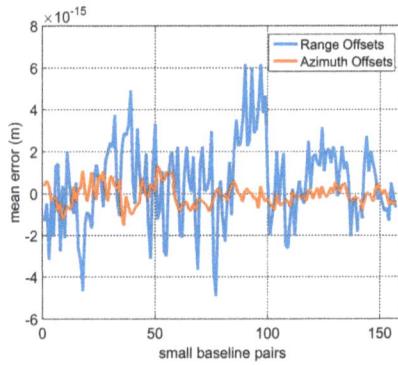

Figure 5. Mean errors of the range and azimuth offsets measured from the stable area of all 157 pairs after residual removal.

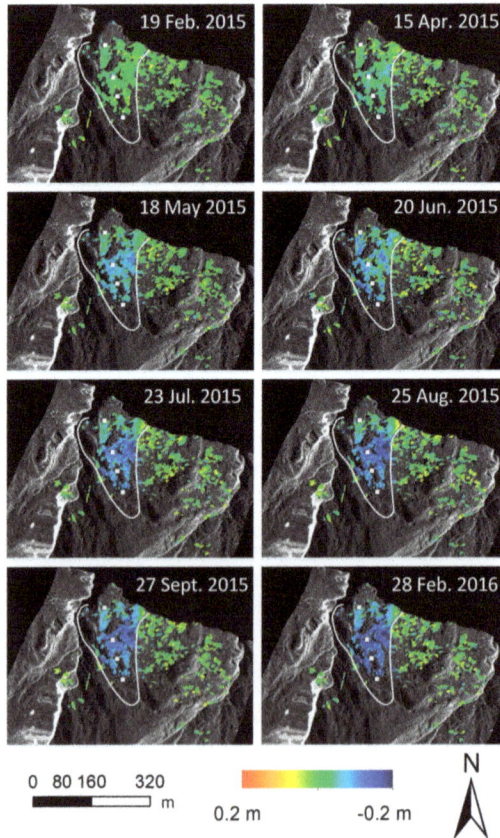

Figure 6. Accumulated azimuth displacement of the Tanjiahe landslide on different acquisition dates derived by the small baseline subset (SBAS) offset tracking approach, superimposed on TSX-ST amplitude, with landslide boundary plotted in white line, and Global Positioning System (GPS) stations marked in white squares. Data source: TerraSAR-X Staring Spotlight © DLR <2015>.

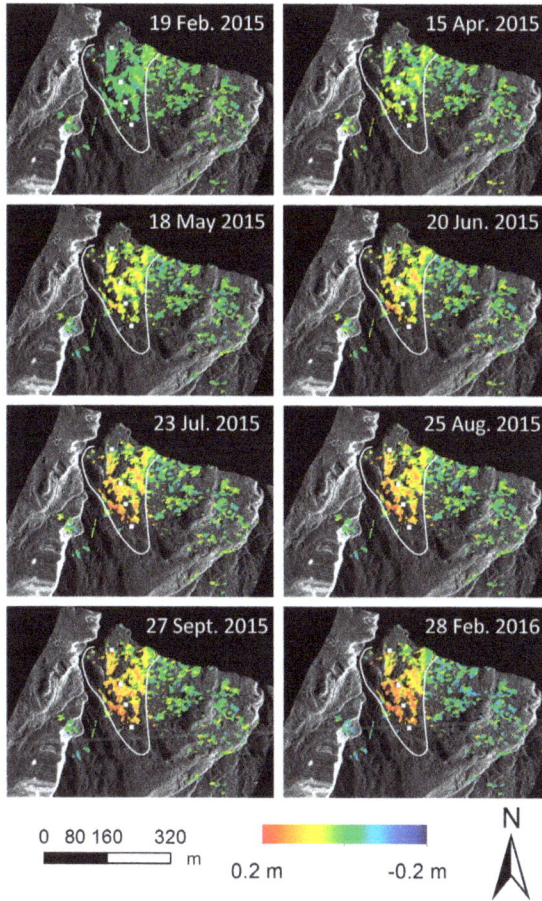

Figure 7. Accumulated range displacement of the Tanjiahe landslide on different acquisition dates derived by the SBAS offset tracking approach, superimposed on TSX-ST amplitude, with landslide boundary plotted in white line, and GPS stations marked in white squares. Data source: TerraSAR-X Staring Spotlight © DLR <2015>.

From both the azimuth and range offset maps shown in Figures 6 and 7, the whole landslide body shows consistent pattern of temporal evolution, i.e., the slope experienced biggest displacement rate in April–August and tend to be stable in the following months. The spatial distribution of the landslide is also clear. The upper part of the slope shows bigger deformation magnitude, whilst smaller displacement rate is observed in the lower part of the landslide.

Three pixels on the landslide body are selected for time series analysis, with their locations displayed in Figure 8 and corresponding offset measurements plotted in Figure 9.

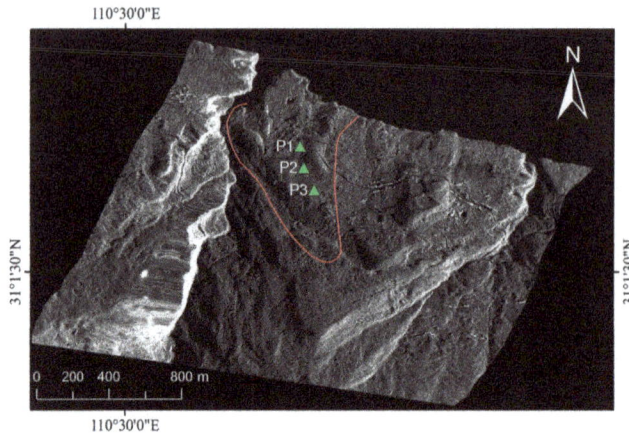

Figure 8. Location of the three pixels 'P1', 'P2', and 'P3' selected for time series analysis. The three pixels marked by green triangles and the landslide boundary in red are superimposed on the SAR amplitude image over the Tanjiahe landslide site. Data source: TerraSAR-X Staring Spotlight © DLR <2015>.

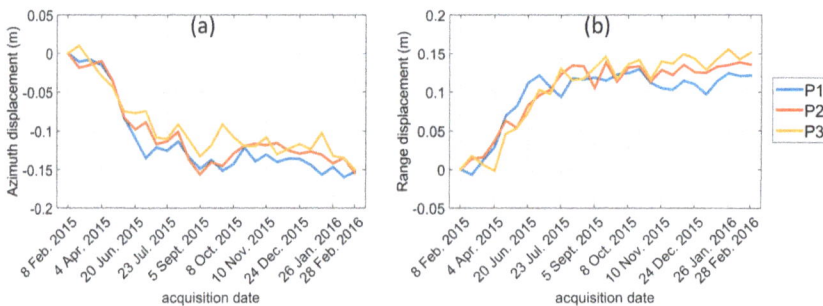

Figure 9. Time series offset measurements of selected pixels in the Tanjiahe landslide area. (**a**) Accumulated azimuth displacement on each acquisition date; (**b**) accumulated range displacement on each acquisition date.

In Figure 9, we can see that the four pixels show quite similar magnitudes of deformation, with maximum azimuth displacement around −0.15 m, and slant range displacement ranging from 0.1 m to 0.15 m. As all these data were acquired with right-looking SAR in a descending orbit, the negative magnitude of azimuth displacement corresponds to the reverse along-track direction (predominantly to the North) and the positive magnitude of range displacement represents the movement away from the sensor. In the time series analysis, the Tanjiahe landslide shows a seasonal pattern with a big increase in the displacement rate in April–August, which slows down in the remainder of the year.

3.4. Precision Assessment and Comparison with GPS Measurements

The precision of offset measurements is assessed as follows: as the accumulated 2D displacements on each acquisition date have been retrieved, for each valid pixel in the stable area, the standard deviation errors over time of accumulated azimuth and range offset measurements are calculated respectively. The overall error level is estimated by spatially analysed statistics in terms of 'mean ± STD', as shown in Table 3.

Table 3. Overall precision assessment based on the standard deviation errors of the azimuth and range offset measurements derived from the stable ground, calculated along the temporal baseline.

	Azimuth Offset (m)	Range Offset (m)
Standard deviation errors	0.025 ± 0.011	0.027 ± 0.009

Due to the absence of direct geodetic measurements of the landslide, we compare the deformation magnitudes of the selected pixels derived by the offset tracking approach with GPS measurements found in literature. Four GPS stations were installed along the longitudinal section of the Tanjiahe slope (ZG287, ZG288, ZG289, and ZG290) and surveyed monthly [29]. The paper presented a schematic view of the GPS stations, a table showing annual displacement magnitudes from 2007 to 2015, as well as a plot of GPS displacement curves. The displacement time series of the four GPS stations show a very consistent pattern of accumulated displacement. Smaller deformation magnitude is observed from the ZG290 station located at the lower part of the slope, whist higher displacement rate is found on the three GPS stations installed on the upper part of the slope (i.e., ZG287, ZG288, ZG289). This distribution is identical to the spatial distribution revealed by the offset maps in Figures 6 and 7.

We have no access to the coordinates and actual measurements of the GPS time series used in the abovementioned publication [29] (the plot in Zhang et al. [29] does not give the digit corresponding to each point of the GPS time series). Nevertheless, we manually aligned the schematic view of the GPS locations to the geocoded SAR amplitude over the Tanjiahe landslide area, using the river shoreline as the matching features. In this way, the sketch maps of GPS stations are coarsely co-registered with the offset tacking results. Then the pixels in the same area of GPS stations are extracted for a comparison. The GPS time series plot was taken from the publication [29] and digitized, in order to obtain the GPS measurements corresponding to the curves in the plot. The GPS measurements span the time period from December 2014 to December 2015, overlapping 10 months in time with the offset measurements (February 2015–February 2016).

Prior to the comparison, the GPS time series measurements are projected onto the azimuth and slant range directions, based on the knowledge of the main sliding direction of the landslide (340 degrees clock-wise from the North), and slope degrees of each position derived from the reference SRTM DEM. The projected GPS monthly measurements are interpolated to the acquisition dates of each image in the TSX data stack, in order to make the time series comparison on a one-to-one basis.

The time series of pixels located in the same area of the GPS stations are plotted against the annual displacements of individual GPS stations, as shown in Figure 10. As ZG290 and ZG287 are located out of the mask of valid pixels, the other two stations, ZG288 and ZG289, are used in the following analysis.

In Figure 10, we can see the time series offsets measured from pixels located on the positions of ZG288 and ZG289 closely follow the corresponding GPS measurements; the differences between the offsets and GPS time series are estimated by RMSE, and summarized in Table 4. Considering the precision of SBAS offset tracking (as estimated in Table 3) of 2.5 ± 1.1 cm in the azimuth direction and 2.7 ± 0.9 cm in the range direction, the RMSE between the offset time series and GPS data are not statistically significant.

Table 4. The root mean square error (RMSE) between the time series displacements measured by SBAS offset tracking and GPS stations.

	ZG288	ZG289
Azimuth RMSE (cm)	1.89	1.20
Range RMSE (cm)	3.38	1.80

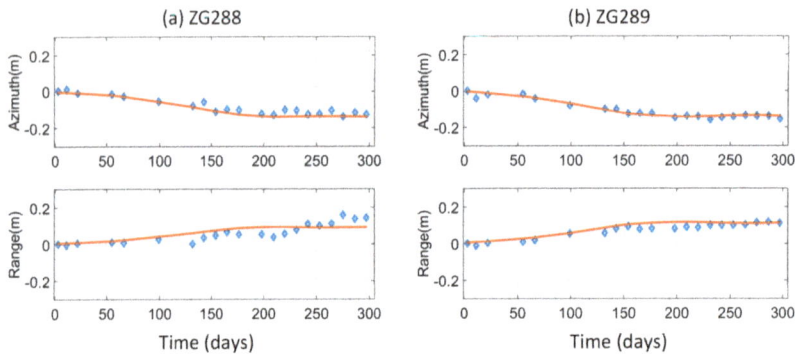

Figure 10. Time series azimuth and range displacement (marked in blue diamonds) measured from pixels located in the same area of the GPS stations, plotted versus the GPS time series over the time period from 19 February 2015 to 2 December 2015. (**a**) Comparison with the GPS time series measured from ZG288 station. (**b**) Comparison with the GPS time series measured from ZG289 station. The *x* axis represents the days counted from 19 February 2015 (red lines). As some TSX images supposed to be acquired in a 11-day repeat cycle are absent in the original data stack, the offset time series measurements in this figure are not evenly spaced in the *x* axis.

4. Discussion

4.1. The Relationship between the Landslide and Water Level Variations of the Three Gorges Reservoir

The construction of the Three Gorges Dam was started in 2003 and completed in 2009. The reservoir level rose from 66 m to 135 m in 2003, then to 156 m in 2006, and finally to 175 m above sea level after three impoundments. After reaching the designed maximum height of 175 m in 2010, the water level experiences 30 m of fluctuation between 145–175 m every year. A drawdown-filling cycle is repeated every year at almost exactly the same time.

As is well known [31,32], the ground water table or pore-water pressure within the soil layers of the landslide body are affected by the reservoir surface fluctuation and local precipitation, which decreases the effective normal stress leading to a decrease in shear strength of the soils. In previous studies, the fluctuation of the reservoir water level and seasonal rainfall are found as the two main triggering factors for landslides along the Yangtze River banks [31,33–36]. Two studies of the Tanjiahe landslide show controversial results on whether the local rainfall is the key driving factor of the slope movements in this area [29,37]. However, previous studies of the Ivancich landslide in central Italy point out there is a lack of correlation between the rainfall and the extremely slow landslide displacement (<16 mm·year^{-1}) [38–40]. Thus, in this section, the landslide displacement is respectively compared with reservoir water level measurements and daily rainfall data. The rainfall data was measured from a gauge station in Badong County, 16 km upstream from the Tanjiahe landslide site.

It should be noted that the time series measurements of azimuth and range offset are not evenly spaced in the time domain, as some acquisition dates are missing. Thus, prior to the analysis, a cubic spline interpolation is used to interpolate across the missing dates for the every-11-day measurements. The interpolated displacements of the three selected pixels 'P1', 'P2', and 'P3'are displayed in Figure 11 as follows.

As we can see from Figure 11, the measurements from all three pixels show a consistent pattern of displacements in both the azimuth and range directions. The azimuth offset measurements of 'P1' are then selected for subsequent analysis in Section 4.1.

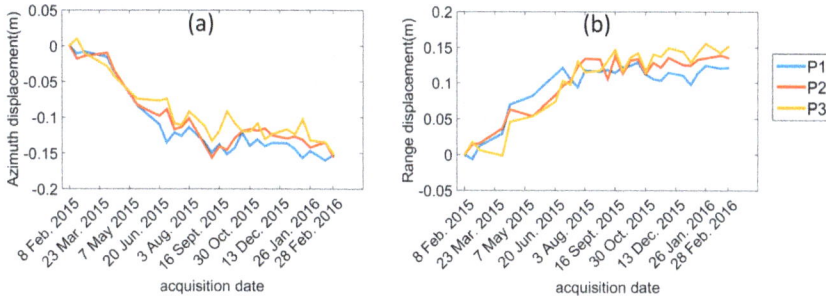

Figure 11. Interpolated time series measurements of (**a**) azimuth and (**b**) range displacements of the Tanjiahe landslide derived by SBAS offset tracking.

The every-11-day offset measurements with a corresponding trend line are plotted against the water level measurements (Figure 12) and local rainfall data (Figure 13).

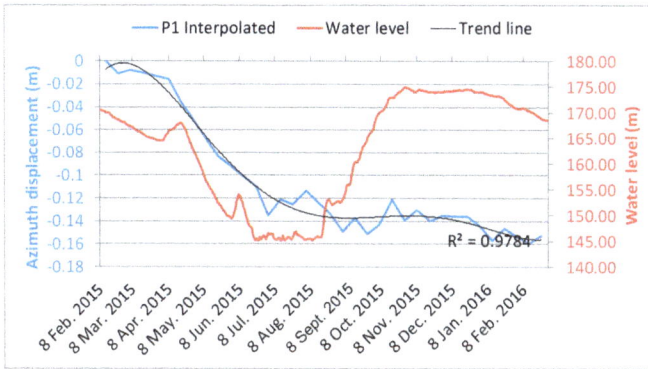

Figure 12. Interpolated time series azimuth displacement of P1 vs. water level measurements of the Three Gorges Reservoir in the period from 8 February 2015 to 28 February 2016.

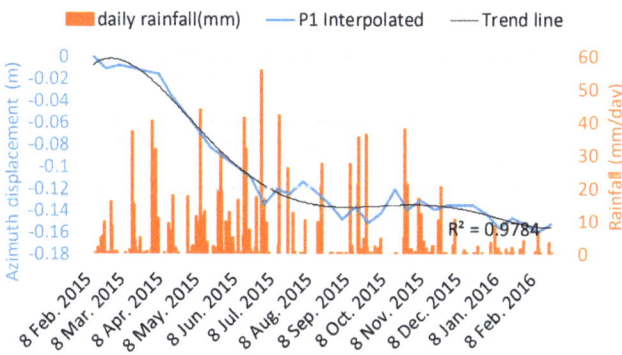

Figure 13. Interpolated time series azimuth displacement of P1 vs. local rainfall data in the period from 8 February 2015 to 28 February 2016.

In Figure 12, a significant and abrupt increase in deformation magnitude can be observed, synchronized with the sharp reservoir drawdown in April–July 2015, and non-significant displacement (with regard to the 2–3 cm precision of azimuth offset measurements) over the following months. It is evident that the active landslide period coincides with the fast drawdown of the reservoir water level. No noticeable correlation is found between the dramatic raising of water level in September–October and the landslide displacement.

In Figure 13, heavy and intense rainfall is observed from March to October, covering the period of the greatest displacement rate (April–July 2015). However, in the period of July–October 2015 with intensity of rainfall second only to the previous months, no noticeable deformation can be observed.

The above results suggest that rainfall does not play a key role in triggering the landslide in the observation year, but we cannot rule out the possibly that rainfall has combining effects with the water level variations due to the overlapped period of reservoir fast drawdown and heavy rainfall. Reservoir drawdown is the key driving factor of the landslide. It appears that the slope stability decreases with the fast drawdown of the reservoir water level and increases with the big rise of water level.

4.2. Potential and Limitations of the SBAS Offset Tracking Approach in Comparison with InSAR

Sub-pixel offset tracking techniques only utilize intensity bands of the satellite imagery to retrieve 2D ground deformation. It is less sensitive to low coherence and does not require phase-unwrapping, which leads to most of the failures in time series InSAR due to the low density of valid pixels. As a method free of phase-unwrapping, offset tracking has no limitation in the maximum detectable displacement gradient (DDG). Thus, offset tracking techniques potentially have the capability and advantage to measure slope movements with the speed exceeding the maximum detectable displacement of DInSAR or map deformation in challenging areas such as densely vegetated and steeply sloped terrain.

For a single pair of SAR images, the accuracy of offset tracking is jointly determined by the deformation rates in the area of interest, image resolution, and the correlation coefficient of scatterers in the target area [41]. The presence of high-contrast surface features does help to improve the accuracy. With the availability of high-resolution SAR imagery, offset tracking is able to monitor very slow-moving landslides (16 mm·year^{-1}–1.6 m·year^{-1} as defined in Cruden and Varnes [4], Hungr et al. [5]) and complement the applications of DInSAR. This has been demonstrated in our previous study of the Shuping landslide [3].

Given an area of interest and the same dataset, the accuracy of offset tracking is mainly determined by the correlation coefficient. Low correlation leads to a large uncertainty in cross-correlation and eventually low accuracy of measurements. For this reason, temporal and spatial baseline screening is necessary prior to analysis. If there is a significant dependence between the number of reliable measurements and the temporal–spatial baseline, it is beneficial to create a small baseline network of offset pairs, in order to increase the density and coverage of observation.

The proposed SBAS offset tracking approach is demonstrated of being capable of measuring centimetre-level landslide rates in densely vegetated terrain. Instead of only measuring the deformation of only a few sparsely distributed strong scatterers, the proposed approach provides a synoptic overview of the landslide by constructing a small baseline network and time series inversion of redundant connection of offset pairs. According to the results of our preliminary experiment in conventional offset tracking, to achieve the same precision, the number of reliable measurements derived by SBAS offset tracking is more than 15 times the conventional offset tracking method.

Offset tracking has the advantage of obtaining 2D measurements using data from a single orbit. It should be noted that the azimuth and range offsets only measure the projection of the real displacement on the slant range plane due to the radar geometry. The displacement component perpendicular to the slant range plane, if there was any, would not be detected. To measure 3D displacement, images from at least two different orbits are needed to solve the least squares functions.

Data from descending and ascending orbits are preferred to improve the robustness of the estimation, which should be considered when selecting new data in future studies.

East-West (E-W)-oriented landslides provide a better geometry for line-of-sight (LOS) measurement from sun-synchronous SAR imaging instruments, as the E-W and downward component of the sliding vector can both be captured by satellite LOS measurements. For North-South (N-S)-oriented landslides, as most of the landslides on the banks of Yangtze River appear to be, Offset Tracking is of great importance to provide measurements in the azimuth direction (approximately N-W) when repeat data are only available from a single orbit.

InSAR techniques have been widely used for displacement monitoring in many areas including the Three Gorges Region, with success demonstrated through measurements of much smaller displacement rates in urban areas using time series InSAR approaches [42,43]. This case study, shows results of so-called 'very slow-moving' landslide (16 mm·year^{-1} to 1.6 m·year^{-1} as defined in Cruden and Varnes [4]) with annual displacement rates up to 20 cm. As calculated in Table 2 using a multi-looking factor of 2, the upper limit of measurable displacement in one repeat cycle (11 days), is 0.59 cm, over a ground distance of 1 m. For the Tanjiahe landslide showing a dramatic increase in deformation over a short period, there is a high probability of underestimation by InSAR based techniques, especially on the landslide boundary.

In addition to the limitation of maximum detectable displacement gradient (DDG), the rapid loss of phase coherence is a major issue in densely vegetated areas. In our preliminary work, using the TerraSAR-X Staring Spotlight data with an 11-day repeat cycle and 0.23 m × 0.84 m resolution, the coherence loss is still a problem, resulting in too low a redundancy of data connections to apply time series InSAR. Satellite data with a shorter re-visit cycle and high resolution (1–3 m at least) is expected can help to address this issue, which should be exploited in future work. A shorter repeat interval is much-needed by DInSAR in this kind of study area. The improvement of the re-visit time will also increase the maximum measurable displacement of DInSAR.

Geometric distortion is also a key factor affecting the quality of interferograms, especially in an area characterised by many steep slopes. DEM products often exhibit higher height errors in areas with rugged topography due to geometric distortions. Low resolution, inaccuracies, or both in the reference DEM can lead to large residual errors of topographic phase, which further decrease the quality of interferograms.

Theoretically, time series InSAR has higher accuracy (millimetre-level) compared with Offset Tracking (centimetre-level). In other types of terrain (e.g., with more man-made structures, less vegetation, or lower slopes), the use of InSAR techniques may have more advantages for displacement monitoring, especially for slower displacement.

For slow-moving landslides, the use of offset tracking is recommended to assess if the assumption of maximum displacement gradient of InSAR can be fulfilled. Offset Tracking is less sensitive to low coherence and is able to derive 2D displacement using data from a single orbit, whilst time series InSAR can help to detect the smaller magnitude of deformation (e.g., during a less active period of the landslide). Thus, the two techniques are complementary to each other and there appears to be hope of more improvements with the availability of satellite data of shorter revisit cycle and higher resolution.

5. Conclusions

This work demonstrates the capability of the SBAS Offset Tracking approach to monitor centimetre-level landslide displacement in a challenging area characterised by dense vegetation cover and steep slopes. In the case study of the Tanjiahe landslide, as significance is found between the number of reliable measurements and the temporal baseline, a small baseline network of offset pairs is created to minimize temporal decorrelation, and increase the density and coverage of the offset measurements at the end. Considering the centimetre-level displacement rate, an extra step is taken to remove the residual offsets due to co-registration errors and orbit inaccuracies before the SVD inversion of all offset pairs. Taking advantage of the sub-metre resolution of the TSX Staring Spotlight data, the proposed SBAS offset tracking approach has been shown of being capable of measuring centimetre-level landslide rates with an average precision of 2–3 cm, with point density more than 15 times of the conventional offset tracking approach. The offset results have been validated of good agreement with published GPS measurements. This approach is of particular interests for deformation monitoring in many rural areas lack of high contrast surface features, especially over densely vegetated and steep terrain.

In the case study, the relationship between the landslide and local rainfall, as well as the water level changes of the Three Gorges Reservoir has been assessed. The reservoir fast drawdown is identified as a major triggering factor of the landslide, and rainfall does not appear to be a key triggering factor in the observation period.

Acknowledgments: This research has been supported by the National Key Research and Development Program of China (Project No. 2017YFB0504200), the UCL Dean's Prize, and China Scholarship Council Scholarship. This research is linked to the ESA-MOST DRAGON-3 Project #10665: Monitoring ground surface displacements in China from EO through case studies of landslides in the Three Gorges Area, crustal tectonic movement in Tibet, and subsidence in South China. The TerraSAR-X data employed in this study was provided by the German Aerospace Centre (DLR) under data grant GEO2630. We greatly thank J. Zhang, Q. Jiao, and T. Xue from the China Earthquake Administration for support on our fieldwork. The water level measurements used in this research were accessed from the Three Gorges Corporation website: http://www.ctg.com.cn/.

Author Contributions: Experiment design, data processing and analysing, and draft of the manuscript was done by Luyi Sun. Supervision and selection of areas as well as editing of the original manuscript prior to submission was done by Jan-Peter Muller. Supervision, advice on data processing and interpretation, as well as editing of the revised manuscript was done by Jinsong Chen.

Conflicts of Interest: The authors declare no conflict of interest. The founding sponsors had no role in the design of the study; in the collection, analyses, or interpretation of data; in the writing of the manuscript; and in the decision to publish the results.

Appendix A

In this section, we firstly provide the details on parameter selection for cross-correlation, followed by an example showing 2D image of the cross-correlation for a case with good correlation and another with bad results.

The general performance of sub-pixel cross-correlation is assessed through cumulative histograms of the azimuth and range deformation fields [44] derived from the stable area (Figure 2) surrounding the landslide body. Bigger discrepancies from the centre of the Cumulative Distribution Function (CDF) suggest higher error level in the stable area, indicating lower accuracy of cross-correlation. CDF of azimuth and range displacements are plotted for different correlation window sizes using the same oversampling factor of 16, as displayed in Figure A1. The time consumption of different parameter settings is summarised in Table A1.

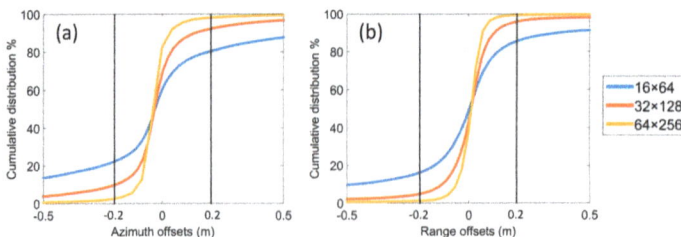

Figure A1. (**a**) Cumulative histograms of azimuth offsets derived from the stable area surrounding the landslide body; (**b**) Cumulative histograms of range offsets derived from the stable area surrounding the landslide body. This is plotted for different correlation window sizes of 16 × 64 pixels, 32 × 128 pixels, and 64 × 256 pixels.

Table A1. Processing time corresponding to different window sizes of cross-correlation, taking into account the time consumption of image co-registration.

Correlation Window Size (in Pixels)	Elapsed Time
16 × 64	1 h 25 min
32 × 128	5 h 40 min
64 × 256	25 h 41 min

From Figure A1 and Table A1, we can see that a larger window size improves the accuracy but dramatically increases the processing time. In experiments, we also found that larger window sizes increase artifacts and reduce the resolution of the output deformation fields. In the case study, the window size of 32 × 128 pixels was selected for cross-correlation, as a trade-off between the correlation accuracy, time consumption, and output resolution. Using the window size of 32 × 128 pixels, over 80% of pixels in the azimuth CDF are characterised by offsets around zero and within ±0.2 m. About 90% of pixels in the range CDF are centred on zero and within ±0.2 m of offsets. For the case study, we found that this performance is good enough as the correlation output of individual offset pairs.

In Figure A2, we present a comparison of a case with good correlation from high-contrast features and another of bad correlation from vegetated surface.

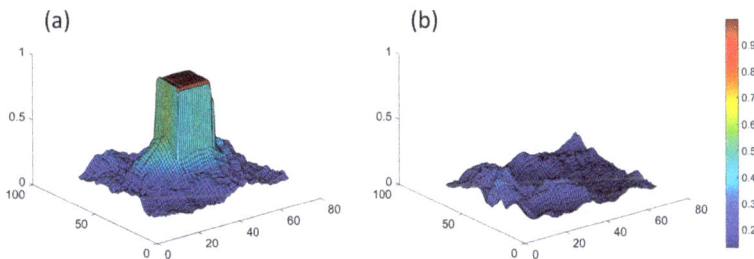

Figure A2. (**a**) Correlation peaks of high-contrast surface features; (**b**) correlation peaks of pixels on the vegetated surface. The correlation peaks in (**a,b**) are both plotted from the 8 Feburary 2015 and 28 Feburary 2016 image pair by extracting a window of 64 × 64 pixels centred by the targeted pixel. The colour bar represents the correlation coefficient ranging from 0 to 1. In (**a**) there is a square bar in the middle instead of a single peak. This is because the cross-correlation used a step size of 2 pixels in range direction and 8 pixels in azimuth directions with regard to 32 × 128 pixels of correlation window size.

References

1. Li, X.; Muller, J.-P.; Fang, C.; Zhao, Y. Measuring displacement field from terrasar-x amplitude images by sub-pixel correlation: An application to the landslide in shuping, three gorges area. *Acta Petrol. Sin.* **2011**, *27*, 3843–3850.
2. Singleton, A.; Li, Z.; Hoey, T.; Muller, J.P. Evaluating sub-pixel offset techniques as an alternative to d-insar for monitoring episodic landslide movements in vegetated terrain. *Remote Sens. Environ.* **2014**, *147*, 133–144. [CrossRef]
3. Sun, L.; Muller, J.-P. Evaluation of the use of sub-pixel offset tracking techniques to monitor landslides in densely vegetated steeply sloped areas. *Remote Sens.* **2016**, *8*, 659. [CrossRef]
4. Cruden, D.M.; Varnes, D.J. *Landslide Types and Processes*; Special Report; National Research Council, Transportation Research Board: Washington, DC, USA, 1996.
5. Hungr, O.; Leroueil, S.; Picarelli, L. The varnes classification of landslide types, an update. *Landslides* **2014**, *11*, 167–194. [CrossRef]
6. Chen, C.W.; Zebker, H.A. Network approaches to two-dimensional phase unwrapping: Intractability and two new algorithms. *J. Opt. Soc. Am. A* **2000**, *17*, 401–414. [CrossRef]

7. Michel, R.; Avouac, J.P.; Taboury, J. Measuring ground displacements from sar amplitude images: Application to the landers earthquake. *Geophys. Res. Lett.* **1999**, *26*, 875–878. [CrossRef]
8. Jiang, M.; Li, Z.; Ding, X.; Zhu, J.-J.; Feng, G. Modeling minimum and maximum detectable deformation gradients of interferometric sar measurements. *Int. J. Appl. Earth Obs. Geoinf.* **2011**, *13*, 766–777. [CrossRef]
9. Strozzi, T.; Luckman, A.; Murray, T.; Wegmuller, U.; Werner, C.L. Glacier motion estimation using sar offset-tracking procedures. *IEEE Trans. Geosci. Remote Sens.* **2002**, *40*, 2384–2391. [CrossRef]
10. Kääb, A. Monitoring high-mountain terrain deformation from repeated air- and spaceborne optical data: Examples using digital aerial imagery and aster data. *ISPRS J. Photogramm. Remote Sens.* **2002**, *57*, 39–52. [CrossRef]
11. Yamaguchi, Y.; Tanaka, S.; Odajima, T.; Kamai, T.; Tsuchida, S. Detection of a landslide movement as geometric misregistration in image matching of spot hrv data of two different dates. *Int. J. Remote Sens.* **2003**, *24*, 3523–3534. [CrossRef]
12. Delacourt, C.; Allemand, P.; Casson, B.; Vadon, H. Velocity field of the "la clapière" landslide measured by the correlation of aerial and quickbird satellite images. *Geophys. Res. Lett.* **2004**, *31*. [CrossRef]
13. Wangensteen, B.; Guðmundsson, Á.; Eiken, T.; Kääb, A.; Farbrot, H.; Etzelmüller, B. Surface displacements and surface age estimates for creeping slope landforms in northern and eastern iceland using digital photogrammetry. *Geomorphology* **2006**, *80*, 59–79. [CrossRef]
14. Debella-Gilo, M.; Kääb, A. Sub-pixel precision image matching for measuring surface displacements on mass movements using normalized cross-correlation. *Remote Sens. Environ.* **2011**, *115*, 130–142. [CrossRef]
15. Lacroix, P.; Berthier, E.; Maquerhua, E.T. Earthquake-driven acceleration of slow-moving landslides in the colca valley, Peru, detected from pléiades images. *Remote Sens. Environ.* **2015**, *165*, 148–158. [CrossRef]
16. Stumpf, A.; Malet, J.P.; Delacourt, C. Correlation of satellite image time-series for the detection and monitoring of slow-moving landslides. *Remote Sens. Environ.* **2017**, *189*, 40–55. [CrossRef]
17. Yonezawa, C.; Takeuchi, S. Decorrelation of sar data by urban damages caused by the 1995 hyogoken-nanbu earthquake. *Int. J. Remote Sens.* **2001**, *22*, 1585–1600. [CrossRef]
18. De Zan, F. Accuracy of incoherent speckle tracking for circular gaussian signals. *IEEE Geosci. Remote Sens. Lett.* **2014**, *11*, 264–267. [CrossRef]
19. Casu, F.; Manconi, A.; Pepe, A.; Lanari, R. Deformation time-series generation in areas characterized by large displacement dynamics: The sar amplitude pixel-offset sbas technique. *IEEE Trans. Geosci. Remote Sens.* **2011**, *49*, 2752–2763. [CrossRef]
20. Manconi, A.; Casu, F.; Ardizzone, F.; Bonano, M.; Cardinali, M.; De Luca, C.; Gueguen, E.; Marchesini, I.; Parise, M.; Vennari, C.; et al. Brief communication: Rapid mapping of landslide events: The 3 december 2013 montescaglioso landslide, Italy. *Nat. Hazards Earth Syst. Sci.* **2014**, *14*, 1835–1841. [CrossRef]
21. Pathier, E.; Fielding, E.J.; Wright, T.J.; Walker, R.; Parsons, B.E.; Hensley, S. Displacement field and slip distribution of the 2005 kashmir earthquake from sar imagery. *Geophys. Res. Lett.* **2006**, *33*. [CrossRef]
22. Casu, F.; Manconi, A. Four-dimensional surface evolution of active rifting from spaceborne sar data. *Geosphere* **2016**, *12*, 697–705. [CrossRef]
23. Raucoules, D.; de Michele, M.; Malet, J.P.; Ulrich, P. Time-variable 3d ground displacements from high-resolution synthetic aperture radar (sar). Application to la valette landslide (South French Alps). *Remote Sens. Environ.* **2013**, *139*, 198–204. [CrossRef]
24. Wu, J.; Huang, J.; Han, X.; Xie, Z.; Gao, X. Three-gorges dam—Experiment in habitat fragmentation? *Science* **2003**, *300*, 1239–1240. [CrossRef] [PubMed]
25. Wang, H.; Harvey, A.M.; Xie, S.; Kuang, M.; Chen, Z. Tributary-junction fans of China's Yangtze Three-Gorges Valley: Morphological implications. *Geomorphology* **2008**, *100*, 131–139. [CrossRef]
26. Qin, H. The Mechanisam and Reactivation Criteria of Landslides Induced by Water Level Fluctuation and Rainfall in the Three Gorges Region. Ph.D. Thesis, Three Gorges University, Yichang, China, 2011.
27. Fan, J.; Xia, Y.; Zhao, H.; Li, M.; Wang, Y.; Guo, X.; Tu, P.; Liu, G.; Lin, H. Monitoring of Landslide deformation based on the coherent targets of high resolution insar data. In Proceedings of the Remote Sensing of the Environment: 18th National Symposium on Remote Sensing of China, Wuhan, China, 20–23 October 2012.
28. Ministry of Environmental Protection of the People's Republic of China. Three Gorges Bulletin in 2007. Available online: http://english.mep.gov.cn/standards_reports/threegorgesbulletin/Bulletin_2007/ (accessed on 19 October 2017).

29. Zhang, G.; Tan, T.; Xu, Z.; Qiu, C.; Li, X.; Lu, S. Anlaysis of the deformation monitoring results of the tanjiahe landslide in the three gorges region. *J. Nat. Disasters* **2017**, 185–192. [CrossRef]

30. Berardino, P.; Fornaro, G.; Lanari, R.; Sansosti, E. A new algorithm for surface deformation monitoring based on small baseline differential sar interferograms. *IEEE Trans. Geosci. Remote Sens.* **2002**, *40*, 2375–2383. [CrossRef]

31. Luo, X.; Wang, F.; Zhang, Z.; Che, A. Establishing a monitoring network for an impoundment-induced landslide in three gorges reservoir area, China. *Landslides* **2009**, *6*, 27–37. [CrossRef]

32. Miao, H.; Wang, G.; Yin, K.; Kamai, T.; Li, Y. Mechanism of the slow-moving landslides in jurassic red-strata in the three gorges reservoir, china. *Eng. Geol.* **2014**, *171*, 59–69. [CrossRef]

33. Wang, F.; Zhang, Y.; Huo, Z.; Peng, X.; Araiba, K.; Wang, G. Movement of the shuping landslide in the first four years after the initial impoundment of the three gorges dam reservoir, China. *Landslides* **2008**, *5*, 321–329. [CrossRef]

34. Wang, F.; Yin, Y.; Huo, Z.; Zhang, Y.; Wang, G.; Ding, R. Slope deformation caused by water-level variation in the three gorges reservoir, china. In *Landslides: Global Risk Preparedness*; Sassa, K., Rouhban, B., Briceño, S., McSaveney, M., He, B., Eds.; Springer: Berlin/Heidelberg, Germany, 2013; pp. 227–237.

35. Liu, J.G.; Mason, P.J.; Clerici, N.; Chen, S.; Davis, A.; Miao, F.; Deng, H.; Liang, L. Landslide hazard assessment in the three gorges area of the yangtze river using aster imagery: Zigui–badong. *Geomorphology* **2004**, *61*, 171–187. [CrossRef]

36. He, K.; Li, X.; Yan, X.; Guo, D. The landslides in the three gorges reservoir region, China and the effects of water storage and rain on their stability. *Environ. Geol.* **2008**, *55*, 55–63.

37. Wang, S.; Liu, J.; Wang, L.; Yang, Q. Analysis on influence of tgp reservoir water level fluctuation on stability of tanjiahe landslide. *Yangtze River* **2015**, *46*, 83–86.

38. Ardizzone, F.; Rossi, M.; Calo, F.; Paglia, L.; Manunta, M.; Mondini, A.C.; Zeni, G.; Reichenbach, P.; Lanari, R.; Guzzetti, F. Preliminary analysis of a correlation between ground deformations and rainfall: The ivancich landslide, central Italy. *Proc. SPIE* **2011**. [CrossRef]

39. Calò, F.; Ardizzone, F.; Castaldo, R.; Lollino, P.; Tizzani, P.; Guzzetti, F.; Lanari, R.; Angeli, M.G.; Pontoni, F.; Manunta, M. Enhanced landslide investigations through advanced dinsar techniques: The ivancich case study, assisi, italy. *Remote Sens. Environ.* **2014**, *142*, 69–82. [CrossRef]

40. De Novellis, V.; Castaldo, R.; Lollino, P.; Manunta, M.; Tizzani, P. Advanced three-dimensional finite element modeling of a slow landslide through the exploitation of dinsar measurements and in situ surveys. *Remote Sens.* **2016**, *8*, 670. [CrossRef]

41. De Zan, F. Coherent shift estimation for stacks of sar images. *Geosci. IEEE Remote Sens. Lett.* **2011**, *8*, 1095–1099. [CrossRef]

42. Perissin, D.; Wang, T. Repeat-pass sar interferometry with partially coherent targets. *IEEE Trans. Geosci. Remote Sens.* **2012**, *50*, 271–280. [CrossRef]

43. Liu, P.; Li, Z.; Hoey, T.; Kincal, C.; Zhang, J.; Zeng, Q.; Muller, J.-P. Using advanced insar time series techniques to monitor landslide movements in badong of the three gorges region, china. *Int. J. Appl. Earth Obs. Geoinf.* **2013**, *21*, 253–264. [CrossRef]

44. Yun, S.H.; Zebker, H.; Segall, P.; Hooper, A.; Poland, M. Interferogram formation in the presence of complex and large deformation. *Geophys. Res. Lett.* **2007**, *34*. [CrossRef]

remote sensing

MDPI

Letter

Landslide Displacement Monitoring with Split-Bandwidth Interferometry: A Case Study of the Shuping Landslide in the Three Gorges Area

Xuguo Shi [1], Houjun Jiang [2,*], Lu Zhang [3,4,*] and Mingsheng Liao [3,4]

[1] Faculty of Information Engineering, China University of Geosciences, 388 Lumo Road,
    Wuhan 430074, China; shixg@cug.edu.cn
[2] Department of Surveying and Geoinformatics, Nanjing University of Posts and Telecommunications,
    9 Wenyuan Road, Nanjing 210023, China
[3] State Key Laboratory of Information Engineering in Surveying, Mapping and Remote Sensing,
    Wuhan University, 129 Luoyu Road, Wuhan 430079, China; liao@whu.edu.cn
[4] Collaborative Innovation Center for Geospatial Technology, 129 Luoyu Road, Wuhan 430079, China
*   Correspondence: jianghouj@njupt.edu.cn (H.J.); luzhang@whu.edu.cn (L.Z.)

Academic Editors: Salvatore Stramondo and Prasad S. Thenkabail
Received: 3 August 2017; Accepted: 8 September 2017; Published: 10 September 2017

Abstract: Landslides constitute a major threat to people's lives and property in mountainous regions such, as in the Three Gorges area in China. Synthetic Aperture Radar Interferometry (InSAR) with its wide coverage and unprecedented displacement measuring capabilities has been widely used in landslide monitoring. However, it is difficult to apply traditional InSAR techniques to investigate landslides having large deformation gradients or moving primarily in north-south direction. In this study, we propose a time series split-bandwidth interferometry (SBI) procedure to measure two dimensional (azimuth and range) displacements of the Shuping landslide in the Three Gorges area with 36 TerraSAR-X high resolution spotlight (HS) images acquired from February 2009 to April 2010. Since the phase based SBI procedure is sensitive to noise, we focused on extracting displacements of corner reflectors (CRs) installed on or surrounding the Shuping landslide. Our results agreed well with measurements obtained by the point-like targets offset tracking (PTOT) technique and in-situ GPS stations. Centimeter level accuracy could be achieved with SBI on CRs which shows great potential in futures studies on fast moving geohazards.

Keywords: landslide; the Three Gorges; corner reflector; split-bandwidth interferometry

1. Introduction

A landslide refers to the movement of rock, earth, and debris downward upon a slope which may be caused by natural factors or human activities [1]. As a prevalent geohazard in mountainous areas, it poses a serious danger to local residents. Thus, continuous monitoring of landslide activity in these areas is essential to ensure public safety.

According to investigations, the Three Gorges area is frequently affected by landslide hazards, and there have been more than 3800 reported landslides [2]. Serving the largest hydro-power station in China, the Three Gorges Dam began construction in 1994 and it was completed in 2006. Presently, the water level varies between 145 m and 175 m according to the dam's operation scheme. The annual water level fluctuation contributed much to the destabilization of bank slopes along the Yangtze River which can inevitably aggravate geohazards such as landslides and rock falls.

SAR interferometry (InSAR) can be used as an earth displacement monitoring tool, having the advantages of wide area coverage and high accuracy; it has been widely used in previous landslide monitoring [3–7] studies. The dense vegetation in the Three Gorges area causes serious decorrelation which is unfavorable for InSAR analysis [8]. To overcome this problem, several corner reflectors (CRs) were installed on a few well-known landslides to improve the performance of differential InSAR [8,9]. Incidentally, it should be noted that the applicability of differential InSAR are seriously limited for landslides without CRs installed. Thus, advanced InSAR methods such as persistent scatters SAR interferometry (PSI) [1,10–12], Quasi Permanent Scatterers (QPS) InSAR [13] and Small baselines subset (SBAS) InSAR [3,14] which mainly make use of point-like scatterers have also been applied to identify and monitor active landslides in this area. Generally, if displacement between neighboring pixels exceed 1/4 of the wavelength, an unavoidable phase unwrapping error will cause underestimation [6], especially in densely vegetated areas, and this error should be mitigated.

In the Three Gorges area, there are many active landslides with annual displacement rates of meter-level [15]. Thus, the effectiveness of the InSAR method when applied to monitoring these landslides with large deformation gradients will be significantly reduced. In such cases, one solution is to employ the pixel offset tracking method to track the displacements that have occurred in the azimuth and range directions using amplitude information of the SAR images. This technique has been successfully applied to monitor the Shuping landslide [16–19] and the Kaziwan landslide [19] in the Three Gorges area. Meanwhile, the split-bandwidth interferometry (SBI), which is also known as multi-aperture interferometry (MAI), can also measure large azimuth displacements by analyzing forward and backward single look complex SAR images with centimeter-level accuracy [20–22]. Usually, this method is applied to obtain the displacements in the azimuthal direction. However, SBI can also be used in the range direction although this is comparatively rare because of the low sensitivity to displacements compared with standard InSAR methods, as well as a high sensitivity to pixel's signal to cluster ratio (SCR) [23]. Nevertheless, phase unwrapping errors will occur with standard InSAR methods, especially when measuring rapid movements on sparse points. In contrast, the characteristic of having low sensitivity to displacements for SBI becomes an advantage when measuring landslide deformations of large gradient [23]. As a result, unwrapping errors in the traditional InSAR method might be greatly suppressed.

In this study, using high resolution TerraSAR-X SAR images, split-bandwidth interferometry in the azimuth and range directions were applied to retrieve two-dimensional time series displacements of CRs installed on the Shuping landslide, located in the Three Gorges area. The effectiveness of our method was validated against results obtained from point-like targets offset tracking in a previous study [16] and Global Positioning System (GPS) measurements. This study is organized as follows: Section 2 describes our study area and datasets, and the principles of our method are elaborated in Section 3. Our results were given and evaluated in Section 4, followed by study conclusions.

2. Study Areas and Datasets

2.1. The Shuping Landslide

The first impoundment of the Three Gorges dam in 2003 activated many ancient landslides such as the Qianjiangping landslide [24], the Shuping landslide [25] etc. The Shuping landslide is located at the south bank of Yangtze River as shown in Figure 1a. The possible catastrophic failure of the Shuping landslide may pose great threats to the safety of local marine transportation and infrastructure. Thus, the movement status of the slope has drawn significant concern. GPS stations were promptly installed on the Shuping landslide for ongoing monitoring of the slope evolution [25]. InSAR methods have also been employed to monitor the displacement of the Shuping landslide [26]. As we can see from Figure 1b, the Shuping landslide is covered by dense vegetation. This results in a significant loss of coherence hindering the application of InSAR. Thus, triangular trihedral corner reflectors (CRs) with edges of 1 meter were installed on the Shuping landslide to assist InSAR-based landslide displacement monitoring [8,9]. Readers may refer to [8] for detailed design specifications of the CRs in the Three Gorges area. Generally, the radar cross section of the CR can reach more than 20 dB, which can be easily identified in vegetated areas [8]. The locations of CRs installed on or surrounding the Shuping landslide are marked with red dots in Figure 1b.

Previous studies with standard InSAR failed because of the rapid displacement of the Shuping landslide [3]. As an alternative method, SAR pixel offset tracking analysis has been successfully employed to obtain the displacements of the Shuping landslide using high resolution TerraSAR-X datasets [16,17]. Here, we explored the feasibility of phase based SBI for monitoring rapid movements in azimuth and range directions.

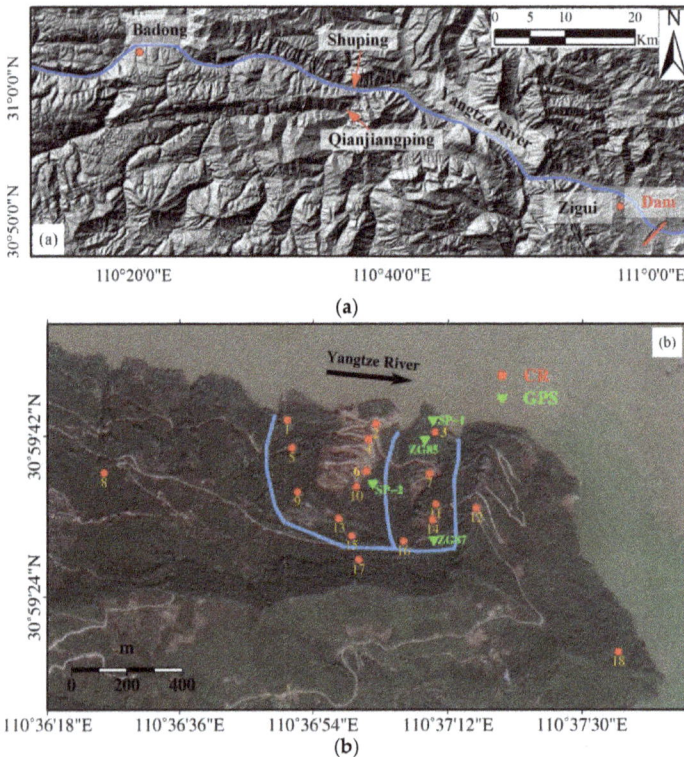

(a)

(b)

Figure 1. (a) Location of the Shuping landslide; (b) Location of corner reflectors and GPS stations installed on the Shuping landslide overlaid on ZY-3 satellite multispectral optical image.

2.2. Datasets

There were 36 TerraSAR-X High resolution Spotlight (HS) images from February 2009 to April 2010 that were acquired to monitor the movement of shuping landslide. Basic parameters of the dataset are summarized in Table 1. According to our previous study, very minor displacements of Shuping landslide occurred during winter when the water level of the Three Gorges Reservoir remained relatively stable and precipitation was low [16]. Thus, the image acquired in 12 November 2009 during this period with maximum correlation with other SAR images was selected as a common master for time series analysis. The temporal and perpendicular baseline distributions are given in Figure 2. A SRTM DEM with approximately 30 m resolution was used for DEM assisted registration and geocoding. Horizontal displacements measured by 4 GPS stations, namely SP-1, SP-2, ZG85 and ZG87, are collected for validation. The locations of these GPS stations are marked by green triangles in Figure 1b.

Table 1. Basic information for high resolution spotlight TerraSAR-X datasets.

Parameters	Values
Orbit direction	Descending
Look angle (°)	39
Heading (°)	189.6
Polarization	HH
Azimuth spacing (m)	0.87
Range spacing (m)	0.45
Range Bandwidth (MHz)	300
Azimuth Bandwidth (Hz)	7277

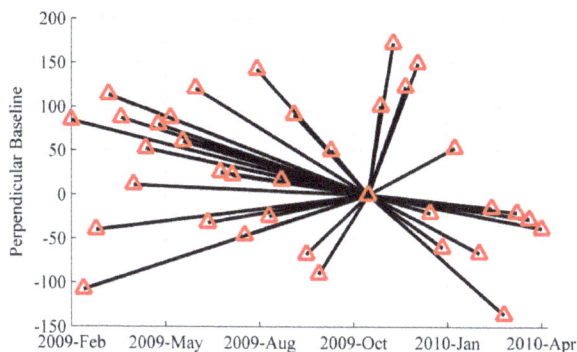

Figure 2. Temporal and perpendicular baseline distributions.

3. Time Series Displacement Retrieval Methods

3.1. Split-Bandwidth Interferometry

Split-bandwidth interferometry estimates pixel shifts between two single-look complex (SLC) SAR images by measuring phase differences of two interferograms formed by using lower and upper portions of imaging spectrum separately. Since the impulse response of a SAR system has a linear phase variation within its main lobe depending on the signal's center frequency f_c, a pixel offset Δt in seconds will lead to an additional phase term $2\pi f_c \Delta t$ in standard InSAR, mixed with interferometric phases containing topography, atmospheric delay, and displacement [27]. The pixel shift can be extracted by performing a phase differential operation between two interferograms from the same image pair with different azimuth or range center frequencies but identical interferometric phases.

In this study, we implement SBI as follows: a pair of TerraSAR-X HS images m and s are split into four low-resolution sublook images m_l, m_u, s_l, and s_u by filtering out the lower and upper bands in the frequency domain. Then, by combining the sublook images with common spectra, two interferograms are formed and their phase difference ϕ_{split} is derived from $(m_u \cdot s_u^*)(m_l \cdot s_l^*)^*$, where * indicates conjugate multiplication. Finally, the pixel shift Δt in units of time can be retrieved from the phase difference ϕ_{split}, given the spectral separation Δf_c between the two sublook images [27]:

$$\Delta t = \frac{\phi_{split}}{2\pi \Delta f_c} \tag{1}$$

In the case of range SBI, the phase difference ϕ_{split} contains a component corresponding to the range offsets produced by topography and the InSAR baseline. This component should be removed by using the DEM-assisted co-registration [28,29]. DEM and orbit ephemerides are used in conjunction to calculate the topographic offsets. In the azimuth case, little topographic information remains in ϕ_{split} because the azimuth offsets are insensitive to the cross-track baseline.

For the TerraSAR-X HS image, a larger spectral separation of sublooks can be achieved compared to the image acquired in Stripmap Mode (SM), dramatically improving the sensitivity and accuracy of SBI. According to Equation (1), the sensitivities to ground displacements for one cycle of ϕ_{split} are ~0.75 m and ~1.46 m for range and azimuth split-bandwidth interferometry respectively. Because of the finer bandwidth increment, the sensitivity of using TerraSAR-X HS image is approximately 3 times that of the SM image.

SBI is equivalent to coherent cross correlation (CCC), the maximum likelihood estimator for the pixel offset between distributed Gaussian targets. Since the requirement of removing interferometric phase complicates the implementation of CCC, SBI is often used as a replacement in practice. In theory, the standard deviation in displacement estimation from the SBI is given by [30]:

$$\sigma = \frac{1}{2\pi\Delta f_c} \frac{1}{\sqrt{N}} \sqrt{\frac{B}{b}} \frac{\sqrt{1-\gamma^2}}{\gamma} \frac{p_{spa}}{\Delta t_s} \tag{2}$$

where $\Delta f_c = B - b$, B is the processed bandwidth of a single target, b is the sublook bandwidth which is often selected as one third of the bandwidth [30,31], N is the number of independent samples averaged, γ is the interferometric coherence, Δt_s is the image sampling in seconds, and p_{spa} is the pixel spacing. In order to improve the reliability of SBI measurements, pixels exhibiting the same point-like behaviors as the center pixel were selected first using a two-sample KS test within an estimation window of 64 pixels in azimuth and 32 pixels in range. Then, adaptive multi-looking was carried out on these pixels. In our case, it is easy to find more than 10 point-like behaved pixels since the side lobe is very obvious for such high resolution TerraSAR-X HS images. For a TerraSAR-X HS image, B is 300 MHz in range and 7277 Hz in azimuth. Assuming $\gamma = 0.95$ on CRs, then $N = 10$ and $b = B/3$, σ is ~2.1 cm and ~3.8 cm in range and azimuth directions respectively.

The workflow of our process is shown in Figure 3. Split-bandwidth interferograms were first generated with the method described above. After pairwise azimuth and range SBI were performed, the corresponding phases of the CRs can then be extracted. As mentioned before, the phase components from topography and baseline have been removed. Thus, three dimensional phase unwrapping [32] can be carried out directly to retrieve the time series displacements.

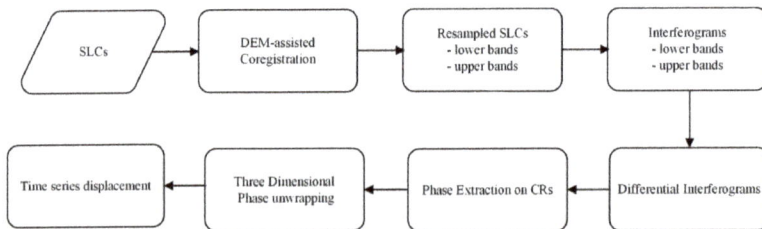

Figure 3. Diagram of Split-Bandwidth Interferometry process.

3.2. Point-Like Targets Offset Tracking

Rapidly moving ground targets might lead to loss of coherence or phase unwrapping problems which greatly affect the applicability of standard InSAR. In such situations, a pixel offset tracking method can also be used to extract azimuth and range displacement from the amplitude information using high resolution SAR images. In our study, point-like targets offset tracking (PTOT) making use of pixels with high amplitude values are employed to track the movements of the CRs. PTOT mainly makes use of pixels with high amplitude values. Usually, high SCRs will be maintained on these pixels to ensure greater accuracy of our measurements. Readers can find a detailed workflow of PTOT in our previous study [16]. Here, the PTOT results are used for cross validation with SBI. The theoretical accuracy of amplitude pixel offset tracking on point-like targets can be expressed as [30,33]:

$$\sigma = \frac{\sqrt{3}}{\pi} \frac{1}{\sqrt{SCR}} p_{spa} \tag{3}$$

where SCR is the signal to cluster ratio of output correlation for point-like targets and p_{spa} is the pixel spacing. In our case, the typical value of SCR for CRs are more than 150 (corresponding to 21 dB). Thus, the achievable accuracy can be better than 0.045 pixels corresponding to ~2 cm and ~4 cm in the range and azimuth directions separately for high resolution TerraSAR-X spotlight datasets, which is very close to the theoretical accuracy of SBI, as we discussed above.

4. Results and Discussions

4.1. Comparsion of Time Series Displacements from SBI and PTOT

As shown in Figure 1b, there were four CRs installed surrounding Shuping landslide. In order to mitigate systematic biases, CR12 was selected as a reference to calibrate the measurements of other CRs. Time series displacements measured by the SBI and PTOT methods in both azimuth and range directions for the other three CRs, namely CR8, CR17 and CR18, are given in Figure 4. Corresponding statistics of these three CRs are given in Table 2. Overall, good agreement was achieved for the measurements from both methods. As expected, all three of these CRs were stable during the period of more than one year.

Since CRs are ideal point-like targets, very high SCRs can be maintained over a long period. As mentioned in Section 3, the theoretical achievable accuracy from the PTOT method is almost the same as that from the SBI method in both azimuth and range directions on CRs. As expected, the statistics in Table 2 suggest that comparable accuracy was achieved by the SBI and PTOT in both azimuth and range direction. Statistics on these data indicate that the achievable accuracy can reach centimeter-level on point-like targets. At the same time, the mean and standard deviation of measurements in the range direction are lower than in the azimuth direction for both methods. This can be explained by the fact that the pixel spacing in the range direction are much higher than that in the azimuth direction. The standard deviations from SBI measurements are relatively higher than that from PTOT measurements, which means the consistency of PTOT measurements are relatively better than that of SBI measurements. This could be attributed to the higher sensitivity of noise for SBI. In our case, noise induced by vegetation or geometrical distortions in the mountainous setting can be very serious.

Furthermore, we notice that the differences between SBI and PTOT measurements on a few CRs, e.g., CR18 in Figure 4 and CR7, CR11 in Figures 5 and 6, are higher than other CRs. Signals from the target's surroundings such as vegetation in the resolution cell, e.g., CR18 lying in layover areas, might affect the accuracy of SBI measurements. Multi-looking of the SBI process also plays an important role in the measurement accuracy. Since CRs are ideal point-like targets, it is difficult to find enough point-like targets for adaptive multi-looking. Time series displacements of 14 CRs installed on Shuping landslide measured with SBI and PTOT in azimuth and range directions were shown in Figures 5 and 6, respectively. As expected, very good agreements were achieved at all CRs in both azimuth and range directions with only minor disparities between SBI measurements and PTOT measurements identified. Due to the higher sensitivity in the range direction compared with the azimuth direction for both methods, a higher consistency of time series measurements in the range direction was achieved. Similar displacement patterns with different magnitudes were identified on CRs which is mainly caused by the decline of water level in the Three Gorges area [16]. The most significant displacements in azimuth direction and range direction reached nearly 0.9 m on CR11 and nearly 0.7 m on CR14, respectively, during a period of over one year. Both CRs are located on the top of the Shuping landslide. The large displacements are mainly concentrated in the eastern and central part of Shuping landslide. Very small displacements detected on the western part of Shuping landslide where CR1, CR5 and CR9 are located indicate that this part might be stable.

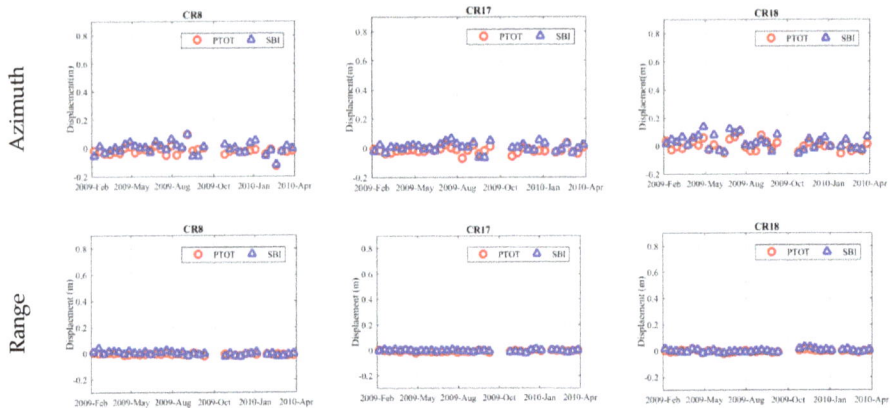

Figure 4. Comparison of displacement measurements at stable Corner reflectors (CRs) by Split-bandwidth interferometry (SBI) and point-like targets offset tracking (PTOT) in azimuth and range directions.

Table 2. Basic statistics of stable CRs outside Shuping landslide (unit: cm).

CR	SBI				PTOT			
	Azimuth		Range		Azimuth		Range	
	Mean	STD	Mean	STD	Mean	STD	Mean	STD
CR8	−0.4	4.1	−0.1	1.4	−1.9	3.3	−0.8	0.7
CR17	0.5	2.3	0.6	0.8	−1.5	2.1	−0.4	0.6
CR18	2.5	5.0	0.0	1.2	0.2	3.8	−0.8	0.9

4.2. Valadation with GPS Measurments

To further validate our results, comparisons between GPS observations and SAR measurements from SBI and PTOT were also carried out. As mentioned in Section 2, only the easting and northing components of GPS measurements are available. Thus, GPS measurements were converted into SAR azimuth geometry which is not considered sensitive to displacements in the vertical direction for the simplicity of validation.

$$D_{AZ} = D_N \cos \alpha + D_E \sin \alpha \tag{4}$$

where D_{AZ}, D_N and D_E are displacements in SAR azimuth direction, northing and easting directions respectively. α is the heading angle of the satellite at the target point. All the measurements from 4 GPS stations were initially calibrated with respect to the measurement obtained at 10 November 2009 that is closest to the master image of TerraSAR-X dataset.

The locations of CRs and GPS stations are not identical, as shown in Figure 1b. According to the first law of geography, the variation of displacement magnitude will be generally limited within a small area. Thus, validations between measurements from GPS stations and nearby CRs were carried out. Comparisons of CR and GPS pairs CR3 and SP-1, CR3 and ZG85, CR6 and SP2, CR10 and SP-2 as well as CR16 and ZG87 were given in Figure 7. The length of error bar for the GPS measurements is 10 cm. The displacement pattern of the GPS and CR time series measurements are shown to be similar. Specific magnitude of time series displacements for CR3 and SP-1, CR3 and ZG85 as well as CR6 and SP-2 agreed quite well. All the measurements from SBI and PTOT method are distributed within the error bars. However, disparities between GPS stations and SAR measurements were more apparent with the increased distance as shown for CR10 and SP-2, CR16 and ZG87. These disparities are unavoidable for the Shuping landslide with non-uniform displacements. Nevertheless, we can conclude that both the SBI and PTOT methods can achieve centimeter-level accuracy on point-like targets.

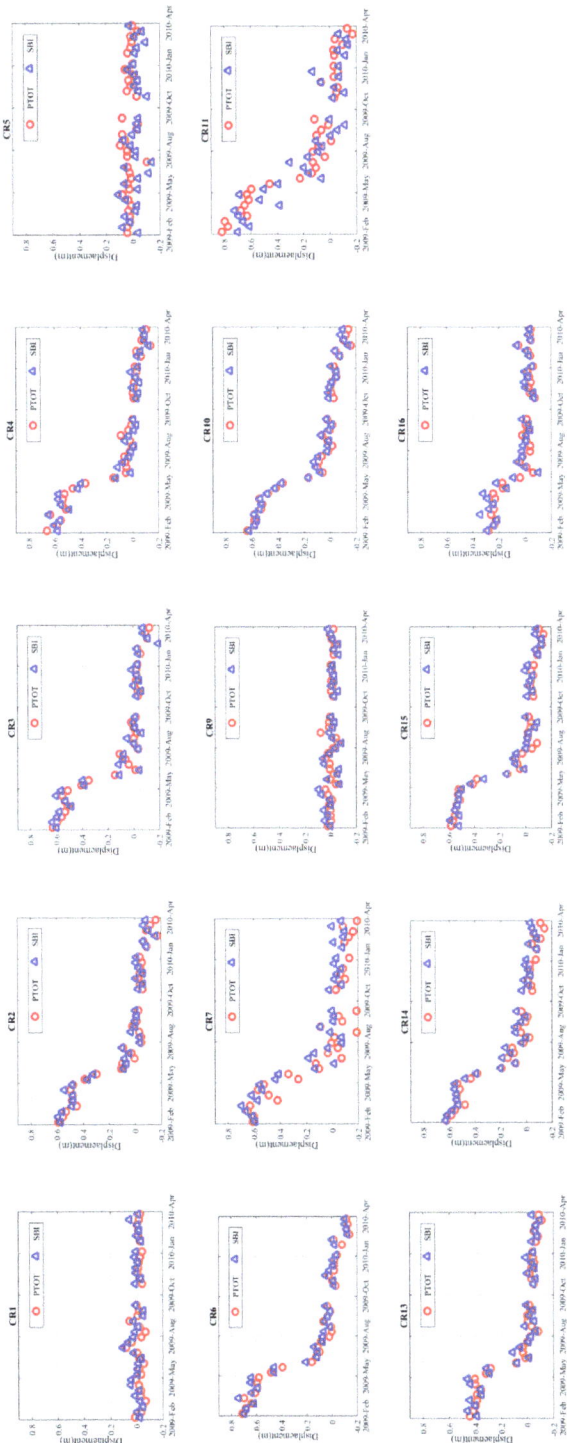

Figure 5. Comparison of time series displacements of CRs at Shuping landslide in azimuth direction obtained by PTOT and SBI method.

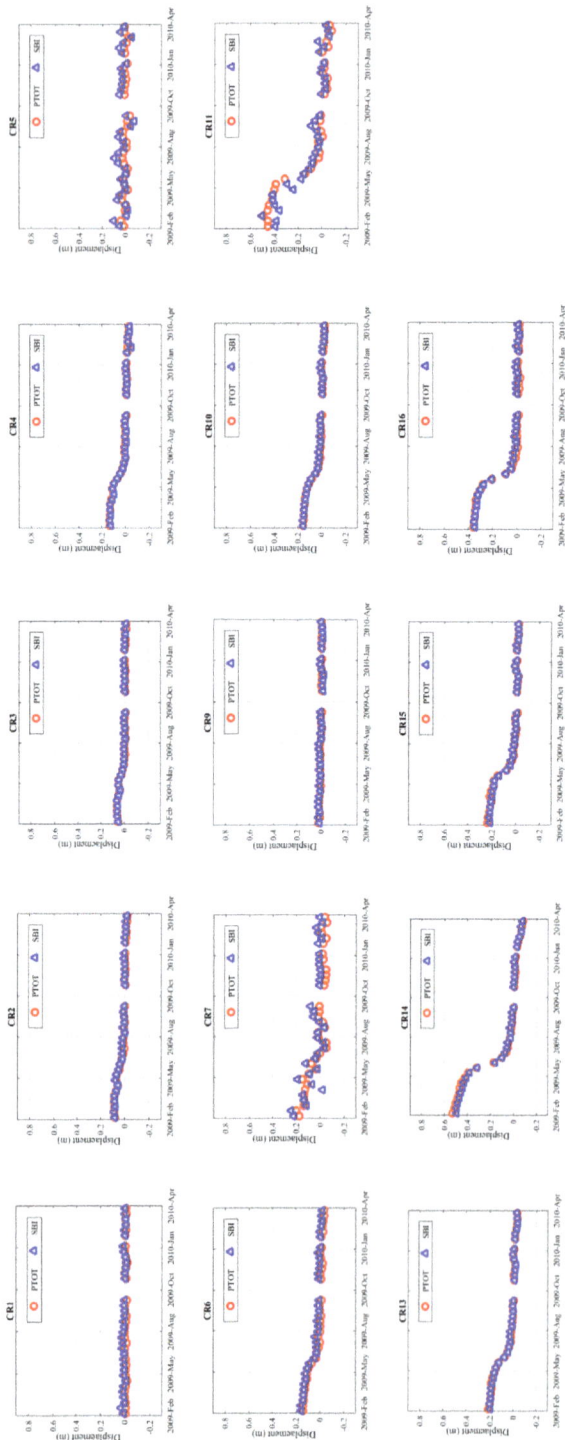

Figure 6. Comparison of time series displacements of CRs on Shuping landslide in range direction obtained by PTOT and SBI method.

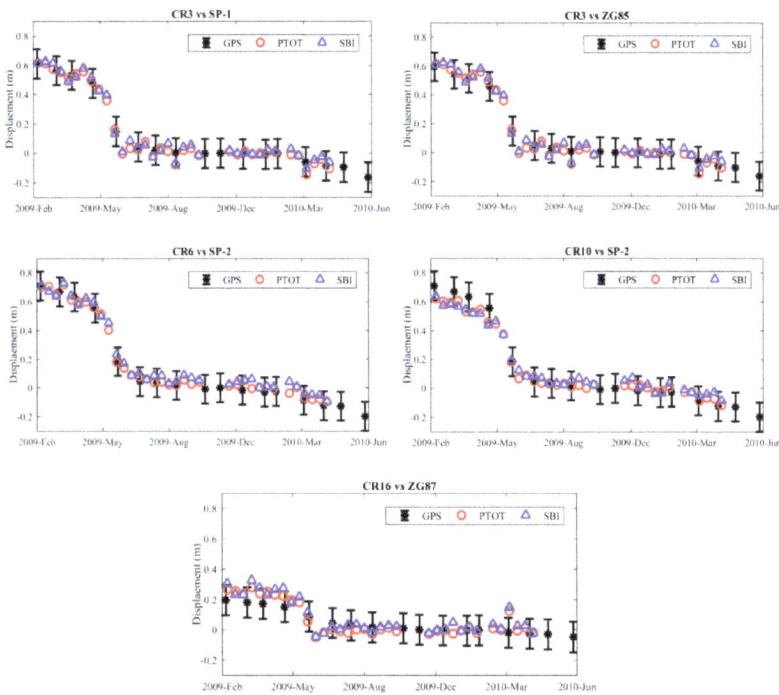

Figure 7. Comparison of measured displacements from SBI, PTOT and adjacent GPS stations in the azimuth direction. The length of the error bar is 10 cm.

5. Conclusions

This study successfully employed the azimuth and range split-bandwidth interferometry to derive the two-dimensional time series displacements of CRs installed on or surrounding the Shuping landslide in the Three Gorges area. The results of our method was first validated with point-like targets offset tracking (PTOT) results from our previous study [16]. Then, comparisons with GPS measurements were carried out to evaluate the effectiveness of both the SBI and PTOT methodologies. Both methods can achieve centimeter-level accuracy on corner reflectors. Our study indicates that it is promising to employ both an azimuth and range SBI method to monitor rapid movement of ground targets. Meanwhile, the phase unwrapping problem for standard InSAR methodologies can also be significantly reduced. It is also worth noting that the multi-look process should be carried out for natural targets, such as exposed rocks, to improve the performance of SBI. Thus, the theoretical achievable accuracy for the SBI method can demonstrate a better performance than the pixel offset tracking method on natural targets [21]. This method can be employed to monitor fast moving landslides with sparse vegetation coverage, such as the Guobu slope in Qinghai Province, China [34].

Acknowledgments: This work was financially supported by the National Key R&D Program of China (Grant No. 2017YFB0502700), the National Natural Science Foundation of China (Grant No. 61331016, 41702376, 41774006 and 41501497), Fundamental Research Funds for the Central Universities, China University of Geosciences (Wuhan) (Grant No. CUG170634), NUPTSF (Grant No. NY214197). The TerraSAR-X datasets were provided by German Aerospace Center (DLR) through the TSX-Archive-2012 AO project (GEO1856).

Author Contributions: X.S., H.J., L.Z. and M.L. conceived and designed the experiments; X.S. performed the experiments; X.S., H.J. and L.Z. analyzed the results; X.S. and H.J. wrote original manuscript. L.Z. edited the manuscript. All authors have read and approved the final manuscript.

Conflicts of Interest: The authors declare no conflict of interest.

References

1. Liao, M.; Tang, J.; Wang, T.; Balz, T.; Zhang, L. Landslide monitoring with high-resolution SAR data in the Three Gorges region. *Sci. China Earth Sci.* **2012**, *55*, 590–601. [CrossRef]
2. Liu, C.; Liu, Y.; Wen, M.; Li, T.; Lian, J.; Qin, S. Geo-hazard Initiation and Assessment in the Three Gorges Reservoir. In *Landslide Disaster Mitigation in Three Gorges Reservoir, China*; Wang, F., Li, T., Eds.; Springer: Berlin/Heidelberg, Germany, 2009; pp. 3–40.
3. Shi, X.; Liao, M.; Li, M.; Zhang, L.; Cunningham, C. Wide-Area Landslide Deformation Mapping with Multi-Path ALOS PALSAR Data Stacks: A Case Study of Three Gorges Area, China. *Remote Sens.* **2016**, *8*, 136. [CrossRef]
4. Zhao, C.; Lu, Z.; Zhang, Q.; de la Fuente, J. Large-area landslide detection and monitoring with ALOS/PALSAR imagery data over Northern California and Southern Oregon, USA. *Remote Sens. Environ.* **2012**, *124*, 348–359. [CrossRef]
5. Sun, Q.; Zhang, L.; Ding, X.L.; Hu, J.; Li, Z.W.; Zhu, J.J. Slope deformation prior to Zhouqu, China landslide from InSAR time series analysis. *Remote Sens. Environ.* **2015**, *156*, 45–57. [CrossRef]
6. Wasowski, J.; Bovenga, F. Investigating landslides and unstable slopes with satellite Multi Temporal Interferometry: Current issues and future perspectives. *Eng. Geol.* **2014**, *174*, 103–138. [CrossRef]
7. Abdolmaleki, N.; Motagh, M.; Bahroudi, A.; Sharifi, M.A.; Haghshenas Haghighi, M. Using Envisat InSAR time-series to investigate the surface kinematics of an active salt extrusion near Qum, Iran. *J. Geodyn.* **2014**, *81*, 56–66. [CrossRef]
8. Xia, Y.; Kaufmann, H.; Guo, X. Landslide monitoring in the Three Gorges area using D-InSAR and corner reflectors. *Photogramm. Eng. Remote Sens.* **2004**, *70*, 1167–1172.
9. Xia, Y.; Kaufmann, H.; Guo, X. Differential SAR interferometry using corner reflectors. In Proceedings of the IEEE International Geoscience and Remote Sensing Symposium, Toronto, ON, Canada, 24–28 June 2002; Volume 2, pp. 1243–1246.
10. Shi, X.; Zhang, L.; Liao, M.; Balz, T. Deformation monitoring of slow-moving landslide with L- and C-band SAR interferometry. *Remote Sens. Lett.* **2014**, *5*, 951–960. [CrossRef]
11. Wang, T.; Perissin, D.; Liao, M.; Rocca, F. Deformation monitoring by long term D-InSAR analysis in Three Gorges area, China. In Proceedings of the IEEE International Geoscience and Remote Sensing Symposium, Boston, MA, USA, 7–11 July 2008; pp. 65–68.
12. Tantianuparp, P.; Shi, X.; Zhang, L.; Balz, T.; Liao, M. Characterization of Landslide Deformations in Three Gorges Area Using Multiple InSAR Data Stacks. *Remote Sens.* **2013**, *5*, 2704–2719. [CrossRef]
13. Perissin, D.; Teng, W. Repeat-Pass SAR Interferometry with Partially Coherent Targets. *IEEE Trans. Geosci. Remote Sens.* **2012**, *50*, 271–280. [CrossRef]
14. Liu, P.; Li, Z.; Hoey, T.; Kincal, C.; Zhang, J.; Zeng, Q.; Muller, J.-P. Using advanced InSAR time series techniques to monitor landslide movements in Badong of the Three Gorges region, China. *Int. J. Appl. Earth Obs. Geoinf.* **2013**, *21*, 253–264. [CrossRef]
15. Miao, H.; Wang, G.; Yin, K.; Kamai, T.; Li, Y. Mechanism of the slow-moving landslides in Jurassic red-strata in the Three Gorges Reservoir, China. *Eng. Geol.* **2014**, *171*, 59–69. [CrossRef]
16. Shi, X.; Zhang, L.; Balz, T.; Liao, M. Landslide deformation monitoring using point-like target offset tracking with multi-mode high-resolution TerraSAR-X data. *ISPRS J. Photogramm. Remote Sens.* **2015**, *105*, 128–140. [CrossRef]
17. Singleton, A.; Li, Z.; Hoey, T.; Muller, J.P. Evaluating sub-pixel offset techniques as an alternative to D-InSAR for monitoring episodic landslide movements in vegetated terrain. *Remote Sens. Environ.* **2014**, *147*, 133–144. [CrossRef]
18. Li, X.; Muller, J.; Chen, F.; Zhang, Y. Measuring displacement field from TerraSAR-X amplitude images by subpixel correlation: An application to the landslide in Shuping, Three Gorges Area. *Acta Petrol. Sin.* **2011**, *27*, 3843–3850.
19. Shi, X.; Liao, M.; Zhang, L.; Balz, T. Landslide stability evaluation using high-resolution satellite SAR data in the Three Gorges area. *Q. J. Eng. Geol. Hydrogeol.* **2016**, *49*, 203–211. [CrossRef]

20. Hyung-Sup, J.; Joong-Sun, W.; Sang-Wan, K. An Improvement of the Performance of Multiple-Aperture SAR Interferometry (MAI). *IEEE Trans. Geosci. Remote Sens.* **2009**, *47*, 2859–2869. [CrossRef]
21. Bechor, N.B.; Zebker, H.A. Measuring two-dimensional movements using a single InSAR pair. *Geophys. Res. Lett.* **2006**, *33*, L16311. [CrossRef]
22. Jung, H.S.; Lu, Z.; Won, J.S.; Poland, M.P.; Miklius, A. Mapping Three-Dimensional Surface Deformation by Combining Multiple-Aperture Interferometry and Conventional Interferometry: Application to the June 2007 Eruption of Kilauea Volcano, Hawaii. *Geosci. Remote Sens. Lett.* **2011**, *8*, 34–38. [CrossRef]
23. Jiang, H.; Feng, G.; Wang, T.; Bürgmann, R. Toward full exploitation of coherent and incoherent information in Sentinel-1 TOPS data for retrieving surface displacement: Application to the 2016 Kumamoto (Japan) earthquake. *Geophys. Res. Lett.* **2017**, *44*. [CrossRef]
24. Wang, F.; Zhang, Y.; Huo, Z.; Peng, X.; Wang, S.; Yamasaki, S. Mechanism for the rapid motion of the Qianjiangping landslide during reactivation by the first impoundment of the Three Gorges Dam reservoir, China. *Landslides* **2008**, *5*, 379–386. [CrossRef]
25. Wang, F.; Zhang, Y.; Huo, Z.; Peng, X.; Araiba, K.; Wang, G. Movement of the Shuping landslide in the first four years after the initial impoundment of the Three Gorges Dam Reservoir, China. *Landslides* **2008**, *5*, 321–329. [CrossRef]
26. Xia, Y. CR-Based SAR-Interferometry for landslide monitoring. In Proceedings of the IEEE International Geoscience and Remote Sensing Symposium, Boston, MA, USA, 7–11 July 2008; Volume 2, pp. 1239–1242.
27. Scheiber, R.; Moreira, A. Coregistration of interferometric SAR images using spectral diversity. *IEEE Trans. Geosci. Remote Sens.* **2000**, *38*, 2179–2191. [CrossRef]
28. Sansosti, E.; Berardino, P.; Manunta, M.; Serafino, F.; Fornaro, G. Geometrical SAR image registration. *IEEE Trans. Geosci. Remote Sens.* **2006**, *44*, 2861–2870. [CrossRef]
29. Wang, T.; Jonsson, S.; Hanssen, R.F. Improved SAR Image Coregistration Using Pixel-Offset Series. *IEEE Geosci. Remote Sens. Lett.* **2014**, *11*, 1465–1469. [CrossRef]
30. Bamler, R.; Eineder, M. Accuracy of differential shift estimation by correlation and split-bandwidth interferometry for wideband and delta-k SAR systems. *IEEE Geosci. Remote Sens. Lett.* **2005**, *2*, 151–155. [CrossRef]
31. Zan, F.D. Coherent Shift Estimation for Stacks of SAR Images. *IEEE Geosci. Remote Sens. Lett.* **2011**, *8*, 1095–1099. [CrossRef]
32. Hooper, A.; Zebker, H.A. Phase unwrapping in three dimensions with application to InSAR time series. *J. Opt. Soc. Am. A* **2007**, *24*, 2737–2747. [CrossRef]
33. Stein, S. Algorithms for ambiguity function processing. *IEEE Trans. Acoust. Speech Signal Process.* **1981**, *29*, 588–599. [CrossRef]
34. Shi, X.; Zhang, L.; Tang, M.; Li, M.; Liao, M. Investigating a reservoir bank slope displacement history with multi-frequency satellite SAR data. *Landslides* **2017**. [CrossRef]

remote sensing

MDPI

Article

Split-Band Interferometry-Assisted Phase Unwrapping for the Phase Ambiguities Correction

Ludivine Libert [1,*], Dominique Derauw [1], Nicolas d'Oreye [2,3], Christian Barbier [1] and Anne Orban [1]

[1]   Centre Spatial de Liège, Université de Liège, Avenue du Pré-Aily, B-4031 Angleur, Belgium;
      dderauw@ulg.ac.be (D.D.); cbarbier@ulg.ac.be (C.B.); aorban@ulg.ac.be (A.O.)
[2]   European Centre for Geodynamics and Seismology, Rue Josy Welter 19, L-7256 Walferdange,
      Grand-Duchy of Luxembourg; ndo@ecgs.lu
[3]   National Museum of Natural History, Rue de Munster 25, L-2160 Luxembourg,
      Grand-Duchy of Luxembourg
*   Correspondence: llibert@ulg.ac.be; Tel.: +32-4-372-47-01

Received: 20 July 2017; Accepted: 19 August 2017; Published: 23 August 2017

Abstract: Split-Band Interferometry (SBInSAR) exploits the large range bandwidth of the new generation of synthetic aperture radar (SAR) sensors to process images at subrange bandwidth. Its application to an interferometric pair leads to several lower resolution interferograms of the same scene with slightly shifted central frequencies. When SBInSAR is applied to frequency-persistent scatterers, the linear trend of the phase through the stack of interferograms can be used to perform absolute and spatially independent phase unwrapping. While the height computation has been the main concern of studies on SBInSAR so far, we propose instead to use it to assist conventional phase unwrapping. During phase unwrapping, phase ambiguities are introduced when parts of the interferogram are separately unwrapped. The proposed method reduces the phase ambiguities so that the phase can be connected between separately unwrapped regions. The approach is tested on a pair of TerraSAR-X spotlight images of Copahue volcano, Argentina. In this framework, we propose two new criteria for the frequency-persistent scatterers detection, based respectively on the standard deviation of the slope of the linear regression and on the phase variance stability, and we compare them to the multifrequency phase error. Both new criteria appear to be more suited to our approach than the multifrequency phase error. We validate the SBInSAR-assisted phase unwrapping method by artificially splitting a continuous phase region into disconnected subzones. Despite the decorrelation and the steep topography affecting the volcanic test region, the expected phase ambiguities are successfully recovered whatever the chosen criterion to detect the frequency-persistent scatterers. Comparing the aspect ratio of the distributions of the computed phase ambiguities, the analysis shows that the phase variance stability is the most efficient criterion to select stable targets and the slope standard deviation gives satisfactory results.

Keywords: synthetic aperture radar; interferometry; phase unwrapping; split-band; multichromatic analysis

1. Introduction

Over the years, performances of Synthetic Aperture Radar (SAR) sensors have been improved to finally reach the metric resolution by combining the synthetic aperture principle in the azimuth direction with an increase of the radar signal bandwidth in the range direction. Using such data, the well-known SAR Interferometry (InSAR) solves the relationship between the phase and the optical path difference to retrieve the topography. However, the spectral information of the range component is rarely exploited. Split-Band Interferometry (SBInSAR), also known as Multichromatic

Analysis, exploits the information contained in the frequency domain to put an added value to SAR Interferometry. It applies InSAR to subrange images obtained by splitting the large available range bandwidth of recent sensors and explores the phase trend through the partial interferograms in order to provide pointwise absolute phase measurements. This process is equivalent to an absolute and spatially independent phase unwrapping, as long as it is performed on scatterers with a stable behaviour across the spectral domain. Such targets are called frequency-persistent scatterers (PS$_f$) [1]. The theoretical applicability of Split-Band Interferometry and its performance regarding the spectral decomposition parameters are discussed in [2]. This work showed that the quality of the split-band phase is the result of a trade-off between increasing the number of subbands of the spectral decomposition and preserving a sufficient resolution for the subrange images [2]. The practical feasibility of topographic measurements has been reported in [3] for airborne data in X-band with a total bandwidth of 400 MHz. The study in [1] applied the technique to spaceborne TerraSAR-X data in spotlight mode (300 MHz) over the Uluru monolith in Australia. The same test site has been considered by [4] using Cosmo-SkyMed images with a total bandwidth of 325 MHz. Another study [5] also demonstrated the potential of SBInSAR for height retrieval using a TanDEM-X bistatic pair of images of 100 MHz bandwidth over Nyiragongo volcano, but stressed the need for a larger value of the initial range bandwidth. So far, the frequency-persistent scatterers have been selected using the multifrequency phase error [6] that is an estimator of the coherence from one spectral subband to another and that quantifies the quality of the phase measurements. In [7], the scattering properties of the frequency-persistent scatterers and "temporally" coherent scatterers (PS) from the Permanent Scatterers Interferometry (PSI) [8] are investigated. Among the potential applications of SBInSAR, let us mention absolute height retrieval, change detection [9], ionospheric correction [10,11] and urban monitoring. The spectral diversity of the range bandwidth can also be used to improve coregistration as discussed in [12] or to estimate high-gradient surface displacements, such as earthquake ruptures [13].

Most InSAR phase unwrapping algorithms are aimed to perform relative measurements and determine the phase of a pixel with respect to the phase of its neighbours, rather than the absolute phase. It is a real issue in practice because noncoherent patches due to geometrical distortions (layover, shadowing) or time changes can isolate coherent regions from each other and cause a separate phase unwrapping from one region of the interferogram to another. This introduces unknown phase ambiguities and prevents from comparing the phase between two separated regions. Because SBInSAR provides absolute phase measurements, it can potentially solve these phase ambiguities and reconnect the phase of distinct regions. Since we only need to know the integer number of cycles that must be added to the wrapped phase to solve the phase ambiguities, the accuracy requirements regarding the split-band phase are less demanding than in the case of height retrieval.

In this study, we propose an approach based on SBInSAR to complement the InSAR phase unwrapping that estimates and corrects the phase ambiguities, and we demonstrate its efficiency. In Section 2, we set the basic notions and equations of Split-Band Interferometry and we present the method for the phase ambiguities correction. We also propose new methods to select robust PS$_f$. In Section 3, the SBInSAR-assisted phase unwrapping is tested on spotlight images acquired over Copahue volcano. The test site is described, as well as the data set and the processing. An indirect validation procedure is also presented, along with an indicator to compare the precision of the results.

2. Methods

Split-Band Interferometry is a three-step process derived from classical SAR Interferometry. It takes advantage of the large range bandwidth of recent SAR sensors to work the absolute interferometric phase out. The splitting of the range bandwidth of a SAR scene into several narrower subbands produces lower resolution images of this scene, each one with a frequency slightly shifted with respect to the initial one. During the first step of the SBInSAR process, the same spectral decomposition is applied to the already coregistered master and slave images of a given interferometric pair. In a second step, interferometry is performed on each pair of master and slave subimages. It yields a set of interferograms where the pointwise phase evolves linearly across the spectral domain. The slope of the final pixel-by-pixel linear regression of the phase is proportional to the absolute optical path, and it therefore enables performing absolute phase measurements on the points considered as spectrally stable targets.

The operating and rationale of the SBInSAR processor have already been presented in [5]. In this section, we will first outline the basic principles of Split-Band Interferometry and the corresponding equations. We will then define the estimators for the characterization and detection of frequency-persistent scatterers. Finally, we will present our approach for the phase-offset determination.

2.1. Rationale of Split-Band Interferometry

Let us consider an interferometric pair of coregistered images with a bandwidth B and a carrier frequency v_0. They are spectrally decomposed into N subbands of partial bandwidth B_N centered at frequencies v_i $(i = 1, 2, \ldots N)$, N being odd. Frequencies of adjacent subbands are shifted of Δv. As stated in [5], the interferometric phase of the coregistered and spectrally decomposed images in the ith partial interferogram is expressed by:

$$\Delta\phi_i = \frac{4\pi}{c}(r_s - r_m - e_c)\,v_0 + \frac{4\pi}{c}e_c\,v_i, \tag{1}$$

where r_m and r_s are the range coordinates in master and slave images, respectively, e_c is the coregistration error and the dependence on the coordinates of the pixel is implicit for the sake of clarity. The phase behavior of a point across the N partial interferograms is fitted by a simple linear function:

$$p\,(v_i) = s\,v_i + u, \tag{2}$$

where s and u are the fit parameters. The slope of this linear regression is given by:

$$s = \frac{\partial(\Delta\phi_i)}{\partial v_i} = \frac{4\pi}{c}e_c. \tag{3}$$

In this case, the absolute optical path difference is the sum of the registration Δr applied on the range coordinate and the coregistration error e_c. The phase issued by the split-band process, called the split-band phase, is therefore computed as:

$$
\begin{aligned}
\Delta\varphi &= \Delta\varphi_{reg} + \Delta\varphi_{e_c} \\
&= \frac{4\pi}{c}v_0\,\Delta r + \frac{4\pi}{c}v_0\,e_c.
\end{aligned}
\tag{4}
$$

Nevertheless, the split-band phase measurement can only be considered as absolute if it is known with a sufficient accuracy. Since the second term of Equation (4) is obtained by multiplying the slope of the linear regression by the initial carrier frequency, the accuracy of the split-band phase is directly related to the accuracy of slope through:

$$\sigma_{\Delta\varphi} = v_0\,\sigma_s, \tag{5}$$

with $\sigma_{\Delta\varphi}$ being the standard deviation of the split-band phase and σ_s the standard deviation of the slope coefficient s of the linear regression. Considering the chi-square fitting of a straight line, the latter can be expressed as follows:

$$\sigma_s = \frac{1}{\Delta \nu} \sqrt{\frac{\sum\limits_{i=1}^{N} \frac{1}{\sigma_{\phi_i}^2}}{\sum\limits_{i=1}^{N} \frac{1}{\sigma_{\phi_i}^2} \sum\limits_{i=1}^{N} \frac{x_i^2}{\sigma_{\phi_i}^2} - \left(\sum\limits_{i=1}^{N} \frac{x_i}{\sigma_{\phi_i}^2}\right)^2}}, \tag{6}$$

where x_i is the subband index ranging from $-\frac{N-1}{2}$ to $\frac{N-1}{2}$ and $\sigma_{\phi_i}^2$ is the phase variance in the ith partial interferogram. This expression holds for independent data points, i.e., nonoverlapping subbands.

2.2. Detection of Frequency-Persistent Scatterers

The linearity of the phase assumed in Equation (1) holds only for targets with a coherent behavior across the spectral domain, i.e., for frequency-persistent scatterers. The consequence is that the stable nature of a frequency-persistent scatterer insures the accuracy of the split-band phase measurement, and it is therefore fundamental to correctly detect PS_f.

In the following, we define the multifrequency phase error that is the commonly used criterion to detect spectrally stable targets, and we propose two new detection criteria: the slope standard deviation and the phase variance stability. These new criteria are meant to improve the selection of the PS_f population and their efficiency will be compared to the one of the multifrequency phase error when applied to the test case.

2.2.1. Multifrequency Phase Error

So far, most studies have exploited the multifrequency phase error σ_ν to detect stable targets. This estimator of the split-band phase quality is basically the a posteriori uncertainty of the phase value in the partial interferograms, and it is mathematically defined as:

$$\sigma_\nu = \sqrt{\frac{1}{N-2} \sum_{i=1}^{N} (\Delta\phi_i - p(\nu_i))^2}. \tag{7}$$

2.2.2. Slope Standard Deviation

In the framework of the phase ambiguities retrieval, the split-band phase must be measured with accuracy better than a cycle, and we use this one-cycle accuracy to characterize frequency-persistent scatterers. Setting a threshold of 2π in Equation (5) leads us to a first criterion of selection based on the standard deviation of the slope of the linear regression:

$$\sigma_s < \frac{2\pi}{\nu_0}. \tag{8}$$

This criterion only depends on the initial carrier frequency. In X-band, the upper limit of the phase slope standard deviation has a typical value of 0.65 rad/GHz.

2.2.3. Phase Variance Stability

In order to establish the second criterion, which will be referred to as the phase variance stability, let us assume that the spectral decomposition is symmetrical with respect to ν_0 and that the phase variance $\sigma_{\phi_i}^2$ does not vary much from one subband to another. In this case, the squared sum in Equation (6) can be neglected and we can introduce an upper bound $\sigma_{\phi,max}^2$ on the partial phase variance:

$$\sigma_s = \frac{1}{\Delta \nu} \sqrt{\frac{1}{\sum\limits_{i=1}^{N} \frac{x_i^2}{\sigma_{\phi_i}^2}}}$$

$$\leq \frac{\sigma_{\phi,max}}{\Delta \nu} \sqrt{\frac{1}{\sum\limits_{i=1}^{N} x_i^2}}. \tag{9}$$

Given the symmetry of the x_i values, the remaining sum in Equation (9) can be developed as the double sum of the $\frac{N-1}{2}$ first squared integers. Inserting the relation (9) into Equation (5) and setting once again a 2π threshold on $\sigma_{\Delta \varphi}$, we finally obtain:

$$\sigma_{\phi,max}^2 < \left(2\pi \frac{\Delta \nu}{\nu_0}\right)^2 \frac{N(N+1)(N-1)}{12}. \tag{10}$$

For a given pixel in the stack, if the value of the phase variance is lower than this limit in every partial interferogram, i.e.,:

$$\sigma_{\phi_i}^2 < \sigma_{\phi,max}^2 \quad \forall i = 1, 2, \dots N, \tag{11}$$

then the required accuracy should be insured and the point is considered as being a PS$_f$. In practice, the spectral decomposition is always symmetrical with respect to the central carrier frequency. For the second assumption, we consider it as verified when the value of $\sigma_{\Delta \varphi}$ varies of less than 5% when the squared sum is neglected.

Let us note that the upper bound given by Equation (10) increases as N^3. This can be interpreted in the following way: for a large number of bands, the partial phase variance can be important as long as its value from one band to another remains consistent. In this case, the split-band phase will be measured accurately anyway. However, increasing the number of subbands reduces the resolution. The key point of the spectral decomposition will be to determine the number of subbands so as to find a trade-off between accuracy and resolution.

It is important to note that the phase variance stability criterion for the selection of frequency-persistent scatterers is quite stringent: due to the assumption that must be satisfied, the number of selected points will be low. Some valid PS$_f$ may even be missed during the selection. However, the accuracy for the selected points is guaranteed, as we will demonstrate with the test case.

2.3. SBInSAR-Assisted Phase Unwrapping

In a conventional InSAR process, phase unwrapping algorithms generally provide relative measurements of the phase. However, due to decorrelation, they frequently fail to unwrap the interferogram as a whole and parts of the image are separately processed, introducing phase ambiguities that prevent from comparing the phase from one region to another.

We present here an approach based on Split-Band Interferometry to determine and correct the local phase ambiguities of the unwrapped phase. We consider the general case where the phase ambiguity is an unknown number of cycles $2\pi n$, with n being an integer. The phase ambiguity has the same value for all the points through a given region of continuously unwrapped phase. In theory, the presence of a single stable target per independent area is sufficient to solve the phase ambiguities. In practice, however, we have to deal with the phase noise and the uncertainties, and it is not possible to determine which scatterer is the most stable. For this reason, we adopt a statistical analysis of the PS$_f$ to derive this integer number of cycles. Let us specify that the phase-offset, or phase ambiguity, denotes the $2\pi n$-discrepancy. However, we will largely use these terms in the following sections to refer to the number of cycles n alone.

Let us note $\Delta\phi$ the unwrapped phase obtained with the classical phase unwrapping process. For a given PS$_f$ at pixel's coordinates (k,l) in the image, neglecting the noise and the phase unwrapping errors, the unknown number of cycles can be computed as:

$$n(k,l) = \frac{\Delta\varphi(k,l) - \Delta\phi(k,l)}{2\pi}. \tag{12}$$

The SBInSAR-assisted phase unwrapping will consider one region of the phase unwrapping at a time. In a first time, it will select the PS$_f$ of this region based on one of the criteria presented in the previous section. It will then estimate the phase ambiguities using Equation (12) for all the selected pixels and round each of these values to the nearest integer. The rounded value with the largest number of occurrences, i.e., the mode of the distribution, is assumed to be the phase ambiguity we are looking for. Finally, this value is multiplied by 2π and added to the unwrapped phase of all the pixels of the region in order to correct the phase ambiguity. This procedure is repeated for each region separately unwrapped. When the distribution of the rounded phase ambiguities has multiple modes, no correction is applied. Regions with a population below 10 PS$_f$ are not corrected either, since they frequently show multiple modes. The algorithm steps are presented in Figure 1. In the end of the process, an image of the leveled unwrapped phase is provided, showing only the shifted areas. Let us stress that, despite the loss of resolution in the subproducts, this final unwrapped phase image preserves the initial range resolution.

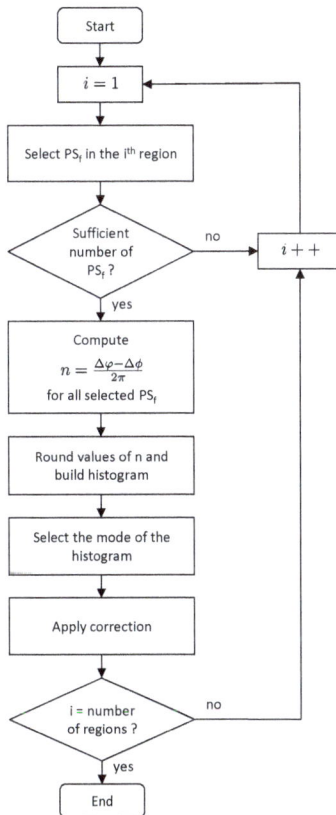

Figure 1. Flow chart of the Split-Band Interferometry-assisted phase unwrapping algorithm.

The routine supports topographic and deformation modes. For both modes, the selection of the stable targets population will be identical, but, in deformation mode, the DEM contribution will be removed from both the unwrapped phase and the split-band phase. Since the offset estimation is based on the difference between these two phases, no difference of performance is expected between topographic and deformation modes, except due to the quality of the unwrapped phase. In the following, we will only consider the topographic mode and no subtraction of the DEM will be made.

Besides, in practice, the rounding of the phase ambiguities to the nearest integer mitigates the influence of the noise and the phase unwrapping error. Although infinitely accurate phase measurements would be necessary in theory, we will show that the one-cycle accuracy assumed for PS_f is enough to determine the phase-offset.

3. Copahue Test Case

3.1. Test Site

Copahue is an active strato-volcano located in the northwest of the Argentinean province of Neuquén in the Andes, at the border between Argentina and Chile (Figure 2a). This volcano has an elongated elliptical shape (22 km × 8 km) oriented in the SW–NE direction and reaches a maximum elevation of 2997 m. It has nine craters clustered along the N 60° E direction, but only the eastern-most one is active. The active crater is about 300 m deep and contains an acid lake created by abundant precipitations and ice melting [14]. Recent eruptions have been reported in 2000, 2012 and 2014 and the last eruption was accompanied by a degassing unrest [15]. Deformations observed over Copahue volcano using InSAR are discussed in [16,17].

A Google Earth™ (Copahue, Neuquén, Argentina) optical view of the area is given in Figure 2b. The area of interest shows steep topography as well as moderate slopes, little vegetation and a snow cover that can vary over the year. Change in snow cover and frequent precipitations can cause local loss of coherence in InSAR products. Variety of topography, of slope orientations and of geometrical distortions, along with the presence of natural scatterers make Copahue volcano an interesting site to apply SBInSAR-assisted phase unwrapping.

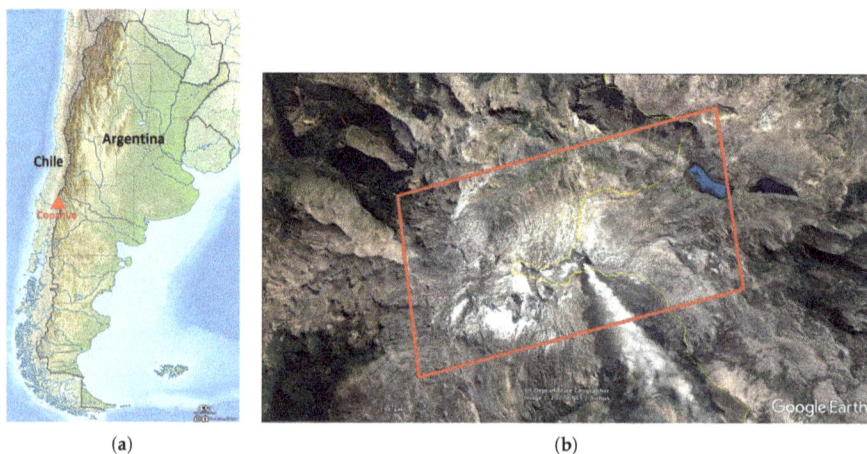

(a)   **(b)**

Figure 2. (**a**) Location of Copahue volcano (red triangle) on the border between Chile and Argentina. (**b**) Google Earth™ image of Copahue volcano in 2017. The footprint of the InSAR pair is drawn in red.

3.2. Data Set and Processing

The data set used to test the SBInSAR-assisted phase unwrapping method consists in two spotlight images acquired on ascending orbits by TerraSAR-X on 15 and 26 December 2014. Each image is the master image of a pair acquired in pursuit monostatic mode, and the ensemble of both constitutes therefore a standard spotlight interferometric pair. They are acquired in VV-polarization with a look angle ranging from approximately 32.8° to 33.8° and a range bandwidth of 300 MHz. The interferometric pair has a perpendicular baseline of about 32 m that corresponds to a height of ambiguity of 163 m, and a temporal baseline of 11 days, which minimizes the effects of temporal decorrelation.

Studying the deformations over Copahue volcano is outside the scope of the present study, as we wish to make a methodological demonstration only. Therefore, no DEM is used to remove topographic information of the phase during the InSAR processing. Hence, the analyzed signal contains the topography but also possible deformations or artefacts from atmospheric origin.

A multilooking of 5 pixels × 5 pixels is applied to the images and a coherence threshold of 0.5 is chosen, above which the phase is considered for unwrapping. A branch-cut algorithm is used to unwrap the phase [18]. The coherence map and the corresponding unwrapped InSAR phase are presented in Figure 3. In the coherence map shown in Figure 3a, large decorrelated areas are present in regions corresponding mainly to the snow cover. In Figure 3b, we observe a smooth phase gradient on the main part of the interferogram and numerous smaller regions phase-shifted with respect to the main area.

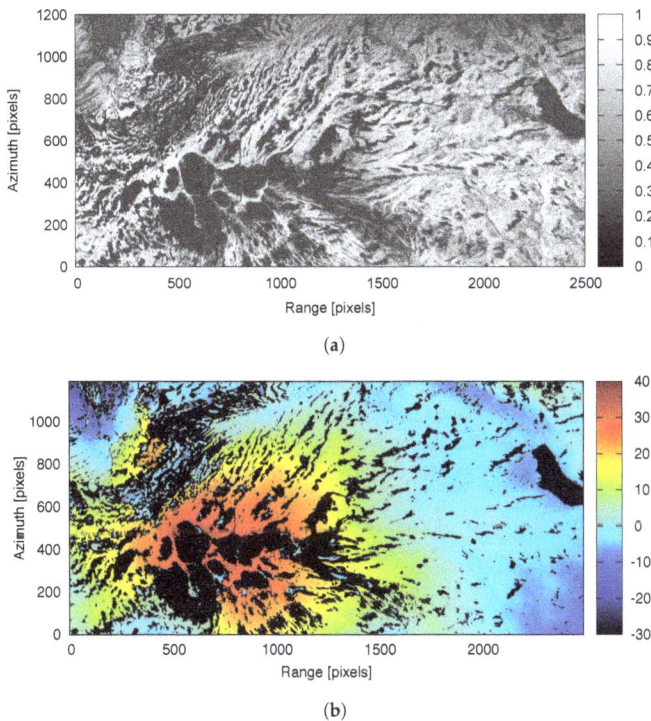

(a)

(b)

Figure 3. (**a**) coherence image of the test pair over the Copahue volcano. (**b**) fully connected unwrapped phase. Color chart values are given in radians.

The SBInSAR processing is applied with a spectral decomposition into five subbands of 60 MHz. Given the 300 MHz initial range bandwidth, the subbands do not overlap and the rationale presented in Section 2 is valid. Moreover, this large partial bandwidth mitigates the resolution loss and therefore insures the quality of the split-band measurements. In the split-band phase image (Figure 4), we observe large patches of noisy phase corresponding mainly to noncoherent areas in the InSAR phase. These noisy patches with high dispersion of values illustrate clearly that all the scatterers are not stable regarding the SBInSAR processing and that the adequate pixels must be selected somehow.

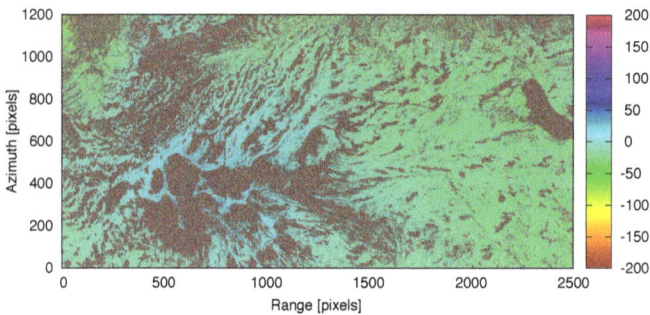

Figure 4. Phase measured with Split-Band Interferometry over Copahue volcano. Color chart values are given in radians.

3.3. Validation Procedure

In order to demonstrate the applicability of the SBInSAR-assisted phase unwrapping without additional measurements, we propose an indirect validation strategy. It consists of disconnecting some regions of the interferogram by introducing artificial cuts during the phase unwrapping process, which is based on a branch-cut method. Knowing the unwrapped phase $\Delta\phi_c$ of the "fully connected" version of the interferogram and the unwrapped phase $\Delta\phi_a$ of the "artificially disconnected" version, and knowing that their values only differ by an entire number of cycles $2\pi m$ with m integer, one can obviously state:

$$m(k,l) = \frac{\Delta\phi_c(k,l) - \Delta\phi_a(k,l)}{2\pi}. \tag{13}$$

This phase-offset will not necessarily be the same as the correction n computed with the SBInSAR-assisted phase unwrapping, but the phase-offset difference between two pixels of coordinates (k_1, l_1) and (k_2, l_2) should be the same for both m and n:

$$n(k_1, l_1) - n(k_2, l_2) = m(k_1, l_1) - m(k_2, l_2). \tag{14}$$

If we focus on pixels in two separate regions, then the relative offset values can be used to validate the results, as illustrated in Figure 5. The artificially disconnected regions of the test pair are shown in Figure 6a. We cut three areas from the main coherent area. The corresponding unwrapped phase is given in Figure 6b. It is obvious from this figure that a phase shift has been introduced for region 2. The relative phase shifts between these regions are listed in Table 1. This will be used in the next section to validate our results.

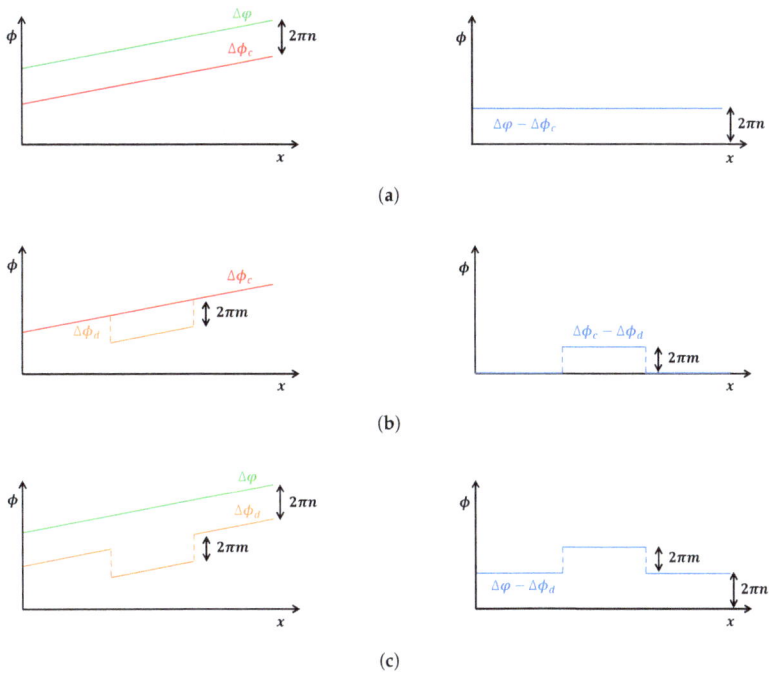

(a)

(b)

(c)

Figure 5. Diagrams of the validation procedure for a one-dimensional simplified interferogram. (**a**) If the connected InSAR phase $\Delta\phi_c$ is subtracted from the split-band phase $\Delta\varphi$, the difference gives an offset $2\pi n$. (**b**) If a region of the one-dimensional interferogram is disconnected, an offset $2\pi m$ is introduced between the connected phase $\Delta\phi_c$ and the disconnected InSAR phase $\Delta\phi_d$. (**c**) If the disconnected InSAR phase is subtracted from the split-band phase, the relative offset $2\pi m$ between the disconnected regions remains the same as in case (**b**).

Table 1. Relative phase-offset values of artificially disconnected regions.

Relative Phase-Offset	Cycles
$m_1 - m_2$	-1
$m_1 - m_3$	0
$m_1 - m_4$	-1
$m_2 - m_3$	1
$m_2 - m_4$	0
$m_3 - m_4$	-1

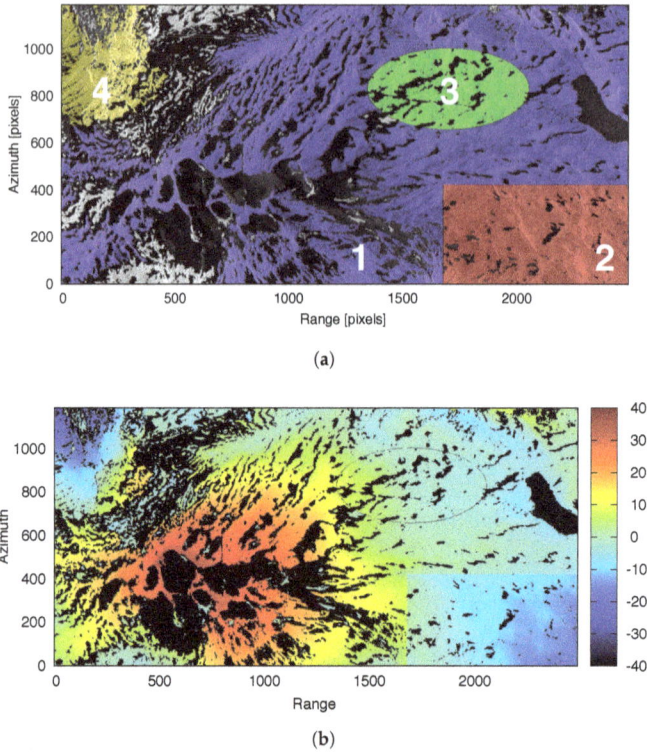

Figure 6. (a) map of the disconnected regions. The main coherent region from which areas have been artificially cut is represented in blue. We refer to the blue region, the red rectangle, the green ellipse and the yellow area, respectively, as the regions 1, 2, 3 and 4. Regions in black are regions with coherence lower than the threshold applied during phase unwrapping. White areas are naturally disconnected parts of the unwrapped phase and they are not considered for the validation of the SBInSAR-assisted phase unwrapping. **(b)** artificially disconnected version of the unwrapped phase over Copahue volcano. Color chart values are given in radians.

3.4. Indicator of Quality

In the proposed approach for the SBInSAR-assisted phase unwrapping, the phase ambiguity is chosen as the mode of the phase-offsets distribution. The statistical nature of the method makes two requirements necessary in order to have confidence in the correction: first, the probability associated to the mode value must be high; second, the dispersion of the distribution must be low. These two conditions are summarized by a low W/H ratio, W and H being, respectively, the width and the height of the normalized distribution. The normalized distribution of the rounded phase-offsets can be fitted by a normal law:

$$f(n) = \frac{1}{\sigma\sqrt{2\pi}} \exp\left(-\frac{1}{2}\left(\frac{n-\mu}{\sigma}\right)^2\right) \qquad (15)$$

with the expectation value μ and the standard deviation σ being the parameters of the fit. The height of the distribution is defined as the maximum of the fitted normal law and the width is characterized by the half width at half maximum. The ratio is then given by:

$$\frac{W}{H} = \frac{\sigma\sqrt{2\ln(2)}}{f(\mu).} \tag{16}$$

The aspect ratio of the distribution is an indication regarding the precision of the measurement, not its accuracy. It will therefore allow to compare the precision of two different estimations, but it will not assess if the measurement is correct or not.

4. Results and Discussion

In a first time, the SBInSAR-assisted phase unwrapping is applied on the four artificially disconnected areas of the Copahue test case in order to validate the approach for the phase ambiguities correction and determine which detection criterion is the best. Four situations are considered from the PS_f selection point of view: in the initial situation, we do not discriminate the PS_f and keep all the pixels. In the other cases, the PS_f are selected using either the multifrequency phase error σ_V, the threshold on standard deviation of the slope σ_s of the linear regression or the stability of the phase variance $\sigma^2_{\phi_i}$. We set a threshold of 0.5 on the multifrequency phase error. The number of selected pixels according to the region and the detection criterion is shown in Figure 7. Since the first region is noticeably larger than the three others, it shows therefore a larger population of selected pixels for any detection criterion. The number of detected pixels is much higher in the case of the multifrequency phase error than for the other two criteria. The phase variance stability classifies approximately 2–3% of the initial population as PS_f while the proportion is about ten times higher for the standard deviation of the slope.

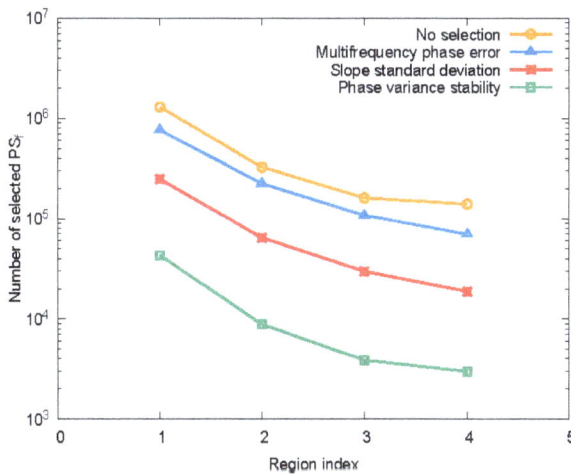

Figure 7. Number of selected PS_f for regions 1 to 4 using the different detection criteria. The y-axis is on a logarithmic scale.

All four of the selection methods provide the same phase ambiguities corrections of the disconnected regions. These corrections are listed in Table 2. As indicated before, the expected relative phase-offsets are given in Table 1. Based on the validation strategy proposed in the previous section, we verify that our results are consistent with these values. Even though the four situations give similar measurements, the precision of the result is not the same. Looking at the histogram of the normalized distribution of the phase-offset estimates in region 3 (Figure 8), we observe that the dispersion of the distribution varies from one criterion to the other, but the modes of the distributions are indeed the same. When no selection of PS$_f$ is applied, the distribution is spread over a large range of values with a low probability for the mode bin. This behavior is similar for histograms over all the other regions. The quality of the results without selection criterion applied is quantified by a W/H ratio of about 45–50 for most regions, with the highest value of 85 for region 4 (see Table 3). The larger dispersion in region 4 is probably due to the large uncorrelated patch present in the split-band phase. Those targets are most probably unstable and they are not discriminated in this case. The W/H aspect ratio is lowered when the multifrequency phase error criterion is applied to select stable pixels. In this case, the contrast between the dispersion in region 4 and the three others is significantly reduced.

Table 2. Phase-offset corrections for artificially disconnected regions.

Computed Phase-Offset	Cycles
n_1	-3
n_2	-2
n_3	-3
n_4	-2

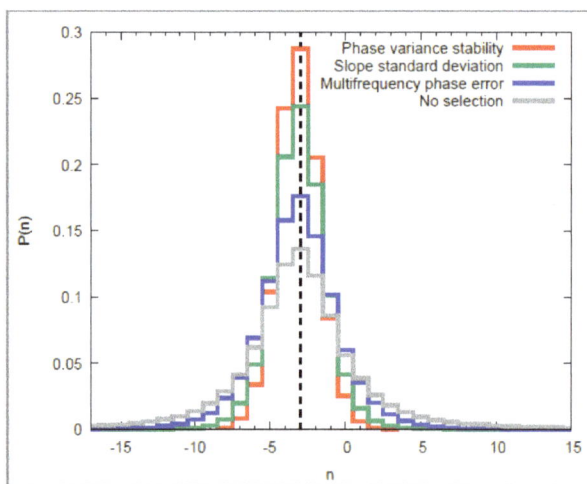

Figure 8. Normalized histograms of the estimated phase-offset values n for region 3. The gray line histogram represents the case where no selection of PS$_f$ is applied. For other cases, the PS$_f$ population is selected using three different criteria. The vertical dashed line indicates the expected phase ambiguity. Similar figures are obtained for the three other regions.

Table 3. Ratio W/H of the phase-offsets distributions of artificially disconnected regions.

PS$_f$ Selector	Region 1 W/H	Region 2 W/H	Region 3 W/H	Region 4 W/H
None	54	43	43	85
σ_v	25	26	25	30
σ_s	12	14	12	13
$\sigma_{\phi_i}^2$	8	9	8	9

Let us now consider our two new selection criteria: with the standard deviation of the slope, the aspect ratio is still improved by a factor 2. The best ratio is, however, obtained by using the phase variance stability criteria. Similar results are found for the other areas. The W/H ratio seems to be correlated with the population of frequency-persistent scatterers: the smallest the number of selected pixels, the lowest the aspect ratio. Among the PS_f selection criteria considered in this study, the phase variance stability appears to be the most efficient. However, when applied to a less favorable case, e.g., to images with a smaller bandwidth or small disconnected areas, this criterion can be too restrictive and the population of selected PS_f will be too limited to estimate the phase ambiguity reliably. In such cases, the standard devation of the slope is a satisfactory alternative.

After validation of the procedure, the SBInSAR-assisted phase unwrapping is applied to the ensemble of the naturally and artificially disconnected regions of the unwrapped phase. Using the phase variance stability to detect stable targets, 33 regions out of 1796 are corrected while, using the standard deviation of the slope, we reach an amount of 74 regions. Let us remind that a region will be corrected if it holds at least 10 PS_f and if the phase-offset mode is unique. During the test on the artificially disconnected areas, we observed that the phase variance stability selected fewer targets than the standard deviation of the slope in a given region. We reach a similar conclusion for the naturally disconnected areas (Figure 9). It is interesting to note that the majority of the pixels identified by the phase variance stability are also identified by the standard deviation of the slope. Less than 1% of the PS_f population is selected by the phase variance stability only.

For regions with a PS_f population larger than 30 pixels, we observe in Figure 10 that the mode of the phase-offset distribution represents approximately 20–30% of the detected PS_f for the standard deviation of the slope and a slightly higher value of 25–35% for the phase variance stability. Regions with smaller population of stable targets can exhibit even larger values. The regions with an occurrence of the mode below 20% of the PSf_f population are an exception, whatever the detection criterion.

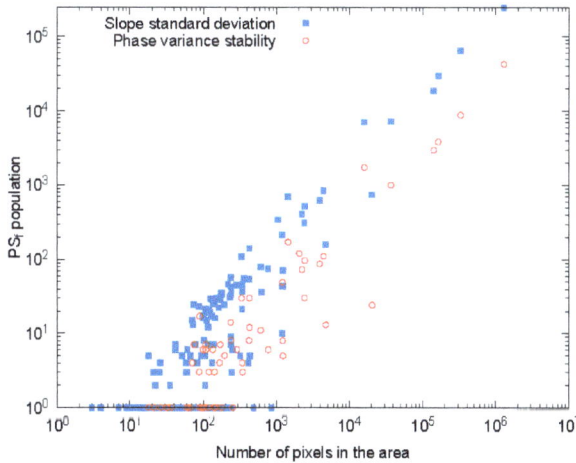

Figure 9. PS_f population as a function of the size of the area, for both artificially and naturally disconnected areas. Regions with no PS_f are not represented. Axes are in logarithmic scale.

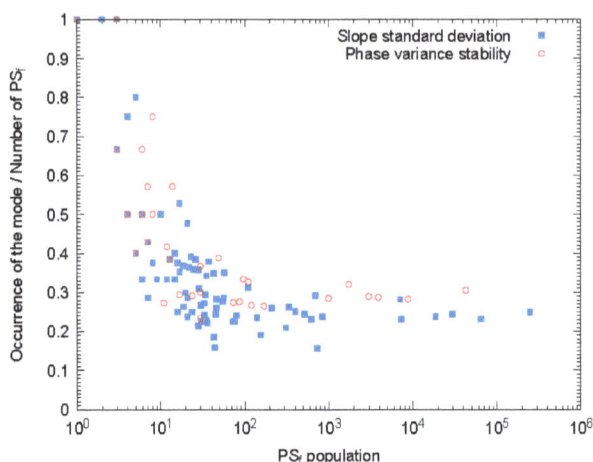

Figure 10. Probability of the mode of the phase-offset distribution as a function of the amount of selected PS_f in a given region, for both artificially and naturally disconnected areas. Regions with multiple modes are not represented. The *x*-axis is on the logarithmic scale.

As it can be seen in Figure 9, whose axes are in logarithmic scale, there seems to be no relationship between the size of a patch and the PS_f population. The stable nature of a target is probably related to its intrinsic characteristics and/or the geometry of observation, which causes a heterogeneous distribution of the targets across the scene. Since the population of stable targets is the key point to determine the phase ambiguity, and it cannot be related to the initial population of a given region, it is not possible to define the minimum size of an area where the SBInSAR-assisted phase unwrapping can be applied. However, we observe that when the standard deviation of the slope is considered, the largest region with no PS_f at all is made of less than 400 pixels. For the phase variance stability criterion, the largest region has a size of 3665 pixels, but most of them contain less than 500 pixels.

5. Conclusions

This study has presented a probabilistic approach based on Split-Band Interferometry to correct the local phase ambiguities introduced during phase unwrapping of classical InSAR process. The applicability and potential of the proposed approach have been demonstrated on a TerraSAR-X pair of images over Copahue volcano, showing steep topography and local loss of coherence. The corrections have been computed for artificially disconnected areas, allowing an indirect validation of the results. We have shown that the SBInSAR-assisted phase unwrapping is efficient for X-band images with a large range bandwidth like spotlight images.

We suggested two new criteria to select frequency-persistent scatterers, one based on the phase variance stability and the other on the slope standard deviation. Both were tested and compared to the multifrequency phase error based on the aspect ratio of the offsets distribution. With each one of the three selectors, the expected corrections were retrieved. However, the phase variance stability showed more precise estimates, though the standard deviation of the slope gave satisfactory results. Both appeared to be more efficient than the multifrequency phase error.

When we applied the SBInSAR-assisted phase unwrapping to naturally disconnected regions as well, without validation of the phase-offset correction, we noted a factor larger than two between the amount of regions corrected using the stability of the phase variance and using the standard deviation of the slope. The phase variance stability appears to be too demanding for hlsmall regions. So far, no correlation has been observed between the size of a patch and the number of stable targets detected in this patch. As a consequence, the minimum size of a corrigible region could not be determined.

Conventional phase unwrapping algorithms, such as SNAPHU [19–21], propose deformation mode to handle abrupt deformations or normalization options to level the phase of close areas. In this case, the consequence may be that the phase ambiguity is not an integer number of cycles anymore and an a priori estimation of the deformation gradient might be necessary. With the SBInSAR-assisted phase unwrapping, disconnected regions are reconnected by correcting an offset that is an integer number of cycles, preserving thus phase information integrity. Moreover, this leveling of the phase is done by exploiting only the spectral information of SAR images and does not require additional data, like on-ground measurements (e.g., GPS). The drawback of this approach is that it will not reconnect small regions because of the need for a sufficient population of stable targets. Additionally, spectral decomposition is demanding regarding memory and computing time. However, this technique keeps all its interest for practical cases with steep topography, local coherence losses, geometrical distortions or high-gradient deformations leading to phase unwrapping issues.

Future work will focus on the applicability of the method to images with smaller bandwidth and the definition of the best selector of PS_f in a standard case. Decreasing the bandwidth to 100–150 MHz, we expect a reduced population of PS_f due to the loss of resolution but still reasonable results. In addition, the method has been validated on disconnected areas with an important initial population (>105 pixels) and consequently with a higher probability to include stable targets. The next step will be to apply and validate it for smaller regions. Finally, due to the dependency of the split-band phase accuracy on the frequency, we expect better results with C-band or L-band data. In the future, the SBInSAR-assisted phase unwrapping will be tested on Sentinel-1 data.

Acknowledgments: This work was carried out in the framework of the MUZUBI (MUlti-Zone phase Unwrapping using advanced split-Band Interferometry) and RESIST (REmote Sensing and In Situ detection and Tracking of geohazards) projects funded respectively by the Belgian Science Policy contracts Nos. SR/00/324 and SR/00/305. Test data were provided by the DLR in the framework of the TanDEM-X Science Phase announcement of opportunity for the project "Split-Band InSAR monitoring of Virunga and Copahue volcanoes using TanDEM-X."

Author Contributions: Ludivine Libert developed the theoretical aspects of the frequency-persistent scatterers selection and the SBInSAR-assisted phase unwrapping approach, and processed the data. Dominique Derauw developed the split-band processor and provided technical support for its use. Nicolas d'Oreye, Christian Barbier and Anne Orban participated to the data analysis and interpretation. All of the authors participated in editing and reviewing the manuscript.

Conflicts of Interest: The authors declare no conflict of interest.

Abbreviations

The following abbreviations are used in this manuscript:

PS_f	Frequency-Persistent Scatterer(s)
SAR	Synthetic Aperture Radar
InSAR	Synthetic Aperture Radar Interferometry
SBInSAR	Split-Band Interferometry
DEM	Digital Elevation Model

References

1. Bovenga, F.; Giacovazzo, V.M.; Refice, A.; Nitti, D.O.; Veneziani, N.V. Interferometric Multi-Chromatic Analysis of High Resolution X-Band Data. In Proceedings of the Fringe 2011 Workshop, Frascati, Italy, 19–23 September 2011.
2. Veneziani, N.; Bovenga, F.; Refice, A. A Wide-Band Approach to the Absolute Phase Retrieval in SAR Interferometry. *Multidimens. Syst. Signal Process.* **2003**, *14*, 183–205.
3. Bovenga, F.; Giacovazzo, V.M.; Refice, A.; Veneziani, N. Multichromatic Analysis of InSAR Data. *IEEE Trans. Geosci. Remote Sens.* **2013**, *51*, 4790–4799.
4. Bovenga, F.; Rana, F.M.; Refice, A.; Veneziani, N. Multichromatic Analysis of Satellite Wideband SAR Data. *IEEE Geosci. Remote Sens. Lett.* **2014**, *11*, 1767–1771.

5. De Rauw, D.; Kervyn, F.; d'Oreye, N.; Albino, F.; Barbier, C. Split-Band Interferometric SAR Processing Using TanDEM-X Data. In Proceedings of the FRINGE'15: Advances in the Science and Applications of SAR Interferometry and Sentinel-1 InSAR Workshop, Frascati, Italy, 23–27 March 2015.

6. Bovenga, F.; Giacovazzo, V.M.; Refice, A.; Veneziani, N.; Vitulli, R. Multi-Chromatic Analysis of InSAR Data: Validation and Potential. In Proceedings of the Fringe 2009, Frascati, Italy, 30 November–4 December 2009.

7. Bovenga, F.; Derauw, D.; Rana, F.M.; Barbier, C.; Refice, A.; Veneziani, N.; Vitulli, R. Multi-Chromatic Analysis of SAR Images for Coherent Target Detection. *Remote Sens.* **2014**, *6*, 8822–8843.

8. Ferretti, A.; Prati, C.; Rocca, F. Permanent Scatterers in SAR Interferometry. *IEEE Trans. Geosci. Remote Sens.* **2001**, *39*, 8–20.

9. Derauw, D.; Orban, A.; Barbier, C. Wide Band SAR Sub-Band Splitting and Inter-Band Coherence Meaurements. *Remote Sens. Lett.* **2010**, *1*, 133–140.

10. Rosen, P.A.; Hensley, S.; Chen, C. Measurement and mitigation of the ionosphere in L-band Interferometric SAR data. In Proceedings of the 2010 IEEE Radar Conference, Arlington, VA, USA, 10–14 May 2010; pp. 1459–1463.

11. Furuya, M.; Suzuki, T.; Derauw, D. A step-by-step recipe of band splitting technique for isolation of ionospheric signal in L-band InSAR data. In Proceedings of the AGU Fall Meeting, San Francisco, CA, USA, 12–16 December 2016.

12. Schreiber, R.; Moreira, A. Coregistration of Interferometric SAR Images Using Spectral Diversity. *IEEE Trans. Geosci. Remote Sens.* **2000**, *38*, 2179–2191.

13. Jiang, H.; Feng, G.; Wang, T.; Bürgmann, R. Toward full exploitation of coherent and incoherent information in Sentinel-1 TOPS data for retrieving surface displacement: Application to the 2016 Kumamoto (Japan) earthquake. *Geophys. Res. Lett.* **2017**, *44*, 1758–1767.

14. Naranjo, J.A.; Polanco, E. The 2000 AD eruption of Copahue Volcano, Southern Andes. *Rev. Geol. Chile* **2004**, *31*, 279–292.

15. Tamburello, G.; Agusto, M.; Caselli, A.; Tassi, F.; Vaselli, O.; Calabrese, S.; Rouwet, D.; Capaccioni, B.; Napoli, R.D.; Cardellini, C.; et al. Intense magmatic degassing through the lake of Copahue volcano, 2013–2014. *J. Geophys. Res. Solid Earth* **2015**, *120*, 6071–6084, doi:10.1002/2015JB012160.

16. Fournier, T.J.; Pritchard, M.E.; Riddick, S.N. Duration, magnitude, and frequency of subaerial volcano deformation events: New results from Latin America using InSAR and a global synthesis. *Geochem. Geophys. Geosystems* **2010**, *11*, doi:10.1029/2009GC002558.

17. Velez, M.L.; Euillades, P.; Caselli, A.; Blanco, M.; Díaz, J.M. Deformation of Copahue volcano: Inversion of InSAR data using a genetic algorithm. *J. Volcanol. Geotherm. Res.* **2011**, *202*, 117–126.

18. Goldstein, R.M.; Zebker, H.A.; Werner, C.L. Satellite radar interferometry: Two-dimensional phase unwrapping. *Radio Sci.* **1988**, *23*, 713–720.

19. Chen, C.W.; Zebker, H.A. Network approaches to two-dimensional phase unwrapping: intractability and two new algorithms. *J. Opt. Soc. Am. A* **2000**, *17*, 401–414.

20. Chen, C.W.; Zebker, H.A. Two-dimensional phase unwrapping with use of statistical models for cost functions in nonlinear optimization. *J. Opt. Soc. Am. A* **2001**, *18*, 338–351.

21. Chen, C.W.; Zebker, H.A. Phase unwrapping for large SAR interferograms: Statistical segmentation and generalized network models. *IEEE Trans. Geosci. Remote Sens.* **2002**, *40*, 1709–1719.

remote sensing

MDPI

Article

Better Estimated IEM Input Parameters Using Random Fractal Geometry Applied on Multi-Frequency SAR Data

Ali Ghafouri [1], Jalal Amini [1,*], Mojtaba Dehmollaian [2] and Mohammad Ali Kavoosi [3]

[1] School of Surveying and Geospatial Engineering, Faculty of Engineering, University of Tehran, Tehran 1439957131, Iran; ali.ghafouri@ut.ac.ir
[2] Center of Excellence on Applied Electromagnetic Systems, School of Electrical and Computer Engineering, University of Tehran, Tehran 1439957131, Iran; m.dehmollaian@ut.ac.ir
[3] Department of Geology, Exploration Directorate of National Iranian Oil Company, Tehran 1994814695, Iran; m.kavoosi@niocexp.ir
* Correspondence: jamini@ut.ac.ir; Tel.: +98-912-459-4685; Fax: +98-21-8860-4534

Academic Editors: Timo Balz, Uwe Soergel, Mattia Crespi, Batuhan Osmanoglu, Nicolas Baghdadi and Prasad Thenkabail
Received: 26 February 2017; Accepted: 4 May 2017; Published: 5 May 2017

Abstract: Microwave remote sensing can measure surface geometry. Via the processing of the Synthetic Aperture Radar (SAR) data, the earth surface geometric parameters can be provided for geoscientific studies, especially in geological mapping. For this purpose, it is necessary to model the surface roughness against microwave signal backscattering. Of the available models, the Integral Equation Model (IEM) for co-polarized data has been the most frequently used model. Therefore, by the processing of the SAR data using this model, the surface geometry can be studied. In the IEM, the surface roughness geometry is calculable via the height statistical parameter, the *rms-height*. However, this parameter is not capable enough to represent surface morphology, since it only measures the surface roughness in the vertical direction, while the roughness dispersion on the surface is not included. In this paper, using the random fractal geometry capability, via the implementation of the power-law roughness spectrum, the precision and correctness of the surface roughness estimation has been improved by up to 10%. Therefore, the random fractal geometry is implemented through the calculation of the input geometric parameters of the IEM using the power-law surface spectrum and the spectral slope. In this paper, the in situ roughness measurement data, as well as SAR images at frequencies of L, C, and X, have been used to implement and evaluate the proposed method. Surface roughness, according to the operational frequencies, exhibits a fractal or a diffractal behavior.

Keywords: Synthetic Aperture Radar (SAR); Integral Equation Model (IEM); random fractal geometry

1. Introduction

Studying the surface roughness geometry, especially studying the natural surfaces, requires appropriate satellite data and processing methodologies [1–3]. The Synthetic Aperture Radar (SAR) data acquired by the airborne and space-borne sensors has made it possible to examine the surface roughness, which provides useful information for geoscientists and geologists. The backscattered signal in all polarizations is affected by surface roughness, and contains the surface geometry information [4–6]. To measure the surface geometry using the SAR data, the surface geometric parameter(s) must be modeled against the backscattering coefficient on each polarization. Generally speaking, to describe and differentiate the patterns and the geometric surface texture, it is indispensable to model the interaction between the backscattered signals and the surface properties. The radar signal is sensitive to certain surface roughness measures, which is determined by the operating radar signal frequency [7–9].

Of the available models, the standard theoretical models of backscattering include: (1) the Kirchhoff Approximation (KA), which encompasses Geometric Optics (GO) and Physical Optics (PO); and (2) the Small Perturbation Model (SPM) [7,10].

The Geometric Optics model for very rough surfaces, the Physical Optics model for moderate roughness, and the Small Perturbation Model for approximately smooth surfaces, have been used. Fung et al. (1992) have developed the Integral Equation Model (IEM) as a physically-based electromagnetic transfer model which combines the Kirchhoff models and the SPM, and which constructs a more applicable model that can theoretically tolerate a really wide range of roughness dimensions. It is of note that the IEM is not restricted to any special frequency range or roughness measures [11].

The IEM exploits the rms-height parameter to characterize the surface geometry [12]. In this paper, the statistical rms-height and the Gaussian auto-correlation function (ACF) have been implemented as the *conventional* IEM. In the conventional method, the surface roughness geometry is only considered in the vertical direction, and only the Euclidean geometry is applied.

However, using the so called *fractal* IEM, which considers roughness in the horizontal direction and its dispersion on the surface, it can be possible to improve the surface morphology estimation and soar the precision of the microwave discrimination by means of the power-law surface spectrum and the spectral slope parameter [13].

In this paper, as depicted in Figure 1, to provide the surface geometric characteristics for the IEM computation, the in situ surface roughness measurement has been performed by the field surveying operation. The dielectric constants have been extracted from the tables produced by Martinez et al. (2000) [14]. Furthermore, the SAR satellite data of ALOS PALSAR, Sentinel-C, and TerraSAR, respectively, in the bands L, C, and X, have been used to compute the surface roughness, i.e., the geological morphology.

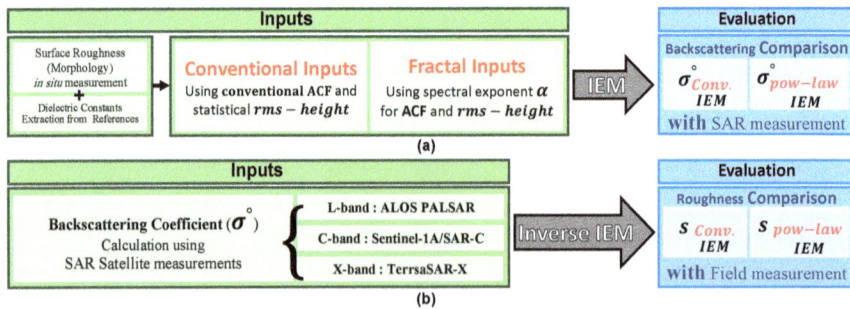

Figure 1. Flowchart of the study and results evaluation. (**a**) Direct model evaluation: computation of backscattering using the in situ measured data; (**b**) inverse model evaluation: calculation of surface roughness using the backscattering coefficient.

In this paper, the IEM geometric input parameters have been calculated using three different methods for two types of surface assumptions; two conventional geometry methods (stationary surface assumption) and the random fractal geometry (for power-law surface assumption). After running the model with each of the inputs, the backscattering coefficient at each point corresponding to each pixel of the SAR measurement data is computed (Figure 1a). In addition, through the IEM inversion method, the surface parameters are calculated, corresponding to the in situ measurements (Figure 1b). The IEM computation results, as well as the inverse IEM compared with the measurements, can show the model's level of efficacy.

The IEM calculated and the SAR measured backscattering coefficients are mutually compared to evaluate the methodology (Figure 1a). Additionally, comparing the in situ rms-height with the inverse IEM results, the level of efficacy when using the power-law inputs versus the conventional ones can be cleared [12,15] (Figure 1b).

In Section 2, the IEM backscattering model with the conventional inputs will be addressed. The methodology of random fractal geometry will be developed in Section 3, in order to determine the power-law surface parameters as the input of the IEM. Finally, the implementation and validation of the fractal IEM will be discussed in Section 4 for the multi-frequency SAR data. Section 5 presents the conclusion.

2. The IEM Backscattering Model

Backscattering is the amount of the scattered signal per surface unit in the scattering angle from 0 to 180 degrees [16]. The calculation of the backscattering coefficient ($\sigma°$) process depends on the satellite antenna specifications. As defined by Fung et al. (1992) and explained by themselves (1994), the IEM model describes the relation of the backscattering coefficient to the surface roughness parameters, as well as the dielectric constant and the incidence angle. The IEM is characterized for the co-polarized and cross-polarized backscattering calculations [8,17]. Yet, in this paper, just the co-polarized equation is implemented. The co-polarized backscattering coefficient equation according to Fung et al. (2004), which is termed I^{2}EM by Ulaby (2014), is as follows [7,8,10,11,16]:

$$\sigma_{pp}° = \frac{k^2}{4\pi} e^{-2k^2 s^2 \cos^2\theta} \sum_{n=1}^{+\infty} |I_{pp}^n|^2 \frac{W^{(n)}(2k\sin\theta, 0)}{n!} \tag{1}$$

where:

$$I_{pp}^n = (2k \, s \, \cos\theta) f_{pp} exp\left(-k^2 s^2 \cos^2\theta\right) + (k \, s \, \cos\theta)^n F_{pp} \tag{2}$$

and *pp* is either the *hh* or *vv* polarizations; *k* stands for the radar wavenumber ($k = \frac{2\pi}{\lambda}$, λ wavelength); *s* is the rms-height; θ denotes the incidence angle; and $W^{(n)}$ is the Fourier transform of *n*th power of the ACF. f_{hh}, f_{vv}, F_{hh}, and F_{vv} are approximated by the following equations [10,16]:

$$f_{hh} = \frac{-2R_h}{\cos\theta} \tag{3a}$$

$$f_{vv} = \frac{2R_v}{\cos\theta} \tag{3b}$$

$$F_{hh} = 2\frac{\sin^2\theta}{\cos\theta}\left[4R_h - \left(1 - \frac{1}{\varepsilon}\right)(1 + R_h)^2\right] \tag{3c}$$

$$F_{vv} = 2\frac{\sin^2\theta}{\cos\theta}\left[\left(1 - \frac{\varepsilon \cos^2\theta}{\varepsilon - \sin^2\theta}\right)(1 - R_v)^2 - \left(1 - \frac{1}{\varepsilon}\right)(1 + R_v)^2\right] \tag{3d}$$

where the horizontally and vertically polarized Fresnel reflection coefficients, R_h and R_v, are given by the following:

$$R_h = \frac{\cos\theta - \sqrt{\varepsilon - \sin^2\theta}}{\cos\theta + \sqrt{\varepsilon - \sin^2\theta}} \tag{4a}$$

$$R_v = \frac{\varepsilon \cos\theta - \sqrt{\varepsilon - \sin^2\theta}}{\varepsilon \cos\theta + \sqrt{\varepsilon - \sin^2\theta}} \tag{4b}$$

where ε is the ground relative dielectric constant; a complex number concerns the storage and dissipation of electricity.

In the IEM, *s* and the surface ACF are the two geometric elements whose calculations are the main issue considered in this paper.

Due to the nonlinearity of (1), the model inversion, i.e., solving the IEM for its parameters analytically, is almost impossible. There are a few arithmetic methods which can be employed to calculate the surface parameters knowing the backscattering coefficient and the imaging parameters [12]. In spite of the development of intelligent computation methods such as neural networks and the Bayesian method, one of the best and most direct methods in this respect is the Look-up table (LUT) [18,19]. In the LUT method, the backscattering coefficient values for different values of the surface roughness parameters and dielectric constant are calculated using (1), and then, the surface parameters corresponding to the backscattering coefficients can be calculated by interpolation and reversed matching.

Section 2.1 describes the validity range and Section 2.2 exhibits the conventional input parameters calculation of this model.

2.1. The IEM Validity Region

The IEM is applicable on a wide range of surfaces, from smooth to rough. The IEM's validity range, given by Fung et al. (1994), is:

$$ks < 3 \tag{5a}$$

$$(kl)(ks) < \mu\sqrt{|\varepsilon|} \tag{5b}$$

$$cos^2\theta \frac{(ks)^2}{\sqrt{0.46kl}} exp\left[-\sqrt{2 \times 0.46\, kl\, (1 - sin\theta)}\right] \ll 1 \tag{5c}$$

where μ is a constant and its Gaussian value, and the exponential ACFs are 1.6 and 1.2, respectively [10,20].

Dierking (1999) has also estimated the validity range of the IEM for different frequencies, similar to the ones specified by Fung et al. [21]. Furthermore, different studies have concerned the real validity range of this model and it is mostly shown that the applicable validity domain of this model is more limited [18,22,23].

2.2. Conventional Inputs for IEM

The rms-height is calculated using the following formula (6):

$$s = \sqrt{\frac{1}{N}\left[\left(\sum_{i=1}^{N} z_i^2\right) - N\bar{z}^2\right]} \;\; \forall\, \bar{z} = \frac{1}{N}\sum_{i=1}^{N} z_i \tag{6}$$

Moreover, $W^{(n)}$ is another geometric term in the IEM, i.e., the Fourier transform of the *n*th power of the surface ACF, $A(x)$, either its exponential, or Gaussian regression [24,25] (i.e., respectively (7)–(9)):

$$A(x) = \frac{\sum_{i=1}^{N-j} z_i z_{i+j}}{\sum_{i=1}^{N} z_i^2} \tag{7}$$

$$A(x) = e^{\frac{-|x|}{l}} \tag{8}$$

$$A(x) = e^{\frac{-x^2}{l^2}} \tag{9}$$

where l is the correlation length. Because of its dependence on the profile length, as well as the rms-height, the calculation of the correlation length has been considered as a difficult problem [9,26]. In this paper, both exponential and Gaussian ACFs are used for conventional implementation, which have presented a better estimation and have been frequently recommended [3,12,15]. In the calculation of the ACFs, the correlation length is considered as one third of the semivariogram range [27].

3. Power-Law Inputs for the IEM

Functions, like y, proportional to some power of the input x (i.e., $y = x^p$), are power-law functions. Power-law surfaces have the spectrum in the form of $S(f) = c/f^\alpha$ (α: spectral slope). In natural surface modeling, unlike the conventional geometry, the random fractal geometry provides better results. The term "fractal", in many cases, is considered as the same as "self-similar". Natural phenomena are statistically self-similar; i.e., every part of their structure has statistical properties (the mean and the standard deviation) similar to those of the whole structure [28].

The surfaces with a power-law roughness spectrum over the interval $0 \le f \le \infty$ are ideal random fractals. For such surfaces, the rms-height s, the correlation length l, and the ACF do not exist [21,29] Thus, the surface modeling must be realized within the limits of the spatial frequencies (i.e., sampling rates) $f_{min} \le f \le f_{max}$. This is a semi-self-affine surface [13,30]. In other words, using the fractal geometry, natural surfaces can be modeled through the power-law form within f_{min} and f_{max}.

Yordanov et al. (2002) present a general expression for the ACF valid for the arbitrary topological dimension [31]. Since after the roughness measurement of the study sites, the geometric parameters are considered to be calculated on some arbitrary linear profiles, the ACF for a linear profile of the samples can be realised in the form of [32]:

$$A(x) = 2\pi^{D-1} \int_{k_{min}}^{k_{max}} k^{D-1-\alpha} \cos(kx)dk \tag{10}$$

where k is the wavenumber and α has a limited amount; $D < \alpha < D+2$, D is the topological dimension, and for the linear profile, it is considered one ($D = 1$) [29,33], representing the slope of the linear best-fit of the power spectral density (PSD) on a logarithmic scale [29,32].

It is noted that this linear best-fit must be applied to the trendless profile. In other words, the spectral slope (α) might not include the trend part of the PSD, the early part of the PSD [29].

According to Dierking (1999), in addition to the ACF, two other regular geometric surface parameters, the power-law rms-height and the correlation length for a linear profile, can also be calculated using the spectral slope parameter [21]:

$$s = \sqrt{cL^{\alpha-1}/(\alpha-1)} \tag{11}$$

$$l = \frac{(\alpha-1)^2 L}{2(2\alpha-1)} \tag{12}$$

where L is the profile length, and c is the spectral constant or the spectral offset [21].

Hence, in order to evaluate the role of the random fractals for improving the IEM results, having ascertained the surface morphology, using (8) and (9), the ACF for conventional-1 and conventional-2 methods, respectively, is achieved. The rms-height (s) can be calculated using (6) for these two conventional methods.

For the power-law calculation of the input parameters, the welch method is applied to the in situ measured data to calculate the spectral slope, α. Having obtained α and using (11), the rms-height (s) is computed. Using (10), the ACF is computed.

4. Implementation and Evaluation

Three different methods for the calculation of the input parameters of the IEM are implemented for the in situ field measurement to apply the methodology of the flowchart (Figure 1a). Afterward, following the second flowchart (Figure 1b), the LUT can be formed to determine the surface roughness of the terrestrial points corresponding to each measured SAR backscattering (on L, C, and X-bands) pixels. The case study is located in western Iran, between the cities of Ilam and Dehloran. From the view point of geology, this region is known as the "Anaran" anticline, located in the Zagros fold-thrust belt. Figure 2 depicts the geographic location of the case study and the 10 measurement sites. Figure 2b depicts the main geological formations of the region; however, the morphology of two of them (Gurpi Fm and Quaternary Sediments) is not considered to have been studied in this paper.

The field measurement is planned in the form of a mesh of points; thus, the geometric parameters for any arbitrary profiles on the surface can be calculated. The in situ surface roughness measurement is performed on ten different sites with different morphologies using the programmable surveying Total Station (Trimble™5600). This instrument is used to measure a 25 × 25 mesh with a 50 cm interval on each measurement site. The measurement instrument has a better than one centimeter precision. Being measured as a mesh of points allows for the calculation of roughness parameters along any arbitrary profile and certainly increases the precision of the parameters estimation. Generally, there are three formations belonging to different geological periods. Table 1 shows the details of these geological formations.

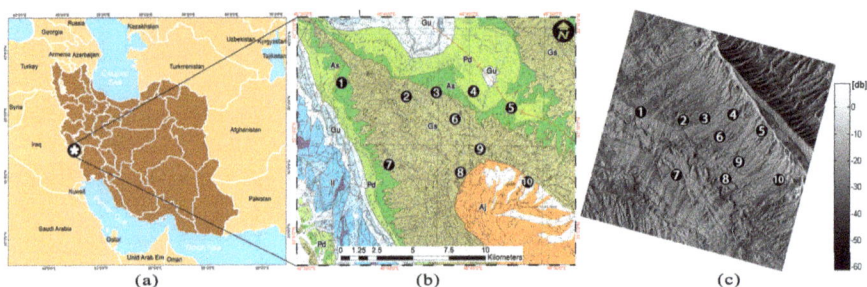

Figure 2. (**a**) The case study location on a map of Iran; (**b**) Surface roughness measurement sites position on the geological map (*As*, Asmari Fm; *Gs*, Gachsaran Fm; *Pd*, Pubdeh Fm; *Gu*, Gurpi Fm; *Qu*, Quaternary Sediments); (**c**) Measurement sites position on the X-Band SAR backscattering image; the color-bar of the values of backscattering is beside the image in db.

Table 1. Geological age and general specification of the morphology of the study area formations.

Geological Formation	Geological Period	General Morphology Appearance
Asmari (As)	Oligocene to Lower Miocene	Rough
Gachsaran (Gs)	early Miocene	Moderate
Pabdeh (Pd)	Paleocene to early Miocene	Smooth

Furthermore, the SAR measured data are described in Table 2. Three SAR measurement data are used in three microwave bands. The acquisition modes are also specified.

Table 2. Specifications of the SAR measurement data in three different frequencies for the study.

Satellite/Antenna	Acquisition Mode	Band	Freq. (GHz)	Incidence Angle/Pass/Look
ALOS–Palsar-2	Spot-light	L	1.200	32.3°/Ascending/Left
Sentinel-1A	Strip Map Mode	C	5.405	38.1°/Ascending/Right
TerraSAR-X	Staring Spot-Light	X	9.650	22.7°/Descending/Left

The earth surface roughness of a single study area in each of the radar bands is seen differently, due to the difference in the operational frequency (or wavelength). Figure 3 depicts the face of the formations surface and shows the extent to which they are affected by weathering and erosion. Obviously, this variety of roughness must not be seen as the same in different wavelengths. The smoother surfaces are more susceptible to erosion, whereas the rougher ones are more resistant against weathering. The level to which a lithology gets eroded is dependent on its material and its age.

Figure 3. The geological formations surface. (**a**) Asmari (As) fm., Oligocene to Lower Miocene; (**b**) Gachsaran (Gs) fm., early Miocene; (**c**) Pabdeh (Pd) fm., Paleocene to early Miocene. The extent to which they are affected by weathering and erosion can be seen. The photos are taken from ~1 m above the surface.

It must be considered that prior to any backscattering processing, as well as results assessment, the SAR imagery data introduced in Table 2, which are contaminated by the speckle noise, were subjected to a pre-processing stage, i.e., despeckling using the wavelet transform method [34].

Figure 4 illustrates how the roughness of the formations appears in the SAR images of different frequencies. The backscattering in the images is shown in the equalized black and white images, but the side color bars show the pixel values in db. The whiter the images, the rougher the face because of double, triple, and in general, multi-scattering signals. The blacker pixel values are because of the smoothness of the surface.

Figure 4. Backscattering on the SAR measurement, as equalized black and white images; (**a**) the L-Band SAR data, the *ALOS-PALSAR* satellite; (**b**) the C-Band SAR data, the *Sentinel-1A* satellite; (**c**) the X-Band SAR data, the *TerraSAR-X* satellite; the value of backscattering is depicted beside each image on the color-bar in db. Lower values, i.e., dark pixels, show smoothness and large db values are because of the surface roughness.

The numerical geometric characteristics on each site surface are tabulated in Table 3. The rms-height and the correlation length are calculated using the conventional geometry; the α and the fractal dimension using the power-law geometry.

Since there was a considerably dry climatic condition when the satellite data was being acquired, and the field measurement was performed in the same seasonal conditions, it was obviously perceived that the moisture until 5 cm below the surface was absolutely 0%; that is why the dielectric constant values were used from the references who presented these values for the dry climate [14]. The values for the three geological formations of the study area, i.e., Asmari, Gachsaran, and Pabdeh, are 3.6, 4.0, and 4.1, respectively.

Figure 5 illustrates how the spectral slope (α) is calculated using the in situ measurements of the PSD of the 10 selected study sites. The PSDs are estimated using the Welch's power spectral density [35,36]. According to the previous section, the data trend or the result of large topography must not be considered in the measurement of the spectral slope (α). For this aim, the red line (i.e., a linear regression) in the plots of Figure 3 is drawn on a polynomial regression of the PSD, notwithstanding the general trend part of the PSD. The experimental analysis obviously shows that the regression of the PSD for frequencies between 0.85 and 3.15 (Hz) results in the best values for the spectral slope. This empirically decided frequency range is somehow equal for different sites.

Table 3. Morphological and surface geometric characteristics of the study sites based on the site visit and the in situ measurement. The roughness parameters are calculated using the equations and methods presented in Section 3. Figure 5 shows how α is estimated for each site surface. The Rms-height and the correlation length are the conventional ones and the average of the multiple arbitrary profiles on each of the site surfaces.

Site No.	Dominant Geological Formation	rms-Height (cm)	Corr. Length (cm)	α
Site 1	As	6.02	81.07	1.5345
Site 2	Gs	2.2	45.3	1.7112
Site 3	As-Gs	4.64	54.07	1.6865
Site 4	Pd	1.11	14.9	1.7923
Site 5	Pd	1.21	18.03	1.8825
Site 6	Gs	1.98	40.71	1.8817
Site 7	As	4.97	74.38	1.5349
Site 8	Gs	1.66	19.31	1.7501
Site 9	Gs	2.86	50.9	1.7709
Site 10	Qu	1.76	31.31	1.7472

Table 4 shows the capability of each radar frequency for the discrimination of each formation. The conformity in this table is determined according to the contents of Tables 2 and 3.

Table 4. The status of the validity of the IEM based on (5a,b,c) for each geological formation on each of the study bands, based on Fung et al. (1994) [10]. Check marks show the conformity of the formation roughness parameters (on Table 3) to the IEM constraints at each microwave band and the X marks show the non-conformity.

Geological Formation	L-Band			C-Band			X-Band		
	(5a)	(5b)	(5c)	(5a)	(5b)	(5c)	(5a)	(5b)	(5c)
Asmari Fm.	✓	✓	✓	✓	✓	✓	×	×	×
Gachsaran Fm.	✓	×	✓	✓	✓	✓	✓	×	✓
Pabdeh Fm.	✓	×	✓	✓	✓	✓	✓	✓	✓

In this table, it becomes clear which of the three constraints determined by Fung et al. (1994) via (5a,b,c) are approved or rejected in order to morphologically differentiate the formations. In this table, the approbation (✓) of an equation indicates the validity of the constraint for the relevant band and the rejection (×) of the equation means the invalidity of the constraint.

4.1. Geometric Parameters Analysis

The microwave signal, depending on the radar frequency, is sensitive to a limited range of the roughness spectrum; thus, a surface seen as being rough in one frequency measurement may be seen as completely smooth in another. Figure 6 represents the backscattering of the incident radar signal in the radar bands L, C, and X on *hh*, *vv*, and *hv* polarizations for the earth surface of the main geological formations present in the field measured sites. As the frequency or the local incident angle increases, the smaller roughness scale becomes significant. The graphs confirm the different capability of the operational frequencies for the surfaces discrimination, presented in Table 4.

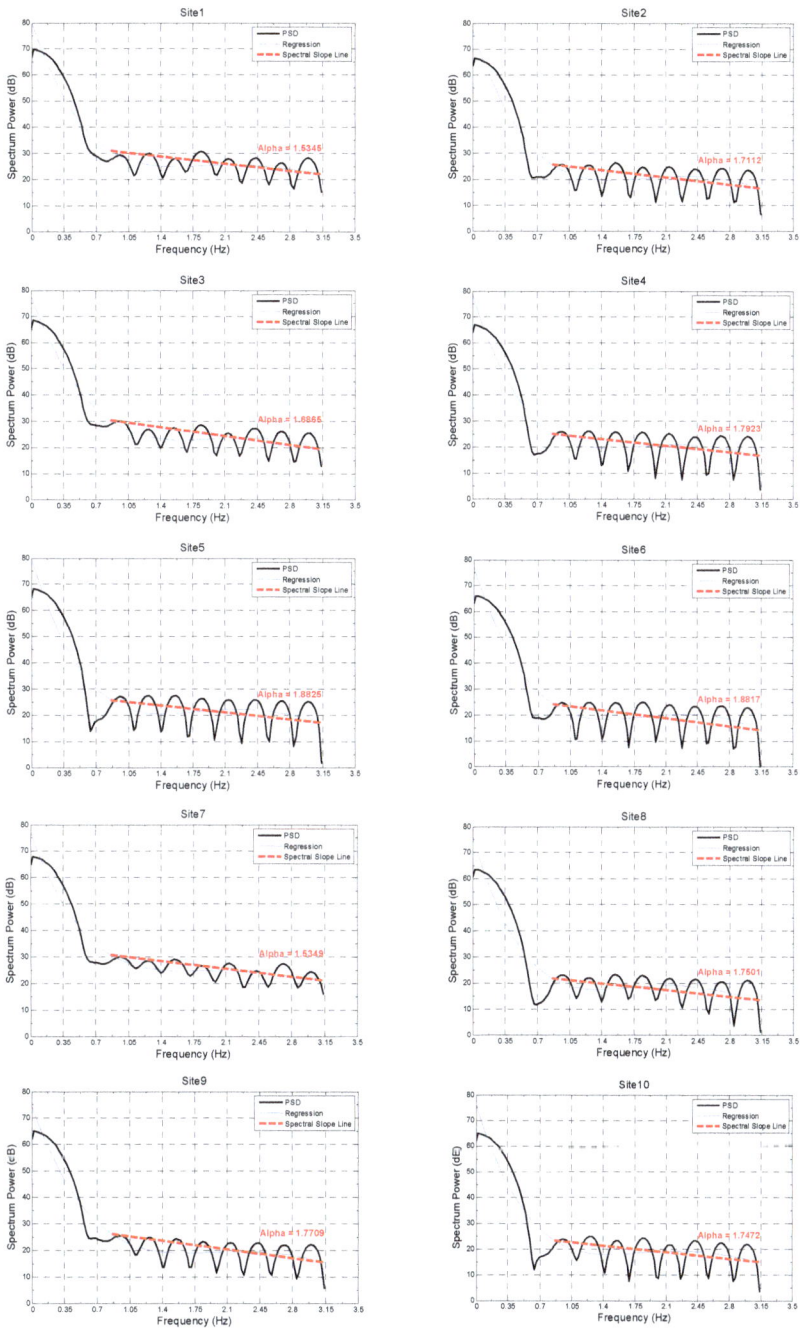

Figure 5. Calculation of the spectral slope (α) via the linear regression of a polynomial regression of the PSD, calculated from the in situ measurements of the 10 selected study sites using the Welch's method. Experimentally, the trend of the data is between 0 and 0.85 Hz, which has not been considered in the calculation of the spectral slope.

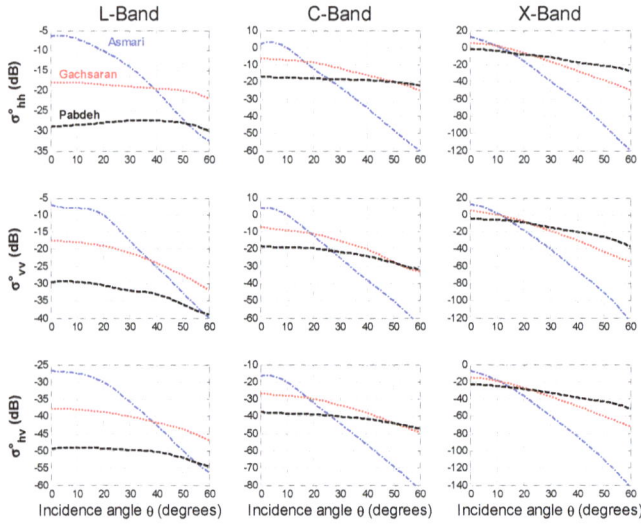

Figure 6. Backscattering (σ°) for three bands L, C, and X on *hh*, *vv*, and *hv* polarizations for the main geological formations of the study area.

4.2. Evaluation of Backscattering and Surface Roughness Simulation

What is needed to discriminate geological formations according to their morphology is the surface roughness [25]. In this section, the results of the IEM implementation to simulate the backscattering coefficient (σ°) using the surface data and a comparison with the SAR measured ones in the radar bands L, C, and X for the surface roughness computation is described (Figure 1a). Via inverse IEM using the LUT, having ascertained the SAR measurements, the rms-height of the surface is simulated. An evaluation of the simulated surface roughness can be performed using the rms-height of the field measurements (Figure 1b).

Having obtained the surface geometric characteristics for a 25 × 25 points mesh and then, taking arbitrary profiles according to geological maps and the site visit, a matrix 25 × 25 of σ° is computed for each site. Moreover, using the IEM inversion, by the values of σ° on a 25 × 25 pixels SAR image, a 25 × 25 matrix of the rms-height can be calculated for each of the sites.

The statistical rms-height, as well as the ACF, are calculated beside the extracted dielectric constant of each site from the published measured values to compute the conventional IEM backscattering coefficient, i.e., the values of σ°_{hh} and σ°_{vv}. The power-law IEM is computed using the ACF and the rms-height is calculated using the spectral slope parameter.

Figures 7–9, respectively, for the bands L, C, and X, present the comparisons between the backscattering coefficients calculated through the three methods of calculation of the IEM input parameters, namely two "Conventional" methods (exponential and Gaussian ACFs) and a "Power-law" method, in comparison with the measured SAR backscattering coefficient, referred to as an "SAR Measurement". This comparison is performed on 30 randomly selected pixels for 10 measured sites. Precisely being a point on the diagonal line of each graph indicates that the IEM simulated value on the corresponding pixel is exactly equal to the measured backscattering on that pixel. Therefore, in these graphs, the farness of the diagonal line shows the simulation error [25,37].

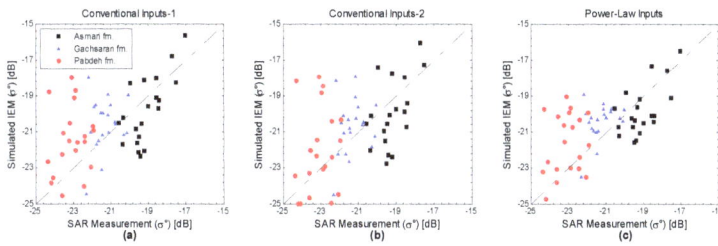

Figure 7. Backscattering simulation accuracy via the IEM for the geological formations on the L-band, (**a**) using the conventional geometry and the exponential ACF; (**b**) using the conventional geometry and the Gaussian ACF; (**c**) using the Power-Law geometry for the calculation of the IEM input parameters.

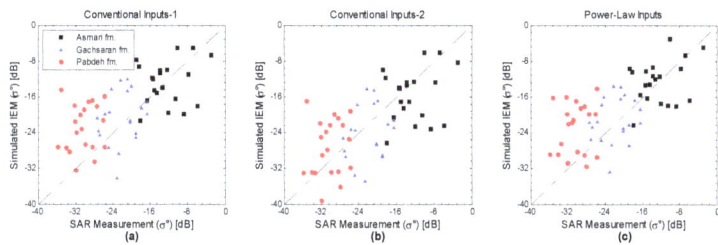

Figure 8. Backscattering simulation accuracy via the IEM for the geological formations on the C-band, (**a**) using the conventional geometry and the exponential ACF; (**b**) using the conventional geometry and the Gaussian ACF; (**c**) using the Power-Law geometry for the calculation of the IEM input parameters.

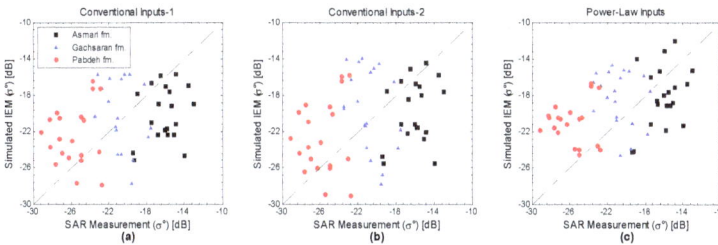

Figure 9. Backscattering simulation accuracy via the IEM for the geological formations on the X-band, (**a**) using the conventional geometry and the exponential ACF; (**b**) using the conventional geometry and the Gaussian ACF; (**c**) using the Power-Law geometry for the calculation of the IEM input parameters.

For an assessment of the aforementioned calculation methods, it is necessary to compare the simulated surface roughness parameter (rms-height) resulting from the inverse IEM (i.e., extracted from Look-up table) with the in situ measured parameter for the ten sites. For this comparison, the point graphs of Figures 10–12 depict the comparison of the computed rms-height of 30 selected pixels on bands L, C, and X, and the measured values on their corresponding field points.

Precisely being a point on the diagonal line of each graph indicates that the simulated rms-height using the inverse IEM on the corresponding pixel is exactly equal to the field measured rms-height. Evidently, the proximity of the points to the diagonal line announces the simulation correctness [37].

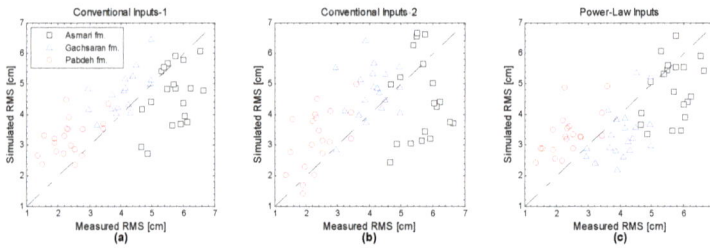

Figure 10. The Rms-height simulation accuracy via the inverse IEM for the geological formations on the L-band, (**a**) Using the conventional geometry and the exponential ACF; (**b**) using the conventional geometry and the Gaussian ACF; (**c**) using the Power-Law geometry for the calculation of the IEM input parameters.

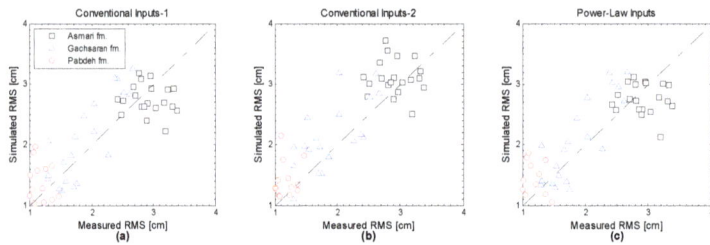

Figure 11. The Rms-height simulation accuracy via the inverse IEM for the geological formations on the C-band, (**a**) Using the conventional geometry and the exponential ACF; (**b**) using the conventional geometry and the Gaussian ACF; (**c**) using the Power-Law geometry for the calculation of the IEM input parameters.

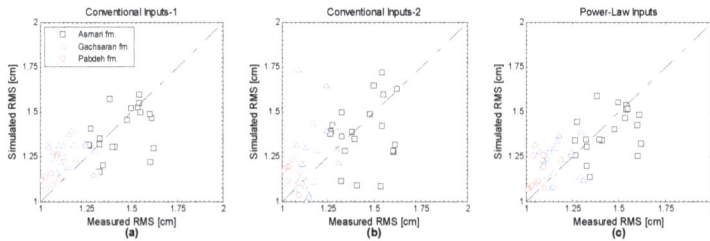

Figure 12. The Rms-height simulation accuracy via the inverse IEM for the geological formations on the X-band, (**a**) using the conventional geometry and the exponential ACF; (**b**) using the conventional geometry and the Gaussian ACF; (**c**) using the Power-Law geometry for the calculation of the IEM input parameters.

In Table 5, the standard deviation of the calculated rms-height in the methods of IEM implementation are tabulated. Therefore, the statistical dispersion of the calculated results on each site for each polarization versus the field measured values are comparable, in order to evaluate the efficiency of the proposed power-law calculation of the inputs instead of the conventional one. Evidently, the smaller the standard deviation, the higher the accuracy of the simulated data.

Table 5. The standard deviation of the rms-height, calculated using the inverse IEM for the geological formations on bands L, C, and X, using three methods of inputs calculation, two conventional (1: The exponential ACF; 2: The Gaussian ACF) and the power-law geometry. The bold values show a decrease of the standard deviation of using the power-law inputs versus both the conventional ones.

Geological Formation	L-Band			C-Band			X-Band		
	As.	Gs.	Pd.	As.	Gs.	Pd.	As.	Gs.	Pd.
Inputs using the Conventional Geometry1	1.049	0.919	0.873	0.352	0.378	0.505	0.207	0.239	0.341
Inputs using the Conventional Geometry2	0.985	0.799	0.834	0.375	0.309	0.506	0.232	0.283	0.355
Inputs using the Power-law Geometry	**0.839**	**0.721**	0.970	**0.326**	0.370	0.512	0.224	0.240	**0.285**

Comparing the values of the standard deviation in Table 5 shows that the use of random fractals geometry has improved the results by more than 10% on average, i.e., dividing the largest standard deviation values of the two conventional methods by the standard deviation of the Power-law method can give such a value. The bold numbers are used to indicate that the Power-law geometry overcomes both of the conventional methods. For both co-polarizations, the same improvement is achieved.

Figure 13 illustrates the standard deviation values on the bar-charts to easily compare the different frequencies, as well as the different formations.

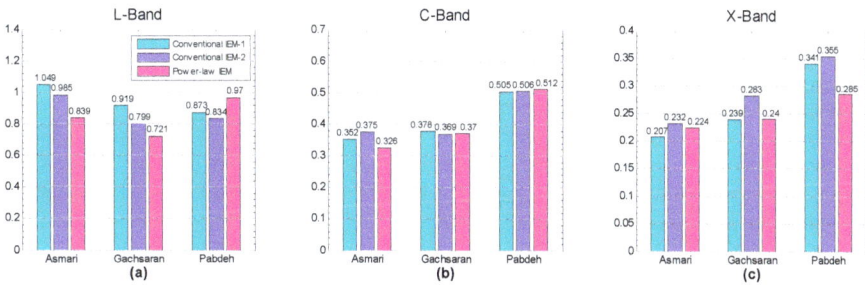

Figure 13. The Standard Deviation of the calculated rms-height values using the IEM Look-up table calculated via three methods of inputs calculation, conventional-1: using the exponential ACF, the conventional-2: using the Gaussian ACF, and the power-law geometry for the geological formations of the study sites, using the backscattering coefficient on (**a**) L-band; (**b**) C-band; and (**c**) X-band.

5. Discussion

In this paper, the Integral Equation Model (IEM) backscattering model, as well as its inverse (using the Look-up table method), are implemented to measure the geometry of the earth surface roughness or morphology. For such implementation, the input parameters are calculated using three different methods: two conventional geometries (the exponential and the Gaussian correlation functions, the ACFs) and one power-law method. The implementation of the fractal geometry in this paper is performed through the spectral slope parameter (α), which shows the coincidence of the power-law ACF with the earth surface. The IEM results (i.e., $\sigma°$, backscattering coefficients) are compared with the Synthetic Aperture Radar (SAR) measurement (i.e., *TerraSAR-X*) and the inverse IEM results (i.e., rms-height) were compared with the in situ measurement.

The fractal nature of the roughness causes a higher efficiency of the power-law geometry versus the conventional geometry when modeling the surface; Figures 7–12 show better results for the fractal modeling and eliminate more uncertainties, affirming [21,38] studies. Comparing the two conventional geometry implementations, apart from the C-band as the moderate wavelength, the Gaussian ACF causes less deviation for higher wavelengths, and the exponential ACF results are better for lower wavelengths; this supports the deductions presented in [12,37,39]. The values of the standard deviation in Table 5 show a general improvement, but when considering the behavior of different geological

Remote Sens. **2017**, *9*, 445

surfaces with various levels of roughness on each of the SAR frequency image, we deduce that for lower frequencies (e.g., L-band), the rougher surfaces are in the fractal regime, confirming the conclusions presented in [21,40]; and for the higher frequencies (e.g., X-band), the smoother ones are in accordance with [6]. Therefore, the rougher surfaces in high frequencies and the smoother ones in low frequencies display an obvious diffractal behavior. In general, having the in situ measurements, applying the most famous backscattering model, and completing an evaluation with the SAR measured data, the results of significant studies such as [17,38,41,42] are somehow endorsed with comparable numerical results, showing the efficiency of applying the power-law methods for IEM implementation for the SAR backscattering studies.

6. Conclusions

Using the random fractal geometry via the implementation of the power-law model in the backscattering modeling offers results which are up to 10% better for the calculation of the surface geometry. This methodology can improve the results of the spectral processings in the morphological mapping of the formations. Additionally, using the achieved information of morphology, we improved the scale of the geological maps.

For each radar frequency, the specific size of the surface roughness can be measured via signal backscattering. Using the three SAR bands L, C, and X for the earth surface of three known-geological surfaces shows the capability of this technology in geological morphology mapping.

The selection of the sites in this paper was undertaken considering that no large-scale topography has been inside the sampling mesh. Moreover, for the calculation of the spectral slope parameter (α), the trend part of the Power Spectral Density (PSD) graph has been ignored, and the spectral slope line was drawn on a polynomial regression of the PSD, which experimentally offered more accurate results.

A proper estimation of the surface roughness spatial frequency and consequently the spectral slope α, using the power-law ACF, offered appropriate IEM input parameters computation; therefore, after running the model, a more precise backscattering coefficient which exhibited a lower standard deviation was achieved. Also, using the random fractal geometry, the model inversion with the Look-up table method offered better roughness approximation. However, the geometric behavior of the surface roughness against the SAR frequencies is not constant, i.e., as the micro-topography decreases, the fractal regim exists in higher frequencies, and vice versa.

Acknowledgments: The authors would like to thank the University of Tehran Vice Chancellor for the Research and also the Research and Development department of NIOC Exploration Directorate for the support of this paper. All the respectable reviewers are acknowledged for their fruitful comments and suggestions about the paper.

Author Contributions: A.G., J.A., M.D., and M.A.K. contributed to conception. A.G. and M.A.K. carried out the acquisition, analysis, and interpretation of data. A.G. drafted the manuscript. All authors read and approved the final manuscript.

Conflicts of Interest: The authors declare that they have no competing interests.

References

1. Panciera, R.; Tanase, M.A.; Lowell, K.; Walker, J.P. Evaluation of iem, dubois, and oh radar backscatter models using airborne l-band sar. *IEEE Trans. Geosci. Remote Sens.* **2014**, *52*, 4966–4979. [CrossRef]
2. Rodrigues, F.Á.; Neto, J.R.; Marques, R.P.; de Medeiros, F.S.; Nobre, J.S. Sar image segmentation using the roughness information. *IEEE Geosci. Remote Sens. Lett.* **2016**, *13*, 132–136. [CrossRef]
3. Zhu, L.; Walker, J.P.; Ye, N.; Rudiger, C. The effect of radar configuration on effective correlation length. In Proceedings of the 2016 International Conference on Electromagnetics in Advanced Applications (ICEAA), Cairns, Australia, 19–23 September 2016; pp. 820–823.
4. Baghdadi, N.; Zribi, M.; Paloscia, S.; Verhoest, N.E.; Lievens, H.; Baup, F.; Mattia, F. Semi-empirical calibration of the integral equation model for co-polarized l-band backscattering. *Remote Sens.* **2015**, *7*, 13626–13640. [CrossRef]

5. Gorrab, A.; Zribi, M.; Baghdadi, N.; Mougenot, B.; Chabaane, Z.L. Potential of x-band terrasar-x and cosmo-skymed sar data for the assessment of physical soil parameters. *Remote Sens.* **2015**, *7*, 747–766. [CrossRef]
6. Martínez-Agirre, A.; Álvarez-Mozos, J.; Lievens, H.; Verhoest, N.E.; Giménez, R. Sensitivity of c-band backscatter to surface roughness parameters measured at different scales. In Proceedings of the 2015 IEEE International Geoscience and Remote Sensing Symposium (IGARSS), Milan, Italy, 26–31 July 2015; pp. 700–703.
7. Ulaby, F.T.; Long, D.G. *Microwave Radar and Radiometric Remote Sensing*; University of Michigan Press: Ann Arbor, MI, USA, 2014; pp. 425–445.
8. Baghdadi, N.; Cresson, R.; Pottier, E.; Aubert, M.; Mehrez, M.; Jacome, A.; Benabdallah, S. A potential use for the c-band polarimetric sar parameters to characterize the soil surface over bare agriculture fields. *IEEE Trans. Geosci. Remote Sens.* **2012**, *50*, 3844–3858. [CrossRef]
9. Zribi, M.; Gorrab, A.; Baghdadi, N. A new soil roughness parameter for the modelling of radar backscattering over bare soil. *Remote Sens. Environ.* **2014**, *152*, 62–73. [CrossRef]
10. Fung, A.K. *Microwave Scattering and Emission Models and Their Applications*; Artech House: Norwood, MA, USA, 1994.
11. Fung, A.K.; Li, Z.; Chen, K. Backscattering from a randomly rough dielectric surface. *IEEE Trans. Geosci. Remote Sens.* **1992**, *30*, 356–369. [CrossRef]
12. Hajnsek, I. Inversion of Surface Parameters Using Polarimetric Sar. Ph.D. Thesis, Friedrich-Schiller-Universität Jena, Jena, Germany, 2001.
13. Franceschetti, G.; Riccio, D. *Scattering, Natural Surfaces, and Fractals*; Academic Press: Cambridge, MA, USA, 2006.
14. Martinez, A.; Byrnes, A.P. *Modeling Dielectric-Constant Values of Geologic Materials: An Aid to Ground-Penetrating Radar Data Collection and Interpretation*; Kansas Geological Survey, University of Kansas: Lawrence, KS, USA, 2001.
15. Mazaheri Tehrani, H. Soil Moisture Estimation with Polarimetric Sar Data. Ph.D. Thesis, University of Calgary, Calgary, AL, Canada, 2014.
16. Fung, A.K.; Chen, K.S. An update on the iem surface backscattering model. *IEEE Geosci. Remote Sens. Lett.* **2004**, *1*, 75–77. [CrossRef]
17. Baghdadi, N.; Holah, N.; Zribi, M. Calibration of the integral equation model for sar data in c-band and hh and vv polarizations. *Int. J. Remote Sens.* **2006**, *27*, 805–816. [CrossRef]
18. Barrett, B.W.; Dwyer, E.; Whelan, P. Soil moisture retrieval from active spaceborne microwave observations: An evaluation of current techniques. *Remote Sens.* **2009**, *1*, 210–242. [CrossRef]
19. Chen, K.; Kao, W.; Tzeng, Y. Retrieval of surface parameters using dynamic learning neural network. *Remote Sens.* **1995**, *16*, 801–809. [CrossRef]
20. Sahebi, M.R.; Bonn, F.; Bénié, G.B. Neural networks for the inversion of soil surface parameters from synthetic aperture radar satellite data. *Can. J. Civ. Eng.* **2004**, *31*, 95–108. [CrossRef]
21. Dierking, W. Quantitative roughness characterization of geological surfaces and implications for radar signature analysis. *IEEE Trans. Geosci. Remote Sens.* **1999**, *37*, 2397–2412. [CrossRef]
22. Álvarez-Mozos, J.; Verhoest, N.E.; Larrañaga, A.; Casalí, J.; González-Audícana, M. Influence of surface roughness spatial variability and temporal dynamics on the retrieval of soil moisture from sar observations. *Sensors* **2009**, *9*, 463–489. [CrossRef] [PubMed]
23. Choker, M.; Baghdadi, N.; Zribi, M.; El Hajj, M.; Paloscia, S.; Verhoest, N.E.; Lievens, H.; Mattia, F. Evaluation of the oh, dubois and iem backscatter models using a large dataset of sar data and experimental soil measurements. *Water* **2017**, *9*, 38. [CrossRef]
24. Verhoest, N.E.; Lievens, H.; Wagner, W.; Álvarez-Mozos, J.; Moran, M.S.; Mattia, F. On the soil roughness parameterization problem in soil moisture retrieval of bare surfaces from synthetic aperture radar. *Sensors* **2008**, *8*, 4213–4248. [CrossRef] [PubMed]
25. Ghafouri, A.; Amini, J.; Dehmollaian, M.; Kavoosi, M. Measuring surface roughness of geological rock surfaces in sar data using fractal geometry. *C. R. Geosci.* **2017**, *11*, 327–338.
26. Baghdadi, N.; Chaaya, J.A.; Zribi, M. Semiempirical calibration of the integral equation model for sar data in c-band and cross polarization using radar images and field measurements. *IEEE Geosci. Remote Sens. Lett.* **2011**, *8*, 14–18. [CrossRef]

27. Western, A.W.; Bloschl, G.; Grayson, R.B. How well do indicator variograms capture the spatial connectivity of soil moisture? *Hydrol. Process.* **1998**, *12*, 1851–1868. [CrossRef]

28. Mandelbrot, B.B. *The Fractal Geometry of Nature*; Macmillan Publishers: London, UK, 1983.

29. Durst, P.J.; Mason, G.L.; McKinley, B.; Baylot, A. Predicting rms surface roughness using fractal dimension and psd parameters. *J. Terramech.* **2011**, *48*, 105–111. [CrossRef]

30. Martino, G.D.; Franceschetti, G.; Riccio, D.; Zinno, I. Spectral processing for the extraction of fractal parameters from sar data. In Proceedings of the 2011 17th International Conference on Digital Signal Processing (DSP), Corfu, Greece, 6–8 July 2011; pp. 1–7.

31. Yordanov, O.; Atanasov, I. Self-affine random surfaces. *Eur. Phys. J. B* **2002**, *29*, 211–215. [CrossRef]

32. Yordanov, O.; Ivanova, K. Description of surface roughness as an approximate self-affine random structure. *Surf. Sci.* **1995**, *331*, 1043–1049. [CrossRef]

33. Rasigni, G.; Llebaria, A.; Lafraxo, M.; Buat, V.; Rasigni, M.; Abdellani, F. Autoregressive process for characterizing statistically rough surfaces. *J. Opt. Soc. Am. A* **1993**, *10*, 1257–1262. [CrossRef]

34. Roomi, S.; Kalaiyarasi, D.; Rangan, N.K. Discrete wavelet transform based despeckling for sar images. In Proceedings of the 2011 World Congress on Information and Communication Technologies, Mumbai, India, 11–14 December 2011; pp. 373–378.

35. Wang, F.; Zhang, B.; Yang, D.; Li, W.; Zhu, Y. Sea-state observation using reflected beidou geo signals in frequency domain. *IEEE Geosci. Remote Sens. Lett.* **2016**, *13*, 1656–1660. [CrossRef]

36. Fadzal, C.C.W.; Mansor, W.; Khuan, L.; Mohamad, N.; Mahmoodin, Z.; Mohamad, S.; Amirin, S. Welch power spectral density of eeg signal generated from dyslexic children. In Proceedings of the 2014 IEEE Region 10 Symposium, Kuala Lumpur, Malaysia, 14–16 April 2014; pp. 560–562.

37. Ghafouri, A.; Amini, J.; Dehmollaian, M.; Kavoosi, M. Random fractals geometry in surface roughness modeling of geological formations using synthetic aperture radar images. *J. Geomat. Sci. Technol. Iran. Soc. Surv. Geomat. Eng.* **2015**, *5*, 97–108.

38. Di Martino, G.; Iodice, A.; Riccio, D.; Ruello, G.; Zinno, I. On the fractal nature of volcano morphology detected via sar image analysis: The case of somma-vesuvius volcanic complex. *Eur. J. Remote Sens.* **2012**, *45*, 177–187. [CrossRef]

39. Zribi, M. Développement de Nouvelles Méthodes de Modélisation de la Rugosité Pour la Rétrodiffusion Hyperfréquence de la Surface Du Sol. Ph.D. Thesis, Université Paul Sabatier, Toulouse, France, 1998.

40. Franceschetti, G.; Iodice, A.; Maddaluno, S.; Riccio, D. A fractal-based theoretical framework for retrieval of surface parameters from electromagnetic backscattering data. *IEEE Trans. Geosci. Remote Sens.* **2000**, *38*, 641–650. [CrossRef]

41. Franceschetti, G.; Iodice, A.; Migliaccio, M.; Riccio, D. Scattering from natural rough surfaces modeled by fractional brownian motion two-dimensional processes. *IEEE Trans. Antennas Propag.* **1999**, *47*, 1405–1415. [CrossRef]

42. Zribi, M.; Ciarletti, V.; Taconet, O.; Paillé, J.; Boissard, P. Characterisation of the soil structure and microwave backscattering based on numerical three-dimensional surface representation: Analysis with a fractional brownian model. *Remote Sens. Environ.* **2000**, *72*, 159–169. [CrossRef]

remote sensing

MDPI

Article

The Role of Resolution in the Estimation of Fractal Dimension Maps From SAR Data

Gerardo Di Martino [1], Antonio Iodice [1], Daniele Riccio [1,*], Giuseppe Ruello [1] and Ivana Zinno [2]

[1] Department of Electrical Engineering and Information Technology, University of Naples Federico II, Via Claudio 21, 80125 Napoli, Italy; gerardo.dimartino@unina.it (G.D.M.); iodice@unina.it (A.I.); giuseppe.ruello@unina.it (G.R.)
[2] Institute for Electromagnetic Sensing of the Environment, Research National Council (CNR), Via Diocleziano 328, 80124 Napoli, Italy; zinno.i@irea.cnr.it
*   Correspondence: Daniele.Riccio@unina.it; Tel.: +39-081-7683114

Received: 29 September 2017; Accepted: 20 December 2017; Published: 22 December 2017

Abstract: This work is aimed at investigating the role of resolution in fractal dimension map estimation, analyzing the role of the different surface spatial scales involved in the considered estimation process. The study is performed using a data set of actual Cosmo/SkyMed Synthetic Aperture Radar (SAR) images relevant to two different areas, the region of Bidi in Burkina Faso and the city of Naples in Italy, acquired in stripmap and enhanced spotlight modes. The behavior of fractal dimension maps in the presence of areas with distinctive characteristics from the viewpoint of land-cover and surface features is discussed. Significant differences among the estimated maps are obtained in the presence of fine textural details, which significantly affect the fractal dimension estimation for the higher resolution spotlight images. The obtained results show that if we are interested in obtaining a reliable estimate of the fractal dimension of the observed natural scene, stripmap images should be chosen in view of both economic and computational considerations. In turn, the combination of fractal dimension maps obtained from stripmap and spotlight images can be used to identify areas on the scene presenting non-fractal behavior (e.g., urban areas). Along this guideline, a simple example of stripmap-spotlight data fusion is also presented.

Keywords: synthetic aperture radar; rough surfaces; fractals

1. Introduction

Cosmo/SkyMed synthetic aperture radar (SAR) multi-operational capabilities allow the observation of a scene at different spatial scales, i.e., with different levels of detail. Usually, the choice of the operational mode better fitting a specific application is dictated by a trade-off between resolution and coverage. In particular, in the case of Cosmo/SkyMed the range of attainable resolutions spans from about 1 m in enhanced spotlight mode, with a coverage of 10 km, to about 100 m in scansar huge region mode, with a coverage of about 200 km, whereas the stripmap mode achieves a resolution of 3 m, with a coverage of 40 km. In this context, stripmap images, or even better, scansar images, are well suited for large-scale applications, whereas spotlight products can be fruitfully used for local-scale analyses, able to provide more detailed information on the observed scene. The combined use of multi-operational data, therefore, can be set in the framework of a multi-scale approach, in which coarse resolution images can provide the necessary regional-scale survey of the area of interest, while high resolution images can be used to refine the analysis on a local scale. Regarding the analysis of natural surfaces, the mentioned multi-scale approach can be used to gain the required information investigating different spatial scales. In fact, the physical and geometrical information regarding the scene under survey can present strong variations, which cannot be fully appreciated through low-resolution SAR data, which, in turn, allow only the estimation of somehow averaged quantities.

When the focus is on surface roughness, fractal models represent the best way to model the behavior of natural surfaces [1–12]. The fractal dimension is a concise and meaningful entity, bearing meaningful physical information regarding the geometrical and geophysical characterization of a natural surface [5–9]. In particular, geologists usually use the fractal dimension to model the roughness of natural surfaces, because it is not dependent on the size of the observed surface. In fact, a typical problem of classical statistical roughness descriptors, such as the height standard deviation and correlation length, lies in their dependence on the observation scale and on the extension of the surface from which they are estimated [5–9].

The use of fractal concepts in the remote-sensing community is a long-standing topic, testified by applications regarding image analysis [13], and, in particular, SAR image segmentation [14,15] and feature detection [16,17]. However, all the mentioned works are based on the estimation of fractal parameters of image texture, leaving aside the problem of investigating how the fractal dimension of the observed surfaces is related to the amplitude values of image pixels. This kind of analysis was first considered in Pentland [18], and Kube and Pentland [19] for the case of optical images of Lambertian natural surfaces; this pioneering work was extended to non-Lambertian fractal surfaces in Korvin [20]. More recently this analysis has been performed for the case of SAR imaging: in particular, in Di Martino et al. [21,22] the authors introduced appropriate models and algorithms, allowing for the estimation of the fractal dimension of an observed surface from a SAR amplitude image. Therefore, while works [14–17] were based on the introduction of convenient fractal-based parameters to perform image processing tasks, without entering the problem of associating to them a clear physical meaning, the estimation framework of Di Martino et al. [21,22] provides estimates of the fractal dimension of the observed surface, i.e., a parameter with a clear physical meaning that can be easily managed by physicists and geologists for the characterization of natural phenomena. Moreover, works [21,22] forerun (and partly stimulated) the development of SAR processing techniques based on the use of fractal dimension: indeed, recently fractal dimension has been fruitfully used in SAR interferometry, to support coregistration [23], regularization [24], and phase unwrapping [25], and in SAR speckle filtering [26,27]. All these techniques benefit from the availability of accurate estimates of the surface fractal dimension.

In more detail, in Di Martino et al. [21,22] the estimation of the surface fractal dimension from a SAR amplitude image is performed exploiting a sliding window scouring the whole image, thus allowing the generation, as output, of a new value-added SAR product, the *fractal dimension map*, i.e., a point-by-point map of the estimated fractal dimension of the imaged surface, accounting for local variations of the surface fractal dimension. Some properties of the technique we propose are remarkable [28]. The fractal dimension clearly depends on a single feature of the observed scene, namely its roughness. From this viewpoint, the fractal dimension is one of the few parameters retrievable from SAR data that allows the separation of the influence of a single physical parameter (in particular, the roughness) from the others involved in SAR image formation. Actually, in many practical situations, the post-processing of SAR images provides products showing a significant dependence on the acquisition geometry of the employed SAR image, and this is the major disadvantage from the end-user viewpoint. For instance, the severe dependence on the SAR acquisition geometry leads to classification maps that are not easily comparable and cannot be straightforwardly managed by non-expert SAR users, whenever they are obtained from SAR images acquired from different satellite tracks. Conversely, the estimation of the fractal dimension maps is almost independent of the acquisition geometry. In fact, in Di Martino et al. [28] we analyzed a wide set of multi-angle SAR images of the same area and verified that the estimated fractal dimension maps are dependent neither on the sensor look angle nor on the local incidence angle, at least in the hypothesis of validity of the theoretical model (i.e., on natural surfaces). Furthermore, in Di Martino et al. [29] we demonstrated that the maps are not dependent on polarization, at least when co-polarized channels are considered.

Due to the above-mentioned properties, the fractal dimension maps are value-added SAR products that can have a significant impact on the end users' community. For this reason, in the present paper we investigate the role of resolution, in order to unveil possible dependencies on the surface spatial scales involved in the estimation. For purely fractal surfaces, the fractal dimension keeps constant over all the spatial scales; however, it must be noted that, since the fractional Brownian motion (fBm) surface root mean square (rms) slope decreases as the observation scale increases, at larger observation scales the small-slope hypothesis is more easily satisfied: from this viewpoint, stripmap images are more convenient than spotlight images. In conclusion, we expect that in natural, homogeneous (at the scale of window size) areas the estimated fractal dimension is the same on spotlight and stripmap images, except in the uncommon cases in which the small-slope assumption is satisfied for the stripmap image and not satisfied for the spotlight image. It must be also noted that, sometimes, actual natural surfaces present multifractal behavior, i.e., their fractal dimension changes according to the spatial scale [5,8,9]. This may be another source of difference between the fractal dimension maps obtained from spotlight and stripmap images. Finally, in areas containing man-made objects the employed fractal model does not hold, and since such objects are usually strongly scale-dependent, we expect that the estimated fractal dimension obtained from stripmap and spotlight images can be different.

For the analysis, we use two sets of COSMO-SkyMed stripmap and enhanced spotlight SAR images relevant to the areas of Bidi in Burkina Faso and Naples in Italy. All the images were acquired with similar look angles, between 31° and 33°. Two subsets were cropped from each image as representative of different kind of land-cover and texture properties: they span from a homogeneous bare-soil scenario to a very heterogeneous urban scenario. In this way, we are able to investigate the behavior of the algorithm according to different regimes of validity of the theoretical model. In fact, the theoretical model was originally developed for bare-soil natural surfaces, which can be conveniently modeled through fractals. For such natural surfaces, we expect that the estimated fractal dimension maps do not significantly depend on the employed resolution, due to the scale-invariance property of fractal objects. Accordingly, the same results should be obtained from stripmap and spotlight images. This is the first property that we want to verify in this paper. However, in the presence of man-made objects, or other non-fractal objects, even if the theoretical model does not hold and the algorithm does not provide a true fractal dimension, the estimated meta-parameter can provide meaningful information, which can be used for the characterization of the area of interest [30]. In this case we cannot a priori expect that the estimated fractal dimension does not depend on resolution, and it is of interest to verify the behavior of the obtained results at different resolutions. The analysis is performed within a statistical framework, along the same guidelines of Di Martino et al. [28]. However, we also present results regarding the joint use (on a pixel basis) of the fractal dimension maps obtained from the stripmap and the spotlight images, where we try to identify fine textural details related to specific land-cover classes.

The paper is organized as follows. In Section 2 the theoretical and methodological background is summarized and the remarkable properties presented by the fractal dimension maps are detailed. In Section 3 we present the experimental setup and the statistical analysis of the obtained maps, along with an example of stripmap–spotlight data fusion. Finally, conclusions and relevant suggestions are reported in Section 4.

2. Materials and Methods

In the present section, we summarize the main theoretical and implementation aspects regarding the generation of the fractal dimension map from a single SAR image relevant to a natural surface.

2.1. Basic Theory

With regard to the surface model, we consider here the fractional Brownian motion (fBm) model, which is an everywhere continuous, nowhere differentiable process. It can be conveniently described in terms of its increment probability density function (pdf) [2–4]; in particular, a stochastic process $z(x,y)$ is an (isotropic) fBm surface if, for every x, x', y, y' it satisfies the following relation:

$$\Pr\{z(x,y) - z(x',y') < \bar{\zeta}\} = \frac{1}{\sqrt{2\pi s\tau^H}} \int_{-\infty}^{\bar{\zeta}} \exp\left(-\frac{\zeta^2}{2s^2\tau^{2H}}\right) d\zeta \tag{1}$$

where $\Pr\{\cdot\}$ stands for "probability", τ is the distance between the points (x,y) and (x',y'), s [m^{1-H}] is the *incremental standard deviation*, i.e., the standard deviation of the surface increments at unitary distance, and H is the *Hurst coefficient* ($0 < H < 1$), related to the fractal dimension D through the relationship $D = 3 - H$. In the frequency domain, the power spectral density (PSD) of the isotropic two-dimensional fBm process exhibits an appropriate power-law behavior [1–4]:

$$S(k) = S_0 k^{-\alpha} \tag{2}$$

where in S_0 and α are the fBm spectral parameters, related to the previously introduced spatial parameters by the following relationships [2]:

$$S_0 = 2^{H+1}\Gamma^2(1+H)\sin(\pi H)s^2 \tag{3}$$

$$\alpha = 2 + 2H = 8 - 2D \tag{4}$$

$\Gamma(\cdot)$ being the Euler Gamma function.

In Di Martino et al. [22] the authors presented a forward model linking the stochastic characterization of an SAR image to the fractal parameters of the observed surface. In particular, the authors demonstrated that—in the hypothesis of a small slope regime for the surface roughness—the modulus of the reflectivity function $|\gamma(x,y)|$ (where x and y represent azimuth and ground-range, respectively) depends, to the first order, on the partial derivative $p(x,y)$ of the surface height in the range direction. Hence, in a first order approximation, $|\gamma(x,y)|$ does not depend on the partial derivative of surface height in the azimuth direction $q(x,y)$ [22]:

$$|\gamma(x,y)| = a_0 + a_1 p(x,y) + o(p,q) \tag{5}$$

where a_0 and a_1 are the coefficients of the Mc Laurin series expansion, depending on the considered scattering model, and $o(\cdot)$ indicates terms that are infinitesimal of an order higher than one. The coefficients a_0 and a_1, and, in turn, the validity limits of the proposed model, depend on the considered look-angle, the fractal parameters of the observed surface, the scattering model, and the considered resolution.

In Di Martino et al. [22], a closed form expression for the PSD of the azimuth and range cuts of the derivative process $p(x,y)$ has been evaluated via appropriate Fourier transforms of their autocorrelations. The expression obtained for the PSD of the range cuts is

$$S_p(k_y; \Delta y) = 2s^2 \Delta y^{-1+2H} \Gamma(1+2H)\sin(\pi H)\left[1 - \cos\left(|k_y|\Delta y\right)\right]\frac{1}{\left(|k_y|\Delta y\right)^{1+2H}} \tag{6}$$

where k_y is the spatial wavenumber associated with the range direction and Δy is the ground range resolution of the image [21,22]. Moreover, it was demonstrated that for small wavenumbers the range cut PSD holds a linear behavior in a log-log plane, while the azimuth-cut PSD does not present this remarkable property [22]. In particular, when $k_y \Delta y \to 0$ we obtain the following asymptotic expression for the range-cut PSD, $\tilde{S}_p(k_y)$:

$$\tilde{S}_p(k_y) = s^2 \Gamma(1+2H)\sin(\pi H)\frac{1}{|k_y|^{2H-1}} \tag{7}$$

Hence, taking into account the relation between the reflectivity and the derivative process p reported in Equation (5), and the band-limiting effect of the SAR impulse response [22], we obtain for S_i the following expression:

$$S_i(k_y; \Delta y) = a_1^2 S_p(k_y; \Delta y) \text{Rect}[\frac{\Delta y \sin^2 \vartheta_0 \, k_y}{\pi}] \tag{8}$$

where Rect[·] stands for the rectangular function and, in a wide range of small wavenumbers, the asymptotic expression in Equation (7) can be assumed for S_p.

2.2. Estimation Method

Actually, the asymptotic spectrum reported in Equation (7) exhibits linear behavior in a log-log plane: in particular, looking at Equation (7), we can conclude that its slope is equal to $1 - 2H$. Therefore, a straightforward way to estimate the value of the fractal dimension of an observed scene directly from its corresponding SAR image can be based on the use of linear regression techniques. Based on this observation, an algorithm has been developed allowing the retrieval of the point-by-point fractal dimension map of a natural surface starting from its single look amplitude SAR image [22]. The algorithm evaluates the fractal dimension D associated with each pixel using the information relevant to neighboring pixels enclosed in a sliding window which, spanning the whole image, generates the fractal dimension map. The estimation of the PSD of the range cuts enclosed in each processing window is performed using the Capon estimator [22,31,32]. This requires the preliminary estimation of the autocorrelation matrix, which we perform using the modified covariance method [18,19]. Then, a linear regression step on the obtained PSD allows the evaluation of the fractal dimension D. The Capon estimator has been chosen because it overcomes the leakage and high variance problems arising when facing the estimation of power-law spectra, as detailed in [10,31,32]. In Figure 1 we report a flow chart of the processing steps performed in each instance of the sliding window.

Figure 1. Flow chart of the processing used for the estimation of the fractal dimension within the sliding window.

The key parameter to be set for the elaboration is the size of the sliding window. Its choice results from a trade-off between the accuracy and resolution of the output map [22]. It was verified that a good choice in most cases is a window size of about 50×50 pixels [22]. Moreover, the size of the sliding window also dictates the maximum spatial scale involved in the estimation. In fact, the spatial scales involved in the estimation are dictated on the one hand by the resolution, which is related to the minimum spatial scale, and on the other hand by the size of the sliding window, which is related to the maximum scale. Therefore, if we analyze the behavior of the algorithm on images presenting different resolutions using the same window size (in order to obtain a comparable estimation accuracy for the two maps), different spatial scales will be involved in the estimation. This fact, in principle, can lead to different values of the fractal dimension on the two maps, for the reasons discussed in the Introduction and experimentally explored in Section 3.

3. Results and Discussion

The data set used in the experimental framework consists of Cosmo/SkyMed images acquired in two different areas, which present very different characteristics with respect to climate, geology and land-cover. In particular, we consider images obtained in stripmap and enhanced spotlight operational modes, so that the area covered by the spotlight images is a subset of the area imaged in stripmap mode. The data set consists of:

- A first couple of images relevant to the area of Bidi in the Yatenga district of Burkina Faso: the stripmap image was acquired on 18 August 2011 and the enhanced spotlight image on 22 August 2011. The data are in Single look Complex Slant Balanced (SCS_B format and acquired in Horizontal-Horizontal (HH) polarization with a look angle of about 33°. The guaranteed resolution of the stripmap image is 3×3 m^2 in the azimuth-ground range, while for the enhanced spotlight it is 1×1 m^2; the pixel spacing is 1.9×2.1 m^2 for the stripmap and 0.7×0.6 m^2 for the enhanced spotlight.
- A second couple of images relevant to the city of Naples in Italy: in this case, the stripmap image was acquired on 1 August 2011 and the enhanced spotlight image on 29 August 2011. The images are in Single look Complex Slant Unbalanced (SCS_U) data format (i.e., unbalanced data with no weighting applied to the processed bandwidth), with a look angle of about 31° and HH polarization. The guaranteed resolution is 2.5×2.5 m^2 in azimuth-ground range for the stripmap case and 0.85×0.85 m^2 for the enhanced spotlight one; the pixel spacing is 2.1×2.1 m^2 for the stripmap and 0.7×0.6 m^2 for the enhanced spotlight.

For each couple of images, we selected two subsets presenting different characteristics from the viewpoint of land-cover and scene distribution. To each subset, we applied the estimation algorithm described in the previous section: we set the elaboration window size at 51×51 pixels for all the examined cases. In this way, we obtained fractal estimates presenting similar accuracies and, hence, the values of the fractal dimension maps obtained from the stripmap and spotlight data will be comparable for homogeneous areas. This corresponds to about 102×107 m^2 in the stripmap case and 36×31 m^2 in the spotlight case. The images and the maps were geocoded using a standard technique based on the Shuttle Radar Topography Mission (SRTM) Digital Elevation Model (DEM) of the area of interest and were interpolated in order to obtain a pixel spacing of 5×5 m^2. Actually, the geocoding step is necessary to allow an effective comparison of data acquired with very different acquisition parameters; furthermore, the use of geocoded products is of fundamental importance for the applicative community (e.g., geologists, geophysicists), which usually works with geo-referenced data, rather than with data in SAR native reference systems [28]. However, with regard to the fractal dimension maps, the geocoding step cannot be applied to SAR images prior to the fractal dimension estimation step, since it significantly degrades the fractal characteristics present in the image [28]: therefore, we first evaluated the fractal dimension maps on non-geocoded images and then applied the geocoding directly to the obtained maps.

Once the geocoded fractal dimension maps are available, we can analyze their statistical behavior. The proposed statistical analysis is based on the evaluation of the first order statistics of the fractal dimension maps of the subsets. This kind of statistical analysis is also justified by the fact that the proposed product is devoted to the applicative community, which is typically interested in the geophysical characteristics of homogeneous areas, rather than in the punctual characteristics related to the statistical behavior of single image pixels. Note that we did not need to implement any calibration steps, because the fractal dimension maps are not dependent on the absolute image calibration [28]. In the following subsections, we discuss in detail the results obtained for each processed subset.

3.1. Burkina 1

In Figure 2a we show a Google Earth image of the first area of interest, 2.8×2.4 km^2. The available optical Satellites Pour l'Observation de la Terre (SPOT) image was acquired on 5 May 2013, i.e., during the Burkina Faso dry season. In fact, the area of interest is located in the Sahelian zone of Burkina Faso, where the climate is characterized by two main seasons, a long dry season from October to May and an extremely rainy season from June to September. Due to these extreme climate conditions, strong inter-seasonal variability occurs in the land-cover of the area, which at the end of the dry season is almost free of vegetation: this is the case in the reported Google Earth image. A detailed knowledge of the characteristics of this area has been gained in the frame of several projects regarding the use of SAR data for monitoring the environment [33,34].

In Figure 2b,c we present the geocoded stripmap SAR image and the corresponding fractal dimension map respectively, while in Figure 2d,e the enhanced spotlight SAR image and the corresponding fractal dimension map are respectively shown. Since the geocoded images were interpolated to the same pixel spacing of 5×5 m^2, the difference between the stripmap and the spotlight acquisitions can be mainly appreciated noting the presence of a stronger speckle reduction on the spotlight image. In fact, before the geocoding step, we applied a 3×3 spatial multilook on the spotlight image, thus obtaining almost the same pixel spacing of the single-look stripmap image. It is worth noting that the fractal maps shown in Figure 2c,e have been evaluated starting respectively from the stripmap and the enhanced spotlight single look SAR images, without performing any type of transformation that could alter the fractal dimension estimation. This also explains the evident different resolutions of the two fractal dimension maps. The spatial multilook operation was carried out on the spotlight image only for a better visual comparison with the stripmap one.

Looking at the images in Figure 2a,b,d, we can note that the area of interest consists essentially of bare soil, with the presence of a small village in the upper part of the scene. Actually, the images in Figure 2b,d were acquired during the rainy season and witness a larger presence of vegetation with respect to the Google image; moreover, some small water basins, which during the dry season are completely empty and cannot be located on the optical image, are present close to the village. The dark spots on the images in Figure 2b,d are due to the presence of eroded soil, which is unable to keep a water-content sufficient to allow the growth of even small vegetation [33,34]. We chose this kind of scene because it represents a natural almost-canonical case, where the linear imaging model should be valid over almost the entire scene. Actually, the soil is bare almost everywhere, with minor presence of vegetation, and no significant geophysical features can be appreciated on the scene. In this situation, we expect the fractal dimension estimation to be almost independent of the considered spatial scales.

In Figure 3 we show the histograms of the fractal dimension maps shown in Figure 2c,e. The values of the mean and the standard deviation of the fractal dimension maps are reported in Table 1. From the histograms in Figure 3 and the statistics in Table 1, it is evident that the fractal dimension maps estimated on the stripmap and spotlight images share very similar behavior, presenting a difference in the average estimated values of the fractal dimension of 0.02, i.e., significantly smaller than the standard deviations of both the maps. The slightly lower mean value obtained in the spotlight case can be ascribed to the fine-scale details present over the spotlight image. In fact, in the presence of strong point-like scatterers and, more in general, of discontinuities (i.e., resolution-scale features or man-made objects) the employed fractal model does not hold, and the algorithm provides an estimate of the fractal dimension that may be not enclosed in the range of allowed fractal values [30], implying the presence of dark square spots whose size is related to the used processing window size. It is well known that in enhanced spotlight mode the number of bright points can be much larger with respect to the stripmap one, because, due to the higher resolution, small objects can also easily act as corner reflectors.

Table 1. Statistics of the subsets.

Subset	Mode	D_{mean}	D_{stdev}
Burkina 1	Strip	2.17	0.07
	Spot	2.15	0.08
Burkina 2	Strip	2.21	0.06
	Spot	2.15	0.08
Naples 1	Strip	2.17	0.05
	Spot	2.18	0.06
Naples 2	Strip	2.14	0.09
	Spot	2.08	0.11

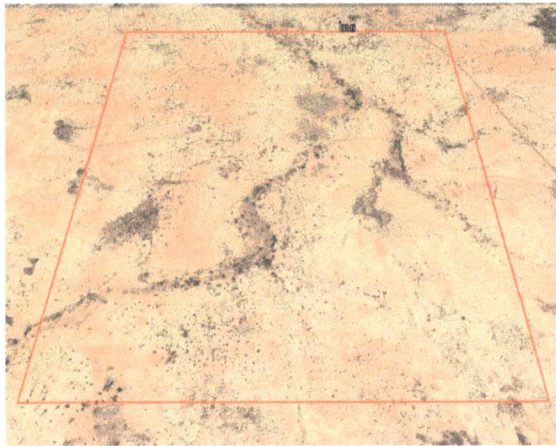

(a)

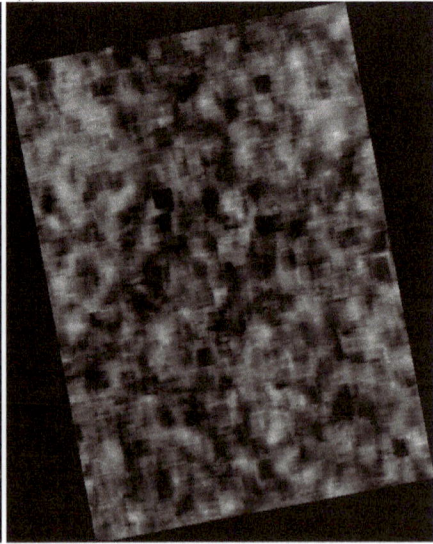

(b)   (c)

Figure 2. *Cont.*

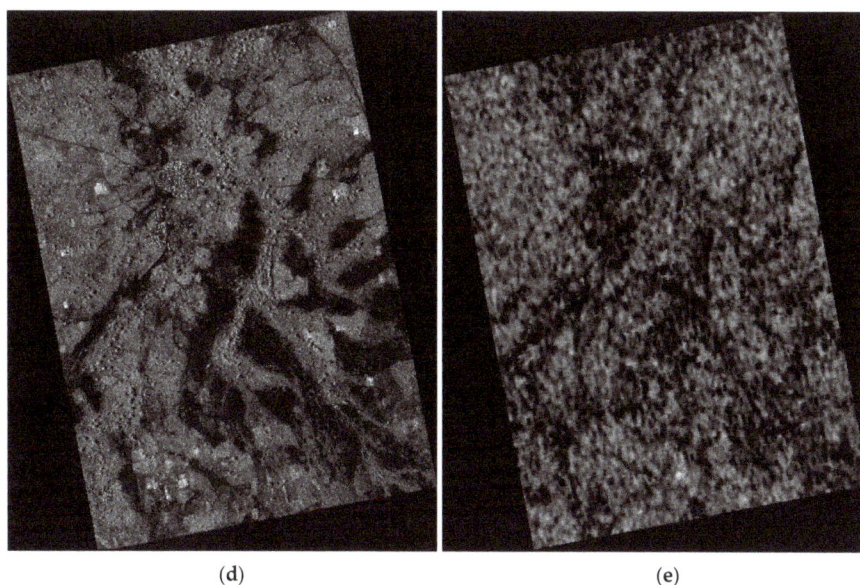

(d)   (e)

Figure 2. (**a**) Google Earth view of the subset Burkina 1, with the footprint of the imaged area of interest marked in red; (**b**) geocoded stripmap acquisition; (**c**) fractal dimension map corresponding to (**b**); (**d**) geocoded enhanced spotlight image; (**e**) fractal dimension map corresponding to (**d**).

Until now we have not mentioned the presence of speckle. Indeed, as discussed in Section 2.2, the considered estimation technique is based on the Capon estimator, which performs an intrinsic filtering of speckle, discarding the high-wavenumber region of the image spectrum, i.e., the part mostly affected by speckle [22]. However, in addition to the comparison between single-look stripmap and spotlight images sharing similar speckle levels, a multilook can be applied to the high-resolution spotlight image, so that an image with approximately the same resolution of the stripmap one, but with significantly lower levels of speckle, can be obtained. In Figure 4 we show a comparison of the histograms obtained on the stripmap (i.e., the same as in Figure 3) and on the spotlight image after a multilook of 3×3 pixels was applied. The average fractal dimension obtained for the multilook spotlight image was 2.1, with a standard deviation of 0.1: significant differences between the histograms can now be observed. The value of the average fractal dimension decreased. This is coherent with the obtained reduction of speckle power: indeed, the effect of speckle is to flatten the high-frequency portion of the spectrum, thus reducing the spectral slope and increasing the retrieved fractal dimension. However, the decrease of the fractal dimension is associated with an increase in the standard deviation and, more importantly, with a significantly larger number of pixels presenting values of the fractal dimension lower than two, i.e., outside the fractal range. This is probably related to the fact that the application of speckle filtering on areas where a texture due to topography is present results in texture distortion [35]. Regarding the estimation of fractal dimension maps, this issue was analyzed in Di Martino et al. [36], where the applied despeckling filters significantly impaired fractal dimension estimation. In conclusion, the role of speckle and speckle filtering in fractal dimension estimation remains an open issue: however, a comprehensive discussion of this topic is beyond the scope of this paper.

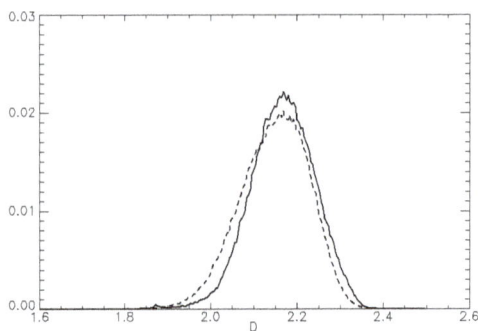

Figure 3. Histogram of the fractal dimension maps of Figure 2c,e: solid line for the stripmap and dashed line for the enhanced spotlight.

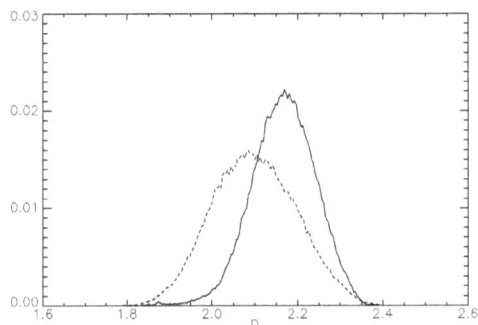

Figure 4. Histogram of the fractal dimension map of Figure 2c compared to that of the fractal dimension map obtained from enhanced spotlight 3 × 3 multilook data: solid line for the stripmap and dashed line for the multilook enhanced spotlight.

In summary, the land-cover in the examined case can be considered substantially homogeneous, and the fractal dimension maps estimated from the stripmap and enhanced spotlight images share very similar mean values. Therefore, when we are interested in the evaluation of the fractal dimension of natural surfaces not presenting relevant fine-scale details, the use of the less expensive stripmap data does not imply any information loss and is, therefore, advisable. Moreover, due to the lower number of samples for equally-sized areas, the estimation of the fractal dimension from stripmap images is also more convenient from a computational viewpoint.

3.2. Burkina 2

In Figure 5a a Google Earth image of the area of interest is shown (2.3 × 2.3 km²). Also in this case, the optical SPOT image was acquired at the end of Burkina Faso dry season. In Figure 5b,c we present the geocoded stripmap SAR image and the corresponding fractal dimension map, respectively, while in Figure 5d,e the enhanced spotlight SAR image and the corresponding fractal dimension map are shown. Note that, as in the previous case, we applied the geocoding and multilooking procedure described above to the SAR images, and we estimated the fractal maps from the single look SAR images. The same considerations on seasonal variability and on the comparison between the Google Earth and the SAR image hold in this case.

With respect to the images discussed in the previous case, here we can appreciate an increase in vegetated areas and, more in general, the presence of more resolution-scale details. In addition, the image is characterized by the juxtaposition of homogeneous areas of different types (e.g., vegetated and eroded-soil areas), which are responsible for the presence of many edges, i.e., regions of separation between the different areas. The presence of edges allows the testing of the algorithm in the presence of discontinuities, which are one of the critical features for the model validity: in particular, local fractal dimension estimation algorithms behave as edge detectors [17,30], due to the presence within the estimation window of a non-homogeneous texture. Comparing the fractal dimension maps in Figure 5c,e, we can note that the impact of this kind of texture on the map estimated from the spotlight image is significant and it is witnessed by the presence of large areas presenting low values of the fractal dimension (see Figure 5e). This is confirmed by the histograms of the maps in Figure 6 and by the statistics reported in Table 1. In fact, the difference in the average value of the fractal dimension is equal to 0.06, thus being comparable to the standard deviations of both fractal dimension maps.

Very interestingly, the discussed behavior can also be observed on a pixel basis, investigating the joint behavior of the two maps via a feature-based data-fusion technique [37]. In particular, we consider the image obtained by taking the difference of the two geocoded fractal dimension maps, namely the spotlight-based map minus the stripmap-based one. In order to highlight the areas where the two maps present significantly different behavior, in Figure 5f we show a classification map obtained by thresholding the difference image with two thresholds (±0.25): the areas with a difference lower than −0.25 are highlighted in green, while those with a difference greater than 0.25 are red. The areas with a difference in the range [−0.25,0.25] are represented in black. Remembering that very low values of the fractal dimension are obtained in areas close to bright spots, whose presence influences the estimates obtained in an area comparable to the estimation window [30]; we can identify two different mechanisms responsible for the appearance of the two classes. First, the presence of bright spots due to strong trihedral scatterers that are observable on both spot and strip images imply a low fractal dimension on both maps. However, due to the fact that the elaboration windows used to estimate the two maps present the same number of pixels and, hence, a different size in meters, the extension of the low-fractal-dimension area is different in the two maps. This can be observed looking at the red areas in Figure 5f: they usually frame a black area, which is the area interested by the presence of the bright spot, where the fractal dimension is almost the same on the two images. Conversely, the green areas are related to the presence of fine textural details, which are only observable on spotlight images. In particular, in the scene of interest these details are related to the presence of isolated trees.

In summary, as discussed above, the effect of fine-scale (non-fractal) details is more significant on the fractal dimension map obtained from the spotlight image than on the one estimated from the stripmap image and, in this case, this difference emerges in a more evident way.

(a)

Figure 5. *Cont.*

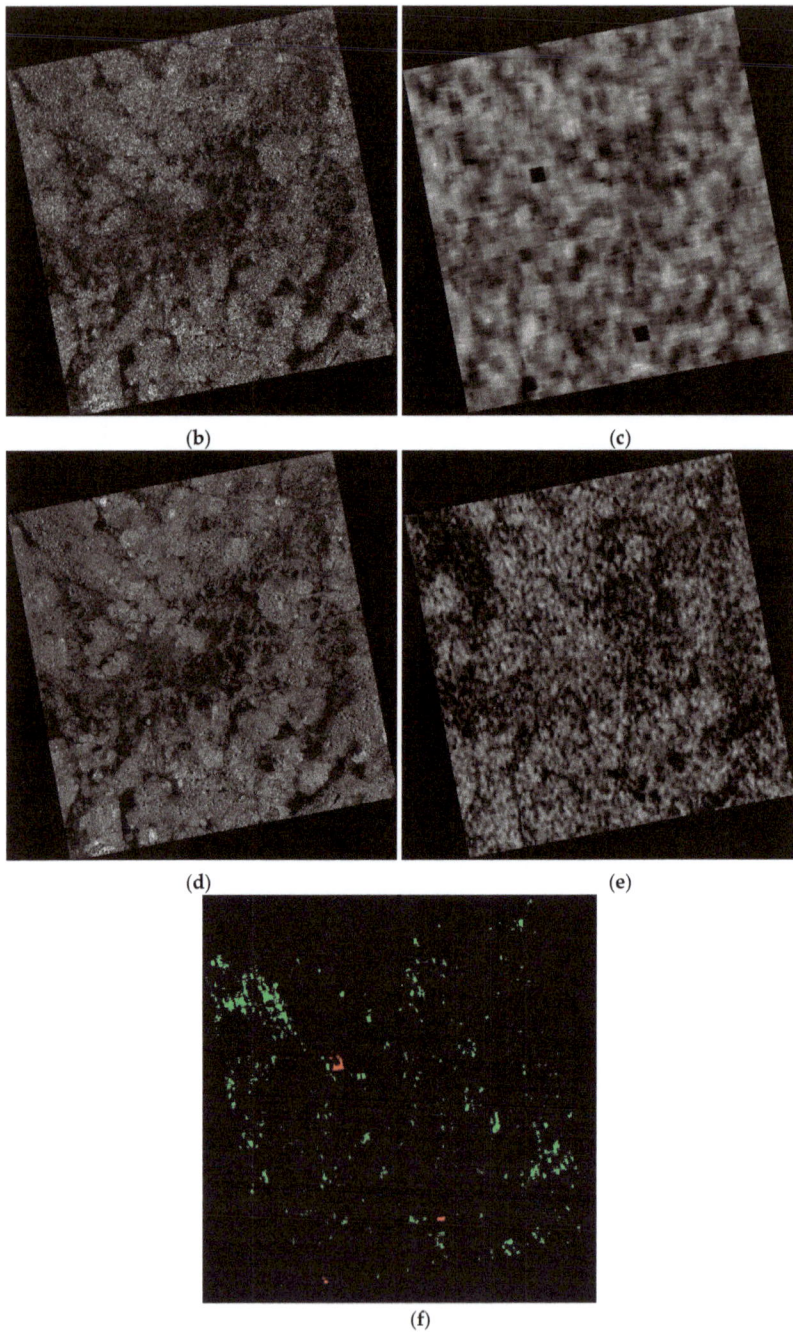

Figure 5. (**a**) Google Earth view of the subset Burkina 2, with the footprint of the imaged area of interest marked in red; (**b**) geocoded stripmap acquisition; (**c**) fractal dimension map corresponding to (**b**); (**d**) geocoded enhanced spotlight image; (**e**) fractal dimension map corresponding to (**d**); (**f**) thresholded difference map.

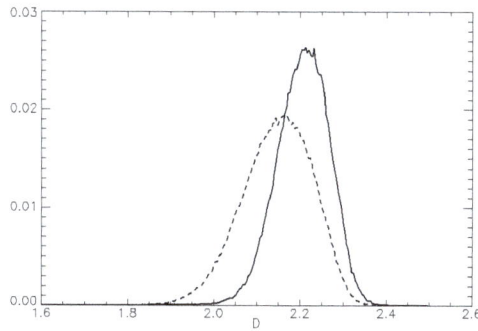

Figure 6. Histogram of the fractal dimension maps of Figure 4c,e: solid line for the stripmap and dashed line for the enhanced spotlight.

3.3. Naples 1

The area of interest in this case is the Bosco di Capodimonte, an urban wood area in the city of Naples, Italy. The Google Earth optical image of the area (2×2.7 km^2) is shown in Figure 7a. After the application of the same processing used in the previous cases, we obtained the stripmap geocoded image and its corresponding fractal dimension presented in Figures 7b and 7c, respectively, and the enhanced spotlight geocoded image and its associated fractal dimension map shown in Figure 7d,e. Forested areas are frequently used as calibration sites in SAR processing, since they provide very stable values of the reflectivity [38]. We chose this area to test the algorithm behavior over a very homogeneous natural area consisting of closely spaced tall vegetation, rather than of bare soil. Since we are dealing with X-band images, the penetration of the transmitted field under the forest upper-level structure will be very low and the algorithm is supposed to estimate the fractal dimension of the envelope of the treetops.

Looking at the fractal dimension maps in Figure 7c,e, it is evident that the estimated fractal dimension is uniform (i.e., spatially homogeneous) over the forested area, while it presents significant spatial variations in its surroundings. This is due to the presence of man-made objects, whose effect on the estimation algorithm [30] will be better highlighted in the last case study. To avoid the influence of these objects, a subset of the images was considered in the statistical analysis: the considered area is marked in red in Figure 7b–e. In Figure 8 we show the histograms of the fractal dimension maps, while their statistics are reported in Table 1. Actually, the statistical behavior of the two maps is very similar: in particular, the difference in the average values of the fractal dimension is very low with respect to the map's standard deviations. Therefore, we can conclude that in this case the estimated fractal dimension maps bear essentially the same information.

This can be observed also on the thresholded fractal dimension difference map in Figure 7f (obtained along the same guidelines of the one presented in Figure 4f), where the only relevant differences can be appreciated in the urban areas present on the image. In particular, no significant difference is present on the forested area enclosed in the red box.

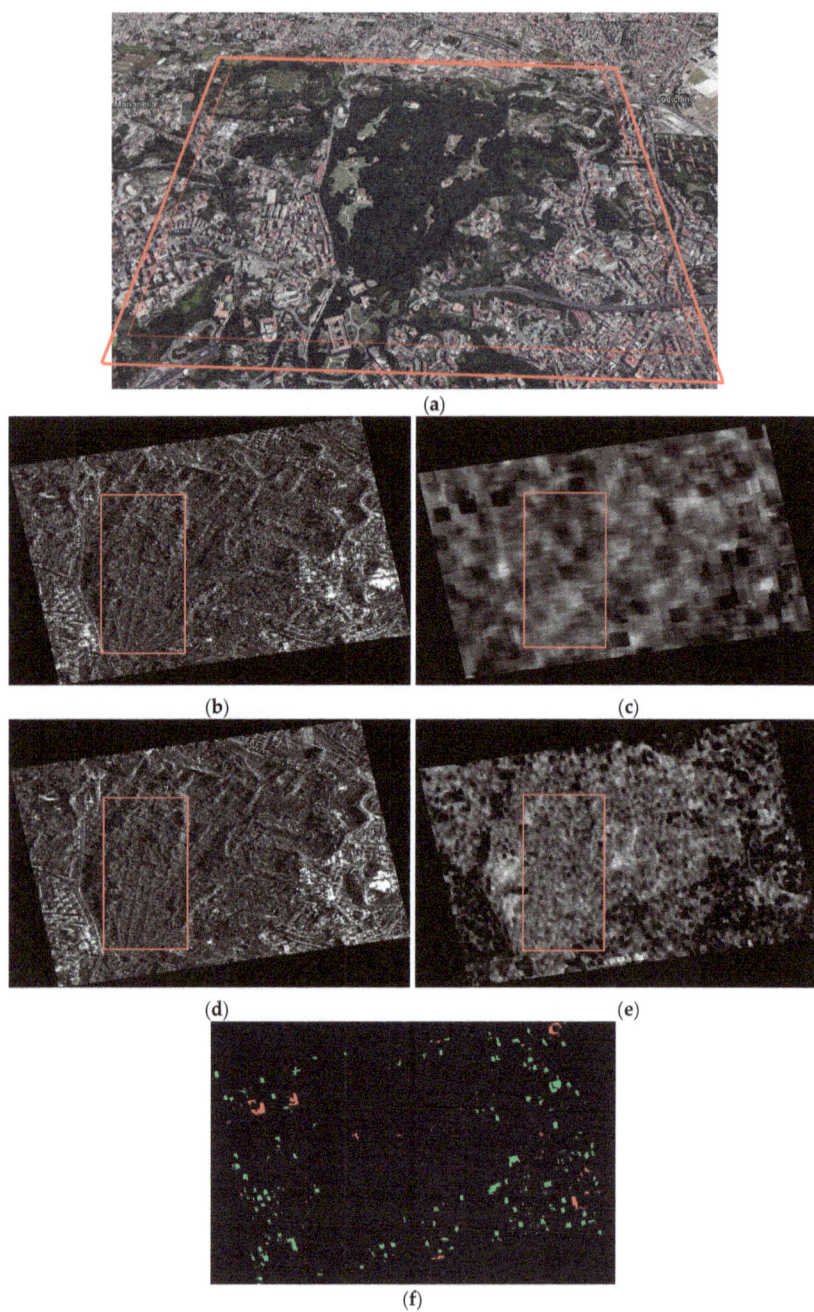

Figure 7. (**a**) Google Earth view of the subset Naples 1, with the footprint of the imaged area of interest marked in red; (**b**) geocoded stripmap acquisition; (**c**) fractal dimension map corresponding to (**b**); (**d**) geocoded enhanced spotlight image; (**e**) fractal dimension map corresponding to (**d**); (**f**) thresholded difference map.

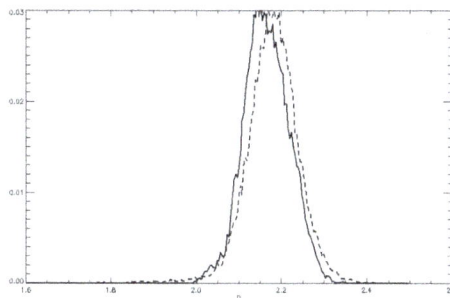

Figure 8. Histogram of the fractal dimension maps of Figure 7c,e: solid line for the stripmap and dashed line for the enhanced spotlight.

3.4. Naples 2

For the last case study, we selected an image of the city of Naples, Italy, covering an area of 2.7 × 2.7 km^2 close to the business district and the central station. The Google Earth image of the area is reported in Figure 9a, while in Figure 9b–e the geocoded stripmap and spotlight images and their corresponding fractal dimension maps are shown. In this case, we are interested in studying the behavior of the algorithm over a non-natural area, consisting essentially of buildings and other man-made objects. In this kind of scenario, the fractal models do not hold and the obtained fractal dimension maps present quite large spatial variations.

This situation can be considered as a reference for the assessment of similarity of the fractal dimension maps obtained in the previous case studies. In fact, in this case the spotlight acquisition detects much more details than the stripmap, as can be appreciated looking at the high density of dark spots present on the fractal dimension map shown in Figure 9e with respect to the one in Figure 9c. This is also confirmed by the histograms presented in Figure 10 and by the statistics of the maps reported in Table 1. With respect to the previous cases, the standard deviations of the maps are significantly higher and the difference in the average values of the fractal dimension is almost comparable with the standard deviation. In this situation, the thresholded difference mask is not statistically significant, and bears no useful information. Obviously, when applied to man-made areas, that cannot be described through fractal models, the algorithm does not provide values that can be treated as a true fractal dimension of the imaged area. Anyway, these values, though not bearing a precise physical meaning, could provide a valuable support for the identification and characterization of urban areas [30].

(a)

Figure 9. *Cont.*

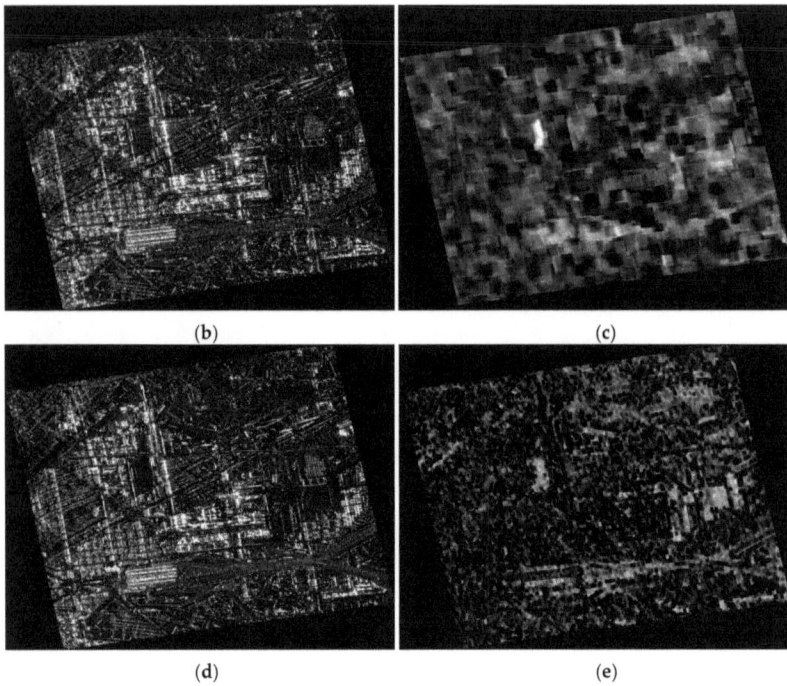

Figure 9. (a) Google Earth view of the subset Naples 2, with the footprint of the imaged area of interest marked in red; (b) geocoded stripmap acquisition; (c) fractal dimension map corresponding to (b); (d) geocoded enhanced spotlight image; (e) fractal dimension map corresponding to (d).

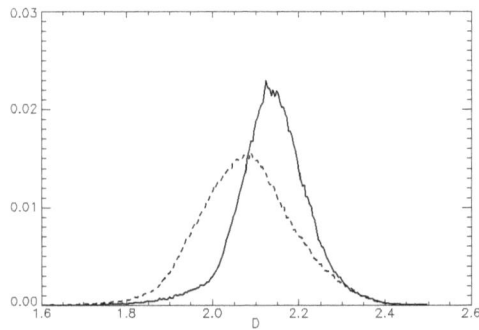

Figure 10. Histogram of the fractal dimension maps of Figure 9c,e: solid line for the stripmap and dashed line for the enhanced spotlight.

4. Conclusions

In this paper we analyzed the behavior of the fractal dimension maps estimated from SAR data acquired using different operational modes. The fractal dimension maps represent the point-by-point fractal dimension of a surface and are estimated from SAR amplitude images. The used data set consists of stripmap and enhanced spotlight Cosmo/SkyMed images collected in the area of Bidi in the North of Burkina Faso and Naples in Italy.

The objective of the work was to highlight potential dependencies of the estimation method on the resolution of the input SAR data. The results of our analysis show that for natural homogeneous areas (e.g., on scarcely vegetated areas and woods) the fractal dimensions estimated from enhanced spotlight and stripmap images are substantially the same; however, estimations from spotlight images are more sensitive to the presence of small objects and sharp edges, which may impair the retrieval of the fractal dimensions performed over areas that include them. Accordingly, when evaluating the fractal dimension of the topography of natural areas is of exclusive interest, it may be more convenient to use stripmap SAR data.

Conversely, in areas containing man-made objects or, more in general, many resolution-scale features (e.g., urban areas), fractal dimension estimates evaluated starting from enhanced spotlight and stripmap images may significantly differ, and, this time, the higher sensitivity to small objects and to man-made typical structures (i.e., buildings, streets, etc.) of fractal dimension maps estimated from spotlight acquisitions is a convenient feature to distinguish and identify such structures. This discussion was also confirmed from the pixel-wise analysis of the image obtained from the difference of the two maps. In this case, we implemented a feature-based data fusion, showing how it is possible to identify meaningful land-cover classes (e.g., trees).

The presented results provide a new basis for the development of further data fusion techniques based on the combined use of multi-operational data. In fact, the combination of fractal dimension values obtained from stripmap and spotlight data is not trivial, and should be based not simply on the classical trade-off between coverage and resolution, but also on the specific application. In general, if we are interested in obtaining a reliable estimate of the fractal dimension of the observed natural scene, stripmap images should be chosen in view of economic and computational considerations (stripmap data are significantly less expensive than spotlight data and the number of samples for equal sized areas is lower). In turn, the combination of fractal dimension maps obtained from stripmap and spotlight images can be used to identify areas on the scene presenting non-fractal behavior (e.g., urban areas).

Acknowledgments: The images were provided by ASI in the framework of two Cosmo-SkyMed 2007 AO projects: Project 1200 "Exploitation of fractal scattering models for COSMO-SkyMed images interpretation," and Project 2218 "Use of high resolution SAR data for water resource management in semi-arid regions."

Author Contributions: All the authors equally contributed to the conceiving, writing, and experiments of the work.

Conflicts of Interest: The authors declare no conflict of interest.

References

1. Mandelbrot, B.B. *The Fractal Geometry of Nature*; Freeman: New York, NY, USA, 1983.
2. Franceschetti, G.; Riccio, D. *Scattering, Natural Surfaces and Fractals*; Academic Press: Burlington, VT, USA, 2007.
3. Falconer, K. *Fractal Geometry*; Wiley: Chichester, UK, 1989.
4. Feder, J.S. *Fractals*; Plenum: New York, NY, USA, 1988.
5. Turcotte, D.L. *Fractals and Chaos in Geology and Geophysics*; Cambridge University Press: Cambridge, UK, 1997.
6. Campbell, B.A. Scale-dependent surface roughness behavior and its impact on empirical models for radar backscatter. *IEEE Trans. Geosci. Remote Sens.* **2009**, *47*, 3480–3488. [CrossRef]
7. Dierking, W. Quantitative Roughness Characterization of Geological Surfaces and Implications for Radar Signature Analysis. *IEEE Trans. Geosci. Remote Sens.* **1999**, *37*, 2397–2412. [CrossRef]
8. Shepard, M.K.; Campbell, B.A.; Bulmer, M.H.; Farr, T.G.; Gaddis, L.R.; Plaut, J.J. The roughness of natural terrain: A planetary and remote sensing perspective. *J. Geophys. Res.* **2001**, *106*, 32777–32795. [CrossRef]
9. Gaddis, L.R.; Mouginis-Mark, P.J.; Hayashi, J.N. Lava flow surface textures: SIR-B radar image texture, field observations, and terrain measurements. *Photogramm. Eng. Remote Sens.* **1990**, *56*, 211–224.
10. Austin, T.; England, A.W.; Wakefield, G.H. Special problems in the estimation of power-law spectra as applied to topographical modeling. *IEEE Trans. Geosci. Remote Sens.* **1994**, *32*, 928–939. [CrossRef]

11. Di Martino, G.; Iodice, A.; Riccio, D.; Ruello, G. Equivalent number of scatterers for SAR speckle modeling. *IEEE Trans. Geosci. Remote Sens.* **2014**, *52*, 2555–2564. [CrossRef]

12. Franceschetti, G.; Callahan, P.S.; Iodice, A.; Riccio, D.; Wall, S.D. Titan, Fractals, and Filtering of Cassini Altimeter Data. *IEEE Trans. Geosci. Remote Sens.* **2006**, *44*, 2055–2062. [CrossRef]

13. Pesquet-Popescu, B.; Véhel, J.L. Stochastic fractal models for image processing. *IEEE Signal Process. Mag.* **2002**, *19*, 48–62. [CrossRef]

14. Stewart, C.V.; Moghaddam, B.; Hintz, K.J.; Novak, L.M. Fractional Brownian Motion Models for Synthetic Aperture Radar Imagery Scene Segmentation. *Proc. IEEE* **1993**, *81*, 1511–1522. [CrossRef]

15. Kaplan, L.M. Extended Fractal Analysis for Texture Classification and Segmentation. *IEEE Trans. Image Process.* **1999**, *8*, 1572–1585. [CrossRef] [PubMed]

16. Jansing, E.D.; Chenoweth, D.L.; Knecht, J. Feature detection in synthetic aperture radar images using fractal error. In Proceedings of the IEEE Aerospace Conference, Aspen, CO, USA, 13 February 1997; pp. 187–195.

17. Jansing, E.D.; Allen, B.; Chenoweth, D.L. Edge enhancement using the fractal error metric. In Proceedings of the 1st International Conference on Engineering Design and Automation, Bangkok, Thailand, 18–21 March 1997.

18. Pentland, A.P. Fractal-Based Description of Natural Scenes. *IEEE Trans. Pattern Anal. Mach. Intell.* **1984**, *6*, 661–674. [CrossRef] [PubMed]

19. Kube, P.; Pentland, A. On the imaging of fractal surfaces. *IEEE Trans. Pattern Anal. Mach. Intell.* **1988**, *10*, 704–707. [CrossRef]

20. Korvin, G. Is the optical image of a non-Lambertian fractal surface fractal? *IEEE Geosci. Remote Sens. Lett.* **2005**, *2*, 380–383. [CrossRef]

21. Di Martino, G.; Iodice, A.; Riccio, D.; Ruello, G. Imaging of Fractal Profiles. *IEEE Trans. Geosci. Remote Sens.* **2010**, *48*, 3280–3289. [CrossRef]

22. Di Martino, G.; Riccio, D.; Zinno, I. SAR Imaging of Fractal Surfaces. *IEEE Trans. Geosci. Remote Sens.* **2012**, *50*, 630–644. [CrossRef]

23. Danudirdjo, D.; Hirose, A. Local Subpixel Coregistration of Interferometric Synthetic Aperture Radar Images Based on Fractal Models. *IEEE Trans. Geosci. Remote Sens.* **2013**, *51*, 4292–4301. [CrossRef]

24. Danudirdjo, D.; Hirose, A. InSAR Image Regularization and DEM Error Correction with Fractal Surface Scattering Model. *IEEE Trans. Geosci. Remote Sens.* **2015**, *53*, 1427–1439. [CrossRef]

25. Danudirdjo, D.; Hirose, A. Anisotropic Phase Unwrapping for Synthetic Aperture Radar Interferometry. *IEEE Trans. Geosci. Remote Sens.* **2015**, *53*, 4116–4126. [CrossRef]

26. Di Martino, G.; Di Simone, A.; Iodice, A.; Riccio, D. Scattering-Based Nonlocal Means SAR Despeckling. *IEEE Trans. Geosci. Remote Sens.* **2016**, *54*, 3574–3588. [CrossRef]

27. Di Martino, G.; Di Simone, A.; Iodice, A.; Poggi, G.; Riccio, D.; Verdoliva, L. Scattering-Based SARBM3D. *IEEE J. Sel. Top. Appl. Earth Obs. Remote Sens.* **2016**, *9*, 2131–2144. [CrossRef]

28. Di Martino, G.; Iodice, A.; Riccio, D.; Ruello, G.; Zinno, I. Angle Independence Properties of Fractal Dimension Maps Estimated from SAR Data. *IEEE J. Sel. Top. Appl. Earth Obs. Remote Sens.* **2013**, *6*, 1242–1253. [CrossRef]

29. Di Martino, G.; Iodice, A.; Riccio, D.; Ruello, G.; Zinno, I. The effects of polarization on fractal dimension maps estimated from SAR data. In Proceedings of the PolInSAR 2013 (ESA SP-713), Frascati, Italy, 28 January–1 February 2013.

30. Di Martino, G.; Iodice, A.; Riccio, D.; Ruello, G.; Zinno, I. Fractal Filtering Applied to SAR Images of Urban Areas. In Proceedings of the Urban Remote Sensing Event JURSE, Munich, Germany, 11–13 April 2011.

31. Capon, J. High-Resolution Frequency-Wavenumber Spectrum Analysis. *Proc. IEEE* **1969**, *57*, 1408–1418. [CrossRef]

32. Kay, S.M. *Modern Spectral Analysis*; Patience Hall: Englewood Cliffs, NJ, USA, 1999.

33. Amitrano, D.; Di Martino, G.; Iodice, A.; Riccio, D.; Ruello, G.; Ciervo, F.; Papa, M.N.; Koussoube, Y. Effectiveness of high-resolution SAR for water resource management in low-income semi-arid countries. *Int. J. Remote Sens.* **2014**, *35*, 70–88. [CrossRef]

34. Amitrano, D.; Di Martino, G.; Iodice, A.; Riccio, D.; Ruello, G. A New Framework for SAR Multitemporal Data RGB Representation: Rationale and Products. *IEEE Trans. Geosci. Remote Sens.* **2015**, *53*, 117–133. [CrossRef]

Remote Sens. **2018**, *10*, 9

35. Di Martino, G.; Poderico, M.; Poggi, G.; Riccio, D.; Verdoliva, L. Benchmarking Framework for SAR Despeckling. *IEEE Trans. Geosci. Remote Sens.* **2014**, *52*, 1596–1615. [CrossRef]

36. Di Martino, G.; Poggi, G.; Riccio, D.; Verdoliva, L. Effects of Despeckling on the Estimation of Fractal Dimension from SAR Images. In Proceedings of the IEEE International Geoscience and Remote Sensing Symposium, Melbourne, Australia, 21–26 July 2013; pp. 3950–3953.

37. Pohl, C.; Van Genderen, J.L. Multisensor image fusion in remote sensing: Concepts, methods and applications. *Int. J. Remote Sens.* **1998**, *19*, 823–854. [CrossRef]

38. Freeman, A. SAR calibration: An overview. *IEEE Trans. Geosci. Remote Sens.* **1992**, *30*, 1107–1121. [CrossRef]

remote sensing

MDPI

Review

Statistical Modeling of Polarimetric SAR Data: A Survey and Challenges

Xinping Deng [1], Carlos López-Martínez [2], Jinsong Chen [1],* and Pengpeng Han [1]

[1] Shenzhen Institutes of Advanced Technology, Chinese Academy of Sciences, Shenzhen 518055, China; xp.deng1@siat.ac.cn (X.D.); pp.han@siat.ac.cn (P.H.)
[2] Remote Sensing Laboratory, Signal Theory and Communications Department, Universitat Politècnica de Catalunya, 08034 Barcelona, Spain; carlos.lopez@tsc.upc.edu
* Correspondence: js.chen@siat.ac.cn; Tel.: +86-755-86392370

Academic Editors: Timo Balz, Uwe Soergel, Mattia Crespi, Batuhan Osmanoglu, Zhenhong Li and Prasad S. Thenkabail
Received: 6 March 2017; Accepted: 2 April 2017; Published: 5 April 2017

Abstract: Knowledge of the exact statistical properties of the signal plays an important role in the applications of Polarimetric Synthetic Aperture Radar (PolSAR) data. In the last three decades, a considerable research effort has been devoted to finding accurate statistical models for PolSAR data, and a number of distributions have been proposed. In order to see the differences of various models and to make a comparison among them, a survey is provided in this paper. Texture models, which could capture the non-Gaussian behavior observed in high resolution data, and yet keep a compact mathematical form, are mainly explained. Probability density functions for the single look data and the multilook data are reviewed, as well as the advantages and applicable context of those models. As a summary, challenges in the area of statistical analysis of PolSAR data are also discussed.

Keywords: statistical modeling; polarimetric SAR; texture models; finite mixture models; copulas

---

1. Introduction

Synthetic Aperture Radar (SAR) and Polarimetric SAR (PolSAR) are widely used for observation of natural scenes. In most SAR or PolSAR systems, the size of a resolution cell is much larger than the wavelength. The measured signal is then a coherent addition of the echoes from all individual targets within that cell. Depending on the relative phases of each scattered wave, the coherent addition may be constructive or destructive, and it produces a salt-and-pepper appearance known as speckle over SAR images [1]. The target information, therefore, should be extracted through statistical analysis of the data. Hence, an accurate statistical model to describe the data becomes very important for the extraction of ground target properties [2–6].

Gaussian statistics for the radar returns have been frequently assumed when the spatial resolution of PolSAR images is moderate and the speckle is fully developed [1,7,8]. Actually, the number of targets in a resolution cell of low or medium resolution data is large. According to the Central Limit Theorem (CLT), Gaussian statistics could give a proper approximation to the data distribution. The Gaussian distribution is both mathematically tractable and efficient, making it very useful in specific applications. For SAR or PolSAR data, the mean value of the complex echo is generally assumed to be zero, and all the statistical properties are determined by the covariance matrix (or the coherency matrix) under the Gaussian assumption.

As the image resolution increases, analysis of real PolSAR data, however, reveals that non-Gaussian models give a more accurate representation of the data. The change of the observing surface could also give rise to non-Gaussian distributed data. Applications based on such models have better performance [2,4,9,10]. A common way to introduce non-Gaussianity is to divide the randomness of the radar images into two unrelated factors, texture and speckle. The texture models the natural spatial variation of the radar cross section, whereas the speckle, following a Gaussian distribution, conveys the polarimetric information. The texture and the speckle are incorporated with a product operation which leads to a doubly stochastic model called product model [11]. In the last two decades, a considerable research effort has been dedicated to investigate accurate product models for PolSAR data [12–16].

Another way to model the non-Gaussian behavior of PolSAR data is the so called finite mixture model [17–19], which assumes the data under analysis is a discrete mixture of different targets. This makes sense in certain scenes such as urban areas, which usually consist of coherent targets like houses and roads, as well as distributed targets like trees and grass. The backscattering from the urban area is a combination of different scattering mechanisms. It has been shown that for complex regions with irregular histograms, multimodal or spiky for example, the finite mixture model is more accurate than a single distribution [17].

As summarized in [20], there are many non-Gaussian distributions, including the Weibull distribution, the lognormal distribution, and the α-stable distribution, suggested for the one-dimensional SAR data. However, these distributions are difficult to generalize to the multidimensional PolSAR data. A possible solution to this problem is to consider the idea of copulas [21]. First, we can use various non-Gaussian distributions to model the data of each polarimetric channel separately (called marginal distribution), and then introduce some common multivariate distributions to model the dependence of these marginal distributions. With the copulas, different marginal distributions and simple correlation structure can make up complex distributions for the PolSAR data [22,23].

As we can see, there are many statistical models proposed for the PolSAR data from different aspects. In this paper, a survey of these models is provided. PDFs for the single look data and the multilook data are mainly reviewed, as well as the advantages of those models. Analysis of real PolSAR data are performed using different statistical methods to evaluate the models.

The remainder of this paper is organized as follows. First, a few basic concepts of the polarimetric SAR are introduced, especially the notation employed in this paper. Then, statistics of the fully developed speckle will be discussed. Properties of the single look data and the multilook data are studied under the Gaussian assumption. The introduction of texture is followed, along with the widely studied texture models, including both the scalar texture models and the multi-texture models. Finally, finite mixture models as well as copula based models are detailed. Several experiments to validate applicability of different models are also given. Challenges in statistical modeling is summarized at the end.

2. Polarimetric SAR

PolSAR systems measure the properties of a distant target by detecting the change of the polarization state that the target induces to the incident wave. Let the polarized incident wave $\mathbf{E}^i$ and scattered wave $\mathbf{E}^s$ be expressed as the Jones Vectors [24]

$$\mathbf{E}^i = \begin{bmatrix} E_h^i \\ E_v^i \end{bmatrix} \quad \mathbf{E}^s = \begin{bmatrix} E_h^s \\ E_v^s \end{bmatrix} \tag{1}$$

where h represents the horizontal polarization state and v the vertical polarization state. It is possible to relate the incident and the scattered waves by means of a 2×2 complex matrix [24]

$$\begin{bmatrix} E_h^s \\ E_v^s \end{bmatrix} = \frac{e^{-jkz}}{z} \begin{bmatrix} S_{hh} & S_{hv} \\ S_{vh} & S_{vv} \end{bmatrix} \begin{bmatrix} E_h^i \\ E_v^i \end{bmatrix} \tag{2}$$

where z is the distance between the target and the receiving antenna, and k is the wave number of the illuminating wave. The 2×2 transformation matrix is generally referred to as scattering matrix and denoted by **S**. It characterizes the target under observation with four complex-valued scattering coefficients. The diagonal elements of the scattering matrix receive the name "co-pol", since they relate the same polarization for the incident and the scattered waves. The off-diagonal elements are known as "cross-pol" terms as they relate orthogonal polarization states [24].

The definition of **S** depends on the coordinate systems. There are two principal conventions concerning the coordinate systems where the polarimetric scattering process can be considered: Forward Scattering Alignment (FSA) and Back Scattering Alignment (BSA) [24]. The difference lies in the way the coordinate system is selected to describe the polarization state of the scattered wave. The FSA is usually used when the transmitter and the receiver are not placed at the same spatial location, for example, for bistatic radar measurements. In contrast, the BSA is often adopted in monostatic radar measurements, in which the transmitting and receiving antennas are collocated in space. In this paper, we assume that the BSA convention is employed.

The interaction between the electromagnetic waves with a reciprocal medium follows the vector reciprocity theorem, which states that if we transmit a polarization state P_A from position A, then the component polarized in the P_B direction at position B is equal to the P_A component of the scattered radiation when we illuminate the same object from B with polarization P_B [25]. The reciprocity theorem applies to ground targets generally [25]. In the BSA coordinate system, the reciprocity theorem says that the cross-pol channels of the scattering matrix are equal, that is $S_{hv} = S_{vh}$. Therefore, there are only three independent complex coefficients required to characterize the scatterer under observation.

In many cases, it is more flexible to represent the scattering matrix **S** as a vector which is known as scattering vector. The vectorization can be performed through [26]

$$\mathbf{k} = \frac{1}{2} \operatorname{Tr}(\mathbf{S\Psi}) \tag{3}$$

where $\operatorname{Tr}(\cdot)$ is the matrix trace and $\mathbf{\Psi}$ is a 2×2 complex matrix from a basis set which are constructed as an orthonormal set under the Hermitian inner product. The lexicographic basis and the Pauli basis are the most common ones in the context of radar polarimetry. The selection of the basis to vectorize the scattering matrix depends on the final purpose of the vectorization itself. When studying the statistical behavior of the PolSAR data, the lexicographic basis is more convenient due to its simplicity. The lexicographic basis set consists of the straightforward lexicographic ordering of the elements of the scattering matrix. For a reciprocal target, the scattering vector in this case can be expressed as

$$\mathbf{k} = \begin{bmatrix} S_{hh} \\ \sqrt{2}S_{hv} \\ S_{vv} \end{bmatrix} \tag{4}$$

Targets under observation are commonly situated in a dynamically changing environment and are subjected to spatial and temporal variations. Despite the radar system transmits a perfectly polarized wave, the wave scattered by the target is partially polarized [25]. Such scatterers are called distributed targets. The analysis of this type of targets can not be performed exactly by one target but a population of targets. More precisely, they are analyzed by introducing the concept of space and time varying stochastic processes, where the targets are described by the second order moments such as the polarimetric coherency or covariance matrices.

The covariance matrix is defined as the expectation of the outer product of the target vector with its transpose conjugate

$$\mathbf{\Sigma} = \mathrm{E}\{\mathbf{k}\mathbf{k}^\dagger\} = \begin{bmatrix} \mathrm{E}\{S_{hh}S_{hh}^*\} & \sqrt{2}\,\mathrm{E}\{S_{hh}S_{hv}^*\} & \mathrm{E}\{S_{hh}S_{vv}^*\} \\ \sqrt{2}\,\mathrm{E}\{S_{hv}S_{hh}^*\} & 2\,\mathrm{E}\{S_{hv}S_{hv}^*\} & \sqrt{2}\,\mathrm{E}\{S_{hv}S_{vv}^*\} \\ \mathrm{E}\{S_{vv}S_{hh}^*\} & \sqrt{2}\,\mathrm{E}\{S_{vv}S_{hv}^*\} & \mathrm{E}\{S_{vv}S_{vv}^*\} \end{bmatrix} \tag{5}$$

where $(\cdot)^{\dagger}$ and $(\cdot)^{*}$ denote the transpose conjugate and conjugate, respectively. In practice, the number of scattering vectors used to calculate the expectation is limited. Let L denote the number of pixels to compute the average, the PolSAR data are then represented by the so-called sample covariance matrix

$$C_L = \frac{1}{L} \sum_{i=1}^{L} \mathbf{k}_i \mathbf{k}_i^{\dagger} \tag{6}$$

where $\mathbf{k}_i$ is the ith scattering vector. The averaging is also called multilook processing which can be employed to reduce the speckle of PolSAR data, with L referring to the number of looks.

3. Gaussian Statistics

Under the assumption that the speckle is fully developed, it has been experimentally verified that the Gaussian statistics generally provide a good fit to SAR data, especially in homogeneous natural areas [7,27–30]. The multivariate Gaussian distribution, which is both mathematically tractable and efficient, is proper to model the scattering vectors when the surface roughness is relatively low, the spatial resolution is moderate, and a large number of scatterers are present [1,24]. The Gaussian assumption indicates that the statistical properties of the data are determined by the covariance matrix. The sample covariance matrix in this case follows a complex Wishart distribution, which is widely used in the applications of PolSAR data. There exist also some variations of the Wishart distribution that are shown to be more accurate in certain circumstances.

3.1. Gaussian Distribution

When a radar illuminates an area of a random surface containing many elementary scatterers, the scattering vector, $\mathbf{z}$, can be modeled as having a d-dimensional complex Gaussian distribution with zero mean. The Probability Density Function (PDF) is given by [31]

$$p(\mathbf{z}; \mathbf{\Sigma}) = \frac{1}{\pi^d |\mathbf{\Sigma}|} \exp(-\mathbf{z}^{\dagger} \mathbf{\Sigma}^{-1} \mathbf{z}) \tag{7}$$

where $|\cdot|$ is the determinant operation. The complex Gaussian distribution is denoted by $\mathbf{z} \sim \mathcal{CN}(0, \mathbf{\Sigma})$ for brevity. The real and imaginary parts of any complex element of $\mathbf{z}$ are assumed to follow a circular Gaussian distribution. Consider the ith element $\mathbf{z}_i = x_i + jy_i$ for example, the joint PDF of the real and the imaginary parts can be written as

$$p(x_i, y_i; \sigma_i) = \frac{1}{\pi \sigma_i^2} \exp\left(-\frac{x_i^2 + y_i^2}{\sigma_i^2}\right) \tag{8}$$

where $\sigma_i^2 = \Sigma_{ii}$. Let r_i be the amplitude and θ_i be the phase of a complex value, then the real part of $\mathbf{z}_i$ can be written as $x_i = r_i \cos \theta_i$, and the imaginary part as $y_i = r_i \sin \theta_i$. The Jacobian determinant of the transform from (x_i, y_i) to (r_i, θ_i) is given by

$$J = \begin{vmatrix} \cos \theta_i & -r_i \sin \theta_i \\ \sin \theta_i & r_i \cos \theta_i \end{vmatrix} = r_i \tag{9}$$

Subsequently, the joint PDF of the amplitude and the phase can be obtained from (8) after changing variables, giving

$$p(r_i, \theta_i; \sigma_i) = \frac{r_i}{\pi \sigma_i^2} \exp\left(-\frac{r_i^2}{\sigma_i^2}\right) \tag{10}$$

The circular Gaussian assumption implies that the phase θ_i is uniformly distributed over $(-\pi, \pi]$, and independent from the amplitude. Averaging over the phase, therefore, gives the PDF of the amplitude

$$p(r_i; \sigma_i) = \frac{2r_i}{\sigma_i^2} \exp\left(-\frac{r_i^2}{\sigma_i^2}\right) \tag{11}$$

Equation (11) is known as the Rayleigh distribution, with mean value $\sigma_i \sqrt{\pi}/2$. The intensity of the ith channel, $I_i = x_i^2 + y_i^2 = r_i^2$, can be easily proved to have a negative exponential distribution

$$p(I_i; \sigma_i) = \frac{1}{\sigma_i^2} \exp\left(-\frac{I_i}{\sigma_i^2}\right) \tag{12}$$

with mean value $E\{I_i\} = \sigma_i^2$ and variance $Var\{I_i\} = \sigma_i^4$. This distribution shows that the useful information is described by a single degree of freedom, corresponding to the mean intensity.

Besides the intensity, the joint properties of two different polarimetric channels are of great interest. Considering two polarimetric channels $z_i = x_i + jy_i$ and $z_k = x_k + jy_k$, the complex correlation coefficient is determined by

$$\rho e^{j\varphi} = \frac{\Sigma_{ik}}{\sqrt{\Sigma_{ii}\Sigma_{kk}}} \tag{13}$$

where ρ and φ are, respectively, the amplitude and the phase of the complex correlation coefficient. The joint PDF of the real part and the imaginary part can be derived from (7), which is given as follows [30,32,33]

$$p(x_i, y_i, x_k, y_k) = \frac{1}{\pi^2 \psi^2 (1 - \rho^2)} \exp\left(-\frac{\sigma_k^2(x_i^2 + y_i^2) + \sigma_i^2(x_k^2 + y_k^2)}{\psi^2(1 - \rho^2)}\right. $$
$$\left. + \frac{2\psi\rho[(x_ix_k + y_iy_k)\cos\varphi + (x_ky_i - x_iy_k)\sin\varphi]}{\psi^2(1 - \rho^2)}\right) \tag{14}$$

where $\sigma_i^2 = \Sigma_{ii}$, $\sigma_k^2 = \Sigma_{kk}$ and $\psi = \sigma_i\sigma_k$. Write the complex values in the polar form, i.e., $r_ie^{j\theta_i} = x_i + jy_i$ and $r_ke^{j\theta_k} = x_k + jy_k$, by changing variables from (x_i, y_i, x_k, y_k) to $(r_i, \theta_i, r_k, \theta_k)$, the previous distribution becomes

$$p(r_i, \theta_i, r_k, \theta_k) = \frac{r_ir_k}{\pi^2\psi^2(1 - \rho^2)} \exp\left(-\frac{\sigma_k^2 r_i^2 + \sigma_i^2 r_k^2 - 2\psi r_ir_k\rho\cos(\theta_i - \theta_k - \varphi)}{\psi^2(1 - \rho^2)}\right) \tag{15}$$

We are interested in the distributions of the product of the two amplitudes $z = r_ir_k$, and the phase difference $\phi = \theta_i - \theta_k$, since their values reflect the correlation between different polarimetric channels. It can be shown that the Jacobian determinant of the transform from $(r_i, r_k, \theta_i, \theta_k)$ to (r_i, z, θ_i, ϕ) is $-1/r_i$. Thus the following distribution can be obtained after changing variables

$$p(r_i, z, \theta_i, \phi) = \frac{z}{\pi^2\psi^2(1 - \rho^2)}\frac{1}{r_i} \exp\left(-\frac{\sigma_k^2 r_i^2 + \frac{\sigma_i^2 z^2}{r_i^2} - 2\psi\rho z\cos(\phi - \varphi)}{\psi^2(1 - \rho^2)}\right) \tag{16}$$

from which the joint PDF of z and ϕ can be further derived by integrating over θ_i and r_i and employing the equality (A1)

$$p(z, \phi) = \frac{2z}{\pi\psi^2(1 - \rho^2)} \exp\left(\frac{2\rho z\cos(\phi - \varphi)}{\psi(1 - \rho^2)}\right) K_0\left(\frac{2z}{\psi(1 - \rho^2)}\right) \tag{17}$$

Here K_v is the modified Bessel function of the second kind of order v [34]. The marginal distribution of the product of the amplitudes, subsequently, is found to be

$$p(z) = \frac{4z}{\psi^2(1-\rho^2)} I_0 \left(\frac{2\rho z}{\psi(1-\rho^2)} \right) K_0 \left(\frac{2z}{\psi(1-\rho^2)} \right) \tag{18}$$

where $I_0(z)$ is the modified Bessel function of the first kind [34] resulting from the integral identity (A2). Similarly, integrating (17) over the amplitudes and following the identity (A3) gives the marginal distribution of the phase difference

$$p(\phi) = \frac{1-\rho^2}{2\pi(1-\beta^2)} \left\{ \frac{\beta}{\sqrt{\beta^2-1}} \ln(-\beta + \sqrt{\beta^2-1}) + 1 \right\} \tag{19}$$

with $\beta = \rho \cos(\phi - \varphi)$. Note that $-\beta + \sqrt{\beta^2 - 1}$ is a complex number since β is less than 1. Therefore, it can be represented in the polar form, e.g., $-\beta + \sqrt{\beta^2 - 1} = \exp(j(\pi - \arccos \beta))$, and as a result, (19) becomes

$$p(\phi) = \frac{1-\rho^2}{2\pi(1-\beta^2)} \left\{ \frac{\beta(\pi - \arccos \beta)}{\sqrt{1-\beta^2}} + 1 \right\} \tag{20}$$

The PDFs shown in (18) and (20) can be also found in [32,33]. The Gaussian assumption implies that the statistics of the PolSAR data is completely determined by the covariance matrix. The properties described by the multivariate distribution (7) can be analyzed separately by the intensity (12), the product of amplitudes (18) and the phase difference (20).

3.2. Wishart Distribution

SAR data are frequently multilook processed for speckle reduction. Under the Gaussian assumption, the sample covariance matrix $\mathbf{C}_L$ follows a complex Wishart distribution, $\mathbf{C}_L \sim \mathcal{CW}(L, \mathbf{\Sigma})$, with PDF given by [31]

$$p(\mathbf{C}_L; L, \mathbf{\Sigma}) = \frac{L^{Ld}|\mathbf{C}_L|^{L-d} \exp(-L\, \mathrm{Tr}(\mathbf{\Sigma}^{-1}\mathbf{C}_L))}{\Gamma_d(L)|\mathbf{\Sigma}|^L} \tag{21}$$

where the normalization factor $\Gamma_d(L)$ is defined as

$$\Gamma_d(L) = \pi^{\frac{d(d-1)}{2}} \prod_{i=1}^{d} \Gamma(L-i+1) \tag{22}$$

with $\Gamma(\cdot)$ referring to the gamma function. The Wishart distribution is valid only if $L \geq d$. The random variables of this distribution are the diagonal terms of $\mathbf{C}_L$ as well as the real and imaginary parts of the upper (or lower) off-diagonal terms. For a d-dimensional radar signal, the total number of independent variables is d^2.

Considering only one polarimetric channel, from (21), we have the distribution of the intensity as

$$p(I_i; L, \sigma_i) = \frac{1}{\Gamma(L)} \left(\frac{L}{\sigma_i^2} \right)^L I_i^{L-1} \exp\left(-\frac{L}{\sigma_i^2} I_i \right) \tag{23}$$

It is known as the gamma distribution with mean value $\mathrm{E}\{I_i\} = \sigma_i^2$ and variance $\mathrm{Var}\{I_i\} = \sigma_i^4/L$ [35]. The number of looks can be estimated using the mean and the variance of the intensity

$$\hat{L} = \frac{\mathrm{E}^2\{I_i\}}{\mathrm{Var}\{I_i\}}. \tag{24}$$

When L is equal to 1, the gamma distribution reduces to the exponential distribution (12). The variances of the two different distributions show that the multilook process reduces the speckle by scaling down the fluctuation magnitude with a factor $1/L$.

For two polarimetric channels, saying channel i and channel k, the sample covariance matrix can be written as

$$\mathbf{C}_L = \begin{bmatrix} I_i & R_{ik} + jI_{ik} \\ R_{ik} - jI_{ik} & I_k \end{bmatrix}. \tag{25}$$

Let $\rho e^{j\varphi}$ represent the complex correlation coefficient, the joint distribution of I_i, I_k, R_{ik} and I_{ik} can be derived from (21), giving

$$p(I_i, I_k, R_{ik}, I_{ik}) = \frac{L^{2L}(I_i I_k - R_{ik}^2 - I_{ik}^2)^{L-2}}{\pi \Gamma(L)\Gamma(L-1)\psi^{2L}(1-\rho^2)^L} \times \tag{26}$$
$$\exp\left(-L\frac{\sigma_i^2 I_k + \sigma_k^2 I_i - 2\rho\psi(R_{ik}\cos\varphi - jI_{ik}\sin\varphi)}{\psi^2(1-\rho^2)}\right)$$

where $\sigma_i^2 = \Sigma_{ii}$, $\sigma_k^2 = \Sigma_{kk}$, and $\psi = \sigma_i\sigma_k$. Write the off-diagonal element in the polar form, $ze^{j\phi} = R_{ik} + jI_{ik}$, by changing variables from $(I_i, I_k, R_{ik}, I_{ik})$ to (I_i, I_k, z, ϕ), the following result can be obtained

$$p(I_i, I_k, z, \phi) = \frac{zL^{2L}(I_i I_k - z^2)^{L-2}}{\pi\Gamma(L)\Gamma(L-1)\psi^{2L}(1-\rho^2)^L}\exp\left(-L\frac{\sigma_i^2 I_k + \sigma_k^2 I_i - 2z\rho\psi\cos(\phi - \varphi)}{\psi^2(1-\rho^2)}\right) \tag{27}$$

The determinant of $\mathbf{C}_L$ must be greater than 0, therefore, we have $I_i I_k - z^2 > 0$. Integrating I_i over $(z^2/I_k, \infty)$ using (A4) and then I_k over $(0, \infty)$ using (A1) gives

$$p(z, \phi) = \frac{2L^{L+1}z^L}{\pi\Gamma(L)\psi^{L+1}(1-\rho^2)}\exp\left(\frac{2Lz\rho\cos(\phi - \varphi)}{\psi(1-\rho^2)}\right)K_{L-1}\left(\frac{2Lz}{\psi(1-\rho^2)}\right) \tag{28}$$

Subsequently, the marginal distribution of the amplitude can be obtained following the integral identity (A2)

$$p(z) = \frac{4L^{L+1}z^L}{\Gamma(L)\psi^{L+1}(1-\rho^2)}I_0\left(\frac{2Lz\rho}{\psi(1-\rho^2)}\right)K_{L-1}\left(\frac{2Lz}{\psi(1-\rho^2)}\right) \tag{29}$$

and the distribution of the phase difference by identity (A5)

$$p(\phi) = \frac{(1-\rho^2)^L}{2\sqrt{\pi}(1-\beta)^{2L}}\frac{\Gamma(2L)}{\Gamma(L)\Gamma(L+\frac{3}{2})}{}_2F_1\left(2L, L - \frac{1}{2}, L + \frac{3}{2}, \frac{\beta+1}{\beta-1}\right). \tag{30}$$

where $\beta = \rho\cos(\phi - \varphi)$, and ${}_2F_1(a, b; c; z)$ is the Gauss hypergeometric function [34]. Again, the statistical properties of the multilook data can be analyzed separately using (23), (29) and (30). The Wishart distribution is widely used in the modeling of PolSAR data [7,36–38], and there are several variations that make the model more accurate or efficient.

3.2.1. Relaxed Wishart Model

Compared with the multivariate complex Gaussian distribution, the Wishart distribution depends on an additional parameter, L, the number of looks. Assume that the multilook processing has different contributions to different types of targets, Anfinsen et al. proposed a refined model called relaxed Wishart distribution [39], in which the number of looks L is treated as a variable shape parameter. In other words, the number of looks is assumed to be distinct in different areas. It is observed that varying L gives a better representation of the data than using a constant L over all regions [39].

3.2.2. Wishart-Kotz Distribution

Another variation of the Wishart distribution is the Wishart-Kotz model [40,41], which exhibits the heavy tails needed to fit the data found in high resolution PolSAR images. In this model, there are no special mathematical functions involved that limit the usefulness by inflicting high computational cost and numerical instability. The sample covariance matrix in the Wishart-Kotz model is assumed to follow a Wishart-Kotz type I distribution with PDF defined as [40]

$$p(\mathbf{C}_L; L, \mathbf{\Sigma}, \rho, \beta) = \frac{c|\mathbf{C}_L|^{L-d}}{|\mathbf{\Sigma}|^L}(\text{Tr}(\mathbf{\Sigma}^{-1}\mathbf{C}_L))^{\beta-1}\exp(-[L\ \text{Tr}(\mathbf{\Sigma}^{-1}\mathbf{C}_L)]^\rho) \tag{31}$$

with additional parameters ρ and β, and a normalization constant factor c

$$c = \frac{\rho L^{\beta+Ld-1}\Gamma(Ld)}{\Gamma_d(L)\Gamma(\frac{\beta+Ld-1}{\rho})} \tag{32}$$

Here $\Gamma_d(L)$ is the same as that in Wishart model, see (22). The Wishart-Kotz distribution is a generalization of the Wishart distribution, which reduces to the latter when $\rho = 1$ and $\beta = 1$.

4. Texture Model

The properties of the fully developed speckle are detailed in the previous section. This section illustrates how to model the texture statistically. There are two main manners to manage this: (1) consider the texture as a scalar random variable, or (2) consider it as a vector having the same dimension as the speckle component. They lead to the so called scalar texture model and multi-texture model, respectively. The texture random variable is generally assumed to be positive with unity mean. Therefore, it models the variation of the radar cross section only, leaving the intensities to the speckle component [7,42]. The statistical properties could be described by a certain distribution, or just a stochastic process without a specific PDF.

4.1. Scalar Texture Model

The scalar texture model assumes that the texture component in the product model is a positive scalar random variable. The scattering vector in this case can be written as [7,11,43,44]

$$\mathbf{k} = \sqrt{\tau}\mathbf{z} \tag{33}$$

where τ is the texture parameter with mean value equal to 1, and $\mathbf{z}$ is the speckle vector, following a multivariate Gaussian distribution (7). The scalar texture model is also referred to as scale mixture of Gaussian [4], or Sphereically Invariant Random Vector (SIRV) [45–47]. For the multilook data, the sample covariance matrix can be expressed as

$$\mathbf{C}_L = \frac{1}{L}\sum_{i=1}^{L}\tau_i\mathbf{z}_i\mathbf{z}_i^\dagger = \frac{\tau}{L}\sum_{i=1}^{L}\mathbf{z}_i\mathbf{z}_i^\dagger \tag{34}$$

under the assumption that the texture has a higher spatial correlation than the speckle and the texture parameter is constant over the multilook processing window [13].

For a known τ, (33) implies that the scattering vector $\mathbf{k}$ follows a complex Gaussian distribution (see Section 3.1) with PDF given by

$$p(\mathbf{k}|\tau; \mathbf{\Sigma}) = \frac{1}{\pi^d|\mathbf{\Sigma}|}\frac{1}{\tau^d}\exp\left(-\frac{\mathbf{k}^\dagger\mathbf{\Sigma}^{-1}\mathbf{k}}{\tau}\right) \tag{35}$$

And the distribution of the sample covariance matrix is given by

$$p(\mathbf{C}_L|\tau; L, \boldsymbol{\Sigma}) = \frac{L^{Ld}|\mathbf{C}_L|^{L-d}}{\Gamma_d(L)|\boldsymbol{\Sigma}|^L} \frac{1}{\tau^{Ld}} \exp\left(-\frac{L \operatorname{Tr}(\boldsymbol{\Sigma}^{-1}\mathbf{C}_L)}{\tau}\right) \tag{36}$$

which is known as the Wishart distribution detailed in Section 3.2.

If the PDF of the texture random variable is not explicitly specified, τ can be viewed as an unknown deterministic parameter from pixel to pixel [47]. According to the concept of SIRV, an approximate maximum likelihood estimator for the texture parameter of each pixel is found to be [45,47]

$$\hat{\tau}_i = \frac{\mathbf{k}_i^\dagger \hat{\boldsymbol{\Sigma}}^{-1} \mathbf{k}_i}{d}$$

$$\hat{\boldsymbol{\Sigma}} = \frac{1}{N} \sum_{i=1}^{N} \frac{\mathbf{k}_i \mathbf{k}_i^\dagger}{\hat{\tau}_i} \tag{37}$$

where $\hat{\tau}_i$ is the texture parameter of the ith pixel, d is the dimension of the target vector, and N is the number of pixels in the neighborhood. The estimators of the texture parameter and the covariance matrix depend on each other. They can be decoupled using a recursive process. Inserting $\hat{\tau}_i$ into the second line of the above equation, the covariance matrix in the $(k+1)$th iteration can be estimated by [45–47]

$$\hat{\boldsymbol{\Sigma}}_{k+1} = \frac{d}{N} \sum_{i=1}^{N} \frac{\mathbf{k}_i \mathbf{k}_i^\dagger}{\mathbf{k}_i^\dagger \hat{\boldsymbol{\Sigma}}_k^{-1} \mathbf{k}_i} \tag{38}$$

The process can be initialized by any matrix, even an identity matrix [47], and it is stopped when the Frobenius distance between two consecutive estimated matrices reaches some limit. More details about the existence as well as the convergence can be found in [46]. This estimator is referred to as fixed point estimator [47].

On the contrary, if the texture random variable is specified by a distribution, averaging all possible τ gives the unconditional or marginal PDF of the scattering vector

$$p(\mathbf{k}; \boldsymbol{\Sigma}) = \int_0^\infty p(\mathbf{k}|\tau; \boldsymbol{\Sigma}) p(\tau) d\tau \tag{39}$$

which is analytically solvable for some choices of $p(\tau)$. The PDF of the sample covariance matrix can be obtained similarly by

$$p(\mathbf{C}_L; L, \boldsymbol{\Sigma}) = \int_0^\infty p(\mathbf{C}_L|\tau; L, \boldsymbol{\Sigma}) p(\tau) d\tau \tag{40}$$

A number of models have been proposed in the literature by introducing different distributions for the texture component, including the $\mathcal{K}$ distribution [13], the $\mathcal{G}^0$ distribution [14,15], the Kummer-$\mathcal{U}$ distribution [16], the $\mathcal{W}$, and the $\mathcal{M}$ distribution [48], to represent different scenes of PolSAR data. They are explained in the following subsections.

4.1.1. $\mathcal{K}$ Distribution

The $\mathcal{K}$ distribution, assuming that the texture is gamma distributed, is widely used to model forests and the sea surface, and it can be unarguably regarded as one of the most successful radars models [4,10,12,13]. The gamma distribution is given by [35]

$$p(x; \alpha, \theta) = \frac{1}{\Gamma(\alpha)\theta^\alpha} x^{\alpha-1} \exp\left(-\frac{x}{\theta}\right) \tag{41}$$

with shape parameter α and scale parameter θ. The mean value is $\mu = \alpha\theta$. Let $\tau = \frac{x}{\mu}$ to ensure the mean value of the texture is equal to 1, the distribution can be written as

$$p(\tau;\alpha) = \frac{\alpha^\alpha}{\Gamma(\alpha)} \tau^{\alpha-1} \exp(-\alpha\tau) \tag{42}$$

The PDF of the scattering vector **k** can be obtained by substituting the texture distribution into (39) and employing the integral equality (A1)

$$p(\mathbf{k};\alpha,\mathbf{\Sigma}) = \frac{1}{\pi^d|\mathbf{\Sigma|}} \frac{2\alpha^{\frac{\alpha+d}{2}}}{\Gamma(\alpha)} (\mathbf{k}^\dagger\mathbf{\Sigma}^{-1}\mathbf{k})^{\frac{\alpha-d}{2}} K_{\alpha-d}\left(2\sqrt{\alpha\mathbf{k}^\dagger\mathbf{\Sigma}^{-1}\mathbf{k}}\right) \tag{43}$$

By the same procedure, inserting (42) into (40), we have the PDF of the sample covariance matrix as follows

$$p(\mathbf{C}_L;\alpha,L,\mathbf{\Sigma}) = \frac{L^{Ld}|\mathbf{C}_L|^{L-d}}{\Gamma_d(L)|\mathbf{\Sigma}|^L} \frac{2\alpha^{\frac{\alpha+Ld}{2}}}{\Gamma(\alpha)} \left(L\,\mathrm{Tr}(\mathbf{\Sigma}^{-1}\mathbf{C}_L)\right)^{\frac{\alpha-Ld}{2}} K_{\alpha-Ld}\left(2\sqrt{\alpha L\,\mathrm{Tr}(\mathbf{\Sigma}^{-1}\mathbf{C}_L)}\right) \tag{44}$$

4.1.2. Normal Inverse Gaussian (NIG)

The Normal Inverse Gaussian (NIG) distribution assumes that the texture follows an inverse Gaussian distribution [49,50]. The PDF of the inverse Gaussian distribution is given by

$$p(x;\mu,\gamma) = \left(\frac{\gamma}{2\pi x^3}\right)^{1/2} \exp\left(\frac{-\gamma(x-\mu)^2}{2\mu^2 x}\right) \tag{45}$$

where μ is the mean value. By letting μ equal to 1 and replacing the random variable x with τ, the texture distribution becomes

$$p(\tau;\gamma) = \left(\frac{\gamma}{2\pi}\right)^{1/2} \tau^{-3/2} e^\gamma \exp\left(-\frac{1}{2}\left(\frac{\gamma}{\tau} + \gamma\tau\right)\right) \tag{46}$$

Subsequently, the PDF of the scattering vector and the sample covariance matrix can be obtained by following the integral Equation (A1), giving

$$p(\mathbf{k};\gamma,\mathbf{\Sigma}) = \frac{1}{\pi^d|\mathbf{\Sigma}|} \frac{\sqrt{2\gamma}e^\gamma}{\sqrt{\pi}} \left(\frac{\gamma}{2\mathbf{k}^\dagger\mathbf{\Sigma}^{-1}\mathbf{k}+\gamma}\right)^{\frac{1+2d}{4}} K_{d+\frac{1}{2}}\left(\sqrt{\gamma(\gamma+2\mathbf{k}^\dagger\mathbf{\Sigma}^{-1}\mathbf{k})}\right) \tag{47}$$

and

$$\begin{aligned}
p(\mathbf{C}_L;\gamma,L,\mathbf{\Sigma}) = {} & \frac{L^{Ld}|\mathbf{C}_L|^{L-d}}{\Gamma_d(L)|\mathbf{\Sigma}|^L} \frac{\sqrt{2\gamma}e^\gamma}{\sqrt{\pi}} \left(\frac{\gamma}{2L\,\mathrm{Tr}(\mathbf{\Sigma}^{-1}\mathbf{C}_L)+\gamma}\right)^{\frac{1+2Ld}{4}} \\
& \times K_{Ld+\frac{1}{2}}\left(\sqrt{\gamma(\gamma+2L\,\mathrm{Tr}(\mathbf{\Sigma}^{-1}\mathbf{C}_L))}\right)
\end{aligned} \tag{48}$$

The NIG distribution has strong theoretical grounds derived from Brownian motion theory. Experiments demonstrate that it usually gives a better representation of the data than the Wishart distribution or the $\mathcal{K}$ distribution does, because the inverse Gaussian distribution captures larger distribution shape variations than the gamma distribution [50]. In addition, the NIG distribution has less trouble at boundary mixtures than the $\mathcal{K}$ distribution [50].

4.1.3. $\mathcal{G}$ and $\mathcal{G}^0$ Distributions

It is shown that the $\mathcal{G}$ distribution and the $\mathcal{G}^0$ distribution have a good representation of extremely heterogeneous regions such as urban areas [15]. Especially, the $\mathcal{G}^0$ distribution has the same number of parameters as the $\mathcal{K}$ distribution, but without complex special functions like the Bessel function which requires intensive computations [14,15].

The $\mathcal{G}$ distribution assumes that the texture parameter obeys the Generalized Inverse Gaussian (GIG) law which is characterized by the PDF [14,51]

$$p(x; a, b, p) = \frac{1}{2K_p(\sqrt{ab})} \left(\frac{a}{b}\right)^{\frac{p}{2}} x^{p-1} \exp\left(-\frac{1}{2}\left(\frac{b}{x} + ax\right)\right) \tag{49}$$

where $a > 0$, $b > 0$ and p is a real parameter. The mean value of this distribution is $\mu = \sqrt{\frac{b}{a}} \frac{K_{p+1}(\sqrt{ab})}{K_p(\sqrt{ab})}$. Letting $\tau = \frac{x}{\mu}$ gives

$$p(\tau; a, b, p) = \frac{1}{2} \frac{K_{p+1}^p(\sqrt{ab})}{K_p^{p+1}(\sqrt{ab})} \tau^{p-1} \exp\left(-\frac{\sqrt{ab}}{2}\left(\frac{K_p(\sqrt{ab})}{K_{p+1}(\sqrt{ab})} \frac{1}{\tau} + \frac{K_{p+1}(\sqrt{ab})}{K_p(\sqrt{ab})}\tau\right)\right) \tag{50}$$

which can be further rewritten as follows by replacing $\sqrt{ab}$ with ω to reduce the number of parameters

$$p(\tau; \omega, p) = \frac{1}{2} \frac{K_{p+1}^p(\omega)}{K_p^{p+1}(\omega)} \tau^{p-1} \exp\left(-\frac{\omega}{2}\left(\frac{K_p(\omega)}{K_{p+1}(\omega)} \frac{1}{\tau} + \frac{K_{p+1}(\omega)}{K_p(\omega)}\tau\right)\right) \tag{51}$$

Substituting (51) into (39) and (40), and calculating the integral using (A1) leads to

$$p(\mathbf{k}; \omega, p, \mathbf{\Sigma}) = \frac{1}{\pi^d|\mathbf{\Sigma}|} \frac{1}{\eta^p K_p(\omega)} \left(\eta^2 + \frac{2\eta}{\omega}\mathbf{k}^\dagger \mathbf{\Sigma}^{-1}\mathbf{k}\right)^{\frac{p-d}{2}} K_{p-d}\left(\sqrt{\omega^2 + \frac{2\omega}{\eta}\mathbf{k}^\dagger \mathbf{\Sigma}^{-1}\mathbf{k}}\right) \tag{52}$$

and

$$p(\mathbf{C}_L; \omega, p, L, \mathbf{\Sigma}) = \frac{L^{Ld}|\mathbf{C}_L|^{L-d}}{\Gamma_d(L)|\mathbf{\Sigma}|^L} \frac{1}{\eta^p K_p(\omega)} \left(\eta^2 + \frac{2\eta}{\omega}L\,\mathrm{Tr}(\mathbf{\Sigma}^{-1}\mathbf{C}_L)\right)^{\frac{p-Ld}{2}}$$
$$\times K_{p-Ld}\left(\sqrt{\omega^2 + \frac{2\omega}{\eta}L\,\mathrm{Tr}(\mathbf{\Sigma}^{-1}\mathbf{C}_L)}\right) \tag{53}$$

where $\eta = \frac{K_p(\omega)}{K_{p+1}(\omega)}$. The above expressions are the PDFs of the scattering vector and the sample covariance matrix following $\mathcal{G}$ distributions [14,52].

The $\mathcal{G}^0$ distribution can be obtained from the $\mathcal{G}$ distribution by letting $a \to 0$. Representing the modified Bessel function $K_v(z)$ using (A6), Equation (49) becomes

$$p(x; a, b, p) = \frac{2^{p-1}\Gamma\left(p + \frac{1}{2}\right)}{b^p\sqrt{\pi}} x^{p-1} \exp\left(-\frac{1}{2}\left(\frac{b}{x} + ax\right)\right) \times \left(\int_1^\infty e^{-\sqrt{ab}t}(t^2 - 1)^{p-\frac{1}{2}}dt\right)^{-1} \tag{54}$$

If $a \to 0$, $p = -\lambda$, $b = 2\beta$, then after calculating the integral via (A7), the PDF of the GIG distribution is reduced to

$$p(x; \lambda, \beta) = \frac{\beta^\lambda}{\Gamma(\lambda)} x^{-\lambda-1} \exp\left(-\frac{\beta}{x}\right) \tag{55}$$

Equation (55) is known as the inverse gamma distribution, or the reciprocal of the gamma distribution, with mean value $\mu = \frac{\beta}{\lambda - 1}$. Let $\tau = \frac{x}{\mu}$ to ensure the mean value of the texture τ is equal to 1, the PDF becomes

$$p(\tau; \lambda) = \frac{(\lambda - 1)^{\lambda}}{\Gamma(\lambda)} \tau^{-\lambda - 1} \exp\left(-\frac{\lambda - 1}{\tau}\right) \tag{56}$$

The PDFs of the scattering vector and the sample covariance matrix of the $\mathcal{G}^0$ distribution can be obtained by plugging the texture distribution into (39) and (40) respectively, and calculating the integral by (A9), giving

$$p(\mathbf{k}; \lambda, \mathbf{\Sigma}) = \frac{1}{\pi^d |\mathbf{\Sigma}|} \frac{\Gamma(\lambda + d)(\lambda - 1)^{\lambda}}{\Gamma(\lambda)} \left(\lambda - 1 + \mathbf{k}^{\dagger} \mathbf{\Sigma}^{-1} \mathbf{k}\right)^{-\lambda - d} \tag{57}$$

and

$$p(\mathbf{C}_L; \lambda, L, \mathbf{\Sigma}) = \frac{L^{Ld} |\mathbf{C}_L|^{L-d}}{\Gamma_d(L) |\mathbf{\Sigma}|^L} \frac{\Gamma(\lambda + Ld)(\lambda - 1)^{\lambda}}{\Gamma(\lambda)} \left(\lambda - 1 + L \operatorname{Tr}(\mathbf{\Sigma}^{-1} \mathbf{C}_L)\right)^{-\lambda - Ld} \tag{58}$$

Another extreme case of the GIG distribution is the gamma distribution when $b \to 0$, which leads to the $\mathcal{K}$ distribution [14].

4.1.4. Kummer-$\mathcal{U}$ Distribution

Assuming that the texture parameter follows a Fisher distribution, also known as the F-distribution or the Fisher-Snedecor distribution, with PDF given by [35]

$$p(x; d_1, d_2) = \frac{1}{B\left(\frac{d_1}{2}, \frac{d_2}{2}\right)} \left(\frac{d_1}{d_2}\right)^{\frac{d_1}{2}} x^{\frac{d_1}{2} - 1} \left(1 + \frac{d_1}{d_2} x\right)^{-\frac{d_1 + d_2}{2}} \tag{59}$$

where $d_1 > 0$ and $d_2 > 0$, the scattering vector or the sample covariance matrix are Kummer-$\mathcal{U}$ distributed, with the ability to model different types of textures, because the Fisher distribution covers a large range of distributions [16,53]. The mean value of the Fisher distribution is $\mu = \frac{d_2}{d_2 - 2}$. Let $\tau = \frac{x}{\mu}$ to ensure the mean value of the texture variable equal to 1, we have the distribution for the texture as

$$p(\tau; \xi, \zeta) = \frac{\Gamma(\xi + \zeta)}{\Gamma(\xi)\Gamma(\zeta)} \frac{\xi}{\zeta - 1} \left(\frac{\xi}{\zeta - 1}\tau\right)^{\xi - 1} \left(\frac{\xi}{\zeta - 1}\tau + 1\right)^{-\xi - \zeta} \tag{60}$$

Here parameters d_1 and d_2 are replaced by $\xi = d_1/2$ and $\zeta = d_2/2$ to make the expression more concise. Inserting the texture distribution into (39), the PDF of the scattering vector can be calculated by

$$p(\mathbf{k}; \xi, \zeta, \mathbf{\Sigma}) = \frac{\Gamma(\xi + \zeta)}{\Gamma(\xi)\Gamma(\zeta)\pi^d |\mathbf{\Sigma}|} \left(\frac{\xi}{\zeta - 1}\right)^{\xi} \times$$
$$\int_0^{\infty} \tau^{\xi - 1 - d} \left(\frac{\xi}{\zeta - 1}\tau + 1\right)^{-\xi - \zeta} \exp\left(-\frac{\mathbf{k}^{\dagger} \mathbf{\Sigma}^{-1} \mathbf{k}}{\tau}\right) d\tau \tag{61}$$

Replacing τ by $\frac{\zeta - 1}{\xi} t^{-1}$, and using (A10) to calculate the integral results into the distribution of the scattering vector

$$p(\mathbf{k}; \xi, \zeta, \mathbf{\Sigma}) = \frac{1}{\pi^d |\mathbf{\Sigma}|} \frac{\Gamma(\xi + \zeta)\Gamma(\zeta + d)}{\Gamma(\xi)\Gamma(\zeta)} \left(\frac{\xi}{\zeta - 1}\right)^d$$
$$\times U\left(d + \zeta, d - \xi + 1, \frac{\xi}{\zeta - 1} \mathbf{k}^{\dagger} \mathbf{\Sigma}^{-1} \mathbf{k}\right) \tag{62}$$

where $U(a, b, z)$ is the hyper-geometric function of the second kind [34]. By the same procedure, the distribution of the sample covariance matrix can be obtained as

$$
\begin{aligned}
p(\mathbf{C}_L;\xi,\zeta,L,\mathbf{\Sigma}) =& \frac{L^{Ld}|\mathbf{C}_L|^{L-d}}{\Gamma_d(L)|\mathbf{\Sigma}|^L}\frac{\Gamma(\xi+\zeta)\Gamma(\zeta+Ld)}{\Gamma(\xi)\Gamma(\zeta)}\left(\frac{\xi}{\zeta-1}\right)^{Ld} \\
& \times U\left(Ld+\zeta, Ld-\xi+1, \frac{\xi}{\zeta-1}L\ \mathrm{Tr}(\mathbf{\Sigma}^{-1}\mathbf{C}_L)\right)
\end{aligned}
\tag{63}
$$

As a matter of fact, Fisher distributions are the Pearson VI solutions and cover a large range of distributions. It is not only confined to urban scenes, but also fits reasonably in forest and agricultural fields [16,53]. The behavior of the head and tail of the distribution can be controlled by the two parameters ξ and ζ.

4.1.5. $\mathcal{W}$ Distribution

The $\mathcal{W}$ distribution assumes the texture to follow a beta distribution [48], which is given by [35]

$$
p(x;\alpha,\beta) = \frac{1}{B(\alpha,\beta)}x^{\alpha-1}(1-x)^{\beta-1}, \quad x\in[0,1]
\tag{64}
$$

The mean value of the beta distribution is $\mu = \frac{\alpha}{\alpha+\beta}$. Let $\tau = \frac{x}{\mu}$, $\xi = \alpha$, $\zeta = \alpha+\beta$, the distribution of the normalized texture can be written as

$$
p(\tau;\xi,\zeta) = \frac{\Gamma(\zeta)}{\Gamma(\xi)\Gamma(\zeta-\xi)}\frac{\xi}{\zeta}\left(\frac{\xi}{\zeta}\tau\right)^{\xi-1}\left(1-\frac{\xi}{\zeta}\tau\right)^{\zeta-\xi-1}, \quad \tau\in[0,\frac{\zeta}{\xi}]
\tag{65}
$$

The distribution of the scattering vector in this case can be calculated by

$$
\begin{aligned}
p(\mathbf{k};\xi,\zeta,\mathbf{\Sigma}) =& \frac{\Gamma(\zeta)}{\Gamma(\xi)\Gamma(\zeta-\xi)\pi^d|\mathbf{\Sigma}|}\left(\frac{\xi}{\zeta}\right)^{\xi-1}\times \\
& \int_0^{\frac{\zeta}{\xi}}\tau^{\xi-1-d}\left(\frac{\zeta}{\xi}-\tau\right)^{\zeta-\xi-1}\exp\left(-\frac{\mathbf{k}^\dagger\mathbf{\Sigma}^{-1}\mathbf{k}}{\tau}\right)d\tau
\end{aligned}
\tag{66}
$$

which leads to the following result according to the integral identity (A11)

$$
\begin{aligned}
p(\mathbf{k};\xi,\zeta,\mathbf{\Sigma}) =& \frac{1}{\pi^d|\mathbf{\Sigma}|}\frac{\Gamma(\zeta)}{\Gamma(\xi)}\left(\frac{\xi}{\zeta}\right)^{\frac{\xi+d-1}{2}}\left(\mathbf{k}^\dagger\mathbf{\Sigma}^{-1}\mathbf{k}\right)^{\frac{\xi-d-1}{2}}\times \\
& \exp\left(-\frac{\xi}{2\zeta}\mathbf{k}^\dagger\mathbf{\Sigma}^{-1}\mathbf{k}\right)W_{\frac{d+1+\xi-2\zeta}{2},\frac{\xi-d}{2}}\left(\frac{\xi}{\zeta}\mathbf{k}^\dagger\mathbf{\Sigma}^{-1}\mathbf{k}\right)
\end{aligned}
\tag{67}
$$

where $W_{a,b}(z)$ is Whittaker W function [34]. The distribution of the sample covariance matrix can be obtained by the same way

$$
\begin{aligned}
p(\mathbf{C}_L;\xi,\zeta,L,\mathbf{\Sigma}) =& \frac{L^{Ld}|\mathbf{C}_L|^{L-d}}{\Gamma_d(L)|\mathbf{\Sigma}|^L}\frac{\Gamma(\zeta)}{\Gamma(\xi)}\left(\frac{\xi}{\zeta}\right)^{\frac{\xi+Ld-1}{2}}\left(L\ \mathrm{Tr}(\mathbf{\Sigma}^{-1}\mathbf{C}_L)\right)^{\frac{\xi-Ld-1}{2}}\times \\
& \exp\left(-\frac{\xi}{2\zeta}L\ \mathrm{Tr}(\mathbf{\Sigma}^{-1}\mathbf{C}_L)\right)W_{\frac{Ld+1+\xi-2\zeta}{2},\frac{\xi-Ld}{2}}\left(\frac{\xi}{\zeta}L\ \mathrm{Tr}(\mathbf{\Sigma}^{-1}\mathbf{C}_L)\right)
\end{aligned}
\tag{68}
$$

4.1.6. $\mathcal{M}$ Distribution

Another possible distribution for the texture is the beta prime distribution, also known as inverted beta distribution, with PDF given by [35]

$$
p(x;\alpha,\beta) = \frac{1}{B(\alpha,\beta)}x^{\alpha-1}(1+x)^{-\alpha-\beta}, \quad x>0
\tag{69}
$$

The mean value can be calculated by $\mu = \frac{\alpha}{\beta-1}$. Again, scale the random variable to ensure the mean value is equal to 1 by letting $\tau = \frac{\beta-1}{\alpha+\beta-1}(1+x)$, the above distribution becomes

$$p(\tau;\xi,\zeta) = \frac{\Gamma(\zeta)}{\Gamma(\xi)\Gamma(\zeta-\xi)}\frac{\zeta-1}{\xi-1}\left(\frac{\zeta-1}{\xi-1}\tau\right)^{-\zeta}\left(\frac{\zeta-1}{\xi-1}\tau-1\right)^{\zeta-\xi-1}, \quad \tau > \frac{\xi-1}{\zeta-1} \tag{70}$$

where the parameters are changed to $\zeta = \alpha + \beta$, $\xi = \beta$ to make the expression brief. Equation (70) is the texture distribution of the $\mathcal{M}$ distribution [48]. According to the product model, the distribution of the scattering vector can be calculated by

$$p(\mathbf{k};\xi,\zeta,\boldsymbol{\Sigma}) = \frac{\Gamma(\zeta)}{\Gamma(\xi)\Gamma(\zeta-\xi)\pi^d|\boldsymbol{\Sigma}|}\left(\frac{\xi-1}{\zeta-1}\right)^{\xi} \times$$
$$\int_{\frac{\xi-1}{\zeta-1}}^{\infty} \tau^{-\zeta-d}\left(\tau - \frac{\xi-1}{\zeta-1}\right)^{\zeta-\xi-1}\exp\left(-\frac{\mathbf{k}^{\dagger}\boldsymbol{\Sigma}^{-1}\mathbf{k}}{\tau}\right)d\tau \tag{71}$$

Employing the integral identity (A12), we have the PDF of the scattering vector as

$$p(\mathbf{k};\xi,\zeta,\boldsymbol{\Sigma}) = \frac{1}{\pi^d|\boldsymbol{\Sigma}|}\frac{\Gamma(\zeta)\Gamma(\xi+d)}{\Gamma(\xi)\Gamma(\zeta+d)}\left(\frac{\zeta-1}{\xi-1}\right)^d M\left(\xi+d,\zeta+d,-\frac{\zeta-1}{\xi-1}\mathbf{k}^{\dagger}\boldsymbol{\Sigma}^{-1}\mathbf{k}\right) \tag{72}$$

and the PDF of the sample covariance matrix as

$$p(\mathbf{C}_L;\xi,\zeta,L,\boldsymbol{\Sigma}) = \frac{L^{Ld}|\mathbf{C}_L|^{L-d}}{\Gamma_d(L)|\boldsymbol{\Sigma}|^L}\frac{\Gamma(\zeta)\Gamma(\xi+Ld)}{\Gamma(\xi)\Gamma(\zeta+Ld)}\left(\frac{\zeta-1}{\xi-1}\right)^{Ld}$$
$$\times M\left(\xi+Ld,\zeta+Ld,-\frac{\zeta-1}{\xi-1}L\,\mathrm{Tr}(\boldsymbol{\Sigma}^{-1}\mathbf{C}_L)\right) \tag{73}$$

Here $M(a,b,z)$ is the confluent hypergeometric function of the first kind, also known as the KummerM function [34]. The $\mathcal{W}$ distribution and the $\mathcal{M}$ distribution are able to model data with low variance but extreme skewness, which is particularly relevant to data with textural variability after a speckle filtering [48].

4.1.7. Wishart-Generalized Gamma Distribution

The Wishart-Generalized Gamma (WGΓ) distribution employs the generalized gamma distribution to model the texture. The generalized gamma distribution has a more compact form and a larger variety of alternative distributions, with the gamma, the Weibull, the Rayleigh, and the exponential distributions being its special cases. Thus it is of greater flexibility in the statistical modelling [54]. The PDF of the generalized gamma distribution is given by [35]

$$p(x;v,\theta,k) = \frac{v}{\theta\Gamma(k)}\left(\frac{x}{\theta}\right)^{kv-1}\exp\left(-\left(\frac{x}{\theta}\right)^v\right), \quad v > 0, \theta > 0, k > 0 \tag{74}$$

which reduces to the gamma distribution (41) when $v = 1$. The mean value is given by $\mu = \theta\Gamma(k+\frac{1}{v})/\Gamma(k)$. Scaling the mean value to 1, the PDF for the texture is obtained as

$$p(\tau;v,k) = \frac{v\beta^{kv}}{\Gamma(k)}\tau^{kv-1}e^{-(\beta\tau)^v} \tag{75}$$

where $\beta = \Gamma(k+\frac{1}{v})/\Gamma(k)$. The distribution of the scattering vector $\mathbf{k}$ then can be calculated by

$$p(\mathbf{k};v,k,\boldsymbol{\Sigma}) = \frac{v\beta^{kv}}{\Gamma(k)\pi^d|\boldsymbol{\Sigma}|}\int_0^{\infty}\tau^{kv-d-1}\exp\left(-(\beta\tau)^v - \frac{\mathbf{k}^{\dagger}\boldsymbol{\Sigma}^{-1}\mathbf{k}}{\tau}\right)d\tau \tag{76}$$

There is no closed form expression for the above equation, but it can be solved numerically [54]. The distribution of the sample covariance matrix can be calculated by

$$p(\mathbf{C}_L; v, k, L, \mathbf{\Sigma}) = \frac{v\beta^{kv}L^{LD}|\mathbf{C}_L|^{L-d}}{\Gamma(k)I(L,d)|\mathbf{\Sigma}|^L} \int_0^\infty \tau^{kv-Ld-1} \exp\left(-(\beta\tau)^v - \frac{L\,\mathrm{Tr}(\mathbf{\Sigma}^{-1}\mathbf{C}_L)}{\tau}\right) d\tau \qquad (77)$$

It is reported that the WGΓ distribution could provide better fitness than the $\mathcal{K}$ and Kummer-$\mathcal{U}$ distributions for different land cover types of homogeneous, heterogeneous, and extremely heterogeneous terrains [54].

4.1.8. Generalized $\mathcal{K}$ Distribution

The well-known gamma distribution sometimes cannot fit the texture distribution accurately in very heterogeneous areas. In order to improve the flexibility of the model, it is assumed that the texture follows a Laguerre expansion of the gamma distribution [55], with its PDF given by

$$p(\tau; \alpha, \mu) = \frac{\tau^{\alpha-1}}{\Gamma(\alpha)}\left(\frac{\alpha}{\mu}\right)^\alpha \exp\left(-\frac{\alpha\tau}{\mu}\right) \sum_{u=0}^\infty \zeta_u \frac{\Gamma(\alpha)u!}{\Gamma(u+\alpha)} L_u^{\alpha-1}\left(\frac{\alpha\tau}{\mu}\right) \qquad (78)$$

where μ, the mean value, is normally assumed to be equal to 1, and

$$\zeta_u = \sum_{k=0}^u (-1)^k \binom{u+\alpha-1}{u-k} \frac{1}{k!}\left(\frac{\alpha}{\mu}\right)^k E\{x^k\} \qquad (79)$$

The Laguerre polynomial $L_u^{\alpha-1}(x)$ is given by

$$L_u^{\alpha-1}(x) = \sum_{k=0}^u (-1)^k \binom{u+\alpha-1}{u-k} \frac{x^k}{k!} \qquad (80)$$

The PDF of the sample covariance matrix in this case can be expressed as [55]

$$p(\mathbf{C}_L; \alpha, \mu, L, \mathbf{\Sigma}) = \frac{L^{Ld}|\mathbf{C}_L|^{L-d}}{\Gamma_d(L)|\mathbf{\Sigma}|^L} \frac{\alpha^\alpha}{\Gamma(\alpha)\mu^\alpha} \times \sum_{u=0}^\infty \zeta_u \frac{\Gamma(\alpha)u!}{\Gamma(u+\alpha)} \sum_{k=0}^u (-1)^k$$
$$\frac{2}{k!} \frac{(u+\alpha-1)!}{(u-k)!(\alpha-1+k)!}\left(\frac{\alpha}{\mu}\right)^k \left(\frac{L\mu\,\mathrm{Tr}(\mathbf{\Sigma}^{-1}\mathbf{C}_L)}{\alpha}\right)^{\frac{\alpha+k-Ld}{2}} \qquad (81)$$
$$K_{\alpha+k-Ld}\left(2\sqrt{\frac{\alpha}{\mu}L\,\mathrm{Tr}(\mathbf{\Sigma}^{-1}\mathbf{C}_L)}\right)$$

which is a weighted combination of a series of $\mathcal{K}$ distributions based on a Laguerre polynomial expansion. It shows that the generalized $\mathcal{K}$ distribution gives a better approximation than the $\mathcal{K}$ distribution when there exist strong scatterers in the scene [55].

4.2. Multi-Texture Model

In the scalar texture model, different polarimetric channels are assumed to have a common texture variable. However, if the electromagnetic wave sees different geometrical or dielectric properties of the target, and if those properties are spatially modulated, then the texture of each channel should be different [56]. For example, in scattering from forest areas, volume scattering will affect the cross-pol component stronger than the co-pol channels, whereas surface scattering will have the opposite effect [57]. The scalar texture model must, therefore, be extended to take into consideration the different radar cross section modulations in polarimetric channels. One solution is to allow for a vector component of the texture in the product model. This type of models are called multi-texture models.

Under the assumption of reciprocity, there are only three independent complex coefficients required to characterize the scatterer under observation. The multi-texture model then can be formulated as [57–60]

$$\mathbf{k} = \mathbf{\Lambda}^{1/2}\mathbf{z} \tag{82}$$

where $\mathbf{z}$ represents the speckle, following a multivariate Gaussian distribution (see Section 3.1), and $\mathbf{\Lambda}$ is a diagonal matrix containing texture variables for each channel

$$\mathbf{\Lambda} = \begin{bmatrix} \tau_{hh} & 0 & 0 \\ 0 & \tau_{hv} & 0 \\ 0 & 0 & \tau_{vv} \end{bmatrix} \tag{83}$$

The texture parameters are assumed to be positive, and we have $E\{\mathbf{\Lambda}\}$ equal to $\mathbf{I}$, the identity matrix. Assuming that the texture variables are constant on the scale of the multilook processing window, the sample covariance matrix can be written as

$$\mathbf{C}_L = \frac{1}{L}\sum_{i=1}^{L}\mathbf{k}_i\mathbf{k}_i^T = \mathbf{\Lambda}^{1/2}\mathbf{W}\mathbf{\Lambda}^{1/2} \tag{84}$$

where $\mathbf{W}$ is Wishart distributed, see Section 3.2.

Provided that the distributions of the texture variables are known, the PDF of the scattering vector can be calculated using

$$p(\mathbf{k};\mathbf{\Sigma}) = \int_{\Omega^+} p(\mathbf{k}|\mathbf{\Lambda};\mathbf{\Sigma})p(\mathbf{\Lambda})d\mathbf{\Lambda} \tag{85}$$

where Ω^+ is the set of all diagonal matrices with non-negative entries. After changing variable by $\mathbf{z} = \mathbf{\Lambda}^{-1/2}\mathbf{k}$, the conditional distribution of $\mathbf{k}$ on $\mathbf{\Lambda}$ can be obtained from (7), giving

$$p(\mathbf{k}|\mathbf{\Lambda};\mathbf{\Sigma}) = \frac{1}{\pi^d|\mathbf{\Sigma}||\mathbf{\Lambda}|}\exp\left(-\mathbf{k}^\dagger\mathbf{\Lambda}^{-1/2}\mathbf{\Sigma}^{-1}\mathbf{\Lambda}^{-1/2}\mathbf{k}\right) \tag{86}$$

By the similar way, we have the distribution of the sample covariance matrix as [57,59]

$$p(\mathbf{C}_L;L,\mathbf{\Sigma}) = \int_{\Omega^+} p(\mathbf{C}_L|\mathbf{\Lambda};L,\mathbf{\Sigma})p(\mathbf{\Lambda})d\mathbf{\Lambda} \tag{87}$$

where

$$p(\mathbf{C}_L|\mathbf{\Lambda};L,\mathbf{\Sigma}) = \frac{L^{Ld}|\mathbf{C}_L|^{L-d}}{\Gamma_d(L)|\mathbf{\Sigma}|^L|\mathbf{\Lambda}|^L}\exp\left(-L\,\mathrm{Tr}(\mathbf{\Sigma}^{-1}\mathbf{\Lambda}^{-1/2}\mathbf{C}_L\mathbf{\Lambda}^{-1/2})\right) \tag{88}$$

Different texture variables for the multi-texture model can be: (1) totally dependent, in which case it reduces to the scalar texture model, (2) independent from each other, that is, texture variables follow different distributions with different parameters, or (3) partially correlated [58,61]. In many cases, it is reasonable to assume co-pol channels have the same texture but different from that of the cross-pol channels. This type of models is usually referred to as dual-texture model [57,59,62]. For reciprocal media with reflection symmetry for example, the PDF of the sample covariance matrix can be expanded as [59]

$$\begin{aligned}
p(\mathbf{C}_L;L,\mathbf{\Sigma}) =\ & \frac{L^{3L}|\mathbf{C}_L|^{L-3}}{I(L,3)|\mathbf{\Sigma}|^L}\int_0^\infty \exp\left(-L\frac{q_{22}c_{22}}{T_x}\right)\frac{p(T_x)}{T_x^L}dT_x\times \\
& \int_0^\infty \exp\left(-L\frac{q_{11}c_{11}+q_{13}c_{31}+q_{31}c_{13}+q_{33}c_{33}}{T_{co}}\right)\frac{p(T_{co})}{T_{co}^{2L}}dT_{co}
\end{aligned} \tag{89}$$

where q_{ij} and c_{ij} denote the (i,j)th entry of matrix $\mathbf{C}_L$ and Σ respectively. The texture of the co-pol channels is represented by T_{co} and that of the cross-pol channel by T_x.

4.2.1. Correlated $\mathcal{K}$ Distribution

The correlated $\mathcal{K}$ distribution assumes that the texture variables of different polarimetric channels are partially correlated, each following a gamma distribution [58,61]. Unfortunately, there is no explicit expression for the joint distribution of the texture variables, or the correlated gamma distribution. In this model, the texture of polarimetric channel i, specified by the PDF (42) with parameter α, is given by [61]

$$\tau_i = \frac{1}{2\alpha} \sum_{k=1}^{2\alpha} [g_i^{(k)}]^2 \tag{90}$$

where $g_i^{(k)}$ is the ith element of the vector $\mathbf{g}^{(k)}, k = 1, \cdots, 2\alpha$, which is Gaussian distributed with zero mean, variance one, and correlation matrix $\mathbf{T}$. The correlation properties of the texture variables is also specified by $\mathbf{T}$. The characteristic function of the vector containing all texture variables is [61]

$$C(\boldsymbol{\omega}) = \frac{1}{|\mathbf{I} + j(1/\alpha)\mathbf{T}\mathbf{W}|^\alpha} \tag{91}$$

where $\mathbf{W}$ is a diagonal matrix having the entry (i,i) equal to the ith element of the characteristic function variable $\boldsymbol{\omega}$. This model requires that all polarimetric channels have the same half-integer distribution parameter α, e.g., 0.5, 1.5, 2.5 and so on.

4.2.2. Dual-Texture $\mathcal{G}$ Distribution

The dual-texture $\mathcal{G}$ distribution is derived by considering different texture variables for co-pol and cross-pol channels. Both the co-pol and the cross-pol texture variables are modelled by the GIG distributions (49), which yields a more flexible multivariate distribution [62]. Under the assumption of reciprocity and reflection symmetry, the statistical properties of the single look complex data is characterized by the distribution [62]

$$p(\mathbf{k}; \Sigma, \boldsymbol{\theta}) = \frac{1}{\pi^d |\Sigma|} \prod_{i=1}^{2} \frac{(\eta_i^2 + 2\eta_i s_i/\omega_i)^{\frac{p_i-d+i}{2}}}{\eta_i^{p_i} K_{p_i}(\omega_i)} K_{p_i-d+i}\left(\sqrt{\omega_i^2 + 2\omega_i s_i/\eta_i}\right) \tag{92}$$

where $\boldsymbol{\theta} = \{\omega, p_i, \eta_i\}$ consists of all parameters for the GIG texture distributions (see Section 4.1.3), $s_1 = z_{11}c_{11} + z_{13}c_{31} + z_{31}c_{13} + z_{33}c_{33}$, and $s_2 = z_{22}c_{22}$, with z_{ij} and c_{ij} indexing entries of $\mathbf{Z} = \mathbf{k}\mathbf{k}^\dagger$ and Σ respectively.

5. Other Models

To model a complex scene using texture models, we often need to introduce complicated distributions with many parameters to describe the statistical behavior of the texture component. However, having more parameters requires a more complicated estimation process by considering higher order statistics. In addition, higher order moment estimators are known to have higher variance. With the limited sample sizes used in the modelling, such complicated modelling may be very inefficient [50]. To overcome this problem, some researchers try to divide a complex model into multiple simple components and then find a way to combine these components together. The finite mixture model and copula based model detailed as follows are based on this idea.

5.1. Finite Mixture Model

The heterogeneity that appears in PolSAR data may result from the mixture of different targets. For instance, from an urban area which usually consists of different objects like houses, trees and roads, the backscattering is a combination of different scattering mechanisms. The forest areas sometimes can be treated as a mixture of bright clutters and dark ones, corresponding to the strong returns from the crowns of trees and the shadows behind them. To represent this type of data, a simple model would be inappropriate. Finite mixture models, instead, could achieve reasonable level of accuracy [17–19,63,64].

Assume that the region under analysis can be modeled by a mixture of K components, then the overall PDF of the data can be written as a weighted sum of the probabilities of each component [65]

$$p(\mathbf{x}; \boldsymbol{\theta}) = \sum_{k=1}^{K} w_k p_k(\mathbf{x}; \boldsymbol{\theta}_k) \tag{93}$$

where $\boldsymbol{\theta}$ is a vector containing all the parameters of the distribution and the mixing proportions obey

$$\sum_{k=1}^{K} w_k = 1, \ w_k \geq 0 \tag{94}$$

It has been shown that for complicated regions with more irregular histograms (multimodal, spiky), the finite mixture model is more accurate than a single distribution [17–19].

There are many options for the distributions of the mixing components, but here we mainly focus on the mixture of Wishart distributed components. For different mixing components, the number of looks are the same. The PDF, therefore, can be written as

$$p(\mathbf{C}_L; L, \boldsymbol{\theta}) = \frac{L^{Ld} |\mathbf{C}_L|^{L-d}}{\Gamma_d(L)} \sum_{k=1}^{K} \frac{w_k \exp(-L \operatorname{Tr}(\boldsymbol{\Sigma}_k^{-1} \mathbf{C}_L))}{|\boldsymbol{\Sigma}_k|^L} \tag{95}$$

where $\boldsymbol{\theta} = \{\boldsymbol{\Sigma}_k, k = 1, \cdots, K\}$ and $\Gamma_d(L)$ is given by (22). The PDF of the ith channel intensity, which is also a finite mixture, is found to be

$$p(I_i; L, \boldsymbol{\theta}) = \frac{I_i^{L-1}}{\Gamma(L)} \sum_{k=1}^{K} \left(\frac{L}{\sigma_{k,i}^2}\right)^L \exp\left(-\frac{L}{\sigma_{k,i}^2} I_i\right) \tag{96}$$

where $\sigma_{k,i}^2 = [\boldsymbol{\Sigma}_k]_{ii}$. The most interesting property of a mixture density is that the shape of the density is extremely flexible. A mixture density may be multimodal, or even if it is unimodal, may exhibit considerable skewness or additional humps. For this reason, finite mixture distributions offer a flexible way to describe rather heterogeneous data by summarizing the characteristics of the data in terms of the number and the spread of the mixture components [65].

5.2. Copula Based Model

Copulas are popular in high-dimensional statistical applications as they allow one to easily model and estimate the distribution of random vectors by estimating marginals and dependence separately [21]. They are of great interest for two main reasons: (1) as a way to study scale-free measures of dependence; and (2) as a starting point for constructing families of multivariate distributions [21]. For the PolSAR data, we often have a much better idea about the marginal behaviour of individual polarimetric channels than we do about their dependence structure. The copula approach allows us to combine our more developed marginal models with a variety of possible dependence models to investigate the statistical behavior of the data.

A d-dimensional copula, denoted by $C(\mathbf{u}) = C(u_1, \ldots, u_d)$, is a joint Cumulative Distribution Function (CDF) of a d-dimensional random vector on the unit hypercube $[0, 1]^d$ with uniform marginals. More specifically, a copula is a function C from $[0, 1]^d$ to $[0, 1]$ with the following properties [21,66]:

1. $C(u_1, \ldots, u_{i-1}, 0, u_{i+1}, \ldots, u_d) = 0$, the copula is equal to 0 if at least one parameter is 0.
2. $C(1, \ldots, 1, u_i, 1, \ldots, 1) = u_i$, the copula is equal to u_i if all parameters are 1 except u_i.
3. For each hyperrectangle $B = \prod_{i=1}^{d}[x_i, y_i] \subseteq [0, 1]^d$ where $x_i \leq y_i$, the C-volume of B is non-negative

$$\sum_{\mathbf{z} \in \times_{i=1}^{d}\{x_i, y_i\}} (-1)^{N(\mathbf{z})} C(\mathbf{z}) \geq 0 \tag{97}$$

where $\mathbf{z}$ represents the corners of the hyperrectangle, and $N(\mathbf{z}) = \#\{k : z_k = x_k\}$ is the number of elements in $\mathbf{z}$ reaching the lower bound of the hyperrectangle.

According to Sklar's Theorem, any multivariate joint distribution can be written in terms of univariate marginal distribution functions and a copula which describes the dependence structure between the variables [21]

$$H(x_1, \ldots, x_d) = C(F_1(x_1), \ldots, F_d(x_d)) \tag{98}$$

where F_i is the continuous marginal CDF $F_i(x) = P(X_i \leq x)$. The copula C contains all information about the dependence structure whereas the marginal cumulative distribution functions F_i contains all information about an individual random variable.

There are many parametric copula families available, which usually have one or more parameters controlling the strength of dependence. The most popular ones include the elliptical copulas (such as the Gaussian copula and the student t copula), and the Archimedian copulas. In the context of PolSAR data modeling, the Ali-Mikhail-Haq copula which belongs to the Archimedian family is demonstrated to be appropriate [22,23,67]. The Gaussian copula is also found to be proper to model the wavelet coefficients [68]. Though it is a hot topic, the study of copulas and the role they play in statistics and stochastic processes is a subject still in its infancy. There are many open problems and much work to be done.

6. Model Analysis

In the previous sections, the statistical models proposed for the PolSAR data are reviewed, with an emphasis on the derivation of PDFs for the scattering vector and the sample covariance matrix. The models are categorized into three groups: (1) Gaussian Models, (2) Texture Models, and (3) The Others. Table 1 shows a summary of all these models. As we can see, texture models are still the main focus in statistical modeling of PolSAR data. Several examples of the texture distributions with different distribution parameters are plotted in Figure 1.

In the remaining of this section, we will show some experimental results on the applicability of different statistical models.

First of all, two homogeneous Regions Of Interest (ROI) over the farmland of a RADARSAT-2 image are analyzed, as shown in Figure 2. The data, in single look complex format, has a spatial resolution of 11.1 m $\times$ 7.6 m (Range $\times$ Azimuth). It was acquired over Flevoland (The Netherlands) with the Fine Quad-Pol mode during the ESA-led AgriSAR 2009 campaign. Statistical properties are analyzed separately by the histograms of the intensity, the product of amplitudes and the phase difference between two polarimetric channels. To tell whether Gaussian distributions are proper or not, the histograms are compared with the PDFs defined by (12), (18) and (20). The covariance matrices of the Gaussian distributions are estimated using the simple mean estimator.

Figure 2 shows the fit of the HH intensity, and the fit of the product of HH Channel and HV channel. It demonstrates that the histograms conform to the corresponding PDFs, implying that Gaussian distributions are suitable for these crops areas. Though it could work, the comparison of histogram and PDF is not visually effective, see Figure 2f,g for example. So in the next experiments we will try different methods to validate the applicability of statistical models.

As mentioned in the previous sections, the spatial resolution of PolSAR images is one of the most important factors that have strong impact on the data statistics. To demonstrate this, real SAR data including a RADARSAT-2 Fine Quad-Pol data (RST2) as well as a F-SAR X-band full-pol data (FSAR) are analyzed. The two data have quite different spatial resolutions, 11.1 m $\times$ 7.6 m for the RST2 data, and 0.25 m $\times$ 0.25 m for the FSAR data. Three ROIs over the crops area from each data are tested, see Figure 3a,b. For the RST2 data, each ROI covers 50 $\times$ 50 pixels. The ROIs in the FSAR data are much larger thanks to a higher spatial resolution, each covering 200 $\times$ 200 pixels. The Pauli decomposition shows that the ROIs in both images are very homogeneous, no appreciable texture is observed.

Table 1. Summary of statistical PolSAR data models.

Category	Model	PDF	References	Summary
Gaussian	Gaussian	(7)	[31,33]	Simple, high mathematical tractability, suitable for data of low or moderate spatial resolution.
	Wishart	(21)	[31–33]	
	Relaxed Wishart	(21)	[39]	More flexible than the Wishart distribution, but assigning different values to the number of looks L is not so convincing.
	Wishart-Kotz	(31)	[40,41]	With ability to model heavy tail behaviors, computationally efficient and numerically stable, but at the expense of adding two more parameters.
Texture Models	$\mathcal{K}$	(43), (44)	[4,7,10]	Suitable for non-Gaussian data, widely used to model forest, ocean and so on, strong physical background.
	NIG	(47), (48)	[49,50]	Large shape variations, strong theoretical grounds derived from Brownian motion.
	$\mathcal{G}$	(52), (53)	[14,15,52]	Able to model different types of texture, but requires more parameters (two parameters).
	$\mathcal{G}^0$	(57), (58)	[14,15]	Suitable for extremely heterogeneous data, no complex special function involved.
	Kummer-$\mathcal{U}$	(62), (63)	[16,53]	Able to model different types of texture, but requires more parameters (two parameters), texture distribution belongs to Pearson family.
	$\mathcal{W}$	(67), (68)	[5]	Able to model data with low variance but extreme skewness, e.g., textured data after speckle filtering.
	$\mathcal{M}$	(72), (73)	[5]	
	WGΓ	No Explicit	[54]	Of great flexibility (generalization of many other distributions), but the PDF needs to be calculated numerically.
	Generalized $\mathcal{K}$	(81)	[55]	Good approximation of data when there exist strong scatterers, very complex PDF with polynomial expansions.
	Correlated $\mathcal{K}$	No Explicit	[58,61]	Able to model texture correlations of different channels, no explicit expression for the texture variables, distribution parameters are limited to specific values.
	Dual-Texture $\mathcal{G}$	(92)	[62]	Different texture distributions for the co-pol and the cross-pol channels.
Others	Finite Mixture	(93)	[17–19]	Extremely flexible (covering both unimodal and multimodal distributions), able to model data with considerable skewness, suitable for rather heterogeneous data.
	Copula Based	No Explicit	[22,67]	Divides complex multivariate distributions into marginal distributions and dependence structure, and analyze them separately, but it is not very straightforward to choose the best copulas.

Normalized Intensity Moments (NIM) are employed to determine whether the Gaussian distributions are suitable for the test ROIs or not [9,69]. Let I denote the intensity of a polarimetric channel, the NIM of the vth order is defined as

$$NIM_v = \frac{\mathrm{E}\{I^v\}}{\mathrm{E}^v\{I\}} \tag{99}$$

For Gaussian distributed data, the intensity will follow an exponential distribution as defined in (12). The NIMs in this case are independent of the data, which can be calculated by

$$NIM_v^* = \Gamma(v+1) \tag{100}$$

By comparing the estimated values from the data with those of Gaussian distributions, we can easily make conclusions on the applicability of Gaussian distributions.

The HH channel is analyzed for both the RST2 data and the FSAR data. Results are shown in Figure 3c,d, where black lines represent theoretical values of the exponential distribution and different markers represent values estimated from the test ROIs. As it can be seen, the NIMs estimated from the RST2 data fit those calculated from the exponential distribution very well. Same results can be obtained for the HV channel and the VV channel. It is rational to conclude that these ROIs can be modeled by Gaussian distributions. In contrast, the result on the FSAR data shows different behaviors. There are large discrepancies between the estimated values and the theoretical values for all ROIs. Apparently, Gaussian distributions are not accurate any more.

A further validation on the FSAR data is performed. Assuming that the intensity of each ROI can be modeled by a Weibull distribution, then the distribution parameter, denoted by γ, can be estimated using the first order moment. Furthermore, the NIM of the vth order can be computed by

$$NIM_v^\dagger = \frac{\Gamma\left(1 + \frac{v}{\gamma}\right)}{\Gamma^v\left(1 + \frac{1}{\gamma}\right)} \tag{101}$$

In Figure 3e, the estimated NIMs (markers) and those calculated using the above equation (lines) are plotted for each ROI. The Weibull distribution seems to be applicable in ROI 2 and ROI 3. Compared with the exponential distribution, the Weibull distribution could capture larger variance by introducing an additional distribution parameter. However, even the Weibull distribution could not give an accurate representation for ROI 1. Complex distributions with more parameters may achieve reasonable fit.

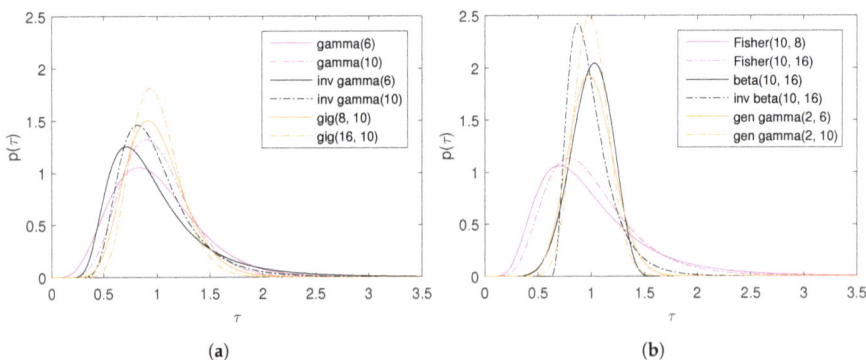

Figure 1. Examples of different texture distributions. (**a**) PDFs of the gamma (gamma(6) and gamma(10)), the inverse gamma (inv gamma(6) and inv gamma(10)) and the GIG (gig(8, 10) and gig(16, 10)) distributions. (**b**) PDFs of the Fisher (Fisher(10, 8) and Fisher(10, 16)), the beta (beta(10, 16)), the inverted beta (inv beta(10, 16)), and the generalized gamma (gen gamma(2, 6) and gen gamma(2, 10)) distributions.

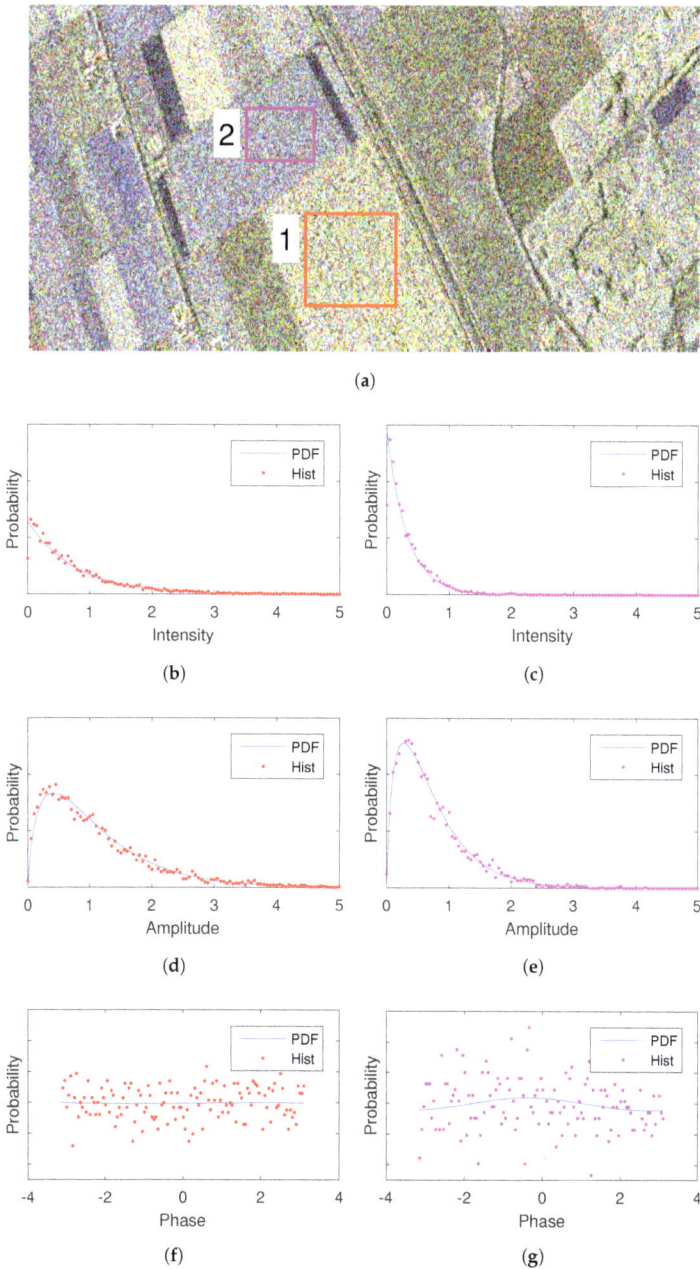

Figure 2. Histograms of two homogeneous areas in a RADARSAT-2 image and the PDFs under Gaussian assumption. Parameters of the Gaussian distributions are estimated using moments. (**a**) Pauli decomposition of the RADARSAT-2 data and two ROIs. (**b,c**) Intensity of the S_{hh}. (**d,e**) Amplitude of $S_{hh}S_{hv}$. (**f,g**) Phase of $S_{hh}S_{hv}$.

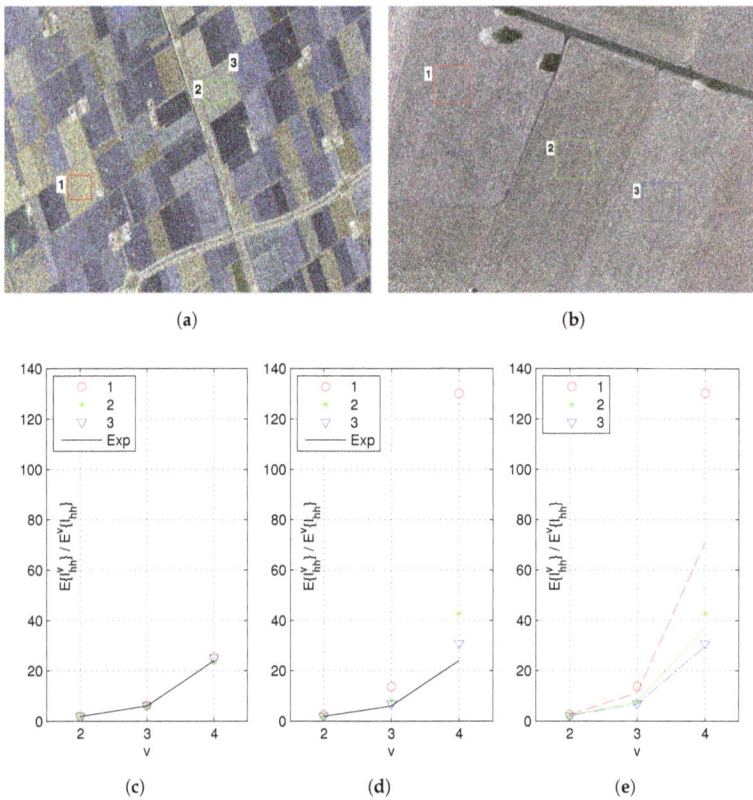

Figure 3. NIMs of the 2nd, the 3rd and the 4th order estimated from three crops areas in a RADARSAT-2 data and a F-SAR data. (**a**) Pauli decomposition and ROIs of the RADARSAT-2 image. (**b**) Pauli decomposition and ROIs of the F-SAR image. (**c**) NIMs of the ROIs in the RADARSAT-2 data and the exponential distribution. (**d**) NIMs of the ROIs in the F-SAR data and the exponential distribution. (**e**) NIMs of the ROIs in the F-SAR data and Weibull distributions.

In general, the Weibull distribution is advisable to model the intensity of high resolution single channel data. However, for PolSAR data, the correlations between different polarimetric channels convey useful information, besides the intensities. In order to describe the statistical behavior correctly, copulas (introduced in Section 5.2) can be adopted. By modeling the dependence structure between polarimetric channels using copulas, and the intensities by Weibull distributions, a good approximation of the data could be expected. However, how to choose the proper copulas needs to be investigated intensively. We haven't found a copula capturing the dependency properly for the testing ROIs.

Another aspect that causes non-gaussianity in PolSAR data is the fluctuation of radar cross section due to the change of surface properties, e.g., height of the observing surface. Usually, this type of data should be modeled using texture distributions. To validate the applicability of texture models, two PolSAR images, the RADARSAT-2 Fine Quad-Pol data (RST2) and the ALOS-2 level 1.1 Full-Pol data (ALOS2), are analyzed. Both images were acquired over Barcelona (Spain) with similar incidence angles. The spatial resolution is different, 11.1 m × 7.6 m (Range × Azimuth) for the RST2 data and 3.49 m × 3.84 m for the ALOS2 data, respectively. Original data are in the single look complex format, from which the sample covariance matrices are obtained after applying a multilook processing.

We have selected ROIs locating at similar positions in the urban area, in the agriculture area, in the sea and the forest areas. Pauli decomposition of the test data and ROIs are shown in Figure 4.

Matrix variate log-cumulants [42] are used to examine the suitability of texture models. The matrix variate log-cumulants are of great value for the analysis of sample covariance matrix, and that they can be employed to derive estimators for the distribution parameters with low bias and variance [42]. Different from the NIMs, there is no need to study each polarimetric channel separately with matrix variate log-cumulants. Define the Mellin kind matrix-variate characteristic function as the Mellin Transform of the PDF

$$\phi(s) = \int_{\Omega_+} |\mathbf{Z}|^{s-d} p(\mathbf{Z}) d\mathbf{Z} \tag{102}$$

then, the vth-order log-cumulant, or Mellin kind cumulant, is given by

$$\kappa_v = \frac{d^v}{ds^v} \ln \phi(s) \bigg|_{s=d} \tag{103}$$

Meanwhile, the sample log-cumulants can be estimated from the data using

$$\hat{\kappa}_v = \hat{\mu}_v - \sum_{i=1}^{v-1} \binom{v-1}{i-1} \hat{\kappa}_i \hat{\mu}_{v-i} \tag{104}$$

where $\hat{\mu}_v$ is the estimated log-moments

$$\hat{\mu}_v = \frac{1}{M} \sum_{i=1}^{M} (\ln |\mathbf{C}_i|)^v \tag{105}$$

with M denoting the number of samples and $\mathbf{C}_i$ the ith sample covariance matrix.

To see if a texture model is applicable, we can compare the log-cumulants calculated from the PDF (κ_v) and those estimated from the sample data ($\hat{\kappa}_v$). In [42], a diagram is proposed to visualize the comparison by plotting the second order log-cumulants κ_2 against the third order log-cumulants κ_3 in a plane, where different distributions place in different regions, as shown in Figure 4c,d. In this diagram, estimated log-cumulants are represented by the "+" markers (values from different ROIs are distinguished by various colors), and theoretical values of different texture distributions are represented by curves (the $\mathcal{K}$ and the $\mathcal{G}^0$ distributions) as well as regions (the Kummer-$\mathcal{U}$, the $\mathcal{M}$ and the $\mathcal{W}$ distributions).

From Figure 4, we can see that the urban areas (red and green rectangles) can be modeled by the $\mathcal{G}^0$ or the Kummer-$\mathcal{U}$ distributions, which have the capability to model heterogeneous areas. The two ROIs in urban area represent two different urban structures, one is of tall and densely distributed apartments, the other is of short and sparse houses. This may be an explanation as to why different statistics, the $\mathcal{G}^0$ vs the Kummer-$\mathcal{U}$, are obtained. In agriculture areas (cyan and yellow rectangles), $\mathcal{K}$ distribution is shown to be the most suitable model. The forest area (black rectangle) shows weak texture in the RST2 data. In the ALOS2 data, there is a strong fluctuation in the backscattering due to the radar foreshortening. To eliminate the effect of radar image distortions, another forest region (purple rectangle) is analyzed, which is found to follow a $\mathcal{K}$ distribution. In most cases, texture is not observed in the sea areas.

As explained in Section 5.1, the finite mixtures could also give rise to non-gaussian statistics. To further distinguish textures from mixtures, higher order log-cumulants are required [70]. A large number of samples are demanded in order to estimate the higher order log-cumulants correctly. There are only 20 × 20 pixels in each of the previous ROIs, not enough to obtain a satisfying estimation of the fourth order log-cumulants. So another experiment is carried out on an airborne SAR data, a UAVSAR image.

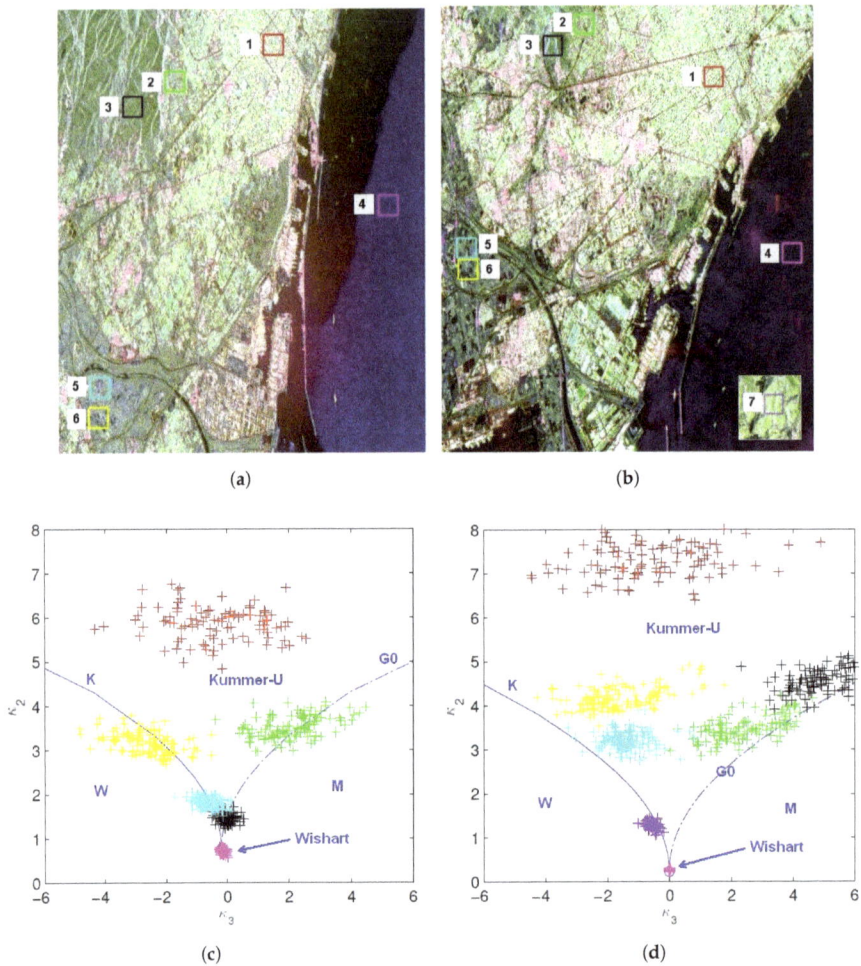

Figure 4. Matrix variate log-cumulants of the 2nd and the 3rd order estimated from a RADARSAT-2 data and an ALOS-2 data. Theoretical values calculated from the $\mathcal{K}$, the $\mathcal{G}^0$, the Kummer-$\mathcal{U}$, the $\mathcal{W}$, the $\mathcal{M}$ and the Wishart distributions are also plotted as references. (**a**) ROIs of the RADARSAT-2 data. (**b**) ROIs of the ALOS-2 data. (**c**) Matrix variate log-cumulants of the RADARSAT-2 data. (**d**) Matrix variate log-cumulants of the ALOS-2 data.

The test site is in the West Panhandle of Florida (USA), and the data is in the multilook cross-product slant range format, with number of looks in the range dimension and azimuth dimension equal to 3 and 12 respectively. The ENL is estimated as 12.73 over a homogeneous sea area. Four ROIs covering land types of ocean area (ROI 1), forest (ROI 2), wetland (ROI 3), and urban area (ROI 4), are tested, see Figure 5. Thanks to a higher spatial resolution, 1.67 m × 0.8 m (Range × Azimuth), each ROI contains 90 × 70 pixels, much more samples than those in the previous experiment.

Figure 5. Test regions on the UAVSAR data. Four ROIs over different land types are tested, including the sea, the forest, the wetland and the urban area.

The log-cumulants of the second order, the third order and the fourth order are calculated. From the log-cumulant diagrams (Figure 6a,b), we can see that different ROIs show different statistical behaviors. The ocean area can be modeled by a Wishart distribution, and the forest by a $\mathcal{K}$ distribution. The wetland and the urban area are very heterogeneous, especially the urban area, which has a very small κ_3. The point clouds representing estimated statistics are less widely spread than those in Figure 4c,d. This is because more samples are used to estimate the values.

The fourth order log-cumulant is considered to further discriminate the texture from mixture. As shown in Figure 6c,d, the log-cumulants of major texture models can construct a smooth surface, while those of the finite mixture model will lie below this surface. The results show that texture models are proper for the sea area and forest area, while a finite mixture model make a better representation than a texture model for the wetland area and the urban area, because the point clouds estimated from ROI 1 and ROI 2 are on the product model surface, whereas those from ROI 3 and ROI 4 are below it. Actually, the Pauli decomposition in Figure 5 shows that the first two ROIs are very homogeneous and ROI 3 consists of different targets. Urban area, made up of distributed targets and point targets usually, has very large variance. This can be verified by the log-cumulant cube in Figure 6d, where both of the absolute values of κ_3 and κ_4 are very large. The estimation of the number as well as the weights of mixing components needs to be further studied.

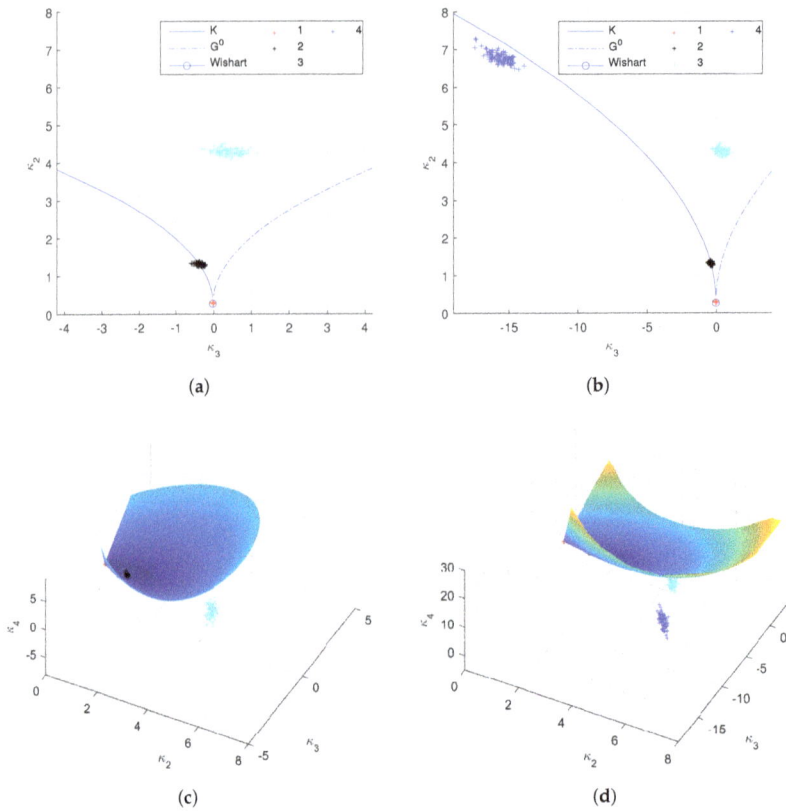

Figure 6. Log-cumulants of the 2nd, the 3rd and the 4th orders on the UAVSAR data. The right column and the left column are the same results but with different axes limits. (**a**,**b**) Log-cumulants of the 2nd and the 3rd order. ROIs over different ground targets show different statistics. (**c**,**d**) Log-cumulant up to the 4th order. It shows some ROIs can be modeled by texture models, while others should be represented using finite mixture model.

At last, statistical properties of the sea area at two different conditions are examined. One is with smooth surface, and the other with waves, as shown in Figure 7. Both data are acquired by RADARSAT-2 at C-band, and they have similar spatial resolutions. To study the textures of different polarimetric channels, the intensity of each channel is checked separately using the scalar log-cumulants [42,71]. The second order and the third order log-cumulants are employed as before. From Figure 7, we can see that in the first case, no texture is observed in all polarimetric channels. This can be also validated by the Pauli decomposition. When there exist sea waves, however, the log-cumulants are quite different. The HH channel and the VV channel have similar statistics, but different from those of the HV channel. In other words, multi-texture is observed in the test area. The result supposes that we can model the data using a dual-texture model in which the co-pol channels share a same texture distribution and the cross-pol channel with another one. The correlation between different textures needs to be further investigated to see if a partially correlated texture distribution is required.

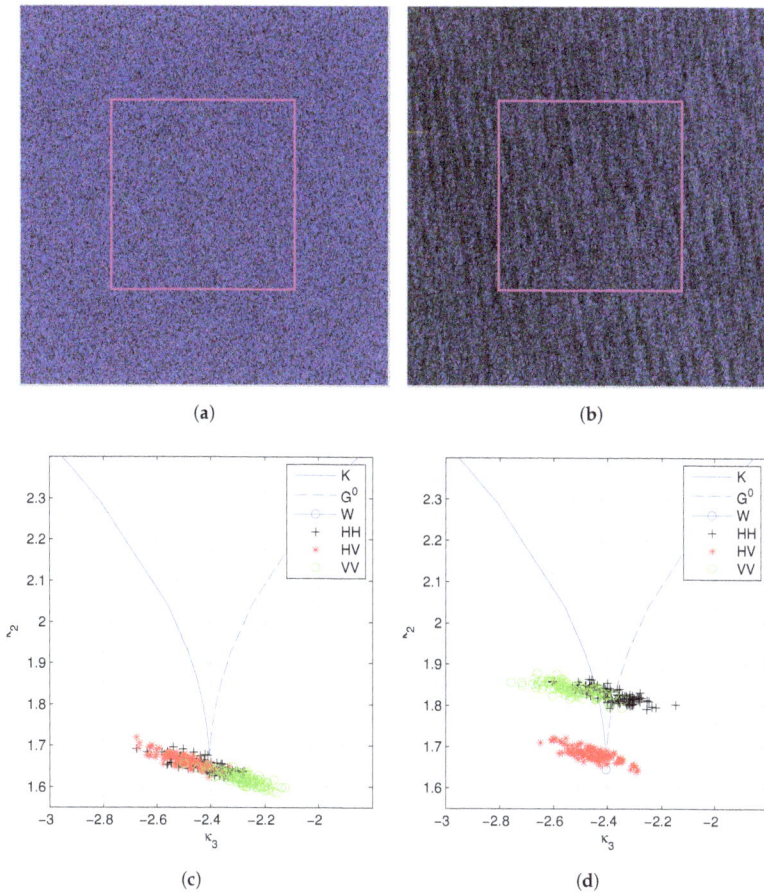

Figure 7. Scalar log-cumulants of different polarimetric channels (the HH, the HV and the VV channels) over sea areas. (**a**) Test region over smooth sea surface. (**b**) Test region over sea surface with waves. (**c**) Log-cumulants of the smooth sea surface. (**d**) Log-cumulants of the sea surface with waves.

7. Challenges

When the spatial resolution is not very high and the data is very homogeneous, the Gaussian distributions could provide a good representation of the data. As the spatial resolution increases, PolSAR data usually show non-Gaussian behaviour, e.g., exhibiting heavy tails. The texture models, which adopt additional random variables to model the spatial variation of the radar cross section, are found to be accurate for this kind of data. Texture models could model the non-Gaussian behavior observed in high resolution data, and yet keep a compact mathematical form. However, to model textures over complex scenes, sophisticated distributions are generally required. In addition, they are known to present problems in estimating parameters accurately. General distributions that cover a wide range of other distributions are suggested by many researchers. However, they usually have a complex form. Using several simple distributions from a certain family, Pearson family for example, could be a better idea. The distributions from a same family often have similar behaviors, but can be further distinguished by statistics like higher order log-cumulants [70].

In the product model as shown in (33) or (34), positive random variable following any distribution can be employed to model the texture. Additionally, PDFs of the scattering vector or the sample covariance matrix can be obtained subsequently after mathematical calculations. The PDFs, however, give no information about why the data following a specific distribution is obtained. Most of the texture models lack a physical explanation of the underlying scattering process. A possible way to solve this problem could be the random walk model, which treats the received signal as an addition of responses from all the scatterers in the same resolution cell [9,12,72]. The random walk model can be interpreted as a discrete analog of the SAR focusing process.

Texture information has been used in optical image processing for a long time. In SAR or PolSAR images, it is also found to be useful to distinguish different target types. For example, trees of different heights can be distinguished by texture information [73]. However, currently the most common way to make use of texture models is to design probability based algorithms (e.g., classification and segmentation) by replacing the Gaussian distribution or the Wishart distribution [4–6]. How to extract texture information and let PolSAR applications benefit from it is not involved. Apparently, combining polarimetric information and texture information could improve the performance of applications since more knowledge is introduced. Therefore, a further study in this aspect will be of great value.

Besides texture models, there are non-Gaussian models subdividing complicated distributions into components, each with a simple distribution. For example, the finite mixture model treats a distribution as a weighted sum of those of different target types. In addition, the copula based model divides a multivariate distribution into marginal distributions and a general dependence structure. In the finite mixture model, robust algorithms to estimate the mixing number and the mixing weights are in urgent need. For the copula based method, we have many options for the marginal distributions. However, for the dependence structure, not many experiments were implemented to show which copula is the best. Additionally, it is a big challenge to extend the bivariate copulas which are intensively investigated in the field of statistical analysis to multivariate ones to fit the PolSAR data.

The statistical properties of PolSAR data are characterized totally by the PDFs of the scattering vectors or the sample covariance matrices. However, it is difficult to use these PDFs directly because they are multivariate ones. Normally, the statistics of each polarimetric channel are studied separately, and the correlation between different polarimetric channels are neglected. Another way is to analyze the determinants of sample covariance matrices. The widely used matrix variate log-cumulant is an example. However, we need to filter the data (the multilook process) to obtain the sample covariance matrices, which could change the actual statistical properties of the data. To overcome these problems, the l_2-norms of the scattering vectors can be employed, and they are found to be a useful tool for texture analysis of PolSAR data [73]. However, there are also limitations, e.g., the difference between models are not very large.

In summary, statistical modeling and texture analysis of PolSAR data covers a wide range of topics. To make a better understanding of texture and to make good use of it, there is still have a lot of work to do.

Acknowledgments: This work has been financed by: the Shenzhen Science & Technology Program with code JSGG20150512145714247, the Spanish Science, Research and Innovation Plan (Ministerio de Economía y Competitividad) with project code TIN2014-55413-C2-1-P, and the National Key Research Plan with code 2016YFC0500201-07. In addition, the authors would like to acknowledge ESA (European Space Agency) for the RADARSAT-2 data provided in the frame of the AgriSAR 2009 campaign, JAXA (Japan Aerospace Exploration Agency) for the ALOS-2 data provided in the framework of the 4th ALOS Research Announcement for ALOS-2, NASA (National Aeronautics and Space Administration) for the UAVSAR data, and Andreas Reigber from DLR for the F-SAR data.

Author Contributions: Xinping Deng designed and performed the experiments, analyzed the data, and wrote the paper; Carlos López-Martínez and Jinsong Chen provided valuable suggestions to write the paper, and revised the paper; Pengpeng Han prepared some of the data used in the experiments.

Abbreviations

The following abbreviations are used in this manuscript:

BSA   Back Scattering Alignment
CDF   Cumulative Distribution Function
CLT   Central Limit Theorem
FSA   Forward Scattering Alignment
GIG   Generalized Inverse Gaussian
NIG   Normal Inverse Gaussian
NIM   Normalized Intensity Gaussian
PDF   Probability Density Function
PolSAR   Polarimetric SAR
ROI   Region Of Interest
SAR   Synthetic Aperture Radar
SIRV   Spherically Invariant Random Vector

Appendix A

Some integral identities used in this paper are listed out here.

1. ([74] p. 368, Equation (3.471-9))

$$\int_0^\infty x^{v-1} \exp\left(-\frac{\beta}{x} - \alpha x\right) dx = 2 \left(\frac{\beta}{\alpha}\right)^{v/2} K_v\left(2\sqrt{\beta\alpha}\right) \tag{A1}$$

$$\text{Re}\,\beta > 0, \text{Re}\,\alpha > 0$$

K_v is the modified Bessel function of the second kind of order v.

2. ([74] p. 340, Equation (3.339))

$$\int_0^\pi \exp(z\cos x)dx = \pi I_0(z) \tag{A2}$$

$I_0(z)$ is the modified Bessel function of the first kind.

3. ([74] p. 702, Equation (6.624-1))

$$\int_0^\infty xe^{-\alpha x}K_0(\beta x)dx = \frac{1}{\alpha^2 - \beta^2}$$

$$\times \left\{ \frac{\alpha}{\sqrt{\alpha^2 - \beta^2}} \ln\left(\frac{\alpha}{\beta} + \sqrt{\left(\frac{\alpha}{\beta}\right)^2 - 1}\right) - 1 \right\} \tag{A3}$$

4. ([74] p. 347, Equation (3.382-2))

$$\int_u^\infty (x-u)^v e^{-\mu x}dx = \mu^{-v-1}e^{-u\mu}\Gamma(v+1), \quad u > 1, \text{Re}\,v > -1, \text{Re}\,\mu > 0 \tag{A4}$$

5. ([74] p. 700, Equation (6.621-3))

$$\int_0^\infty x^{\mu-1}e^{-\alpha x}K_v(\beta x)dx = \frac{\sqrt{\pi}(2\beta)^v}{(\alpha+\beta)^{\mu+v}} \frac{\Gamma(\mu+v)\Gamma(\mu-v)}{\Gamma(\mu+1/2)}$$

$$\times {}_2F_1\left(\mu+v, v+\frac{1}{2}; \mu+\frac{1}{2}; \frac{\alpha-\beta}{\alpha+\beta}\right) \tag{A5}$$

$$\text{Re}\,\mu > |\text{Re}\,v|, \text{Re}\,(\alpha+\beta) > 0$$

${}_2F_1(a, b; c; z)$ is the Gauss hypergeometric function.

6. ([74] p. 917, Equation (8.432-3))

$$K_v(z) = \frac{\left(\frac{z}{2}\right)^v \Gamma\left(\frac{1}{2}\right)}{\Gamma\left(v + \frac{1}{2}\right)} \int_1^\infty e^{-zt}(t^2 - 1)^{v - \frac{1}{2}} dt, \quad \mathrm{Re}\left(v + \frac{1}{2}\right) > 0, |\arg z| < \frac{\pi}{2} \tag{A6}$$

7. ([74] p. 325, Equation (3.252-3))

$$\int_1^\infty x^{\mu - 1}(x^p - 1)^{v - 1} = \frac{1}{p} B\left(1 - v - \frac{\mu}{p}, v\right) \tag{A7}$$

$$p > 0, \mathrm{Re}\, v > 0, \mathrm{Re}\, \mu < p(1 - \mathrm{Re}\, v)$$

8. The gamma function is defined as

$$\Gamma(t) = \int_0^\infty x^{t-1} e^{-x} dx. \tag{A8}$$

Let $x = \frac{\beta}{y}$ where $\beta > 0$, we have the following equation after changing variables

$$\int_0^\infty y^{-t-1} \exp\left(-\frac{\beta}{y}\right) dy = \Gamma(t)\beta^{-t} \tag{A9}$$

9. ([34] p. 505, Equation (13.2.5))

$$\int_0^\infty e^{-zt} t^{a-1}(1 + t)^{b-a-1} dt = \Gamma(a)U(a, b, z) \tag{A10}$$

U is the confluent hypergeometric function of the second kind, or KummerU function.

10. ([74] p. 367, Equation (3.471-2))

$$\int_0^u x^{v-1}(u - x)^{\mu - 1} \exp\left(-\frac{\beta}{x}\right) dx = \beta^{\frac{v-1}{2}} u^{\frac{2\mu + v - 1}{2}} \Gamma(\mu)$$

$$\times \exp\left(-\frac{\beta}{2u}\right) W_{\frac{1 - 2\mu - v}{2}, \frac{v}{2}}\left(\frac{\beta}{u}\right) \tag{A11}$$

$$\mathrm{Re}\, \mu > 0, \mathrm{Re}\, \beta > 0, \mu > 0$$

W is Whittaker W function.

11. ([74] p. 368, Equation (3.471-5))

$$\int_u^\infty x^{v-1}(x - u)^{\mu - 1} \exp\left(\frac{\beta}{x}\right) dx = B(1 - \mu - v, \mu) u^{\mu + v - 1}$$

$$\times M\left(1 - \mu - v, 1 - v, \frac{\beta}{u}\right) \tag{A12}$$

$$0 < \mathrm{Re}\, \mu < \mathrm{Re}\,(1 - v), u > 0$$

M is the confluent hypergeometric function of the first kind, also known as the KummerM function.

References

1. Goodman, J.W. Some fundamental properties of speckle. *JOSA* **1976**, *66*, 1145–1150.
2. Lee, J.S.; Grunes, M.R.; De Grandi, G. Polarimetric SAR speckle filtering and its implication for classification. *IEEE Trans. Geosci. Remote Sens.* **1999**, *37*, 2363–2373.
3. Kersten, P.R.; Lee, J.S.; Ainsworth, T.L. Unsupervised classification of polarimetric synthetic aperture radar images using fuzzy clustering and EM clustering. *IEEE Trans. Geosci. Remote Sens.* **2005**, *43*, 519–527.

4. Doulgeris, A.P.; Anfinsen, S.N.; Eltoft, T. Classification with a non-Gaussian model for PolSAR data. *IEEE Trans. Geosci. Remote Sens.* **2008**, *46*, 2999–3009.

5. Bombrun, L.; Vasile, G.; Gay, M.; Totir, F. Hierarchical segmentation of polarimetric SAR images using heterogeneous clutter models. *IEEE Trans. Geosci. Remote Sens.* **2011**, *49*, 726–737.

6. Doulgeris, A.P.; Akbari, V.; Eltoft, T. Automatic PolSAR segmentation with the U-distribution and Markov random fields. In Proceedings of the 2012 EUSAR—9th European Conference on Synthetic Aperture Radar, Nuremberg, Germany, 23–26 April 2012; pp. 183–186.

7. Lee, J.S.; Mitchell, R.G.; Kwok, R. Classification of multi-look polarimetric SAR imagery based on complex Wishart distribution. *Int. J. Remote Sens.* **1994**, *15*, 2299–2311.

8. Lee, J.S.; Grunes, M.; Ainsworth, T.; Du, L.J.; Schuler, D.; Cloude, S. Unsupervised classification using polarimetric decomposition and the complex Wishart classifier. *IEEE Trans. Geosci. Remote Sens.* **1999**, *37*, 2249–2258.

9. Jakeman, E.; Pusey, P.N. A model for non-Rayleigh sea echo. *IEEE Trans. Antennas Propag.* **1976**, *24*, 806–814.

10. Jakeman, E.; Pusey, P. Significance of K distributions in scattering experiments. *Phys. Rev. Lett.* **1978**, *40*, 546–550.

11. Ward, K. Compound representation of high resolution sea clutter. *Electron. Lett.* **1981**, *17*, 561–563.

12. Yueh, S.H.; Kong, J.A.; Jao, J.K.; Shin, R.T.; Novak, L.M. K-distribution and polarimetric terrain radar clutter. *J. Electromagn. Waves Appl.* **1989**, *3*, 747–768.

13. Lee, J.S.; Schuler, D.L.; Lang, R.H.; Ranson, K.J. K-distribution for multi-look processed polarimetric SAR imagery. In Proceedings of the International Geoscience and Remote Sensing Symposium, Surface and Atmospheric Remote Sensing: Technologies, Data Analysis and Interpretation, Pasadena, CA, USA, 8–12 August 1994; Volume 4, pp. 2179–2181.

14. Frery, A.C.; Muller, H.J.; Yanasse, C.C.F.; Sant'Anna, S.J.S. A model for extremely heterogeneous clutter. *IEEE Trans. Geosci. Remote Sens.* **1997**, *35*, 648–659.

15. Freitas, C.C.; Frery, A.C.; Correia, A.H. The polarimetric G distribution for SAR data analysis. *Environmetrics* **2005**, *16*, 13–31.

16. Bombrun, L.; Beaulieu, J.M. Fisher distribution for texture modeling of polarimetric SAR data. *IEEE Geosci. Remote Sens. Lett.* **2008**, *5*, 512–516.

17. Moser, G.; Zerubia, J.; Serpico, S.B. Dictionary-based stochastic expectation-maximization for SAR amplitude probability density function estimation. *IEEE Trans. Geosci. Remote Sens.* **2006**, *44*, 188–200.

18. Krylov, V.; Moser, G.; Serpico, S.B.; Zerubia, J. *Modeling the Statistics of High Resolution SAR Images*; Technical Report; Institut National de Recherche en Informatique et en Automatique, INRIA: Rocquencourt, France, 2008.

19. Wang, Y.; Ainsworth, T.L.; Lee, J. On Characterizing High-Resolution SAR Imagery Using Kernel-Based Mixture Speckle Models. *IEEE Geosci. Remote Sens. Lett.* **2015**, *12*, 968–972.

20. Gao, G. Statistical modeling of SAR images: A survey. *Sensors* **2010**, *10*, 775–795.

21. Nelsen, R.B. *An Introduction to Copulas*; Springer Science & Business Media: New York, NY, USA, 2007.

22. Mercier, G.; Bouchemakh, L.; Smara, Y. The use of multidimensional copulas to describe amplitude distribution of polarimetric SAR data. In Proceedings of the IEEE International Geoscience and Remote Sensing Symposium, Barcelona, Spain, 23–28 July 2007; pp. 2236–2239.

23. Voisin, A.; Krylov, V.A.; Moser, G.; Serpico, S.B.; Zerubia, J. Supervised Classification of Multisensor and Multiresolution Remote Sensing Images With a Hierarchical Copula-Based Approach. *IEEE Trans. Geosci. Remote Sens.* **2014**, *52*, 3346–3358.

24. Lee, J.S.; Pottier, E. *Polarimetric Radar Imaging: From Basics to Applications*; CRC Press: Boca Raton, FL, USA, 2009.

25. Cloude, S. *Polarisation: Applications in Remote Sensing*; Oxford University Press: Oxford, UK, 2009.

26. Cloude, S.R.; Pottier, E. A review of target decomposition theorems in radar polarimetry. *IEEE Trans. Geosci. Remote Sens.* **1996**, *34*, 498–518.

27. Frost, V.S.; Stiles, J.A.; Shanmugan, K.S.; Holtzman, J.C. A Model for Radar Images and Its Application to Adaptive Digital Filtering of Multiplicative Noise. *IEEE Trans. Pattern Anal. Mach. Intell.* **1982**, *PAMI-4*, 157–166.

28. Kong, J.; Yueh, S.; Lim, H.; Shin, R.; Van Zyl, J. Classification of earth terrain using polarimetric synthetic aperture radar images. *Prog. Electromagn. Res.* **1990**, *3*, 327–370.

29. Sarabandi, K. Derivation of phase statistics from the Mueller matrix. *Radio Sci.* **1992**, *27*, 553.
30. Ulaby, F.; Sarabandi, K.; Nashashibi, A. Statistical properties of the Mueller matrix of distributed targets. In Proceedings of the IEE Proceedings F—Radar and Signal Processing, Barcelona, Spain, 6 August 1992; Volume 139, pp. 136–146.
31. Goodman, N.R. Statistical analysis based on a certain multivariate complex Gaussian distribution (An Introduction). *Ann. Math. Stat.* **1963**, *34*, 152–177.
32. Lee, J.S.; Hoppel, K.W.; Mango, S.A.; Miller, A.R. Intensity and phase statistics of multilook polarimetric and interferometric SAR imagery. *IEEE Trans. Geosci. Remote Sens.* **1994**, *32*, 1017–1028.
33. Tough, R.J.A.; Blacknell, D.; Quegan, S. A statistical description of polarimetric and interferometric synthetic aperture radar data. *Proc. R. Soc. Lond. A Math. Phys. Sci.* **1995**, *449*, 567–589.
34. Abramowitz, M.; Stegun, I.A. *Handbook of Mathematical Functions: With Formulas, Graphs, and Mathematical Tables*; Courier Corporation: North Chelmsford, MA , USA, 1964; Volume 55.
35. Walck, C. *Hand-Book on Statistical Distributions for Experimentalists*; University of Stockholm: Stockholm, Sweden, 2007.
36. López-Martínez, C.; Fabregas, X. Polarimetric SAR speckle noise model. *IEEE Trans. Geosci. Remote Sens.* **2003**, *41*, 2232–2242.
37. López-Martínez, C.; Pottier, E.; Cloude, S.R. Statistical assessment of eigenvector-based target decomposition theorems in radar polarimetry. *IEEE Trans. Geosci. Remote Sens.* **2005**, *43*, 2058–2074.
38. Alonso-González, A.; López-Martínez, C.; Salembier, P. Filtering and segmentation of polarimetric SAR data based on binary partition trees. *IEEE Trans. Geosci. Remote Sens.* **2012**, *50*, 593–605.
39. Anfinsen, S.N.; Eltoft, T.; Doulgeris, A.P. A relaxed Wishart model for polarimetric SAR data. In *Proceedings of the Fourth International Workshop on Science and Applications of SAR Polarimetry and Polarimetric Interferometry PolInSAR*; European Space Agency: Noordwijk, The Netherlands, 2009; pp. 26–30.
40. Kersten, P.R.; Anfinsen, S.N. A flexible and computationally efficient density model for the multilook polarimetric covariance matrix. In Proceedings of the 9th European Conference on Synthetic Aperture Radar, Nuremberg, Germany, 23–26 April 2012; pp. 760–763.
41. Kersten, P.R.; Anfinsen, S.N.; Doulgeris, A.P. The Wishart-Kotz classifier for multilook polarimetric SAR data. In Proceedings of the IEEE International Geoscience and Remote Sensing Symposium, Munich, Germany, 22–27 July 2012; pp. 3146–3149.
42. Anfinsen, S.N.; Eltoft, T. Application of the matrix-variate Mellin transform to analysis of polarimetric radar images. *IEEE Trans. Geosci. Remote Sens.* **2011**, *49*, 2281–2295.
43. Novak, L.M.; Burl, M.C. Optimal speckle reduction in polarimetric SAR imagery. *IEEE Trans. Aerosp. Electron. Syst.* **1990**, *26*, 293–305.
44. Sheen, D.R.; Johnston, L.P. Statistical and spatial properties of forest clutter measured with polarimetric synthetic aperture radar (SAR). *IEEE Trans. Geosci. Remote Sens.* **1992**, *30*, 578–588.
45. Gini, F.; Greco, M. Covariance matrix estimation for CFAR detection in correlated heavy tailed clutter. *Signal Process.* **2002**, *82*, 1847–1859.
46. Pascal, F.; Forster, P.; Ovarlez, J.; Larzabal, P. Performance analysis of covariance matrix estimates in impulsive noise. *IEEE Trans. Signal Process.* **2008**, *56*, 2206–2217.
47. Vasile, G.; Ovarlez, J.P.; Pascal, F.; Tison, C. Coherency matrix estimation of heterogeneous clutter in high-resolution polarimetric SAR images. *IEEE Trans. Geosci. Remote Sens.* **2010**, *48*, 1809–1826.
48. Bombrun, L.; Anfinsen, S.N.; Harant, O. A complete coverage of log-cumulant space in terms of distributions for Polarimetric SAR data. In Proceedings of the 5th International Workshop on Science and Applications of SAR Polarimetry and Polarimetric Interferometry (POLinSAR 2011), Friscati, Italy, 24–28 January 2011; pp. 1–8.
49. Doulgeris, A.; Anfinsen, S.N.; Eltoft, T. Analysis of non-Gaussian POLSAR data. In Proceedings of the IEEE International Geoscience and Remote Sensing Symposiumm, Barcelona, Spain, 23–28 July 2007, pp. 160–163.
50. Doulgeris, A.P.; Eltoft, T. Scale Mixture of Gaussian Modelling of Polarimetric SAR Data. *EURASIP J. Adv. Signal Process.* **2009**, *2010*, 874592.
51. Koudou, A.E.; Ley, C. Characterizations of GIG laws: A survey. *Probab. Surv.* **2014**, *11*, 161–176.
52. Khan, S.; Guida, R. Application of Mellin-Kind Statistics to Polarimetric G Distribution for SAR Data. *IEEE Trans. Geosci. Remote Sens.* **2014**, *52*, 3513–3528.

53. Tison, C.; Nicolas, J.M.; Tupin, F.; Maître, H. A new statistical model for Markovian classification of urban areas in high-resolution SAR images. *IEEE Trans. Geosci. Remote Sens.* **2004**, *42*, 2046–2057.

54. Song, W.; Li, M.; Zhang, P.; Wu, Y.; Jia, L.; An, L. The WGΓ Distribution for Multilook Polarimetric SAR Data and Its Application. *IEEE Geosci. Remote Sens. Lett.* **2015**, *12*, 2056–2060.

55. Bian, Y.; Mercer, B. Multilook polarimetric SAR data probability density function estimation using a generalized form of multivariate K-distribution. *Remote Sens. Lett.* **2014**, *5*, 682–691.

56. De Grandi, G.; Lee, J.S.; Schuler, D.; Nezry, E. Texture and speckle statistics in polarimetric SAR synthesized images. *IEEE Trans. Geosci. Remote Sens.* **2003**, *41*, 2070–2088.

57. Eltoft, T.; Anfinsen, S.N.; Doulgeris, A.P. A multitexture model for multilook polarimetric radar data. In Proceedings of the IEEE International Geoscience and Remote Sensing Symposium, Vancouver, BC, Canada, 24–29 July 2011; pp. 1048–1051.

58. Yu, Y. Textural-partially correlated polarimetric K-distribution. In Proceedings of the IEEE International Geoscience and Remote Sensing Symposium, Seattle, WA, USA, 6–10 July 1998; Volume 1, pp. 60–62.

59. Eltoft, T.; Anfinsen, S.N.; Doulgeris, A.P. A multitexture model for multilook polarimetric synthetic aperture radar data. *IEEE Trans. Geosci. Remote Sens.* **2014**, *52*, 2910–2919.

60. Doulgeris, A.P.; Anfinsen, S.N.; Eltoft, T. Segmentation of polarimetric SAR data with a multi-texture product model. In Proceedings of the IEEE International Geoscience and Remote Sensing Symposium, Nuremberg, Germany, 22–27 July 2012; pp. 1437–1440.

61. Lombardo, P.; Farina, A. Coherent radar detection against K-distributed clutter with partially correlated texture. *Signal Process.* **1996**, *48*, 1–15.

62. Khan, S.; Guida, R. The new dual-texture G distribution for single-look PolSAR data. In Proceedings of the IEEE International Geoscience and Remote Sensing Symposium, Nuremberg, Germany, 22–27 July 2012; pp. 1469–1472.

63. Anfinsen, S.N. *Statistical Unmixing of SAR Images*; Technical Report; Munin Open Research Archive, University of Tromsø—The Arctic University of Norway: Tromsø, Norway, 2016.

64. Wang, Y.; Ainsworth, T.L.; Lee, J.S. Application of Mixture Regression for Improved Polarimetric SAR Speckle Filtering. *IEEE Trans. Geosci. Remote Sens.* **2017**, *55*, 453–467.

65. Frühwirth-Schnatter, S. *Finite Mixture and Markov Switching Models*; Springer: Berlin, Germany, 2006.

66. McNeil, A.J.; Frey, R.; Embrechts, P. *Quantitative Risk Management: Concepts, Techniques and Tools*; Princeton University Press: Princeton, NJ, USA, 2005.

67. Mercier, G.; Moser, G.; Serpico, S.B. Conditional copulas for change detection in heterogeneous remote sensing images. *IEEE Trans. Geosci. Remote Sens.* **2008**, *46*, 1428–1441.

68. Regniers, O.; Bombrun, L.; Guyon, D.; Samalens, J.C.; Germain, C. Wavelet-based texture features for the classification of age classes in a maritime pine forest. *IEEE Geosci. Remote Sens. Lett.* **2015**, *12*, 621–625.

69. Oliver, C. Fundamental properties of high-resolution sideways-looking radar. *IEEE F-Commun. Radar Sig.* **1982**, *129*, 385–402.

70. Deng, X.; López-Martínez, C. Higher Order Statistics for Texture Analysis and Physical Interpretation of Polarimetric SAR Data. *IEEE Geosci. Remote Sens. Lett.* **2016**, *13*, 912–916.

71. Nicolas, J.M. Introduction aux statistiques de deuxième espèce: Applications des logs-moments et des logs-cumulants à l'analyse des lois d'images radar. *TS. Traitement du Signal* **2002**, *19*, 139–167.

72. Deng, X.; López-Martínez, C.; Varona, E.M. A Physical Analysis of Polarimetric SAR Data Statistical Models. *IEEE Trans. Geosci. Remote Sens.* **2016**, *54*, 3035–3048.

73. Deng, X.; López-Martínez, C. On the Use of the l_2-Norm for Texture Analysis of Polarimetric SAR Data. *IEEE Trans. Geosci. Remote Sens.* **2016**, *54*, 6385–6398.

74. Jeffrey, A.; Zwillinger, D. *Table of Integrals, Series, and Products*, 7th ed.; Academic Press: Boston, MA, USA, 2007.

remote sensing

MDPI

Article

Multi-Feature Segmentation for High-Resolution Polarimetric SAR Data Based on Fractal Net Evolution Approach

Qihao Chen, Linlin Li, Qiao Xu, Shuai Yang, Xuguo Shi and Xiuguo Liu *

Faculty of Information Engineering, China University of Geosciences (Wuhan), Wuhan 430074, China; chenqihao@cug.edu.cn (Q.C.); lllcug@163.com (L.L.); xu_qiao_cug@126.com (Q.X.); cug_ys@163.com (S.Y.); shixg@cug.edu.cn (X.S.)

* Correspondence: liuxg318@hotmail.com; Tel.: +86-27-6788-3728; Fax: +86-27-6788-3809

Academic Editors: Timo Balz, Uwe Soergel, Mattia Crespi, Batuhan Osmanoglu, Zhenhong Li and Prasad S. Thenkabail
Received: 24 February 2017; Accepted: 4 June 2017; Published: 6 June 2017

Abstract: Segmentation techniques play an important role in understanding high-resolution polarimetric synthetic aperture radar (PolSAR) images. PolSAR image segmentation is widely used as a preprocessing step for subsequent classification, scene interpretation and extraction of surface parameters. However, speckle noise and rich spatial features of heterogeneous regions lead to blurred boundaries of high-resolution PolSAR image segmentation. A novel segmentation algorithm is proposed in this study in order to address the problem and to obtain accurate and precise segmentation results. This method integrates statistical features into a fractal net evolution algorithm (FNEA) framework, and incorporates polarimetric features into a simple linear iterative clustering (SLIC) superpixel generation algorithm. First, spectral heterogeneity in the traditional FNEA is substituted by the G^0 distribution statistical heterogeneity in order to combine the shape and statistical features of PolSAR data. The statistical heterogeneity between two adjacent image objects is measured using a log likelihood function. Second, a modified SLIC algorithm is utilized to generate compact superpixels as the initial samples for the G^0 statistical model, which substitutes the polarimetric distance of the Pauli RGB composition for the CIELAB color distance. The segmentation results were obtained by weighting the G^0 statistical feature and the shape features, based on the FNEA framework. The validity and applicability of the proposed method was verified with extensive experiments on simulated data and three real-world high-resolution PolSAR images from airborne multi-look ESAR, spaceborne single-look RADARSAT-2, and multi-look TerraSAR-X data sets. The experimental results indicate that the proposed method obtains more accurate and precise segmentation results than the other methods for high-resolution PolSAR images.

Keywords: polarimetric synthetic aperture radar (PolSAR); segmentation; high-resolution; fractal net evolution approach (FNEA); G^0 distribution; simple linear iterative clustering (SLIC); multi-feature; superpixels

1. Introduction

1.1. Background

Synthetic aperture radar (SAR) has been widely accepted as an indispensable method for Earth monitoring due to its all-day/all-weather capacity and penetration capability [1–4]. Fully polarimetric SAR (PolSAR) emits or receives two orthogonal polarized radar waves, and allows the discrimination of different scattering mechanisms. PolSAR image segmentation is able to obtain distinct and self-similar pixel groups that depict homogeneous regions, with virtually no speckle noise [5]. Since accurate segmentation is important for subsequent classification and extraction of surface parameters [6,7], PolSAR image segmentation has been increasingly used for land use and land cover classification [8], land development detection [9], and oil seep detection [10].

With a new generation of advanced SAR sensors, higher-resolution PolSAR images of the Earth's surface have been acquired. In addition to being affected by speckle, the high-resolution PolSAR images show the following characteristics:

(1) Spatial characteristics: The decrease in the resolution cell provides richer spatial details of ground objects [11], such as significant geometric shape features and texture information.
(2) Statistical characteristics: The scattering vectors from the homogeneous regions of medium- or low-resolution PolSAR data can be modeled using Gaussian distributions. The corresponding coherency matrices have a complex Wishart distribution [12]. However, in high-resolution PolSAR data, a significantly reduced number of sub-scatterers within a resolution cell leads to a greater heterogeneity [13], particularly in urban areas, where clusters can no longer be modeled using a Gaussian process.

In short, high-resolution PolSAR images usually contain speckle noise, and many heterogeneous regions, with rich spatial features. The complexity makes segmentation of high-resolution PolSAR images a very challenging task. In this paper, research on segmentation for high-resolution PolSAR images is reported.

1.2. Related Work

Some classic segmentation algorithms for PolSAR images have been proposed, including the Markov random field (MRF) [14,15], statistical region merging (SRM) [5], hierarchical segmentation [16,17], and superpixel segmentation [18–20]. Liu et al. [15] proposed a spatially adaptive segmentation method to keep each segment at an appropriate size and shape based on multiscale wedgelet analyses and Wishart MRF (MW-WMRF). Lang et al. [5] utilized the generalized statistical region merging (GSRM) algorithm, based on the product model, which shows improved robustness and anti-noise performance without any assumption on PolSAR data distributions. Alonso-González et al. [17] used binary partition trees (BPT) to develop a novel region-based and multi-scale PolSAR data representation for speckle noise filtering and segmentation, on the basis of the Gaussian hypothesis. Liu et al. [18] oversegmented PolSAR images into many local and coherent regions using the normalized cuts algorithm to improve the classification accuracy by adding inherent statistical characteristics to the contour information. Ersahin et al. [21] segmented PolSAR data with contour information and spatial proximity, based on spectral graph partitioning for object oriented classification. In summary, an increasing number of methods are combining spatial features and statistical properties of PolSAR images to obtain useful segmentation results.

Segmentation using spatial features is a common method to model a labeling process as MRF [14,15], which includes the spatial relationships between pixels. In contrast, the fractal net evolution algorithm (FNEA) makes good use of the geometric shape features and spectral information of targets [22]. This was successfully used in high spatial resolution, optical image segmentation [23,24]. FNEA is a bottom-up region merging technique with a fractal iterative heuristic optimization procedure. It starts with a single pixel and a pairwise comparison of its neighbors, with the aim of minimizing the resulting merged heterogeneity. The heterogeneity is determined using geometric shapes and the standard deviation of spectral properties as its basis. By replacing spectral information with the parameters of $H/\alpha/A$ decomposition [25], Freeman decomposition [26], or Pauli decomposition [8,9], researchers have introduced FNEA into PolSAR image segmentation for object-oriented classification. Benz and Pottier [25] first used FNEA to segment filtered PolSAR images by employing shape features, $H/\alpha/A$, and the total scattered power span. Qi et al. [8] implemented FNEA segmentation on a Pauli RGB composition image of filtered PolSAR data, and successfully applied it to land-use and land-cover classification. However, the segmentation results of these methods were easily influenced by speckle noise, and the results were too fragmented to represent unbroken land parcels [8] as the statistical characteristics of PolSAR data were not used.

Segmentation, using statistical features uses the classical complex Wishart distribution, which has been widely used with PolSAR data [15,17,19–21,27]. As the Gaussian or Wishart model does not agree well with heterogeneous scenes of high-resolution images, heterogeneity is usually modeled as the product of the square root of a textured random variable and an independent, zero-mean, complex circular Gaussian speckle random vector [28]. If the random texture variable is Gamma, inverse Gamma, or Fisher distributed, the target coherency matrices follow multivariate K distribution [29], G^0 distribution [30], or Kummer$U$ distribution [31], respectively. Beaulieu and Touzi [16] presented a hierarchical segmentation method using the K-distribution model, and verified its effectiveness for textured forested areas. Bombrun et al. [32] proposed a hierarchical maximum likelihood segmentation for high-resolution PolSAR images using Kummer$U$ distribution heterogeneous clutter models, which provided a better performance compared to the classical Gaussian criterion. However, it is essential to robustly estimate the parameters of these multiplicative models with enough samples. Generally, the segmentation results have obvious dentate boundaries, as the initial samples are usually collected from image blocks within square windows [32]. Moreover, the accuracy of parameter estimation decreases due to the differences between the square blocks and the actual boundaries of targets in high-resolution PolSAR data.

1.3. The Proposed Approach

Superpixel algorithms group pixels into meaningful atomic regions, which are roughly homogeneous in size and shape, and can be used to replace the rigid structure of the pixel grid or the square block. Since the utilization of superpixels helps to overcome the influence of speckle noise and to preserve statistical characteristics [19,33,34], it has gradually become an important preprocessing step for segmentation or classification. Recently, Achanta et al. proposed a new superpixel algorithm, simple linear iterative clustering (SLIC), which adapted a k-means clustering approach to efficiently generate superpixels [35]. Considering the simplicity, fast processing and excellent boundary adherence of SLIC, Qin et al. [19] introduced the superpixel algorithm into PolSAR image segmentation by combining it with the Wishart hypothesis test distance and the spatial distance.

We propose a novel segmentation algorithm for high-resolution PolSAR data by combining spatial, statistical, and polarimetric features; this algorithm integrates statistical features into a FNEA framework, and the polarimetric features with SLIC, in order to generate pre-segments. Given that the G^0 distribution has been shown to be flexible, computationally inexpensive, and capable of modeling varying degrees of texture [30,36], we substitute the G^0 distribution of statistical heterogeneity for the spectral heterogeneity in the traditional FNEA. Furthermore, we also utilize a modified SLIC algorithm to generate compact, approximately homogeneous superpixels as initial samples for the statistical model, which utilizes the polarimetric distance of Pauli RGB composition instead of the CIELAB color distance.

The remainder of this paper is organized as follows; Section 2 describes the proposed segmentation method for high-resolution PolSAR data. The employed PolSAR images and the experimental and evaluation results are reported in Section 3. The discussion of the results is presented in Section 4. Conclusions are given in Section 5.

2. Methodology

The proposed approach is based on the FNEA framework, and can be divided into three main parts: (1) a statistical heterogeneity measure using the G^0 distribution model for high-resolution PolSAR data; (2) initial sample generation for the statistical model using the SLIC algorithm with polarimetric features; and (3) segmentation with the G^0 statistical and shape features, based on the FNEA framework. The details of these are explained in subsequent subsections.

2.1. FNEA

FNEA merges two adjacent objects with a fractal iterative heuristic optimization procedure, which starts with single-pixel objects, and satisfies the condition of minimizing the resulting merged heterogeneity [22,23].

The heterogeneity between two adjacent image objects is defined by integrating the change of spectral heterogeneity Δh_{spc} with the change of shape heterogeneity Δh_{shp} in a virtual merge [22], as follows:

$$\Delta h = w_{shp} \Delta h_{shp} + \left(1 - w_{shp}\right) \Delta h_{spc} \tag{1}$$

where w_{shp} is the weight of the shape feature, and $w_{shp} \epsilon [0, 1]$. Adjacent objects i and j are merged when the smallest growth in heterogeneity occurs.

For multispectral remote sensing images, Δh_{spc} can be described by adding weight w_c to image channels c [22],

$$\Delta h_{spc} = \sum_c w_c [(n_i + n_j)\delta_c^{i \cup j} - (n_i \delta_c^i + n_j \delta_c^j)] \tag{2}$$

where n denotes the objects size, which is the number of pixels in an image object; and δ_c denotes the spectral heterogeneity of the image object, which is the standard deviation within the objects of channel c.

The change of shape heterogeneity Δh_{shp} can be expressed as

$$\Delta h_{shp} = (n_i + n_j)h_{shp}^{i \cup j} - (n_i h_{shp}^i + n_j h_{shp}^j) \tag{3}$$

where h_{shp} denotes the shape heterogeneity of the image object, which is described with regard to smoothness and compactness. It is described by

$$h_{shp} = w_{smth}\frac{p}{b} + (1 - w_{smth})\frac{p}{\sqrt{n}} \tag{4}$$

where w_{smth} is the weight of smoothness, and $w_{smth} \epsilon [0, 1]$. The smoothness heterogeneity is defined as the ratio of factual edge length p and border length b; b is given by the bounding box of an image object parallel to the raster while the compactness heterogeneity is defined as the ratio of factual edge length p and the square root of n [22].

At each iteration in FNEA, an image object is merged into its adjacent image object with the minimum heterogeneity, and when the heterogeneity is less than threshold t (i.e., scale parameter). If all increases exceed the scale parameter, no further merging occurs and the segmentation stops. The larger scale parameter t is, the more image objects can be merged and the larger the image objects grow.

2.2. Statistical Heterogeneity Measure by the G^0 Model

PolSAR data are mainly provided in two forms: The single-look scattering matrix and the multi-look polarimetric coherency (or covariance) matrix. Each pixel of PolSAR data can be described by a 2×2 complex scattering matrix S [2]:

$$S = \begin{bmatrix} S_{HH} & S_{HV} \\ S_{VH} & S_{VV} \end{bmatrix} \tag{5}$$

where H and V represent the horizontal and vertical polarization directions, respectively.

In a reciprocal medium, $S_{HV} = S_{VH}$ and S can be transformed into a three-dimensional single-look scattering vector using the complex Pauli spin matrix basis set [2,27]:

$$k = \frac{1}{\sqrt{2}} \begin{bmatrix} S_{HH} + S_{VV} & S_{HH} - S_{VV} & 2S_{HV} \end{bmatrix}^T \tag{6}$$

where $[\cdot]^T$ denotes the matrix transpose. In this paper, the dimension of the scattering vector is denoted by d ($d = 3$ for the reciprocal case).

Usually, L-look coherency matrix T is computed to suppress speckle using the average of k of the surrounding pixels, as follows [2,27]:

$$T = \frac{1}{L} \sum_{i=1}^{L} k_i k_i^{H} \tag{7}$$

where k_i^{H} is the conjugate transpose of scattering vector k_i, and L is the number of looks.

2.2.1. G^0 Model for High-Resolution PolSAR Data

High-resolution PolSAR images are greatly affected by heterogeneity due to the significantly reduced number of scatterers within a resolution cell [11]. The heterogeneity is usually modeled with the multiplicative models [28].

For single-look complex PolSAR data, the multiplicative model is given by

$$k = \sqrt{\tau} x \tag{8}$$

where τ is a texture random variable, and x is a d-dimensional speckle vector, which follows an independent zero-mean multivariate complex Gaussian distribution.

Assume that τ in Equation (8) obeys the inverse Gamma distribution, in which the probability density function (PDF) is given by

$$p_\tau(\tau) = \frac{\tau^{\alpha-1}}{(-\alpha-1)\Gamma(-\alpha)} exp\left(\frac{\alpha+1}{\alpha}\right), \ -\alpha, \ \tau > 0 \tag{9}$$

where α is the shape parameter and $\Gamma(\cdot)$ is the standard Euler Gamma function.

In this case, the target scattering vector k follows the G^0 distribution, which is characterized by the following PDF [37]:

$$p_k(k) = \frac{\Gamma(d-\alpha)}{\pi^d |\Sigma| \Gamma(-\alpha)\Gamma(-\alpha-1)^\alpha} \left(k^H \Sigma^{-1} k - \alpha - 1\right)^{\alpha-d} \tag{10}$$

where $\Sigma = E[kk^H]$ is the covariance matrix of k, $E[\cdot]$ denotes the mathematical expectation, and $|\cdot|$ represents the determinant, while $(\cdot)^{-1}$ denotes the inverse.

For multi-look PolSAR data, coherency matrix T is modeled as the product of random variable τ and independent random matrix X:

$$T = \tau X \tag{11}$$

where X obeys a Wishart distribution.

For an inverse gamma distributed texture, target coherency matrix T follows the G^0 distribution [30], which is characterized by the PDF, as follows:

$$p_T(T) = \frac{L^{Ld} |T|^{L-d} \Gamma(Ld-\alpha)}{\Gamma_d(L) |\Sigma|^L \Gamma(-\alpha)\Gamma(-\alpha-1)^\alpha} \left(Ltr\left(\Sigma^{-1}T\right) - \alpha - 1\right)^{\alpha-Ld}, \ \alpha < -1 \tag{12}$$

where $\Sigma = E[T]$, $tr(\cdot)$ is the trace operator and L is number of looks. $\Gamma_d(L)$ represents the multivariate gamma function, defined as

$$\Gamma_d(L) = \pi^{d(d-1)/2} \prod_{i=0}^{d-1} \Gamma(L-i). \tag{13}$$

The G^0 model is particularly suitable for high-resolution PolSAR image description, which is a flexible model for different texture classes of SAR images [30,36].

2.2.2. Statistical Heterogeneity Measure

Since the log likelihood function can be used to measure the similarity between two segments in a hierarchical segmentation of PolSAR Images [16,32], it is utilized to measure the statistical heterogeneity between two image objects. At each iteration, merging two image objects using statistical features yields a decrease in the log likelihood function. Thus, the two adjacent image objects, i and j, should be merged to produce the smallest decrease of the log likelihood function. The change of statistical heterogeneity Δh_{stt} can be expressed as

$$\Delta h_{stt} = h_{stt}^i + h_{stt}^j - h_{stt}^{i \cup j} \tag{14}$$

where h_{stt} denotes the statistical heterogeneity of image object O, which is the maximum log likelihood value of the image object.

For single-look complex PolSAR data, the statistical heterogeneity of image object O is given by

$$h_{stt}^O = \sum_{k \in O} \ln[p_k(k)] \tag{15}$$

where pixels in image object O are considered independent realizations, with respect to the assumed PDF in Equation (10).

According to Equations (10), (14), and (15), the statistical heterogeneity of image object O can be simplified to

$$h_{stt}^O \cong -n \ln|\hat{\Sigma}| - n\hat{\alpha} \ln(-\hat{\alpha} - 1) - n \ln \left[\frac{\Gamma(-\hat{\alpha})}{\Gamma(d - \hat{\alpha})} \right] - (d - \hat{\alpha}) \sum_{k \in O} \ln \left(k^H \hat{\Sigma}^{-1} - \hat{\alpha} - 1 \right) \tag{16}$$

where n is the number of pixels in image object O, $\hat{\alpha}$, and $\hat{\Sigma}$ are the best likelihood estimates of α, and Σ for this image object.

For multi-look PolSAR data, with respect to the assumed PDF in Equation (12), the statistical heterogeneity of image object O is

$$h_{stt}^O = \sum_{i \in O} \ln[p_T(T_i)]. \tag{17}$$

Similarly, the statistical heterogeneity of image object O can be simplified by Equations (12), (14), and (17) to

$$h_{stt}^O \cong -nL \ln|\hat{\Sigma}| - n\hat{\alpha} \ln(-\hat{\alpha} - 1) - n \ln \left[\frac{\Gamma(-\hat{\alpha})}{\Gamma(Ld - \hat{\alpha})} \right] - (Ld - \hat{\alpha}) \sum_{i \in O} \ln \left(Ltr(\hat{\Sigma}^{-1}T_i) - \hat{\alpha} - 1 \right). \tag{18}$$

It can be concluded that the statistical heterogeneity from Equation (18) is equivalent to Equation (16) when $L = 1$, which means that the change in statistical heterogeneity can be computed by Equations (14) and (18), for both single-look and multi-look data.

2.2.3. Parameter Estimation

The number of looks, L, is generally an integer provided by the SAR sensor. Statistical heterogeneity measured by the G^0 distribution model is parameterized by scale matrix Σ and shape parameter α. The correct and reasonable merging objects are based on the proper estimation of the involved parameters.

Scale matrix Σ is the mathematical expectation of the coherency matrix, which can be calculated using the classical sample covariance matrix estimator [38], as follows:

$$\hat{\Sigma} = <T>_N \tag{19}$$

where $\langle \cdot \rangle$ denotes sample averaging, and N is the number of samples.

Shape parameter α can be estimated using the method of matrix log-cumulants, which has been proposed in the literature [39]. However, given that this method is more complex, needing a large amount of computation, and that this calculation is performed for each image object in each iteration of the segmentation, we estimate shape parameter α using the method of Doulgeris, provided in the literature [40]. This estimator is given by

$$\hat{\alpha} = \frac{2L\mathit{Var}\{M\} + d(Ld - 1)}{d - L\mathit{Var}\{M\}} \tag{20}$$

where $M = tr(\hat{\Sigma}^{-1}T)$, and $Var\{\cdot\}$ is the statistical variance.

2.3. Initial Samples Generation for Statistical Model

Accurate and robust parameter estimation of the statistical models requires sufficient samples. Since FNEA starts region merging from a single pixel, the number of samples (i.e., pixels in each image object) is too small to accurately estimate parameters at the beginning of the iterations. Generally, the image blocks within square windows are used as samples. However, there are obvious differences between the square blocks and the actual boundaries of targets in high-resolution PolSAR data. Therefore, in this work, the SLIC superpixel algorithm with polarimetric and spatial features was used to produce suitable initial samples before the utilization of statistical heterogeneity.

The SLIC algorithm incorporates k-means clustering to efficiently produce superpixels for images in the CIELAB color space [35]. It includes two main steps: Initializing m cluster centers and assigning each pixel to the nearest cluster center in a local search region. This algorithm has a speed advantage over traditional k-means clustering by limiting the size of the search area to reduce the number of distance calculations. In general, a key parameter, m, is the desired size of the superpixels, with approximately equal pixels [35].

Pauli RGB images can be obtained by using them as blue, red, and green channels, respectively. This has become the standard display mode of PolSAR data [41]. Hence, the polarimetric feature space of Pauli RGB composition was used to replace the CIELAB color space for the polarimetric SAR images.

The method of initial samples generation for statistical model using SLIC algorithm includes the following three main steps:

(1) Initializing cluster centers. The algorithm begins by initializing m cluster centers $C_m = [R_m\ G_m\ B_m\ x_m\ y_m]^T$ by sampling pixels in the Pauli RGB images at regular grid steps g, and the grid interval is $g = \sqrt{N/m}$, where N is the number of pixels of the Pauli RGB image [35]. Then the centers are adjusted to seed locations where the lowest gradient meets in a 3×3 neighborhood [42]. This procedure is important as it avoids centering a superpixel on an edge, and reduces the probability of seeding a superpixel with a noisy pixel.

(2) Associating each pixel with the nearest cluster center. The distance between the superpixel center, C_m, and each pixel, i, is calculated in region $2g \times 2g$ around the C_m [42]. Then, all the pixels can be assigned to the nearest cluster center, and the superpixels with the approximate size of $g \times g$ are finally obtained [35]. The distance measure D combines the polarimetric distance of the Pauli RGB composition and the spatial distance and is described by

$$d_p = \sqrt{(R_i - R_m)^2 + (G_i - G_m)^2 + (B_i - B_m)^2}$$
$$d_s = \sqrt{(x_i - x_m)^2 + (y_i - y_m)^2} \tag{21}$$
$$D = \sqrt{\left(\frac{d_p}{\max(d_p)}\right)^2 + \left(\frac{d_s}{g}\right)^2}$$

where polarimetric distance d_p and spatial distance d_s are normalized by their respective maximum distances within a cluster, max (d_p) and g The equivalent weight between d_p and d_s is utilized to calculate final distance D [35,43]. Once each pixel has been associated to the nearest cluster center, the cluster centers adjust to be the mean $[R\ G\ B\ x\ y]^T$ vector of all the pixels belonging to the cluster. This step can be repeated for 10 iterations, which is enough for most images [35].

(3) In order to provide sufficient numbers of initial samples for the statistical model, the superpixels, of which sizes are less than the specific threshold g^2, are merged with the nearest neighbors according to the distance measure. Then, the final superpixels of the PolSAR images are obtained.

2.4. Segmentation with Statistical and Shape Features

Traditional FNEA starts with single pixel objects and segments an image by integrating spectral feature and shape features. It is difficult to use the statistical feature of PolSAR data based on pixels. Therefore, based on the initial objects generated by SLIC, we substitute the G^0 distribution statistical heterogeneity for spectral heterogeneity in the original FNEA, and then combine the G^0 statistical and shape features to segment PolSAR data.

Similar to Equation (1), the change in total heterogeneity including the G^0 statistical and shape features can be obtained by weighting as follows:

$$\Delta h' = w_{shp}\Delta h_{shp} + \left(1 - w_{shp}\right)\Delta h_{stt} \tag{22}$$

where shape heterogeneity is calculated using Equation (4) by averaging the smoothness and compactness.

Using Equations (3), (4), (14), (18), and (22), the proposed segmentation method starts from superpixels with a fractal iterative procedure, and satisfies the condition of minimizing the resulting merged heterogeneity.

The proposed method consists of two main procedures: Generating superpixels as initial objects using SLIC and FNEA segmentation based on superpixels. The former utilizes Pauli RGB and spatial information for segmentation, while the latter employs G^0 statistics information and shape features. The value of g^2 is related to the start time using statistics information. A greater value of g^2 delays the use of statistical information.

Each iteration of superpixel-based FNEA needs to traverse all the objects, and then the object information is updated after the iteration. In order to ensure the accuracy of the boundary, each object is merged once, at most, in each iteration, similar to the region growing algorithm.

In summary, the details of proposed segmentation method are presented in Algorithm 1.

Algorithm 1. FNEA-based Multi-Feature PolSAR Segmentation.

1: **INPUT:** PolSAR data, samples number g^2, shape weight w_{shp}, scale parameter t.
2: **OUTPUT:** segmentation result.
3: Generate superpixels of PolSAR data using new SLIC algorithm by Equation (21).
4: Produce initial image objects with the superpixels.
5: **do**
6:　Get the number of image objects N_O, $a = N_O$.
7:　**for** each image object i **do**
8:　**for** each adjacent object j of image object i **do**
9:　　Estimate scale matrix Σ, shape parameter α using pixels in object by Equations (19) and (20).
10:　　Compute the change of heterogeneity $\Delta h'$ by Equations (3), (4), (14), (18), and (22).
11:　　**end for**
12:　　Compare all the $\Delta h'$ to obtain the minimum and set it as Δh_{min}
13:　　**if** $\Delta h_{min} \leq t$ **then**
14:　　Merge image objects i and j, create a new image object $i \cup j$, and delete objects i, j.
15:　　$a = a - 1$.
16:　**end if**
17:　**end for**
18: **while** $(a < N_O)$
19: Produce segmentation image and the vector of objects boundaries.

---

3. Experiment and Results

To verify the proposed method, one simulated data set, and three different real-world PolSAR data sets are used in the rest of this section. Moreover, different segmentation methods were adopted for comparison. This section is divided into two subsections to describe the experimental data sets and report on the experimental details of the simulated PolSAR data, RADARSAT-2 image, ESAR image, and TerraSAR-X image.

3.1. Description of the Experimental Data Sets

The first data set is a simulated single-look PolSAR image, 400 × 400 pixels in size, and contains eight different classes: Building areas, forest, bush land, grass land, two different types of crops, road, and a water body. To better reflect the ground reality, the Wishart distribution was adopted to generate the water body, while the G^0 distribution data were adopted to generate the other classes. The initial scattering vectors and distribution parameters were estimated from a real data set. The Pauli RGB image is shown in Figure 1a, and the corresponding reference map is shown in Figure 1b. Figure 2 depicts the theoretical PDF of each class.

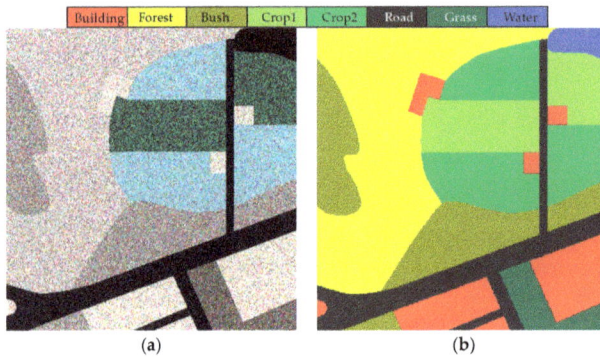

Figure 1. Simulated single-look polarimetric synthetic aperture radar (PolSAR) data as the first data set: (**a**) Pauli RGB image and (**b**) the object spatial distribution reference map of the simulated image.

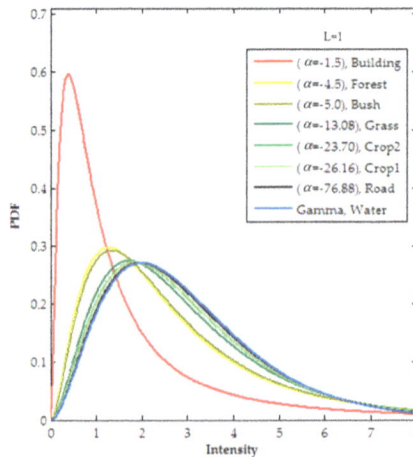

Figure 2. The theoretical probability density function (PDF) of each simulated class.

The second data set is a section of the single-look C-band RADARSAT-2 PolSAR image of Northern Flevoland, Netherlands, and has a spatial resolution of 4.7 m × 4.8 m (range × azimuth). The experimental image is 1400 × 1400 pixels in size and is shown in Figure 3a. The major land cover classes include homogeneous areas (such as water bodies and farmlands), and heterogeneous areas (such as forest and urban areas). A manual interpretation using nine categories was used as the ground truth map, which is shown in Figure 3b.

The third data set is a section of the two-look processed L-band ESAR PolSAR image of Oberpfaffenhofen, Germany, and has a spatial resolution of 1.5 m × 1.8 m (range × azimuth). The experimental image is 800 × 800 pixels in size and is shown in Figure 4a. The major land cover classes include homogeneous areas (such as roads, grasslands, and farmlands) and heterogeneous areas (such as forests and urban areas). A manual interpretation using 16 categories was used as the ground truth map, which is shown in Figure 4b.

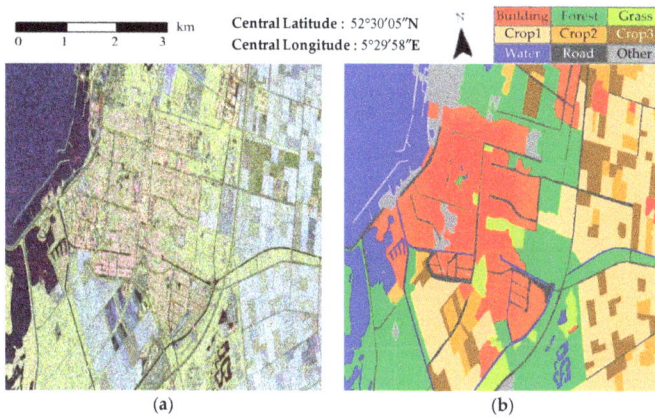

Figure 3. C-band, single-look RADARSAT-2 PolSAR image in Flevoland as the second data set: (**a**) Pauli RGB image and (**b**) the ground truth map of (**a**).

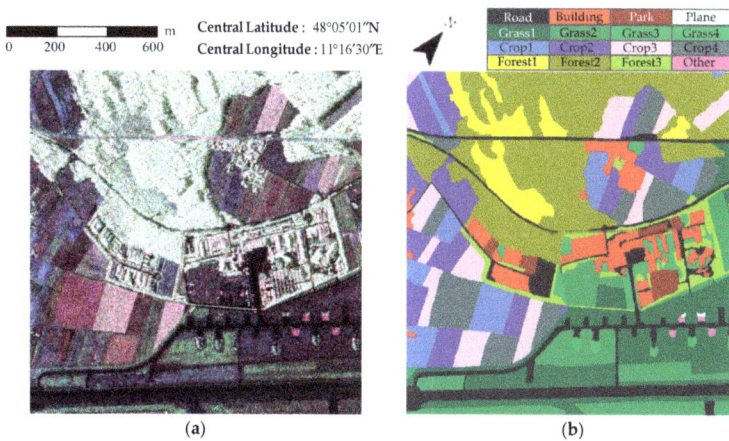

Figure 4. L-band, two-look ESAR PolSAR image in Oberpfaffenhofen as the third data set: (**a**) Pauli RGB image and (**b**) the ground truth map of (**a**).

The fourth data set is obtained from a subset of an X-band TerraSAR-X PolSAR image of Deggendorf, German, which is six-look, with a spatial resolution of about 5.0 m × 4.8 m (range × azimuth). The experimental image, with a size of 541 × 541 pixels, is shown in Figure 5a. The major land cover classes include homogeneous areas (such as river and farmlands) and heterogeneous areas (such as forests and building areas). The corresponding optical image is shown in Figure 5b.

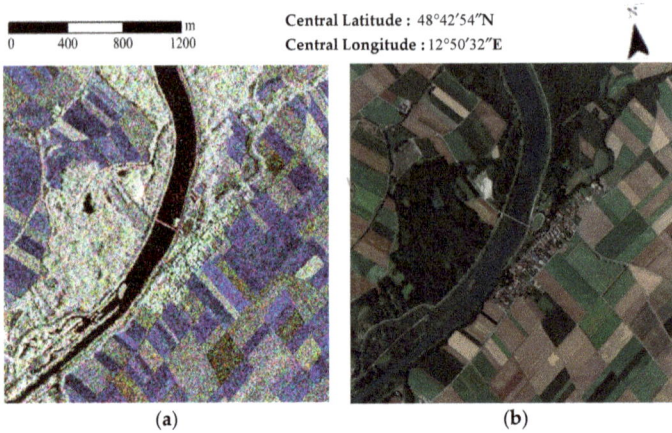

Figure 5. X-band, six-look TerraSAR-X PolSAR image of Deggendorf as the fourth data set: (**a**) Pauli RGB image and (**b**) the reference image from Google Earth.

3.2. Evaluation and Comparison

To verify the improvement in segmentation accuracy by integrating the G^0 statistical features into the FNEA framework and pre-segmenting using SLIC, segmentation experiments were performed using different methods, namely: (a) FNEA segmentation based on Freeman decomposition without SAR statistical features (FFD); (b) FNEA segmentation with Pauli RGB image without SAR statistical features (FPD); (c) improved FNEA with G^0 statistical features start from square blocks (IFGB); (d) improved FNEA using Wishart statistical features with pre-segmenting by SLIC (IFWS); (e) improved FNEA using K statistical features with pre-segmenting by SLIC (IFKS); (f) improved FNEA using the G^0 statistical features with pre-segmenting by SLIC (IFGS); and (g) segmentation using the G^0 statistical features without shape features based on SLIC pre-segmenting (IFGS-S).

In addition to qualitative visual assessment, the quality of segmentation results requires an evaluation criterion. Various accuracy metrics describe the similar aspects of the correspondence between reference objects and segments [44,45], such as the difference in area between reference objects and the segments they intersect as well as the positional difference between reference objects and segments. In this paper, area-based measures were used to evaluate the accuracy of the segmentation results. Let R denote the reference segments that consists of regions representing ground objects, and S denote segmentation result from the processed SAR image. Two area-based metrics are defined as follows:

$$\rho_d = \frac{R \cap S}{R}, \ \rho_q = \frac{R \cap S}{R \cup S} \tag{23}$$

where ρ_d is the area rate of correct segmentation, i.e., detection rate, and ρ_q is the degree of overlap between R and S (i.e., quality rate), which takes the false positive rate into consideration. These metrics are continuous in [0, 1], and the higher values of these metrics mean better segmentation results. Given that serious over-segmentation may also result in a high segmentation accuracy, the total number of objects (TN) was introduced as an auxiliary evaluation criterion. In general, the number of objects should be as small as possible, in the case of satisfying accuracy requirements.

The segmentation results and accuracy measures of the simulated images using the different methods are shown in Figure 6 and Table 1, respectively. In Figure 6, the blue lines represent the boundaries of the segmentation results, and the background is the Pauli RGB image. Figure 6d shows the local details in the lower left part of the superpixel map, which was produced using SLIC with 4×4 pixels, in the desired size of the superpixel.

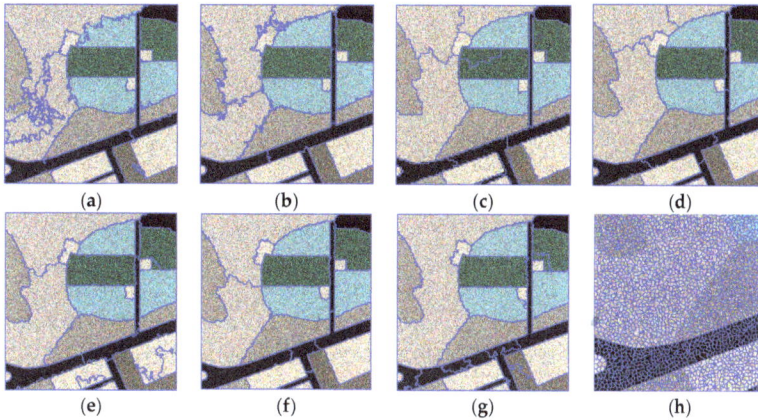

Figure 6. Segmentation results of the simulated image using different methods: (**a**) FNEA segmentation based on Freeman decomposition without SAR statistical features (FFD); (**b**) FNEA segmentation with Pauli RGB image without SAR statistical features (FPD); (**c**) improved FNEA with G^0 statistical features start from square blocks (IFGB); (**d**) improved FNEA using Wishart statistical features with pre-segmenting by simple linear iterative clustering (SLIC) (IFWS); (**e**) improved FNEA using K statistical features with pre-segmenting by SLIC (IFKS); (**f**) improved FNEA using the G^0 statistical features with pre-segmenting by SLIC (IFGS); (**g**) segmentation using the G^0 statistical features without shape features based on SLIC pre-segmenting (IFGS-S); and (**h**) SLIC (local enlarged drawing of lower-left part of the superpixel map).

Table 1. Segmentation accuracy measures of the simulated image.

Method	ρ_d (%)	ρ_q (%)	TN
FFD	94.33	89.27	26
FPD	98.22	96.50	26
IFGB	97.12	94.40	26
IFWS	98.47	96.98	26
IFKS	98.11	96.29	27
IFGS	98.77	97.57	25
IFGS-S	98.70	97.44	29

As shown in Table 1, the proposed IFGS method obtained the best detection and quality rates with the least number of generated objects, while the FFD method had the worst segmentation results. For the FFD method, inaccurate segmentation boundaries appeared, especially for classes with similar polarimetric features (Figure 6a). The results of the Pauli-based FPD method is greatly affected by speckle noise, which causes blurred segmentation boundaries between classes of crops, bush land, forest, and building areas (Figure 6b). As shown in Figure 6c–f, the class boundaries became more accurate when statistical information was utilized. However, the Wishart-based IFWS method is not applicable to heterogeneous areas like forests and urban areas. The K-based IFKS method had inaccurate segmentation results in extreme heterogeneous building areas, as shown in the bottom part of Figure 6e. In contrast, the G^0-based IFGS method obtained the best segmentation results for the areas with different degrees of heterogeneity, with a 98.77% detection rate and a 97.57% quality rate. The contrast between Figure 6c,f demonstrates that dentate boundaries appear when square blocks are taken as the initial samples for the statistical model. The boundaries of straight roads or other regular areas deviated. Compared to the IFGS method, the IFGS-S method obtained the wrong segmentation boundaries in partial regular building areas, due to the absence of utilizing the shape features as shown in Figure 6g. In conclusion, the proposed superpixel-based IFGS method, utilizing the G^0 distribution and shape features, obtained accurate and precise segmentation boundaries for the different areas.

The segmentation results and superpixel map of single-look RADARSAT-2 and two-look ESAR PolSAR images, using these different methods are shown in Figures 7 and 8, respectively. In Figures 7 and 8, the blue lines represent the boundaries of the segmentation results, and the background is a Pauli RGB image. The superpixel maps of the RADARSAT-2 and ESAR images were produced using SLIC, with 4 × 4 pixels, in the desired size of the superpixel. The segmentation accuracies are presented in Tables 2 and 3, respectively.

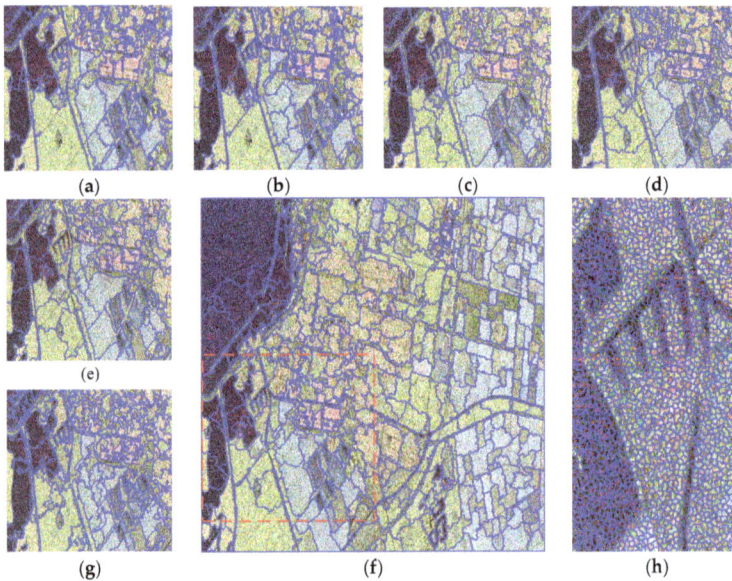

Figure 7. Segmentation results of the single-look RADARSAT-2 PolSAR image using different methods: (a) FFD; (b) FPD; (c) IFGB; (d) IFWS; (e) IFKS; (f) IFGS; (g) IFGS-S; and (h) SLIC (local enlarged drawing of central left part of the superpixel map). (a–e,g) show screenshots of the results in the red dashed rectangle of (f).

Table 2. Segmentation accuracy measures of the single-look RADARSAT-2 image.

Method	ρ_d (%)	ρ_q (%)	TN
FFD	89.30	78.26	558
FPD	88.49	77.07	553
IFGB	90.35	80.10	546
IFWS	89.14	78.17	564
IFKS	89.17	78.03	543
IFGS	91.33	81.53	541
IFGS-S	89.53	78.68	590

For the single-look RADARSAT-2 image, the proposed IFGS method also obtained optimal segmentation results. The contrast between Figure 7a,b and Figure 7c–f demonstrated that the utilization of statistical information helped to suppress the influence of the speckle noise, and generated accurate class boundaries. The segmentation results in Figure 7c,f certify that the superpixel-based method could avoid serrated boundaries. Moreover, the detection rate and quality rate of the IFGS method increased by approximately 1% and 1.4%, compared to the IFGB method, according to Table 2, which validated the effectiveness of the superpixels. As shown in Figure 7d–f and Table 2, the Wishart-based IFWS method obtained seriously fragmented results in areas of high heterogeneity, like the city areas located in the upper right of Figure 7d, and the K-based IFKS method had difficulty segmenting the small rivers accurately. Compared to the IFGS method, the IFGS-S method achieved blurred segmentation boundaries in urban areas, forest, and water areas due to the absence the shape feature. In summary, employing shape features and statistics information comprehensively contributed to generating better segmentation results. According to Table 2, the proposed superpixel-based IFGS method obtained the highest accuracy with the least number of segmentation objects, and the detection rate and quality rate were 91.33% and 81.53%, respectively. The above-mentioned results indicate that the proposed method is applicable to single-look high-resolution PolSAR images.

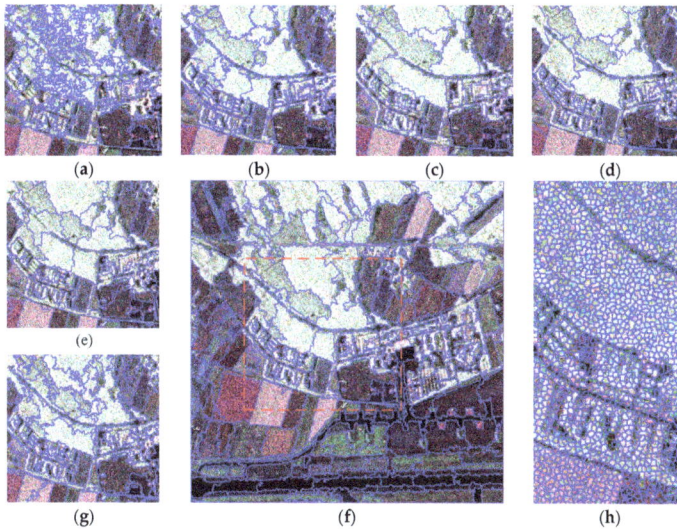

Figure 8. Segmentation results of the two-look ESAR PolSAR image using different methods: (**a**) FFD; (**b**) FPD; (**c**) IFGB; (**d**) IFWS; (**e**) IFKS; (**f**) IFGS; (**g**) IFGS-S; and (**h**) SLIC (local enlarged drawing of central left part of the superpixel map). (**a–e,g**) show screenshots of the results in the red dashed rectangle of (**f**).

For the multi-look ESAR image, the proposed IFGS method still obtained optimal segmentation results. As shown in Figure 8a,b, the segmentation results of the FFD and FPD methods, which utilized polarimetric decomposition features, were greatly affected by the speckle noise, resulting in blurred segmentation boundaries. In contrast, the statistics-based segmentation method suppressed the influence of speckle noise and achieved more accurate segmentation boundaries between the different classes, as shown in Figure 8c–f. However, when the square blocks were taken as initial samples for the statistical models, serrated and inaccurate segmentation boundaries occurred, as shown in Figure 8c. The contrast between Figure 8d–f certifies that the G^0-based IFGS method obtained better segmentation results (such as for roads in the forest) than the Wishart-based IFWS or the K-based IFKS method. For the IFGS-S method, the blurred segmentation boundaries occurred in areas with little change in statistics and polarimetric information, especially for the inner areas of forest, urban, and farmland in Figure 8g. According to Table 3, the proposed IFGS method obtained the highest accuracy with the least number of segmentation objects, similar to that of the single-look RADARSAT-2 image. Specifically, the detection and quality rates of the IFGS method were 88.46% and 72.07%, respectively. This demonstrates the effectiveness of the proposed method for multi-look high-resolution PolSAR images.

In order to further validate the effectiveness of the proposed method, the segmentation results of the third set of ESAR data, generated using the GSRM method [5] and the MW-WMRF method [15] were used as comparison. Figure 9a,b demonstrates the representation maps of the GSRM method and the MW-WMRF method, respectively. The Pauli RGB of each pixel was replaced by the average Pauli RGB of the segment that the pixel belonged to. Similarly, Figure 9c,d gives the representation maps of the proposed method at different scales. Specifically, a number of isolated small segments occurred in heterogeneous areas like forest (area A of Figure 9a) and urban areas (area B of Figure 9a) for the GSRM method, which decreased the visibility and accuracy of the representation map. As for the MW-WMRF method, the segmentation results of urban areas were broken due to the utilization of the Wishart distribution, which is shown in area B of Figure 9b. The boundaries between the different types of farmlands for these two methods were inaccurate (area C of Figure 9a,b). The boundaries for forests of different height were not accurate enough (area A of Figure 9a,b). In contrast, accurate boundaries of farmlands and forest were obtained, with different segmentation scales, for the proposed method (area A and area C of Figure 9c,d), and the urban areas were entirely segmented (area B of Figure 9c,d). Furthermore, the detection rates of the GSRM, MW-WMRF, and IFGS ($t = 13$) method were 87.86%, 90.20%, and 90.60%, respectively. In summary, the proposed method obtained better segmentation results compared to the GSRM method and the MW-WMRF method.

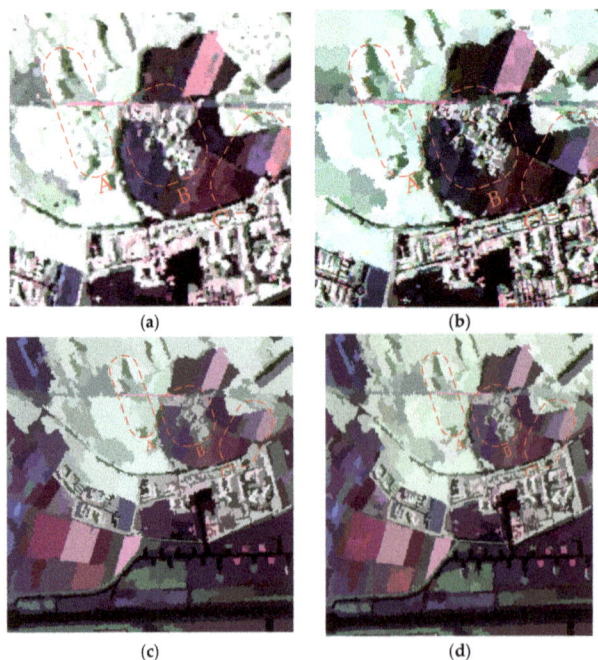

(a)

(b)

(c)

(d)

Figure 9. Representation maps (Pauli RGB images) of segmentation results for the ESAR data using different methods: (**a**) GSRM; (**b**) MW-WMRF; (**c**) IFGS, $t = 17$, $w_{shp} = 0.05$, $g^2 = 16$; and (**d**) IFGS, $t = 13$, $w_{shp} = 0.05$, $g^2 = 16$.

Table 3. Segmentation accuracy measures of the two-look ESAR image.

Method	ρ_d (%)	ρ_q (%)	TN
FFD	83.31	65.13	348
FPD	86.65	69.57	335
IFGB	87.90	71.29	335
IFWS	87.94	71.34	342
IFKS	87.57	70.82	334
IFGS	88.46	72.07	334
IFGS-S	85.38	67.85	343

For the X-band multi-look TerraSAR-X image, the proposed IFGS method also performed well. As shown in Figure 10, the river, bridge, lakes, and building areas in the southeastern area of the river were all accurately segmented. Accurate and delicate boundaries of different types of farmlands and of the inner area of the forest were achieved. Compared to Figure 10a, the segmentation results in Figure 10b were more precise and obtained a higher quantity of objects when the segmentation scale decreased to 12. This experiment further demonstrated the applicability of the proposed method for high-resolution PolSAR images.

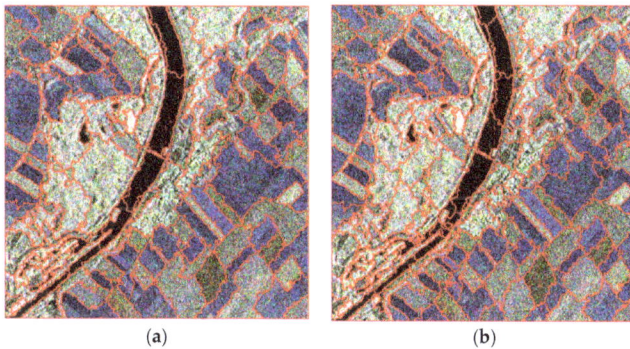

(a) (b)

Figure 10. Segmentation results of the six-look TerraSAR-X PolSAR image using the proposed method: (a) IFGS, $t = 16$, $w_{shp} = 0.05$, $g^2 = 16$; (b) IFGS, $t = 12$, $w_{shp} = 0.05$, $g^2 = 16$.

4. Discussion

4.1. Main Features of the Proposed Method

A superpixel-based FNEA segmentation method for high-resolution PolSAR data, using Pauli polarimetric, spatial, and G^0 distribution statistical features, is proposed. The proposed method was successfully applied to simulated and real-world PolSAR data sets.

The main feature of the method is the comprehensive utilization of G^0 distribution and shape information, based on FNEA. In related studies, traditional FNEA used polarimetric and geometric shape features for PolSAR image segmentation [8,9,25,26], which is easily influenced by speckle noise. Given the absence of statistical characteristics for PolSAR data, the statistical feature is introduced into the FNEA framework in the proposed approach. Many other methods use the classical complex Wishart distribution in order to represent scattering matrix statistics for PolSAR image segmentation [15,17,19–21,27]. Considering the ability to modeling varying degrees of texture [30,36], G^0 distribution is more suitable for heterogeneous or homogeneous areas in high-resolution PolSAR data compared to the Wishart or K distribution. Thus, the proposed method adopts the G^0 distribution to suppress speckle noise and obtains consistent segmentation results compared with the traditional FNEA method.

Another feature of our method is adding pre-segmentation using SLIC with the polarimetric features in order to generate initial samples for the application of the G^0 statistical model. It is essential to robustly estimate model parameters with enough samples. Most of the previous work collected initial samples from image blocks within square windows [32], which led to segmentation results with obvious dentate boundaries. To handle this problem, superpixels were introduced for pre-segmentation in the proposed approach. Given the excellent boundary adherence, SLIC was used in our approach, which combined the polarimetric distance of Pauli RGB compositions and spatial distance. This approach is capable of achieving accurate segmentation results with precise boundaries between the different areas.

4.2. Sensitivity Analysis of the Parameters

According to the experiments using simulated PolSAR data and real-world PolSAR images (include RADARSAT-2 and ESAR data), the desired number of initial samples (g^2), shape weight w_{shp}, and scale parameter t affected the segmentation accuracy; a detailed analysis is as follows.

4.2.1. Number of Initial Samples

As we know, an appropriate number of samples is essential for parameter estimation, namely, the size of superpixels, g^2, affects the use of the G^0 distribution in our method. When the size of the superpixels was too small, parameter estimation of the G^0 distribution became unstable, which made the calculation of statistical heterogeneity inaccurate. On the other hand, statistical features were not adapted for superpixel generation. The time utilizing this statistic's feature for the superpixel-based FNEA can be delayed when the size of superpixels g^2 become too big, which could affect the subsequent segmentation accuracy. Therefore, the proper size of a superpixel is one of key issues for the segmentation experiment.

An additional experiment was conducted to explore the minimum size of the superpixels. Abstractly, enough samples ensure stable parameter estimation of the G^0 distribution, and there was little in terms of statistical heterogeneity differences between adjacent objects in one class. Three different simulated PolSAR images were used in the experiment, which only contained forest, crops, and roads, and is mentioned here in Section 3.1. The three simulated PolSAR images were divided into different sizes of blocks to calculate the standard deviation of the normalized G^0 heterogeneity ($\Delta h_{stt}/g^2$) between adjacent objects [32]. Figure 11 shows the changes of $\Delta h_{stt}/g^2$ with different sizes of blocks. As observed in Figure 11, the standard deviation became stable when the size of the blocks was large enough. In contrast, the standard deviation increased sharply when the size of the blocks was less than 16 pixels, which means that there were large statistical heterogeneity differences, due to the unstable parameter estimation using a small number of samples. Consequently, the superpixels should contain at least 16 pixels in order to ensure the stable calculation for G^0 heterogeneity.

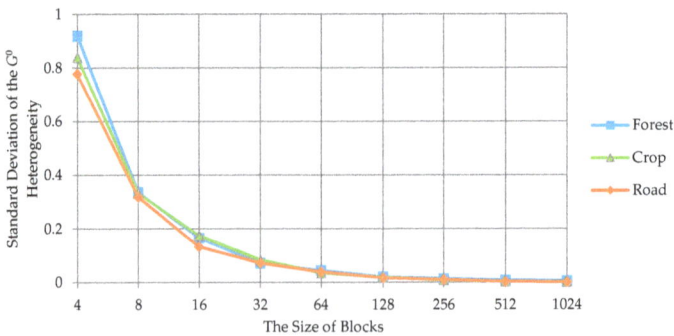

Figure 11. Standard deviation of the normalized G^0 heterogeneity, as a function of block size.

Another experiment was conducted to analyze the effect of size of superpixels on the segmentation accuracy. In the process of generating superpixels, different desired sizes of superpixels were set, and the proposed IFGS method was used in the segmentation experiment. The changes in the detection and quality rates using the different desired sizes of superpixels for the simulated data, ESAR, and RADARSAT-2 images are shown in Figure 12. In practice, all the superpixels of a small size were merged with neighboring, larger superpixels in the generation process. The initial sizes of the superpixels were generally larger than the desired size of the superpixels. The desired sizes of the superpixels were set from 3 pixels × 3 pixels to 10 pixels × 10 pixels in this experiment.

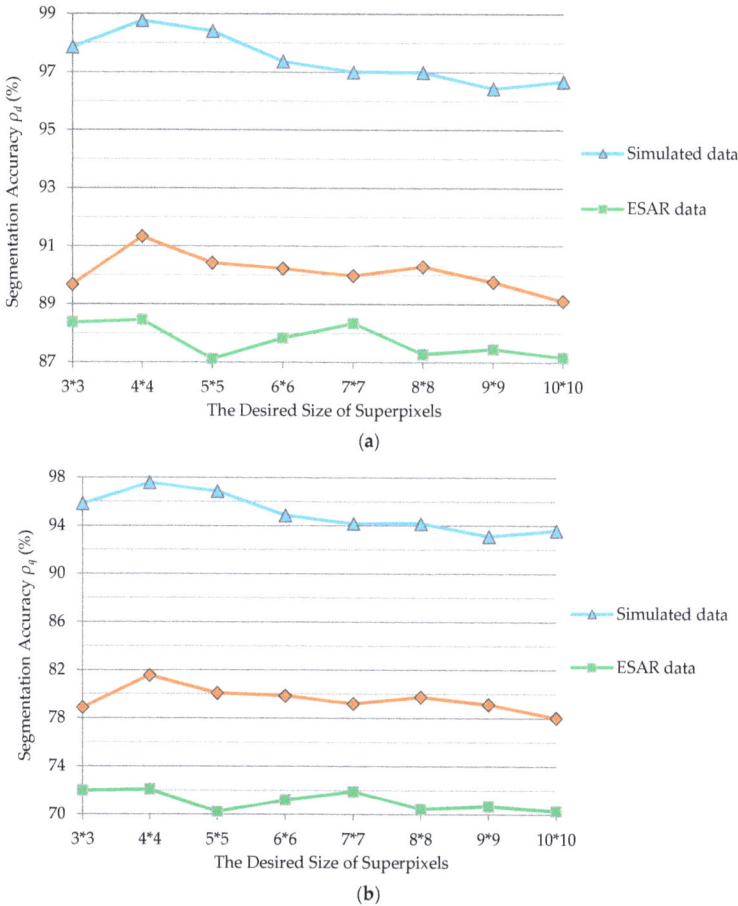

Figure 12. Segmentation accuracy obtained using the IFGS method for the simulated data, ESAR, and RADARSAT-2 data, with different desired sizes of superpixels: (**a**) detection rate; (**b**) quality rate.

As shown in Figure 12, the segmentation accuracy fluctuated and decreased slowly when the desired size of the superpixels increased. For the simulated image, the detection rate and quality rate decreased when the desired size of the superpixels increased from 4 pixels × 4 pixels to 10 pixels × 10 pixels. Specifically, the larger superpixels lead to delayed utilization of statistical information, and then caused a decrease in the segmentation accuracy. On the other hand, when the desired size of the superpixels was 3 pixels × 3 pixels, the number of samples could not ensure stable parameter estimation and generated inaccurate segmentation results. For the ESAR and RADARSAT-2 images, the detection rate and quality rate fluctuated when the desired size of superpixels increased due to the complexity of the real ground objects. Similarly, when the desired size of superpixels was 4 pixels × 4 pixels, the segmentation of the ESAR and RADARSAT-2 image obtained the highest detection and quality rates. Thus, superpixels were generated with a desired size of 4 pixels × 4 pixels for these three images. Sixteen pixels were used for parameter estimation of the G^0 distribution for the proposed method.

4.2.2. Weight of Features

According to the previously presented segmentation experiments, shape features improved the segmentation performance for the three different images. Shape weight w_{shp} affected the use of statistical and shape features, according to Equation (22). In order to set an appropriate feature weight, the ratios of the average changes in shape heterogeneity and statistical heterogeneity $(\overline{\Delta h_{shp}/\Delta h_{stt}})$ were calculated for the simulated data, ESAR data and RADARSAT-2 date, when the shape feature weight was 0 and 1, respectively. The ratios shown in Table 4 indicate that the shape heterogeneity was apparently larger than the statistical heterogeneity in the proposed IFGS segmentation. A relatively small shape weight should be set to balance the shape features and statistical features.

Table 4. Ratios $(\overline{\Delta h_{shp}/\Delta h_{stt}})$ calculated for the simulated data, ESAR, and RADARSAT-2 image in segmentation with the shape weight of 0 and 1.

PolSAR Data	$\overline{\Delta h_{shp}/\Delta h_{stt}}$	
	$w_{shp} = 0$	$w_{shp} = 1$
Simulated image	1.48	2.07
ESAR image	2.09	2.25
RADARSAT-2 image	6.55	8.25

Further experiments were conducted to analyze the effects of weight of shape features on segmentation accuracy. In the process of the proposed IFGS segmentation using the desired superpixel size of 4 pixels × 4 pixels, different shape weights, in the 0–0.3 range, were set. The changes in detection and quality rates, using different shape weights for the simulated data, ESAR, and RADARSAT-2 images, are shown in Figure 13.

As shown in Figure 13, the weight of the shape also had a significant effect on the proposed IFGS segmentation method. For the simulated image, the detection rate and quality rate improved when weight of shape features increased from 0 to 0.05, and decreased when the weight of the shape features exceeded 0.05. Specifically, as the shape weight increased to more than 0.15, the relatively small statistical heterogeneity was not fully utilized, and the segmentation accuracy sharply decreased. When the shape weight varied from 0.02 to 0.08, the integrated utilization of the statistical features and shape features obtained a higher accuracy than single use of the statistical features. For the ESAR image, when the shape weight exceeded 0.1, the segmentation accuracy of the IFGS method was lower than that of the IFGS-S method. However, the same situation occurred for the RADARSAT-2 image when the shape weight exceeded 0.2, which coincided with the case where the ratios $(\overline{\Delta h_{shp}/\Delta h_{stt}})$ of RADARSAT-2 image were larger than those of the ESAR image. Similar to the simulated data, when the weight of shape feature was 0.05, the segmentation of ESAR and RADARSAT-2 images obtained the highest detection and quality rates. Shape features with a weight of 0.05 were used for comprehensive utilization of shape and statistical features in the image segmentation experiment. Moreover, the detection rate of the proposed IFGS method improved by approximately 1%, 3%, and 3% compared to the IFGS-S method and the quality rate improved by approximately 2%, 4%, and 4%, respectively.

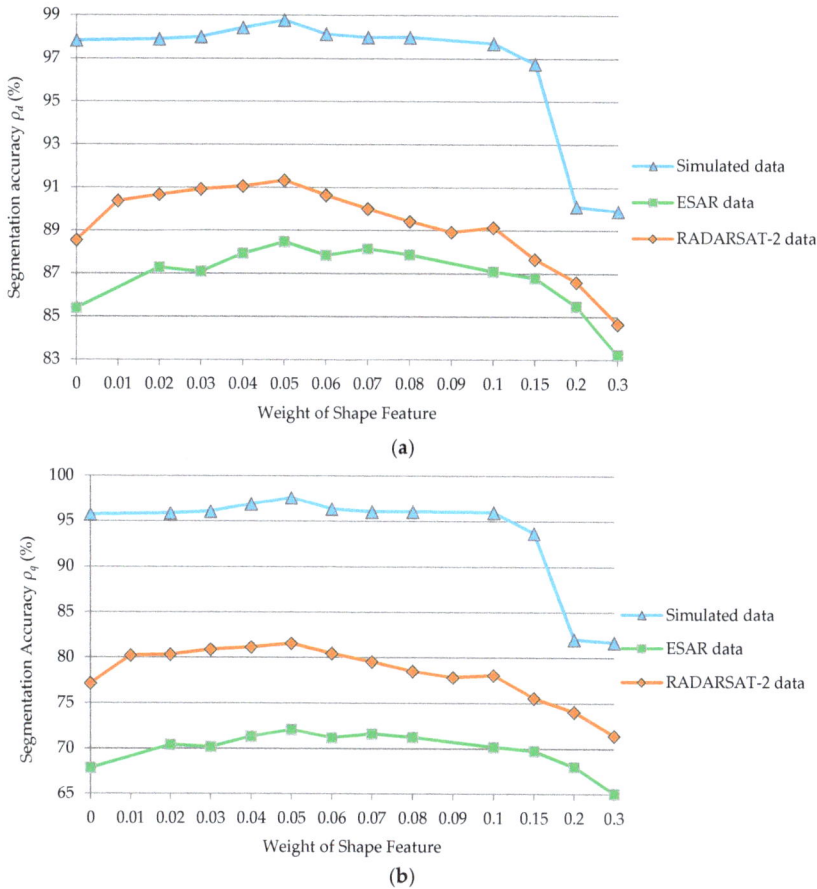

Figure 13. Segmentation accuracy obtained using the IFGS method for the simulated data, ESAR, and RADARSAT-2 data with different shape weights: (**a**) detection rate; (**b**) quality rate.

4.2.3. Scale of Segmentation

Among the three main parameters of the proposed method, scale parameter t is a relative threshold that determines the average size of objects, or the number of objects within a scene. It can be flexibly adjusted, depending on the desired number of objects in the segmentation. An extra experiment was conducted to analyze the effects of scale parameter on segmentation accuracy. In this segmentation experiment, the desired superpixel size was set as 4 pixels × 4 pixels, and the shape weight was 0.05. Different scales were set, and the detection rate and the number of result objects were calculated. Figure 14 shows the changes in the detection rate and the number of result objects with different segmentation scales for the simulated data, ESAR, and RADARSAT-2 images.

As shown in Figure 14, the scale had a consistent effect on the proposed IFGS segmentation method for the three images. As the scale of the segmentation increased, more adjacent objects were merged and the number of final objects became smaller. This means that adjacent objects with a minimal feature difference were not reasonably divided, and resulted in a decrease in segmentation accuracy. On the other hand, the number of objects and segmentation accuracy increased when the scale became smaller. However, the increased number of objects resulted in broken segmentation results. The appropriate number of objects should consider as the true application scene.

For the single-look RADARSAT-2 image, the final segmentation results using four different scales (scale = 12, 16, 20, 25) are shown in Figure 15. As the segmentation scale increased, the detail boundary information decreased and the size of single object became larger. Moreover, targets with larger areas were more intact when they were segmented. As shown in Figure 15a, the narrow rivers and roads were segmented correctly, while the heterogeneous urban areas and larger water bodies were over-segmented, causing broken segmentation results. In Figure 15d, the large water bodies were entirely segmented, but the urban areas, forest areas, and farmland were under-segmented. The boundaries of narrow roads, rivers, and small lakes were inaccurate. Incorrect boundaries occurred between adjacent objects with minimal feature differences, especially for different types of farmlands and forests. In summary, it is necessary to consider the practical use of scale-setting. Specifically, a larger scale is essential for main category classification, and larger target detection. A small scale is applicable to focusing on the details of ground objects.

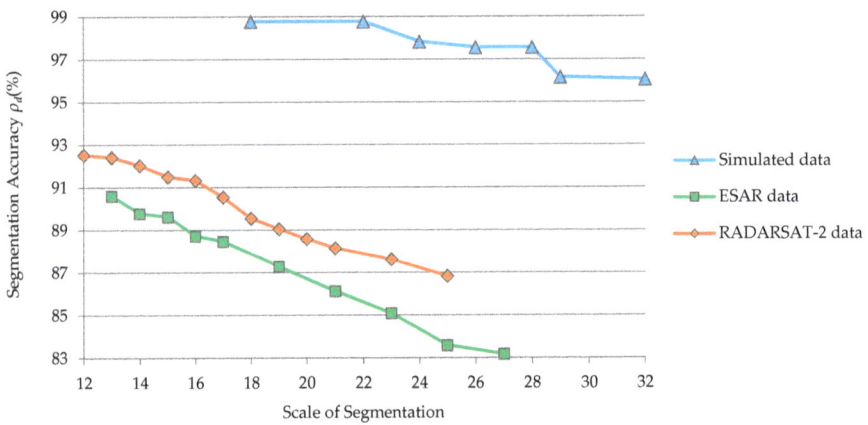

(a)

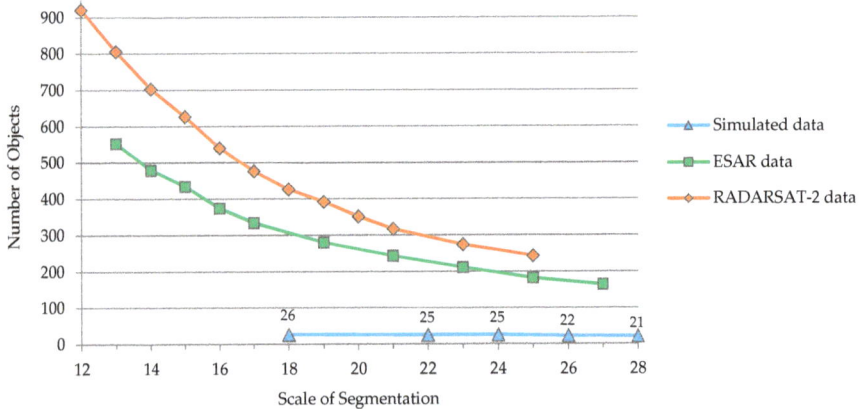

(b)

Figure 14. Segmentation accuracy obtained using the IFGS method for the simulated data, ESAR, and RADARSAT-2 data with different scales: (**a**) detection rate; (**b**) the number of result objects.

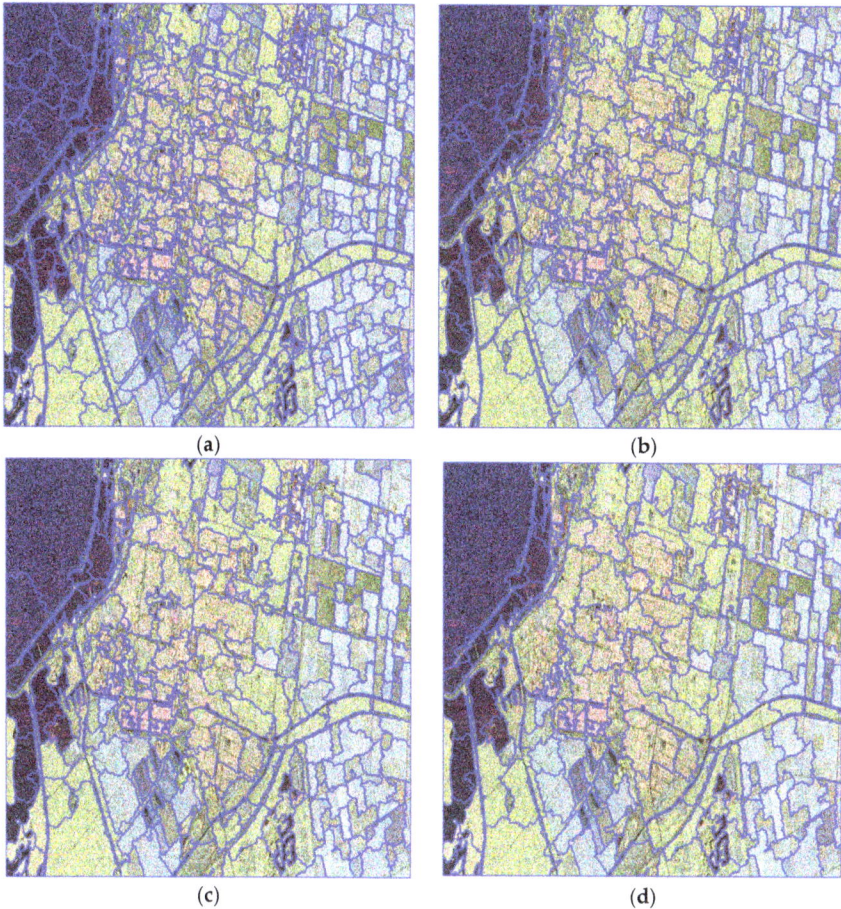

Figure 15. Segmentation results of the single-look RADARSAT-2 PolSAR image with different scales: (**a**) scale = 12; (**b**) scale = 16; (**c**) scale = 20; (**d**) scale = 25.

4.3. Time Performance Analysis

The analysis of the proposed method consists two parts: Initial segmentation (SLIC) and superpixel-based FNEA segmentation (SP-FNEA). The time complexity of SLIC and SP-FNEA are $O(10n)$ and $O(kn)$, respectively, where n is the number of pixels and k is the number of iterations in SP-FNEA. The time complexity of the proposed method is $O(10n) + O(kn)$.

To further analyze the time efficiency of the algorithm, the runtime of SLIC and SP-FNEA for the different PolSAR images with specific segmentation scales was calculated by averaging multiple runtimes. A laptop using the 64-bit Windows 10 operating system, a quad-core Intel i5-4210U, 2.40 GHz, and 8 GB memory was utilized for the segmentation experiments. Table 5 shows the runtimes of the proposed algorithm for different PolSAR images. As shown in Table 5, the proposed algorithm has a good time efficiency. Specifically, the runtime of SLIC is linear with data size, meaning that SLIC has an excellent computational efficiency. For the SP-FNEA algorithm, its runtime efficiency is associated with the number of iterations and the data size. Generally, the larger scale resulted in the merger of adjacent objects, and increased the runtime. Moreover, the efficient SLIC avoids SP-FNEA segmentation, starting from a single pixel, and saving considerable runtime.

Table 5. Runtimes of the proposed algorithm for different PolSAR images.

PolSAR Data	Size	Scale t	Time (s)		
			SLIC	SP-FNEA	Total
Simulated Data	400×400	22	22.52	12.32	34.84
RADARSAT-2	1400×1400	16	267.85	238.60	506.45
ESAR	800×800	17	88.04	64.66	152.70
TerraSAR-X	541×541	16	40.07	27.20	67.27

4.4. Accuracies, Errors, and Uncertainties

We adopted the detection and quality rates to evaluate the performance of the proposed segmentation, which were calculated according to the ground truth data. These two measures are widely considered in the field of image segmentation of remote sensing. It is clear that our method obtained more accurate segmentation results than the other methods, and this advantage was obtained with the least number of result objects.

According to the previous experiments, several factors affect the segmentation accuracy of the proposed method. The parameter estimation accuracy of the G^0 distribution plays an important role in the statistical feature-based segmentation method. For the proposed approach, measurement of statistical heterogeneity and correct object merging depend on the proper estimation of the involved parameters. The determination of the scale parameters is another factor that causes segmentation errors for the proposed method. The scales of different types of targets exhibit differences due to their inequitable heterogeneities. An inappropriate scale parameter leads to under-segmentation in homogeneous regions or over-segmentation in heterogeneous regions, reducing the segmentation accuracy of the target of interest. Future development of this approach should include an accurate parameter estimation method of the G^0 distribution and the determination of the segmentation scale.

The improvement of the proposed method was verified using simulated data, and real-world RADARSAT-2, ESAR, TerraSAR-X images, which mainly cover farmlands, forests, and urban areas. The experimental results show that the proposed method performed well for targets in these images. However, the segmentation results may vary for other application scenarios. In-depth experiments and analyses for other application scenarios, such as forest species classification, building collapse assessment, and oil spill extraction, have not been conducted in the present study. The applicability of the proposed method for specific applications of high-resolution PolSAR image remains uncertain.

5. Conclusions

In high-resolution fully polarimetric Synthetic Aperture Radar (PolSAR) images, speckle noise and heterogeneous regions with rich spatial features makes segmentation a challenging task. In this study, a novel segmentation algorithm for high-resolution PolSAR data has been developed by combining spatial, statistical, and polarimetric features. This integrates the statistical features into a fractal net evolution algorithm (FNEA) framework, and polarimetric features into simple linear iterative clustering (SLIC) for generating pre-segments. The main improvements are as follows: First, spectral heterogeneity in the traditional FNEA was substituted by the G^0 distribution statistical heterogeneity to combine shape features and statistical features of PolSAR data. Second, a modified SLIC algorithm was utilized to generate compact, approximately homogeneous superpixels as the initial samples for the G^0 statistical model, which substituted the polarimetric distance of Pauli RGB composition for the CIELAB color distance.

Several datasets were utilized in the experiments to verify the validity and applicability of the proposed method, including a simulated PolSAR data, a spaceborne single-look RADARSAT-2 image, an airborne multi-look ESAR image, and a spaceborne multi-look TerraSAR-X image. The highest accuracy for each data set was obtained using the proposed approach with the least number of generated objects, i.e., 98.77% on simulated data, 88.46% on ESAR image, and 91.33% on RADARSAT-2 image, respectively. It can thus be concluded that the proposed method achieved accurate and precise segmentation results for high-resolution PolSAR images.

Nevertheless, the performance of the proposed method could be further improved. For instance, the parameter estimation of the statistical model, the initial sample generation, the setting of the weight of features, and the strategy of determining the segmentation scale can be optimized. Moreover, the information included in the polarimetric decomposition parameters was not fully utilized in our method, except in the SLIC pre-segmentation. Hence, combining polarimetric features adequately in the superpixel-based FNEA to improve the performance of high-resolution PolSAR image segmentation is a promising prospect.

Acknowledgments: This work is supported in part by the National Natural Science Foundation of China under Grant No.41301477 and 41471355. We also thank Qi Chen and Qikai Shen for their support during the entire course of this study.

Author Contributions: Qihao Chen drafted the manuscript and was responsible for the research design, experiment; Linlin Li, Qiao Xu, Shuai Yang and Xuguo Shi reviewed the manuscript and were responsible for the analysis; Xiuguo Liu supervised the research and contributed to the editing and review of the manuscript.

Conflicts of Interest: The authors declare no conflict of interest.

References

1. Jiao, L.; Liu, F. Wishart deep stacking network for fast POLSAR image classification. *IEEE Trans. Image Process.* **2016**, *25*, 3273–3286. [CrossRef] [PubMed]
2. Yang, S.; Chen, Q.; Yuan, X.; Liu, X. Adaptive coherency matrix estimation for polarimetric SAR imagery based on local heterogeneity coefficients. *IEEE Trans. Geosci. Remote Sens.* **2016**, *54*, 6732–6745. [CrossRef]
3. Ressel, R.; Singha, S. Comparing near coincident space borne C and X band fully polarimetric SAR data for arctic sea ice classification. *Remote Sens.* **2016**, *8*, 198. [CrossRef]
4. Wei, J.; Zhang, J.; Huang, G.; Zhao, Z. On the use of cross-correlation between volume scattering and helix scattering from polarimetric SAR data for the improvement of ship detection. *Remote Sens.* **2016**, *8*, 74. [CrossRef]
5. Lang, F.; Yang, J.; Li, D.; Zhao, L.; Shi, L. Polarimetric SAR image segmentation using statistical region merging. *IEEE Geosci. Remote Sens. Lett.* **2014**, *11*, 509–513. [CrossRef]
6. Liu, H.; Wang, Y.; Yang, S.; Wang, S.; Feng, J.; Jiao, L. Large polarimetric SAR data semi-supervised classification with spatial-anchor graph. *IEEE J. Sel. Top. Appl. Earth Obs. Remote Sens.* **2016**, *9*, 1439–1458. [CrossRef]
7. Cheng, J.; Ji, Y.; Liu, H. Segmentation-based PolSAR image classification using visual features: RHLBP and color features. *Remote Sens.* **2015**, *7*, 6079–6106. [CrossRef]
8. Qi, Z.; Yeh, A.G.; Li, X.; Lin, Z. A novel algorithm for land use and land cover classification using RADARSAT-2 polarimetric SAR data. *Remote Sens Environ.* **2012**, *118*, 21–39. [CrossRef]
9. Qi, Z.; Yeh, A.G.; Li, X.; Xian, S.; Zhang, X. Monthly short-term detection of land development using RADARSAT-2 polarimetric SAR imagery. *Remote Sens Environ.* **2015**, *164*, 179–196. [CrossRef]
10. Suresh, G.; Melsheimer, C.; Koerber, J.; Bohrmann, G. Automatic estimation of oil seep locations in synthetic aperture radar images. *IEEE Trans. Geosci. Remote Sens.* **2015**, *53*, 4218–4230. [CrossRef]
11. Vasile, G.; Ovarlez, J.; Pascal, F.; Tison, C. Coherency matrix estimation of heterogeneous clutter in high-resolution polarimetric SAR images. *IEEE Trans. Geosci. Remote Sens.* **2010**, *48*, 1809–1826. [CrossRef]
12. Goodman, N.R. Statistical analysis based on a certain multivariate complex Gaussian distribution (an introduction). *Ann. Math. Stat.* **1963**, *34*, 152–177. [CrossRef]
13. Tison, C.; Nicolas, J.M.; Tupin, F.; Maitre, H. A new statistical model for Markovian classification of urban areas in high-resolution SAR images. *IEEE Trans. Geosci. Remote Sens.* **2004**, *42*, 2046–2057. [CrossRef]
14. Rignot, E.; Chellappa, R. Segmentation of polarimetric synthetic aperture radar data. *IEEE Trans. Image Process.* **1992**, *1*, 281–300. [CrossRef] [PubMed]
15. Liu, B.; Zhang, Z.; Liu, X.; Yu, W. Representation and spatially adaptive segmentation for PolSAR images based on wedgelet analysis. *IEEE Trans. Geosci. Remote Sens.* **2015**, *53*, 1–13. [CrossRef]

16. Beaulieu, J.M.; Touzi, R. Segmentation of textured polarimetric SAR scenes by likelihood approximation. *IEEE Trans. Geosci. Remote Sens.* **2004**, *42*, 2063–2072. [CrossRef]
17. Alonso-Gonzalez, A.; Lopez-Martinez, C.; Salembier, P. Filtering and segmentation of polarimetric SAR data based on binary partition Trees. *IEEE Trans. Geosci. Remote Sens.* **2012**, *50*, 593–605. [CrossRef]
18. Liu, B.; Hu, H.; Wang, H.; Wang, K.; Liu, X.; Yu, W. Superpixel-based classification with an adaptive number of classes for polarimetric SAR images. *IEEE Trans. Geosci. Remote Sens.* **2013**, *51*, 907–924. [CrossRef]
19. Qin, F.; Guo, J.; Lang, F. Superpixel segmentation for polarimetric SAR imagery using local iterative clustering. *IEEE Geosci. Remote Sens. Lett.* **2015**, *12*, 13–17.
20. Zhang, Y.; Zou, H.; Luo, T.; Qin, X.; Zhou, S.; Ji, K. A fast superpixel segmentation algorithm for PolSAR images based on edge refinement and revised Wishart distance. *Sensors* **2016**, *16*, 1687. [CrossRef] [PubMed]
21. Ersahin, K.; Cumming, I.G.; Ward, R.K. Segmentation and classification of polarimetric SAR data using spectral graph partitioning. *IEEE Trans. Geosci. Remote Sens.* **2010**, *48*, 164–174. [CrossRef]
22. Benz, U.C.; Hofmann, P.; Willhauck, G.; Lingenfelder, I.; Heynen, M. Multi-resolution, object-oriented fuzzy analysis of remote sensing data for GIS-ready information. *ISPRS J. Photogramm. Remote Sens.* **2004**, *58*, 239–258. [CrossRef]
23. Hay, G.J.; Blaschke, T.; Marceau, D.J.; Bouchard, A. A comparison of three image-object methods for the multiscale analysis of landscape structure. *ISPRS J. Photogramm. Remote Sens.* **2003**, *57*, 327–345. [CrossRef]
24. Burnett, C.; Blaschke, T. A multi-scale segmentation/object relationship modelling methodolgy for landscape analysis. *Ecol. Model.* **2003**, *168*, 233–249. [CrossRef]
25. Benz, U.; Pottier, E. Object based analysis of polarimetric SAR data in alpha-entropy-anisotropy decomposition using fuzzy classification by eCognition. In Proceedings of the IEEE International Geoscience and Remote Sensing Symposium, Sydney, Australia, 9–13 July 2001; pp. 1427–1429.
26. Gao, H.; Yang, K.; Jia, Y.L. Segmentation of polarimetric SAR image using object-oriented strategy. In Proceedings of the 2nd International Conference on Remote Sensing, Environment and Transportation Engineering, Nanjing, China, 1–3 June 2012; pp. 1–5.
27. Cao, F.; Hong, W. An unsupervised segmentation with an adaptive number of clusters using the SPAN/H/α/A space and the complex Wishart clustering for fully polarimetric SAR data analysis. *IEEE Trans. Geosci. Remote Sens.* **2007**, *45*, 3454–3467. [CrossRef]
28. Ulaby, F.T.; Kouyate, F.; Brisco, B.; Williams, T.H.L. Textural information in SAR images. *IEEE Trans. Geosci. Remote Sens.* **1986**, *GE-24*, 235–245. [CrossRef]
29. Quegan, S.; Rhodes, I.; Caves, R. Statistical models for polarimetric SAR data. In Proceedings of the IEEE International Geoscience and Remote Sensing Symposium, Pasadena, CA, USA, 8–12 August 1994; pp. 1371–1373.
30. Freitas, C.; Frery, A.; Correia, A. The polarimetric G distribution for SAR data analysis. *Environmetrics* **2005**, *16*, 13–31. [CrossRef]
31. Bombrun, L.; Beaulieu, J.M. Fisher distribution for texture modeling of polarimetric SAR data. *IEEE Geosci. Remote Sens. Lett.* **2008**, *5*, 512–516. [CrossRef]
32. Bombrun, L.; Vasile, G.; Gay, M.; Totir, F. Hierarchical segmentation of polarimetric SAR images using heterogeneous clutter models. *IEEE Trans. Geosci. Remote Sens.* **2011**, *49*, 726–737. [CrossRef]
33. Salembier, P.; Foucher, S. Optimum graph cuts for pruning binary partition trees of polarimetric SAR images. *IEEE Trans. Geosci. Remote Sens.* **2016**, *54*, 5493–5502. [CrossRef]
34. Feng, J.; Cao, Z.; Pi, Y. Polarimetric contextual classification of PolSAR images using sparse representation and superpixels. *Remote Sens.* **2014**, *6*, 7158–7181. [CrossRef]
35. Achanta, R.; Shaji, A.; Smith, K.; Lucchi, A.; Fua, P.; Süsstrunk, S. SLIC superpixels compared to state-of-the-art superpixel methods. *IEEE Trans. Pattern Anal. Mach. Intell.* **2012**, *34*, 2274–2282. [CrossRef] [PubMed]
36. Khan, S.; Guida, R. Application of Mellin-kind statistics to polarimetric G distribution for SAR data. *IEEE Trans. Geosci. Remote Sens.* **2014**, *52*, 3513–3528. [CrossRef]
37. Khan, S.; Guida, R. On single-look multivariate G distribution for PolSAR data. *IEEE J. Sel. Top. Appl. Earth Obs. Remote Sens.* **2012**, *5*, 1149–1163. [CrossRef]
38. Gini, F.; Greco, M. Covariance matrix estimation for CFAR detection in correlated heavy tailed clutter. *Signal Process.* **2002**, *82*, 1847–1859. [CrossRef]

39. Anfinsen, S.N.; Eltoft, T. Application of the matrix-variate Mellin transform to analysis of polarimetric radar images. *IEEE Trans. Geosci. Remote Sens.* **2011**, *49*, 2281–2295. [CrossRef]

40. Doulgeris, A.P.; Anfinsen, S.N.; Eltoft, T. Classification with a non-Gaussian model for PolSAR data. *IEEE Trans. Geosci. Remote Sens.* **2008**, *46*, 2999–3009. [CrossRef]

41. Cloude, S.R.; Pottier, E. A review of target decomposition theorems in radar polarimetry. *IEEE Trans. Geosci. Remote Sens.* **1996**, *34*, 498–518. [CrossRef]

42. Jiao, H.; Luo, Y.; Wang, N.; Qi, L.; Dong, J.; Lei, H. Underwater multi-spectral photometric stereo reconstruction from a single RGBD image. In Proceedings of the Asia-Pacific Signal and Information Processing Association Annual Summit and Conference (APSIPA), Jeju, South Korea, 13–16 December 2016; pp. 1–4.

43. Xu, Q.; Chen, Q.; Yang, S.; Liu, X. Superpixel-based classification using K distribution and spatial context for polarimetric SAR images. *Remote Sens.* **2016**, *8*, 619. [CrossRef]

44. Möller, M.; Lymburner, L.; Volk, M. The comparison index: A tool for assessing the accuracy of image segmentation. *IEEE J. Sel. Top. Appl. Earth Obs.* **2007**, *9*, 311–321. [CrossRef]

45. Clinton, N.; Holt, A.; Scarborough, J.; Yan, L.; Gong, P. Accuracy assessment measures for object-based image segmentation goodness. *Photogramm. Eng. Remote Sens.* **2010**, *76*, 289–299. [CrossRef]

remote sensing

MDPI

Article

PolSAR Land Cover Classification Based on Roll-Invariant and Selected Hidden Polarimetric Features in the Rotation Domain

Chensong Tao, Siwei Chen *, Yongzhen Li and Shunping Xiao

The State Key Laboratory of Complex Electromagnetic Environment Effects on Electronics and Information System, School of Electronic Science and Engineering, National University of Defense Technology, Changsha 410073, China; taochensongnudt@163.com (C.T.); e0061@sina.com (Y.L.); qwertmingx@tom.com (S.X.)
* Correspondence: chenswnudt@163.com; Tel.: +86-731-8457-3487

Academic Editors: Timo Balz, Uwe Soergel, Mattia Crespi, Batuhan Osmanoglu and Prasad S. Thenkabail
Received: 8 May 2017; Accepted: 15 June 2017; Published: 1 July 2017

Abstract: Land cover classification is an important application for polarimetric synthetic aperture radar (PolSAR). Target polarimetric response is strongly dependent on its orientation. Backscattering responses of the same target with different orientations to the SAR flight path may be quite different. This target orientation diversity effect hinders PolSAR image understanding and interpretation. Roll-invariant polarimetric features such as entropy, anisotropy, mean alpha angle, and total scattering power are independent of the target orientation and are commonly adopted for PolSAR image classification. On the other aspect, target orientation diversity also contains rich information which may not be sensed by roll-invariant polarimetric features. In this vein, only using the roll-invariant polarimetric features may limit the final classification accuracy. To address this problem, this work uses the recently reported uniform polarimetric matrix rotation theory and a visualization and characterization tool of polarimetric coherence pattern to investigate hidden polarimetric features in the rotation domain along the radar line of sight. Then, a feature selection scheme is established and a set of hidden polarimetric features are selected in the rotation domain. Finally, a classification method is developed using the complementary information between roll-invariant and selected hidden polarimetric features with a support vector machine (SVM)/decision tree (DT) classifier. Comparison experiments are carried out with NASA/JPL AIRSAR and multi-temporal UAVSAR data. For AIRSAR data, the overall classification accuracy of the proposed classification method is 95.37% (with SVM)/96.38% (with DT), while that of the conventional classification method is 93.87% (with SVM)/94.12% (with DT), respectively. Meanwhile, for multi-temporal UAVSAR data, the mean overall classification accuracy of the proposed method is up to 97.47% (with SVM)/99.39% (with DT), which is also higher than the mean accuracy of 89.59% (with SVM)/97.55% (with DT) from the conventional method. The comparison studies clearly demonstrate the efficiency and advantage of the proposed classification methodology. In addition, the proposed classification method achieves better robustness for the multi-temporal PolSAR data. This work also further validates that added benefits can be gained for PolSAR data investigation by mining and utilization of hidden polarimetric information in the rotation domain.

Keywords: polarimetric synthetic aperture radar (PolSAR); polarimetric feature; polarimetric matrix rotation; polarimetric coherence pattern; rotation domain; feature selection; classification

---

1. Introduction

With the ability to work day and night under almost all weather conditions and to acquire full polarization information, polarimetric synthetic aperture radar (PolSAR) has become one of the most important microwave remote sensors [1]. Plenty of successful applications have been developed [1–5]. Among them, land cover classification is an important application for PolSAR data utilization. It is able to provide information support to many fields such as general survey of crops, appraisal of cultivated and urban land occupation, environment monitoring, etc.

Plenty of approaches have been proposed to enhance the classification performance from aspects of polarimetric features, classifiers, and both. On one hand, some approaches focused on polarimetric features with better discriminate performance among different land covers through target scattering mechanism understanding and interpretation. The commonly used polarimetric target decomposition techniques can be divided into two categories: eigenvalue-eigenvector-based decomposition and model-based decomposition [5,6]. For eigenvalue-eigenvector-based decomposition, entropy H, anisotropy Ani, and mean alpha angle $\bar{\alpha}$ derived from Cloude-Pottier decomposition are frequently used and an entropy based PolSAR land classification scheme was established thereafter [7]. Lee et al. also used Cloude-Pottier decomposition with Wishart classifier to PolSAR image classification [8]. For model-based decomposition, the derived polarimetric features are the energy contributions of some typical scattering mechanisms from Freeman-Durden three-component decomposition [9], Yamaguchi four-component decomposition [10], and the recently reported generalized model-based decomposition techniques [11,12]. Lee et al. applied Freeman-Durden three-component decomposition with the Wishart classifier to classify PolSAR data [13]. Wang et al. adopted the non-negative eigenvalue decomposition for terrain and land-use classification [14]. Hong et al. proposed a four-component decomposition and applied it to classify wetland vegetation types [15].

On the other hand, other approaches to improve PolSAR classification accuracy aim at designing or choosing the classifier with the better classification performance to take full advantage of the available polarimetric features. Specifically, Pajares et al. proposed an optimization relaxation approach based on the analogue Hopfield Neural Network for cluster refinement of pre-classified results from the Wishart classification [16]. Attarchi and Gloaguen compared the performances of the support vector machine (SVM) classifier, neural networks classifier, and random forest classifier for classifying complex mountainous forests with SAR and other source data [17]. Zhou et al. applied the deep convolutional neural networks (CNN) classifier for PolSAR classification and obtained improved results [18]. In addition, considering both polarimetric feature and classifier at the same time is also an effective way to improve the classification accuracy. Deng et al. used both polarimetric decomposition and time-frequency decomposition to mine the hidden information of objects in PolSAR images and applied a C5.0 decision tree (DT) algorithm for optimal feature selection and final classification [19]. They also proposed an approach to classify the PolSAR data by integrating polarimetric decomposition, sub-aperture decomposition, and DT algorithm [20]. Cheng et al. designed and implemented a segmentation-based PolSAR image classification method incorporating texture features, color features and SVM classifier [21]. Wang et al. proposed a PolSAR classification method based on the generalized polarimetric decomposition of the Mueller matrix and SVM classifier [22].

Among the aforementioned PolSAR classification methods based on polarimetric features, roll-invariant polarimetric features are commonly adopted. An important reason is that polarimetric response of a target is strongly dependent on its orientation [23]. On one hand, the backscattering responses of the same target with different orientations to the PolSAR flight path are significantly various. On the other hand, the backscattering responses of different targets with some specific orientations to the flight path may be similar to each other. This target orientation diversity effect frequently induces scattering mechanism ambiguity and hinders the correct understanding and interpretation of PolSAR data [23]. As a result, roll-invariant polarimetric features which are independent of the target orientation diversity are preferred in many applications. However, roll-invariant polarimetric features may not completely represent target scattering properties. Target orientation diversity also contains rich information which is not sensed by roll-invariant polarimetric features [23]. To further improve the PolSAR classification accuracy, proper exploration of the target orientation diversity is an effective way and is able to provide valuable hidden information for physical parameter retrieval. In this vein, uniform polarimetric matrix rotation theory [23] and a visualization and characterization tool of polarimetric coherence pattern [24] were respectively proposed to extract the hidden polarimetric features in the rotation domain along the radar line of sight for hidden scattering information mining. Parts of these new features achieved successful applications including crop discrimination [25], target enhancement [23], and manmade target extraction [26], etc.

Since these hidden polarimetric features contain rich hidden scattering information of targets in the rotation domain, this work aims to utilize them to enhance PolSAR classification accuracy. Specifically, we firstly propose a polarimetric feature selection scheme to select suitable hidden polarimetric features derived from the rotation domain. Then, the selected hidden polarimetric features and the commonly used roll-invariant polarimetric features of $H/Ani/\bar{\alpha}/Span$ are combined as the discriminant feature set. Finally, a classification method using the combined feature set and the SVM/DT classifier [27,28] is developed.

This work is organized as follows. In Section 2, the two novel schemes for hidden polarimetric information mining in the rotation domain and their corresponding hidden polarimetric features are reviewed. The proposed polarimetric feature selection scheme and classification method are described in Section 3. Comparison experiments with NASA/JPL AIRSAR and multi-temporal UAVSAR datasets are carried out and investigated in Section 4. Section 5 provides the final conclusions and outlook for future work.

2. Hidden Polarimetric Feature Extraction in the Rotation Domain

2.1. Polarimetric Matrices and Their Rotation

For PolSAR, in the horizontal and vertical polarization basis (H, V), the acquired full polarization information can form a scattering matrix **S** with the representation as

$$\mathbf{S} = \begin{bmatrix} S_{HH} & S_{HV} \\ S_{VH} & S_{VV} \end{bmatrix} \tag{1}$$

where S_{HV} is the backscattered coefficient from vertical transmitting and horizontal receiving polarization. Other terms in **S** are defined similarly.

Subject to the reciprocity condition $(S_{HV} = S_{VH})$, the coherency matrix **T** is

$$\mathbf{T} = \left\langle \mathbf{k}_P \mathbf{k}_P^H \right\rangle = \begin{bmatrix} T_{11} & T_{12} & T_{13} \\ T_{21} & T_{22} & T_{23} \\ T_{31} & T_{32} & T_{33} \end{bmatrix} \tag{2}$$

where $\mathbf{k}_P = \frac{1}{\sqrt{2}} \begin{bmatrix} S_{HH} + S_{VV} & S_{HH} - S_{VV} & 2S_{HV} \end{bmatrix}^T$ is the Pauli scattering vector, $\langle \cdot \rangle$ denotes the sample average, the superscript T and H denote the transpose and conjugate transpose respectively, and T_{ij} is the (i, j) entry of the coherency matrix **T**.

With a rotation angle θ along the radar line of sight, the rotated scattering matrix $\mathbf{S}(\theta)$ and coherency matrix $\mathbf{T}(\theta)$ respectively become

$$\mathbf{S}(\theta) = \mathbf{R}_2(\theta)\mathbf{S}\mathbf{R}_2^H(\theta) \tag{3}$$

$$\mathbf{T}(\theta) = \mathbf{k}_P(\theta)\mathbf{k}_P^H(\theta) = \mathbf{R}_3(\theta)\mathbf{T}\mathbf{R}_3^H(\theta) \tag{4}$$

where the rotation matrixes are $\mathbf{R}_2(\theta) = \begin{bmatrix} \cos\theta & \sin\theta \\ -\sin\theta & \cos\theta \end{bmatrix}, \mathbf{R}_3(\theta) = \begin{bmatrix} 1 & 0 & 0 \\ 0 & \cos 2\theta & \sin 2\theta \\ 0 & -\sin 2\theta & \cos 2\theta \end{bmatrix}.$

2.2. Polarimetric Features Derived from Uniform Polarimetric Matrix Rotation Theory

In order to explore the target orientation diversity and mine embedded hidden information, uniform polarimetric matrix rotation theory was proposed [23]. It rotates the acquired polarimetric matrix along the radar line of sight and fully describes the rotation characteristics of each entry of the matrix. Taking the coherency matrix for example, with mathematic transformations, all the elements and powers of the off-diagonal terms of a rotated coherency matrix $\mathbf{T}(\theta)$ can be represented as a uniform sinusoidal function [23]

$$f(\theta) = A \sin[\omega(\theta + \theta_0)] + B \tag{5}$$

where A is the oscillation amplitude, B is the oscillation center, ω is the angular frequency, and θ_0 is the initial angle. Therefore, the new polarimetric feature parameter set $\{A, B, \omega, \theta_0\}$ named as the oscillation parameter set is able to completely characterize the rotation properties of all the elements and powers of the off-diagonal terms of $\mathbf{T}(\theta)$.

Series of new polarimetric features are derived in [23] to describe the hidden scattering information of the target in the rotation domain. Among them, there are eleven independent hidden features as: $\theta_0\_\mathrm{Re}[T_{12}(\theta)]$, $\theta_0\_\mathrm{Im}[T_{12}(\theta)]$, $\theta_0\_\mathrm{Re}[T_{23}(\theta)]$, $\theta_0\_|T_{12}(\theta)|^2$, $\theta_0\_|T_{23}(\theta)|^2$, $A\_\mathrm{Re}[T_{12}(\theta)]$, $A\_\mathrm{Im}[T_{12}(\theta)]$, $A\_|T_{12}(\theta)|^2$, $A\_|T_{23}(\theta)|^2$, $B\_T_{22}(\theta)$, and $B\_|T_{23}(\theta)|^2$. where $\mathrm{Re}[T_{ij}]$ and $\mathrm{Im}[T_{ij}]$ are the real and imaginary parts of T_{ij} respectively, and $\theta_0\_T_{ij}(\theta)$ denotes the initial angle θ_0 of $T_{ij}(\theta)$. The other terms of $A\_T_{ij}(\theta)$ and $B\_T_{ij}(\theta)$ are defined similarly.

2.3. Polarimetric Features Derived from Polarimetric Coherence Pattern

Polarimetric coherence between two polarization channels s_1 and s_2 is also used for target detection and classification. It can be estimated in practice with the sample average of sufficient samples with similar properties [29] as

$$|\gamma_{1-2}| = \frac{|\langle s_1 s_2^* \rangle|}{\sqrt{\langle |s_1|^2 \rangle \langle |s_2|^2 \rangle}} \tag{6}$$

where the superscript $*$ denotes the conjugate, and the value of $|\gamma_{1-2}|$ is within the range of $[0, 1]$.

Due to the sensitivity of polarimetric coherence to the target's orientation to the PolSAR flight path, a visualization and characterization tool of polarimetric coherence pattern [24] was proposed to extend the original polarimetric coherence at a given rotation angle ($\theta = 0$) to the whole rotation domain. It covers all rotation angles ($\theta \in [-\pi, \pi)$) along the radar line of sight for the exploration of complete interpretation of the target's polarimetric coherence as

$$|\gamma_{1-2}(\theta)| = \frac{|\langle s_1(\theta) s_2^*(\theta) \rangle|}{\sqrt{\langle |s_1(\theta)|^2 \rangle \langle |s_2(\theta)|^2 \rangle}} \tag{7}$$

With this approach, a set of new polarimetric features were proposed to quantitatively characterize a polarimetric coherence pattern's variation along the radar line of sight [24]. These derived polarimetric features include: original coherence $\gamma_{-\mathrm{org}}$, coherence degree $\gamma_{-\mathrm{mean}}$, coherence fluctuation $\gamma_{-\mathrm{std}}$, maximum and minimum coherences $\gamma_{-\mathrm{max}}$ and $\gamma_{-\mathrm{min}}$, coherence contrast $\gamma_{-\mathrm{contrast}}$, coherence beamwidth $\gamma_{-\mathrm{bw}}$, maximum and minimum rotation angles $\theta_{\gamma-\mathrm{max}}$ and $\theta_{\gamma-\mathrm{min}}$. The detailed definitions are given in [24].

With (H, V) polarization basis, four independent polarimetric coherence patterns can be obtained [24] as $|\gamma_{HH-VV}(\theta)|$, $|\gamma_{HH-HV}(\theta)|$, $\left|\gamma_{(HH+VV)-HV}(\theta)\right|$, and $\left|\gamma_{(HH-VV)-HV}(\theta)\right|$. For each of them, the aforementioned nine hidden polarimetric features can be extracted. Therefore, there are a total of thirty-six hidden features derived from the polarimetric coherence patterns.

3. Proposed Polarimetric Feature Selection Scheme and Classification Method

3.1. Proposed Polarimetric Feature Selection Scheme

Based on the aforementioned hidden polarimetric features derived in the rotation domain, there are eleven features derived from the uniform polarimetric matrix rotation theory and thirty-six features derived from the polarimetric coherence pattern. So we need to select suitable features among them to avoid information redundancy which may decrease the accuracy of the final land cover classification. Since γ_{-bw} of $\left|\gamma_{(HH-VV)-HV}(\theta)\right|$ is almost invariant for different land covers [24], it is not considered. Then, a polarimetric feature selection scheme is proposed for the other forty-six hidden polarimetric features. The flowchart of it is shown in Figure 1.

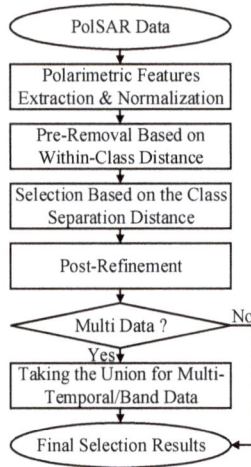

Figure 1. Flowchart of the proposed polarimetric feature selection scheme.

The steps of the proposed polarimetric feature selection scheme are as follow:

(1) The first step is polarimetric features extraction and normalization. Based on the filtered PolSAR data, independent hidden polarimetric features are extracted and normalized to the range of $[0,1]$. The total normalized feature set is $F^{all} = \{f_i, i = 1, ..., I\}$, I is the number of hidden polarimetric features.

(2) Pre-removal is done to the F^{all} based on the within-class distance, which is a measure of the disperse degree of samples within the same class. From ground-truth map, there are X known land covers $C_x, x = 1, ..., X$. For feature f_i, the samples from each land cover C_x can be represented as $C_x^i = \left\{ f_i^{(x,k)}, k = 1, 2, ..., N_x \right\}$. Where $f_i^{(x,k)}$ is the feature value of the kth sample in land cover C_x, N_x is the total sample number of land cover C_x. The within-class distance of land cover C_x with feature f_i is $d_{within}(C_x^i)$ as

$$d_{within}\left(C_x^i\right) = \sqrt{\frac{1}{N_x}\sum_{k=1}^{N_x}\left(f_i^{(x,k)} - center_x^i\right)^2} \tag{8}$$

where $center_x$ is the center of land cover C_x. With feature f_i, $center_x^i = \frac{1}{N_x}\sum_{k=1}^{N_x} f_i^{(x,k)}$. Based on different features, the within-class distances of land cover C_i are $d_{within}(C_x^i)$, $i = 1, ..., I$. The three largest within-class distances of each land covers are all chosen. The features which produce them form the removal feature set $F^{removal}$ which needs to be removed from F^{all} as $F^{preremoval} = F^{all} - F^{removal}$. $F^{preremoval} = \{\tilde{f}_j, j = 1, ..., J\}$, J is the feature number of $F^{preremoval}$ and $J \leq I$.

(3) The preliminary selection is carried out based on the class separation distance defined as the distance between two classes plus the distance between two class centers. A corresponding amount of land cover pairs are constructed by combining each two land covers. There are Y land cover pairs as P_y, $y = 1, ..., Y$. For each land cover pair P_y, two land covers of it are C_{y_1} and C_{y_2}, where y_1 and y_2 are the land cover labels. With feature $\tilde{f}_j$, the samples of land covers C_{y_1} and C_{y_2} can be represented as $C_{y_1}^j = \{\tilde{f}_j^{(y_1,l)}, l = 1, 2, ..., N_{y_1}\}$ and $C_{y_2}^j = \{\tilde{f}_j^{(y_2,m)}, m = 1, 2, ..., N_{y_2}\}$, respectively. The distance between land covers C_{y_1} and C_{y_2} with feature $\tilde{f}_j$ $d_{class}\left(C_{y_1}^j, C_{y_2}^j\right)$ and the distance between their centers $center_{y_1}$ and $center_{y_2}$ with feature $\tilde{f}_j$ $d_{center}\left(center_{y_1}^j, center_{y_2}^j\right)$ are respectively as

$$d_{class}\left(C_{y_1}^j, C_{y_2}^j\right) = \sqrt{\frac{1}{N_{y_1} N_{y_2}} \sum_{l=1}^{N_{y_1}} \sum_{m=1}^{N_{y_2}} \left(\tilde{f}_j^{(y_1,l)} - \tilde{f}_j^{(y_2,m)}\right)^2} \tag{9}$$

$$d_{center}\left(center_{y_1}^j, center_{y_2}^j\right) = \left|\frac{1}{N_{y_1}} \sum_{l=1}^{N_{y_1}} \tilde{f}_j^{(y_1,l)} - \frac{1}{N_{y_2}} \sum_{m=1}^{N_{y_2}} \tilde{f}_j^{(y_2,m)}\right| \tag{10}$$

The class separation distance of land cover pair P_y with feature $\tilde{f}_j$ is proposed as $d_{separation}\left(P_y^j\right) = d_{class}\left(C_{y_1}^j, C_{y_2}^j\right) + d_{center}\left(center_{y_1}^j, center_{y_2}^j\right)$ and is able to measure the land cover separation of land cover pair P_y. Based on different features of $F^{preremoval}$, the class separation distances of land cover pair P_y are $d_{separation}\left(P_y^j\right)$, $j = 1, ..., J$. The selected hidden polarimetric feature of land cover pair P_y is $f_y^{ss} = \underset{\tilde{f}_j \in F^{preremoval}}{argmax} \left\{d_{separation}\left(P_y^j\right)\right\}$. The preliminary selected feature set for all the land cover pairs is $F^{preselection} = f_1^{ss} \cup f_2^{ss} \cup ... \cup f_Y^{ss}$.

(4) After the preliminary selection, post-refinement is implemented. The idea of post-refinement is to determine the features with relatively higher discriminant performance in $F^{preselection}$. For each land cover pairs, the features which lead to the maximum class separation distances are recorded and accounted. Features with appearance higher than a predefined threshold r are all determined as the optimal feature candidates. Then, the final selected feature set can be formed as $F^{selection}$ with these optimal features from $F^{preselection}$.

(5) Finally, if the PolSAR data is a single-temporal/band, the $F^{selection}$ will be the final selection results direct. Or, if it is one of the multi-temporal/band PolSAR data, the union for all the $F^{selection}$ of different temporal/band data will be the final selection results.

This work uses the basic and commonly adopted criterions of within-class distance, distance between two classes, and distance between two class centers to select suitable hidden polarimetric features derived in the rotation domain. Certainly, other feature selection schemes can also be suitable.

3.2. Proposed Classification Method

The main idea of the proposed classification method is to utilize the complementary information between the roll-invariant polarimetric features and the selected hidden polarimetric features in the rotation domain. The combination of them will be used as the classifier input. In order to validate the performance of the proposed classification method, both the SVM and DT classifiers [27,28] are used in this work. The flowchart of the proposed classification method is illustrated in Figure 2 and the corresponding steps are as follows.

(1) In order to extract the roll-invariant polarimetric features of $H/Ani/\bar{\alpha}$, the original PolSAR data is speckle filtered. The recently reported adaptive SimiTest speckle filter [29] is adopted.

(2) Based on the filtered coherency matrix, total scattering power *Span* can be calculated by $Span = T_{11} + T_{22} + T_{33}$.

(3) Roll-invariant polarimetric features of entropy H, mean alpha angle $\bar{\alpha}$ and anisotropy *Ani* can be extracted by Cloude-Pottier decomposition [6].

(4) The selected hidden polarimetric features are extracted using the uniform polarimetric matrix rotation theory [23] and the visualization and characterization tool of polarimetric coherence pattern [24].

(5) Each of these polarimetric features is normalized to the range of $[0, 1]$. And the combination of these normalized features is formed as the classifier input.

(6) Through training and validation processing of the SVM/DT classifier, the final classification results and corresponding accuracies of each land cover and the overall can be obtained.

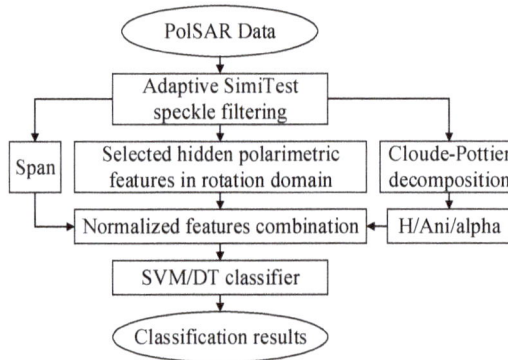

Figure 2. Flowchart of the proposed classification method.

4. Comparison Experiments

4.1. Data Description

First, NASA/JPL AIRSAR L-band PolSAR data collected over Flevoland, the Netherlands, is adopted. The range and azimuth pixel resolutions are 6.6 m and 12.1 m respectively. The data is speckle filtered by the adaptive SimiTest approach with a 15 × 15 sliding window [29] and is shown in Figure 3a. The filter sliding window size of 15 × 15 is recommended in [29], which makes a tradeoff for the filter performance and computational cost. Besides, the filter sliding window size effect will be investigated in Section 4.3. This study area contains various land covers and a ground-truth map for eleven known land covers (including water, rapeseed, grasses, bare soil, potatoes, beet, wheat, lucerne, forest, peas, and stembeans) is shown in Figure 3b.

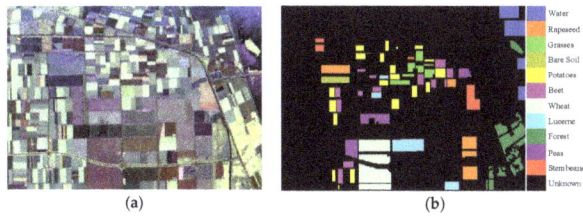

Figure 3. Study area. (**a**) RGB composite image of the filtered AIRSAR data with Pauli basis; (**b**) Ground-truth map for eleven known land covers.

Secondly, NASA/JPL UAVSAR L-band multi-temporal PolSAR data collected over Manitoba, Canada, are also adopted. The range and azimuth pixel resolutions are 5 m and 7 m respectively. Four temporal data are used in the comparison. The acquisition dates are 17 June, 22 June, 5 July, and 17 July in 2012, respectively. With Pauli basis, the RGB composite images of the filtered multi-temporal UAVSAR data are shown in Figure 4a–d. Also, the adaptive SimiTest speckle filter with the recommended 15×15 sliding window [29] is adopted. This study area also contains various land covers and a ground-truth map for seven known land covers (including oats, rapeseed, wheat, corn, soybeans, forage crops, and broadleaf) is shown in Figure 4e.

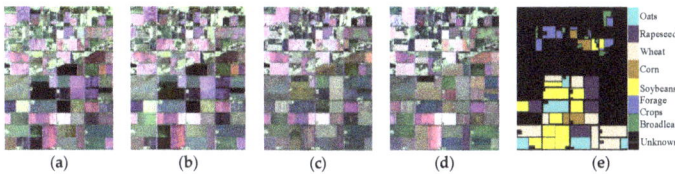

Figure 4. Study area. (**a**–**d**) RGB composite images of the filtered multi-temporal UAVSAR data (17 June, 22 June, 5 July, and 17 July in 2012 respectively) with Pauli basis; (**e**) Ground-truth map for seven known land covers.

4.2. Selected Hidden Polarimetric Features of Different PolSAR Data

For each land cover class, 1000 random samples are used to represent the class in the feature selection processing. For the AIRSAR data, $X = 11$ denotes eleven known land covers and the corresponding number of land cover pairs is $Y = 55$. Meanwhile, for the multi-temporal UAVSAR data, $X = 7$ and $Y = 21$. The preliminary selected feature sets $F^{preselection}$ for different PolSAR data are shown in Table 1. The numbers in brackets indicate the appearance number that this feature leads to the maximum class separation distances. For example, the selected hidden polarimetric feature $\gamma_{-\min}$ of $|\gamma_{HH-VV}(\theta)|$ can maximize the class separation distances within fourteen land cover pairs of the fifty-five pairs of the filtered AIRSAR data.

Table 1. Preliminary selected feature sets for different polarimetric synthetic aperture radar (PolSAR) data.

PolSAR Data		Preliminary Selected Feature Set
AIRSAR		γ_{-min} of $\left|\gamma_{HH-VV}(\theta)\right|$ (14), γ_{-max} of $\left|\gamma_{(HH-VV)-HV}(\theta)\right|$ (11), $\theta_0\_Re[T_{12}(\theta)]$ (9), $\theta_0\_Im[T_{12}(\theta)]$ (6), γ_{-org} of $\left|\gamma_{(HH+VV)-HV}(\theta)\right|$ (3), $\gamma_{-contrast}$ of $\left|\gamma_{(HH-VV)-HV}(\theta)\right|$ (3), γ_{-max} of $\left|\gamma_{(HH+VV)-HV}(\theta)\right|$ (2), γ_{-org} of $\left|\gamma_{HH-HV}(\theta)\right|$ (2), γ_{-org} of $\left|\gamma_{HH-VV}(\theta)\right|$ (2), γ_{-max} of $\left|\gamma_{HH-VV}(\theta)\right|$ (2), γ_{-max} of $\left|\gamma_{HH-HV}(\theta)\right|$ (1)
UAVSAR	17 June	$\theta_0\_Im[T_{12}(\theta)]$ (9), γ_{-org} of $\left|\gamma_{(HH-VV)-HV}(\theta)\right|$ (4), $\theta_0\_Re[T_{12}(\theta)]$ (2), γ_{-max} of $\left|\gamma_{(HH-VV)-HV}(\theta)\right|$ (2), γ_{-mean} of $\left|\gamma_{(HH+VV)-HV}(\theta)\right|$ (1), $\gamma_{-contrast}$ of $\left|\gamma_{(HH+VV)-HV}(\theta)\right|$ (1), γ_{-min} of $\left|\gamma_{(HH-VV)-HV}(\theta)\right|$ (1), γ_{-org} of $\left|\gamma_{HH-VV}(\theta)\right|$ (1)
	22 June	$\theta_0\_Im[T_{12}(\theta)]$ (9), γ_{-org} of $\left|\gamma_{(HH-VV)-HV}(\theta)\right|$ (4), γ_{-min} of $\left|\gamma_{(HH-VV)-HV}(\theta)\right|$ (3), $\theta_0\_Re[T_{12}(\theta)]$ (1), γ_{-mean} of $\left|\gamma_{(HH+VV)-HV}(\theta)\right|$ (1), γ_{-org} of $\left|\gamma_{(HH+VV)-HV}(\theta)\right|$ (1), γ_{-org} of $\left|\gamma_{HH-VV}(\theta)\right|$ (1), γ_{-min} of $\left|\gamma_{HH-VV}(\theta)\right|$ (1)
	5 July	$\theta_0\_Im[T_{12}(\theta)]$ (7), $\theta_0\_Re[T_{12}(\theta)]$ (6), γ_{-min} of $\left|\gamma_{HH-VV}(\theta)\right|$ (3), γ_{-org} of $\left|\gamma_{(HH+VV)-HV}(\theta)\right|$ (3), $\gamma_{-contrast}$ of $\left|\gamma_{(HH+VV)-HV}(\theta)\right|$ (1), γ_{-org} of $\left|\gamma_{HH-VV}(\theta)\right|$ (1)
	17 July	$\theta_0\_Im[T_{12}(\theta)]$ (6), $\theta_0\_Re[T_{12}(\theta)]$ (3), γ_{-org} of $\left|\gamma_{(HH+VV)-HV}(\theta)\right|$ (3), γ_{-min} of $\left|\gamma_{HH-VV}(\theta)\right|$ (3), γ_{-org} of $\left|\gamma_{(HH-VV)-HV}(\theta)\right|$ (2), γ_{-max} of $\left|\gamma_{(HH-VV)-HV}(\theta)\right|$ (1), γ_{-org} of $\left|\gamma_{HH-HV}(\theta)\right|$ (1), γ_{-max} of $\left|\gamma_{HH-HV}(\theta)\right|$ (1), γ_{-max} of $\left|\gamma_{(HH+VV)-HV}(\theta)\right|$ (1)

Based on the preliminary selected feature sets in Table 1, we set $r = 3$ in the followed refinement processing. In this vein, features which have only one or two corresponding land cover pairs are not taken into consideration. As a result, for AIRSAR data, the final selected features are $\theta_0\_Re[T_{12}(\theta)]$, $\theta_0\_Im[T_{12}(\theta)]$, γ_{-org} of $\left|\gamma_{(HH+VV)-HV}(\theta)\right|$, γ_{-max} of $\left|\gamma_{(HH-VV)-HV}(\theta)\right|$, $\gamma_{-contrast}$ of $\left|\gamma_{(HH-VV)-HV}(\theta)\right|$ and γ_{-min} of $\left|\gamma_{HH-VV}(\theta)\right|$. For multi-temporal UAVSAR data, the union of the selected features of different temporal data are the final selection results, which include $\theta_0\_Re[T_{12}(\theta)]$, $\theta_0\_Im[T_{12}(\theta)]$, γ_{-org} of $\left|\gamma_{(HH+VV)-HV}(\theta)\right|$, γ_{-org} of $\left|\gamma_{(HH-VV)-HV}(\theta)\right|$, γ_{-min} of $\left|\gamma_{(HH-VV)-HV}(\theta)\right|$ and γ_{-min} of $\left|\gamma_{HH-VV}(\theta)\right|$.

For AIRSAR data, four commonly adopted roll-invariant polarimetric features of $H/Ani/\bar{\alpha}/Span$ are calculated and shown in Figure 5a–d. Then, the six selected hidden polarimetric features derived in the rotation domain are also calculated and shown in Figure 5e–j. In order to compare the land cover discrimination abilities of the six selected hidden polarimetric features and the four roll-invariant polarimetric features, means and standard deviations of different features in terms of each known land covers are shown in Figure 6. Based on the four roll-invariant polarimetric features of $H/Ani/\bar{\alpha}/Span$ only, land cover 3 (grasses) and 7 (wheat) cannot be successfully discriminated. The discriminations between land cover 3 (grasses) and 8 (lucerne), and land cover 5 (potatoes) and 9 (forest) are also limited. In comparison, with $\theta_0\_Re[T_{12}(\theta)]$, land cover 3 (grasses) and 7 (wheat) can be discriminated successfully. Furthermore, with γ_{-min} of $\left|\gamma_{HH-VV}(\theta)\right|$, better discriminations are achieved for land cover 3 (grasses) and 8 (lucerne), and land cover 5 (potatoes) and 9 (forest). Other selected hidden polarimetric features are also able to enhance the discriminations for some land cover pairs. Thereby, selected hidden polarimetric features can further enhance the land cover discrimination abilities and have the potentials to improve the land cover classification accuracy.

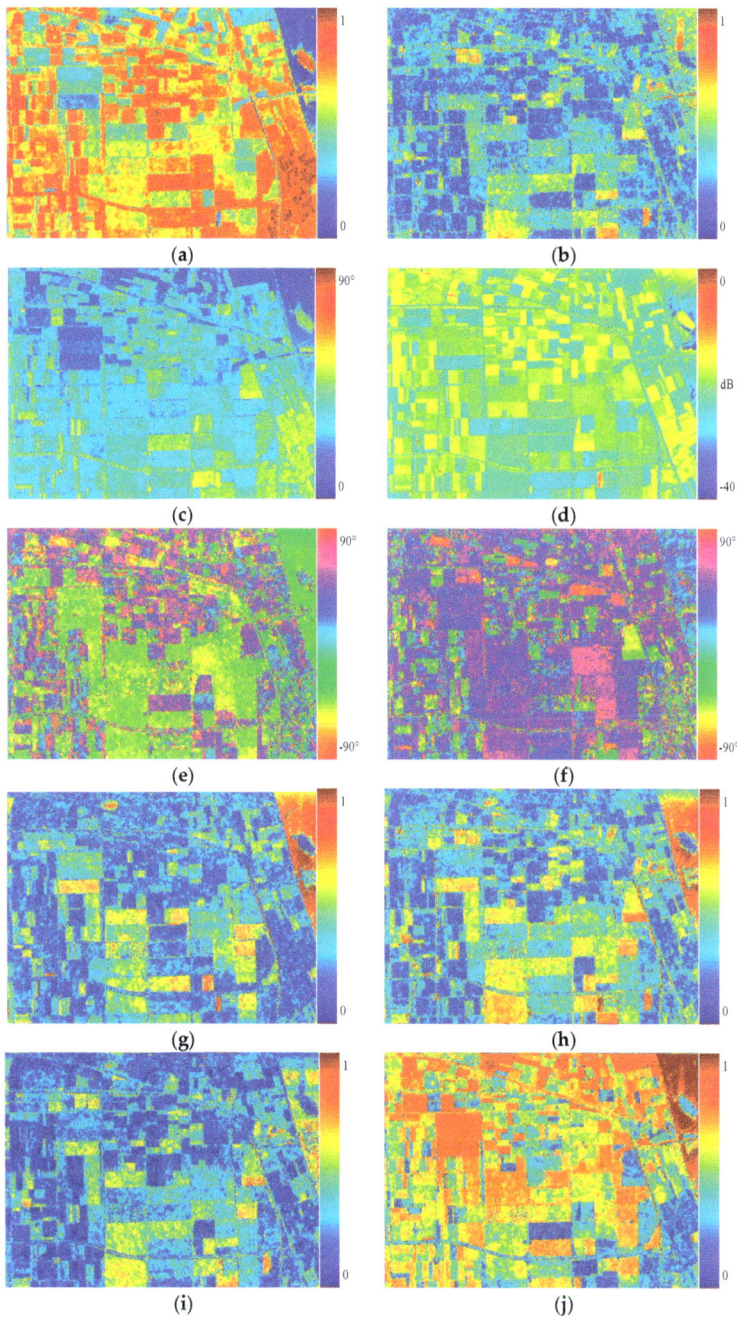

Figure 5. Roll-invariant polarimetric features and selected hidden polarimetric features for AIRSAR data. (**a**) H; (**b**) Ani; (**c**) $\bar{\alpha}$; (**d**) $Span$; (**e**) $\theta_0\_\mathrm{Re}[T_{12}(\theta)]$; (**f**) $\theta_0\_\mathrm{Im}[T_{12}(\theta)]$; (**g**) $\gamma_{-\mathrm{org}}$ of $\left|\gamma_{(HH+VV)-HV}(\theta)\right|$; (**h**) $\gamma_{-\max}$ of $\left|\gamma_{(HH-VV)-HV}(\theta)\right|$; (**i**) $\gamma_{-\mathrm{contrast}}$ of $\left|\gamma_{(HH-VV)-HV}(\theta)\right|$; (**j**) $\gamma_{-\min}$ of $\left|\gamma_{HH-VV}(\theta)\right|$.

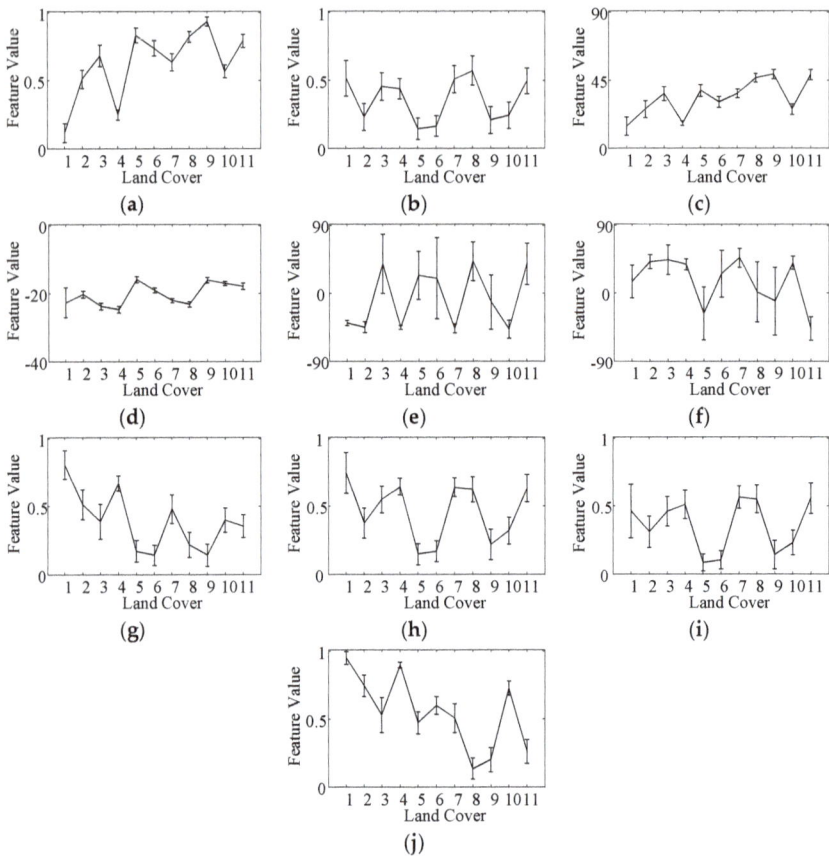

Figure 6. Means and standard deviations comparison for AIRSAR data. Land cover 1–11 indicate water, rapeseed, grasses, bare soil, potatoes, beet, wheat, lucerne, forest, peas, and stembeans respectively. **(a)** H; **(b)** Ani; **(c)** $\bar{\alpha}$; **(d)** $Span$; **(e)** $\theta_0\_Re[T_{12}(\theta)]$; **(f)** $\theta_0\_Im[T_{12}(\theta)]$; **(g)** γ_{-org} of $\left|\gamma_{(HH+VV)-HV}(\theta)\right|$; **(h)** γ_{-max} of $\left|\gamma_{(HH-VV)-HV}(\theta)\right|$; **(i)** $\gamma_{-contrast}$ of $\left|\gamma_{(HH-VV)-HV}(\theta)\right|$; **(j)** γ_{-min} of $\left|\gamma_{HH-VV}(\theta)\right|$.

Similarly, for UAVSAR data (data of 17 June 2012 is used as an example), the four roll-invariant polarimetric features and the six selected hidden polarimetric features are calculated and shown in Figure 7. Means and standard deviations of these features for known land covers are shown in Figure 8. Using $H/Ani/\bar{\alpha}/Span$ only, land cover 1 (oats) and 2 (rapeseed), land cover 1 (oats) and 3 (wheat), land cover 1 (oats) and 5 (soybeans), and land cover 3 (wheat) and 5 (soybeans) may not be successfully discriminated. While they can be discriminated by each hidden polarimetric features of $\theta_0\_Im[T_{12}(\theta)]$, γ_{-org} of $\left|\gamma_{(HH+VV)-HV}(\theta)\right|$, γ_{-min} of $\left|\gamma_{(HH-VV)-HV}(\theta)\right|$, and $\theta_0\_Re[T_{12}(\theta)]$. This further verifies that combining the selected hidden and roll-invariant polarimetric features has better potential to enhance PolSAR classification performance.

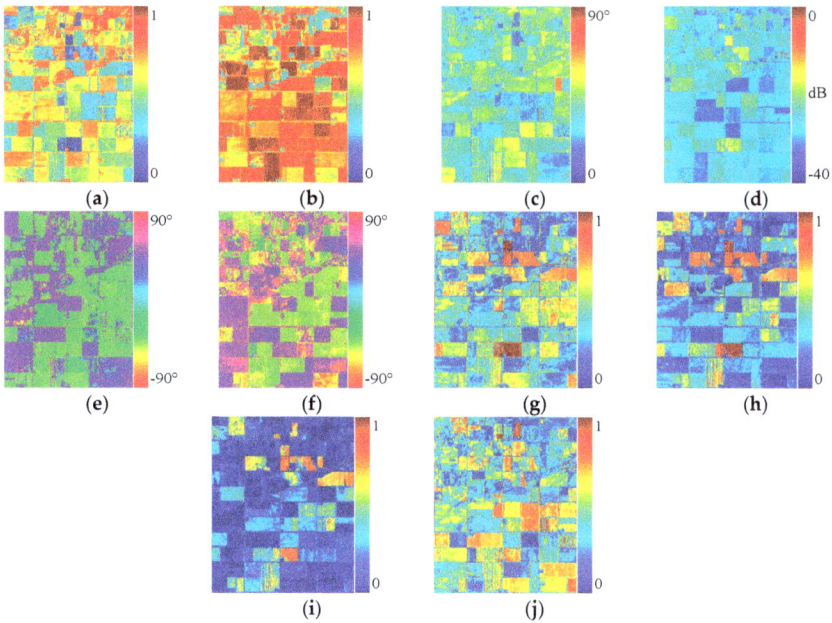

Figure 7. Roll-invariant polarimetric features and selected hidden polarimetric features for UAVSAR data acquired on 17 June 2012. (**a**) H; (**b**) Ani; (**c**) $\bar{\alpha}$; (**d**) $Span$; (**e**) $\theta_0\_\mathrm{Re}[T_{12}(\theta)]$; (**f**) $\theta_0\_\mathrm{Im}[T_{12}(\theta)]$; (**g**) $\gamma_{-\mathrm{org}}$ of $\left|\gamma_{(HH+VV)-HV}(\theta)\right|$; (**h**) $\gamma_{-\mathrm{org}}$ of $\left|\gamma_{(HH-VV)-HV}(\theta)\right|$; (**i**) $\gamma_{-\mathrm{min}}$ of $\left|\gamma_{(HH-VV)-HV}(\theta)\right|$; (**j**) $\gamma_{-\mathrm{min}}$ of $\left|\gamma_{HH-VV}(\theta)\right|$.

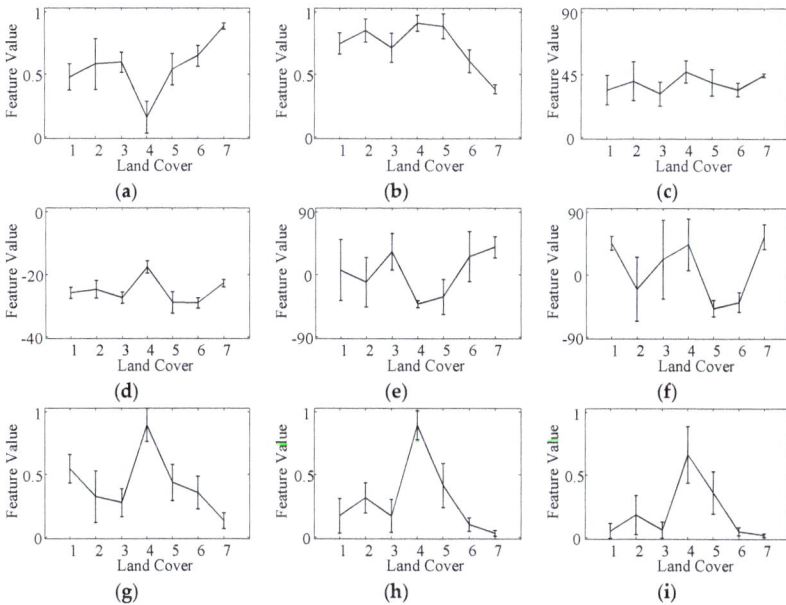

Figure 8. *Cont.*

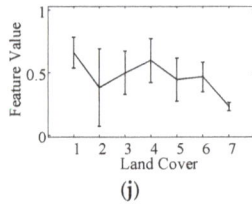

(j)

Figure 8. Means and standard deviations comparison for UAVSAR data acquired on 17 June 2012. Land cover 1–7 indicate oats, rapeseed, wheat, corn, soybeans, forage crops, and broadleaf respectively. (a) H; (b) Ani; (c) $\bar{\alpha}$; (d) $Span$; (e) $\theta_0\_\text{Re}[T_{12}(\theta)]$; (f) $\theta_0\_\text{Im}[T_{12}(\theta)]$; (g) $\gamma_{-\text{org}}$ of $\left|\gamma_{(HH+VV)-HV}(\theta)\right|$; (h) $\gamma_{-\text{org}}$ of $\left|\gamma_{(HH-VV)-HV}(\theta)\right|$; (i) $\gamma_{-\text{min}}$ of $\left|\gamma_{(HH-VV)-HV}(\theta)\right|$; (j) $\gamma_{-\text{min}}$ of $\left|\gamma_{HH-VV}(\theta)\right|$.

4.3. Classification Comparison with AIRSAR Data

In order to demonstrate the added benefits from hidden polarimetric features, the proposed classification method is compared with the conventional classification method which only uses the roll-invariant polarimetric features of $H/Ani/\bar{\alpha}/Span$. For each known land cover in the different PolSAR data, a half of the known samples are randomly selected and used for training the SVM/DT classifier, and the other half of the known samples are used for validation.

First, the AIRSAR data is adopted to compare the classification performance of the conventional and proposed classification methods. Using the SVM classifier, the classification results for the AIRSAR data over eleven known land covers are shown in Figure 9. The classification accuracies are listed in Table 2. It can be observed that the performance of the proposed classification method outperforms that of the conventional one. The overall classification accuracy of the proposed classification method is 95.37%, while that of the conventional classification method is 93.87%. Moreover, for nine of these eleven land covers, the classification accuracies of the proposed classification method are higher than those of the conventional classification method. Especially for grasses, the classification accuracy increase is up to 14.35%, from 66.99% of the conventional method to 81.34% of the proposed method. Besides, the computational costs of the training and validation processing are listed in Table 3. The computational costs are comparable. Finally, the classification results over the full-scene area of this AIRSAR data with the conventional and proposed classification methods respectively are shown in Figure 10.

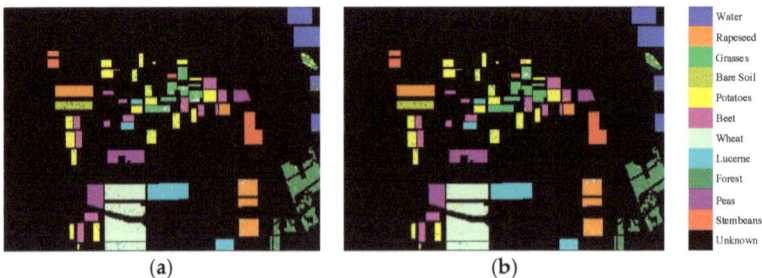

(a)   (b)

Figure 9. Classification results for AIRSAR data over eleven known land covers using support vector machine (SVM) classifier. (a) Conventional classification method; (b) Proposed classification method.

Table 2. Classification accuracies (%) for AIRSAR data using support vector machine (SVM) classifier.

Classification Method	Water	Rapeseed	Grasses	Bare Soil	Potatoes	Beet	Wheat	Lucerne	Forest	Peas	Stembeans	Overall
Conventional	97.65	94.89	66.99	95.84	92.81	94.89	96.12	95.89	92.53	97.85	98.07	93.87
Proposed	**98.39**	**95.38**	**81.34**	**96.75**	**93.42**	**95.76**	**97.58**	**96.33**	**94.08**	97.76	97.44	**95.37**

Table 3. Computational costs (s) of the training and validation processing for AIRSAR data using SVM classifier.

Classification Method	Training	Validation
Conventional	14.3	**38.1**
Proposed	**13.2**	39.7

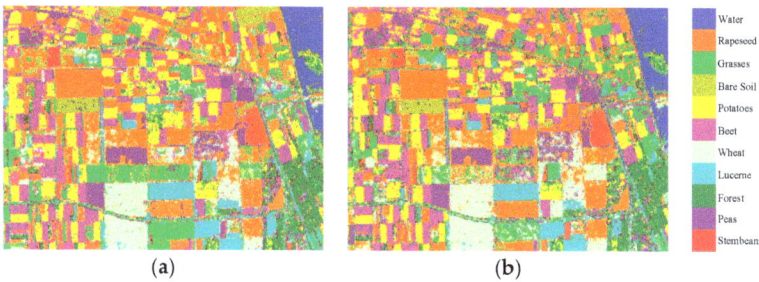

Figure 10. Classification results over the full-scene area of AIRSAR data using SVM classifier. (**a**) Conventional classification method; (**b**) Proposed classification method.

In addition, the DT classifier is used. With the conventional and proposed classification methods, the classification results are shown in Figure 11. The classification accuracies are listed in Table 4. The performance of the proposed classification method still outperforms that of the conventional one. The overall classification accuracy of the proposed classification method is 96.38%, which is higher than the 94.12% of the conventional one. Moreover, for ten of these eleven land covers, the classification accuracies of the proposed classification method are higher than those of the conventional one. Besides, the computational costs of the training and validation processing are listed in Table 5. The computational costs are also comparable. Finally, the classification results over the full-scene area of this AIRSAR data with the conventional and proposed classification methods respectively are shown in Figure 12. The conventional and proposed classification methods both belong to the pixel-based classification methods which are used to deal with the pixels one by one. Besides, in order to compare the performances of the conventional and proposed classification methods, no post-processing is used. Since the misclassification rate is about 5%, these misclassified pixels produce the noisy appearance. Because the DT classifier has a better performance than the SVM classifier and misclassifies less pixels in the full-scene area, the classification results using the DT classifier in Figure 12 look less noisy than those using the SVM classifier in Figure 10. Indeed, some region-based classification methods are suitable to reduce these noisy effects and will be considered in future.

Besides, based on the original AIRSAR data, the overall classification accuracies with different filter sliding window sizes (7×7, 9×9, 11×11, 13×13, 15×15, and 25×25) are examined and listed in Table 6. It is clear that with the same classification method, the larger the filter window size, the higher is the followed classification accuracy. However, the filter window size of 15×15 is chosen based on the tradeoff for both classification accuracy and filter computational cost. In addition, with the same filter window size, the performance of the proposed classification method is still better than that of the conventional one. It verifies the advantage of the proposed classification method further.

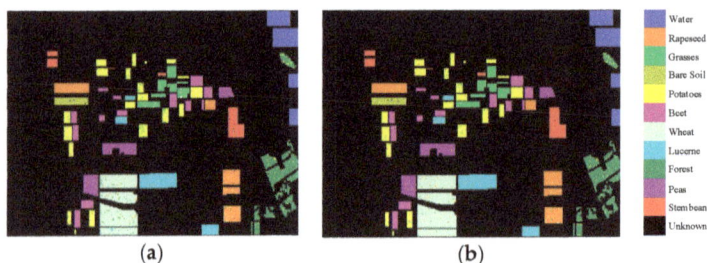

Figure 11. Classification results for AIRSAR data over eleven known land covers using decision tree (DT) classifier. (**a**) Conventional classification method; (**b**) Proposed classification method.

Table 4. Classification accuracies (%) for AIRSAR data using decision tree (DT) classifier.

Classification Method	Water	Rapeseed	Grasses	Bare Soil	Potatoes	Beet	Wheat	Lucerne	Forest	Peas	Stembeans	Overall
Conventional	99.44	94.66	84.56	97.08	91.49	95.64	93.78	94.01	92.16	96.95	96.56	94.12
Proposed	99.39	96.04	93.94	97.09	93.68	96.64	97.49	97.47	94.40	97.52	96.89	96.38

Table 5. Computational costs (s) of the training and validation processing for AIRSAR data using DT classifier.

Classification Method	Training	Validation
Conventional	1.10	0.05
Proposed	2.07	0.04

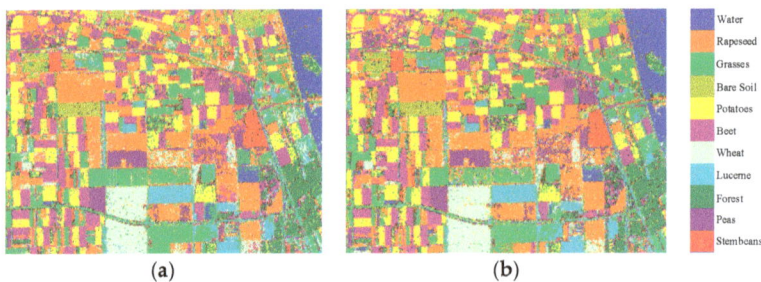

Figure 12. Classification results over the full-scene area of AIRSAR data using DT classifier. (**a**) Conventional classification method; (**b**) Proposed classification method.

Table 6. Overall classification accuracies (%) for AIRSAR data with different filter sliding window sizes.

	7 × 7	9 × 9	11 × 11	13 × 13	15 × 15	25 × 25
Conventional (SVM)	89.07	91.17	92.43	93.21	93.87	95.17
Proposed (SVM)	91.93	93.53	94.41	94.86	95.37	96.63
Conventional (DT)	88.66	91.04	92.39	93.50	94.12	95.57
Proposed (DT)	93.06	94.58	95.32	96.22	96.38	97.28

4.4. Classification Comparison with Multi-Temporal UAVSAR Data

Using the SVM classifier, with the conventional and proposed classification methods, the classification results for the filtered multi-temporal UAVSAR data over seven known land covers are shown in Figure 13. The classification accuracies are listed in Table 7. It is clear that the performance of the proposed classification method is much better than that of the conventional one. The mean overall classification accuracy for four temporal data of the proposed classification method is 97.47%, which is much higher than the 89.59% of the conventional one. Additionally, the overall classification accuracy increments for the four temporal data are 6.45% (17 June: from 90.19% to 96.64%), 6.30% (22 June: from 90.75% to 97.05%), 10.24% (5 July: from 88.03% to 98.27%), and 8.54% (17 July: from 89.39% to 97.93%), respectively. Moreover, the proposed classification method has better robustness for the different temporal PolSAR data. Especially for oats, wheat, and forage crops, the classification accuracy ranges for the four temporal data of the conventional classification method are 77.29–94.61%, 76.85–97.89%, and 56.36–64.51%, while those of the proposed classification method are 94.09–97.39%, 97.79–98.88%, and 83.77–94.16%, respectively. Besides, the computational costs of the training and validation processing are listed in Table 8. It can be seen that the computational costs of the training and validation processing with the proposed classification method are mainly comparable to or less than those with the conventional one. Since the total known samples of each UAVSAR data are about four times as many as those of AIRSAR data, the computational costs with each UAVSAR data are much more than those with AIRSAR data. Finally, the classification results over the full-scene area of these four temporal UAVSAR data with the conventional and proposed classification methods are shown in Figure 14.

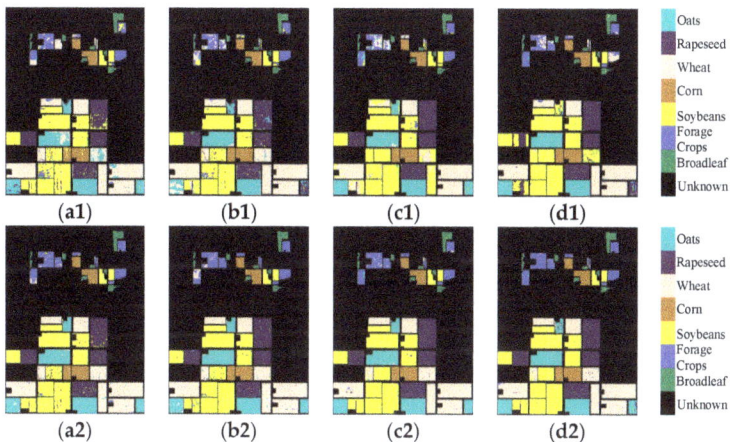

Figure 13. Classification results for multi-temporal UAVSAR data over seven known land covers using SVM classifier. (**a1–d1**) are 17 June, 22 June, 5 July, and 17 July in 2012 with the conventional classification method respectively; (**a2–d2**) are 17 June, 22 June, 5 July, and 17 July in 2012 with the proposed classification method respectively.

In addition, with the DT classifier, the classification results for the multi-temporal UAVSAR data over seven known land covers are shown in Figure 15. The classification accuracies are listed in Table 9. We can see that the performance of the proposed classification method is still better than that of the conventional one. The mean overall classification accuracy for four temporal data of the proposed classification method is 99.39%, which is still higher than the 97.55% of the conventional classification method. In addition, the overall classification accuracy enhancements for the four temporal data are 1.79% (17 June: from 97.48% to 99.27%), 1.65% (22 June: from 97.63% to 99.28%), 1.91% (5 July: from 97.65% to 99.56%), and 2.00% (17 July: from 97.45% to 99.45%), respectively. Moreover, the proposed classification method still has better robustness for the different temporal PolSAR data especially in the areas of oats, rapeseed, and forage crops. Besides, the computational costs of the training and validation processing are listed in Table 10. From it, the computational costs of the validation processing are comparable. Finally, the classification results over the full-scene area of these four temporal UAVSAR data with the conventional and proposed classification methods are shown in Figure 16.

Table 7. Classification accuracies (%) for multi-temporal UAVSAR data using SVM classifier.

	Classification Method	Oats	Rapeseed	Wheat	Corn	Soybeans	Forage Crops	Broadleaf	Overall
17 June	Conventional	86.37	91.70	93.63	96.12	92.64	62.24	98.47	90.19
	Proposed	**96.72**	**96.60**	**98.06**	**98.58**	**96.59**	**88.92**	**98.49**	**96.64**
22 June	Conventional	77.29	93.82	97.89	97.30	94.14	61.38	**98.05**	90.75
	Proposed	**97.21**	**97.76**	**98.88**	**98.93**	**97.68**	**83.77**	97.75	**97.05**
5 July	Conventional	94.61	99.24	76.85	**99.55**	92.31	56.36	**98.63**	88.03
	Proposed	**97.39**	**99.26**	**98.58**	99.45	**99.35**	**90.87**	98.60	**98.27**
17 July	Conventional	82.98	92.19	84.76	**99.78**	97.38	64.51	96.86	89.39
	Proposed	**94.09**	**99.74**	**97.79**	99.75	**99.47**	**94.16**	**97.20**	**97.93**
Mean	Conventional	85.31	94.24	88.28	98.19	94.12	61.12	**98.00**	89.59
	Proposed	**96.35**	**98.34**	**98.33**	**99.18**	**98.27**	**89.43**	98.01	**97.47**

Table 8. Computational costs (s) of the training and validation processing for multi-temporal UAVSAR data using SVM classifier.

Dates	Classification Method	Training	Validation
17 June	Conventional	**610.3**	558.0
	Proposed	699.6	**407.4**
22 June	Conventional	957.0	594.7
	Proposed	**520.1**	**410.5**
5 July	Conventional	784.7	578.6
	Proposed	**633.9**	**285.8**
17 July	Conventional	764.7	435.4
	Proposed	**591.5**	**291.5**

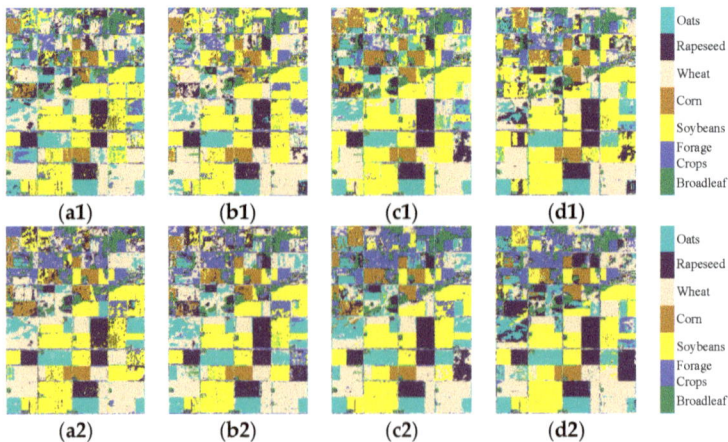

Figure 14. Classification results over the full-scene area of multi-temporal UAVSAR data using SVM classifier. (**a1–d1**) are 17 June, 22 June, 5 July, and 17 July in 2012 with the conventional classification method respectively; (**a2–d2**) are 17 June, 22 June, 5 July, and 17 July in 2012 with the proposed classification method respectively.

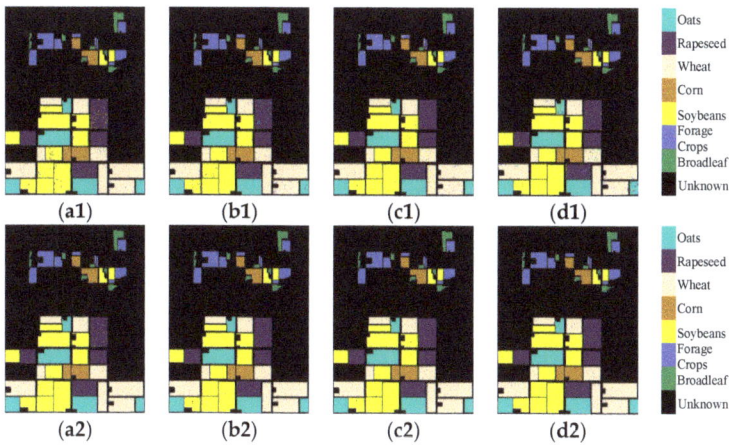

Figure 15. Classification results for multi-temporal UAVSAR data over seven known land covers using DT classifier. (**a1–d1**) are 17 June, 22 June, 5 July, and 17 July in 2012 with the conventional classification method respectively; (**a2–d2**) are 17 June, 22 June, 5 July, and 17 July in 2012 with the proposed classification method respectively.

Table 9. Classification accuracies (%) for multi-temporal UAVSAR data using DT classifier.

	Classification Method	Oats	Rapeseed	Wheat	Corn	Soybeans	Forage Crops	Broadleaf	Overall
17 June	Conventional	98.98	95.79	98.59	98.28	97.09	94.42	98.71	97.48
	Proposed	**99.56**	**98.72**	**99.55**	**99.55**	**99.25**	**98.68**	**99.12**	**99.27**
22 June	Conventional	97.66	97.45	97.87	99.00	98.24	92.72	98.46	97.63
	Proposed	**99.55**	**99.02**	**99.47**	**99.47**	**99.46**	**97.82**	**98.97**	**99.28**
5 July	Conventional	98.39	99.46	97.16	99.80	97.64	91.62	98.64	97.65
	Proposed	**99.46**	**99.59**	**99.66**	**99.81**	**99.78**	**98.46**	**98.77**	**99.56**
17 July	Conventional	94.88	97.77	97.06	99.78	98.61	95.74	97.94	97.45
	Proposed	**99.23**	**99.53**	**99.34**	**99.85**	**99.83**	**98.47**	**98.47**	**99.45**
Mean	Conventional	97.48	97.62	97.67	99.22	97.90	93.63	98.44	97.55
	Proposed	**99.45**	**99.22**	**99.51**	**99.67**	**99.58**	**98.36**	**98.83**	**99.39**

Table 10. Computational costs (s) of the training and validation processing for multi-temporal UAVSAR data using DT classifier.

Dates	Classification Method	Training	Validation
17 June	Conventional	**3.89**	0.13
	Proposed	7.21	**0.12**
22 June	Conventional	**3.67**	0.13
	Proposed	7.37	**0.12**
5 July	Conventional	**3.33**	0.13
	Proposed	7.06	**0.12**
17 July	Conventional	**3.17**	0.13
	Proposed	6.54	**0.11**

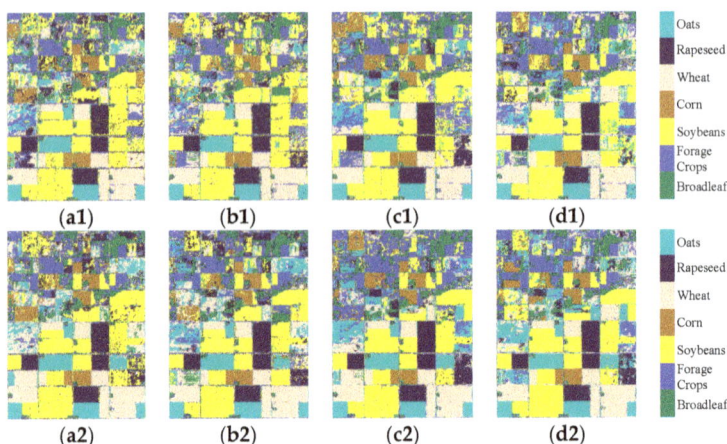

Figure 16. Classification results over the full-scene area of multi-temporal UAVSAR data using DT classifier. (**a1–d1**) are 17 June, 22 June, 5 July, and 17 July in 2012 with the conventional classification method respectively; (**a2–d2**) are 17 June, 22 June, 5 July, and 17 July in 2012 with the proposed classification method respectively.

Besides, based on the original UAVSAR data acquired on June 17, 2012, the overall classification accuracies with different filter sliding window sizes (7×7, 9×9, 11×11, 13×13, 15×15, and 25×25) are investigated and listed in Table 11. Similar conclusion can be obtained as that obtained from AIRSAR data.

Table 11. Overall classification accuracies (%) for UAVSAR data acquired on June 17, 2012 with different filter sliding window sizes.

	7×7	9×9	11×11	13×13	15×15	25×25
Conventional (SVM)	88.23	88.95	89.46	89.86	90.19	91.49
Proposed (SVM)	**94.88**	**95.57**	**96.06**	**96.38**	**96.64**	**97.31**
Conventional (DT)	95.17	96.16	96.76	97.18	97.48	98.04
Proposed (DT)	**98.44**	**98.84**	**99.06**	**99.16**	**99.27**	**99.44**

5. Conclusions and Outlook

This work validates that added benefits can be gained for PolSAR data investigation by mining and utilization of hidden polarimetric information in the rotation domain along the radar line of sight. A PolSAR land cover classification method by combining roll-invariant features and selected hidden features is established. With the added benefits, the land cover discrimination ability is enhanced and the followed classification accuracies are improved significantly. The comparison experiments based on NASA/JPL AIRSAR and multi-temporal UAVSAR data respectively clearly demonstrate the efficiency and advantage of the proposed classification methodology. Moreover, the proposed classification method is also able to achieve better robustness for multi-temporal PolSAR data. Besides, with the SVM/DT classifier, the computational costs of the proposed classification method are always comparable to those of the conventional one. These added benefits are general for the PolSAR land cover classification and the proposed classification technique can be suitable for other kinds of PolSAR data.

This work provides a new vision for PolSAR image interpretation and application. Moreover, other better feature selection schemes, some region-based classification methods, and more advanced classifiers such as CNN classifier are all worth conducting in future.

Acknowledgments: This work was supported in part by the National Natural Science Foundation of China under Grants 41301490, 61490690, and 61490692.

Author Contributions: This work was prepared and accomplished by Chensong Tao, who also wrote the manuscript. Siwei Chen outlined the research and supported the analysis and discussion. He also revised the work in presenting the technical details. Yongzhen Li and Shunping Xiao both suggested the design of comparison experiments and supervised the writing of the manuscript at all stages.

Conflicts of Interest: The authors declare no conflict of interest.

References

1. Lee, J.S.; Pottier, E. *Polarimetric Radar Imaging: From Basics to Applications*; CRC Press: Boca Raton, FL, USA, 2009.
2. Sato, M.; Chen, S.W.; Satake, M. Polarimetric SAR analysis of tsunami damage following the March 11, 2011 East Japan earthquake. *Proc. IEEE* **2012**, *100*, 2861–2875. [CrossRef]
3. Chen, S.W.; Sato, M. Tsunami damage investigation of built-up areas using multitemporal spaceborne full polarimetric SAR images. *IEEE Trans. Geosci. Remote Sens.* **2013**, *51*, 1985–1997. [CrossRef]
4. Chen, S.W.; Wang, X.S.; Sato, M. Urban damage level mapping based on scattering mechanism investigation using fully polarimetric SAR data for the 3.11 East Japan earthquake. *IEEE Trans. Geosci. Remote Sens.* **2016**, *54*, 6919–6929. [CrossRef]
5. Chen, S.W.; Li, Y.Z.; Wang, X.S.; Xiao, S.P.; Sato, M. Modeling and interpretation of scattering mechanisms in polarimetric synthetic aperture radar: Advances and perspectives. *IEEE Signal Process. Mag.* **2014**, *31*, 79–89. [CrossRef]
6. Cloude, S.R.; Pottier, E. A review of target decomposition theorems in radar polarimetry. *IEEE Trans. Geosci. Remote Sens.* **1996**, *34*, 498–518. [CrossRef]
7. Cloude, S.R.; Pottier, E. An entropy based classification scheme for land applications of polarimetric SARs. *IEEE Trans. Geosci. Remote Sens.* **1997**, *35*, 68–78. [CrossRef]
8. Lee, J.S.; Grunes, M.R.; Ainsworth, T.L.; Du, L.; Schuler, D.L.; Cloude, S.R. Unsupervised classification using polarimetric decomposition and the complex Wishart classifier. *IEEE Trans. Geosci. Remote Sens.* **1999**, *37*, 2249–2258. [CrossRef]
9. Freeman, A.; Durden, S.L. A three-component scattering model for polarimetric SAR data. *IEEE Trans. Geosci. Remote Sens.* **1998**, *36*, 963–973. [CrossRef]
10. Yamaguchi, Y.; Moriyama, T.; Ishido, M.; Yamada, H. Four-component scattering model for polarimetric SAR image decomposition. *IEEE Trans. Geosci. Remote Sens.* **2005**, *43*, 1699–1706. [CrossRef]
11. Chen, S.W.; Wang, X.S.; Li, Y.Z.; Sato, M. Adaptive model-based polarimetric decomposition using PolInSAR coherence. *IEEE Trans. Geosci. Remote Sens.* **2014**, *52*, 1705–1718. [CrossRef]
12. Chen, S.W.; Wang, X.S.; Xiao, S.P.; Sato, M. General polarimetric model-based decomposition for coherency matrix. *IEEE Trans. Geosci. Remote Sens.* **2014**, *52*, 1843–1855. [CrossRef]
13. Lee, J.S.; Grunes, M.R.; Pottier, E.; Ferro-Famil, L. Unsupervised terrain classification preserving polarimetric scattering characteristics. *IEEE Trans. Geosci. Remote Sens.* **2004**, *42*, 722–731. [CrossRef]
14. Wang, C.L.; Yu, W.D.; Wang, R.; Deng, Y.K.; Zhao, F.J.; Lu, Y.C. Unsupervised classification based on non-negative eigenvalue decomposition and Wishart classifier. *IET Radar Sonar Navig.* **2014**, *8*, 957–964. [CrossRef]
15. Hong, S.H.; Kim, H.O.; Wdowinski, S.; Feliciano, E. Evaluation of polarimetric SAR decomposition for classifying wetland vegetation types. *Remote Sens.* **2015**, *7*, 8563–8585. [CrossRef]
16. Pajares, G.; Lopez-Martinez, C.; Sanchez-Llado, F.J.; Molina, I. Improving Wishart classification of polarimetric SAR data using the Hopfield Neural Network optimization approach. *Remote Sens.* **2012**, *4*, 3571–3595. [CrossRef]
17. Attarchi, S.; Gloaguen, R. Classifying complex mountainous forests with L-band SAR and Landsat data integration: A comparison among different machine learning methods in the Hyrcanian forest. *Remote Sens.* **2014**, *6*, 3624–3647. [CrossRef]
18. Zhou, Y.; Wang, H.P.; Xu, F.; Jin, Y.Q. Polarimetric SAR images classification using deep convolutional neural networks. *IEEE Geosci. Remote Sens. Lett.* **2016**, *13*, 1935–1939. [CrossRef]

19. Deng, L.; Yan, Y.N.; Wang, C.Z. Improved POLSAR image classification by the use of multi-feature combination. *Remote Sens.* **2015**, *7*, 4157–4177. [CrossRef]
20. Deng, L.; Yan, Y.N.; Sun, C. Use of sub-aperture decomposition for supervised PolSAR classification in urban area. *Remote Sens.* **2015**, *7*, 1380–1396. [CrossRef]
21. Cheng, J.; Ji, Y.Q.; Liu, H.J. Segmentation-based PolSAR image classification using visual features: RHLBP and color features. *Remote Sens.* **2015**, *7*, 6079–6106. [CrossRef]
22. Wang, H.N.; Zhou, Z.M.; Turnbull, J.; Song, Q.; Qi, F. Pol-SAR classification based on generalized polar decomposition of Mueller matrix. *IEEE Geosci. Remote Sens. Lett.* **2016**, *13*, 565–569. [CrossRef]
23. Chen, S.W.; Wang, X.S.; Sato, M. Uniform polarimetric matrix rotation theory and its applications. *IEEE Trans. Geosci. Remote Sens.* **2014**, *52*, 4756–4770. [CrossRef]
24. Chen, S.W.; Wang, X.S. Polarimetric coherence pattern: A visualization tool for PolSAR data investigation. In Proceedings of the IEEE International Geoscience and Remote Sensing Symposium (IGARSS), Beijing, China, 10–15 July 2016; pp. 7509–7512.
25. Chen, S.W.; Li, Y.Z.; Wang, X.S. Crop discrimination based on polarimetric correlation coefficients optimization for PolSAR data. *Int. J. Remote Sens.* **2015**, *36*, 4233–4249. [CrossRef]
26. Xiao, S.P.; Chen, S.W.; Chang, Y.L.; Li, Y.Z.; Sato, M. Polarimetric coherence optimization and its application for manmade target extraction in PolSAR data. *IEICE Trans. Electron.* **2014**, *E97C*, 566–574. [CrossRef]
27. Chang, C.C.; Lin, C.J. LIBSVM: A library for support vector machines. *ACM Trans. Intell. Syst. Technol.* **2011**, *2*, 389–396. [CrossRef]
28. Webb, A.; Copsey, K. *Statistical Pattern Recognition*, 3rd ed.; John Wiley & Sons Ltd.: Chichester, UK, 2011.
29. Chen, S.W.; Wang, X.S.; Sato, M. PolInSAR complex coherence estimation based on covariance matrix similarity test. *IEEE Trans. Geosci. Remote Sens.* **2012**, *50*, 4699–4709. [CrossRef]

remote sensing

MDPI

Article

A SAR-Based Index for Landscape Changes in African Savannas

Andreas Braun * and Volker Hochschild

Institute for Geography, University of Tübingen, Rümelinstraße 19-23, 72070 Tübingen, Germany;
volker.hochschild@uni-tuebingen.de
*   Correspondence: an.braun@uni-tuebingen.de; Tel.: +49-7071-29-78940

Academic Editors: Timo Balz, Uwe Soergel, Mattia Crespi, Batuhan Osmanoglu, Nicolas Baghdadi and
Prasad S. Thenkabail
Received: 24 March 2017; Accepted: 9 April 2017; Published: 11 April 2017

Abstract: Change detection is one of the main applications in earth observation but currently there are only a few approaches based on radar imagery. Available techniques strongly focus on optical data. These techniques are often limited to static analyses of image pairs and are frequently lacking results which address the requirements of the user. Some of these shortcomings include integration of user's expertise, transparency of methods, and communication of results in a comprehensive understandable way. This study introduces an index describing changes in the savanna ecosystem around the refugee camp Djabal, Eastern Chad, based on a time-series of ALOS PALSAR data between 2007 and 2017. Texture based land-use/land cover classifications are transferred to values of natural resources which include comprehensive pertinent expert knowledge about the contributions of the classes to environmental integrity and human security. Changes between the images are analyzed, within grid cells of one kilometer diameter, according to changes of natural resources and the variability of these changes. Our results show the highest resource availability for the year of 2008 but no general decline in natural resources. Largest loss of resources occurred between 2010 and 2011 but regeneration could be observed in the following years. Neither the settlements nor the wadi areas of high ecologic importance underwent significant changes during the last decade.

Keywords: ALOS PALSAR; multi-temporal analysis; resource monitoring; refugee camps; synthetic aperture radar (SAR)

1. Introduction

Detecting and understanding processes at the earth's surface are among the key tasks of spaceborne remote sensing. Thousands of images stored in archives allow for the analysis of dense time-series of nearly every region of the earth. In particular is the Landsat continuity mission, delivering valuable data since the 1970s, which provides an excellent foundation of long term observations [1,2]. Their potential for the mapping of land-use and land cover (LULC) has been demonstrated in numerous studies which exploit the temporal variability of the image information in different methodological frameworks [3–10]. Among others, Song et al. portray the problems regarding atmospheric disturbances in the data and raise the question to what extent these applications are affected by cloud cover [11]. Of primary concern is large-scaled classification of LULC often being constrained by cloud cover. According to the US Department of Energy, about 52% of global land surfaces are covered by clouds on average [12]. In particular, regions within the intertropical convergence zone (ITC) are highly affected by seasonal or full-year cloud coverage [13–15] which hinders proper analysis of LULC based on series of optical satellite imagery [16–18].

To overcome this dependency upon favorable atmospheric conditions, satellites with synthetic aperture radar (SAR), operating at wavelengths which can penetrate cloud cover, are employed.

Their reliability in acquiring usable imagery is one of the main reasons to utilize them in LULC applications. Accordingly, the long-term missions of ERS-1/2 [19] and RADARSAT-1/2 [20], plus the relatively recent Sentinel-1 constellation [21] which was launched in 2014, show high potential for the investigation of changes in land cover over decades. They also serve well as complementary sources when operating at the same wavelength [22,23].

The use of SAR imagery has proven effective for LULC classification in many cases. Early studies mostly applied knowledge-based methods on SAR backscatter in order to separate different classes of land cover [24–26]. In fact, the first article published in Remote Sensing of Environment, back in 1969, was an interpretation key for SAR backscatter at the landscape level [27]. These studies require detailed a-priori knowledge about the study area but offer a high degree of control to the user. Additional features were used to further increase the classification quality in later approaches, mainly interferometric parameters such as coherence [28–30], but also textural information within the intensity values at different levels [31–33]. Besides these technological advances, new methods of supervised classification were developed to assign classes based on sample data. These were required because, in contrast to optical image information, SAR parameters are not necessarily of consistent units and value ranges. Among the most popular methods are Bayesian classifiers [34–36], neural networks [37–39] and random forest classifiers [40–42]. These developments targeting LULC classifications were complemented by progress in SAR polarimetry. Deeper knowledge of the backscattering behavior of surfaces and decompositions into different scattering mechanisms additionally increased the quality of analyses related to land cover [43–46].

Utilizing SAR imagery in a multi-temporal context to monitor changes in land-use or land cover requires particular attention because of the characteristics of image acquisition and signal propagation. While the very identification of areas of change is scientifically well-proven and widely-used [47–49], quantitative and long-term monitoring of distinct classes requires accurate calibration and a robust handling of speckle [50,51].

Studies investigating landscape sensitivity, the severity of disturbances and human impact based on SAR data are rare: Townsend and Foster proposed a statistical indicator for the intensity of flooding for the Roanoke River floodplain (United States) based on eleven RADARSAT-1 scenes acquired over a period of seventeen months [52]. They used the distinctive signature of water in SAR images expressed in usually very low backscatter values. They achieved overall model accuracies of 87.8%, however they only looked at two classes (flooded and non-flooded). Beisl et al. followed a similar approach for the estimation of landscape sensitivity towards floods based on two JERS-1 mosaics in Western Arizona [53]. Their study targeted four LULC classes and five levels of flood hazards but only compared two images (pre- and post-flooding).

Hoffmann et al. derived forest fire damage in East Kalimantan, Indonesia, based on 56 scenes of ERS-2 from 1997 [54]. They discriminated three different classes of burn severity and additionally derived a LULC classification. This study is one of the few making full use of large SAR archives for the assessment of impacts of disturbances on landscapes.

As well, human impacts upon forest systems deforestation of the Amazon rainforests is of special interest in remote sensing studies. Saatchi et al. made use of the polarimetric covariance between channels of SIR-C [55], comparing an intra-annual pair of images—interferometric signatures and coherence [56]. L-band data have been found to be of special importance due to its pronounced interaction with vegetation volumes such as canopy structures [57–59]. Most of the studies, however, do not make use of large sets of SAR images provided by the JAXA archives.

For the African continent, Mitchard et al. have found a robust relationship for the estimation of biomass changes in savanna ecosystems [60,61]. They derived change maps based on 7 biomass classes for an image pair from 1997 (JERS-1) and 2007 (ALOS PALSAR).

Shen et al. used polarimetric RADARSAT-2 data to derive landscape metrics from eight LULC classes in the Nanjishan Wetland Nature Reserve, PR China [62]. Although this approach aims at a deeper understanding of the classified landscape it doesn't make allowance to investigate temporal changes.

Based on the prior developments and findings we identified the following research deficits. While indices of spatial change based on optical data are numerous, there is almost no SAR-based methodology for long-term analysis of landscape developments at a class level. Pixel-based approaches developed for optical data are not applicable for SAR data as they do not allow for smaller misclassifications or isolated pixels arising from speckle effects. Additionally, most studies strongly focus on a stepwise comparison of image pairs without considering the total variation along the full time-series [63]. This is especially a problem in landscapes with high dynamics due to wild-fires, land degradation, vegetation encroachment or strong human impact. For those situations, distinct indices have to be developed, especially when dealing with SAR or very high resolution (VHR) optical data [64]. Other indices are highly elaborate but hard to read for people from outside the field of research. Accordingly, Walker and Peters argue that findings of multi-temporal remote sensing products are often complicated to read or even misleading and susceptible to making false conclusions [65].

In order to perform long-term analyses of landscape change independently from cloud cover, a robust and transparent approach based on radar data has to be developed. As a result, an integrative index should explain both the location and intensity of changes, as well as their implications for environmental integrity and human well-being. This index must be reproducible and adjustable to different needs of the users according to the significance of the targeted classes. Still, it has to be easy to interpret for anyone, especially because stakeholders and decision-makers are often not familiar with or interested in the technical background [66–68].

In this paper, we propose an approach challenging these requirements. For this purpose, we investigate landscape changes in a savannah ecosystem in Eastern Chad. As it is located at the southern border of the Sahel, desertification, impacts of climate change, and limited resources are of main concern [69,70]. These types of savannas are observed to show significant changes regarding vegetation persistence during the last two decades, both positive and negative [71]. The study is used to answer the following questions: How, where, and how much did the semi-arid landscape change during the last ten years? Can an overall decrease of natural resources, such as the availability of fire wood, be observed and to what extent? Do the LULC types show different developments regarding dynamics and variability? Can land degradation be observed in the direct surroundings of the settlements as an indicator of human impact? And finally, can these aspects be visualized in an output map which is easy to read and still includes the severity of changes during the entire period investigated?

2. Materials and Methods

2.1. Study Area

The study area is located in the center of the Sila region of Eastern Chad (Figure 1, upper right map). It hosts a total population of about 470,000 inhabitants, resulting in a moderately low population density of 12.5 inhabitants per square kilometer [72]. It includes the town of Goz Beïda and the refugee camp of Djabal which was opened to Sudanese refugees in 2004 as a consequence of the large influx of people seeking shelter from the Darfur crisis [73]. Djabal is one of numerous refugee settlements (both temporary and permanent) along the 500 km long Sudanese border within Chad's territory which were established for people fleeing from civil war and environmental degradation [74,75].

The region lies at the southern border of the Sahel and is regularly affected by droughts [76] but receives more than 600 mm of annual precipitation on average, allowing for small-scaled agricultural use [77]. A decrease of summer rains and an increase of temperatures by 0.8° Celsius for the past 25 years have been reported, leading to both stronger dependency on crop yields and higher pressure on usable land [78]. Precipitation is mainly limited to the rainy season between June and September while the rest of the year is considered as highly arid. Climate can therefore be described as BSh (Hot semi-arid climate), referring to the Köppen-Geiger scheme.

The area is dominated by Precambrian bedrock and Tertiary sediments as it lies at the transition between the geological Sud province and the Nubian Shield within the Chad Basin [79]. Both camp Djabal and Goz Beïda are located at around 575 m above sea level but they are framed by the Hadjer Arkop massif in the south and northwest reaching up to 900 m [80]. Due to its location at the transition from desert to savannas most of the study area is covered by edaphic grassland, rupicolous shrubs, and scattered dry forest. Generally, tree cover is sparse in this area and mainly consists of single deciduous trees of the Combretaceae family [81]. The ridges of the Hadjer Arkop massif are sparsely covered by various Boswellia trees and shrubs [82]. Wadi Aouada, ranging from West to North across the study area, is accompanied by smaller semi-deciduous riparian forests (Figure 1, main map). Agricultural use of this land lies at around 30% and mainly concentrates around the few settlements. It is composed of small-scaled semi-permanent cultivation and bush fallow.

Goz Beïda lies along an important transport axis of Eastern Chad which connects the large cities of the North (Fada) with the ones in the South (Sarh). A Dutch military base (Ciara) under the mandate of the European Union was temporarily established between 2008 and 2009 as an additional peace force in the region [83]. Today, only the Goz Beïda airport remains at this location.

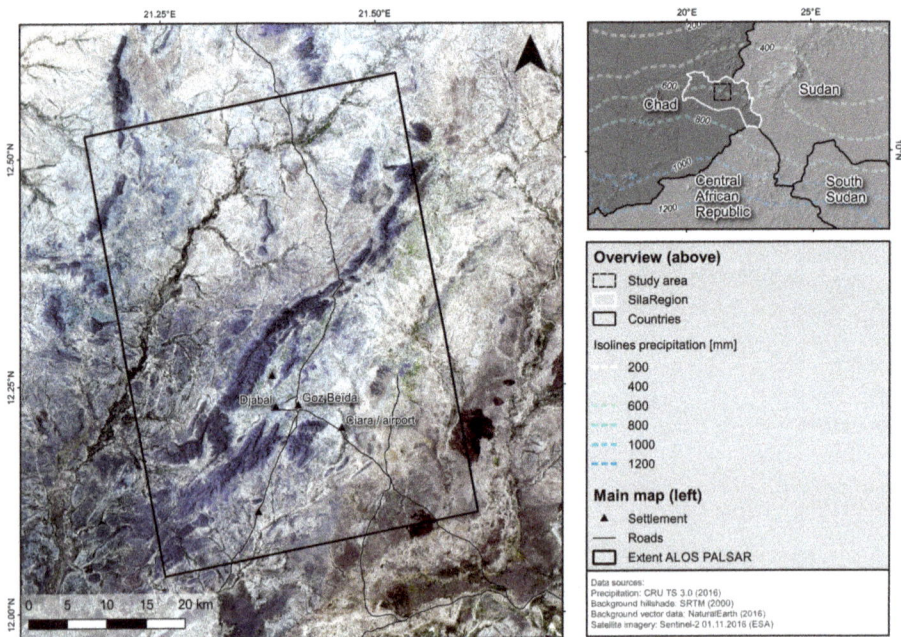

Figure 1. Location and landscape characteristics of the study area.

2.2. Data and Pre-Processing

2.2.1. Satellite Imagery

As the investigated ecosystem shows strong seasonal dynamics, the data to be used should be selected carefully. Therefore, all investigated data were acquired during the dry season which extends from November until March where rainfall below 5 mm is expected. These conditions are favorable because SAR backscatter strongly varies with changing soil moisture. This could cause temporally misleading signatures during the rainy season, e.g., high values at smooth and non-vegetated bare soil.

Additionally, overall vegetation dynamics are smaller during the dry season which increases the inter-annual comparability of the images despite smaller temporal differences.

Table 1 lists all satellite images used in this study. Considering its temporal coverage, wavelength, and high spatial resolution, ALOS PALSAR has been chosen as a main input. Landsat data was additionally used for the collection of reference data. For each year analyzed, a pair of ALOS and Landsat imagery has been found with a temporal difference Δt between 0 and 30 days. There is a gap of observation between 2012 and 2015 because the ALOS PALSAR archive only features data between 2006 and 2011 and ALOS-2 (Daichi) was launched in May 2014. Landsat ETM+ data experienced a failure of its Scan Line Corrector mechanism in May 2003 causing striped gaps in the data (SLC-off) [84]. However, our study area lies in the center of the scene where this effect is minimal, since 2015 data from Landsat 8 (OLI/TIRS) could be used instead.

Table 1. Satellite imagery used in this study.

Year/Δt	Date	Sensor	Comment
2007/18 days	24 December 2006 11 January 2007	ALOS PALSAR Landsat ETM+	SLC-off
2008/2 days	27 December 2007 29 December 2007	ALOS PALSAR Landsat ETM+	SLC-off
2009/18 days	29 December 2008 16 January 2009	ALOS PALSAR Landsat ETM+	SLC-off
2010/30 days	2 December 2009 1 January 2010	Landsat ETM+ ALOS PALSAR	SLC-off
2011/20 days	19 February 2011 11 March 2011	ALOS PALSAR Landsat ETM+	SLC-off
2015/0 days	23 January 2015 23 January 2015	ALOS-2 Landsat OLI/TIRS	
2016/6 days	22 January 2016 28 January 2016	ALOS-2 Landsat OLI/TIRS	
2017/15 days	2 March 2017 15 February 2017	ALOS-2 Landsat OLI/TIRS	

2.2.2. Pre-Processing and Collection of Samples

ALOS data was obtained as Level 1.1 products in Slant Range Complex format [85,86]. Pre-processing included radiometric calibration to radar brightness (Beta Naught, $\beta°$ [87]), multi-looking (n = 2), terrain flattening to normalize radiometric effects caused by different incidence angles (Flattened Gamma Naught, $\gamma°$ [88]) and Range-Doppler terrain correction to adjust topographic distortions using a digital elevation model (1 Arc-Second SRTM) [89]. All rasters were resampled to a common ground resolution of 30 m. No speckle filtering was applied in order to conserve image texture as accuracies of classifications based on SAR texture alone are reported to decrease when speckle was removed [90,91].

In order to derive additional information layers we derived image textures based on the concept of Grey-Level Co-Occurrence Matrix (GLCM) [92] consisting of Cluster Prominence, Cluster Shade, Correlation, Difference Of Entropies, Difference Of Variances, Energy, Entropy, Grey-Level Nonuniformity, HT10, Haralick Correlation, High Grey-Level Run Emphasis, IC1, IC2, Inertia, Inverse Difference Moment, Long Run Low Grey-Level Emphasis, Low Grey-Level Run Emphasis, Mean, Run Length Non-uniformity, Run Percentage, Short Run High Grey-Level Emphasis, Short Run Low Grey-Level Emphasis, Sum Entropy, Sum Variance, and Variance. Each of these 25 textures was calculated for kernels of 3, 9, and 15 pixels in order to extract patterns emerging at different spatial scales, leading to a total of 75 SAR texture layers per analyzed year.

Landsat data was obtained as Level-1T (terrain corrected) products. All rasters were radiometrically corrected by applying conversion top of atmosphere (TOA) reflectance and dark object subtraction (DOS) [93,94].

Six LULC classes were defined for the analysis: (1) urban areas, (2) bare soil, (3) bare rock, (4) grassland, (5) shrubland, and (6) forest. We did not include water as a separate class since the study area does not feature any permanent water bodies and the few remaining temporary flooded areas are covered by the forests of Wadi Aouada.

Each class is represented by training areas which were automatically derived from the Landsat scenes. We utilized indices and threshold values from several studies to define areas which characterize the corresponding class to a best possible degree in each scene. These are listed in Table 2. The selection of training samples underlies two steps. First, representative areas for every class are derived in each year observed (see Table 1). Areas which are assigned to the same class throughout all scenes were then considered as stable over time and transparent according to the corresponding LULC. For example, a Normalized Difference Vegetation Index (NDVI) value greater than 0.2 is reported as high for the dry season in Sudanian savanna as found in the study area [95]. If a pixel fulfills this criterion throughout all images it was considered as forest. Information from near infrared (NIR) and short wave infrared (SWIR) bands was used for criteria of the classes grassland and shrubland but also for thresholds for the abiotic surfaces of bare rock and soil. As reference for urban areas, the extent of camp Djabal in the year 2007 was digitized from the Landsat image. These areas are observed to be stable over time as the camp reached a stable phase.

To prevent misclassifications, we removed areas smaller than 500,000 m^2 from the identified sample areas. Recognizing the these kinds of savanna ecosystems are highly affected by wildfires [96], we calculated the Normalized Burn Ratio (NBR [97]) for all years and scenes. This characterizes burnt areas which were excluded from the identification of sample areas because they no longer reveal which land cover was present before the fire. Furthermore, clouded areas were excluded throughout all scenes.

In a second step, a number of 1600 random points was generated within the remaining areas as sample locations for the SAR classification (see Section 2.3). This technique guarantees that the sample points were chosen both stratified and random while partially respecting the spatial occurrence of each class within the study area. Table 2 lists the criteria used for their identification of the sample areas and the final number of samples per class.

Table 2. Land-use and land cover (LULC) classes and identification of sample areas in the study area.

Class	Samples	Criterion	Source
Urban area	161 (7.2%)	Area of the camp in 2007	-
Bare soil	390 (17.6%)	Red < 0.25 & SWIR1 > 0.35	Drury (2001) [98]
Bare rock	352 (15.9%)	SWIR1/SWIR2 > 1.5	Drury (2001) [98]
Grassland	374 (15.6%)	NDVI > 0.15 & NDVI < 0.25	Forkuor et al. (2012) [95]
Shrubland	126 (5.6%)	Red < 0.2 & NIR > 0.3 & SWIR < 0.3	Liu et al. (2016) [99]
Forest	172 (7.8%)	NDVI > 0.2	Forkuor et al. (2012) [95]
Burnt areas	0 (0%)	NBR > 0.15	López-García & Caselles (1991) [97]

Figure 2 shows the remaining areas which were used for the stratified random sampling. It shows the result of the criteria given in Table 2 which helped to identify stable LULC areas over the investigated period. Clearly visible are camp Djabal in the middle, surrounded by mostly bare soil resulting from less vegetation cover and agricultural use, the bare rocks of Hadjer Arkop and the denser forests of Wadi Aouada in the Northwest. As indicated in Figure 1 grassland and shrubland cover the wide plains in the North and East.

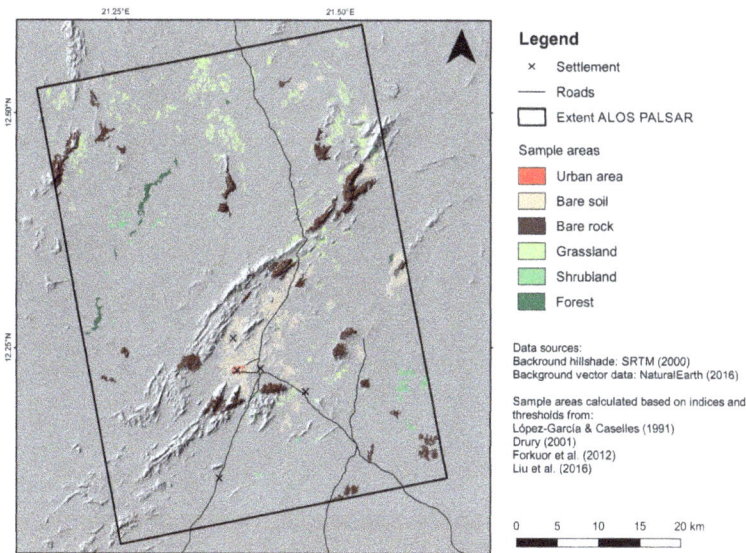

Figure 2. Identified areas used for the stratified random sampling.

2.3. Image Classification

Image classification was performed for each SAR image in Table 1 in order to analyze the temporal developments in the study area. However, as a huge number of SAR textures are used covering many different value ranges in various units, traditional classifiers for pixel-based approaches—such as k-means clustering [100] or Maximum Likelihood estimation [101]—are not suitable. We therefore chose a Random Forest classifier (RF [102]) for our study. It is based on the concept of classification and regression trees (CARTs [103]) and repeatedly uses random subsets of the training data for the modeling of target classes. This automatically makes the best use of the feature data (texture layers in our case) with the best prediction ability to make LULC estimations for the full scene. For each year, we calculated a number of n = 500 classification trees based on random subsets of $\sqrt{n} = 22$ features. Figure 3 shows the outcomes of the classification as an intermediate product of the analysis.

An accuracy assessment has been performed by manually collected sampling points. These points were visually placed on the Landsat image of each year independently from the sampling areas described in Section 2.2 so they represent the real occurrence of each LULC during each time step. A number of 150 points was collected per class and year which were then compared with the results from the SAR classification for the generation of a confusion matrix.

We achieved overall accuracies of 84.4% (2007), 83.3% (2008), 85.7% (2009), 85.0% (2010), 84.0% (2011), 80.3% (2015), 82.3% (2016) and 81.47% (2017) with respective Kappa values of 0.81 (2007), 0.80 (2008), 0.83 (2009), 0.81 (2010), 0.81 (2011), 0.76 (2015), 0.78 (2016) and 0.77 (2017). These may appear low for change detection approaches at first sight, but the proposed evaluation of changes by an aggregated index (see Section 2.4) is not pixel-based and compensates smaller misclassifications. As shown in Tables 3 and 4, urban areas and bare rock show the highest accuracies due to their clear signal in both the optical and the radar images. Bare soil areas also show high accuracies because of its distinct backscatter characteristics during the dry season, but were misclassified as grassland in some areas of transition. The tables also show that grassland was over-estimated throughout all images (low user's accuracies) while shrubland and forest reveal the lowest producer's accuracies due to their similar signature in the SAR data.

Table 3. Producer's accuracy for the texture-based classification of SAR images using the Random Forest classifier.

	2007	2008	2009	2010	2011	2015	2016	2017
Urban area	89.3%	91.3%	90.7%	92.0%	89.3%	86.0%	90.7%	70.53%
Bare soil	92.0%	90.7%	92.0%	93.3%	92.0%	89.3%	86.7%	77.88%
Bare rock	96.0%	95.3%	96.0%	94.0%	96.0%	94.0%	93.3%	90.91%
Grassland	83.3%	80.7%	88.7%	83.3%	82.0%	94.0%	76.7%	82.23%
Shrubland	75.3%	70.7%	76.0%	77.3%	75.3%	68.7%	76.7%	82.64%
Forest	70.7%	71.3%	70.7%	70.0%	69.3%	67.3%	64.0%	66.92%

Table 4. User's accuracy for the texture-based classification of SAR images using the Random Forest classifier.

	2007	2008	2009	2010	2011	2015	2016	2017
Urban area	100.0%	100.0%	100.0%	100.0%	98.5%	99.2%	100.0%	100.0%
Bare soil	82.6%	84.5%	84.7%	85.4%	82.1%	76.1%	79.3%	82.5%
Bare rock	97.3%	95.3%	99.3%	97.2%	94.1%	95.9%	94.0%	95.9%
Grassland	63.1%	61.4%	64.6%	62.8%	64.7%	57.8%	59.3%	57.6%
Shrubland	83.7%	84.8%	85.7%	82.3%	82.5%	80.5%	80.5%	78.4%
Forest	89.8%	82.3%	90.6%	92.9%	89.7%	84.2%	93.2%	87.3%

2.4. Index of Landscape Change

As described in the previous section, pixel-based image classifications based on SAR data can reveal smaller misclassifications due to the lack of spectral diversity. These can, of course, lead to false conclusions in multi-temporal approaches when wrongly classified pixels may appear as change in land-use or land cover. Furthermore, changes from one class into another within single pixels do not allow for implications on the state of the environment, nor do they reveal the consequences of these changes for man. For this reason, Hagenlocher et al. proposed a weighted index which estimates the impact of LULC changes within larger aggregated units in terms of environmental integrity and human security [104]. It is uses expert-based weightings which determine its ecologic and social-economic value of each LULC class so that changes can directly be interpreted as a percentage of resource depletion for different sectors in the study area. This weighted natural resource depletion index (NRD$_w$) was originally designed for very high-resolution data which needed to be aggregated at a larger level in order to overcome the limitations of pixel-based approaches for change detection. We transferred this concept on to the SAR data to compensate the deficiencies of our classification resulting from lack of spectral diversity, as reported in Section 2.3.

The weights were defined by two regional experts with ecological and humanitarian background as shown in Table 5. They refer to environmental integrity (EnvInt) and human security (HumSec) which were then averaged to a mean value of abundant natural resources (NR) of each pixel. As our study addresses the human-related aspects of landscape change, we used a ratio of 0.35/0.65 for the calculation of NR to place emphasis on their socio-economic impact.

Table 5. Importance of the analyzed land-use and land cover classes on environmental integrity (EnvInt) and human security (HumSec) for the calculations of an integrated value of natural resources (NR).

Class	EnvInt 35%	HumSec 65%	NR 100%
Urban area	0.00	0.50	0.313
Bare soil	0.30	0.35	0.331
Bare rock	0.20	0.25	0.231
Grassland	0.50	0.45	0.469
Shrubland	0.70	0.75	0.731
Forest	0.95	0.90	0.919

We chose a hexagon grid with a diameter of 1 km as a unit of analysis which aggregates the NR values of inherent LULC classes according to their spatial proportions within the grid cells. We chose hexagons as spatial units because they are suitable to represent nearest neighbor areas in a regular structure and, compared to rectangles, reduce sampling bias at their edges. Additionally, they are reported to support visual inspection of spatial patterns [105]. These NR values can then pairwise be compared per grid cell in order to retrieve its percentage increase or decrease of resources for different time increments. An example for the calculation of the NRD$_w$ is given in Figure 3.

Figure 4 shows the design of the study: Based on the eight SAR images eight LULC classifications have been derived. The weighted index for Natural Resource Depletion is then calculated per image pair. The result is shown as an example for the first image pair of 2007 and 2008. This map indicates the spatial distribution of changes as well as their estimated impact on natural environment.

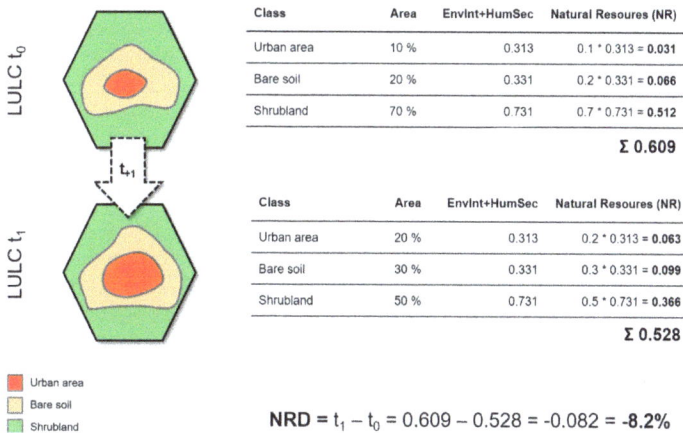

Figure 3. Calculation of the weighted Natural Resource Depletion Index (NRD$_w$).

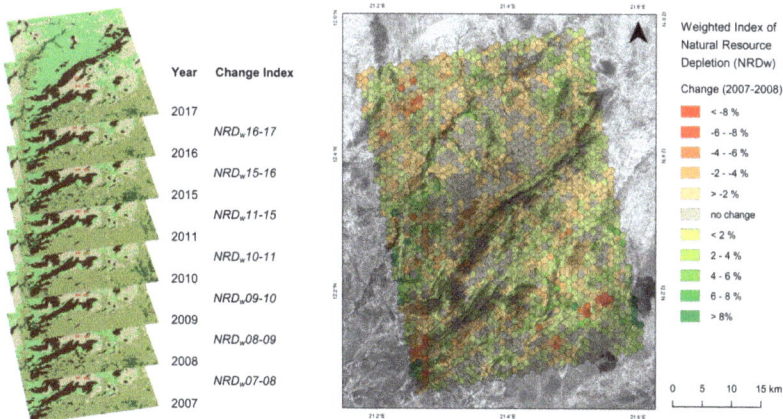

Figure 4. Left: Design of the approach: Each classified image delivers aggregated values of natural resources. Right: Changes between the years are expressed as weighted Natural Resource Depletion (NRD$_w$), demonstrated for the period between 2007 and 2008.

2.5. Index of Landscape Variability

Besides observing inter-annual changes, general conclusions about developments within an area can be derived best based on direct comparisons of conditions at the beginning and the end of the investigated time. However, this excludes two major points in change detection.

Firstly, savanna ecosystems are subject of large biotic and abiotic variation at multiple spatiotemporal scales [106]. It is a key task of remote sensing to reveal these variations to an adequate degree [107]. Besides simply comparing the intensity of changes, information on the temporal variability of regions—or in other words, the sum of all changes along the investigated period—has to be communicated as well. Surely, some areas remain stable after the transition of one LULC class into another while others tend to underlie regular variations. Analyzing changes over multiple images, especially in a region at the border between two ecosystems, surely should include the aspect of resilience towards long-term exterior influences.

Secondly, class-based measurements are always subject to smaller misclassifications, especially if the data source is challenging. It is still widely reported that classifications based on SAR data alone are outperformed by approaches using multispectral optical data [108–110]. Similarly, single misclassifications within a time-series can distort the results towards changes which did not happen. In addition to that, the dates of image acquisitions are highly sensitive towards seasonal variations. Even if all images were taken at the same date of the year, inter-annual shifts in phenology of even a few weeks could cause the erroneous detection of 'pseudo changes'.

Consequently, a second indicator is needed which spatially allocates high temporal variation within the study area and simultaneously quantifies the number of smaller changes which might only be subject to seasonal shifts in phenology.

We therefore decided to apply a post-classification change vector analysis. Vector-based approaches make use of a feature space consisting of spectral or classified information and the temporal dimension [111,112]. In our case, the variable determining the value of each hexagon grid cell is its weighted sum of natural resources as described in the previous section. These values are then plotted against the temporal period which is analyzed as exemplified in Figure 5: In a first step, the length of the vector describing overall change (C_o) is calculated based on the NR values of 2007 and 2017. As a second variable contributing to the variability of an area, the length of the change vectors between each chronological image pairs are summed up to create the sum of annual changes (C_a). For reasons of normalization, a factor of 10 has been applied to NR values for the calculation of vector lengths (e.g., a NR decrease of 0.55 to 0.40 during one year leads to a x-distance of 1.5 instead of 0.15. This is done because the y-distance is 1 except for 2011–2015). The ratio of C_o and C_a is then calculated and describes the variability (v) of the corresponding hexagon cell throughout the observed period. If C_o and C_a are of same length, v gets the value of 1. The longer the vectors of annual changes in relation to the overall change, the higher is the calculated variability v which is therefore a dimensionless value between 0 and 7, as the sum of the maximum NR value of 1 for a period of seven investigated years. This information can then be included in the discussion of the results as an indicator of variability and instability of an area.

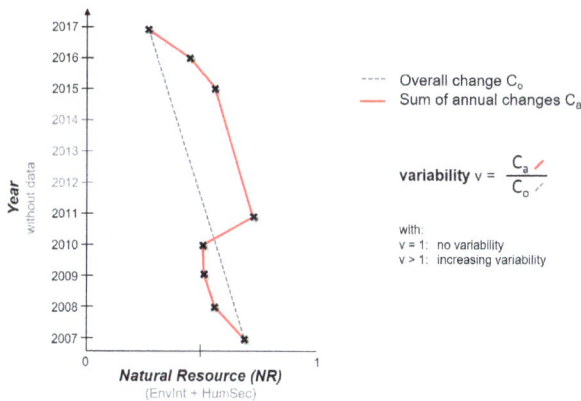

Figure 5. Calculation of the variability v of an area.

3. Results

3.1. Annual Changes

3.1.1. Natural Resources

The results gathered based on the LULC classifications were transferred into NR for a hexagon grid as described in Section 2.4. A summary of different statistics over all 1793 grid cells is given in Table 6. It shows that overall resources did not significantly decline within the study area during the investigated period. Neither their mean (NR_{mean}) nor their median (NR_{median}) shows significant trends. A smaller drop of NR_{median} can be observed for 2015. This indicates that natural resources were slightly the lowest in that year or at least at the time of image acquisition. This is supported by the value of the overall sum of all Natural Resources (NR_{sum}). However, the percentiles of 95% and 5% ($NR_{95\%}$ and $NR_{5\%}$) indicate that this could also have been caused by a smaller proportion of outliers for the year 2015.

Table 6. Development of Natural Resources (NR) and weighted Natural Resource Depletion (NRD_w).

	2007	2008	2009	2010	2011	2015	2016	2017
NR_{max}	91.9%	91.5%	90.7%	90.1%	91.4%	91.7%	89.5%	88.6%
$NR_{95\%}$	73.5%	76.2%	74.7%	74.3%	74.6%	73.9%	73.1%	73.3%
NR_{mean}	46.9%	47.1%	47.0%	47.1%	46.9%	46.5%	46.8%	47.0%
NR_{median}	45.7%	45.4%	45.9%	45.9%	45.8%	44.3%	45.6%	45.1%
$NR_{5\%}$	24.8%	24.4%	25.0%	24.9%	24.2%	25.0%	25.4%	25.6%
NR_{min}	23.1%	23.1%	23.1%	23.1%	23.1%	23.1%	23.1%	23.1%
NR_{sum}	840.8	844.6	843.5	843.8	840.7	833.1	839.9	842.8
$NRD_{w\ mean}$		0.187%	−0.037%	0.016%	−0.175%	−0.421%	0.379%	0.158%
$NRD_{w\ net}$		3.36	−0.65	0.28	−3.14	−7.55	6.79	2.83

A small decline in the maximum NR value (NR_{max}) reached can be observed as it decreases from 91.9% to 88.6% over the investigated period. The minimum (NR_{min}) was constant each year as it represents cells which are completely covered by bare rock—this being the class with the lowest attributed importance (see Table 5).

Overall, no significant change can be observed for the investigated area. A spatially explicit description of changes based on the weighted Natural Resource Depletion is presented in Section 3.1.2.

3.1.2. Natural Resource Depletion

Based on the derivation of NR values for each grid cell their impact upon environmental integrity (EnvInt, 35%) and human security (HumSec, 65%) can be estimated. Their summarized temporal development is demonstrated in Table 6: The mean change of resources ($NRD_{w\ mean}$) ranges between +0.379% (2015–2016) and −0.421% (2011–2015). This means that positive and negative changes widely balance within the study area. At no time are changes dominated in one direction only. This is even more clearly shown by the median, which is not given in the table, as it levels at 0.0 throughout all years. Still, there are some years which show a slightly higher impact than others. If the NRD_w values of all grid cells are added together per year ($NRD_{w\ net}$), values between 6.79 and −7.55 emerge. These reveal a more distinct view on the impact of changes over time. While the overall changes in the study area are around ±3, a more pronounced impact of −7.55 can be assigned to the period between 2011 and 2015 as a consequence of missing data for an annual change detection.

However, the large increase between 2015 and 2016 (+6.79) indicates that the remarkably low value in 2015 of (−7.55) is not only caused by the longer time span but also depicts a local negative peak which needs to be discussed in Section 4.

The main advantage of the NRD_w is that a value of percentaged increase or decrease of landscape quality can be assigned to each grid cell. This allows for the spatial characterization of landscape changes and their impacts. The results are provided as maps for the image pairs throughout the investigated period as demonstrated in Figure 6. It shows where the environment within the study area changed during the investigated years and in which directions changes occured regarding the expert-based weights shown in Table 5. At a first impression, no clear trends can be recognized in the maps. In particular, the regions around Camp Djabal and Goz Beïda show no considerable decrease in environmental resources during the last ten years. Also, the Hadjer Arkop massif is predominantly stable during the investigated period. Variations in NR can well be observed around Wadi Aouada and its areas of higher and more developed vegetation. Whether these are just of seasonal nature or a steady development cannot be clarified by comparing single conditions within time-series. This will be addressed more specifically in Section 3.2. In general, the areas of strongest developments are shrubland in the Southeast of the study area and grassland-covered plains in the North.

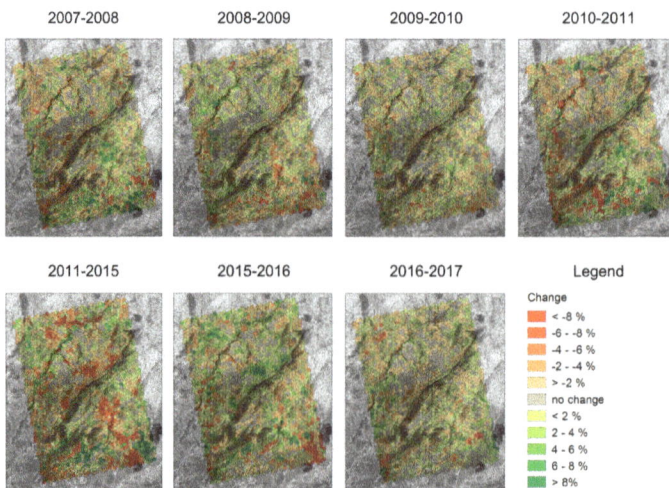

Figure 6. Weighted Natural Resource Depletion Index (NRD_w) for the study area during the investigated period. Note that, due to lack of available L-band SAR data, an annual investigation was not possible between 2011 and 2015.

3.2. Overall Change and Variability of Natural Resources

As indicated in Section 2.5, attentive interpretation of change intensities requires information on the variability of the different grid cells. If an area shows moderate change between 2007 and 2017, but a high variability, as derived according to Figure 5, this area is either subject to higher intra-annual or inter-annual seasonal changes or to generally observable instabilities regarding the interrelations between related types of land cover (such as shrubland or grassland). In turn, if a low variability (smaller than 1.5) can be observed, changes are rather of long-term nature, such as trends in vegetation cover or transitions between types of land-use (bare soil to built-up areas). Figure 7 shows the changes as assessed by the NRD_w between 2007 and 2017 in combination with the variability v per grid cell. While colors indicate the direction and severity of changes, line signatures within the grid cells indicate their variability. Developments can now be interpreted regarding their long-term information content and the affinity of an area towards changes. The indications from Section 3.1.2 can be affirmed and specified: There is no general loss of natural resources in the study area but certain areas are more prone to change than others. Areas around the settlements are almost stable whereas grids of both positive and negative development are accumulated at the South East of the study area. These result from the transition between forest and shrubland. Variability values of mostly above 2.00 indicate that this area is not gradually changing but generally prone to changes. Similar developments can be attached to Wadi Aouada where forest, shrubland, and grassland underlie both seasonal and inter-annual changes.

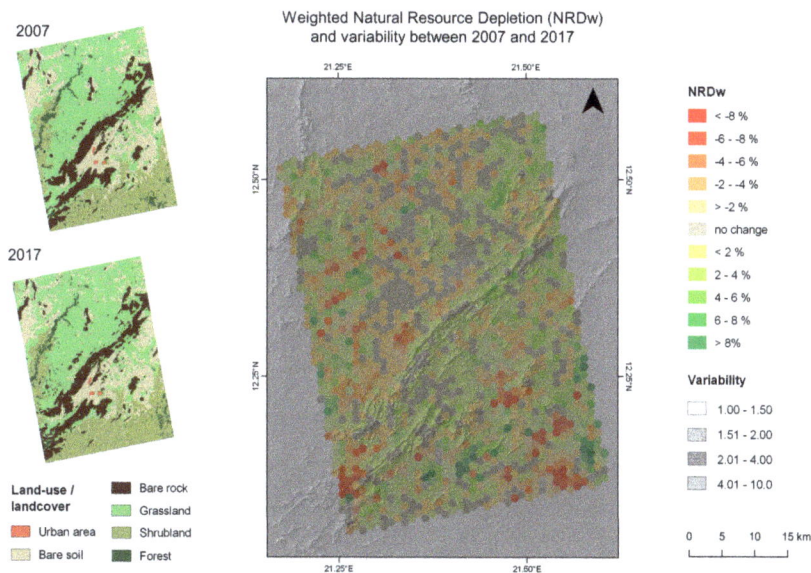

Figure 7. Weighted Natural Resource Depletion Index (NRD_w) and variability for the study area between 2007 and 2017.

General conclusions about the study area can be drawn based on the statistical distribution of the different variability values compared to their NRD_w value as shown in Figure 8. It clearly demonstrates that areas of smaller variability (v < 2.00) form the majority in the study area whereas areas with distinctively high variability (v > 4.00) are very rare. It furthermore shows that areas with slight changes ($\pm 2\%$) are more frequent than areas with no change at all, especially for higher variabilities. This indicates that many areas within the study area are affected by smaller changes and show a relatively unsteady development. It is additionally interesting that areas of highest positive development (>8%) increase with larger variability. This leads to the conclusion that some of them may also result from misclassifications as the classes of high variability no longer show an equal distribution.

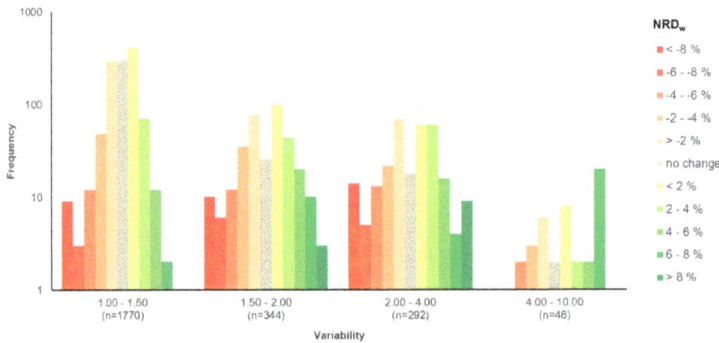

Figure 8. Frequency distribution of the variability of grid cells within the study area compared to their NRD_w index for the time between 2007 and 2017. Note the logarithmic scaling of the y-axis.

4. Discussion

Altogether, the proposed approach allows the derivation of detailed change maps along a series of SAR images which exceed the information content of standard transition matrices between classes. However, there are things to notice in both the ecologic and methodic domain. This section aggregates our findings in the perspective of landscape development and according to the requirements defined in the introduction.

4.1. Landscape Changes

One main point to report is that the landscape in the study area did not underlie large-scale changes, neither positively nor negatively. As shown in Figure 6, there in fact are certain hot spots of accumulated change over the investigated period, however, these did not add up to a gradual overall increase or decrease of landscape quality for larger areas.

Addressing anthropogenic impact on the investigated landscape, neither of the two main settlements of camp Djabal and the city of Goz Beïda, nor the plains within the Hadjer Arkop massif, had noticeable impact on their surrounding landscapes. These findings were expected in terms of no reported growth of these settlements during the investigated period [113–115]. It is still notable that this landscape did not negatively respond to the direct presence of human activities as arid savannas are often observed to be susceptible towards anthropogenic pressure [116]. One reason for the described stability within this area is the high amount of bare soil as a consequence of the relatively intensive agricultural use. Accordingly, their NR values were already quite low and only expansion of human settlements—or the transition from soil areas to bare rock—would have caused negative NRD_w developments. In addition to that, no source of land degradation or desertification, as it might be indicated by transitions from vegetation to bare soil or soil to rock, could be observed among the bare rock areas of the Hadjer Arkop massif.

One region which was prone to many changes during the investigated period is Wadi Aouada, reaching from the West to the North of the study area. As it is the only area with vital vegetation of higher orders during the whole year, it is particularly prone to changes in external circumstances. A direct comparison of the NR of 2007 with 2017 (Figure 7) does not depict a clear image on which parts of the wadi underlay significant positive or negative developments. It can, however, be stated that the core area of the wadi did not change in a critical degree—for example, due to erosion or lesser subsurface drainage. Looking at the annual changes in Figure 6 indicates that there are some years where the wadi is subject to larger changes, notably 2008–2009 (positive) and 2010–2011 (negative). These observations could be linked to climatic variations. However, comparison to records of temperature and precipitation provided by the Climatic Research Unit (CRU) of University of East Anglia (UEA) [117] does not confirm these trends. As Figure 9 shows, the contrary could have been expected as a considerable decrease in precipitation occurred between 2008 and 2009, which in turn shows the most visible positive development within Wadi Aouada. So it can be concluded that climatic variations are most likely not responsible for the smaller changes within the wadi. We expect them to result from overall variation in the vegetation signature within the SAR data.

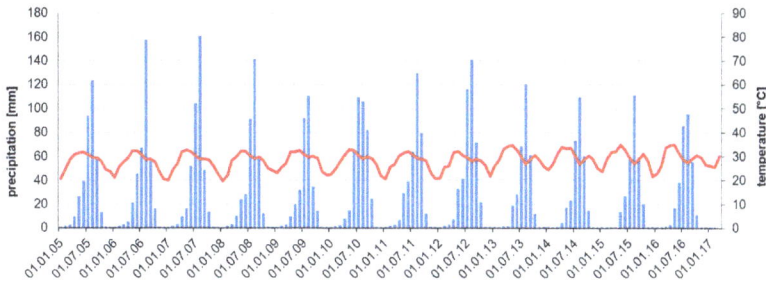

Figure 9. Monthly precipitation and temperature for the study area.

Another mentionable point is the striking accumulation of negative NRD_w values in for 2011–2015 in the North of the Hadjer Arkop massif near the center of the study area. However, the fact that this area is again strikingly positive for the subsequent image pair from 2015–2016 indicates that this anomaly is rather the outcome of a wrong classification. The low variability values in this area (Figure 7) furthermore support this indication and lead to the conclusion that this area is in fact stable regarding LULC and the assessed changes are wrong. Nevertheless, many larger changes for the period from 2011–2015 are realistic results obtained by the observed increment of 4 years instead of one. But this is one example of how investigating variabilities along a time-series can help to detect errors and check the results for plausibility.

To further test the overall condition of the landscape for interrelations with climatic conditions, we compared the sum of all NRD_w values per year, ($NRD_{w\ net}$), with the corresponding change in annual precipitation (Δprec) as shown in Table 7. We found a Pearson correlation coefficient of $r = 0.47$ between both variables which affirms that overall sum of natural resources could in fact fluctuate to a certain degree according to the water availability in the study area. A correlation of $r = -0.61$ was found for the relationship between $NRD_{w\ net}$ and the temporal baseline between the respective SAR images (SAR Δd), defined as the absolute difference in days from an optimum temporal difference between two images of 365 days. It seems obvious that potential changes increase if both images are not taken on the same day of the year. The larger the difference of two images regarding their time of acquisition (see Table 1), the higher is the risk of identifying changes related to phenology instead of identifying changes to overall development of a landscape.

Table 7. NRD$_{w\,net}$ values for the investigated time-steps compared to trends in precipitation (Δprec) and temporal baselines of the used SAR images (SAR Δd).

	2007–2008	2008–2009	2009–2010	2010–2011	2011–2015	2015–2016	2016–2017	r $_{(NRDw)}$
NRD$_{w\,net}$	3.36	−0.66	0.28	−3.14	−7.55	6.80	2.833	1.00
Δprec	−37.7	−55.7	97.7	−49.9	−74.6	15.8	- *	0.47
SAR Δd	12	12	18	88	1078 **	8	12	−0.61

* Changes in precipitation could not be assessed between 2016 and 2017 as no all-season records were available at the time of the study. ** In order to not distort the statistics, the temporal baseline of four years was excluded from the calculation of r.

Lastly, the histogram in Figure 8 demonstrates how changes in the image can be interpreted regarding their long-term predication. Overall changes with low variabilities (between 1.00 and 1.50) make up 72% in our study. If frequencies are not equally distributed in both positive and negative directions, trends can be evaluated regarding their variability. We found that high variability values either indicate single misclassifications between scenes or, when spatially aggregated, characterize areas of constant change. In the latter case, intensities of changes should be interpreted with more care.

4.2. Methods

On the technical side the following points can be highlighted. As the first findings made by Braun et al. [23] indicated, the application of the NRDw concept from Hagenlocher et al. [104] is transferable to radar imagery of medium spatial resolution. This study was able to provide a framework more robust towards smaller misclassifications as they often occur in SAR data which is still flexible enough to be based on different sensors or to be transferred to other regions. The results created reflect different aspects of landscape changes: namely environmental integrity and human security which can be weighted according to the expertise and the needs of the user. The final maps include information on both spatial and temporal variation of landscape changes but are still easy to read and interpret. These maps highlight regions which require special attention and allow a more differentiated dealing with the outcomes.

ALOS PALSAR proved to be suitable for long-term landscape monitoring due to its long wavelength interaction with volume scatterers for the discrimination of different types of vegetation. It would have been interesting to observe how the study area changed annually in this perspective between 2011 and 2015 but no accessible L-band data is available for this period. It is however expected that other SAR sensors with similar archives such as ERS, Radarsat, and Envisat bear the same potential, especially at the temporal scale. Regarding seasonality, additional data at the middle or end of the rainy season would surely have substantially increased the validity of the approach as it could have been used to depict a clearer image of each year's resources. However, this data was not available for all the investigated years. Future approaches surely should consider using a combination of several scenes at crucial points in the study area's phenology which can be determined by the variation of NDVI, for example [118–120]. This would also decrease the weight of textural information in favor of additional seasonal effects.

However, the used textures were able to discriminate all selected LULC classes to a sufficient degree. As Tables 3 and 4 demonstrate, Grassland was the class with lowest accuracies due to its indifferent degree of coverage in the study area and comparably low interaction with the L-band signal. Using these large numbers of input rasters surely requires classifiers based on machine-learning which extract the most valuable information. The random forest classifier performed efficient but other approaches, such as support vector machines (SVM) or artificial neuronal networks (ANN), can be employed as well for these kinds of analyses, especially when phenological variation is additionally introduced as suggested above [121,122].

The automatic and literature-based selection of training areas proved to increase the study's credibility as it reduces the possible bias of manual sampling collection. In our case, field investigations were not possible due to security reasons but using optical data from Landsat grants for independency of the training and validation process from the actual image classification was possible. Also worth noting, optical data does not have to be complete or cloud free. As only the areas which consistently fulfilled the criteria listed in Table 2 throughout all images were used for possible locations of the randomized samples, we could flexibly deal with cloud coverage, burnt areas and missing scan lines in Landsat ETM+ products.

Smaller misclassifications occurred between classes with similar response to the radar signal as shown in Tables 3 and 4. These could, however, be compensated by the creation of hexagon grid cells as a main unit of analysis. Each of them aggregates information of 1280 pixels and represents their percentaged share as one value of natural resources. This reduces the effect of ambiguities at the transition between two LULC classes and facilitates both the visual inspection and interpretation of results. Detailed changes at borders between two classes are no longer visible in this approach but since the fact that savanna ecosystems often consist of a mosaic of continuous LULC this is neither possible nor considered as reasonable. Shifts in phenology could, however, been better handled with an investigation over a longer time span. The longer the time-series of images, the more stable is the approach as it more clearly reveals outliers along the investigated period. ESA's Sentinel-1 mission lays a solid foundation for multi-decade investigations as it was designed to continue the archived data of ERS and Envisat. All of them operate at C-band wavelengths and a comparable spatial resolution.

One crucial point is the definition of investigated LULC classes and their corresponding contribution to both environmental integrity and human security (see Table 5). As this has to be done at the beginning of the study, changes in the numbers of investigated classes can, of course, strongly affect the later results. If interest is placed on the change of a certain type of LULC, the definition of classes should be appropriate and respectively balanced in this context. Also, the expert-based weights have large influence on the evaluation of the changes. It is therefore advisable to include the expertise of as many people as possible, preferably from different domains (local population, humanitarian units, authorities, and other stakeholders). This not only provides for a balanced assessment of weights but also prevents the abuse of weights to intentionally target desired results. As Laczko & Aghazarm argue, research on the impacts of refugee camps upon their environment is needed but cannot act as the only argument for repatriation in political debates [123].

Lastly, the proposed variability can be a valuable measure for both the quality and the long-term information content of results. However, a way of normalization has to be performed when comparing index values along a temporal sequence. We used a rather basic model of change vectors as it met our requirements for the comparison of eight images. If longer or more detailed time-series are employed, the concept of variability has to be refined according to variation at different temporal scales—such as the differentiation between seasonal, inter-annual, and overall variation of a grid cell.

5. Conclusions

Our study showed that landscape changes can be analyzed by radar imagery in a transparent and adaptive way. It is not limited to certain landscape types or specific sensors, but requires a minimum number of consecutive images from the same time of the year. The proposed approach substantially contributes to field of post-classification change analysis as it overcomes limitations regarding the number of investigated scenes, the variation within the investigated time, and the occasionally criticized complexity of high-level approaches which cannot be adapted by users with limited technical or scientific background. It therefore serves as an ideal intermediate between innovative analyses and user-friendly, adaptable frameworks. These will gain further importance with the growing availability of archived and newly acquired SAR data for scientific, organizational, and commercial operational use.

The integration of user expertise about the relevance of land-use and land cover classes hinders the automation of the process but clearly improves the validity of the results. Attaching weights to the selected classes allows for the generation of an index which refines the results according to the knowledge of the user and the information he requires.

Remote Sens. **2017**, *9*, 359

Although it did not reveal large scale trends in our study area, the proposed index is a good approach for a semi-automated evaluation of changes over long time spans and within greater regions. The use of radar data additionally reduces the dependency from atmospherically undisturbed conditions. All parts of the study were performed on free and open source tools (QGIS, ESA SNAP, OTB, Python) which allows for a transfer of the method onto other case areas. We encourage readers to implement and further improve the proposed approach in order to increase the quality of time-series analyses of radar data. This work is a first step towards new standards in change detection applications which are comprehensible and still specified to the users' needs.

Acknowledgments: This study was funded by the Austrian Research Promotion Agency (FFG) under the Austrian Space Applications Programme (ASAP 12, 854041). ALOS data was provided by Japan Aerospace Exploration Agency (JAXA). Landsat and SRTM data was provided by the US Geological Survey (USGS). The authors would like to thank Ted Cahill for proofreading the manuscripts and the anonymous reviewers for their valuable comments and suggestions to improve the manuscript. We acknowledge support by Deutsche Forschungsgemeinschaft and Open Access Publishing Fund of University of Tübingen.

Author Contributions: Andreas Braun and Volker Hochschild designed the study and selected the data; Andreas Braun processed and analyzed the data; Results were interpreted and discussed by Andreas Braun and Volker Hochschild The manuscript was submitted, edited and revised by Andreas Braun

Conflicts of Interest: The authors declare no conflict of interest.

References

1. Irons, J.R.; Dwyer, J.L.; Barsi, J.A. The next Landsat satellite: The Landsat data continuity mission. *Remote Sens. Environ.* **2012**, *122*, 11–21. [CrossRef]
2. Wulder, M.A.; White, J.C.; Goward, S.N.; Masek, J.G.; Irons, J.R.; Herold, M.; Cohen, W.B.; Loveland, T.R.; Woodcock, C.E. Landsat continuity: Issues and opportunities for land cover monitoring. *Remote Sens. Environ.* **2008**, *112*, 955–969. [CrossRef]
3. Byrne, G.F.; Crapper, P.F.; Mayo, K.K. Monitoring land-cover change by principal component analysis of multitemporal landsat data. *Remote Sens. Environ.* **1980**, *10*, 175–184. [CrossRef]
4. Fung, T. An Assessment of TM Imagery for Land-cover Change Detection. *IEEE Trans. Geosci. Remote Sens.* **1990**, *28*, 681–684. [CrossRef]
5. Guerschman, J.P.; Paruelo, J.M.; Di Bella, C.; Giallorenzi, M.C.; Pacin, F. Land cover classification in the Argentine Pampas using multi-temporal Landsat TM data. *Int. J. Remote Sens.* **2003**, *24*, 3381–3402. [CrossRef]
6. Yuan, F.; Sawaya, K.E.; Loeffelholz, B.C.; Bauer, M.E. Land cover classification and change analysis of the Twin Cities (Minnesota) Metropolitan Area by multitemporal Landsat remote sensing. *Remote Sens. Environ.* **2005**, *98*, 317–328. [CrossRef]
7. Bakr, N.; Weindorf, D.C.; Bahnassy, M.H.; Marei, S.M.; El-Badawi, M.M. Monitoring land cover changes in a newly reclaimed area of Egypt using multi-temporal Landsat data. *Appl. Geogr.* **2010**, *30*, 592–605. [CrossRef]
8. Zhu, Z.; Woodcock, C.E.; Holden, C.; Yang, Z. Generating synthetic Landsat images based on all available Landsat data: Predicting Landsat surface reflectance at any given time. *Remote Sens. Environ.* **2015**, *162*, 67–83. [CrossRef]
9. Masek, J.G.; Huang, C.; Wolfe, R.; Cohen, W.; Hall, F.; Kutler, J.; Nelson, P. North American forest disturbance mapped from a decadal Landsat record. *Remote Sens. Environ.* **2008**, *112*, 2914–2926. [CrossRef]
10. He, L.; Chen, J.M.; Zhang, S.; Gomez, G.; Pan, Y.; McCullough, K.; Birdsey, R.; Masek, J.G. Normalized algorithm for mapping and dating forest disturbances and regrowth for the United States. *Int. J. Appl. Earth Obs. Geoinf.* **2011**, *13*, 236–245. [CrossRef]
11. Song, C.; Woodcock, C.E.; Seto, K.C.; Lenney, M.P.; Macomber, S.A. Classification and change detection using Landsat TM data: When and how to correct atmospheric effects? *Remote Sens. Environ.* **2001**, *75*, 230–244. [CrossRef]
12. Warren, S.G.; Hahn, C.J.; London, J.; Chervin, R.; Jenne, R.L. Global Distribution of Total Cloud Cover and Cloud Type Amounts over Land. NCAR Tech. Note TN-273+STR. Available online: http://opensky.ucar.edu/islandora/object/technotes%3A444/datastream/PDF/download/citation.pdf (accessed on 22 November 2016).

13. Gu, G.; Zhang, C. Cloud components of the Intertropical Convergence Zone. *J. Geophys. Res.* **2002**, *107*. [CrossRef]

14. Kummerow, C.; Barnes, W.; Kozu, T.; Shiue, J.; Simpson, J. The Tropical rainfall measuring mission (TRMM) sensor package. *J. Atmos. Ocean. Technol.* **1998**, *15*, 809–817. [CrossRef]

15. Kummerow, C.; Simpson, J.; Thiele, O.; Barnes, W.; Chang, A.T.C.; Stocker, E.; Adler, R.F.; Hou, A.; Kakar, R.; Wentz, F.; et al. The status of the tropical rainfall measuring mission (TRMM) after two years in orbit. *J. Appl. Meteorol.* **2000**, *39*, 1965–1982. [CrossRef]

16. Asner, G.P. Cloud cover in Landsat observations of the Brazilian Amazon. *Int. J. Remote Sens.* **2001**, *22*, 3855–3862. [CrossRef]

17. Mas, J.-F. Monitoring land-cover changes: A comparison of change detection techniques. *Int. J. Remote Sens.* **1999**, *20*, 139–152. [CrossRef]

18. Ju, J.; Roy, D.P. The availability of cloud-free Landsat ETM+ data over the conterminous United States and globally. *Remote Sens. Environ.* **2008**, *112*, 1196–1211. [CrossRef]

19. Fletcher, K. *ERS Missions. 20 Years of Observing Earth*; ESA Communications: Noordwijk, The Netherlands, 2013.

20. Flett, D.; Crevier, Y.; Girard, R. The RADARSAT Constellation Mission: Meeting the government of Canada's needs and requirements. In Proceedings of the 2009 IEEE International Geoscience and Remote Sensing Symposium, Cape Town, South Africa, 12–17 July 2009.

21. Torres, R.; Snoeij, P.; Geudtner, D.; Bibby, D.; Davidson, M.; Attema, E.; Potin, P.; Rommen, B.; Floury, N.; Brown, M.; et al. GMES Sentinel-1 mission. *Remote Sens. Environ.* **2012**, *120*, 9–24. [CrossRef]

22. Liew, S.C.; Kam, S.P.; Tuong, T.P.; Chen, P.; Minh, V.Q.; Lim, H. Landcover classification over the Mekong river delta using ERS and RADARSAT SAR images. In Proceedings of the 1997 IEEE International Geoscience and Remote Sensing Symposium, Remote Sensing—A Scientific Vision for Sustainable Development, Singapore, 3–8 August 1997.

23. Braun, A.; Lang, S.; Hochschild, V. Impact of refugee camps on their environment a case study using multi-temporal SAR data. *J. Geogr. Environ. Earth Sci. Int.* **2016**, *4*, 1–17. [CrossRef]

24. McAvoy, J.G.; Krakowskii, E.M. A Knowledge based system for the interpretation of SAR Images of sea ice. In Proceedings of the 12th Canadian Symposium on Remote Sensing Geoscience and Remote Sensing Symposium, Vancouver, BC, Canada, 10–14 July 1989.

25. Pierce, L.E.; Ulaby, F.T.; Sarabandi, K.; Dobson, M.C. Knowledge-based classification of polarimetric SAR images. *IEEE Trans. Geosci. Remote Sens.* **1994**, *32*, 1081–1086. [CrossRef]

26. Dobson, M.C.; Pierce, L.E.; Ulaby, F.T. Knowledge-based land-cover classification using ERS-1/JERS-1 SAR composites. *IEEE Trans. Geosci. Remote Sens.* **1996**, *34*, 83–99. [CrossRef]

27. Nunnally, N.R. Integrated landscape analysis with radar imagery. *Remote Sens. Environ.* **1969**, *1*, 1–6. [CrossRef]

28. Askne, J.; Hagberg, J.O. Potential of interferometric SAR for classification of land surfaces. In Proceedings of the IEEE International Geoscience and Remote Sensing Symposium, Tokyo, Japan, 18–21 August 1993.

29. Dutra, L.V. Feature extraction and selection for ERS-1/2 InSAR classification. *Int. J. Remote Sens.* **1999**, *20*, 993–1016. [CrossRef]

30. Engdahl, M.E.; Hyyppa, J.M. Land-cover classification using multitemporal ERS-1/2 insar data. *IEEE Trans. Geosci. Remote Sens.* **2003**, *41*, 1620–1628. [CrossRef]

31. Huber, R.; Dutra, L.V. Feature selection for ERS-1/2 InSAR classification: High dimensionality case. In Proceedings of the 1998 IEEE International Geoscience and Remote Sensing, Seattle, WA, USA, 6–10 July 1998.

32. Oliver, C.J. Rain forest classification based on SAR texture. *IEEE Trans. Geosci. Remote Sens.* **2000**, *38*, 1095–1104. [CrossRef]

33. Bruzzone, L.; Marconcini, M.; Wegmuller, U.; Wiesmann, A. An advanced system for the automatic classification of multitemporal SAR images. *IEEE Trans. Geosci. Remote Sens.* **2004**, *42*, 1321–1334. [CrossRef]

34. Van Zyl, J.J.; Burnette, C.F. Bayesian classification of polarimetric SAR images using adaptive a priori probabilities. *Int. J. Remote Sens.* **1992**, *13*, 835–840. [CrossRef]

35. Rignot, E.; Williams, C.L.; Way, J.; Viereck, L.A. Mapping of forest types in Alaskan boreal forests using SAR imagery. *IEEE Trans. Geosci. Remote Sens.* **1994**, *32*, 1051–1059. [CrossRef]

36. Chiang, H.-C.; Moses, R.L.; Potter, L.C. Model-based classification of radar images. *IEEE Trans. Inf. Theory* **2000**, *46*, 1842–1854. [CrossRef]

37. Tzeng, Y.C.; Chen, K.S. A fuzzy neural network to SAR image classification. *IEEE Trans. Geosci. Remote Sens.* **1998**, *36*, 301–307. [CrossRef]

38. Chen, C.-T.; Chen, K.-S.; Lee, J.-S. The use of fully polarimetric information for the fuzzy neural classification of SAR images. *IEEE Trans. Geosci. Remote Sens.* **2003**, *41*, 2089–2100. [CrossRef]

39. Zhang, Y.; Wu, L.; Neggaz, N.; Wang, S.; Wei, G. Remote-sensing image classification based on an improved probabilistic neural network. *Sensors* **2009**, *9*, 7516–7539. [CrossRef] [PubMed]

40. Van Beijma, S.; Comber, A.; Lamb, A. Random forest classification of salt marsh vegetation habitats using quad-polarimetric airborne SAR, elevation and optical RS data. *Remote Sens. Environ.* **2014**, *149*, 118–129. [CrossRef]

41. Du, P.; Samat, A.; Waske, B.; Liu, S.; Li, Z. Random forest and rotation forest for fully polarized SAR image classification using polarimetric and spatial features. *ISPRS J. Photogramm. Remote Sens.* **2015**, *105*, 38–53. [CrossRef]

42. Gupta, S.; Singh, D.; Singh, K.P.; Kumar, S. An efficient use of random forest technique for SAR data classification. In Proceedings of the 2015 IEEE International Geoscience and Remote Sensing Symposium, Milan, Italy, 26–31 July 2015.

43. Lee, J.-S.; Grunes, M.R.; Kwok, R. Classification of multi-look polarimetric SAR imagery based on complex Wishart distribution. *Int. J. Remote Sens.* **1994**, *15*, 2299–2311. [CrossRef]

44. Cloude, S.R.; Pottier, E. An entropy based classification scheme for land applications of polarimetric SAR. *IEEE Trans. Geosci. Remote Sens.* **1997**, *35*, 68–78. [CrossRef]

45. Alberga, V. A study of land cover classification using polarimetric SAR parameters. *Int. J. Remote Sens.* **2007**, *28*, 3851–3870. [CrossRef]

46. Qi, Z.; Yeh, A.G.-O.; Li, X.; Lin, Z. A novel algorithm for land use and land cover classification using RADARSAT-2 polarimetric SAR data. *Remote Sens. Environ.* **2012**, *118*, 21–39. [CrossRef]

47. White, R.G. Change detection in SAR imagery. *Int. J. Remote Sens.* **1991**, *12*, 339–360. [CrossRef]

48. Rignot, E.; van Zyl, J.J. Change detection techniques for ERS-1 SAR data. *IEEE Trans. Geosci. Remote Sens.* **1993**, *31*, 896–906. [CrossRef]

49. Bazi, Y.; Bruzzone, L.; Melgani, F. An unsupervised approach based on the generalized Gaussian model to automatic change detection in multitemporal SAR images. *IEEE Trans. Geosci. Remote Sens.* **2005**, *43*, 874–887. [CrossRef]

50. Dekker, R.J. Speckle filtering in satellite SAR change detection imagery. *Int. J. Remote Sens.* **1998**, *19*, 1133–1146. [CrossRef]

51. Bovenga, F.; Refice, A.; Nutricato, R.; Pasquariello, G.; de Carolis, G. Automated calibration of multi-temporal ERS SAR data. In Proceedings of the IEEE International Geoscience and Remote Sensing Symposium, Toronto, ON, Canada, 24–28 June 2002.

52. Townsend, P.A.; Foster, J.R. A synthetic aperture radar-based model to assess historical changes in lowland floodplain hydroperiod. *Water Resour. Res.* **2002**, *38*, 20-1–20-10. [CrossRef]

53. Beisl, C.H.; de Miranda, F.P.; Evsukoff, A.G.; Pedroso, E.C. Assessment of environmental sensitivity index of flooding areas in western Amazonia using fuzzy logic in the dual season GRFM JERS-1 SAR image mosaics. In Proceedings of the 2003 IEEE International Geoscience and Remote Sensing Symposium, Toulouse, France, 21–25 July 2003.

54. Hoffmann, A.A.; Siegert, F.; Hinrichs, A. *Fire Damage in East Kalimantan in 1997/98 Related to Land Use and Vegetation Classes. Satellite Radar Inventory Results and Proposals for Further Actions*; IFFM/SFMP: Samarinda, Indonesia, 1987.

55. Saatchi, S.S.; Soares, J.V.; Alves, D.S. Mapping deforestation and land use in amazon rainforest by using SIR-C imagery. *Remote Sens. Environ.* **1997**, *59*, 191–202. [CrossRef]

56. Strozzi, T.; Wegmuller, U.; Luckman, A.; Balzter, H. Mapping deforestation in Amazon with ERS SAR interferometry. In Proceedings of the IEEE 1999 International Geoscience and Remote Sensing Symposium, Hamburg, Germany, 28 June–2 July 1999.

57. Almeida-Filho, R.; Rosenqvist, A.; Shimabukuro, Y.E.; Dos Santos, J.R. Evaluation and perspectives of using multitemporal L-Band SAR data to monitor deforestation in the Brazilian Amazonia. *IEEE Geosci. Remote Sens. Lett.* **2005**, *2*, 409–412. [CrossRef]

58. Almeida-Filho, R.; Shimabukuro, Y.E.; Rosenqvist, A.; Sánchez, G.A. Using dual-polarized ALOS PALSAR data for detecting new fronts of deforestation in the Brazilian Amazônia. *Int. J. Remote Sens.* **2009**, *30*, 3735–3743. [CrossRef]

59. Ryan, C.M.; Hill, T.; Woollen, E.; Ghee, C.; Mitchard, E.; Cassells, G.; Grace, J.; Woodhouse, I.H.; Williams, M. Quantifying small-scale deforestation and forest degradation in African woodlands using radar imagery. *Glob. Chang. Biol.* **2012**, *18*, 243–257. [CrossRef]

60. Mitchard, E.T.A.; Saatchi, S.S.; Woodhouse, I.H.; Nangendo, G.; Ribeiro, N.S.; Williams, M.; Ryan, C.M.; Lewis, S.L.; Feldpausch, T.R.; Meir, P. Using satellite radar backscatter to predict above-ground woody biomass: A consistent relationship across four different African landscapes. *Geophys. Res. Lett.* **2009**. [CrossRef]

61. Mitchard, E.; Saatchi, S.S.; Lewis, S.L.; Feldpausch, T.R.; Woodhouse, I.H.; Sonké, B.; Rowland, C.; Meir, P. Measuring biomass changes due to woody encroachment and deforestation/degradation in a forest–savanna boundary region of central Africa using multi-temporal L-band radar backscatter. *Remote Sens. Environ.* **2011**, *115*, 2861–2873. [CrossRef]

62. Shen, G.; Liao, J.; Zhang, L.; Li, X. Wetland landscape analysis using polarimetric RADARSAT-2 data. In Proceedings of the 2014 IEEE International Geoscience and Remote Sensing Symposium, Quebec City, QC, Canada, 13–18 July 2014.

63. Tewkesbury, A.P.; Comber, A.J.; Tate, N.J.; Lamb, A.; Fisher, P.F. A critical synthesis of remotely sensed optical image change detection techniques. *Remote Sens. Environ.* **2015**, *160*, 1–14. [CrossRef]

64. Hussain, M.; Chen, D.; Cheng, A.; Wei, H.; Stanley, D. Change detection from remotely sensed images: From pixel-based to object-based approaches. *ISPRS J. Photogramm. Remote Sens.* **2013**, *80*, 91–106. [CrossRef]

65. Walker, P.A.; Peters, P.E. Making Sense in Time: Remote Sensing and the Challenges of Temporal Heterogeneity in Social Analysis of Environmental Change. *Hum. Ecol.* **2007**, *35*, 69–80. [CrossRef]

66. Hauck, J.; Görg, C.; Varjopuro, R.; Ratamäki, O.; Maes, J.; Wittmer, H.; Jax, K. "Maps have an air of authority": Potential benefits and challenges of ecosystem service maps at different levels of decision making. *Ecosyst. Serv.* **2013**, *4*, 25–32. [CrossRef]

67. Kienberger, S.; Füreder, P.; Hölbling, D.; Tiede, D.; Contreras Mojica, D.M.; Hagenlocher, M.; Zeil, P.; Lang, S. Von Geodaten zu nutzbarer Geoinformation: Entwicklung von und Anforderung an kartografische Produkte im Katastrophenmanagement-Zyklus. In Proceedings of the Workshop "Raum Zeit Risiko", München, Germany, 28 November 2013.

68. Ariti, A.T.; van Vliet, J.; Verburg, P.H. Land-use and land-cover changes in the Central Rift Valley of Ethiopia: Assessment of perception and adaptation of stakeholders. *Appl. Geogr.* **2015**, *65*, 28–37. [CrossRef]

69. Hansen, K.F. Chad. In *Africa Yearbook Volume 11*; Elischer, S., Mehler, A., Hofmeier, R., Melber, H., Eds.; Brill: Leiden, The Netherlands, 2015; pp. 201–208.

70. Rishmawi, K.; Prince, S. Environmental and anthropogenic degradation of vegetation in the Sahel from 1982 to 2006. *Remote Sens.* **2016**. [CrossRef]

71. Southworth, J.; Zhu, L.; Bunting, E.; Ryan, S.J.; Herrero, H.; Waylen, P.R.; Hill, M.J. Changes in vegetation persistence across global savanna landscapes, 1982–2010. *J. Land Use Sci.* **2014**, *11*, 7–32. [CrossRef]

72. Office for the Coordination of Humanitarian Affairs (OCHA). Profil Régional du Sila: Novembre 2012. Available online: https://docs.unocha.org/sites/dms/CHAD/Profil_Dar%20Sila_Novembre%202012.pdf (accessed on 8 December 2016).

73. Clark, J.; Tan, V. Hundreds Flee New Fighting in Darfur; UNHCR Opens 8th Camp in Chad. Available online: http://www.unhcr.org/print/40c081119.html (accessed on 15 November 2016).

74. Bouchardy, J.-Y. Radar Images and Geographic Information Helping Identify Water Resources during Humanitarian Crisis: The Case of Chad/Sudan (Darfur) Emergency. In *Global Monitoring for Sustainability and Security, Proceedings of the 31st International Symposium of Remote Sensing & the Environment, St. Petersburg, Russian, 20–24 June 2005*; Bobylev, L.P., Ed.; International Center for Remote Sensing of Environment: Berlin, Germany, 2005.

75. Humanitarian Information Unit. Sudan (Darfur) and Chad Border Region: Confirmed Damaged and Destroyed Villages. Available online: http://reliefweb.int/map/sudan/sudan-darfur-and-chad-border-region-confirmed-damaged-and-destroyed-villages-2-august-2004 (accessed on 14 November 2016).

76. Middleton, N.; Thomas, D.S.G. *World Atlas of Desertification*, 2nd ed.; Arnold: London, UK, 1997.

77. Harris, I.; Jones, P.D.; Osborn, T.J.; Lister, D.H. Updated high-resolution grids of monthly climatic observations—The CRU TS3.10 Dataset. *Int. J. Climatol.* **2014**, *34*, 623–642. [CrossRef]

78. Funk, C.; Rowland, J.; Eilerts, G.; Adoum, A.; White, L. *A Climate Trend Analysis of Chad: Famine Early Warning Systems Network—Informaing Climate Change Adaption Series*; FEWSNET/US Geological Survey Fact Sheet 2012–3070; U.S. Geological Survey: Reston, VA, USA, 2012.

79. Brownfield, M.E. Assessment of undiscovered oil and gas resources of the sud province, Central East Africa. In *Geologic Assessment of Undiscovered Hydrocarbon Resources of Sub-Saharan Africa*; Brownfield, M.E., Ed.; U.S. Geological Survey: Reston, VA, USA, 2011.

80. Pias, J. Les formations tertiaires et quaternaires de la cuvette tchadienne (République du Tchad). In *Congrès Panafricain de Préhistoire et de L'étude du Quaternaire*; Dakar: Bondy, France, 1976; pp. 425–429.

81. White, F. *Vegetation Map of Africa. A Descriptive Memoir to Accompany the Unesco/AETFAT/UNSO Vegetation Map of Africa*; Unesco: Paris, France, 1981.

82. Wickens, G.E. *The Flora of Jebel Marra (Sudan Republic) and Its Geographical Affinities*; H.M. Stationery Off.: London, UK, 1976.

83. Ministerie van Defensie. European Union Force Chad/CAR. Available online: https://www.defensie.nl/binaries/ defence/documents/leaflets/2013/02/05/european-union-force-chad-car-pdf/european-union-force-chad-car-pdf.pdf (accessed on 8 December 2016).

84. Markham, B.L.; Storey, J.C.; Williams, D.L.; Irons, J.R. Landsat sensor performance: History and current status. *IEEE Trans. Geosci. Remote Sens.* **2004**, *42*, 2691–2694. [CrossRef]

85. JAXA. ALOS/PALSAR Level 1.1/1.5 Product Format Description. NEB-070062B. Available online: http://www. eorc.jaxa.jp/ALOS/en/doc/fdata/PALSAR_x_Format_EL.pdf (accessed on 5 April 2017).

86. JAXA. ALOS-2/PALSAR-2 Level 1.1/1.5 Product Format Description. Ü129-132. Available online: http://www. eorc.jaxa.jp/ALOS-2/en/doc/fdata/PALSAR-2_xx_Format_CEOS_E_r.pdf (accessed on 5 April 2017).

87. Shimada, M.; Isoguchi, O.; Tadono, T.; Isono, K. PALSAR Radiometric and Geometric Calibration. *IEEE Trans. Geosci. Remote Sens.* **2009**, *47*, 3915–3932. [CrossRef]

88. Small, D. Flattening Gamma: Radiometric Terrain Correction for SAR Imagery. *IEEE Trans. Geosci. Remote Sens.* **2011**, *49*, 3081–3093. [CrossRef]

89. Loew, A.; Mauser, W. Generation of geometrically and radiometrically terrain corrected SAR image products. *Remote Sens. Environ.* **2007**, *106*, 337–349. [CrossRef]

90. Collins, M.J.; Wiebe, J.; Clausi, D.A. The effect of speckle filtering on scale-dependent texture estimation of a forested scene. *IEEE Trans. Geosci. Remote Sens.* **2000**, *38*, 1160–1170. [CrossRef]

91. Prasad, T.S.; Gupta, R.K. Texture based classification of multidate SAR images—A case study. *Geocarto Int.* **1998**, *13*, 53–62. [CrossRef]

92. Haralick, R.M.; Shanmugam, K.; Dinstein, I. Textural Features for Image Classification. *IEEE Trans. Syst. Man Cybern.* **1973**, *3*, 610–621. [CrossRef]

93. Chander, G.; Markham, B.L.; Helder, D.L. Summary of current radiometric calibration coefficients for Landsat MSS, TM, ETM+, and EO-1 ALI sensors. *Remote Sens. Environ.* **2009**, *113*, 893–903. [CrossRef]

94. Moran, M.; Jackson, R.D.; Slater, P.N.; Teillet, P.M. Evaluation of simplified procedures for retrieval of land surface reflectance factors from satellite sensor output. *Remote Sens. Environ.* **1992**, *41*, 169–184. [CrossRef]

95. Forkuor, G.; Landmann, T.; Conrad, C.; Dech, S. Agricultural land use mapping in the sudanian savanna of West Africa: Current status and future possibilities. In Proceedings of the 2012 IEEE International Geoscience and Remote Sensing Symposium, Munich, Germany, 22–27 July 2012.

96. Guiguindibaye, M.; Belem, M.O.; Boussim, J.L.; Ndoutorlengar, M. Effect of early fires on the behavior of some perennial woody and herbaceous species in Sudan savanna in chad. *Indian J. Sci. Res. Technol.* **2015**, *3*, 56–65.

97. López-García, M.J.; Caselles, V. Mapping burns and natural reforestation using thematic Mapper data. *Geocarto Int.* **1991**, *6*, 31–37. [CrossRef]

98. Drury, S.A. *Image Interpretation in Geology*, 3rd ed.; Blackwell Science: Malden, MA, USA, 2001.

99. Liu, J.; Heiskanen, J.; Aynekulu, E.; Maeda, E.; Pellikka, P. Land Cover Characterization in West Sudanian Savannas Using Seasonal Features from Annual Landsat Time Series. *Remote Sens.* **2016**. [CrossRef]

100. Hartigan, J.A.; Wong, M.A. Algorithm AS 136: A K-Means Clustering Algorithm. *Appl. Stat.* **1979**, *28*, 100. [CrossRef]

101. Fisher, R.A. On an Absolute Criterion for Fitting Frequency Curves. *Stat. Sci.* **1912**, *12*, 39–41.

102. Breiman, L. Random Forests. *Mach. Learn.* **2001**, *45*, 5–32. [CrossRef]

103. Breiman, L.; Friedman, J.H.; Olshen, R.A.; Stone, C.J. *Classification and Regression Trees*; Belmont: Wadsworth, OH, USA, 1984.

104. Hagenlocher, M.; Lang, S.; Tiede, D. Integrated assessment of the environmental impact of an IDP camp in Sudan based on very high resolution multi-temporal satellite imagery. *Remote Sens. Environ.* **2012**, *126*, 27–38. [CrossRef]

105. Birch, C.P.; Oom, S.P.; Beecham, J.A. Rectangular and hexagonal grids used for observation, experiment and simulation in ecology. *Ecol. Model.* **2007**, *206*, 347–359. [CrossRef]

106. Pickett, S.; Cadenasso, M.L.; Benning, T.L. Biotic and abiotic variability as key determinants of savanna heterogeneity at multiple spatiotemporal scales. In *The Kruger Experience: Ecology and Management of Savanna Heterogeneity*; Biggs, H., du Toit, J.T., Walker, B.H., Rogers, K.H., Sinclair, A.R.E., Eds.; Island Press: Washington, DC, USA, 2003.

107. Hobbs, R.J. Remote sensing of spatial and temporal dynamics of vegetation. In *Remote Sensing of Biosphere Functioning*; Hobbs, R.J., Mooney, H.A., Eds.; Springer: New York, NY, USA, 1990; pp. 203–219.

108. Haack, B.N.; Solomon, E.K.; Bechdol, M.A.; Herold, N.D. Radar and optical data comparison/integration for urban delineation: A case study. *Photogramm. Eng. Remote Sens.* **2002**, *68*, 1289–1296.

109. Le Toan, T.; Mermoz, S.; Fichet, L.V.; Sannier, C.; Bouvet, A. Comparison of optical and SAR data for forest cover mapping. In Proceedings of the 2014 IEEE International Geoscience & Remote Sensing Symposium, Québec City, QC, Canada, 13–18 July 2014; IEEE: Piscataway, NJ, USA, 2014.

110. Hyyppä, J.; Hyyppä, H.; Inkinen, M.; Engdahl, M.; Linko, S.; Zhu, Y.-H. Accuracy comparison of various remote sensing data sources in the retrieval of forest stand attributes. *For. Ecol. Manag.* **2000**, *128*, 109–120. [CrossRef]

111. Malila, W.A. Change Vector Analysis: An Approach for Detecting Forest Changes with Landsat. LARS Symposia. Available online: http://docs.lib.purdue.edu/lars_symp (accessed on 16 December 2016).

112. Serra, P.; Pons, X.; Saurí, D. Post-classification change detection with data from different sensors: Some accuracy considerations. *Int. J. Remote Sens.* **2003**, *24*, 3311–3340. [CrossRef]

113. UNHCR. Sudan/Chad Update: Highlights. Update 15. Available online: http://www.unhcr.org/430f47762.pdf (accessed on 31 January 2017).

114. Farman-Farmaian, M.; Ndakass, V. UNHCR Chad at a Glance. Available online: http://reliefweb.int/sites/reliefweb.int/files/resources/UNHCR%20Chad%20at%20a%20Glance-30April2014.pdf (accessed on 31 January 2017).

115. Monier, C. Goz Beïda: Monthly Report. Available online: https://www.unicef.org/wcaro/wcaro_chad-UNICEFMonthly_Report_Nov_09.pdf (accessed on 31 January 2017).

116. Hirota, M.; Holmgren, M.; van Nes, E.H.; Scheffer, M. Global resilience of tropical forest and savanna to critical transitions. *Science (New York)* **2011**, *334*, 232–235. [CrossRef] [PubMed]

117. Mitchell, T.D.; Carter, T.R.; Jones, P.D.; Hulme, M.; New, M. *A Comprehensive Set of High-Resolution Grids of Monthly Climate for Europe and the Globe: The Observed Record (1901–2000) and 16 Scenarios (2001–2100)*; Tyndall Centre for Climate Change Research Working Paper; Tyndall Centre for Climate Change Research: Norwich, UK, 2004.

118. Heumann, B.W.; Seaquist, J.W.; Eklundh, L.; Jönsson, P. AVHRR derived phenological change in the Sahel and Soudan, Africa, 1982–2005. *Remote Sens. Environ.* **2007**, *108*, 385–392. [CrossRef]

119. Lee, R.; Yu, F.; Price, K.P.; Ellis, J.; Shi, P. Evaluating vegetation phenological patterns in Inner Mongolia using NDVI time-series analysis. *Int. J. Remote Sens.* **2002**, *23*, 2505–2512. [CrossRef]

120. Wagenseil, H.; Samimi, C. Assessing spatio—Temporal variations in plant phenology using Fourier analysis on NDVI time series: Results from a dry savannah environment in Namibia. *Int. J. Remote Sens.* **2006**, *27*, 3455–3471. [CrossRef]

121. Murthy, C.S.; Raju, P.V.; Badrinath, K.V.S. Classification of wheat crop with multi-temporal images: Performance of maximum likelihood and artificial neural networks. *Int. J. Remote Sens.* **2003**, *24*, 4871–4890. [CrossRef]

122. Shao, Y.; Lunetta, R.S. Comparison of support vector machine, neural network, and CART algorithms for the land-cover classification using limited training data points. *ISPRS J. Photogramm. Remote Sens.* **2012**, *70*, 78–87. [CrossRef]

123. Laczko, F.; Aghazarm, C. *Migration, Environment and the Climate Change. Assessng the Evidence*; International Organization for Migration: Geneva, Switzerland, 2009.

remote sensing

MDPI

Technical Note

Semi-Automated Surface Water Detection with Synthetic Aperture Radar Data: A Wetland Case Study

Amir Behnamian [1,*], Sarah Banks [1], Lori White [1], Brian Brisco [2], Koreen Millard [3], Jon Pasher [1], Zhaohua Chen [1], Jason Duffe [1], Laura Bourgeau-Chavez [4] and Michael Battaglia [4]

[1] Environment and Climate Change Canada, National Wildlife Research Centre, 1125 Colonel By Drive, Ottawa, ON K1S 5B6, Canada; sarah.banks@canada.ca (S.B.); lori.white2@canada.ca (L.W.); Jon.Pasher@canada.ca (J.P.); zhaohua.chen@canada.ca (Z.C.); jason.duffe@canada.ca (J.D.)
[2] Canada Centre for Mapping and Earth Observation, Natural Resources Canada, 560 Rochester St., Ottawa, ON K1S 5K2, Canada; brian.brisco@canada.ca
[3] Defence Research and Development Canada (DRDC), Ottawa Research Center, 3701 Carling Ave., Ottawa, ON K2K 2Y7, Canada; Koreen.Millard@drdc-rddc.gc.ca
[4] Michigan Tech Research Institute, Michigan Technological University, Ann Arbor, MI 48105, USA; lchavez@mtu.edu (L.B.-C.); mjbattag@mtu.edu (M.B.)
* Correspondence: abehnamian@uottawa.ca or amir.behnamian@canada.ca; Tel.: +1-613-998-9261

Received: 11 October 2017; Accepted: 20 November 2017; Published: 23 November 2017

Abstract: In this study, a new method is proposed for semi-automated surface water detection using synthetic aperture radar data via a combination of radiometric thresholding and image segmentation based on the simple linear iterative clustering superpixel algorithm. Consistent intensity thresholds are selected by assessing the statistical distribution of backscatter values applied to the mean of each superpixel. Higher-order texture measures, such as variance, are used to improve accuracy by removing false positives via an additional thresholding process used to identify the boundaries of water bodies. Results applied to quad-polarized RADARSAT-2 data show that the threshold value for the variance texture measure can be approximated using a constant value for different scenes, and thus it can be used in a fully automated cleanup procedure. Compared to similar approaches, errors of omission and commission are improved with the proposed method. For example, we observed that a threshold-only approach consistently tends to underestimate the extent of water bodies compared to combined thresholding and segmentation, mainly due to the poor performance of the former at the edges of water bodies. The proposed method can be used for monitoring changes in surface water extent within wetlands or other areas, and while presented for use with radar data, it can also be used to detect surface water in optical images.

Keywords: water mapping; surface water; wetland; SAR; RADARSAT-2; histogram; threshold; segmentation; superpixel

1. Introduction

Wetlands provide a range of ecosystem services, including several that are vital to the health of the environment. Many species rely on wetlands, for example, to provide critical habitat used for procuring food and shelter. Wetlands also improve water quality by naturally filtering toxic substances and sediments. Unfortunately, these qualities were not well recognized in Canada until relatively recently, resulting in numerous wetlands in the south being filled or drained for other uses, including agricultural production [1]. Today, greater efforts are being made to protect these sensitive ecosystems. However, many are still threatened by the effects of anthropogenic disturbance [2], and of particular relevance to this study, climate change [3].

Increased variability and changes to ambient temperature, precipitation, and flow regimes are anticipated [3], and will likely adversely impact wetland ecosystems, which are especially sensitive to changes in the duration of flooding and depth of water [4,5]. Changes in the stability of flow regimes specifically may lead to the loss of suitable habitat for species that are adapted to certain levels of variation, and could lead to increased numbers of invasive and or generalist species [6]. In light of this, there is need for regional monitoring of changes in the extent of surface water within wetlands as this will help identify those areas that are changing. Not only would this inform management strategies, but it would also focus conservation efforts, which would be especially beneficial to those for which wetlands provide habitat and who are already at risk of extinction.

Surface water detection (SWD) using synthetic aperture radar (SAR) data has been the subject of study for many research groups [7–10]. SAR data is a reliable source of information for operational monitoring of water resources since, in contrast to optical data, images can be acquired regardless of cloud cover and haze. Further, due to predominant specular reflection, radar backscatter over non-disturbed water bodies is low relative to most surrounding land and other non-water features. This results in contrasting dark and bright pixels (between water and non-water features), a characteristic that can be used to discriminate both land cover types in SAR images.

Due to their simplicity, threshold-based procedures are widely used for operational SWD, particularly for large datasets [7,8]. Several techniques exist for finding an appropriate threshold value, the simplest of which includes scene-based visual investigation of histograms [11]. This approach is not practical for operational mapping however, as the threshold value differs under different incident angles, wind, and terrain conditions; thus, there is a need for human intervention on a scene-by-scene basis. Alternatively, Bolanos et al. [7] proposed using a normalized threshold value applied to energy texture images to extract surface water extents from RADARSAT-2 images (i.e., $T_e = \frac{t_e - \mu_e}{\sigma_e}$ where T_e is a non-dimensional threshold value standardized by the mean: μ_e, and standard deviation: σ_e, of the energy texture image, and t_e is the threshold value before being normalized).

Statistical modeling of histograms has also been used in several cases to estimate a threshold value for automatic SWD. Matgen et al. [9] estimated the probability density function (PDF) of backscatter values for water using a gamma distribution and adapted an iterative procedure to define an optimal threshold value to separate water and non-water pixels while limiting over-estimation. Schumann et al. [12] computed threshold values for histograms using Otsu's method [13] by applying a criterion to evaluate the between-class variance calculated from a normalized histogram. Notably, one of the limitations of Otsu's method is that when the bimodality of histograms is unbalanced, the threshold value can over or underestimate the extent of water bodies as a result of the dominant mode having a greater effect on the between-class variance, thus resulting in the threshold being drawn closer to its mean [14,15].

To address this, Li and Wang [10] applied a modified version of Otsu's thresholding algorithm to a subset of SAR-based texture (entropy) images. This modified version, called "valley emphasis", attempts to select a threshold that is closer to the valley between the two modes. Li and Wang also subset images so that each contained between 10% and 90% water to ensure that the water and non-water modes of the histogram were more balanced. To do this, an initial water body mask is generated using K-means clustering applied to the SAR intensity image. Despite these advances however, the limitation of Otsu's method remains, especially in cases where the population of water pixels is much lower than the population of non-water pixels.

It is notable that in the literature most SWD methods lack an ancillary process to reinforce the grouping of water pixels, especially along local boundaries between water and non-water features. Attempts to address this common deficiency have focused primarily on making slight adjustments to the threshold value in order to compensate for over or underestimation [9]. As an alternative, we propose the use of image segmentation via the simple linear iterative clustering (SLIC) superpixel method for improved SWD.

Image segmentation has increasingly become a key preprocessing step for computer vision applications such as object class recognition and image segmentation [16,17]. In general, existing algorithms are broadly categorized as graph- or gradient ascent-based approaches [18]. The former are useful for capturing image boundaries, while the latter are beneficial when a regular lattice is required. Achanta et al. [19] suggested three important properties for a desirable segmentation algorithm, including the fact that they are fast and easy to use while maintaining the quality of the results, and also that they adhere to image boundaries. The authors compared five types of superpixel algorithms (two graph-based and three gradient-ascent-based), including their newly developed SLIC superpixel algorithm. Their in-depth comparison showed that their proposed method, which is a gradient-ascent approach based on k-means clustering, outperformed other algorithms on all three aspects (i.e., speed, ease of use, and boundary adherence).

In light of these developments, we have developed a new method for SWD with SAR data that uses both thresholding and image segmentation based on the SLIC superpixel method. In this paper, the method is described in detail and results are demonstrated for multiple RADARSAT-2 images acquired over two study areas. The new thresholding method is based on the statistical characteristics of image histograms (or probability density functions), and select thresholds are applied to image objects. In contrast to previous SWD methods in which the cleanup procedure was implemented implicitly, we also describe a separate and fully-automated cleanup procedure and evaluate its impact on the accuracy of the algorithm.

This paper begins with this introduction followed by the a description of the study area and the methodology. The results are then presented and discussed followed by the conclusion.

2. Study Areas and Data Acquisition

Two study areas with temporally and spatially variable open water bodies are considered in this research (Figure 1). The first site (Figure 1b; referred to hereafter as the Prairie Pothole Region) was also evaluated by Bolanos et al. [7], and thus our newly proposed method was compared to their SWD method. For this site, a Radasat-2 Fine Quad Pol (FQ19) mode image with a nominal ground range resolution of 8.4 m (near-range) covering approximately 25 × 25 km² was acquired on 8 September 2012. A same-day, cloud-free RapidEye image (5-m resolution) over Elk Island National Park, Alberta, was used to extract water polygons for comparison purposes using both methods (Figure 1a). Readers are referred to Bolanos et al. [7] for additional information on this site and dataset, which has not been repeated here for brevity.

The second study area covers the entirety of Prince Edward County and the Bay of Quinte, Ontario (Figure 1c; referred to hereafter as the Bay of Quinte site). This region is located on the Canadian side of Lake Ontario, falling completely within the Mixedwood Plains Ecozone [20]. Here, rainfall generally peaks in September around 90 mm, and temperatures peak in July at around 21 °C (1981 to 2010 Canadian Climate Normals [21]).

With an abundance of fertile soils, the majority of land is used for agricultural production. Wetlands (marsh and swamp) are numerous, and the extent of water bodies varies extensively on an inter and intra-annual basis, both as a result of water level changes, and the emergence of vegetation from the water surface as the growing season progresses. For this site, 11 sets of images of RADARSAT-2 Wide Fine Quad Pol (FQ5W and FQ17W) were acquired between 7 April and 22 September 2016. Each scene covers approximately 50 × 50 km², with a nominal ground range resolution of about 14 (FQ5) and 9 (FQ17) m (Table 1). For accuracy assessment of this site, concurrent and cloud-free WorldView-2 (WVII) images are analyzed. Note that for a meaningful comparison between the results from the two data frames and all dates, we have only focused on analyzing the area covered by all image acquisitions for the Bay of Quinte (see Figure 1c).

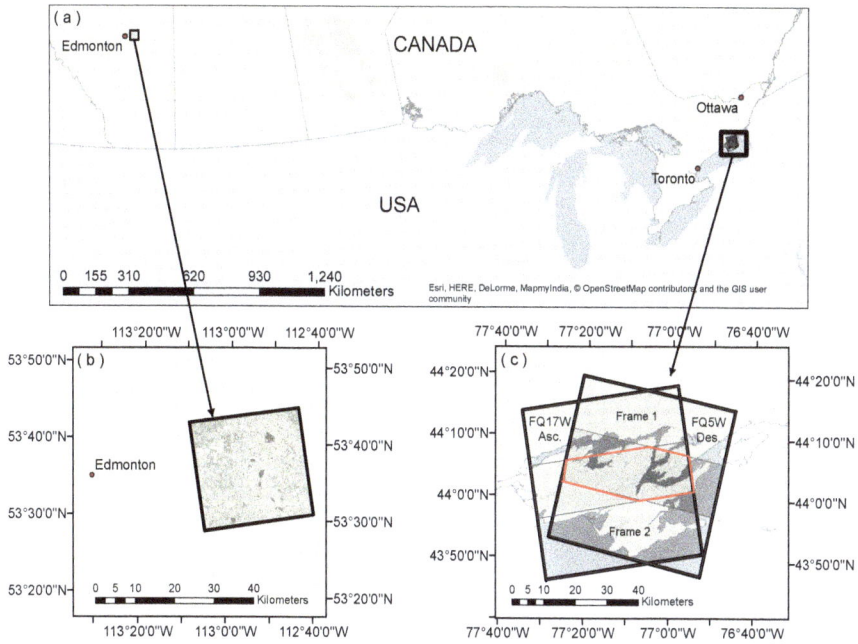

Figure 1. (**a**) Study areas located in Canada, including (**b**) the Prairie Potholes Region in Alberta, and (**c**) the Bay of Quinte in Ontario. Both study areas are covered by concurrent RADARSAT-2 images and high-resolution optical images. The red polygon represents the overlap for two image frames available over the Bay of Quinte site. For the Alberta site only one RADARSAT-2 image was available.

Table 1. RADARSAT-2 images acquired over the Bay of Quinte.

Beam	Incident Angle (°)	Resolution (m)	Orbit Direction	Acquisition Dates in 2016
FQ5W	22.5–26.0	13.6–11.9	Des.	7 April, 1 and 25 May, 18 June, 12 July, 22 September
FQ17W	35.7–38.6	8.9–8.3	Asc.	3 and 27 April, 14 June, 8 July, 25 August

3. Methodology

3.1. Image Processing

Figure 2 summarizes the processing steps applied to available RADARSAT-2 imagery. Image calibration was performed via the extraction of Sigma Nought ($\sigma°$) values using the Constant-Beta look-up tables provided with each scene. A polarimetric Lee filter was then applied to compensate for the effects of speckle. Bolanos et al. [7] discuss the advantages of this filter over others due to its ability to adapt to homogeneous and heterogeneous areas, and its ability to preserve edges. This, in addition to the fact that we wanted to make direct comparison between our method and the method proposed by Bolanos et al., is why we also used the Lee filter in this study. All RADARSAT-2 images were then orthorectified using the rational polynomial coefficients provided with the images and a Shuttle RADAR Topography Mission (SRTM) digital elevation model (DEM). Afterward, all orthorectified images were co-registered using a fast Fourier transform phase matching algorithm with a minimum match score of 0.9 set for automatically selected ground control points (GCPs). These image processing steps were implemented using PCI Geomatics software.

After orthorectification, the calibrated intensity values were then converted to decibels to pronounce the tails of each image's histogram.

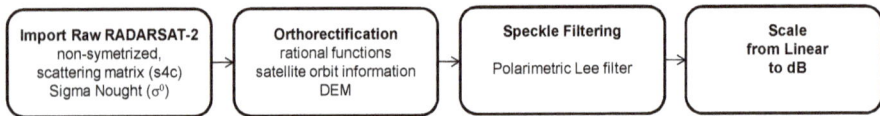

| Import Raw RADARSAT-2 non-symetrized, scattering matrix (s4c) Sigma Nought (σ⁰) | Orthorectification rational functions satellite orbit information DEM | Speckle Filtering Polarimetric Lee filter | Scale from Linear to dB |

Figure 2. The proposed workflow for preparation of RADARSAT-2 imagery. Note that the method could also be adapted to different processing methodologies, SAR data types, and optical imagery. SAR: synthetic aperture radar; DEM: digital elevation model; dB: Decibel.

3.2. Threshold Selection

With the proposed method, two sets of thresholds are extracted from the SAR images for two different purposes: (a) surface water detection; and (b) water boundary detection. For the former, histograms of either HH (Horizontal transmit and Horizontal receive) or HV (Horizontal transmit and Vertical receive) polarizations are used to select a statistically consistent threshold, while for the latter, higher-order texture images, such as variance, are used to select a threshold to detect the boundaries of water bodies. This latter step provides products for an additional cleanup process for the removal of false positives, or areas identified as water bodies, but which are not in fact water bodies. Each threshold type is described in detail in the subsequent section.

3.2.1. Surface Water Detection

Previous studies such as those of Brisco el al. [8] and Manjusree et al. [22] discuss the advantages of using the HH polarization for mapping flooded vegetation due to better canopy penetration, resulting in better contrast between flooded and non-flooded forests (e.g., compared to HV). Over non-disturbed water bodies, co-polarized channels are also used for discriminating land and water, though it is notable that at steep angles backscatter in the HH polarization can be equivalent to the values observed for land, and increased surface roughness (e.g., waves) reduces the separability between water and land [23]. Alternatively, several studies have shown that backscatter values of the cross-polarized channels HV and VH (Vertical transmit and Horizontal receive), are less affected by wind than the co-polarized channels [24,25]. In this study, we are interested in detecting water regardless of the roughness conditions; thus, have chosen to use the HV polarization. Note that this method could be adapted to use any polarization, as it relies solely on the fact that the distribution of values is bimodal (referring to the fact that image histograms show separate modes: one representing water, and one or more others representing non-water pixels).

Bimodality of image pixel values is typically achieved with a sufficient number of water bodies, with low backscatter values, relative to all non-water pixel values. Note that preliminary testing has demonstrated that bimodality can be achieved with as few as 2% of the total pixel population representing water. To define the threshold values, the values are first represented in decibel format (Figure 3a) as this representation is commonly used, and thus users can associate a given value with typically observed responses. This format also reduces noise and improves contrast [7]. In order to better visualize the low probability mode in the histogram, log scaling is applied to the vertical axis (Figure 3b). Then, a high-order polynomial is fitted to the dataset (Figure 3b), and the threshold is found at the local minimum, or valley, between the two modes (i.e., between the first mode, representing lower backscatter values, which typically represent water, and the second mode(s), representing higher backscatter values, which typically represent non-water pixels).

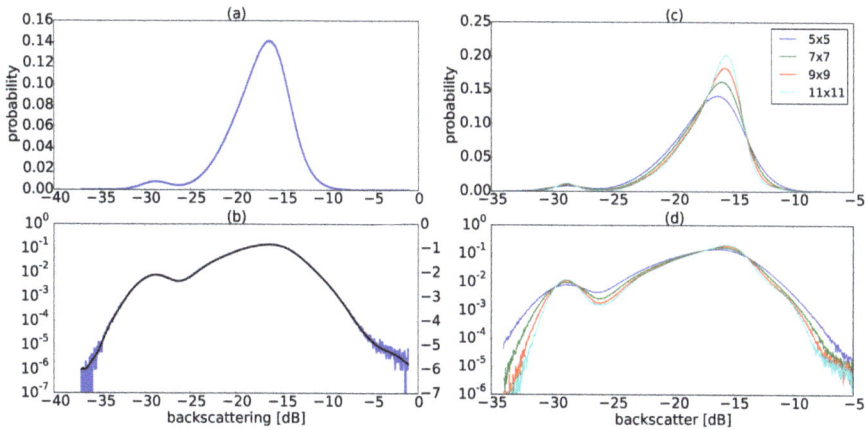

Figure 3. Image histograms generated from the HV intensity values for a Wide Fine Quad Pol image acquired over the Prairie Pothole Region, including (**a**) histogram of backscattering intensity values in dB; (**b**) the same histogram as (**a**) with logarithmic scaling applied to the vertical axis (solid black line represents a 55-order fitted polynomial with its corresponding vertical axis in an arbitrary unit); (**c**) histograms that resulted from applying different window sizes of the polarimetric Lee filter; and (**d**) the same histogram as (**c**) with the vertical axis in logarithmic scale (histograms are plotted using 1000 bins).

Figure 3c,d show the effect of applying speckle filtering with different window sizes on the shape of the image distribution. Speckle filtering consists of reducing the variance of a speckled image to improve the estimation of its mean [26]. In particular, the Lee filter requires the calculation of an adaptive filtering coefficient based on the local statistics defined by a sliding window of N × N pixels, in which homogeneous areas are low-pass filtered but heterogeneous areas that include texture information, such as sharp edges and isolated point targets, are preserved [27]. Table A1 shows threshold values identified via the approach described previously for the image in Figure 4a after having been filtered using various window sizes. Note that the threshold value increases as the window size increases (Table A1). This is mainly due to the average (low-pass) filter that causes an increase in the population of pixels identified as non-water while sliding over regions that are close to water edges. Thus, the thresholds calculated from images filtered with a larger window size can potentially lead to underestimation of the extent of water bodies as a result of increased mixing of land and water pixels. This is also reflected in the number of pixels selected as water, which are listed in Table A1. As such, for all images the polarimetric Lee filter with a 5 × 5 window size is used in this study.

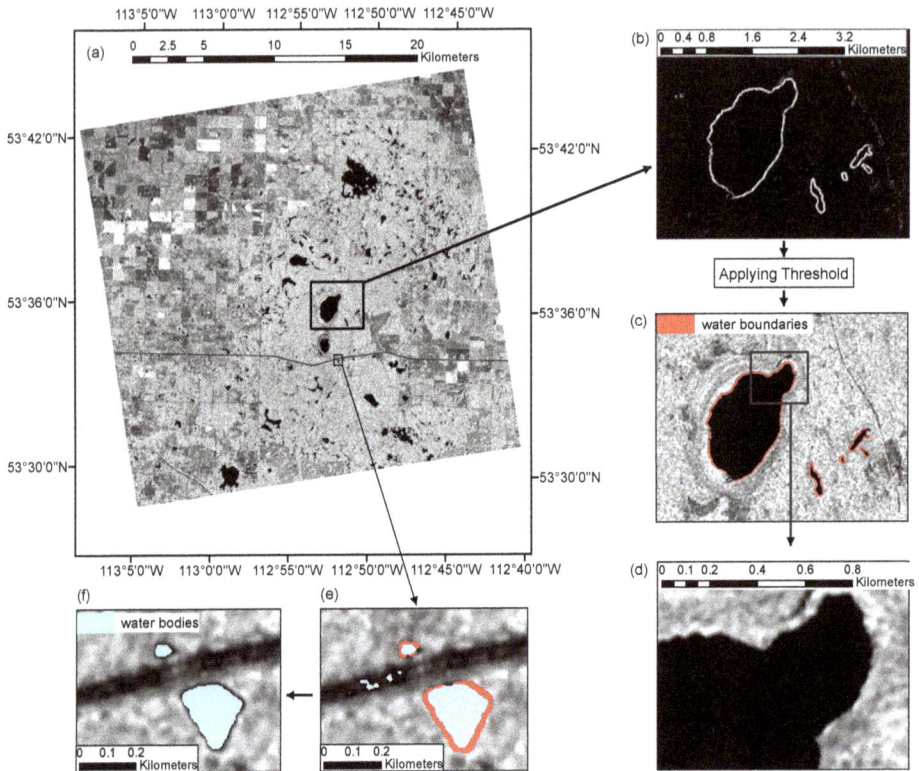

Figure 4. (**a**) HV polarization image from the Fine Quad Pol mode 19 (FQ19) scene acquired over the Prairie Pothole Region; (**b**) variance texture image; (**c**) extracted water boundaries by applying a pixel-based threshold (1.1) on the variance image; (**d**) segmented image generated using the simple linear iterative clustering (SLIC) superpixel algorithm; (**e**) water bodies generated using thresholding and superpixel segmentation; and (**f**) final water bodies after topological intersection of detected water bodies and water boundaries.

3.2.2. Surface Water Boundary

Smooth features such as roads, as well as areas that are affected by shadow, tend to exhibit low backscattering returns, and thus are potentially falsely identified as water via simple thresholding processes. Therefore, we apply a cleanup procedure following the detection of surface water. To do this, we use the boundaries of water bodies that are detected via high-order texture images, specifically the variance of the HV intensity image. Variance is used in this case because backscatter values observed for smooth water bodies are relatively low, and show high homogeneity, whereas backscatter values for land are generally higher. Thus, high variance values are expected when the sliding window within which they are calculated includes both values for water and non-water features (i.e., at the boundary of water bodies).

Variance images were therefore generated from the same cross-polarized channel (i.e., HV) and were similarly log-transformed to pronounce the rare events at the high tail (Figure 5a–c). Given that the histogram of variance image values similarly exhibited a bimodal distribution (Figure 5d) we use the same procedure as described in Section 3.2.1 to define the threshold value (i.e., 1.1) at the local minimum between the two modes. Note that in this case, the mean and standard deviation of the total pixel population are $\mu_v = -0.06$ and $\sigma_v = 0.37$, respectively, and thus the threshold value ($t_v = 1.1$) is approximately three standard deviations from the mean, i.e., $t_v - \mu_v \approx 3\sigma_v$ (or $T_v \approx 3$). Given the availability of a temporal stack of SAR images over the same area acquired at the same incident angle, the variance values of water pixels (and therefore all non-water pixels as well) are expected to be within the same range. Thus, we theorized that a constant value could be used for thresholding variance images at different times, and in cases where histograms of variance image values exhibit weak bimodality (i.e., we theorized that not all variance images need to exhibit strong bimodality to be used in this automated cleanup procedure). We investigated this hypothesis further in the results and discussion section.

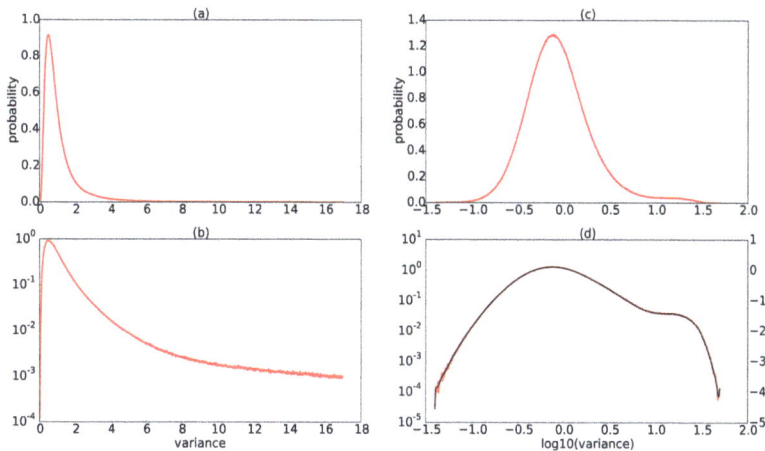

Figure 5. (**a**) Histogram of the variance texture image from the Prairie pothole site, calculated using the HV polarization image from the Wide Fine Quad 19 scene; (**b**) the same histogram with the vertical axis in logarithmic scale; (**c**) log-transformed histogram of the variance image and (**d**) the same histogram as (**c**) with the vertical axis in logarithmic scale (histograms are plotted using 1000 bins, and colors as in Figure 3).

3.3. Superpixel Segmentation

The SLIC superpixel algorithm was used to group adjacent pixels with similar characteristics. Via trial and error, it was found that splitting the image into 1000×1000 pixels blocks with 3600 superpixels in each block worked best for identifying potential water bodies, regardless of their size. In this study, the SLIC superpixel algorithm is implemented using the open-source "skimage" Python package applied to the HV intensity image. The "compactness" and "sigma" parameters in all cases were set to 1, since this tended to produce the best results, and permitted detection of water bodies as small as 25 m^2. The former parameter balances the color and space proximity, with high values giving more weight to the former, resulting in superpixels that are more square. The latter parameter, sigma, is the width of Gaussian smoothing kernel, where a value of zero does not apply any smoothing and higher values apply more smoothing.

3.4. Surface Water Extraction

To generate final surface water products, first, the intensity image is segmented and pixel values are averaged over each superpixel. Then, the mean value is compared with the threshold value defined previously in Section 3.2.1 (to identify surface water). Superpixels with mean values lower than the threshold are selected as water. At this stage, the selection of water pixels is reinforced through two mechanisms: (a) by grouping pixels through the SLIC algorithm to ensure adherence to local boundaries; and (b) by applying the threshold on the average of grouped pixels (i.e., superpixels).

Subsequently, to improve accuracy, water boundaries extracted from thresholding of the variance image are intersected with features identified as water bodies (Figure 4e), and only those that are adjacent to boundaries are included in the final surface water product. In this research, this step is referred to as the cleanup procedure. Figure 4e,f exemplify this process, showing where some of the polygons incorrectly identified as water over a road are excluded from the final surface water product. Figure 6 summarizes the processing steps of the entire SWD workflow. It consists of two separate sub-workflows: the first one (the top line in Figure 6) is the thresholding and segmentation process which results in the generation of water objects; the second one is a boundary detection process (the bottom line in Figure 6) which results in the generation of water boundary objects. The cleanup process is the last step in which a topological intersection between the water objects and boundary objects is performed to remove false positives. In the rest of this paper, the cleanup process is referred to as the process of generating boundary objects and the topological intersection between boundaries and water objects. It is notable that the energy texture image [7] can also be used instead of HH or HV images as an input to the algorithm. Note that that there is no user intervention required in the workflow in Figure 6 since the all workflow can be automated, including the threshold selection process for both surface water and boundary objects. The only user intervention required in the present method is the separation between the single mode and bimodal (or multi-modal) histograms which has to be done after preparation of SAR imagery (i.e., after the process outlined in Figure 2 and before the workflow in Figure 6). As mentioned in Section 3.2.1, a biomodal distribution can be achieved with as few as 2% water pixels in the SAR imagery.

To assess the accuracy of the proposed method, water bodies were manually digitized from a WVII image collected on the same date as one of the RADARSAT-2 scenes (i.e., 27 April 2016). We believe these results to be the most accurate estimate of the true areal extent of surface water and thus results from the proposed method were compared to those values.

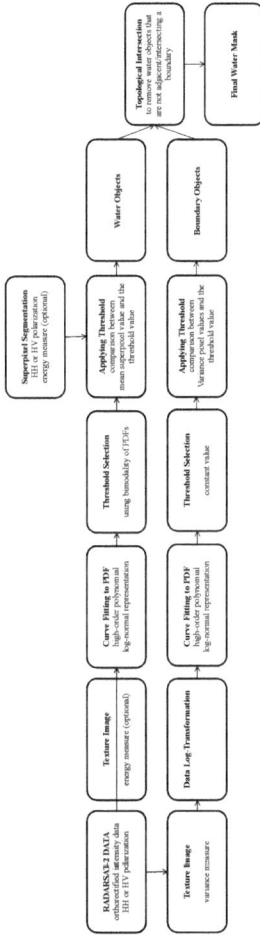

Figure 6. The proposed workflow for surface water detection using RADARSAT-2 imagery. PDF: probability density function.

4. Results and Discussion

4.1. Bay of Quinte

As described previously, the proposed SWD method is a two-step process. First, thresholds for each frame are determined, and second, thresholds are applied to segmented images. The results of the first step, including the normalized threshold values (T_{HV}), are shown in Table 2 and Figure 7 for each frame for the Bay of Quinte. It is notable that in Table 3 the threshold values are quite different from one frame to another, demonstrating that a constant (normalized) threshold value is perhaps less suitable for analysis of temporal data. It is also interesting to note that in Figure 7, the histograms in May and June have an extra mode between water and the dominant non-water mode. This extra mode peaks at approximately at -20 dB, and represents emerging vegetation. Note that this observation was validated via available WorldView data collected throughout the 2016 growing season.

The histograms of the variance texture measure of all frames are plotted in Figure 8. Note that there are many cases in which the distribution of the variance image is not bimodal due to a low population of boundary pixels. Based on these results we investigated whether a constant threshold value could be used for all variance images, given the homogeneity of water bodies and their low backscatter values, relative to the high backscatter from non-water features around water bodies. Four different threshold values were applied on the variance images. Based on the results of the Section 3.2.2 (in which t_v was found 1.1) and the visual investigation of the plots in Figure 8, threshold values of $t_v = 1.0, 1.1, 1.2,$ and 1.3 were evaluated.

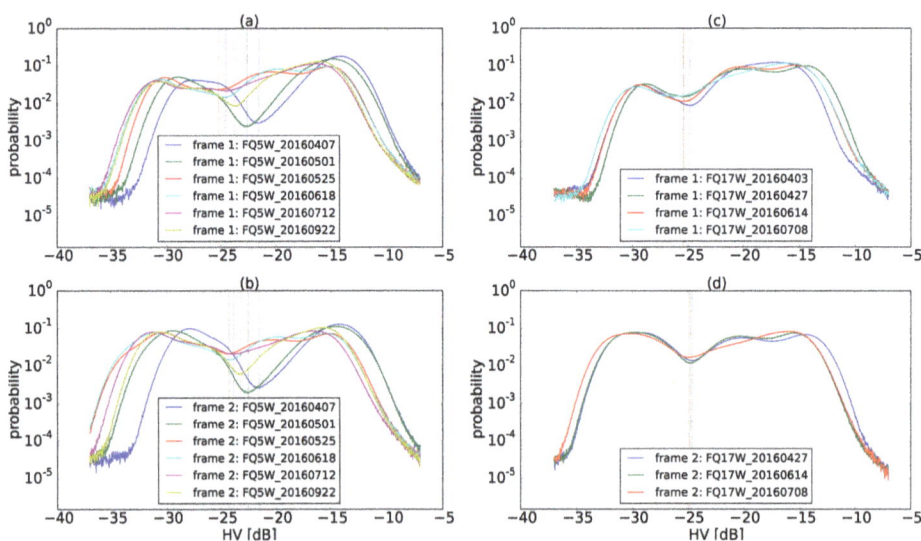

Figure 7. The distribution in log scale of HV backscatter values over the Bay of Quinte on different dates and for each frame for the FQ5 (**a,b**) and FQ17 (**c,d**) modes. Dotted lines represent the local minimum value of the polynomial curve, representing individual threshold values. The threshold values are also listed in Table 2.

Table 2. Threshold values for RADARSAT-2 HV polarization images at different dates and the area of detected surface water extents in the overlap area shown in Figure 1 over the Bay of Quinte (acquisitions in 2016). Digitized water extent in the overlap area is 81.091961 km^2; all areas are in km^2.

	Acquisition Date	FQ5W						FQ17W				
		7 April	1 May	25 May	18 June	12 July	22 September	3 April	27 April	14 June	8 July	25 August
Frame 1	t_{HV} [dB]	−21.71	−22.52	−24.31	−24.15	−24.20	−24.47	−24.57	−25.07	−25.06	−25.84	−24.71
	$T_{HV} = \frac{t_{HV} - \mu_{HV}}{\sigma_{HV}}$	−0.740	−0.730	−0.779	−0.727	−0.671	−0.673	−1.182	−1.266	−1.240	−1.404	−1.323
	Area (only thr.)	79.144050	78.536791	90.325668	81.645851	95.529267	77.592698	80.108943	80.700298	80.798277	83.484192	81.335291
	Area (thr. and seg.; no cleanup)	78.670027	79.046147	92.540965	83.266778	98.375056	78.106496	80.599210	81.973230	81.454872	84.878121	82.414998
	Area (thr. and seg.; cleanup $t_v = 1.1$)	78.396659	78.864226	81.436761	79.157216	83.161622	77.032808	79.841493	79.315489	79.291641	80.771659	79.243893
	Area (thr. and seg.; cleanup manual, A_m)	78.129014	78.634053	77.165192	78.042812	77.489723	76.943957	79.213296	78.602275	78.427127	77.792919	76.951172
Frame 2	t_{HV} [dB]	−21.65	−22.68	−24.10	−24.05	−24.05	−23.25		−24.45	−24.59	−24.73	
	$T_{HV} = \frac{t_{HV} - \mu_{HV}}{\sigma_{HV}}$	−0.208	−0.199	−0.154	−0.138	−0.078	−0.141		−0.302	−0.300	−0.297	
	Area (only thr.)	78.413984	78.405075	90.819369	82.320027	96.334413	78.040422		82.606517	80.857515	85.732206	
	Area (thr. and seg.; no cleanup)	78.854678	79.215934	92.954831	84.087752	96.568122	78.652638		83.322545	82.294253	87.964080	
	Area (thr. and seg.; cleanup $t_v = 1.1$)	78.576186	79.389975	81.886128	79.532122	84.675813	77.146079		79.654046	79.755364	83.387152	
	Area (thr. and seg.; cleanup manual, A_m)	78.257461	78.771559	77.197498	78.124867	77.612345	77.051342		78.790877	78.472412	78.043980	

Table 2 shows the areal extent of water calculated before and after applying the cleanup procedure. Note that before applying the cleanup procedure, the areal extent of water calculated for each image does not show a trend that is consistent with water level gauge information collected at several sites throughout the study area. Water heights increased throughout April, then mostly decreased until reaching September climate normals (ECCC, 2017). This discrepancy is a result of the true extent of surface water being masked by false positives. After applying the cleanup procedure however, accuracy is improved as demonstrated by the fact that the extent of surface water is closer to the values from the manually digitized surface water product, which we believe to be the most accurate estimate of the true areal extent of water. The results of the cleanup procedure are discussed later in this section.

Table 3 shows the percentage of falsely detected water bodies relative to the areal extent of water calculated via manually cleaned up products. For comparison, we use the non-dimensionalized variable $\frac{A - A_m}{A_m}$, where A is area in km^2 and A_m is the best approximation of the true areal extent of water based on thresholding/segmentation, as well as manual cleanup procedures based on visual comparison to the WorldView image acquired on the same day (see Table 2 for A_m values). These false positives are usually generated because of the presence of permanent or non-permanent smooth features (such as roads, airports, agricultural lands, etc.). The results in Table 3 show that by increasing the value of the threshold, the total area of detected water bodies after the automated cleanup procedure is closest to the A_m. However, this improvement comes at the expense of losing some water bodies that were correctly identified. In other words, as we decrease the variance threshold, false positives increase, but if the boundary conditions are too relaxed (i.e., as we increase the variance threshold value) water bodies that do not have very distinct boundaries are also not detected.

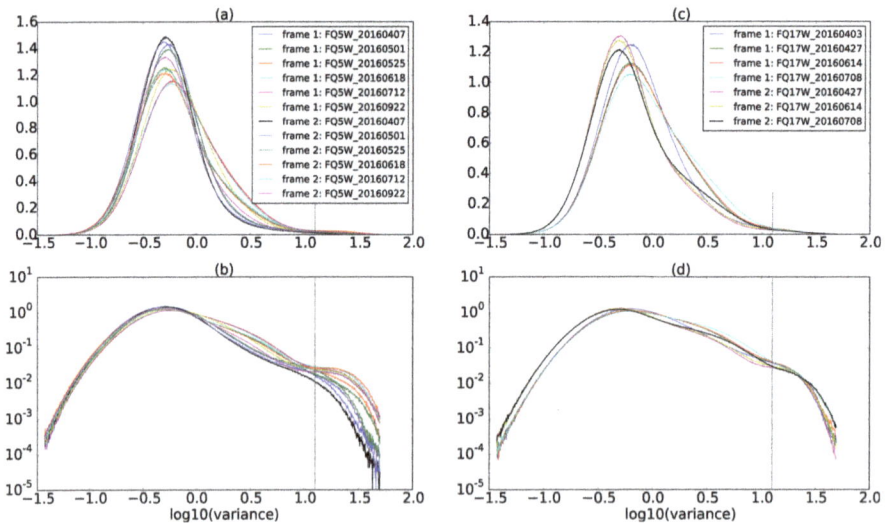

Figure 8. The distribution of variance texture images over the Bay of Quinte at different dates at two different incident angles. (**a**) Wide Fine Quad mode 5; (**b**) the same histogram as (**a**) in log scale; (**c**) Wide Fine Quad mode 17; and (**d**) the same histogram as (**c**) in log scale (dotted lines represent the place of threshold for variance texture images, $t_v = 1.1$).

To determine the appropriate threshold value, we calculated the percent of water losses using a variable $\frac{A_m|_{t_v} - A_m}{A_m}$, where $A_m|_{t_v}$ represents our most accurate estimate of the areal extent of water based on the automated cleanup at t_v, plus a manual cleanup procedure based on visual comparison of the WorldView image acquired on the same day. The difference between the two variables $A_m|_{t_v}$ and A_m represents the area of real water polygons lost due to the cleanup procedure. These results show that the real water extent losses due to an increase in t_v value, for example at $t_v = 1.3$, are as low as 0.03% (Table 4). For a better comparison, the percentage of false positives $\frac{A - A_m}{A_m}$ and the percentage of real water losses $\frac{A_m|t_v - A_m}{A_m}$ are plotted vs. the different variance threshold values in Figure 9. The plot suggests that the slight variation of the threshold value has only a minor effect on removing correctly detected water extents. On the other hand, it illustrates that a slight increase in the value of variance threshold value t_v (as small as 0.1) removes additional false positives at a much greater rate. Thus, this demonstrates that using a constant threshold value for image variance is justifiable for SWD using SAR data, as the increase in the value of the variance threshold removes more false positives, but relatively few correctly-identified water bodies. More details on the accuracy of these results are discussed in the next subsection.

Finally, it is notable that the areal extent of water extracted after the manual cleanup from the two frames shows a less than 0.4% difference for the two different incident angles at different dates (Table 2). This is a satisfactory result that suggests the SWD method proposed in this study can produce accurate and reliable results for images acquired at different incidence angles.

Table 3. Difference in the areal extent of water bodies, in percent, after applying the cleanup procedure and using different variance threshold values. A_m is the area of water extracted following a manual cleanup of results for each date.

				$\frac{A - A_m}{A_m} \times 100$		
Frame No.	Acquisition Date	3 April	27 April	14 June	8 July	25 August
	thr. and seg.; no cleanup	1.7	4.3	3.9	9.1	7.1
	thr. and seg.; cleanup, $t_v = 1.0$	0.9	1.5	1.3	5.6	4.1
Frame 1	thr. and seg.; cleanup, $t_v = 1.1$	0.8	0.9	1.1	3.8	3.0
	thr. and seg.; cleanup, $t_v = 1.2$	0.7	0.5	0.6	2.3	1.9
	thr. and seg.; cleanup, $t_v = 1.3$	0.5	0.2	0.3	1.3	0.8
	thr. and seg.; no cleanup			5.7	4.9	12.7
	thr. and seg.; cleanup, $t_v = 1.0$			1.7	2.0	8.9
Frame 2	thr. and seg.; cleanup, $t_v = 1.1$			1.1	1.6	6.8
	thr. and seg.; cleanup, $t_v = 1.2$			0.4	1.0	3.8
	thr. and seg.; cleanup, $t_v = 1.3$			0.03	0.6	2.4

It is worth mentioning that after visually inspecting WVII images acquired in April, May, June and July, it was observed that floating and/or emerging vegetation was present at several sites beginning in May, but was not detected until June. This is because these features were much smaller in size than the incident microwaves and did not affect backscatter intensity. On 27 April specifically, these features were not visible, and thus did not affect the accuracy assessment completed in this study. Further discussion on floating and or emerging vegetation and its effects on SWD in the context of this study are presented in Section 4.3.

The accuracy of water bodies extracted after the thresholding/segmentation step is affected by two types of errors: false positives and overestimation (or underestimation). The suggested cleanup process was used to improve accuracy by removing false positives. The overall effects of different (variance) thresholds on the removal of false positives have been evaluated and quantified in terms of the percentage of false positives that were detected versus losses of correctly-identified water bodies (see Tables 3 and 4 and Figure 9). To further investigate the details of such losses during the cleanup process, we have listed the number of true water bodies and the area of the largest one lost among those in Table 5. The number of lost polygons consistently increases as the variance threshold increases, but all of the lost polygons are very small water bodies. In many cases, these small water bodies did not exhibit distinct land/water contrast (for SAR data, speckle filtering can reduce the contrast between bright and dark features, especially if they are relatively small in size) and as a result their boundaries were not detected by the variance image. Given the accuracies of water products presented in Tables 3 and 4 and Figure 9, we conclude that a threshold value between 1.1 and 1.2 offers a reasonable balance between the removal of false positives and the loss of real water polygons. However, this can be adjusted for specific applications by users; for example, those that are interested in minimizing the loss of small water bodies can use a lower variance threshold value around 1.0 (though this would be at the expense of more false positives). It is notable that the removal of the false positives, in general, is an easier task to deal with as other ancillary data can be used to mask permanent features (e.g., roads).

Table 4. The variation of the detected area, in percent, after cleanup procedure using different thresholds.

Frame No.	Acquisition Date (27 April)	No. of Polygons	$A_m\|_{t_v}$ (km²)	$\frac{A_m\|_{t_v} - A_m}{A_m} \times 100$	Largest Lost Polyg. (m²)
	thr. and seg.; manual cleanup	81	78.602275	0.0000	
	thr. and seg.; cleanup, $t_v = 1.0$	78	78.601957	−0.0004	149
Frame 1	thr. and seg.; cleanup, $t_v = 1.1$	75	78.600120	−0.0027	1077
	thr. and seg.; cleanup, $t_v = 1.2$	71	78.593149	−0.0116	6232
	thr. and seg.; cleanup, $t_v = 1.3$	66	78.576692	−0.0325	11809
	thr. ; manual cleanup	72	78.052080	−0.7000	1288

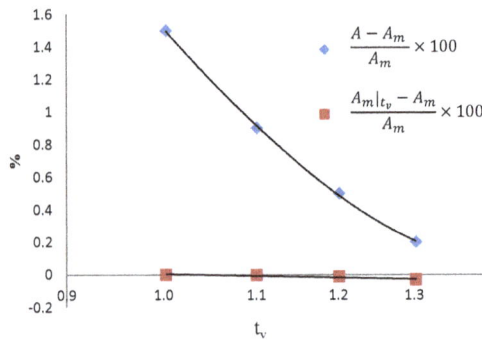

Figure 9. The percentage of false positives (blue diamonds) that were removed and the percentage of true water polygons (red rectangles) that were removed for different variance threshold values. Solid lines represent fitted curves to the data.

Table 4 provides the number of water bodies that were detected after the manual cleanup of the water product generated for the scene acquired on 27 April. Interestingly, the number of detected polygons using thresholding is lower than those detected using thresholding/segmentation (72 and 81, respectively). This is an important observation, as it shows that the thresholding of segmented SAR images improved the number of correctly detected water bodies. Another observation is that the total area of detected water bodies using only thresholding is lower than that detected using thresholding/segmentation. This is in addition to the fact that the largest polygon lost in thresholding had a relatively small area (1288 m²) and the percentage of difference (i.e., $\frac{A_m\|_{t_v} - A_m}{A_m} \times 100$) between the area of the manually cleaned up products from thresholding and thresholding/segmentation approaches is the highest percentage (−0.7) in Table 4.

This clearly suggests that the accuracy of water products generated using thresholding (but not segmentation) is further affected by underestimation of water extents compared to the products generated from thresholding/segmentation. As an example, we have calculated over and underestimated water extents on Fish Lake in the Bay of Quinte (see Figure 10, right). The areal extent of water found using the two approaches are listed in Table 5. The results show that the total area estimated from either approach is less than the true water extent, and that the detected water extent by thresholding is less than that of the thresholding/segmentation approach. We consistently observed this improved performance of the thresholding/segmentation approach compared to the thresholding only approach. The main reason for this improvement is the segmentation process, which is better able to detect the edges of water extents, and which rarely misclassifies the center of water bodies or the edges of scenes which may be noisy or represent slightly higher intensities from waves.

We have also plotted the distribution of the areas of detected water bodies in Figure 11 for the three cases listed in Table 5. These histograms show that both thresholding and thresholding/segmentation approaches are less accurate in detecting water bodies with areal extents between 100 and 1000 m^2, but perform well in detecting water bodies larger than 10,000 m^2. Given the resolution of these data, and the requirement for speckle filtering, this observation is sensible. Specifically, it is reasonable that smaller water bodies cannot be detected as often the filtering processes result in the mixing of adjacent land and water pixels, thus reducing the difference between values, and subsequently the ability to differentiate them.

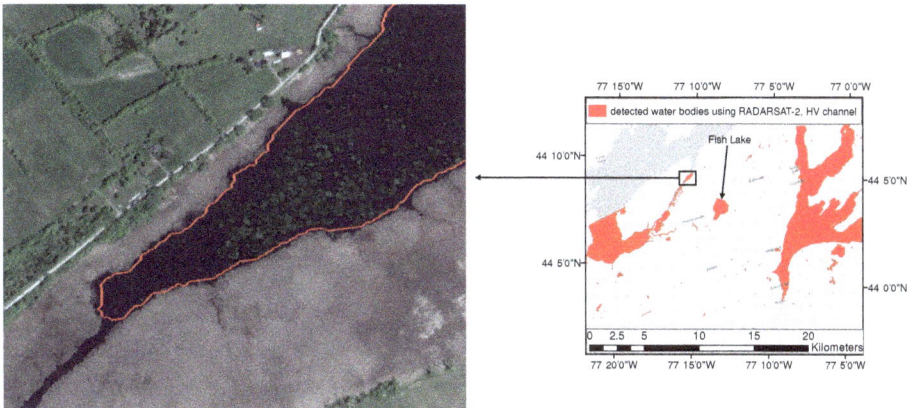

Figure 10. An example of floating and or emerging vegetation from the WorldView-2 (WVII) image acquired on 22 May 2016; the red line represents surface water extent detected using the RADARSAT-2 image acquired on 14 June 2016.

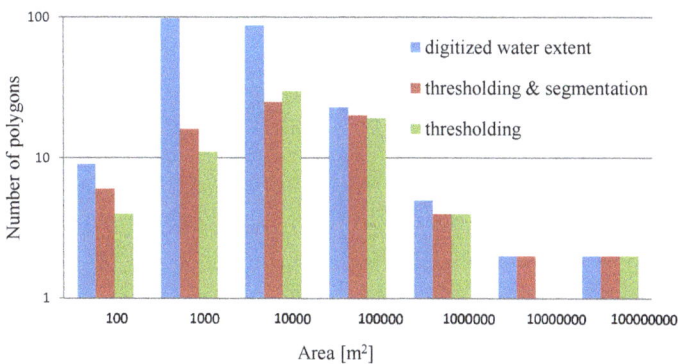

Figure 11. Area distribution of water bodies within the overlap area for the Bay of Quinte site. The total number of polygons for the digitized water extent (blue) is 225, for thresholding and segmentation (red) it is 75, and for threshold only (green) it is 72.

Table 5. Overestimated and underestimated water extents over Fish Lake. All area estimates are in given in square meters.

Frame No.	Fish Lake Acquisition Date (27 April)	Total Area	Underestimated Area	Overestimated Area
Frame 1	digitized water extent	1,608,511	0	0
	thresholding and segmentation	1,574,172	38,447	4192
	thresholding	1,553,469	56,220	1261

4.2. Prairie Pothole Region

Table 6 and Figure 12 summarize results for the Prairie Pothole Region. Similar to what was observed for the Bay of Quinte, the results in Table 6 suggest that the extent of water detected using only thresholding is an underestimation (by about 5% $\simeq \frac{20.07-19.04}{20.07} \times 100$) compared to the products extracted from thresholding and segmentation. As an example, the areas of open water calculated over Astotin lake using thresholding and thresholding/segmentation methods are 3.70 and 3.76 km^2, respectively. This 1.6% underestimation with thresholding only is again due to the poorer performance along the edges of water bodies, and the misclassification of pixels within the center of water bodies (Figure 13).

Table 6. The number of polygons and detected surface water area over the the Prairie Pothole Region using the HV channel.

Region	Acquisition Date (8 September 2012)	No. of Polygons	Detected Surface Water (km^2)
Prairie Potholes	thr. and seg.; no cleanup	785	20.168476
	thr. and seg.; manual cleanup	698	20.100138
	thr. and seg.; cleanup $t_v = 1.1$	669	20.073126
	thr. and seg.; cleanup $t_v = 1.1$ and manual cleanup	662	20.065713
	thr.; manual cleanup	755	19.044606

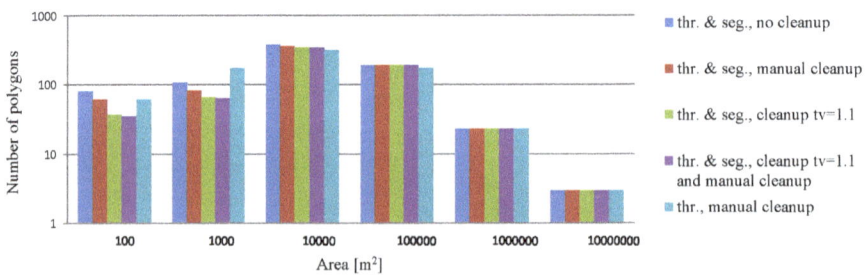

Figure 12. Area distribution of the detected water extents over the Prairie Pothole Region.

In this case, the automated cleanup procedure improved the removal of false positives with the loss of very few true water polygons (29), which were all less than 1000 m^2; making up less than 0.15% of the areal extent of all water bodies. The numbers and sizes of water polygons found using the proposed method are provided in Table 6 and plotted in Figure 12. Two points can be highlighted regarding the histograms in Figure 12. First, the difference between the number of detected water polygons after applying the automated cleanup and automated/manual cleanup is very small, demonstrating that after the automated cleanup most false positives are removed. Second, the thresholding method in this scene shows a slightly better performance for water bodies in the range between 100 and 1000 m^2 compared to thresholding/segmentation method. As mentioned previously, it is also possible to improve the accuracy for these small water bodies by decreasing the variance threshold, though at the cost of an increased number of false positives.

It is notable that the total surface water extent extracted by Bolonos et al. [7] is greater than the one calculated here ($\approx$23 km^2). This difference is a result of Bolanos et al. [7] using a threshold value that is lower (note that the pixel values are all positive in energy texture image; a lower threshold in the energy image is the equivalent of stating a higher threshold in HV channel) than the value at the valley between the two modes of the energy image histogram. Further, the threshold values in their study were normalized values (i.e., subtracted from the mean and divided by standard deviation of the total pixel population), thus a small variation in the normalized threshold value (T_e) is translated to a greater change in non-normalized values (t_e).

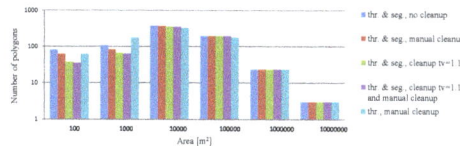

Figure 13. Water extents from the RapidEye image; colored lines represent water extents generated using thresholding and thresholding/segmentation over Astotin lake.

4.3. Limitations and Future Work

The SWD method proposed in this study can be used to generate seasonal, annual, or long-term permanent and non-permanent water masks using SAR data. However, one notable limitation is the inability to detect vegetation in standing water that is much shorter and thinner than the wavelength of incident microwaves. For the Bay of Quinte site specifically, it was observed that when floating and/or emergent vegetation first appears at the beginning of the growing season, it was sometimes still detected as water with the proposed method (Figure 10); however, as both the density and height of features increased through time, they were eventually classified as non-water pixels. Characterization of this type of temporal inaccuracy is complicated using SAR data as the roughness, height, and volume of this floating and/or emerging vegetation vary spatially, especially among different species. We propose that further improvements to this method can be achieved by incorporating information from high-resolution optical images. However, collecting coincident cloud-free imagery may not be possible in all cases.

5. Conclusions

In this study, a new method for semi-automated SWD using SAR data has been described and evaluated. The approach focuses on automatically defining threshold values, identifying water bodies and edge features on a per-object basis, and implementing an automated cleanup procedure which has been demonstrated to improve accuracy compared to thresholding only. This approach is adaptive to images acquired at different incidence angles and dates, removing the need for human intervention on a scene-by-scene basis. Additionally, an independent cleanup procedure is proposed to remove features falsely identified as water. The cleanup procedure is based on the detection of water boundaries extracted by thresholding images generated from high-order texture measures such as variance. The results showed that a constant threshold value can be used for extracting boundaries, thus it can be used in a fully automated cleanup procedure. This method, despite its current and relatively minor limitations, remains an attractive option in cases where there is need for surface water information for a specific time period (e.g., to determine available waterfowl habitat in spring), and for mapping large geographical areas, because results can be generated automatically, and SAR data can be acquired regardless of cloud cover and haze. Environment and Climate Change Canada expects to implement this approach operationally for spatial and temporal detection and monitoring of surface water within wetlands.

Acknowledgments: The authors would like to thank Bhavana Chaudhary for digitizing WVII images to extract water extents over the Bay of Quinte. We would also like to thank Patrick Kirby for his assistance with interface development and Python scripting. The authors would like to thank the Canadian Space Agency for funding in support of the Data Utilization and Applications Plan project (DUAP).

Author Contributions: The initial idea of this method was suggested/developed by White and Behnamian. Banks and Behnamian performed the all analysis, interpreted the results and wrote the first draft of the manuscript. Chen suggested the implementation of the SLIC superpixel algorithm for segmenting SAR images and contributed in developing the segmentation part of the Python script. Banks, White, Pasher, Millard and Behnamian suggested/developed/implemented the validation procedure. Behnamian wrote the Python scripts. All authors advised on the contents and helped edit the original and subsequent versions of the manuscript.

Conflicts of Interest: The authors declare no conflict of interest. The founding sponsors had no role in the design of the study; in the collection, analyses, or interpretation of data; in the writing of the manuscript, and in the decision to publish the results.

Abbreviations

The following abbreviations are used in this manuscript:

DEM	Digital Elevation Model
ECCC	Environment and Climate Change Canada
GCP	Ground Control Point
PDF	Probability Density Function
RGB	Red, Green, Blue
SAR	Synthetic Aperture Radar
SLIC	Simple Linear Iterative Clustering
SRTM	Shuttle RADAR Topography Mission
SWD	Surface Water Detection
WVII	WorldView-2

Appendix A. Lee Filter and Window Size Effects on Threshold Values

Table A1. Threshold values in dB for the HV polarization of the RADARSAT-2 image acquired over the Prairie Pothole Region filtered using a polarimetric Lee filter with different window sizes.

Window Size	Threshold Value (dB)	Number of Pixels Selected as Water
5×5	-26.25	861,052
7×7	-26.02	844,465
9×9	-25.97	806,704
11×11	-25.98	770,204

References

1. Ontario Biodiversity Council (OBC). *State of Ontario's Biodiversity 2010: A Report of the Ontario Biodiversity Council*; Ontario Biodiversity Council: Peterborough, ON, Canada, 2010.
2. Ontario Biodiversity Council (OBC). *State of Ontario's Biodiversity (Web Application)*; Ontario Biodiversity Council: Peterborough, ON, Canada, 2015.
3. Stocker, T.F.; Qin, D.; Plattner, G.K.; Tignor, M.; Allen, S.K.; Boschung, J.; Nauels, A.; Xia, Y.; Bex, B.; Midgley, B.M. *Climate Change 2013: The Physical Science Basis*; Contribution of Working Group I to the Fifth Assessment Report of the Intergovernmental Panel On Climate Change; IPCC: Geneva, Switzerland, 2013.
4. Desgranges, J.L.; Ingram, J.; Drolet, B.; Morin, J.; Savage, C.; Borcard, D. Modelling wetland bird response to water level changes in the Lake Ontario—St. Lawrence River hydrosystem. *Environ. Monit. Assess.* **2006**, *113*, 329–365.
5. Doll, P.; Mueller Schmied, H.; Schuh, C.; Portmann, F.T.; Eicker, A. Global-scale assessment of groundwater depletion and related groundwater abstractions: Combining hydrological modeling with information from well observations and GRACE satellites. *Water Resour. Res.* **2014**, *50*, 5698–5720.

6. Ficke, A.D.; Myrick, C.A.; Hansen, L.J. Potential impacts of global climate change on freshwater fisheries. *Rev. Fish Biol. Fish.* **2007**, *17*, 581–613.

7. Bolanos, S.; Stiff, D.; Brisco, B.; Pietroniro, A. Technical Note: Operational Surface Water Detection and Monitoring Using Radarsat 2. *Remote Sens.* **2016**, *8*, 285.

8. Brisco, B.; Short, N.; van der Sanden, J.; Landry, R.; Raymond, D. Technical Note: A semi-automated tool for surface water mapping with RADARSAT-1. *Can. J. Remote Sens.* **2009**, *40*, 135–151.

9. Matgen, P.; Hostache, R.; Schumann, G.; Pfister, L.; Hoffmann, L.; Savenije, H.H.G. Towards an automated SAR-based flood monitoring system: Lessons learned from two case studies. *Phys. Chem. Earth* **2011**, *36*, 241–252.

10. Li, J.; Wang, S. An automatic method for mapping inland surface waterbodies with Radarsat-2 imagery. *Int. J. Remote Sens.* **2015**, *36*, 1367–1368.

11. White, L.; Brisco, B.; Pregitzer, M.; Tedford, B.; Boychuk, L. Research Note: RADARSAT-2 Beam Mode Selection for Surface Water and Flooded Vegetation Mapping. *Can. J. Remote Sens.* **2014**, *40*, 135–151.

12. Schumann, G.; Baldassarre, G.D.; Alsdorf, D.; Bates, P.D. Near real-time flood wave approximation on large rivers from space: Application to the River Po, Italy. *Water Resour. Res.* **2010**, *46*, doi:10.1029/2008WR007672.

13. Otsu, N. A threshold selection method from gray-level histograms. *IEEE Trans. Syst. Man Cybern.* **1979**, *9*, 62–66.

14. Lee, S.U.; Chung, S.Y.; Park, R.H. A comparative performance study of several global thresholding techniques for segmentation. *Comput. Vis. Graph. Image Process.* **1990**, *52*, 171–190.

15. Vala, M.H.J.; Baxi, A. A review on Otsu image segmentation algorithm. *Int. J. Adv. Res. Comput. Eng. Technol. (IJARCET)* **2013**, *2*, 387–389.

16. Qin, F.; Guo, J.; Lang, F. Superpixel Segmentation for Polarimetric SAR Imagery Using Local Iterative Clustering. *IEEE Geosci. Remote Sens. Lett.* **2015**, *12*, 13–17

17. Ren, X.; Malik, J. Learning a Classification Model for Segmentation. In Proceedings of the Ninth IEEE International Conference on Computer Vision, Nice, France, 13–16 October 2003.

18. Achanta, R.; Shaji, A.; Smith, K.; Lucchi, A.; Fua, P.; Susstrunk, S. *SLIC Superpixels*; Technical report; EPFL 149300; Images and Visual Representation Laboratory: Lausanne, Switzerland, 2010.

19. Achanta, R.; Shaji, A.; Smith, K.; Lucchi, A.; Fua, P.; Susstrunk, S. SLIC Superpixels Compared to State-of-the-Art Superpixel Methods. *IEEE Trans. Pattern Anal. Mach. Intell.* **2012**, *34*, 2274–2281.

20. Ecological Stratification Working Group, Center for Land, Biological Resources Research, State of the Environment Directorate. *A National Ecological Framework for Canada*; Centre for Land and Biological Resources Research, State of the Environment Directorate: Hull, QC, Canada, 1996.

21. Canada, E.A. Canadian Climate Normals. Available online: http://climate.weather.gc.ca/climate_normals/ (accessed on 10 October 2017).

22. Manjusree, P.; Kumar, L.P.; Bhatt, C.M.; Rao, G.S.; Bhanumurthy, V. Optimization of threshold ranges for rapid flood inundation mapping by evaluating backscatter profiles of high incidence angle SAR images. *Int. J. Disaster Risk Sci.* **2012**, *3*, 113–122.

23. Martinis, S. Automatic Near Real-Time Flood Detection in High Resolution X-Band Synthetic Aperture Radar Satellite Data Using Context-Based Classification On Irregular Graphs (Doctoral Dissertation, Lmu). Ph.D. thesis, University of Munchen, Bavaria, Geramany, 2010.

24. Schumann, G.; Hostache, R.; Puech, C.; Hoffmann, L.; Matgen, P.; Pappenberger, F.; Pfister, L. High-resolution 3-D flood information from radar imagery for flood hazard management. *IEEE Trans. Geosci. Remote Sens.* **2007**, *45*, 1715–1725.

25. Henry, J.B.; Chastanet, P.; Fellah, K.; Desnos, Y.L. Envisat multi-polarized ASAR data for flood mapping. *Int. J. Remote Sens.* **2006**, *27*, 1921–1929.

26. Mascarenhas, N.D.A. An overview of speckle noise filtering in SAR images. In Proceedings of the First Latino-American Seminar on Radar Remote Sensing, Buenos Aires, Argentina, 2–4 December 1997; Volume 407, p. 71.

27. Lopes, A.; Touzi, R.; Nezry, E. Adaptive speckle filters and scene heterogeneity. *IEEE Trans. Geosci. Remote Sens.* **1990**, *28*, 992–1000.

Remote Sens. **2017**, *9*, 1209

remote sensing

MDPI

Article

Coherence Change-Detection with Sentinel-1 for Natural and Anthropogenic Disaster Monitoring in Urban Areas

Prosper Washaya [1], **Timo Balz** [1,2,*] **and Bahaa Mohamadi** [1]

1   State Key Laboratory of Information Engineering in Surveying, Mapping and Remote Sensing,
    Wuhan University, Wuhan 430072, China; pwashaya9@gmail.com (P.W.); bh.mo@whu.edu.cn (B.M.)
2   Collaborative Innovation Center for Geospatial Technology, Wuhan 430072, China
*   Correspondence: balz@whu.edu.cn; Tel.: +86-27-6877-9986

Received: 16 May 2018; Accepted: 25 June 2018; Published: 28 June 2018

Abstract: Rapid, reliable, and continuous information is an essential component in disaster monitoring and management. Remote sensing data could be a solution, but often cannot provide continuous data due to an absence of global coverage and weather and daylight dependency. To overcome these challenges, this study makes use of weather and day/light independent Sentinel-1 data with a global coverage to monitor localized effects of different types of disasters using the Coherence Change-Detection (CCD) technique. Coherence maps were generated from Synthetic Aperture Radar (SAR) images and used to classify areas of change and no change in six study areas. These sites are located in Syria, Puerto Rico, California, and Iran. The study areas were divided into street blocks, and the standard deviation was calculated for the coherence images for each street block over entire image stacks. The study areas were classified by land-use type to reveal the spatial variation in coherence loss after a disaster. While temporal decorrelation exhibits a general loss in coherence over time, disaster occurrence, however, indicates a significant loss in coherence after an event. The variations of each street block from the average coherence for the entire image stack, as measured by a high standard deviation after a particular disaster, is an indication of disaster induced building damage.

Keywords: natural disasters; anthropogenic disasters; Sentinel-1; Coherence Change-Detection

---

1. Introduction

Remote sensing data provides a solution for rapid data acquisition during disaster scenarios as well as post disaster information needed for recovery monitoring and management. High-resolution imagery is particularly well suited for identifying changes in landscapes using various change-detection techniques. Moreover, the vast array of techniques that remote sensing technology offers when used alongside Geographical Information System (GIS) software, can reduce uncertainty and serve as a catalyzing agent for analyzing and sharing information [1]. In addition, inaccessible areas can be monitored through remote sensing technologies; for example, areas affected by forest fires can be mapped and monitored, even though they may be difficult to access physically. Efforts have been made to monitor areas affected by disasters using Synthetic Aperture Radar (SAR) imagery [2–6]. These approaches, however, have not considered the globally available data products now realized by the Sentinel-1 mission. The global availability of these data permits disaster monitoring anywhere on the planet, even in areas that lack other types of remote sensing data.

This study demonstrates how changes resulting from anthropogenic and natural disasters in urban areas can be monitored using Sentinel-1 imagery and the Coherence Change-Detection (CCD) technique. Furthermore, this study makes a quantitative comparison of CCD results from several case studies to evaluate the applicability of this approach in varying scenarios. These comparative case studies demonstrate both the efficiency and inefficiencies of remote sensing techniques. There are precedents: Gähler [7] presents remote sensing applications for numerous case studies; flooding in Germany 2013, the Nepal earthquake in 2015, forest fires in Russia 2015, and the search for the missing Malaysian aircraft in 2014. Our study presents a fresh perspective on disaster monitoring, exploring the utility of the recently introduced, globally available Sentinel-1 archive data for detection and quantitative comparison of changes resulting from different disaster scenarios.

1.1. Monitoring Disasters: Anthropogenic and Natural Disasters with Remote Sensing Imagery

In disaster management and monitoring, high-resolution satellite imagery from different dates is especially useful for change-detection [8,9]. Change-detection is defined as a process of identifying differences in the state of an object by observing the pre- and post-event data [10]. Hoque et al. [11] reviewed the various change-detection techniques suitable for the management of tropical cyclone disasters. These events often have devastating impacts on coastal areas across the world. In 2005, Hurricane Katrina wreaked havoc in New Orleans, destroyed critical infrastructure, and damaged the natural environment [12]. Monitoring the recovery of New Orleans using remote sensing techniques, showed that even 10 years after this tropical cyclone, the average vegetation in the affected areas had not fully recovered [13]. In 2017, four major hurricanes—category three or greater—were recorded in the Atlantic by September. These included Irma, Harvey, Jose, and Maria; Maria being regarded as the worst natural disaster in the history of Puerto Rico.

Also, in 2017, California experienced its worst and most expensive wildfire season on record. There were close to 9000 wildfires tearing through the state, burning 1.2 million acres of land, destroying more than 10,800 structures and killing at least 46 people. Forest fires have become a major concern in various areas worldwide. The major fires occurring in the El-Nino year of 1997/1998 burned 25 million hectares of forest area worldwide [14]. Consequently, first responders and decision makers seek to detect and monitor forest fires in a timely way. Optical imagery, such as Landsat imagery, is intuitively understood multispectral data in the visible range of the electromagnetic spectrum, and often employed in land-use classification and disaster monitoring [15]. Landsat imagery has been successfully used to measure fire-induced deforestation and produce burned area maps [15] that support decision-making processes.

Since 2008, the National Aeronautics and Space Administration (NASA) and the United States Geological Survey (USGS) have freely provided an archive of publicly available Landsat imagery data spanning four decades; this data has been used for various applications [16]. This rich body of archival data allows researchers to compare images from different periods to identify and extract areas of changes to understand variations in landscapes over time. Landsat imagery was used to detect urban destruction related to the Syrian conflict [16], as the archive of past images has a temporal fidelity as short as eight days, at high radiometric consistency, with excellent ortho-rectification [17]. However, optical data relies on a passive sensor and, hence, is affected by bad weather conditions; but all-day and all-weather SAR can provide useful information for disaster assessment even under bad weather conditions [3].

Synthetic Aperture Radar (SAR) imagery can detect, extract, and assess disaster-induced damage, such as the destruction caused by earthquakes. SAR is an active sensing system and can overcome the drawbacks of optical imagery [3–6]. The effects of earthquakes can be devastating and may cause significant loss of life and property damage, especially in urban regions. The 2008 Wenchuan earthquake devastated cities in Sichuan province, claiming at least 69,000 lives, and was the most destructive earthquake in China over the last 50 years [18]. In 2017, a magnitude-7.3 earthquake hit the northern border region between Iran and Iraq, the hardest hit town was Sarpol-e Zahab, about 10 miles from the Iraq border [19]. Although TerraSAR-X imagery has been used to detect and assess building damage after an earthquake, this data is sometimes not available for some areas in post-disaster situations. The Sentinel-1 system is a SAR system that, through large-scale mapping capability and high revisiting frequency [20], provides solutions to this drawback.

The Sentinel-1 mission is seen as a potential game changer in operational SAR missions for decades to come [20]. The mission is the first of six Sentinel dedicated missions, introduced as part of the European Copernicus program under the domain of the European Space Agency. The Sentinel-1 system is currently based on a constellation of two SAR satellites (Sentinel-1A and Sentinel-1B) that have on-board C-band sensors. Sentinel-1A was launched on 3 April 2014; the second Sentinel-1 satellite, Sentinel-1B, was launched on 25 April 2016. With a 12-day revisit time, Sentinel-1 operates in four exclusive acquisition modes: Stripmap (SM), Interferometric Wide swath (IW), Extra-Wide swath (EW), and Wave (WV). The introduction of the Sentinel-1 constellation now ensures the timely availability of data through global coverage and the day and night, all-weather availability of data.

1.2. SAR Coherence and Change-Detection

This study employed data products from the Sentinel-1 SAR system with a C-band sensor. Synthetic Aperture Radar (SAR) is a coherent active microwave imaging method in remote sensing used for mapping the scattering properties of the Earth's surface [21]. Unlike down-looking optical sensors, SAR systems are side-looking in nature. This side-looking attribute is responsible for the three main effects inherent in all SAR images: shadow, layover, and foreshortening [22]. Layover occurs when tall objects are displaced towards the sensor when the signal reaches the top before the bottom. Foreshortening is a form of layover where the signal of an illuminated slope is compressed; subsequently appearing shorter in the SAR image than it is in reality. The shadow effect results when a steep slope or a vertical object, like a tall building, causes a radar shadow that appears black in an image. Besides the geometry, SAR signals are affected by a target object, specifically, the dielectric constant and surface roughness and the incident angle of these objects [23]. Buildings are usually structures with rectangular corners made of concrete, stone, metal, and glass where the signal may bounce from the ground to the façade then back to the sensor or from the façade to the ground and back to the sensor. Because of this so-called double-bounce effect, buildings and general infrastructure often appear bright in SAR images, and, therefore, urban built areas are usually represented by bright pixels [24]. Consequently, this study focuses on urban, built-up areas as case studies as they can be detected easily in SAR imagery.

Unlike other SAR techniques that rely on information from a single image, Interferometric Synthetic Aperture Radar (InSAR) exploits the phase differences of at least two complex-valued SAR images acquired from different orbit positions and/or at different times [21]. In addition, unlike SAR, which utilizes the amplitude information of a complex SAR signal, InSAR utilizes phase information. This phase information is used for interferometric products, like coherence images, and permits measurements of change between two images. When interferometric SAR images are not acquired simultaneously, they are affected by different types of noise: atmospheric conditions such as humidity, temperature, and pressure; and change in scatterers, for example, water body scatterers change in just milliseconds. Perpendicular baselines and volume scattering also add noise. The effect of such contributions affects both altitude and terrain deformation measurements [25]. These, therefore, influence the similarity or coherence of the phase signals.

Two waves with a phase difference that remains constant over time are said to be coherent, therefore, the higher the coherence of two waves, the easier it is to predict the properties of one of those waves given knowledge of the other [22]. In this case, coherence estimation becomes essential in generating Digital Elevation Models and measuring deformation [26,27]. Coherence is thus defined as the amplitude of the complex correlation coefficient between two SAR images [28]. The coherence is estimated on a given window size, using Equation (1) below:

$$\gamma = \left| \frac{\frac{1}{N}\Sigma_{i=0}^{N}M_i S_i^*}{\sqrt{\frac{1}{N}\Sigma_{i=0}^{N}M_i M_i^* \frac{1}{N}\Sigma_{i=0}^{N}S_i S_i^*}} \right| \qquad (1)$$

where N is the number of neighboring pixels to be estimated, M and S are the complex master and slave images, respectively, and * denotes the complex conjugate. γ is the resulting coherence [29]. The magnitude used in the equation is so that the values of γ range from 0 (incoherent) to 1 (coherent). The coherence is only equal to 1 when $M = S$, which means the observation is identical in the two images because of stable objects like buildings in the scene. In reality, though, remote sensing measurements cannot be identical over time [22], as a result, values are normally below 1 and often distributed between 0.5 and 0.7, as in the case of urban built-up areas in our study. The high coherence value exhibited by built-up areas is essential as it makes them easy to identify on a coherence image. Therefore, it is essential to exploit the advantages of built-up zones in urban areas (high pixel values and high coherence values) in SAR and coherence images, taking into account the vulnerability of these areas to disasters due to population clustering.

SAR coherence has been used in various applications over the years. Prati and Rocca [30] produced coherence maps for target classification, while Bruzzone et al. [31] proposed a novel system for the classification of SAR images based on concepts of long-term coherence and backscattering temporal variability. Additionally, one well-established application of SAR is the detection of temporal changes in a scene through the Coherence Change-Detection (CCD) [32].

Coherence and intensity characteristics of SAR images have been exploited in techniques to monitor urban activities and changes. Unsupervised thresholding techniques for change-detection using the coherence and intensity characteristics of SAR imagery have been proposed in previous studies [33]. Jendryke et al. [34] combined social media messages with SAR images to express human activities and urban changes in Shanghai. The coherence characteristics of SAR images identified urban areas and the changes occurring therein; linking these data to social media messages permitted the identification of human activity occurring in those areas. However, the SAR coherence techniques applied in these studies face the following setbacks: false positives resulting from the low coherence of vegetation and water bodies and unavailability of post-disaster data.

This study, therefore, sets to solve these problems by using Sentinel-1 imagery, which is globally available, to calculate coherence before and after a disaster and improve the overall accuracy and reduce false positives by calculating the standard deviation of coherence over time and aggregate this into street blocks. Further, the study uses land use classes to measure coherence loss over time and after a disaster.

2. Materials and Methods

2.1. Data Used

Sentinel-1 C band data with VV polarization and baselines of <100 days (temporal) and <150 m (perpendicular) between master and slave images for the six study areas (Aleppo, Damascus, Raqqa, Sarpol Zahab, Santa Rosa, and San Juan) were processed using the SarProZ software [35] for coherence map generation. For Aleppo, 18 Sentinel-1 SLC images were selected. We selected one image as the master image and co-registered it to subsequent slave images, with an image baseline separation of <250 days (temporal) and <100 m perpendicular, and eight coherence maps were generated. The same technique was applied to the other study areas with variations only in the number of images used, the temporal and perpendicular baselines, and the coherence maps generated. For Damascus, 14 images were selected and 13 coherence images produced. For Raqqa, 17 images were selected and 12 coherence maps generated. In the study areas affected by natural disasters, 15 images were selected for Sarpol Zahab and 14 coherence maps generated, for San Juan, Puerto Rico, 13 images were selected and 12 coherence maps generated, and for Santa Rosa, 21 images were selected and 20 coherence maps were generated. The coherence images were integrated to street blocks and land use class polygons.

In this study, the street block was taken as the smallest element in an urban area surrounded by a road at any level of the road hierarchy. Street blocks were generated based on Open Street Map

(OSM) data, which is a form of volunteered geographic information [36]. Street blocks may vary in size, for example, from a block of a few residential apartments to a whole university campus.

Landsat-5 and Landsat-8 images were classified using the supervised classification technique and built-up, vegetated, water bodies, and bare soil were identified as the land-use types in the six study areas using ENVI Software (ENVI, Melbourne, FL, USA). Classification of the optical imagery distinguishes classes in the study areas and permits analysis of the ways in which different disasters affected coherence in the different areas.

2.2. Workflow

The methodology is divided into two parts, which can be broadly categorized as image processing and image analysis (refer to Figure 1). The image processing involves coherence image generation to identify areas and the changes that have taken place therein. The process of coherence map generation included the following steps: image stacking, slave and master selection and extraction, sub-setting the study area, downloading the external Digital Elevation Model (DEM), image filtering, interferogram generation, coherence map generation, and exporting the orthorectified coherence maps in TIFF format. The coherence images were then classified into areas of change and no-change. ArcGIS/ArcMAP software (ArcGIS 10.5, Redlands, CA, USA) was used to perform this task. The images were divided into two classes, and a threshold of 0.6 was applied. The threshold can be calculated using the Renyi's entropy to convert the image into a binary image showing areas of change and no change [33]. However, for urban areas, the 0.6 threshold is considered optimal when identifying built up areas in coherence images [24]. This threshold may vary when longer temporal baselines are considered and also depends on regional factors. In our study, areas with a coherence below the threshold were considered as changed areas and those above were considered as unchanged areas. The results are threshold-based maps classified into change (0) and no change (1) areas.

The study areas were extracted from the coherence images using the shape files to mask out the study area, this was done in order for us to focus on the urban area of the affected regions. The resulting masked coherence images were then converted into point data. This was done to enable the integration of the coherence values to street blocks, which are polygons. The point data was then integrated into the street blocks and the average coherence value per street block was calculated for each image. Standard deviation was then calculated to show which street block deviated from the average over time, indicating change. Standard deviation is, however, not a measure of instability, but a measure of change. For example, coherence in vegetation exhibits instability due to constant de-correlation but has a low standard deviation. A sudden change in a building, however, will result in a high standard deviation as this will show a huge shift from the norm; thus, indicating a change in coherence for buildings. Other types of polygons, for example, grids or hexagons, can be used for this purpose; however, street blocks are more appropriate in an urban set-up as they are relatable objects in the real world [34]. The same technique was applied to the classified polygons by joining them to the coherence point data and calculating the average coherence over time.

To make a comparison of the change-detection results for the different disasters, we extracted the built-up area and vegetation from the land-use classes (a built-up area is less sensitive to coherence loss, thus indicating the coherence loss resulting from a disaster). After calculating the average coherence for the urban classes in the coherence image immediately before a disaster and the image immediately after a disaster, we applied the following change-detection formulae [37]:

$$A = \frac{(T1 - T2)}{T2} * 100\% \tag{2}$$

where A is the percentage of change, $T1$ is the average coherence for the coherence image before the disaster, $T2$ is the average coherence of the image after the disaster, and A is the percentage of change.

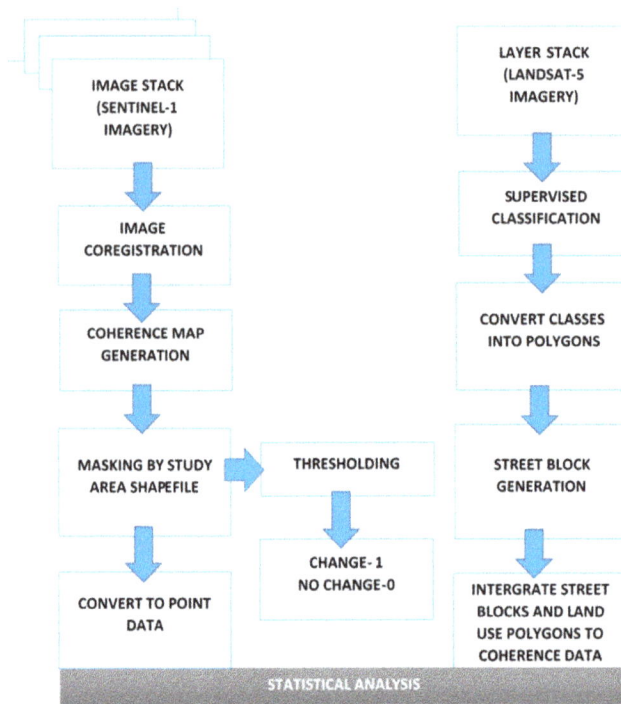

Figure 1. Methodology workflow.

2.3. Study Areas

The study focuses on the following areas affected by different types of disasters (both natural and anthropogenic). For the natural disasters, we selected San Juan, Sarpol Zahab, and Santa Rosa (San Juan) as shown in Figure 2. On 20 September 2017, Hurricane Maria struck Puerto Rico and caused catastrophic damage, which triggered a major humanitarian crisis in San Juan, the capital, and most populous municipality of Puerto Rico. In November 2017, an earthquake measuring 7.3 in magnitude devastated the Iran-Iraq border. It was the strongest on record in the region since 1967. The damage was extensive, and one of the most affected cities was Sarpol Zahab. Another notable natural disaster was the Tubbes fire of October 2017; the fire caused damage estimated at $1.2 billion US dollars, with five percent of Santa Rosa's housing stock destroyed [38].

For anthropogenic disasters, we selected the Civil War in Syria, (focusing on Aleppo, Damascus, and Raqqa) sparked by demonstrations motivated by demands for democratic reforms and release of political prisoners. The Syrian government's response escalated the tensions, which led to the demonstration shifting from their original demands to a demand for the removal of the Assad government [39]; the result was death and destruction in the country's big cities like Aleppo, Raqqa, and Damascus. In 2016, the Syrian Government embarked on a campaign to take back Aleppo city that was in the hands of the Rebels, which resulted in the devastation of the city. In Damascus, war intensified in July 2012 but did not last long as the city became a Syrian government stronghold and witnessed isolated cases of rebel attacks and suicide bombings. In Raqqa, the United States-backed Syrian Defense forces launched an operation in October 2017, which left 80% of the city uninhabitable [39].

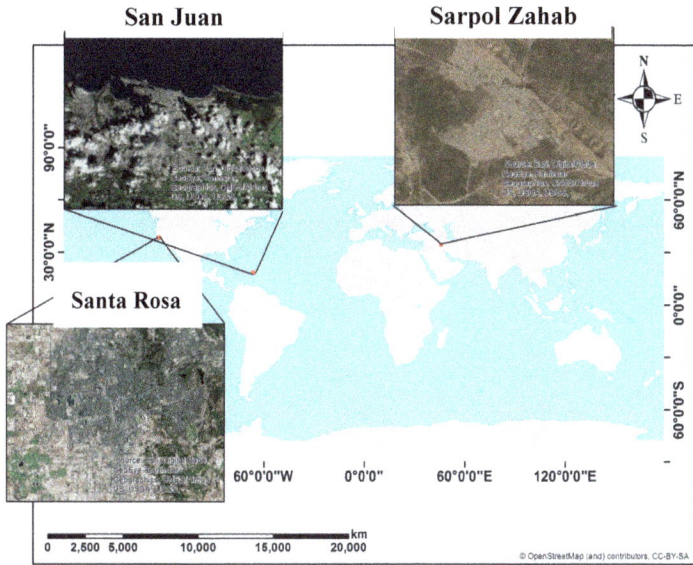

(**a**)

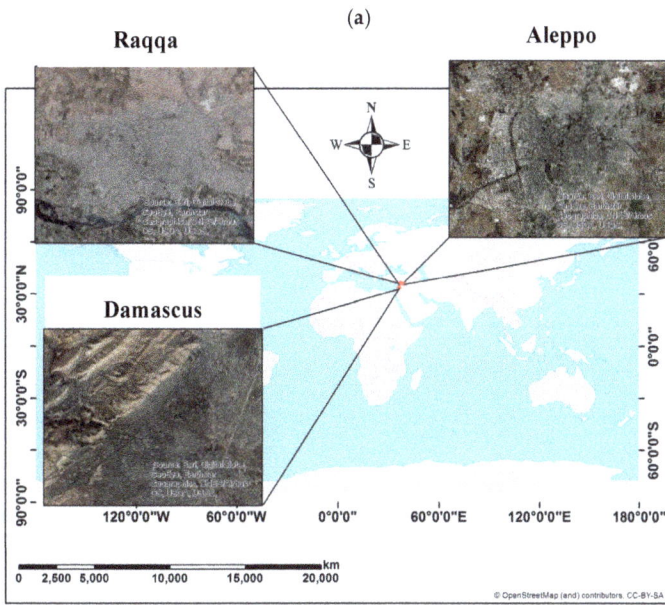

(**b**)

Figure 2. Study areas: (**a**) Areas affected by natural disasters; (**b**) Areas affected by anthropogenic disasters.

3. Results

3.1. Case Studies for Detecting Changes Using CCD

This section presents the results of an analysis of averaging and standard deviation values calculated from the coherence images generated to detect changes in the six study areas. In addition, the supervised classification results permitted an investigation of the response of various land-use classes to coherence loss in the face of disasters.

In general, the results exhibit loss of coherence over time. However, by selecting the image in the middle of the image stack timeline as the master image, coherence matching is done to ensure that coherence loss is distributed evenly in both temporal directions (i.e., pre-master slave images and post-master slave images). For a better visual impression on the coherence images, classification is done on the images into areas of change and no change by applying a threshold of >0.6 [24].

3.1.1. Natural Disasters

Sarpol Zahab, San Juan, and Santa Rosa results are presented in this section. The Sarpol Zahab results visualized in Figure 3 show a substantial loss of coherence in the post-disaster coherence maps compared to the pre-disaster coherence maps. The pre-disaster image immediately before the disaster event (7 November 2017) exhibits coherence loss of 37.8% in the entire area. The post-earthquake image, on the other hand, exhibits a coherence loss of 70.3%. The loss of coherence can be attributed to the destruction of infrastructure (building, roads etc.) by the earthquake, which occurred in November 2017 [40]. However, the southern part of the city also seems to be experiencing coherence loss in both pre- and post-disaster coherence maps, this can be attributed to vegetation, which is highly sensitive to coherence loss as it is constantly changing.

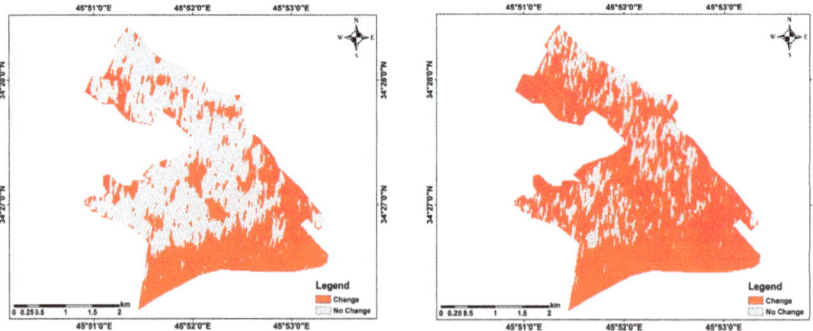

Figure 3. Sarpol Zahab coherence image classified into change and no change showing period before (7 November 2017, **left**) and after (1 December 2017, **right**) the earthquake.

In San Juan, Puerto Rico (Figure 4), results show massive decorrelation, which is an indication of hurricane-induced damages, due to the strong sensibility of interferometric coherence to water and humidity changes. For the pre-disaster coherence map, coherence loss is low but not absent, the results show areas in red (representing coherence loss) that can be attributed to lagoons, rivers, and temporal decorrelation from vegetation. The pre-hurricane coherence image (11 August 2017) shows a coherence loss of 41% while the post-hurricane coherence image (28 September 2017) shows a coherence loss of 64.4%.

In Santa Rosa (Figure 5), the post-fire disaster coherence map shows decorrelation in some parts of the city, especially in the northern part, which is the part of the city that experienced the disaster first. This may be an indication of the route taken by the fire during the disaster [41]. In the southern part, decorrelation can be seen in both pre- and post-fire disaster images. The pre-disaster image

(22 August 2017) shows that the area experienced a coherence loss of 66% while the post-fire coherence image (14 November 2017) shows that the area experienced a total coherence loss of 82%.

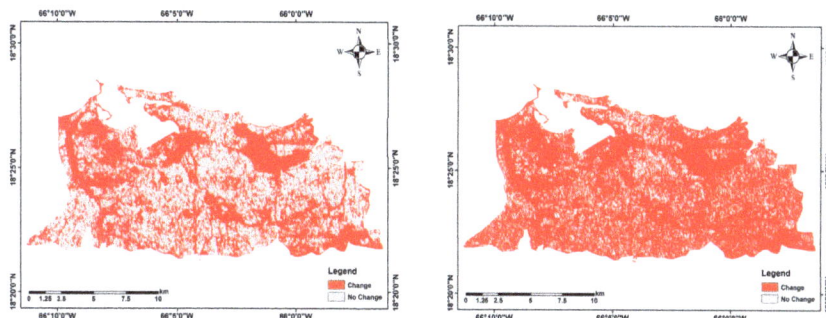

Figure 4. San Juan, Puerto Rico Coherence image classified into change and no change showing images before (11 August 2017, **left**) and after (28 September 2017, **right**) hurricane Maria.

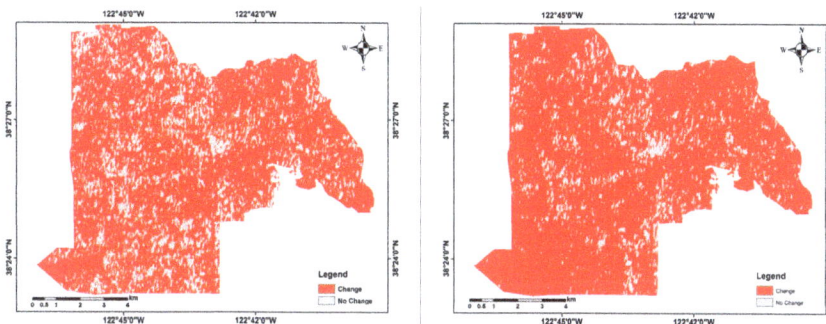

Figure 5. Santa Rosa Coherence image classified into change and no change showing images before (22 August 2017, **left**) and after (14 November 2017, **right**) the fire disaster event.

Natural disasters are usually one-time events that may occur in an area over a short period of time (minutes to hours). Wars and conflicts, on the other hand, involve a series of prolonged events occurring over many months to many years. Areas affected by the war in Syria, therefore, exhibited coherence loss differently from those affected by natural disasters.

3.1.2. Anthropogenic Disasters

For Aleppo, results show that although there was substantial coherence loss in the southeastern part of the city, in the coherence image (30 April 2017), high coherence is generally maintained in the northwestern part of the city (see Figure 6). This can be attributed to the intensification of fighting between the Syrian Government and the rebels between September and December in 2016 [42]. The coherence image (30 April 2017) is from the period after conflict intensification. It shows a 65% loss in total coherence for the city, compared to only 16% loss of coherence in the image for 4 July 2016, collected before the conflict intensified.

In Raqqa, the results in Figure 7 show a significant loss of coherence in the coherence image for 27 September 2017. This loss of coherence can be attributed to the operation launched by the U.S.-backed Syrian Democratic Forces (SDF) to capture Raqqa from the Islamic State of Iraq and Syria (ISIS) [43]. This operation lasted for several months, ending in October 2017. The coherence image representing the period after this operation indicates a loss of 82.5% coherence. In contrast, the coherence image collected before this operation shows a 47.2 loss in coherence.

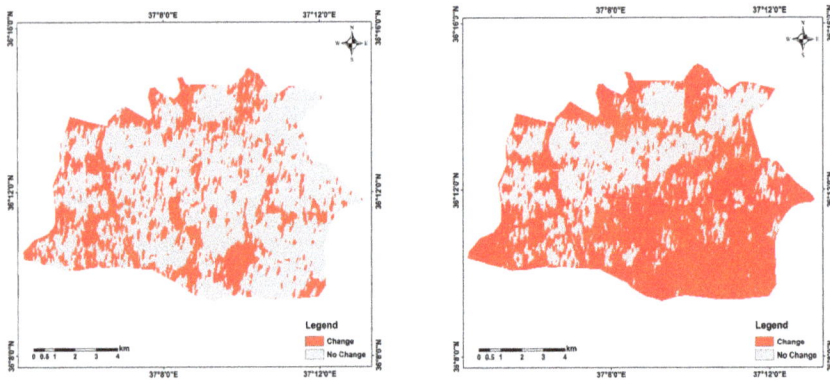

Figure 6. Aleppo Coherence image classified into change and no change showing images before and after the Syrian Government operation. Images before conflict intensification (4 July 2016, **left**) and after conflict intensification (30 April 2017, **right**).

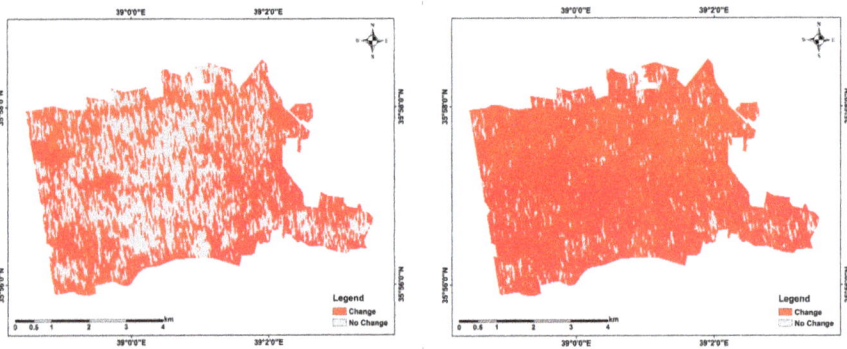

Figure 7. Raqqa Coherence image classified into change and no change showing coherence images of before (14 October 2016, **left**) and after (27 September 2017, **right**) the US-led forces operation against the Islamic State of Iraq and Syria (ISIS).

Considering Damascus, the results in Figure 8 show that towards the northeastern part of the city, a red spot is visible in both coherence maps, this area is a vegetated open space area, which is sensitive to coherence loss and is, therefore, unstable. Furthermore, coherence loss can be seen in the eastern part of the city in coherence maps on 5 April 2017 and is not visible on 31 October 2016, this movement from stability to instability can be an indication of building destruction in the area resulting from intense fighting between Syrian Government forces and rebels [44].

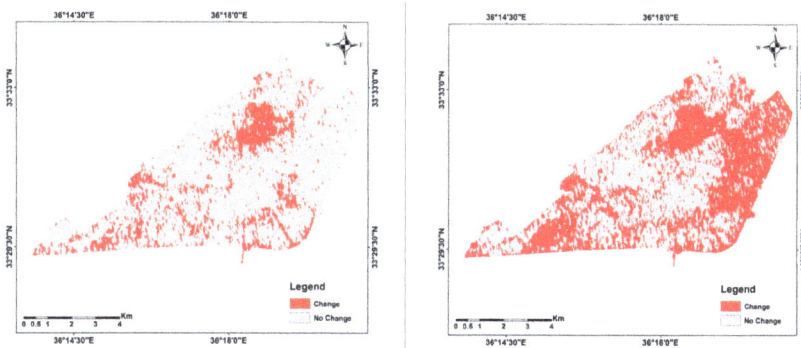

Figure 8. Damascus coherence images classified into change and no change (31 October 2016 on the **left** and 05 April 2017 on the **right**).

3.2. Analysis

3.2.1. Coherence Response to Different Land-Use Classes

Supervised classification results were used for further analysis to get an idea of how each disaster affects each land-use class.

In the study areas, areas classified as built-up, on average, show high coherence as compared to areas classified as vegetation and bare soil. Raqqa was chosen to represent the areas affected by war as it contains the most land use classes of all the war areas in the study, i.e., built-up, vegetated, and bare soil. However, there is a significant drop in built up coherence on 29 July 2017 to the extent that average coherence for the built area almost equals vegetation and bare soil (see Figure 9). This may be a direct result of the Syrian Defense Force (SDF) operation against Islamic State militants (ISIS) in the city around the same period.

For the earthquake disaster, a significant drop in coherence for all classes is observed between 7 November 2017 and 1 December 2017, which is the period after the occurrence of the earthquake. The fire disaster graph shows a dip to an average coherence of below 0.5 for all classes on 14 November 2017, which is after the fire disaster occurred, after that, average coherence remains steady for all three classes although built-up remains slightly above vegetation and bare soil. On the hurricane graph, the average coherence drops to 0.5 for built-up areas below 0.5 for vegetation and water bodies. Throughout the period, the average coherence for water bodies was under 0.5 and was only slightly affected by the hurricane as compared to vegetation and built-up areas. The average coherence increases on 10 October 2017; this could be a result of coherence recovery after floodwater evaporated weeks after the hurricane (see Figure 9).

Figure 9. *Cont.*

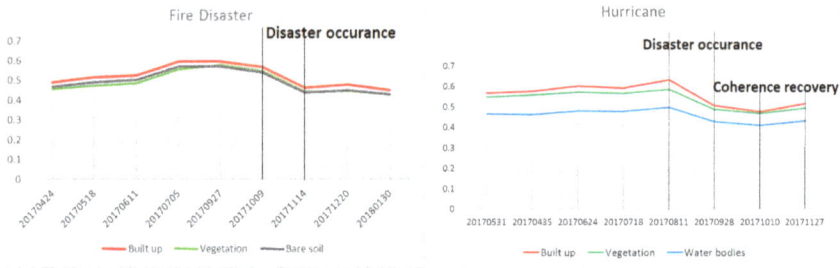

Figure 9. Average coherence graphs for different classes for each disaster.

Comparison of Disasters

Change-detection calculated for the built-up class shows that the war in Raqqa created the highest amount of change (55%) followed by the Sarpol Zahab earthquake (40.5%), which is an indication of significant infrastructure damage. Conversely, change-detection on the built-up classes in the fire and hurricane disaster do not exhibit high levels of change, implying minimal damage on building infrastructure (with 26.4% and 22.3% respectively) as compared to the earthquake and war disasters. The vegetation class, on the other hand, was less affected by the four disasters as compared to the built-up class. In the war disaster, the vegetation class experienced a 14.3% change, in the hurricane disaster, 18% change, and fire and earthquake disasters 23% and 37% change, respectively. The results, as shown in the Figure 10 radar graph, indicate that the war and earthquake disasters have a greater impact on the built-up classes and the hurricane and fire affect the vegetation class more intensely.

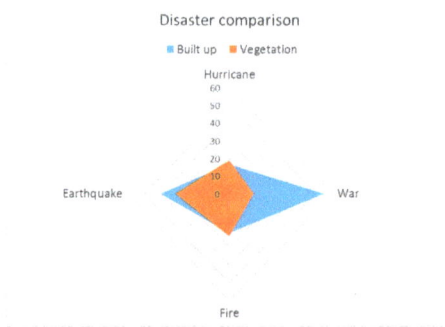

Figure 10. A comparison of built up vegetation land-use class responses to the four disasters.

3.2.2. Standard Deviation Analysis

To understand the way in which coherence in different parts of each study area responds to the various disasters and to reduce false positives from the vegetation and water bodies land use classes, we calculated the standard deviation for every street block over the study periods in the coherence maps. The standard deviation is not a measure of instability, but instead a measure of change. In the standard deviation results, the maps show changes in street blocks. Blocks with a low standard deviation are in light brown, indicating little change, while those that have experienced change (high standard deviation) are in dark brown. The study areas exhibited the following results after standard deviation calculation.

The results in Sarpol Zahab show street blocks with high standard deviation in various parts of the city. A standard deviation map of coherence images before the earthquake shows most of the city with low standard deviation (light brown), however, a comparison with the standard deviation map of all coherence maps (pre- and post-disaster) shows that many street blocks were affected by the earthquake, especially in the northwestern part of the city (refer to Figure 11). This is verified by United Nations Institute for Training and Research (UNITR) that produced a detailed map of damaged structures and a related density map for Sarpol Zahab, seen in Figure 12 [45]. The southern part of the city shows evidence of change indicated by a high standard deviation in the map shown in Figure 11b. Satellite imagery shows this area as agriculture, and the change can be explained by change from different stages of the agricultural season, defined as planting and harvesting times.

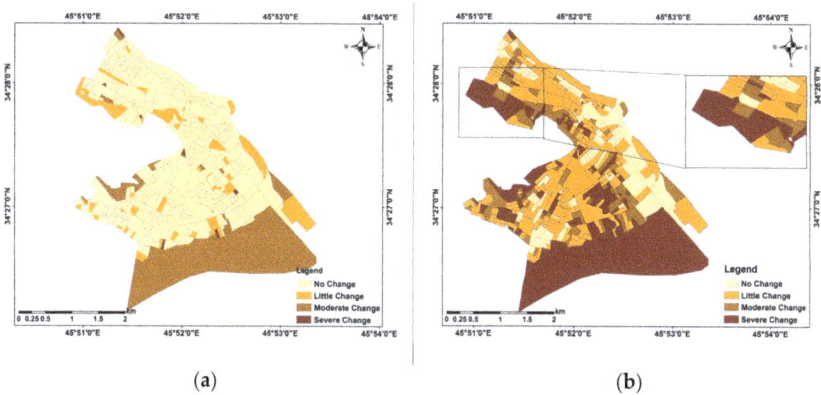

(a) (b)

Figure 11. Sarpol Zahab standard deviation maps, (**a**) shows a lower standard deviation calculated with lighter colored street blocks before the earthquake disaster and (**b**) shows a higher standard deviation calculated with darker street blocks after the earthquake disaster.

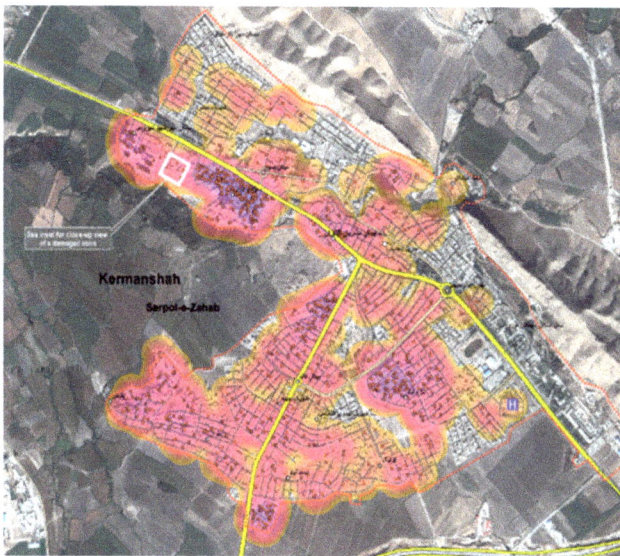

Figure 12. Damage map for Sarpol Zahab. Source: (UNITR) [45].

In San Juan, Puerto Rico, the standard deviation is high throughout the city (see Figure 13). Comparison between the pre- and post-disaster standard deviation maps shows change from light brown street blocks in the pre-disaster map to dark brown street blocks in the post disaster map. This could be an indication of the trail of disaster left by Hurricane Maria. Furthermore, it is interesting to notice that the only parts of the city showing very low standard deviations, or no change, are lagoons and water bodies. In coherence maps, these water bodies and lagoons would normally appear black all the time, indicating instability; this remains the case in the event of a Hurricane.

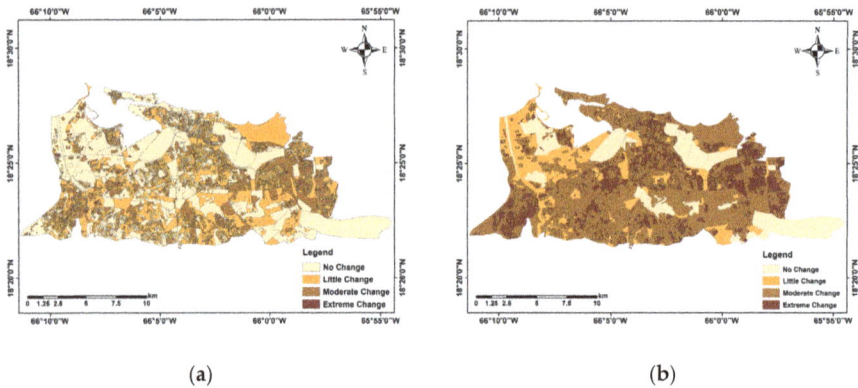

(a)

(b)

Figure 13. San Juan, Puerto Rico standard deviation map (**a**) shows lower standard deviation on street blocks before the hurricane and (**b**) shows higher standard deviation calculated after.

In Santa Rosa, California, the northern part of the city, predominantly residential, shows dark brown street blocks in the post disaster map, which implies change occurring in residential buildings. This area is Coffey Park, and media reports indicate that the fire destroyed many houses in this area [46]. This also shows that the fire came from the north of the city and destroyed houses in the north before it was contained, as shown in Figure 14. Figure 15 shows the fire route according to the New York times.

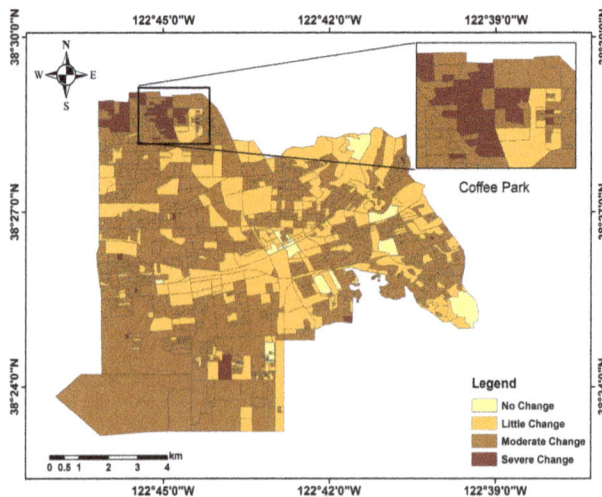

Figure 14. Santa Rosa standard deviation map indicating lower standard deviations on some street blocks in the north before the fire.

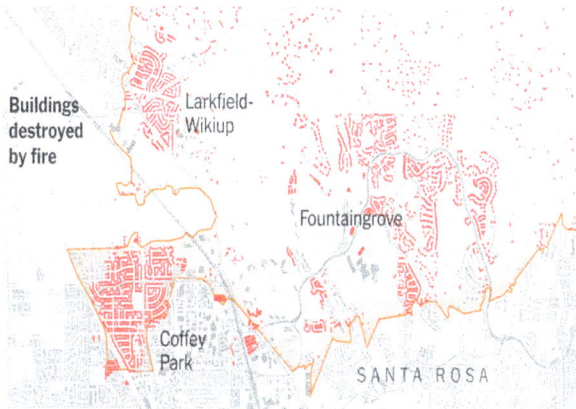

Figure 15. Santa Rosa fire route. Source: New York Times [46].

For areas affected by war, the standard deviation represents changes that occurred at different points in time on different street blocks. This means that the dark brown colors on street blocks represent changes, captured as snapshots of conflict-induced building damages, resulting from airstrikes, suicide bombings, barrel bombings, etc.

In the standard deviation map of Aleppo, the results show that street blocks located at the western side of the city exhibit a low standard deviation (light brown color), an indicator of little to no change. The eastern part of the city, however, shows a high standard deviation showing moderate to extreme change, which may be an indication of building destruction during the war. Maps from the Aljazeera news agency hint this by showing maps of the Syrian government moving to control the eastern part of Aleppo (see Figure 16). The maps show how control of the city in the eastern part changed from rebel to government control, this change indicates that conflict occurred as the government took over from the rebels. Therefore, conflict induced damages resulted in changes in these areas, hence a high standard deviation. A sample of street blocks from Karm Al-Myassar, affected by the civil conflict in the city [47], is shown in the standard deviation map in Figure 17.

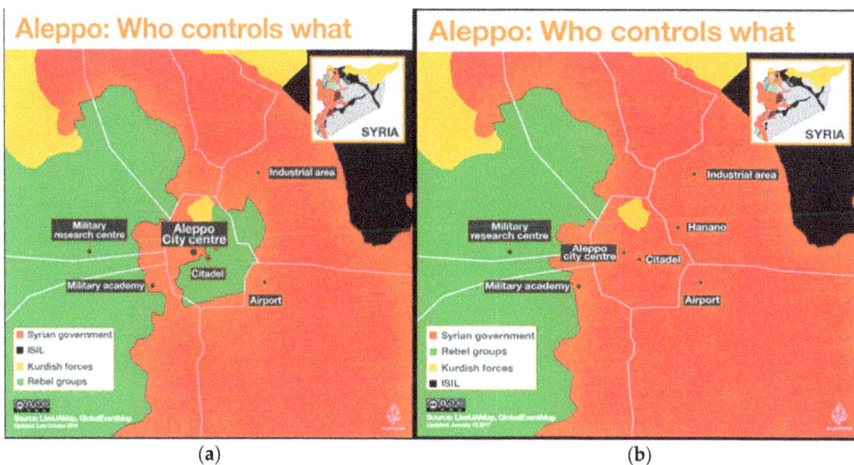

Figure 16. Aleppo Territories. Before Government offensive (**a**) and after Government offensive (**b**). Source: Aljazeera [47].

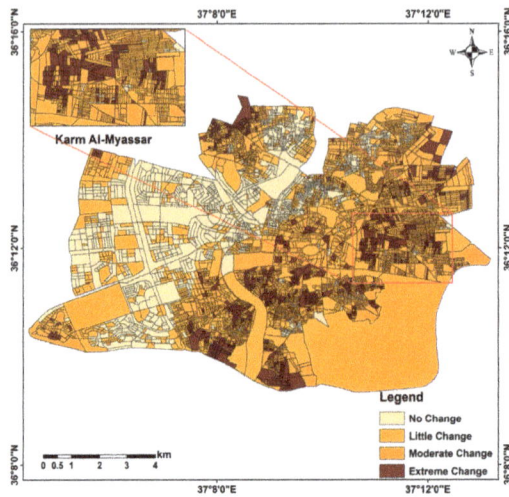

Figure 17. Aleppo standard deviation map shows higher standard deviation on street blocks in the eastern side.

In Raqqa, analysis of the standard deviation map reveals severe change (dark brown) on street blocks in the Central Business District (CBD) (Figure 18a). The high standard deviation may be a result of an offensive assault by the United States backed Syrian Democratic Forces (SDF) [43]. The outskirts of the city, however, have a low standard deviation, hence the street blocks on the fringes of the city experienced less change as compared to the CBD. This was hinted at in maps, indicating how the SDF launched their operation through airstrikes (Figure 18b).

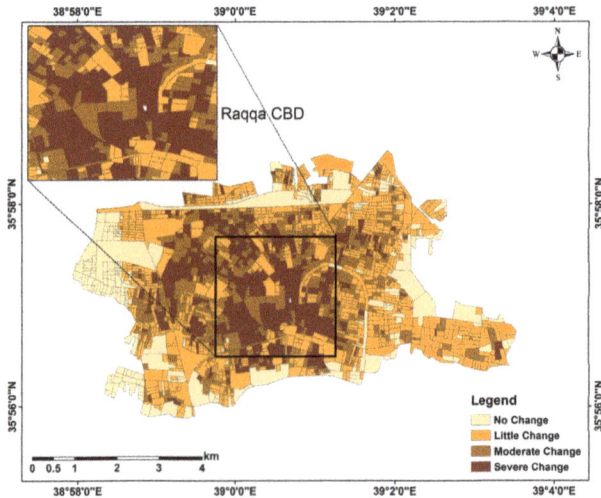

(**a**)

Figure 18. *Cont.*

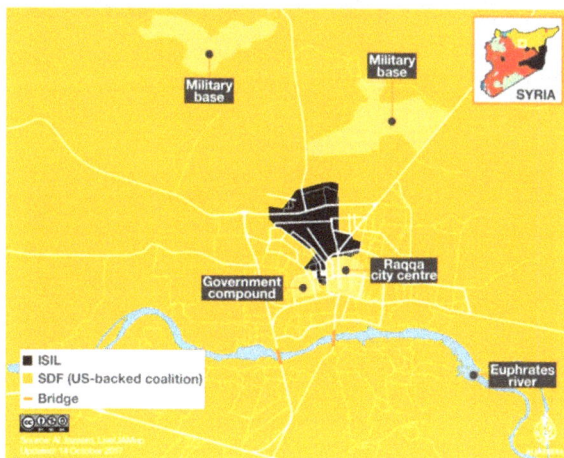

(**b**)

Figure 18. Comparison of Raqqa standard deviation map (**a**) and territory map (**b**) produced by Aljazeera [43].

The standard deviation map for Damascus shown in Figure 19a generally shows low levels of change for most street blocks in the city. The city has been known to be stable and conflict free for most of the civil war period [48], hence most street blocks in the city reflect this by showing little to no change (light brown color). However, the eastern part seems to show street blocks with a very high standard deviation, signifying an extreme change in the area this may have resulted from destruction from air strikes on the rebel-held Jobar district [49]. The northeastern neighborhoods of Jobar and Abbaisid Square are said to have experienced the most intense fighting [50], as illustrated in Figure 19.

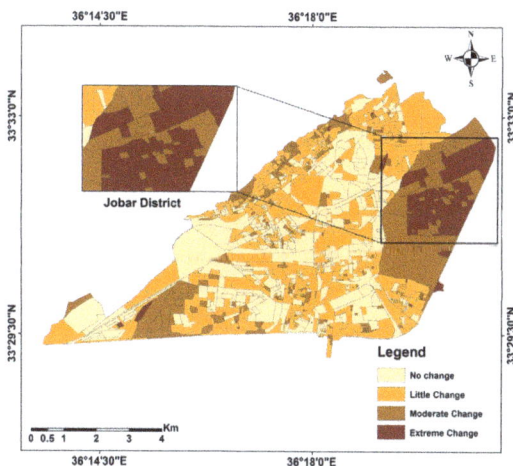

(**a**)

Figure 19. *Cont.*

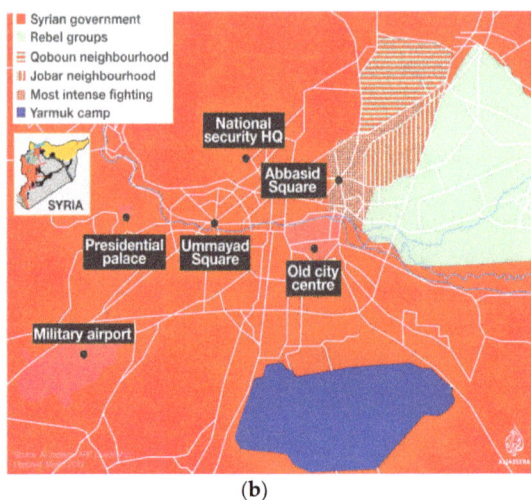

(**b**)

Figure 19. Comparison of Damascus standard deviation map (**a**) and territory map (**b**) produced by Aljazeera [50].

4. Discussion

Apart from the CCD technique, other SAR techniques have been used for change-detection in disaster situations, for example, damage detection techniques based on time-series SAR imagery [4,51]. These methods make use of amplitude and rely on backscatter information from SAR images; in this case, building damage resulting from disaster occurrences can be detected by comparing the backscatter information exhibited by the pre-disaster and post-disaster images [52]. However, since backscatter information is reliant on the dielectric constant, surface roughness, and the incident angle of an object, coherence information becomes essential to detect damage, especially when the backscattering characteristics remain the same between two images. As we have seen from the results, the coherence value does not rely on the backscattering characteristics of the SAR images but instead the similarity/difference of the phase properties of two SAR images. However, other interferometric SAR techniques have been used for damage detection [53]; in this case, the coherence images obtained by interferometric processing are used for damage detection and assessment, as in this study.

The pre-event and post event results in the experiments show rapid changes that occur between these periods, which is indicated by massive loss of coherence, for example, the city of Raqqa shows an 82.5% loss in coherence that is close to the 80% destruction of the city as reported by the media. The CCD technique thus becomes suitable for detection of such changes. In addition to this, we see from the classification results that coherence images can be divided into different classes based on how each class responds to coherence loss. For example, urban areas exhibit predominantly high coherence and generally experience lower coherence loss over time as compared to vegetation and water bodies. This knowledge can be used for classification of SAR images [54], while it is not possible to use amplitude information alone.

The CCD technique is not new and has been widely employed in various remote sensing applications and highlights the capabilities of Sentinel-1 data. This emerging remote sensing resource provides global coverage and archival data, making it ideally suited to support decision-making tasks during and after disaster events. Despite these benefits, however, the Sentinel-1 mission only started in 2014, and as a result, it does not have an archive of images prior to 2014. This is particularly challenging when monitoring the effects of long-term disasters like wars if the period being investigated commenced before the mission began. Processing Sentinel-1 data is also time consuming and user intensive, therefore, limiting the applicability of the CCD technique.

In future work, we will develop an automated CCD technique for global background observation using Sentinel-1 imagery. CCD must be automated for use in urban planning, vegetation monitoring, and forest cover management applications if Sentinel-1 is to be fully exploited [54]. Methods for automatic/unsupervised change-detection using SAR data have been proposed in previous studies [33,35]. An Automatic CCD technique can also provide information on areas affected by a disaster. However, coherence matching requires two images acquired with the same looking angle, and thus a 6 to 12-day revisit time of the Sentinel-1 satellites may not be adequate for immediate intervention and response to disasters. With larger constellations, like the COSMO SkyMed constellation, faster revisit times are possible. However, for the quickest possible response times, images are acquired from different angles, rendering our method unusable under such circumstances. Nevertheless, considering the advantages of Sentinel-1 data, new automatic change-detection methods would facilitate the use of CCD in a wider range of contexts.

As seen from the results, temporal coherence is a product created from repeat-pass InSAR observations and, thus, is susceptible to changes in the scene during the two acquisitions. The effects of temporal decorrelation are evident in the time-series coherence images shown in Figures 3–8 and the classification graphs in Figure 9. Decorrelation increases as temporal distances between the slave and master image increase. In this study, a single master image, selected from the middle of the timeline, was co-registered to subsequent slave images, ensuring that the temporal baseline is minimized by distributing decorrelation in both temporal directions and at the same time maintaining the time separation between pre-event and post-event periods.

In vegetated areas like forests, however, this method may not be effective as coherence is highly sensitive to change. Therefore, Coherence Change-Detection (CCD) is not as effective in highly vegetated areas as it is in urban areas [24]. A possible solution to low coherence resulting from temporal decorrelation would be to apply the short baseline approach that separates the master and slave images by short periods, for example, separating the two images by a period of 12 days to a month, so that high coherence is maintained in each coherence image. This, however, is not very useful for analyzing the effects of disaster events over time. Nevertheless, built up areas in urban settings are generally stable over time. Therefore, temporal decorrelation has little effect on the coherence values representing the built-up area. As a result, selecting one single master image is a more appropriate technique in urban areas. Additionally, the comparative results shown in Figure 10 indicate that as compared to vegetated areas, built up areas are affected more by war and earthquakes. Vegetated areas, however, are more affected by hurricanes and fire disasters. The reasons for this may be that during wartime, infrastructure is targeted for destruction. Earthquakes, however, are characterized by shaking of the ground and result in infrastructure instability leading to destruction. Consequently, the CCD technique may be more useful if applied to areas affected by war and earthquakes rather than by fire and hurricane disasters, as it is a more reliable technique for built-up areas.

Although the standard deviation method, applied in the study, reveals areas that have experienced the most change, it is not effective when quantifying the intensity of damage. Furthermore, coherence is affected by small changes in a scene, hence, at times, it is difficult to distinguish small changes from big changes. This is also problematic when selecting an ideal threshold, especially since a variety of factors might be responsible for coherence loss, for example, new built-up areas cannot be distinguished from some vegetated areas [33]. It is essential, therefore, to find methods to improve detection and assessment of the intensity of damage.

5. Conclusions

This study demonstrates the ability of Sentinel-1 C-band imagery to provide information for disaster monitoring and management. Unlike most imaging remote sensing systems, the Sentinel-1 mission provides continuous, reliable global data, which makes it ideal for monitoring any type of disaster anywhere in the world. Sentinel-1 data from Aleppo, Damascus, and Raqqa in Syria, Sarpol Zahab in Iran, and Puerto Rico, and Santa Rosa CA, USA, were used to generate coherence maps for each of the areas in the periods before, during, and after disasters.

Classifying coherence maps into areas of change and no-change, then applying a threshold of 0.6 and comparing the time series for each period, shows the coherence loss after a particular damaging event. However, differentiating disaster induced coherence loss and temporal coherence loss is not easy on coherence maps; hence, standard deviation analysis was used in combination with street-blocks to identify which street blocks in the areas experienced a sudden loss in coherence, which may be an indication of disaster effects.

Classifying images into built-up, vegetation, bare soil, and water bodies and integrating these land-use classes with the coherence images helped show the effects of coherence loss on each class after each disaster. For future studies, coherence can be used not only as a measure of change in CCD but also for classification, as the study showed that different classes exhibited dissimilar coherence loss over time. Furthermore, future studies will focus on automating the CCD method for global disaster monitoring.

Author Contributions: Methodology, P.W. and T.B.; Supervision, T.B. and B.M.; Writing—original draft, P.W.; Writing—review and editing, T.B. and B.M.

Funding: This work was supported by the Natural Science Foundation of China under the Grant 61331016. Washaya's work was supported by the Chinese Scholarship Council.

Acknowledgments: The authors would like to thank Daniele Perissin for providing the SarProZ software to support this research.

Conflicts of Interest: The authors declare no conflict of interest.

References

1. Huyck, C.; Verrucci, E.; Bevington, J. Remote sensing for disaster response: A rapid, image-based perspective. In *Earthquake Hazard, Risk and Disasters*; Elsevier: Amsterdam, The Netherlands, 2014; pp. 1–24.
2. Ohkura, H.; Jitsufuchi, T.; Matsumoto, T.; Fujinawa, Y. Application of SAR data to monitoring of earthquake disaster. *Adv. Space Res.* **1997**, *19*, 1429–1436. [CrossRef]
3. Dong, Y.; Li, Q.; Dou, A.; Wang, X. Extracting damages caused by the 2008 Ms 8.0 Wenchuan earthquake from SAR remote sensing data. *J. Asian Earth Sci.* **2011**, *40*, 907–914. [CrossRef]
4. Matsuoka, M.; Yamazaki, F. Use of SAR imagery for monitoring areas damaged due to the 2006 Mid Java, Indonesia earthquake. In Proceedings of the 4th International Workshop on Remote Sensing for Post-Disaster Response, Cambridge, UK, 25–26 September 2006.
5. Stramondo, S.; Moro, M.; Tolomei, C.; Cinti, F.; Doumaz, F. InSAR surface displacement field and fault modelling for the 2003 Bam earthquake (southeastern Iran). *J. Geodyn.* **2005**, *40*, 347–353. [CrossRef]
6. Balz, T.; Liao, M. Building damage detection using post-seismic high-resolution SAR satellite data. *Int. J. Remote Sens.* **2010**, *31*, 3369–3391. [CrossRef]
7. Gähler, M. Remote Sensing for Natural or Man-made Disasters and Environmental Changes. *Environ. Appl. Remote Sens.* **2016**, 309–338. [CrossRef]
8. Lu, D.; Mausel, P.; Brondizio, E.; Moran, E. Change detection techniques. *Int. J. Remote Sens.* **2004**, *25*, 2365–2401. [CrossRef]
9. Wang, F.; Xu, Y.J. Comparison of remote sensing change detection techniques for assessing hurricane damage to forests. *Environ. Monit. Assess.* **2010**, *162*, 311–326. [CrossRef] [PubMed]
10. Singh, A. Review article digital change detection techniques using remotely-sensed data. *Int. J. Remote Sens.* **1989**, *10*, 989–1003. [CrossRef]
11. Hoque, M.A.-A.; Phinn, S.; Roelfsema, C.; Childs, I. Tropical cyclone disaster management using remote sensing and spatial analysis: A review. *Int. J. Disaster Risk Reduct.* **2017**, *22*, 345–354. [CrossRef]
12. Ramsey, E., III; Rangoonwala, A.; Middleton, B.; Lu, Z. Satellite optical and radar data used to track wetland forest impact and short-term recovery from Hurricane Katrina. *Wetlands* **2009**, *29*, 66–79. [CrossRef]

13. Li, X.; Yu, L.; Xu, Y.; Yang, J.; Gong, P. Ten years after Hurricane Katrina: Monitoring recovery in New Orleans and the surrounding areas using remote sensing. *Sci. Bull.* **2016**, *61*, 1460–1470. [CrossRef]

14. Rowell, A.; Moore, P.F. *Global Review of Forest Fires*; Citeseer: Forest Grove, OR, USA, 2000.

15. Anggraeni, A.; Lin, C. Application of SAM and SVM Techniques to Burned Area Detection for Landsat TM Images in Forests of South Sumatra. In Proceedings of the International Conference on Environmental Science and Technology, Singapore, 26–28 February 2011; pp. V2160–V2164.

16. Marx, A.J. Detecting urban destruction in Syria: A Landsat-based approach. *Remote Sens. Appl. Soc. Environ.* **2016**, *4*, 30–36. [CrossRef]

17. Woodcock, C.E.; Allen, R.; Anderson, M.; Belward, A.; Bindschadler, R.; Cohen, W.; Gao, F.; Goward, S.N.; Helder, D.; Helmer, E. Free access to Landsat imagery. *Science* **2008**, *320*, 1011–1011. [CrossRef] [PubMed]

18. Liu, C.; Dong, P.; Shi, Y. Recurrence interval of the 2008 Mw 7.9 Wenchuan earthquake inferred from geodynamic modelling stress buildup and release. *J. Geodyn.* **2017**, *110*, 1–11. [CrossRef]

19. CNN, Powerful Iran-Iraq Earthquake Is Deadliest of 2017. Available online: https://edition.cnn.com/2017/11/12/middleeast/iraq-earthquake/index.html (accessed on 27 February 2018).

20. Krassenburg, M. Sentinel-1 mission status. In Proceedings of the 11th European Conference on Synthetic Aperture Radar (EUSAR 2016), Hamburg, Germany, 6–9 June 2016; pp. 1–6.

21. Bamler, R.; Hartl, P. Synthetic aperture radar interferometry. *Inverse Probl.* **1998**, *14*, 53–102. [CrossRef]

22. Woodhouse, I.H. *Introduction to Microwave Remote Sensing*; CRC Press: Boca Raton, FL, USA, 2005.

23. Lusch, D.P. *Introduction to Microwave Remote Sensing*; Center for Remote Sensing and Geographic Information Science Michigan State University: East Lansing, MI, USA, 1999.

24. Jendryke, M.; McClure, S.; Balz, T.; Liao, M. Monitoring the built-up environment of Shanghai on the street-block level using SAR and volunteered geographic information. *Int. J. Digital Earth* **2017**, *10*, 675–686. [CrossRef]

25. Ferrettia, M. *InSAR Principles: Guidelines for SAR Interferometry Processing and Interpretation*; ESTEC: Noordwijk, The Netherlands, 2007.

26. Hagberg, J.O.; Ulander, L.M.; Askne, J. Repeat-pass SAR interferometry over forested terrain. *IEEE Trans. Geosci. Remote Sens.* **1995**, *33*, 331–340. [CrossRef]

27. Li, F.K.; Goldstein, R.M. Studies of multibaseline spaceborne interferometric synthetic aperture radars. *IEEE Trans. Geosci. Remote Sens.* **1990**, *28*, 88–97. [CrossRef]

28. Lopez-Martinez, C.; Fabregas, X.; Pottier, E.; Polarimetrie, E.I.R.-T. In A new alternative for SAR imagery coherence estimation. In Proceedings of the 5th European Conference on Synthetic Aperture Radar (EUSAR'04), Ulm, Germany, 25–27 May 2004.

29. Kampes, B.; Usai, S. In Doris: The delft object-oriented radar interferometric software. In Proceedings of the 2nd International Symposium on Operationalization of Remote Sensing, Enschede, The Netherlands, 16–20 August 1999; p. 20.

30. Prati, C.; Rocca, F. Range resolution enhancement with multiple SAR surveys combination. In Proceedings of the International IEEE Geoscience and Remote Sensing Symposium (IGARSS'92), Houston, TX, USA, 26–29 May 1992; pp. 1576–1578.

31. Bruzzone, L.; Marconcini, M.; Wegmuller, U.; Wiesmann, A. An advanced system for the automatic classification of multitemporal SAR images. *IEEE Trans. Geosci. Remote Sens.* **2004**, *42*, 1321–1334. [CrossRef]

32. Preiss, M.; Stacy, N.J. *Coherent Change Detection: Theoretical Description and Experimental Results*; Defence Science and Technology Organisation Edinburgh: Canberra, Australia, 2006.

33. He, M.; He, X. Urban change detection using coherence and intensity characteristics of multi-temporal SAR imagery. In Proceedings of the 2nd IEEE Asian-Pacific Conference on Synthetic Aperture Radar (APSAR 2009), Xian, China, 26–30 October 2009; pp. 840–843.

34. Jendryke, M.; Balz, T.; McClure, S.C.; Liao, M. Putting people in the picture: Combining big location-based social media data and remote sensing imagery for enhanced contextual urban information in Shanghai. *Comput. Environ. Urban Syst.* **2017**, *62*, 99–112. [CrossRef]

35. Perissin, D.; Wang, Z.Y.; Wang, T. The SarProZ InSAR tool for urban subsidence/manmade structure stabilirt monitoring in China. In Proceedings of the ISRSE, Sidney, Australia, 10–15 April 2011.

36. Mocnik, F.-B.; Zipf, A.; Raifer, M. The OpenStreetMap folksonomy and its evolution. *Geo-Spat. Inf. Sci.* **2017**, *20*, 219–230. [CrossRef]

37. Kafi, K.; Shafri, H.; Shariff, A. An analysis of LULC change detection using remotely sensed data; A Case study of Bauchi City. In *IOP Conference Series: Earth and Environmental Science*; IOP Publishing: Bristol, UK, 2014.

38. LA Times, Santa Rosa Comes to Terms with the Scale of Devastation: 3000 Buildings Lost, Many Dead in Fire. Available online: http://www.latimes.com/local/lanow/la-me-santa-rosa-fire-toll-20171014-story.html (accessed on 15 December 2017).

39. Almasdar News, US-Backed Forces Succeed in Making Raqqa 80 Percent "Uninhabitable". Available online: https://www.almasdarnews.com/article/us-backed-forces-succeed-making-raqqa-80-percent-uninhabitable/ (accessed on 26 October 2017).

40. Reuters, "Strong Earthquake Hits Iraq and Iran, Killing at Least 210". Available online: https://www.reuters.com/article/us-iraq-quake/strong-earthquake-hits-iraq-and-iran-killing-more-than-450-idUSKBN1DC0VZ?il=0 (accessed on 12 November 2017).

41. NY Times, How California's Most Destructive Wildfire Spread, Hour by Hour. Available online: https://www.nytimes.com/interactive/2017/10/21/us/california-fire-damage-map.html (accessed on 3 March 2018).

42. IB Times, The Fall of Aleppo Timeline: How Assad Captured Syria's Biggest City. Available online: http://www.ibtimes.co.uk/fall-aleppo-timeline-how-assad-captured-syrias-biggest-city-1596504 (accessed on 1 November 2017).

43. Reuters, U.S.-Backed Force Launches Assault on Islamic State's 'Capital' in Syria. Available online: https://www.reuters.com/article/us-mideast-crisis-syria-raqqa/u-s-backed-force-launches-assault-on-islamic-states-capital-in-syria-idUSKBN18W29P (accessed on 26 October 2017).

44. Al Jazeera, Fighting Intensifies in Damascus: Rebels Gain Control of Areas Close to Syrian Capital, with Fighting Affecting Operations at International Airport. Available online: https://www.aljazeera.com/news/middleeast/2013/03/2013316133433170554.html (accessed on 6 March 2018).

45. UNITAR, Damaged Structures and Related Density Map in Sarpol-e-Zahab. Available online: http://www.unitar.org/unosat/map/2732 (accessed on 3 March 2018).

46. LA Times, Here's Where More than 7500 Buildings Were Destroyed and Damaged in California's Wine Country Fires. Available online: http://www.latimes.com/projects/la-me-northern-california-fires-structures/ (accessed on 27 February 2018).

47. BBC, Aleppo Battle: Syrian Troops Push Deeper into Rebel-Held East. Available online: http://www.bbc.com/news/world-middle-east-38207231 (accessed on 3 March 2018).

48. Daily Mail, a Tale of Two Cities: How One Side of Damascus Stands Tall BUT the Other Has Been Flattened by Assad in One of the Most Brutal Bombing Campaigns in History. Available online: http://www.dailymail.co.uk/news/article-5456783/A-tale-two-cities-Damascus-brutal-divide.html (accessed on 15 March 2018).

49. Reuters, Syrian Forces and Rebels Fight Fierce Clashes in Northeast Damascus. Available online: https://www.reuters.com/article/us-mideast-crisis-syria-jobar/syrian-forces-and-rebels-fight-fierce-clashes-in-northeast-damascus-idUSKBN16Q09X (accessed on 3 March 2018).

50. The Telegraph, Syrian Rebels Launch Surprise Attack on Damascus. Available online: https://www.telegraph.co.uk/news/2017/03/19/syrian-rebels-launch-surprise-attack-damascus/ (accessed on 3 March 2018).

51. Matsuoka, M.; Yamazaki, F. Use of Interferometric Satellite SAR for Earthquake Damage Detection. In Proceedings of the 6th International Conference on Seismic Zonation, Palm Springs, CA, USA, 12–15 November 2000; EERI: Oakland, CA, USA, 2000.

52. Dong, L.; Shan, J. A comprehensive review of earthquake-induced building damage detection with remote sensing techniques. *ISPRS J. Photogramm. Remote Sens.* **2013**, *84*, 85–99. [CrossRef]

53. Papathanassiou, K.P.; Cloude, S.R. Single-baseline polarimetric SAR interferometry. *IEEE Trans. Geosci. Remote Sens.* **2001**, *39*, 2352–2363. [CrossRef]

54. Santoro, M.; Shvidenko, A.; McCallum, I.; Askne, J.; Schmullius, C. Properties of ERS-1/2 coherence in the Siberian boreal forest and implications for stem volume retrieval. *Remote Sens. Environ.* **2007**, *106*, 154–172. [CrossRef]

remote sensing

MDPI

Letter

Multi-Layer Model Based on Multi-Scale and Multi-Feature Fusion for SAR Images

Aobo Zhai [1,2,*], Xianbin Wen [1,2,*], Haixia Xu [1,2], Liming Yuan [1,2] and Qingxia Meng [3]

1   Key Laboratory of Computer Vision and System, Ministry of Education, Tianjin University of Technology, Tianjin 300384, China; xuhaixiaxhx@163.com (H.X.); yuanleeming@163.com (L.Y.)
2   Tianjin Key Laboratory of Intelligence Computing and Novel Software Technology, Tianjin University of Technology, Tianjin 300384, China
3   School of Computer Science and Technology, Tianjin University, Tianjin 300072, China; 1mengqingxia@163.com
*   Correspondence: aobozhai1@163.com (A.Z.); xbwen317@163.com (X.W.); Tel.: +86-22-6021-6808 (A.Z. & X.W.)

Received: 16 August 2017; Accepted: 18 October 2017; Published: 24 October 2017

Abstract: A multi-layer classification approach based on multi-scales and multi-features (ML–MFM) for synthetic aperture radar (SAR) images is proposed in this paper. Firstly, the SAR image is partitioned into superpixels, which are local, coherent regions that preserve most of the characteristics necessary for extracting image information. Following this, a new sparse representation-based classification is used to express sparse multiple features of the superpixels. Moreover, a multi-scale fusion strategy is introduced into ML–MFM to construct the dictionary, which allows complementation between sample information. Finally, the multi-layer operation is used to refine the classification results of superpixels by adding a threshold decision condition to sparse representation classification (SRC) in an iterative way. Compared with traditional SRC and other existing methods, the experimental results of both synthetic and real SAR images have shown that the proposed method not only shows good performance in quantitative evaluation, but can also obtain satisfactory and cogent visualization of classification results.

Keywords: sparse representation classification (SRC); multi-layer structure; multi-feature fusion; multi-scale; SAR image

1. Introduction

Synthetic aperture radars (SAR) can obtain stable image data as we are observing the planet Earth. It is not affected by light conditions and can be used day and night under various conditions [1,2]. In recent years, SAR image classification has received more attention as an important part of image understanding and interpretation. A considerable number of image classification algorithms have been proposed, such as support vector machine (SVM) [3], neural network (NN) [4], wavelet decomposition, and sparse representation classification (SRC) [5], etc. Among these existing methods, the traditional SVM and NN methods show high reliability in pattern recognition. However, the relevant computation cost is expensive, and they are easily affected by the selection of features. SRC, which is based on sparse representation and was proposed by Mallat and Zhang [6], has been proven to be an extremely powerful tool in image processing and can obtain good performance in the final processing results [7–15].

The basic ideas of SRC are the linear description hypothesis and spatial joint representation mechanism. This is based on the minimum residual between the original and the reconstruction signal. The sparse coefficients associated with the different classes are selected to reconstruct the original signal. Actually, SRC cannot be directly applied to SAR image classification due to the imaging mechanisms of SAR being different to those of nature imagery. However, if an SAR image is transformed into a specific feature space, the SRC can be efficiently used in SAR image classification. A joint sparsity model (JSRM) is proposed based on SRC [16], in which the small neighborhood around the test pixel are represented by linear combinations of a few common training samples. Furthermore, the features cannot be represented well on a single scale, which results in the low accuracy of classification results. Neighboring regions of different scales correspond to the same test pixel and they should offer complementary and correlated information for classification. Different sizes of textures in an image have different performance in different scales. The hierarchical sparse representation-based classification (HSRC) [17] can solve the problem in a previous reference [16] to a certain extent, but the HSRC belongs to classification based on each pixel, which only depends on the selection of features in the spatial domain and the selected scale for each layer. This may lead to a loss or misrepresentation of information, resulting in poor classification accuracy and time-consuming training requirements.

In this paper, aiming to overcome the above-mentioned problems, we proposed a novel approach, which is called the multi-layer with multi-scale and multi-feature fusion model (ML–MFM), for SAR image classification. This maintains high accuracy and robustness in addition to having reduced time requirements. Firstly, in order to fix the deficiency of using a single feature and to provide more textural and gray statistical level information [5,12,16], we extracted three types of features of a SAR image for different classes and different scales, which are respectively the gray-level histogram, gray-level co-occurrence matrix (GLCM), and Gabor filter [18–21]. In other words, a discriminative feature vector is composed of the gray-level histogram, GLCM and Gabor filter for each class, while the feature matrix is constructed by the column vector composed of discriminative feature vectors of all classes and row vectors composed of discriminative feature vectors of all scales. Moreover, motivated by the fusion of characteristics from multiple frames in reference [22], a multi-scale fusion strategy was used to construct the dictionary. Thus, the extracted features under different scales can be merged together to construct the column vectors of the dictionary (see Figure 1), which can allow complementation between sample information and reduce the time complexity. Following this, we should segment an SAR image into a host of homogeneous regions called superpixels, with the structural information captured by a discriminative feature vector extraction for each superpixel. Finally, inspired by the idea of layers in the spatial pyramid in reference [21], the multi-layer operation is utilized to refine the classification results by adding a threshold decision condition to SRC in an iterative way. If a superpixel meets the condition as the new atoms in the dictionary, the category is recorded. Otherwise, it will be used as the testing sample for the next layer (Figure 3 depicts the above-mentioned basic framework). Compared with other methods, the final classification results of the proposed method have higher accuracy.

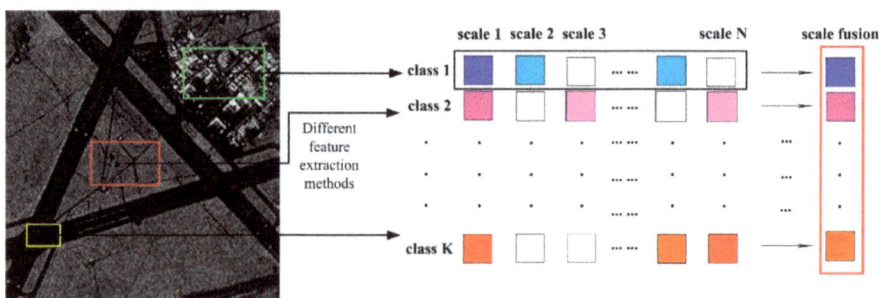

Figure 1. The model of multi-scale fusion strategy.

The remainder of this paper is organized as follows. In Section 2, we briefly review the SRC, while the procedure of our novel model regarding the use of ML–MFM for SAR image classification is explored. The experimental results for synthetic and real SAR images are presented and compared with others in Section 3. Comparison with the HSRC [17] and the major innovation points are provided in Section 4. Finally, conclusions are drawn and future research directions are described in Section 5.

2. Materials and Methods

2.1. SRC

We assume an SAR image contains K classes. D_K is the kth class of the sub-dictionary constructed by concatenating feature vectors of the kth class. We can define a dictionary D constructed by $[D_1, D_2, \ldots, D_K]$ for an SAR image. The testing sample y can be formulated by a series of training samples as follows:

$$y = \psi(y) = Dx \in R^M \tag{1}$$

where $\psi(\cdot)$ is an Eigen function which can be used to realize the transformation from pixel to feature space and $x = [0, \ldots, 0, x_{i,1}, \ldots, x_{i,n_i}, 0, \ldots, 0] \in R^n$ is a sparse coefficient vector whose entries are zeros except those associated with the ith class. A sparse coefficient x indicates that it will be easier to estimate the identity of the testing sample y. A sparse coefficient x can be obtained by solving the following error-constrained Equation (2) or the sparsity-constrained Equation (3):

$$\hat{x} = \mathrm{argmin}\|x\|_0 \text{ subject to } \|y - Dx\|_2 \leq \sigma \tag{2}$$

$$\hat{x} = \mathrm{argmin}\|y - Dx\|_2 \text{ subject to } \|x\|_0 \leq sl \tag{3}$$

where σ is the error tolerant limit and sl is the sparsity level which can represent the maximum number of selected atoms in the dictionary. Moreover, $\|\cdot\|_0$ and $\|\cdot\|_2$ denote l_0 and l_2 norms, respectively. Usually, the problem of solving sparse coefficients can be performed using the orthogonal matching pursuit (OMP) method [23].

After obtaining the sparse coefficient $\hat{x}$, the class label $\hat{k}$ of the test pixel y can be determined by the minimal error between y and its approximation from the sub-dictionary of each class:

$$\hat{k} = \mathrm{argmin}_k \|y - D\hat{x}_k\|_2, \ k = 1, \ldots, K \tag{4}$$

where $\hat{x}_k$ represents the coefficients in $\hat{x}$ belonging to the kth class. In order to demonstrate the drawback of the SRC algorithm clearly, a simple experiment was performed on two real SAR images (the original SAR image is in Figures 10a and 12a), observed in Figure 2. We can see that the final SRC results are unacceptable from Figure 2a,b. This is mainly because the SRC algorithm extracts features of SAR images only by using the pixel-by-pixel method, resulting in a lack of complementation between sample information. Therefore, to solve this problem, superpixels and more complete features need to be taken into account.

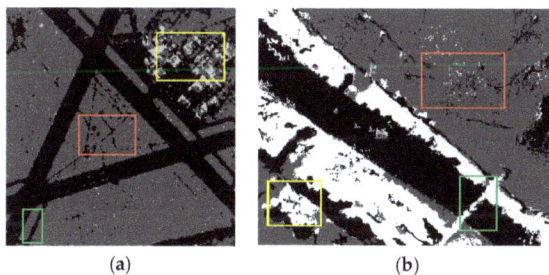

(a)   (b)

Figure 2. Classification results with the sparse representation classification (SRC) algorithm of a previous study [5] on (**a**) SAR1 and (**b**) SAR2.

2.2. Proposed Multi-Layer and Multi-Feature Model (ML–MFM)

In this section, the multi-layers and multi-feature model (ML–MFM) based on the SRC algorithm is proposed. The basic framework of the proposed method is shown in Figure 3.

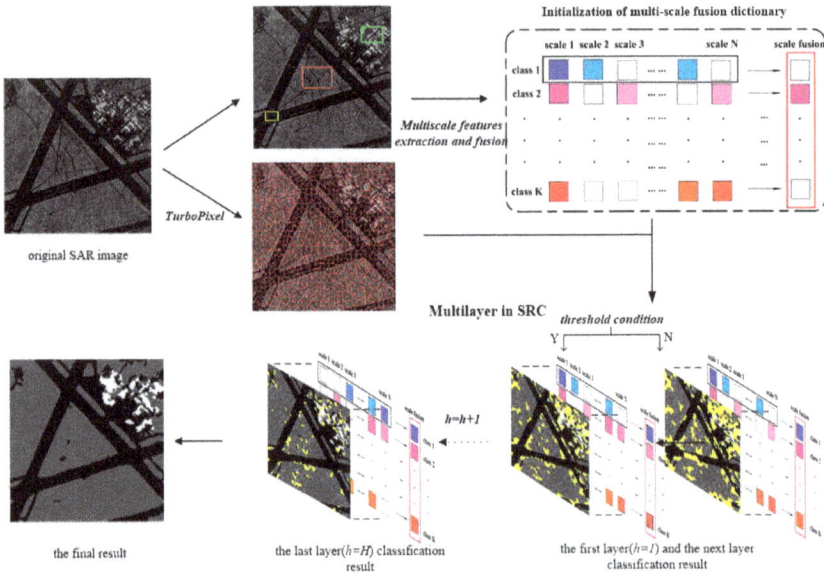

Figure 3. The basic framework of multi-layers and multi-feature model (ML–MFM).

Figure 3 can be understood in four parts: superpixel generation, multi-feature extraction, multi-scale with fusion strategy and multi-layer sparse representation classification. In the first stage, the over-segment algorithm is used. The initialization dictionary is subsequently used in different scales, before being fused by the fusion strategy, which is introduced in Section 2.2.2. The first classification is performed using the initialization dictionary and the superpixels. Finally, SRC is used in multi-layers through iterations to obtain the final result.

2.2.1. Multilayer SRC

There are some difference between SAR image classification and face recognition, the training samples of which can be controlled by a human in a standard data set. SAR images contain various complex terrains. It is difficult to guarantee enough training samples to represent each pixel. To deal with the challenge, reference [17] developed a hierarchical sparse representation classifier to improve the classification of the SAR image, which we called the multi-layer SRC.

In this classifier, h $(1 \leq h \leq H)$ represents the layer of classification. Sparse representation is used for each layer. Thus, we performed sparse representation h times. As the number of layers h $(1 \leq h \leq H)$ increases, the classification map becomes closer to the final result. Finally, the number of h is analyzed in Section 3.1.

2.2.2. Multiscale Fusion-Based Dictionary

In the application of sparse representation, a dictionary is first constructed. To counter the existence of speckle and the complex appearance in the SAR image, we transform the pixel value space into a feature space, which reduces the computational complexity and extracts the discriminative features from the SAR image. The gray-level histogram, GLCM and Gabor filter features are extracted to capture statistical properties in the SAR image. We concentrated on these three types of features to form a feature vector for representing each pixel or superpixel. This non-linear feature will provide competitive performance by representing statistical information and capturing texture information in adjacent areas.

Moreover, different texture features of images show different performances in different scales [24]. In the classification of the image, many experiments have proved that different scales correspond to the same test sample y. The information from different scales complements each other, which is useful to classify each pixel.

We selected n_i vectors of the lth scale training samples from the ith class in the hth layer as columns to construct a matrix $A_{i,l}^h = [f_{i,l,1}^h, f_{i,l,2}^h, \ldots, f_{i,l,n_i}^h] \in R^{m \times n_i}$, where m denotes the dimension of the extracted feature vector. Scale l means the scale sizes $(2l+1) \times (2l+1)$ of the extracted features. Therefore, $l = 1$ is the finest scale and $l = L$ is coarsest scale, which depends on the resolution of the image. This can be calculated by $L = floor(resolution * 3/2)$. We constructed a matrix by the concatenation of the n_i training sample vectors of all K defined classes and all L scales in the fixed hth layer as follows:

$$A^h = \begin{bmatrix} A_{1,1}^h & A_{1,2}^h & \cdots & A_{1,L}^h \\ A_{2,1}^h & A_{2,2}^h & \cdots & A_{2,L}^h \\ \vdots & \vdots & \cdots & \vdots \\ A_{K,1}^h & A_{K,2}^h & \cdots & A_{K,L}^h \end{bmatrix} \tag{5}$$

Following the dictionary D^h in the fixed hth layer is defined by row element using the average fusion strategy in matrix A^h. This is shown as follows:

$$D^h = \begin{bmatrix} D_{1,fusion}^h \\ D_{2,fusion}^h \\ \vdots \\ D_{K,fusion}^h \end{bmatrix} \quad D_{i,fusion}^h = \frac{1}{n_1+n_2+\cdots+n_L} \sum_{l=1}^{L} \sum_{j=1}^{n_L} f_{i,l,j}^h \quad \begin{bmatrix} A_{1,1}^h & A_{1,2}^h & \cdots & A_{1,L}^h \\ A_{2,1}^h & A_{2,2}^h & \cdots & A_{2,L}^h \\ \vdots & \vdots & \cdots & \vdots \\ A_{K,1}^h & A_{K,2}^h & \cdots & A_{K,L}^h \end{bmatrix} = A^h \tag{6}$$

2.2.3. Multi-Layer and Multi-Feature Model

It is well-known that pixels are not natural entities but a result of the discrete representation of an image, with structural information captured in a region rather than a pixel. Furthermore, the computational complexity increases rapidly with an increase in the scale of pixels used. Therefore, we first divided the SAR image into superpixels to integrate the contextual information of neighboring pixels and to reduce computational complexity. In our method, the operation of superpixel generation uses the TurboPixel algorithm [25]. Furthermore, in order to encode gray, textural and spatial information into superpixels, we described each superpixel $sp_t \in sp = \{sp_t | 1 \le t \le N\}$ by a m dimensional feature vector $f_{sp_t} = [f_{t,1}, f_{t,2}, \ldots, f_{t,m}]$, in which N is the maximum number of superpixels.

Unlike other recognition methods, SAR image classification lacks training samples to effectively represent each pixel. To deal with the challenge, inspired by the idea of the hierarchical sparse representation [17], we proposed a multi-layer operation based on SRC and the multi-fusion scale dictionary for SAR image classification.

Based on dictionary construction and the superpixel sp_t, we used Equation (2) with the l_0 norm to ensure a sparse solution. The sparse coefficient can be solved by OMP to obtain $\hat{x}$. We define the hth layer of the minimum residual error and class as Equations (7) and (8), respectively.

$$r^h = res_{min}^h(\psi(sp_t)) = \min_{i=1,\ldots,K} res_i^h(\psi(sp_t)) \tag{7}$$

$$\hat{k}^h = \underset{i}{argmin} \| \psi(sp_t) - D^h \hat{x}_i \|_2, i = 1, 2, \ldots, K \tag{8}$$

where the $res_i^h(\psi(sp_t))$ is the residual error in the hth layer under the fusion scale; r^h is the minimum residual error; and c^h is the category of sp_t. Different from Equation (4), we introduce a parameter ΔT as the threshold value to limit the superpixel and judge whether it belongs to this class instead. If the r^h is within the specified tolerance limit, the pixel belongs to the current class i. Otherwise, the uncertain samples are classified again in the next layer. Following this, the class of superpixels in hth ($1 \leq h < H - 1$, where H is layer number) layer are expressed as:

$$
label(sp_t) = \begin{cases} \underset{i}{argmin}\|\psi(sp_t) - D^h \hat{x}_i\|_2, i = 1, 2, \ldots, K; r^h \leq \Delta T \\ uncertain, \qquad\qquad\qquad otherwise \end{cases} \tag{9}
$$

where $r^h \leq \Delta T$ is a restricted condition to ensure that the superpixel belongs to the category. In fact, the choice of the threshold value ΔT will influence our final results to some extent and it will be further discussed in Section 3.1.

The uncertain superpixels in the $(h + 1)$th $A_i^h = [f_{i,1}^h, f_{i,2}^h, \ldots, f_{i,n_i}^h] \in R^{m \times n_i}$ $(1 \leq h < H)$ layer will be classified by Equations (7)–(9). We selected superpixels from each class, which are labeled as the new training samples in the hth layer. Following this, we extract the feature vector at the hth layer. Arranging these vectors as the columns vector $A_{sp}^h = [A_1^h, A_2^h, \ldots, A_K^h]$, we define the dictionary $D^{(h+1)}$ in the fixed $(h + 1)$th layer based on the dictionary D^h in the fixed hth layer.

$$
D^{(h+1)} = \begin{bmatrix} D_{1,fusion}^{(h+1)} \\ D_{2,fusion}^{(h+1)} \\ \vdots \\ D_{K,fusion}^{(h+1)} \end{bmatrix} = Average \left(\begin{bmatrix} D_{1,fusion}^h \\ D_{2,fusion}^h \\ \vdots \\ D_{K,fusion}^h \end{bmatrix}, \begin{bmatrix} A_1^h \\ A_2^h \\ \vdots \\ A_K^h \end{bmatrix} \right) = Average(D^h, A_{sp}^h) \tag{10}
$$

In fact, the uncertain superpixels decrease with an increase of layers. When $h = H$, there is still a small number of uncertain points (marked by yellow in Figure 3). However, the dictionary D^H is modified by Equation (10), before the remaining pixels will be classified by a traditional sparse repreentation classifier until it outputs the final result. The whole ML–MFM for the SAR image classification Algorithm is summarized as follows:

Algorithm 1. ML–MFM for synthetic aperture radar (SAR) Image Classification

---

Input: SAR image, threshold ΔT, class number K, layer H.
Output: classification map.
1. Segment the SAR image into superpixels by [25].
2. Construct the initial fusion dictionary D^1 by Equations (5) and (6), while the fusion dictionary contains K class, $D^1 = [D_{1,fusion}^1, D_{2,fusion}^1, \ldots, D_{K,fusion}^1]$. Choosing a specified number of pixels n_i from the original SAR image as the samples, each sample can be represented by the m dimension extracted variety of features.
3. Multi-layer SRC and dictionary in layers are constructed.

- Classify all superpixels in the first layer by Equations (7) and (8) with orthogonal matching pursuit (OMP);

- Find the best representative atom's label by Equation (9);

   while $1 \leq h \leq H$
   if $r^h \leq \Delta T$
      $\underset{i}{min} res_i^h(\psi(sp_t))$;
      $label(sp_t) \leftarrow i$;
      updating dictionary by (10)
      $D^{(h+1)} = Average(D^h, A_{sp})$;
   else $uncertain(sp_t^h) \leftarrow sp_t = \psi(sp_t)$;
      $h \leftarrow h + 1$;
   end while

---

3. Results

In this section, the proposed model is now applied in the classification of synthetic and real SAR images. To validate the performance of the proposed method, we use both types of images in quantitative evaluation and visualization results. We mainly compare our results with the results of previous studies [3,5,16,17], in which their parameters are tuned to obtain the best results. Figures 8a, 9a and 10a are the synthetic SAR images, which are from the Brodatz database. These synthetic SAR images have three, four and five types of different textural regions, while the size of each image is 512×512, 335×335, and 512×512, respectively. The test images are named Syn1, Syn2 and Syn3, respectively. In addition, three real SAR images (SAR1, SAR2 and SAR3) were tested in experiments. SAR1 has a size of 256×256, which covers the China Lake Airport, California, with a Ku-Band radar with a 3-m resolution. SAR2 has a size of 321×258, which covers the pipeline over the Rio Grande river near Albuquerque, New Mexico, with a Ku-Band radar with 1-m resolution. SAR3 has a size of 284×284, which covers the X-Band radar with 3-m resolution. The central processing unit time was obtained by running the Matlab code on a DELL computer with Inter (R) Core (TM) i7CPU, 3.4 GHz, 16 GB RAM with MATLAB 2014(a) on Windows 10 (64-bit operating system) in our experiment.

3.1. Experimental Settings

In the experiment, we used the TurboPixel [25] algorithm to over-segment the original image into homogeneous regions and to obtain the superpixels. As each superpixel has different sizes, the features of each superpixel need to be processed so that the fusion features of all superpixels have the same dimensions (i.e., m = 60 in our method). Sixteen effective distribution features and four statistical features suggested by a previous study [17] were used. Thus, the features extracted by gray-level histogram and GLCM have 16 dimensions and four dimensions for each superpixel, respectively. After calculating the convolution of the initial bank of Gabor filters, which consists of 40 filters with five scales and eight orientations, the mean value of the filter response corresponding to each superpixel was computed for every filter. Therefore, the Gabor feature of each superpixel was a 40-dimension vector corresponding to 40 Gabor filters with the total number of dimensions being 60. In addition, the ranges of the radial basis function kernel width and penalty coefficients are (0.0001, 0.001, 0.01, 0.1, 1, 10) and (0.1, 1, 10, 100, 500, 1000) respectively.

3.1.1. Influence of Parameters

It is necessary to set ideal parameters to obtain satisfactory results. There are two main parameters that need to be set in our model, namely H and ΔT (threshold). Based on plenty of experimental data and the analysis of results, each parameter should satisfy the following condition: $1 \leq H \leq 6$, $0.07 \leq \Delta T \leq 0.24$. We noted that the parameter H is influenced by the resolution of the SAR images (similar to a previous reference [17]), as mentioned in Section 2.2. We used an experiment to show the influence of H, which is depicted in Figure 4. The horizontal axis represents the layer h, while the vertical axis represents a certain superpixel number. Here, we artificially set the total superpixel number to 1000 of SAR1. From Figure 4, it is more intuitive to find the most suitable layer range. This is because when $H > 6$, the speed of the growth of certain points slow significantly in histogram and line chart. In many experiments, if we set $H = 6$ as the maximum number of layers, it is the best choice with regards to time and precision. In addition, it is worth noting that ΔT is the threshold to control the categories of accuracy (blue solid line) and the kappa coefficient (green dotted line) with our proposed method. From Figure 5, we can see that when ΔT is too small, there are many uncertain superpixels until $h = H$. However, when ΔT is too large, the finer areas cannot be placed into classes. Therefore, appropriate parameter selection is very important.

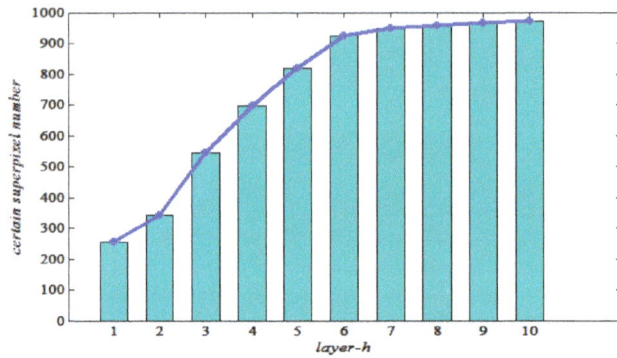

Figure 4. Illustration of certain and uncertain superpixels with different layers corresponding to SAR1 (Figure 7a) in our method (the total number of superpixels is 1000).

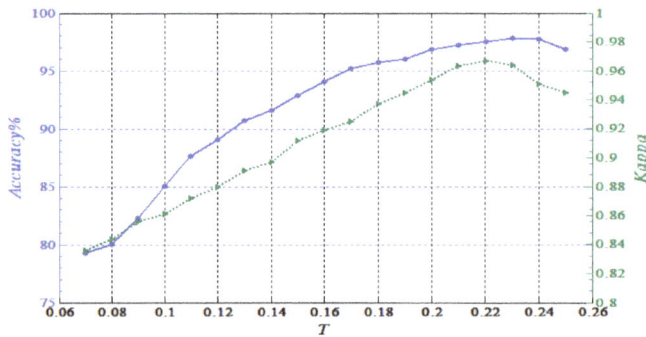

Figure 5. Influence of threshold ΔT in classification accuracy (blue solid line) and Kappa coefficient (green dotted line) corresponding to SAR1 (Figure 7a).

3.1.2. Analysis of Multi-Feature Fusion and Multi-Scale Fusion

Multi-Feature Fusion

In this part, the multi-feature fusion is analyzed to verify its effectiveness in obtaining satisfactory results. In our paper, the fusion strategy is introduced to construct the dictionary. We perform an experiment on the original SAR1 (Figure 6a) to show the influence of multi-feature fusion on the dictionary and the impact of classification results. The rectangular areas of Figure 6a–e are marked by red, yellow and green, respectively. Figure 6b shows the results of the method with the gray-level histogram; Figure 6c shows the results of the GLCM; Figure 6d shows the results of the Gabor method; and Figure 6e shows the results of the multi-features method. We can see that Figure 6e has fewer miscellaneous points than Figure 6b–d. The reason is that the fusion features includes distribution features and four statistical features. Therefore, the dictionary D^h includes more information to obtain better results, which is an advantage that is absent in the method with single features. Therefore, the multi-feature fusion strategy is important.

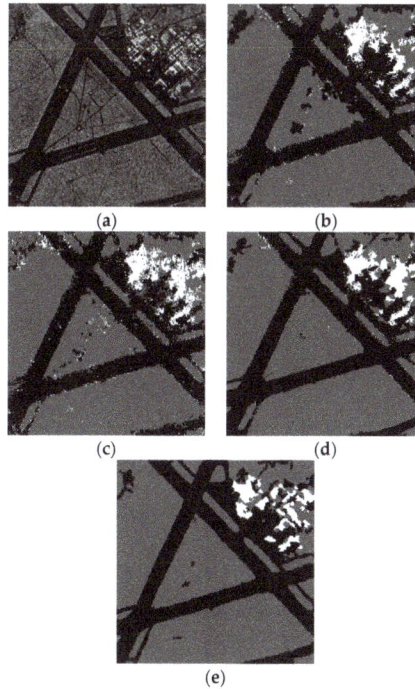

Figure 6. Comparison of (**a**) the original SAR1 with (**b**) gray-level histogram; (**c**) gray-level co-occurrence matrix (GLCM); (**d**) with Gabor; (**e**) with multi-features.

Multi-Scale Fusion

In this part, multi-scale fusion is analyzed to verify its effectiveness in obtaining satisfactory results. In our paper, the fusion strategy is introduced to construct the dictionary. We perform an experiment on the original SAR1 (Figure 7a) to show the influence of the fusion strategy on the dictionary after merging features under different scales. The rectangular areas of Figure 7b,c are both marked by red, yellow and green. Figure 7b shows the results method with the fusion strategy; Figure 7c shows the results method without the fusion strategy. We can see that Figure 7b has fewer miscellaneous points than Figure 7c. The reason for this is that the fusion dictionary D^h includes each scale information (homogeneous and marginal regions). This has similar effects on the dictionary under multi-scales, which are absent in the method without the fusion dictionary and are important to SAR image processing. Therefore, the multi-scale fusion strategy is important.

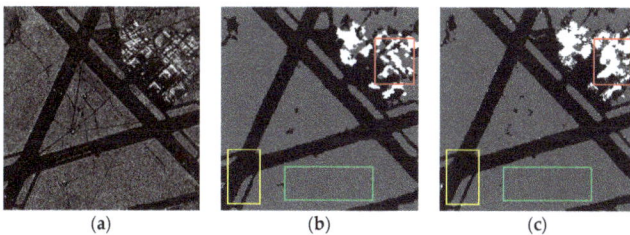

Figure 7. Comparison of (**a**) the original SAR1 (**b**) with fusion strategy; and (**c**) without fusion strategy.

3.2. Results on Synthetic SAR Images

In this section, we test the capability of the proposed algorithm by applying it to the synthetic SAR images Syn1, Syn2, and Syn3. The superpixels of Syn1, Syn2, and Syn3 are 2800, 1500 and 2800, respectively. In our method, $H = 6$ and ΔT is 0.221. The scale (patch) size in the support vector machine (SVM) [3], SRC [5] and JSRM [16] is fixed and we set it to be 3×3. The ground truth was used to calculate the accuracy of the classification results to evaluate the contrast algorithms. We can see that the proposed method can obtain a higher accuracy of classification than the results of previous studies [3,5,16] and can reduce the processing time found in reference [17]. Moreover, as shown Figures 8–10, as well as Table 1, the proposed method can keep the details (edges) in a similar way to reference [17] in the visual representation. However, the results of the other methods in finer textural regions (marked with pink and yellow), such as Figure 9e–g, have significantly different degrees of error, which is caused by the lack of samples. Although our method has no significant improvement in accuracy compared to the method in reference [17], there are benefits to not requiring an extensive amount of time in pixel-by-pixel training and having less miscellaneous points existing in the final classification.

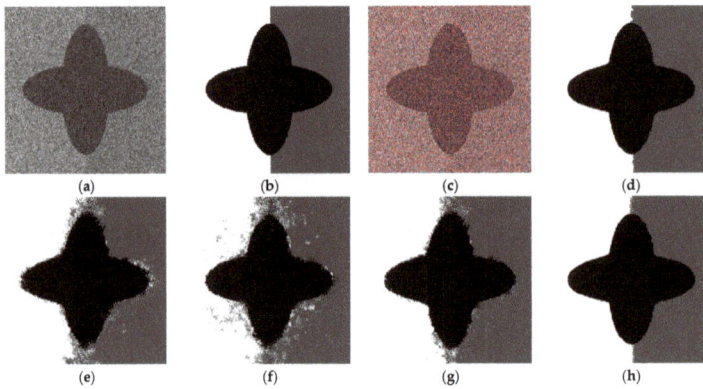

Figure 8. Results of different methods in Syn1: (**a**) Synthetic SAR images; (**b**) Ground truth; (**c**) Superpixels map; (**d**) Proposed method; (**e**) support vector machine (SVM) [3]; (**f**) SRC [5]; (**g**) joint sparsity model (JSRM) [16]; (**h**) hierarchical sparse representation-based classification (HSRC) [17].

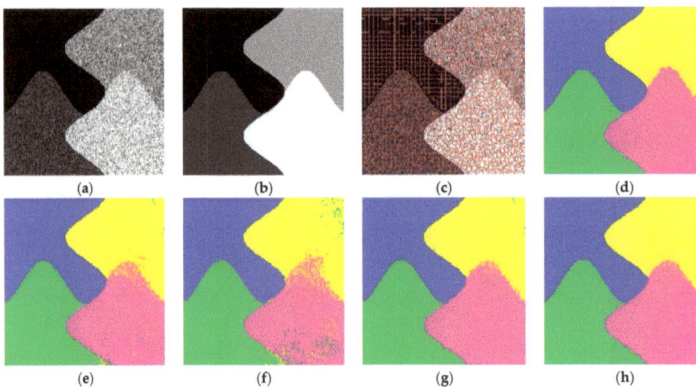

Figure 9. Results of different methods in Syn2: (**a**) Synthetic SAR images; (**b**) Ground truth; (**c**) Superpixels map; (**d**) Proposed method; (**e**) support vector machine (SVM) [3]; (**f**) SRC [5]; (**g**) joint sparsity model (JSRM) [16]; (**h**) hierarchical sparse representation-based classification (HSRC) [17].

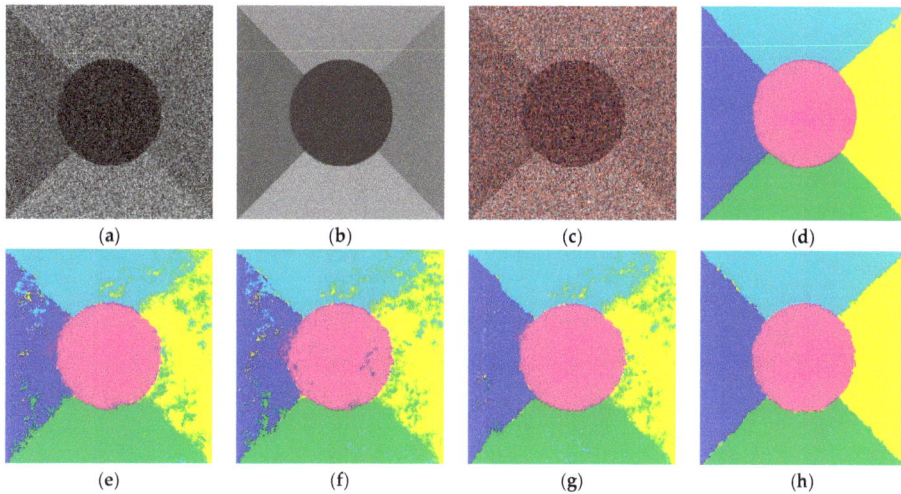

Figure 10. Results of different methods in Syn3: (**a**) Synthetic SAR images; (**b**) Ground truth; (**c**) Superpixels map; (**d**) Proposed method; (**e**) support vector machine (SVM) [3]; (**f**) SRC [5]; (**g**) joint sparsity model (JSRM) [16]; (**h**) hierarchical sparse representation-based classification (HSRC) [17].

Table 1. Comparison of the run times (s) and classification accuracy (%) of different methods.

SAR Image	Proposed Method		SVM [3]		SRC [5]		JSRM [16]		HSRC [17]	
	Accuracy	Time	Accuracy	Time	Accuracy	Time (s)	Accuracy	Time	Accuracy	Time
Syn1	98.79	120.32	80.35	51.88	76.38	101.37	91.73	161.37	98.89	230.59
Syn2	97.76	103.96	88.73	54.89	80.83	85.74	94.78	137.49	98.12	201.14
Syn3	96.04	124.48	73.29	48.14	70.86	106.84	89.14	153.26	96.24	243.32

3.3. Results of Real SAR Images

In this section, three real SAR images are used for further analysis. The compared methods are the same as those used on synthetic SAR images. The results are shown in Figures 11–13. These original real images have three, three and four types of different regions as shown in Figures 11b, 12b and 13b, respectively. The superpixels of SAR1, SAR2 and SAR3 are 1000, 1200 and 1100 as shown in Figures 11b, 12b and 13b, respectively. The evaluation of the classification method is based on the visual inspection of the classification and the run time, accuracy, as well as the kappa coefficient. The scale in SVM, SRC and JSRM is set to 7×7, which represents the best result in the experiments.

From Figure 11c–g, we can see that the proposed method can achieve the classification in different areas and eliminate the influence of shadows, which always leads to categories by mistake. However, when we compare Figure 11c with 11g, it is difficult to know whether our proposed method is better, as it seems that Figure 11g [17] has better visualization results, albeit with some miscellaneous points. Therefore, the accuracy of the quantitative analysis is required for further analysis. From Table 2, it can be seen that the accuracy of the previous study [17] is only slightly higher than our proposed method, but the required running time is too long, as previously seen with synthetic SAR images.

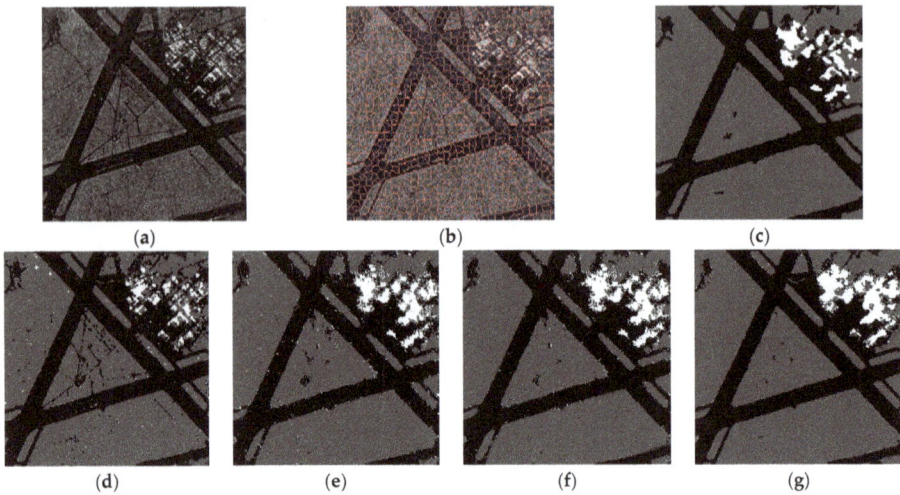

Figure 11. Results of different methods in real SAR1: (**a**) Real SAR images; (**b**) Superpixels map; (**c**) Proposed method; (**d**) support vector machine (SVM) [3]; (**e**) SRC [5]; (**f**) joint sparsity model (JSRM) [16]; and (**g**) hierarchical sparse representation-based classification (HSRC) [17].

From Figure 12c–g, we can see the classification results of different methods, especially in the yellow and red rectangle regions. The yellow and red rectangles of the proposed method in Figure 12c have less miscellaneous points than Figure 12d–f. In general, a smaller number of miscellaneous points indicates a more complete extraction of information and a more stable performance of the algorithm. The different rectangle regions highlight the superiority of the proposed algorithm. However, when we compare Figure 12c with 12g, it is difficult to know whether our proposed method is better, as it seems that Figure 12g [17] has better visualization results, albeit with some miscellaneous points. Therefore, data analysis was used (accuracy, run time and kappa coefficient) for further analysis. From Table 2, it can be seen that the accuracy of the previous method [17] is only slightly higher than our proposed method, but the required running time was too long, as previously seen with synthetic SAR images.

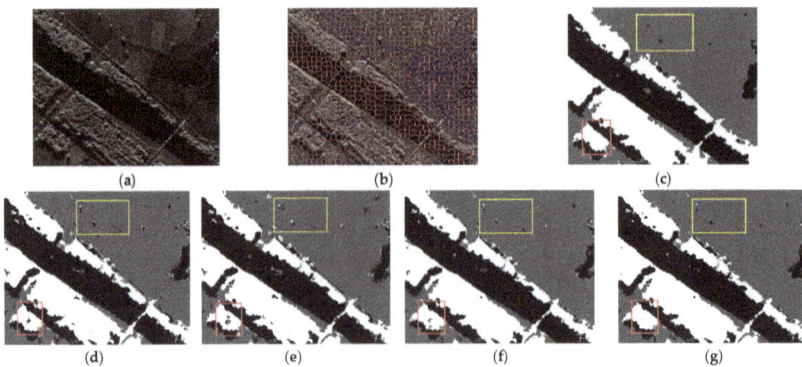

Figure 12. Results of different methods in real SAR2: (**a**) Real SAR images; (**b**) Superpixels map; (**c**) Proposed method; (**d**) support vector machine (SVM) [3]; (**e**) SRC [5]; (**f**) joint sparsity model (JSRM) [16]; and (**g**) hierarchical sparse representation-based classification (HSRC) [17].

The analysis of Figure 13 is similar to Figures 11 and 12. From Figures 11–13, the proposed method is shown to be suitable for the SAR image classification and obtains the optimal results. Table 2 provides the quantitative evaluation of different methods. Although the HSRC obtains higher classification accuracy compared with the others, the running time is too long among the different methods. Our method has the absolute advantage in the running time, with competitive accuracy that is only slightly lower than HSRC. Moreover, our method gets the highest robustness, which is reflected by the kappa coefficient. Above all, our method outperforms the others in terms of time consumption and robustness.

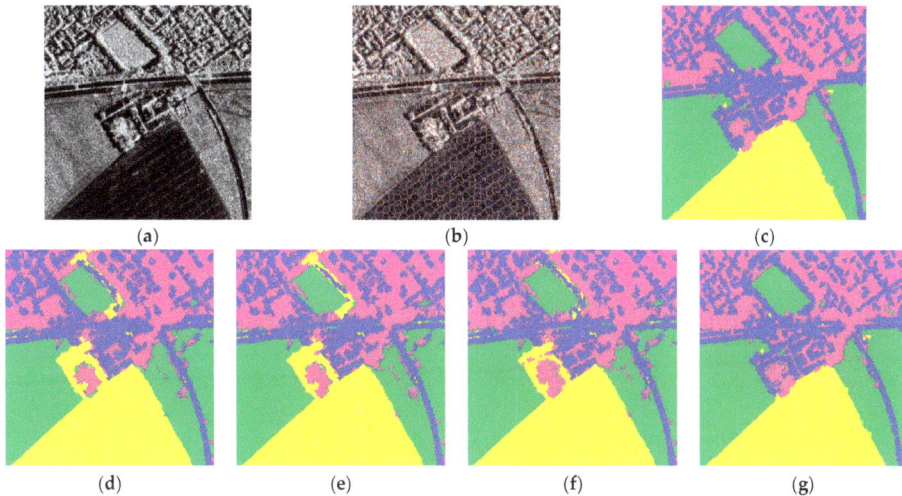

Figure 13. Results of different methods in real SAR3: (**a**) Real SAR images; (**b**) Superpixels map; (**c**) Proposed method; (**d**) support vector machine (SVM) [3]; (**e**) SRC [5]; (**f**) joint sparsity model (JSRM) [16]; and (**g**) hierarchical sparse representation-based classification (HSRC) [17].

Table 2. Comparison of the average criteria and accuracy (%) of different methods.

SAR Image	Proposed Method		SVM [3]		SRC [5]		JSRM [16]		HSRC [17]	
	Accuracy	Time	Accuracy	Time	Accuracy	Time	Accuracy	Time	Accuracy	Time
SAR1	96.18	102.38	89.79	48.68	85.38	98.67	93.67	147.73	96.20	238.95
SAR2	97.57	121.35	87.68	43.19	87.33	99.24	92.48	139.58	97.62	253.48
SAR3	97.52	124.41	83.59	51.94	79.96	102.68	91.04	160.36	96.74	258.72
AA [1]	97.42		87.02		84.22		92.40		97.83	
K [2]	0.961		0.713		0.806		0.941		0.959	

[1] AA is average accuracy; [2] K is kappa coefficient.

4. Discussion

Traditional SVM [3] is limited by lacking samples, which results in low classification accuracy. For instance, the number of training samples is 300, which is 0.46% of the total samples. Fewer samples affect the selection of optimal parameters by SVM for the testing samples, which will decrease the classification accuracy. In the sparse representation method, the HSRC [13] can solve the problem of reference [16] to a certain extent. It introduces the hierarchical concept and multi-size patch feature to solve the problem of lacking samples. Using SRC in SAR classification for both these methods improves the accuracy and stability. However, HSRC classifies images based on each pixel, which only depends on the selection of features in the spatial domain and the selected scale for each layer. This may lead to the loss or misrepresentation of information, which requires a long time for training.

In our paper, we inherit the advantages of reference [17], such as multi-layer. However, the difference is the multi-scale and multi-feature fusion. In the multi-feature fusion stage, we take three different methods to extract the gray and texture characteristics, which are stable in the presence of noise and changes in view, and can enrich the information of images. Moreover, the strategy of the multi-feature fusion was inspired by [23], which fused the different features from multiple layers. We fused the different features from different scales. This reduces the computational time and ensures a rich amount of information. Furthermore, classification based on superpixels can improve the speed of algorithms effectively. Three evaluation metrics (i.e., run time (time), average accuracy (AA) and the Kappa coefficient (K)) are adopted in these experiments to evaluate the quality of classification results. AA represents the mean of the percentage of correctly classified pixels for each class. K estimates the percentage of classified pixels corrected by the number of agreements. We performed comparative experiments with four other methods. The proposed method can solve the time redundancy problem of HSRC, but has its uncertainties. For instance, the uncertain points are always in the process of the algorithm until $h = H$. That is the reason we use the traditional SRC (this step is same as reference [17]) in the last step.

5. Conclusions

In this paper, based on superpixels, we presented a new model of classification of SAR images. It validates that adding multiple features, scales and layers can benefit the results of SRC classification and enrich the information of the images. Furthermore, using multiple layers can decrease the computational time due to the use of superpixels. The fusion strategy was introduced to merge each scale together to form a multi-fusion dictionary. With the added benefits, robustness was enhanced and the classification accuracy was improved significantly. The comparison experiments based on synthetic SAR images and real SAR images clearly demonstrate the efficiency and advantages of the proposed classification method. Moreover, the proposed classification method is also able to achieve lower computational costs. These added benefits are general for SAR image classification, and can be suitable for utility in more applications in the area of SAR image classification, as well as in other areas where the SRC method could be applied.

This method provides a slight improvement in calculation time for SAR image classification and application. Moreover, future research will focus on developing more efficient algorithms to cope with the large-scale SAR images.

Acknowledgments: This work is supported by National Natural Science Foundation of China (No. 61472278 and 61102125). The author would like to thank numerous colleagues for their contribution to this work and three reviewers and editors for improving the manuscript.

Author Contributions: This work was prepared and accomplished by Ao-bo Zhai, who also wrote the manuscript. Xian-bin Wen outlined the research and supported the analysis. He also revised the work in presenting the technical details. Li-ming Yuan, Hai-xia Xu both suggested the design of comparison experiments and supervised the writing of the manuscript at all stages. Qing-xia Meng provided writing suggestions.

Conflicts of Interest: The authors declare no conflict of interest.

References

1. Oliver, C.; Quegan, S. *Understanding Synthetic Aperture Radar Images*; SciTech Publishing: Raleigh, NC, USA, 2004.
2. Adragna, F.; Nicolas, J. *Processing of Synthetic Aperture Radar Images*; Wiley: New York, NY, USA, 2010.
3. Akbarizadeh, G. A new statistical-based kurtosis wavelet energy feature for texture recognition of SAR images. *IEEE Trans. Geosci. Remote Sens.* **2012**, *50*, 4358–4368. [CrossRef]
4. Xue, X.R.; Wang, X.J.; Xiang, F.; Wang, H.F. A new method of SAR image segmentation based on the gray level co-occurrence matrix and fuzzy neural network. In Proceedings of the IEEE 6th International Conference Wireless Communications Networking and Mobile Computing, Chengdu, China, 23–25 September 2010.

5. Wright, J.; Yang, A.Y.; Sastry, S.S.; Ma, Y. Robust face recognition via sparse representation. *IEEE Trans. Pattern Anal. Mach. Intell.* **2009**, *31*, 210–227. [CrossRef] [PubMed]

6. Mallat, S.G.; Zhang, Z. Matching pursuits with time-frequency dictionaries. *IEEE Trans. Signal Process.* **1993**, *41*, 3397–3415. [CrossRef]

7. Elad, M.; Aharon, M. Image Denoising Via Sparse and Redundant Representations over Learned Dictionaries. *IEEE Trans. Image Process.* **2006**, *15*, 3736–3745. [CrossRef] [PubMed]

8. Buades, A.; Coll, B.; Morel, J.M. A Non-Local Algorithm for Image Denoising. In Proceedings of the IEEE Conference Computer Vision Pattern Recognition, San Diego, CA, USA, 20–26 June 2005.

9. Yue, C.; Jiang, W. An algorithm of SAR image denoising in nonsubsampled contourlet transform domain based on maximum a posteriori and non-local restriction. *Remote Sens. Lett.* **2013**, *4*, 270–278. [CrossRef]

10. Fang, L.Y.; Li, S.T.; Mcnabb, R.P.; Nie, Q.; Kuo, A.N.; Toth, C.A.; Izatt, J.A.; Farsiu, S. Fast Acquisition and Reconstruction of Optical Coherence Tomography Images via Sparse Representation. *IEEE Trans. Med. Imaging* **2013**, *32*, 2034–2049. [CrossRef] [PubMed]

11. Li, S.; Yin, H.; Fang, L. Remote Sensing Image Fusion via Sparse Representations over Learned Dictionaries. *IEEE Trans. Geosci. Remote Sens.* **2013**, *51*, 4779–4789. [CrossRef]

12. Xiang, D.; Tang, T.; Hu, C.; Li, Y.; Su, Y. A kernel clustering algorithm with fuzzy factor: Application to SAR image segmentation. *IEEE Geosci. Remote Sens. Lett.* **2011**, *7*, 1290–1294.

13. Wang, W.; Xiang, D.; Ban, Y.; Zhang, J.; Wan, J. Superpixel Segmentation of Polarimetric SAR Images Based on Integrated Distance Measure and Entropy Rate Method. *IEEE J. Sel. Top. Appl. Earth Obs. Remote Sens.* **2017**, *99*, 1–14. [CrossRef]

14. Liu, B.; Hu, H.; Wang, H.; Wang, K.; Liu, X.; Yu, W. Superpixel-Based Classification with an Adaptive Number of Classes for Polarimetric SAR Images. *IEEE Trans. Geosci. Remote Sens.* **2013**, *51*, 907–924. [CrossRef]

15. Zhang, X.; Wen, X.; Xu, H.; Meng, Q. Synthetic aperture radar image segmentation based on edge-region active contour model. *J. Appl. Remote Sens.* **2016**, *10*, 036014. [CrossRef]

16. Chen, Y.; Nasrabadi, N.M.; Tran, T.D. Hyperspectral Image Classification Using Dictionary-Based Sparse Representation. *IEEE Trans. Geosci. Remote Sens.* **2011**, *49*, 3973–3985. [CrossRef]

17. Hou, B.; Ren, B.; Ju, G.; Li, H.; Jiao, L.; Zhao, J. SAR Image Classification via Hierarchical Sparse Representation and Multisize Patch Features. *IEEE Geosci. Remote Sens. Lett.* **2016**, *13*, 33–37. [CrossRef]

18. Swain, M.J.; Ballard, D.H. Color indexing. *Int. J. Comput. Vis.* **1991**, *7*, 11–32. [CrossRef]

19. Haralick, R.M.; Shanmugam, K.; Dinstein, I. Textural Features for Image Classification. *IEEE Trans. Syst. Man Cybern.* **1973**, *3*, 610–621. [CrossRef]

20. Yan, X.; Jiao, L.; Xu, S. SAR image segmentation based on Gabor filters of adaptive window in overcomplete brushlet domain. In Proceedings of the 2nd Asian-Pacific Conference on Synthetic Aperture Radar, Xi'an, China, 26–30 October 2009.

21. Gu, J.; Jiao, L.; Yang, S.; Liu, F.; Hou, B.; Zhao, Z. A Multi-kernel Joint Sparse Graph for SAR Image Segmentation. *IEEE J. Sel. Top. Appl. Earth Obs. Remote Sens.* **2016**, *9*, 1265–1285. [CrossRef]

22. Zhao, S.; Liu, Y.; Han, Y.; Hong, R.; Hu, Q.; Tian, Q. Pooling the Convolutional Layers in Deep ConvNets for Video Action Recognition. *IEEE Trans. Circuits Syst. Video Technol.* **2015**, *1*. [CrossRef]

23. Tan, M.; Tsang, I.W.; Wang, L.; Zhang, X. Convex Matching Pursuit for Large-scale Sparse Coding and Subset Selection. In Proceedings of the AAAI Conference on Artificial Intelligence, Toronto, ON, Canada, 22–26 July 2012.

24. Mo, X.; Monga, V.; Bala, R.; Bala, R.; Fan, Z.G. Adaptive Sparse Representations for Video Anomaly Detection. *IEEE Trans. Circuits Syst. Video Technol.* **2013**, *24*, 631–645.

25. Levinshtein, A.; Stere, A.; Kutulakos, K.N.; Fleet, D.J.; Dickinson, S.J.; Siddiqi, K. TurboPixels: Fast Superpixels Using Geometric Flows. *IEEE Trans. Pattern Anal. Mach. Intell.* **2009**, *31*, 2290–2297. [CrossRef] [PubMed]

remote sensing

Article

L-Band Temporal Coherence Assessment and Modeling Using Amplitude and Snow Depth over Interior Alaska

Yusuf Eshqi Molan [1,*], Jin-Woo Kim [1], Zhong Lu [1] and Piyush Agram [2]

[1]   Roy M. Huffington Department of Earth Sciences, Southern Methodist University, Dallas, TX 75205, USA; jinwook@mail.smu.edu (J.-W.K.); zhonglu@mail.smu.edu (Z.L.)
[2]   Jet Propulsion Laboratory, California Institute of Technology, Pasadena, CA 91109, USA; piyush.agram@jpl.nasa.gov
*   Correspondence: yeshqimolan@smu.edu; Tel.: +1-469-623-0639

Received: 6 December 2017; Accepted: 16 January 2018; Published: 20 January 2018

Abstract: Interferometric synthetic aperture radar (InSAR) provides the capability to detect surface deformation. Numerous processing approaches have been developed to improve InSAR results and overcome its limitations. Regardless of the processing methodology, however, temporal decorrelation is a major obstacle for all InSAR applications, especially over vegetated areas and dynamic environments, such as Interior Alaska. Temporal coherence is usually modeled as a univariate exponential function of temporal baseline. It has been, however, documented that temporal variations in surface backscattering due to the change in surface parameters, i.e., dielectric constant, roughness, and the geometry of scatterers, can result in gradual, seasonal, or sudden decorrelations and loss of InSAR coherence. The coherence models introduced so far have largely neglected the effect of the temporal change in backscattering on InSAR coherence. Here, we introduce a new temporal decorrelation model that considers changes in surface backscattering by utilizing the relative change in SAR intensity between two images as a proxy for the change in surface scattering parameters. The model also takes into account the decorrelation due to the change in snow depth between two images. Using the L-band Advanced Land Observation Satellite (ALOS-2) Phased Array type L-band Synthetic Aperture Radar (PALSAR-2) data, the model has been assessed over forested and shrub landscapes in Delta Junction, Interior Alaska. The model decreases the RMS error of temporal coherence estimation from 0.18 to 0.09 on average. The improvements made by the model have been statistically proved to be significant at the 99% confidence level. Additionally, the model shows that the coherence of forested areas are more prone to changes in backscattering than shrub landscape. The model is based on L-band data and may not be expanded to C-band or X-band InSAR observations.

Keywords: InSAR; temporal coherence modeling; L-band; Interior Alaska

---

1. Introduction

Interferometric synthetic aperture radar (InSAR) provides an all-weather, day-or-night capability to remotely sense mm to cm scale surface deformation with a high spatial resolution of tens of meters or better (e.g., [1–4]). InSAR has been successfully used to detect surface deformation due to various mechanisms, such as volcanism, subsidence, permafrost, and landslides [5–9]. So far numerous methods and approaches have been developed to improve InSAR performance. However, temporal decorrelation, regardless of the processing methodology, is one of the major obstacles for all InSAR applications, especially over vegetated areas. The main sources of the loss of coherence, i.e., decorrelation, are temporal decorrelation, spatial decorrelation, volume decorrelation, thermal decorrelation, and processing errors (e.g., [10,11]). Generally, InSAR coherence decreases with increasing spatial and temporal baselines between two images.

InSAR coherence is sensitive to the changes in surface backscattering, which is dominated by the surface dielectric constant and roughness on the scale of the radar wavelength [12–14]. It has been documented that temporal coherence can be influenced by temporal variations of surface backscattering due to changes in soil moisture, snow depth, surface roughness, and vegetation biomass [12–22]. Simard et al. [12] found precipitation events to be the main cause of temporal decorrelation using fully-polarimetric airborne L-band acquisitions over forested landscapes with up to nine-day temporal baselines. Additionally, they argued that correlation decreases with increasing canopy height regardless of forest type and polarization. Zwieback et al. [15] evaluated soil moisture effects on L-band InSAR and revealed that the phase difference between two SAR images increased with increasing soil moisture difference, whereas the coherence decreased at the same time. Zhang et al. [20], in a case study using C-band ERS SAR data, assessed the relationship between InSAR coherence and soil moisture and inferred that the relation between the two may satisfy an exponential distribution.

Although the effect of the changes in surface backscattering on InSAR coherence has been documented (e.g., [12,16,18]), it has been largely neglected in the coherence models introduced so far. Temporal coherence, in general, is modeled as a univariate exponential function of the temporal baseline [10] with the assumption that the change in the position of scatterers, i.e., mutual displacements of scatterers, is the source of decorrelation [23]. However, we argue that other variables, in addition to the temporal baseline, should be added to the coherence function to compensate for the effect of the temporal variation of surface backscattering on InSAR coherence. In this paper, using ALOS-2 PALSAR-2 images, we analyzed the temporal decorrelation of forested and shrub landscapes in Delta Junction, Alaska, and introduced a new InSAR coherence model, which takes into account the effects of the temporal variations of surface backscattering on InSAR coherence. The model considers the changes in the geometry and dielectric constant of scatterers to be the main sources of decorrelations. The effect of the gradual and natural change in scatterers' geometry has been modeled as a decaying exponential function, which is equivalent to the exponential function of temporal coherence found in the literature [10,13]. The effect of the change in the surface backscattering, mainly due to the change in the dielectric constant of scatterers on InSAR coherence, has been modeled by utilizing the change in InSAR intensity as a proxy for it. The model also takes into account the decorrelation due to the change in snow depth between two images, which induces reversible and seasonal decorrelations. The model, in general, and with different constants, is applicable to model L-band InSAR coherence in other environments and may not be expanded to X-band or C-band SAR observations.

The importance of temporal decorrelation models and their practical use can be better understood by considering the following reasons. Basically, temporal changes of surface parameters describe processes occurring on time scales of the orbit repeat time. In other words, modeling temporal decorrelations provides a means to understand and remotely estimate a wide variety of surficial processes, such as vegetation growth, permafrost freezing and thawing, and soil moisture and vegetation layer induced effects [12].

For instance, it has been shown that both phase and coherence can be used to retrieve soil moisture (e.g., [15]). The coherence, being generally independent of deformation, provides a better means to estimate soil moisture. However, to retrieve soil moisture using temporal coherence, a decorrelation model should be implemented to separate the soil moisture-induced decorrelation, i.e., the change in the dielectric constant, from other decorrelation contributions, such as the decorrelation due to the change in the geometry of scatterers.

The second area of interest is in the design of orbit repeat for new satellite missions, which is driven by considering some important factors, such as tolerable error levels, the attainable baseline, and the expected decorrelation with the time of signals from the regions of interest to be mapped [12]. In this case, temporal decorrelation models can facilitate a priori assessment of the expected coherence levels of interferograms for a new satellite mission designed for a specific application.

Finally, temporal decorrelation models can help better estimate vegetation layer parameters. The total InSAR coherence is the product of spatial, temporal, thermal, volume, and processing coherences (e.g., [23]). Most models used to invert vegetation layer parameters (in PolInSAR studies) only consider the volume decorrelation contribution of the interferometric coherence and ignore other decorrelation contributions. However, leaving non-volumetric decorrelations uncompensated leads to a less accurate parameter estimation. In repeat-pass InSAR systems, the most critical non-volumetric decorrelation contribution is the temporal decorrelation caused by the change of the geometric and/or dielectric properties of the scatterers [23] and its contribution to decorrelation can be quantified using temporal decorrelation models.

The rest of this paper is organized as follows: in Section 2, InSAR coherence estimates are presented over the study area, Delta Junction, Alaska, the temporal decorrelation model and evaluation with real data are described in Section 3, followed by discussions and conclusions in Sections 4 and 5, respectively.

2. Study Area and Data

Our test site, illustrated in Figure 1, is located in Delta Junction, interior Alaska. The area is mostly covered by forest and shrub landscapes [24] and underlain by dis-continuous permafrost. The Alaskan interior between the Alaska and Brooks Mountain Ranges has a strong continental climate with moderate temperatures and precipitation in summer and exceedingly cold and dry weather in winter [25]. The average minimum and maximum annual temperatures in Big Delta station (1937–2005), which is located in the study area, are −6.9 °C and −2.7 °C, respectively. The lowest and highest temperatures occur in January (−23.7 °C) and July (20.8 °C). Average total precipitation, average total snow-fall, and average snow depth are, respectively, about 29, 111.25, and 10.2 centimeters (National Weather Service (http://www.wrcc.dri.edu)).

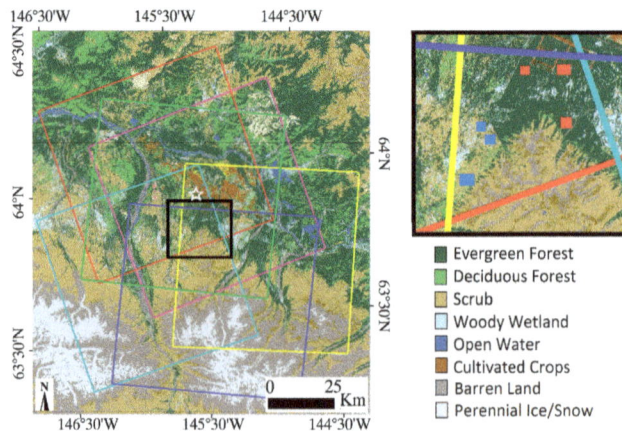

Figure 1. Land cover map of the study area (National Land Cover Database 2011 (NLCD 2011)) [24]. The orbit- frames covering the study area are shown with different colors (explained in Table 1). The black rectangle box shows the overlapping area. The forested and shrub patches are boxed in red and blue, respectively. The location of Snow Telemetry (SNOTEL) site, Granite Creek (963) (Natural Water and Climate Center (https://wcc.sc.egov.usda.gov)), is shown by a white star.

To study the temporal evolution of InSAR coherence, 32 single look complex (SLC) SAR images of L-band ALOS-2 (23.6 cm wavelength) from three ascending and three descending orbital paths in the fine beam and horizontal-horizontal (HH) polarization mode have been used. The data span from August 2014 to March 2017. Each group of the SLC SAR images have been co-registered based on a single master image, which optimizes the geometric and temporal coherence of the interferogram stack.

The SLC images were then used to generate interferograms with a pixel size of about 30 m × 30 m. After removing topographic phase, simulated using the National Elevation Dataset (NED) DEM, and applying range spectral shift and azimuth common band filters, and a linear weighting window size of 5 × 5 (in pixels) was used to estimate correlation. Then, interferograms affected by ionospheric artifacts were excluded and a total number of 75 interferograms with no or very limited effects of ionospheric artifacts have been selected to analyze temporal coherence. Table 1 gives the information of the data and interferograms used in this study.

Table 1. Data used in this study. The letters A and D denote ascending and descending, respectively.

Path-Frame	Orbit Direction	Number of Interferograms	Color of Frame on Figure 1
0040-2330	D	4	Yellow
0041-2330	D	16	Blue
0042-2320	D	10	Green
0137-1280	A	10	Magenta
0138-1280	A	29	Red
0139-1270	A	6	Cyan

For each of the two major land cover types in the study area, forest and shrub, three patches within flat areas with a total number of 1963 and 1729 pixels, respectively, on geo-referenced coherence images that are fully overlapped with all ALOS-2 observations have been selected. Figure 1 shows the selected patches in red (forest) and blue (shrub) boxes. For each of the interferograms, average coherence values of the selected pixels of each of the land cover types have been calculated. Therefore, each interferogram has two coherence values, one for each of the land cover types, evergreen forest and shrub. Figure 2 illustrates the scatter plot of the average coherence versus temporal baseline for the selected patches.

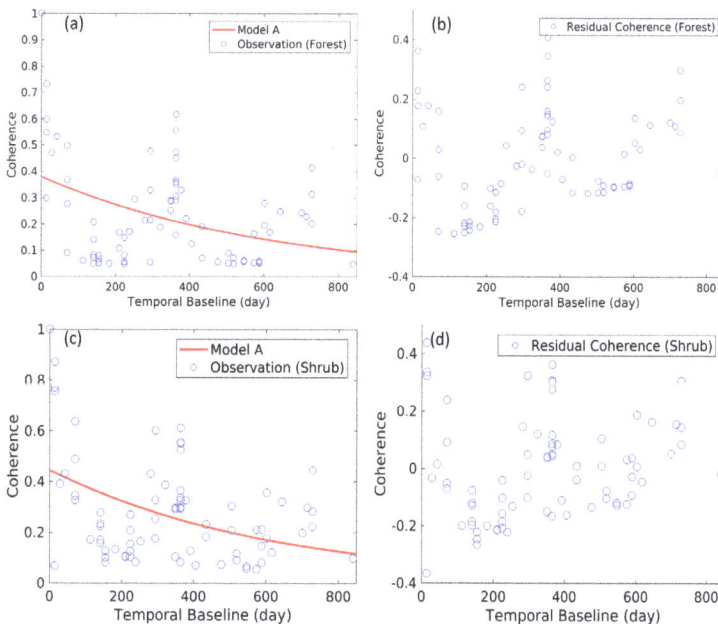

Figure 2. Scatter plots of observations and model A for forested (**a**) and shrub (**c**) land cover types. Scatter plots of residual coherences (observation—model A) for forested (**b**) and shrub (**d**) land cover types.

3. Methods

3.1. InSAR Coherence

An InSAR coherence estimation image is a cross-correlation product of two co-registered complex-valued SAR images (e.g., [10,26,27]) which quantifies radar wavelength-scale changes in backscattering characteristics. Decorrelation, i.e., loss of coherence, is generally increased by increasing spatial and temporal baselines between two image acquisitions [10,27,28]. InSAR coherence assesses the accuracy of the estimated deformation and depends on the amount of phase error in an interferogram [28–30]. Over a small window of pixels, InSAR coherence is estimated by:

$$\gamma = \left| \frac{\sum C_1 C_2^* e^{-j\varnothing}}{\sqrt{\sum |C_1|^2 \sum |C_2|^2}} \right| \tag{1}$$

where C_1 and C_2 are complex-valued backscattering coefficients, C_2^* is the complex conjugate of C_2, $\varnothing$ is the deterministic phase due to baseline error, topography, or large deformation in the correlation window.

The total InSAR coherence is the product of spatial ($\gamma_{spatial}$), temporal ($\gamma_{temporal}$), thermal ($\gamma_{thermal}$), volume (γ_{volume}), and processing ($\gamma_{processing}$) coherences [4,10,28]:

$$\gamma = \gamma_{spatial} \times \gamma_{volume} \times \gamma_{temporal} \times \gamma_{thermal} \times \gamma_{processing}. \tag{2}$$

The spatial (perpendicular) baselines of our dataset, except for two interferograms with spatial baselines of 308 and 347 m, are smaller than 284 m with a mean of ~108 m, whereas the critical baseline of the data is about 11 km. Therefore, a perpendicular baseline of 108 m, i.e., the mean value of the perpendicular baselines, will decrease the coherence by the value of ~0.01 which is negligible. In long wavelength (L-band) SAR sensors, such as ALOS-2 PALSAR-2, the small perpendicular baseline will not affect the variation of spatial decorrelation much. Therefore, we assumed that the spatial decorrelation from the small range of change in the perpendicular baseline is constant. Additionally, with such small perpendicular baselines, the volumetric decorrelation is negligible (e.g., [12]). Here, we focus only on temporal decorrelation by assuming that other decorrelation terms are constant or relatively not significant.

3.2. Temporal Coherence Modeling

The temporal coherence is usually considered as a univariate exponential function of time (e.g., [13,31]) by taking the random motion of scatterers in the resolution cell to be the main source of decorrelation:

$$\gamma_A = \gamma_0 e^{-\frac{t}{\tau}} \tag{3}$$

where subscript A denotes model A, γ_0 is initial coherence, t is the time separation between two SAR images, and $1/\tau$ is its decorrelation rate and is mainly dependent on the wavelength of the radar. Based on model A, the exponential decay of coherence values is expected in general by increasing the temporal baselines. However, the scatter plots of the observed coherence versus temporal baseline and the scatter plots of the residual coherence, i.e., observation—model A, versus the temporal baseline, illustrated in Figure 2, feature strong undulation with local peaks at temporal baselines around one and two years. Model A takes into account decorrelations due to long-term variations of the scatterers' geometry. In the real world, however, in addition to the natural and gradual long-term changes in the scatterers' geometry, seasonal and/or sudden changes in surface backscattering parameters may also contribute to temporal decorrelation. Generally, backscattering is dominated by the surface dielectric constant and roughness among other surface characteristics. Surface parameters, such as soil moisture, vegetation, and temperature alter dielectric constant and roughness and, consequently, backscattering coefficients (e.g., [17,22,32]). Therefore, other term(s) should be added to the temporal coherence function to compensate the effects of changes in surface backscattering between two image acquisitions.

In this paper, we have modified the coherence model to accommodate decorrelations due to the change in surface backscattering parameters. Since the SAR backscatter coefficient and, consequently, SAR intensity, varies as a function of the changes in surface parameters (e.g., [19,33]), here, we use the change in SAR intensity as a proxy for the changes in surface backscattering. Figure 3 shows a semi-logarithmic scatter plot of the coherence ratio, i.e., observation/model A, versus the relative intensity change between two images. Relative intensity is calculated by $r = |10 \log(i_2/i_1)|$, which i_2 and i_1 are SAR intensities of the first and second images, respectively. Considering the linear trend fitted to the semi-logarithmic scatter plots (note R^2 value and very small p-value of the linear regressions), model B is postulated to be:

$$\gamma_B = \gamma_0\, e^{-\left(\frac{t}{\tau}+\frac{r}{\rho}\right)} \tag{4}$$

where r is the relative change in SAR intensity, i.e., the backscattering baseline, and $1/\rho$ is its decorrelation rate. For each of the fitted linear trend, R^2 and p-value of the regression are calculated and shown on the plots. Note that if the p-value of a t-test is smaller than the common alpha values of 0.1, 0.05, and 0.01 (the confidence level of 90, 95, and 99%, respectively), the null-hypothesis is rejected. This means that the additional term related to the relative change in SAR intensity is likely correlated with temporal correlation.

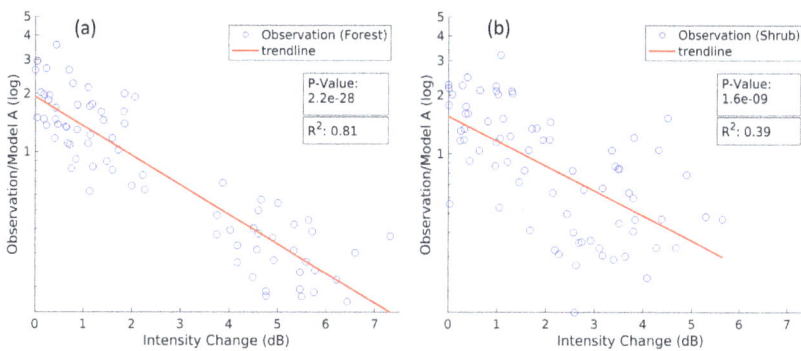

Figure 3. Semi-logarithmic scatter plot of coherence ratio, i.e., observation/model A, versus the relative intensity changes for forested (**a**) and shrub (**b**) land cover types.

The unknown parameters in models A and B, i.e., γ_0, τ, and ρ, can be estimated by solving the equations using known variables, i.e., γ, t, and r. The coherence, γ, is estimated using Equation (1). The temporal baseline of the interferograms, t, ranges between 14 and 840 days. Figure 4 illustrates the scatter plots of model B, observed coherence, and residual coherence values (observation—model B) for the two landscapes. For comparison, the scatter plot of water body's coherence is also shown in the figure. Additionally, Table 2 exhibits the model parameters and RMS error for each model. The RMS error values of model B for the both land cover types are smaller than those of model A, indicating that model B estimates more accurate coherence values. A detailed discussion of how significant the improvement is has been provided in Section 4.

Different snow depths between two images of an interferometric pair is one of the factors that can induce variations in surface scattering behavior, which, in turn, leads to decorrelation. Basically, between two winter images in stable frozen conditions with no change in soil moisture, the change in the dielectric constant is negligible and high coherence values can be expected for open areas [34]. However, the change in snow depth between the two images may change the surface scattering behavior, which, in turn, causes decorrelation. Here, we intend to modify our coherence model by adding the decorrelation term of snow depth changes. The basic assumption here is that the intensity and snow depth changes are independent parameters, i.e., a systematic snow depth change does not produce a systematic intensity change. Figure 5, illustrating the scatter plot of the intensity change versus snow depth changes, shows no trend and indicates that the two parameters are independent.

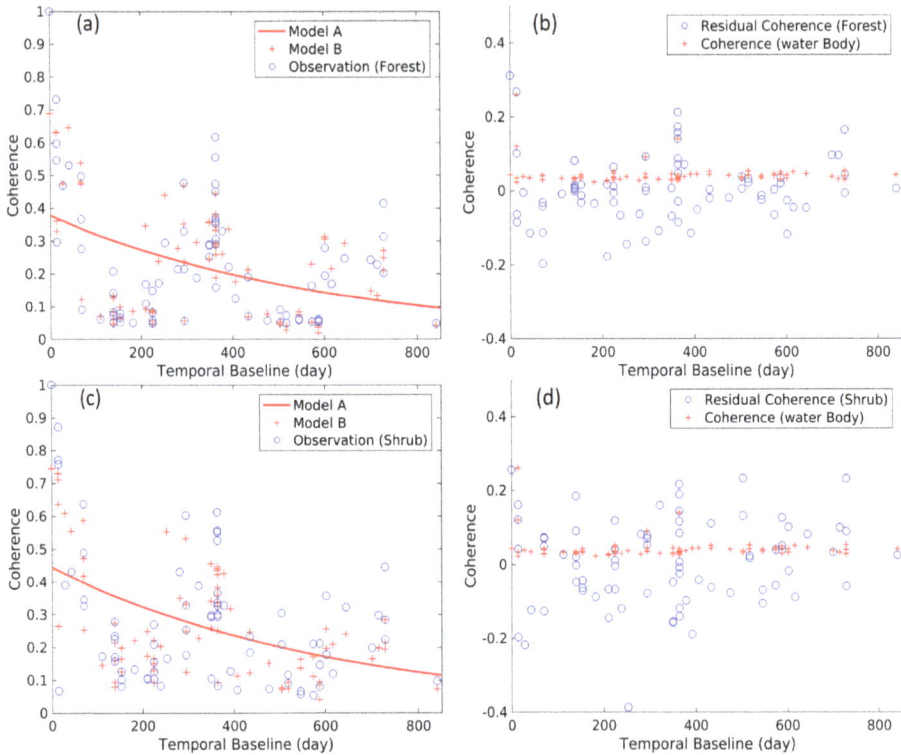

Figure 4. Scatter plots of observations and model B for forested (**a**) and shrub (**c**) land cover types. Scatter plots of residual coherences (observation—model B) for forested (**b**) and shrub (**d**) land cover types. Plot of model A and the scatter plot of coherence of water body are shown for comparison.

Table 2. Model parameters of the two land cover types. Cf is the critical *f*-value.

Model	Land Cover	γ_0	τ (Day)	ρ	σ	RMS	*f*-Test	C_f ($\alpha = 0.01$)
A	Forest	0.37824	616.49	-	-	0.180	-	-
	Shrub	0.444	629.53	-	-	0.186	-	-
B	Forest	0.68885	861.07	2.5406	-	0.092	205.84	6.99
	Shrub	0.74482	879.27	2.5467	-	0.121	102.38	6.99
C	Forest	0.73842	903.7	3.3464	0.62062	0.083	16.23	7.00
	Shrub	0.79153	913.47	5.6462	0.37348	0.101	29.64	7.00

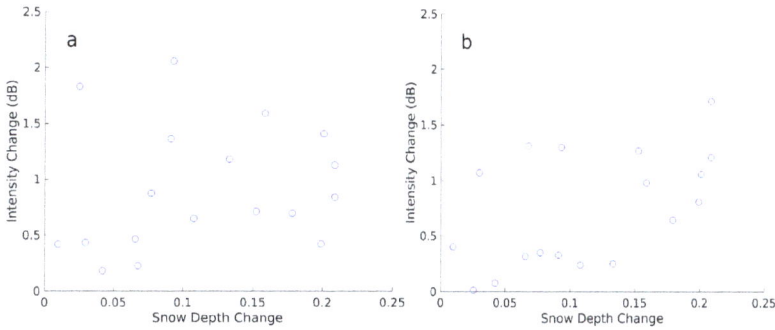

Figure 5. Scatter plot of model intensity changes (dB) versus snow depth changes (meter) for forested (**a**) and shrub (**b**) land cover types.

Figure 6 shows semi-logarithmic scatter plot of coherence ratio, i.e., observation/model B, versus the snow depth change between two images. Considering the linear trend fitted to the semi-logarithmic scatter plots, model C is postulated to be:

$$\gamma_C = \gamma_0 \, e^{-(\frac{t}{\tau} + \frac{r}{\rho} + \frac{s}{\sigma})} \tag{5}$$

where *s* is the snow depth change between images and $1/\sigma$ is its decorrelation rate. The unknown parameters in model C is estimated by solving the equations using known variables. The snow depth values are acquired from SMAP level 4 data (National Snow and Ice Data Center (http://nsidc.org)) and the measurements at the SNOTEL Site Granite Creek (963) (Natural Water and Climate Center (https://wcc.sc.egov.usda.gov)), which is located in our study area (white star in Figure 1). Figure 7 illustrates the scatter plots of model C, observed coherence, and residual coherence values (observation—model C) for the two landscapes. For comparison, the scatter plot of a water body's coherence is also shown. Additionally, Table 2 exhibits the model parameters and RMS error for each model. The RMS error values of model C for the two land cover types are smaller than those of model B. This means that the change in snow depth leads to decorrelation and, taking into account its effect on InSAR coherence improves the model's performance. A statistical analysis of how significant the improvement is has been provided in the next section.

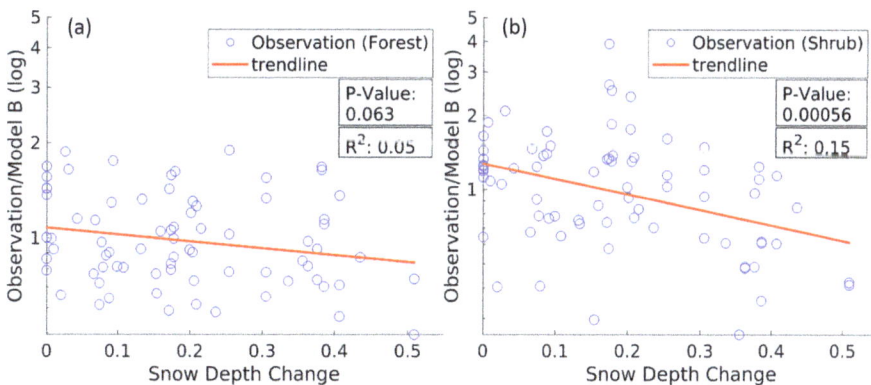

Figure 6. Semi-logarithmic scatter plot of the coherence ratio, i.e., observation/model B, versus snow depth changes (meter) for forested (**a**) and shrub (**b**) land cover types.

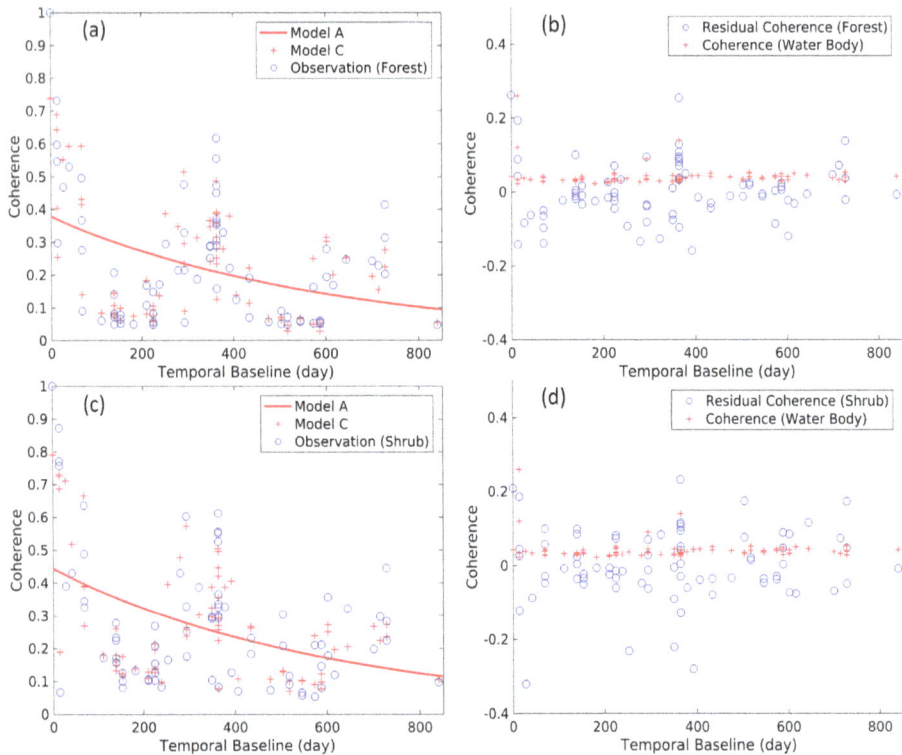

Figure 7. Scatter plots of observations and model C for forested (**a**) and shrub (**c**) land cover types. Scatter plots of residual coherences (observation—model C) for forested (**b**) and shrub (**d**) land cover types. Plot of model A and the scatter plot of coherence of water body are shown for comparison.

4. Discussion

4.1. Scatterers' Type and Decorrelation Sources

Model C has three terms. The first term, $exp\,(-t/\tau)$, is the long-term irreversible decorrelation due to the temporal change in scatterers' geometry. The second term, $exp\,(-r/\rho)$, is the decorrelation due to the changes in backscattering between two images. As stated earlier, the change in SAR intensity was used as a proxy for the change in the surface backscattering. Figure 8 shows the plot of SAR intensity (dB) over forest and shrub landscapes versus the soil moisture measured at the SNOTEL Site Granite Creek (963) (Natural Water and Climate Center (https://wcc.sc.egov.usda.gov)) which is located in our study area and is 12 km away on average from the patches (see Figure 1). The plots demonstrate a general correlation between soil moisture and SAR intensity, indicating that the change in backscattering and SAR intensity is most likely dominated by the change in dielectric constant of scatterers induced by the change in soil and biomass water content.

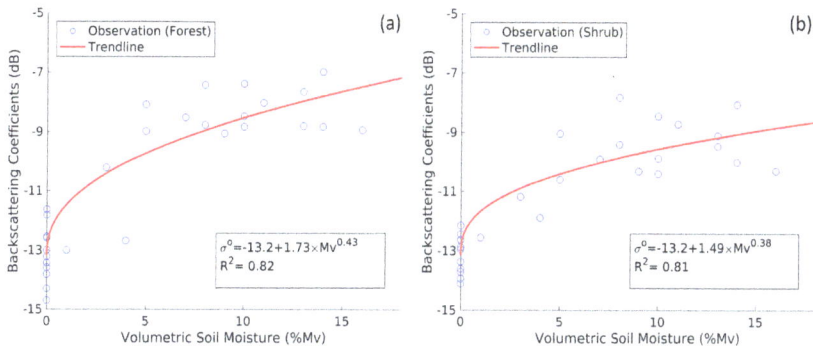

Figure 8. Plot of the change in SAR intensity for forested (**a**) and shrub (**b**) land cover types versus the change in soil moisture measured at the SNOTEL Site Granite Creek (963) (Natural Water and Climate Center (https://wcc.sc.egov.usda.gov)). Data under frozen and unfrozen conditions are included.

The third term, $exp\,(-s/\sigma)$, is the decorrelation due to the change in snow depth between two images. The scatterplot of InSAR intensity change (dB) versus snow depth change for soil moisture-free, i.e., winter-winter, interferograms (Figure 5) show no correlation between the SAR intensity change and the snow depth change over forested and shrub landscapes, indicating that snow depth change-induced decorrelation is most likely dominated by the change in the scatterers' geometry. In Section 4.3, we will perform a statistical assessment to show that the improvement made by considering the snow depth change in the model is statistically significant.

In the case where the temporal evolution of surface parameters such as soil moisture, vegetation layer parameters, and snow depth are known, the general coherence is postulated to be:

$$\gamma = \gamma_0\, e^{-\left(\frac{t}{\tau} + \sum_{i=1}^{n} \frac{p_i}{\mu_i}\right)} \tag{6}$$

where $1/\mu_i$ is the decorrelation rate of the parameter p_i, which is the change in the surface parameters between two images.

In general, shrub landscape is more stable as changes happen in the scatterers' geometry and dielectric constant. Except the decorrelation rate of the change in snow depth, which is lower for forest, the decorrelation rates of long-term and backscattering are lower, i.e., higher τ and ρ values, for shrub landscape compared to the forested landscape (Table 2). The scatterers within a resolution cell are of two types: scatterers associated with the ground surface and scatterers associated with the vegetation layer. Forested landscape (coniferous in this research) possess more backscattering contribution from the vegetation layer compared to the shrub landscape. Since the mutual position of scatterers within a vegetation layer, i.e., the geometry of scatterers, is more likely subject to change than the geometry of the scatterers within a non-vegetated surface, forested areas in the long-term decorrelate faster than non-forested areas as time lapses. Therefore, shrubland is expected to have a lower decorrelation rate associated with the long-term change in the scatterer's mutual position (Table 2). Additionally, the observed higher backscattering decorrelation rate of the forested area is associated, in part, with dielectric variation within the vegetation layer due to, for example, changing water content within the trees. Similarly, the observed higher snow depth decorrelation rate in shrubland is associated with the type of scatterers within each landscape. The lower proportion of snow-covered scatterers within the vegetation layer causes the forested landscape to lose coherence with a lower rate compared to the non-forested landscape as the change in snow depth increases.

The long-term decorrelation rate of the forest is slightly higher than the one of the shrub landscape, whereas the backscattering decorrelation rate of the forest is almost two times greater than that of the shrub landscape. This infers that, with short-baseline datasets, the difference between the temporal decorrelation of snow-free forested and non-forested areas is dominated by the decorrelation induced by the change in the dielectric constant of scatterers within the vegetation layer and not by the change in the geometry of scatterers within the vegetation layer.

4.2. The Effect of Seasonality on Temporal Coherence

Figure 9 depicts coherence as a 3D surface and provides a visual comparison of coherence evolutions of the two land cover types: forested and shrub. The x-axis of the plot is the relative change in SAR intensity and ranges between 0 and 0.65, i.e., the maximum measured relative intensity change, whereas the y-axis is the change in snow depth and ranges between 0 and 0.65 m. It is shown that both land cover types, even with the short temporal baselines, can lose coherence due to the changes in the dielectric constant and snow depth. Additionally, it is illustrated that the forest loses coherence with a higher rate than the shrub landscape with changing backscattering (dielectric constant), whereas the shrub landscape is more prone to decorrelation as the change in snow depth between two images increases.

Table 3 shows statistical properties of the observed coherence values of the two land cover types. The interferograms of each land cover type are subdivided into three sub-groups: summer, winter, and cross-season interferogram categories. It is shown that the shrub landscape has a higher coherence value than forest. Additionally, winter and cross-season interferogram categories possess the highest and lowest coherence values in general.

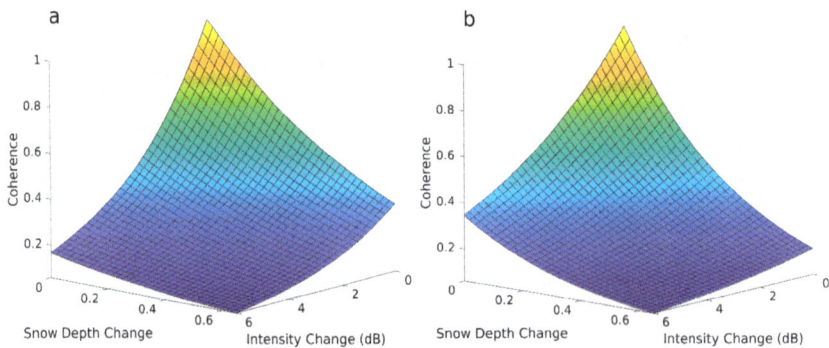

Figure 9. Coherence curves of forested (**a**); and shrub (**b**) land cover types.

Table 3. Statistical properties, i.e., mean and standard deviation (SD), of interferogram categories. The letters S, W, and C denote summer-summer, winter-winter, and cross-season interferogram groups.

Land Cover	Group	Mean	SD	SD/Mean
	S	0.3433	0.0711	0.2070
Forest	W	0.3703	0.0680	0.1835
	C	0.0896	0.0407	0.4546
	S	0.3905	0.0974	0.2495
Shrub	W	0.4074	0.0836	0.2052
	C	0.1522	0.0566	0.3720

Basically, the different scatterers' type and structure within the forested and non-forested resolution cells may result in different decorrelation processes when seasonal or sudden variations in surface parameters and meteorological conditions occur. During a frozen season, a decreased dielectric constant leads to reduced attenuation and a deeper penetration of electromagnetic waves into the forest canopy [34–36]. Consequently, this will cause a decrease in backscatter and influence the polarimetric signature and InSAR coherence [34–38]. In terms of coherence, between two winter images under stable frozen conditions, water content (soil moisture) changes do not occur, leading to low temporal decorrelation for open areas [34]. This means that higher InSAR coherence is expected for winter interferograms. Basically, over frozen forests, compared to the unfrozen condition, more volumetric decorrelation is expected to occur as the perpendicular baseline increases, owing to deeper penetration of electromagnetic waves into the frozen forest canopy [34]. However, with the short-baseline interferograms of L-band ALOS-2 and future datasets with narrower orbital tubes, the volumetric decorrelation induced by the deeper penetration of electromagnetic waves into the frozen soil and the frozen forest canopy become low, resulting in higher winter coherences.

During the unfrozen condition, changing soil moisture, variable water content within the trees, growth-related changes, and wind are among the major sources of temporal decorrelation [39,40]. The variation of the aforementioned surface parameters and meteorological conditions, in turn, results in decreased temporal coherence of the unfrozen condition compared to the frozen condition [34].

4.3. Statistical Assessment on Models' Performance

To statistically assess the improvements of models B and C, which have, respectively, one and two more parameters compared to model A, we used F-test (explained, for example, in Davis, 2002 [41]):

$$F = \frac{SSR_1 - SSR_2}{SSR_2} \frac{n - P_2}{P_2 - P_1} \tag{7}$$

where SSR is the sum of squared residuals of the model, P is the number of free model parameters, and n is the number of observations. If the calculated F-test value is greater than the upper-tailed critical value of the F-distribution, $F_{P2-P1,n-P2,\alpha}$, then, with a $1 - \alpha$ percentage confidence, the null hypothesis is rejected, i.e., the improvement is statistically significant. The calculated F-test values and critical F-distribution values with 99% confidence level (probability level of 0.01) have been presented in Table 2. All the calculated F-test values are greater than the critical values at a 99% confidence level, indicating that the improvements made by the new models are statistically significant.

5. Conclusions

Model C takes into account the long-term irreversible/long-term changes in scatterers' geometry, reversible/seasonal changes in scatterers' dielectric constant, induced mainly by the change in soil and biomass water content, and reversible/seasonal changes in scatterers' geometry, i.e., the third term, due to the change in snow depth. Shrub, in general is more stable than forest, as time lapses and variations happen in the surface backscattering properties. Additionally, the results show high coherence values for winter interferograms compared to summer ones owing to the stable condition of the frozen season. Our model illustrates that snow depth difference between interferogram pairs causes decorrelation, which is shown to mainly result from the change in the scatterers' geometry.

This paper argues that with short-baseline interferograms of L-band ALOS-2 and future data sets with narrower orbital tubes, the differences between temporal decorrelation rates of forest and non-forested areas, in snow-free condition, is dominated by the change in the dielectric constant of scatterers and not by the change in their geometry. It should be noted that the model introduced here is based on L-band data and, therefore, might not be expanded to C-band or X-band InSAR observations. The model provides accurate estimation of InSAR coherence for coniferous forested and shrub land cover types. However, its accuracy over other terrain types should be assessed.

Acknowledgments: ALOS-2 PALSAR data is copyrighted by the Japan Aerospace Exploration Agency (JAXA). We thank Paul Siqueira for facilitating the data access and Scott Hensley for providing several good references. The work was funded by the NASA Earth Surface and Interior Program (NNX16AK56G) and the Shuler-Foscue Endowment at Southern Methodist University. The authors would like to thank the academic editor and three reviewers for their very constructive comments and suggestions.

Author Contributions: Yusuf Eshqi Molan processed the SAR data into In SAR products, conducted modeling, and created the first draft of the manuscript. Jin-Woo Kim, Zhong Lu, and Piyush Agram helped interpret the results. All contributed the writing of this manuscript. Readers who are interested in the data and products from this study can contact directly the corresponding author at the Southern Methodist University (yeshqimolan@smu.edu).

Conflicts of Interest: The authors declare no conflict of interest.

References

1. Massonnet, D.; Feigl, K. Radar interferometry and its application to changes in the Earth's surface. *Rev. Geophys.* **1998**, *36*, 441–500. [CrossRef]
2. Bürgmann, R.; Rosen, P.A.; Fielding, E.J. Synthetic Aperture Radar Interferometry to Measure Earth's Surface Topography and Its Deformation. *Annu. Rev. Earth Planet. Sci.* **2000**, *28*, 169–209. [CrossRef]
3. Simons, M.; Rosen, P. Treatise on Geophysics: Interferometric Synthetic Aperture Radar Geodesy. In *Geodesy*; Schubert, G., Ed.; Elsevier Press: Amsterdam, The Netherlands, 2007; Volume 3, pp. 391–446.
4. Lu, Z.; Dzurisin, D. InSAR imaging of Aleutian volcanoes: Monitoring a volcanic arc from space. In *Geophysical Sciences*; Springer Praxis Books: Chichester, UK, 2014; 390p.
5. Ferretti, A.; Prati, C.; Rocca, F. Permanent scatterers in SAR interferometry. *IEEE Trans. Geosci. Remote Sens.* **2001**, *39*, 8–20. [CrossRef]
6. Kim, J.W.; Lu, Z.; Degrandpre, K. Ongoing Deformation of Sinkholes in Wink, Texas, Observed by Time-Series Sentinel-1A SAR Interferometry (Preliminary Results). *Remote Sens.* **2016**, *8*, 313. [CrossRef]
7. Hu, X.; Wang, T.; Pierson, T.C.; Lu, Z.; Kim, J.; Cecere, T.H. Detecting seasonal landslide movement within the Cascade landslide complex (Washington) using time-series SAR imagery. *Remote Sens. Environ.* **2016**, *187*, 49–61. [CrossRef]
8. Rykhus, R.; Lu, Z. InSAR detects possible thaw settlement in the Alaskan Arctic Coastal Plain. *Can. J. Remote Sens.* **2008**, *34*, 100–112. [CrossRef]
9. Liu, L.; Zhang, T.; Wahr, J. InSAR measurements of surface deformation over permafrost on the North Slope of Alaska. *J. Geophys. Res.* **2010**, *115*, F03023. [CrossRef]
10. Zebker, H.; Villasenor, J. Decorrelation in interferometric radar echoes. *IEEE Trans. Geosci. Remote Sens.* **1992**, *45*, 950–959. [CrossRef]
11. Just, D.; Bamler, R. Phase statistics of interferograms with applications to synthetic aperture radar. *Appl. Opt.* **1994**, *33*, 4361–4368. [CrossRef] [PubMed]
12. Simard, M.; Hensley, S.; Lavalle, M. An empirical assessment of temporal decorrelation using the uninhabited aerial vehicle synthetic aperture radar over forested landscapes. *Remote Sens.* **2012**, *4*, 975–986. [CrossRef]
13. Rocca, F. Modeling Interferograms Stacks. *IEEE Trans. Geosci. Remote Sens.* **2007**, *45*, 3289–3299. [CrossRef]
14. Luo, X.; Askne, J.; Smith, G.; Dammert, P. Coherence characteristics of RADAR signals from rough soil. *Prog. Electromagn. Res.* **2001**, *31*, 68–88. [CrossRef]
15. Zwieback, S.; Hensley, S.; Hajnsek, I. Assessment of soil moisture effects on L-band radar interferometry. *Remote Sens. Environ.* **2015**, *164*, 77–89. [CrossRef]
16. Zwieback, S.; Paulik, C.; Wagner, W. Frozen Soil Detection Based on Advanced Scatterometer Observations and Air Temperature Data as Part of Soil Moisture Retrieval. *Remote Sens.* **2015**, *7*, 3206–3231. [CrossRef]
17. Zwieback, S.; Hensley, S.; Hajnsek, I. A Polarimetric First-Order Model of Soil Moisture Effects on the DInSAR Coherence. *Remote Sens.* **2015**, *7*, 7571–7596. [CrossRef]
18. Lavalle, M.; Simard, M.; Hensley, S. A Temporal Decorrelation Model for Polarimetric Radar Interferometers. *IEEE Trans. Geosci. Remote Sens.* **2012**, *50*, 2880–2888. [CrossRef]
19. Zhang, L.; Shi, J.; Zhang, Z.; Zhao, K. The estimation of dielectric constant of frozen soil-water mixture at microwave bands. In Proceedings of the IEEE International Geoscience and Remote Sensing Symposium, Toulouse, France, 21–25 July 2003; IEEE: Piscataway, NJ, USA, 2003; pp. 2903–2905.
20. Zhang, T.; Zeng, Q.; Li, Y.; Xiang, Y. Study on relation between InSAR coherence and soil moisture. In Proceedings of the ISPRS Congress, Beijing, China, 3–11 July 2008.
21. Borgeaud, M.; Wegmueller, U. On the use of ERS SAR interferometry for the retrieval of geoand bio-physical information. In Proceedings of the 'Fringe 96' Workshop on ERS SAR Interferometry, Zurich, Switzerland, 30 September–2 October 1996; pp. 83–94.

22. Morishita, Y.; Hanssen, R.F. Temporal Decorrelation in L-, C-, and X-band Satellite Radar Interferometry for Pasture on Drained Peat Soils. *IEEE Trans. Geosci. Remote Sens.* **2015**, *53*, 1096–1104. [CrossRef]

23. Lee, S.-K.; Kugler, F.; Papathanassiou, K.P.; Hajnsek, I. Quantification of Temporal Decorrelation Effects at L-Band for Polarimetric SAR Interferometry Applications. *IEEE J. Sel. Top. Appl. Earth Obs. Remote Sens.* **2013**, *6*, 1351–1367. [CrossRef]

24. Homer, C.G.; Dewitz, J.A.; Yang, L.; Jin, S.; Danielson, P.; Xian, G.; Coulston, J.; Herold, N.D.; Wickham, J.D.; Megown, K. Completion of the 2011 National Land Cover Database for the conterminous United States-Representing a decade of land cover change information. *Photogramm. Eng. Remote Sens.* **2015**, *81*, 345–354.

25. O'Neill, K.P.; Kasischke, E.S.; Richter, D.D. Seasonal and decadal patterns of soil carbon uptake and emission along an age-sequence of burned black spruce stands in interior Alaska. *J. Geophys. Res.* **2003**, *108*, 11–15. [CrossRef]

26. Lu, Z.; Freymueller, J. Synthetic aperture radar interferometry coherence analysis over Katmai volcano group, Alaska. *J. Geophys. Res.* **1998**, *103*, 29887–29894. [CrossRef]

27. Dzurisin, D.; Lu, Z. *Interferometric Synthetic Aperture Radar (InSAR) (Chapter 5): Volcano Deformation: Geodetic Monitoring Techniques*; Dzurisin, D., Ed.; Springer-Praxis Publishing Ltd.: Chichester, UK, 2007; pp. 153–194.

28. Hanssen, R. *Radar Interferometry: Data Interpretation and Error Analysis*; Kluwer: Dordrecht, The Netherlands, 2001; Volume 2.

29. Bamler, R.; Hartl, P. Synthetic aperture radar interferometry. *Inverse Prob.* **1998**, *14*, R1–R54. [CrossRef]

30. Touzi, R.; Lopes, A.; Bruniquel, J.; Vachon, P. Coherence estimation for SAR imagery. *IEEE Trans. Geosci. Remote Sens.* **1999**, *37*, 135–149. [CrossRef]

31. Lombardini, F.; Griffiths, H. Effect of temporal decorrelation on 3D SAR imaging using multiple pass beamforming. In Proceedings of the IEE/EUREL Meeting Radar Sonar Signal Processing, Peebles, UK, 5–8 July 1998; pp. 1–4.

32. Ulaby, F.T.; Bradley, G.A.; Dobson, M.C. Microwave backscatter dependence on surface roughness, soil moisture, and soil texture, II, Vegetation covered soil. *IEEE Trans. Geosci. Electron.* **1979**, *17*, 33–40. [CrossRef]

33. Zribi, M.; Gorrab, A.; Baghdadi, N. A new soil roughness parameter for the modelling of radar backscattering over bare soil. *Remote Sens. Environ.* **2014**, *152*, 62–73. [CrossRef]

34. Thiel, C.; Schmullius, C. The potential of ALOS PALSAR backscatter and InSAR coherence for forest growing stock volume estimation in Central Siberia. *Remote Sens. Environ.* **2016**, *173*, 258–273. [CrossRef]

35. Kwok, R.; Rignot, E.J.M.; Way, J.; Freeman, A.; Holt, J. Polarization signatures of frozen and thawed forests of varying environmental state. *IEEE Trans. Geosci. Remote Sens.* **1994**, *32*, 371–381. [CrossRef]

36. Way, J.; Paris, J.; Kasischke, E.; Slaughter, C.; Viereck, L.; Christensen, N.; Weber, J. The effect of changing environmental-conditions on microwave signatures of forest ecosystems—Preliminary-results of the March 1988 Alaskan aircraft SAR experiment. *Int. J. Remote Sens.* **1990**, *11*, 1119–1144. [CrossRef]

37. Dobson, M.G.; McDonald, K.; Ulaby, F.T. Effects of temperature on radar backscatter from boreal forests. In Proceedings of the 10th Annual International Geoscience and Remote Sensing Symposium, College Park, MD, USA, 20–24 May 1990; Mills, R., Ed.; IEEE Publications: College Park, MD, USA, 1990; pp. 2481–2484.

38. Santoro, M.; Eriksson, L.; Fransson, J. Reviewing ALOS PALSAR backscatter observations for stem volume retrieval in Swedish forest. *Remote Sens.* **2015**, *7*, 4290–4317. [CrossRef]

39. Dobson, M.C. Diurnal and seasonal variations in the microwave dielectric constant of selected trees. In Proceedings of the International on Geoscience and Remote Sensing Symposium, Edinburgh, UK, 12–16 September 1988; Guyenne, T.D., Hunt, J.J., Eds.; ESA Publications Devision: Edinburgh, UK, 1988; p. 1754.

40. McDonald, K.C.; Zimmermann, R.; Kimball, J.S. Diurnal and spatial variation of xylem dielectric constant in Norway spruce (*Picea abies* [L.] karst.) as related to microclimate, xylem sap flow, and xylem chemistry. *IEEE Trans. Geosci. Remote Sens.* **2002**, *40*, 2063–2082. [CrossRef]

41. Davis, J.C. *Statistics and Data Analysis in Geology*, 3rd ed.; John Wiley and Sons: Hoboken, NJ, USA, 2002.

MDPI

St. Alban-Anlage 66

4052 Basel

Switzerland

Tel. +41 61 683 77 34

Fax +41 61 302 89 18

www.mdpi.com

Remote Sensing Editorial Office

E-mail: remotesensing@mdpi.com

www.mdpi.com/journal/remotesensing

www.ingramcontent.com/pod-product-compliance
Lightning Source LLC
Chambersburg PA
CBHW051701210326
41597CB00032B/5326